ANNUAL REVIEW OF
NEUROSCIENCE

EDITORIAL COMMITTEE (1988)

ANNUAL REVIEW OF NEUROSCIENCE

VOLUME 11, 1988

W. MAXWELL COWAN, *Editor*
Washington University

ERIC M. SHOOTER, *Associate Editor*
Stanford University School of Medicine

CHARLES F. STEVENS, *Associate Editor*
Yale University School of Medicine

RICHARD F. THOMPSON, *Associate Editor*
University of Southern California

ANNUAL REVIEWS INC. 4139 EL CAMINO WAY P.O. BOX 10139 PALO ALTO, CALIFORNIA 94303-0897

ANNUAL REVIEWS INC.
Palo Alto, California, USA

International Standard Serial Number: 0147-006X
International Standard Book Number: 0-8243-2411-0

Annual Review and publication titles are registered trademarks of Annual Reviews
Inc.

Annual Reviews Inc. and the Editors of its publications assume no responsibility
for the statements expressed by the contributors to this *Review*.

TYPESET BY AUP TYPESETTERS (GLASGOW) LTD., SCOTLAND
PRINTED AND BOUND IN THE UNITED STATES OF AMERICA

Annual Review of Neuroscience
Volume 11, 1988

CONTENTS

(Note: Titles of chapters in Volumes 7–11 are arranged by category on pages 576–580.)

vi   CONTENTS (*Continued*)

SOME RELATED ARTICLES FROM OTHER *ANNUAL REVIEWS*

From the *Annual Review of Biochemistry*, Volume 57 (1988):

> *Molecular and Cellular Biology of Intermediate Filaments*, P. M. Steinert and D. R. Roop

From the *Annual Review of Cell Biology*, Volume 3 (1987):

> *Cell Adhesion in Morphogenesis*, D. R. McClay and C. A. Ettensohn

> *Cell Surface Receptors for Extracellular Matrix Molecules*, C. A. Buck and A. F. Horwitz

> *Laminin and Other Basement Membrane Components*, G. R. Martin and R. Timpl

> *Polypeptide Growth Factors: Roles in Normal and Abnormal Cell Growth*, T. F. Deuel

> *Intracellular Transport Using Microtubule-Based Motors*, R. D. Vale

From the *Annual Review of Entomology*, Volume 33 (1988):

> *Insect Behavioral Ecology: Some Future Paths*, T. Burk

From the *Annual Review of Physiology*, Volume 50 (1988):

> *Prefatory Chapter: Muscular Contraction*, Sir A. Huxley

> *Voltage Dependence of the Na-K Pump*, P. De Weer, D. C. Gadsby, and R. F. Rakowski

> *Site-Directed Mutagenesis and Ion-Gradient Driven Active Transport: On the Path of the Proton*, H. R. Kaback

> *Electroconformational Coupling: How Membrane-Bound ATPase Transduces Energy from Dynamic Electric Fields*, T. Y. Tsong and R. D. Astumian

> *Modulation of the Na,K-ATPase by Ca and Intracellular Proteins*, D. R. Yingst

> *CNS Peptides and Regulation of Gastric Acid Secretion*, Y. Taché

> *Structural and Chemical Organization of the Myenteric Plexus*, C. Sternini

> *Genetic Analysis of Ion Channels in Vertebrates*, W. A. Catterall

> *Ion Channels in* Drosophila, D. M. Papazian, T. L. Schwarz, B. L. Tempel, L. C. Timpe, and L. Y. Jan

> *Gene Transfer Techniques to Study Neuropeptide Processing*, G. Thomas, B. A. Thorne, and D. E. Hruby

From the *Annual Review of Pharmacology and Toxicology*, Volume 28 (1988):

From the *Annual Review of Psychology*, Volume 39 (1988):

ANNUAL REVIEWS INC. is a nonprofit scientific publisher established to promote the advancement of the sciences. Beginning in 1932 with the *Annual Review of Biochemistry*, the Company has pursued as its principal function the publication of high quality, reasonably priced *Annual Review* volumes. The volumes are organized by Editors and Editorial Committees who invite qualified authors to contribute critical articles reviewing significant developments within each major discipline. The Editor-in-Chief invites those interested in serving as future Editorial Committee members to communicate directly with him. Annual Reviews Inc. is administered by a Board of Directors, whose members serve without compensation.

ANNUAL REVIEWS OF
Anthropology
Astronomy and Astrophysics
Biochemistry
Biophysics and Biophysical Chemistry
Cell Biology
Computer Science
Earth and Planetary Sciences
Ecology and Systematics
Energy
Entomology
Fluid Mechanics
Genetics
Immunology

Materials Science
Medicine
Microbiology
Neuroscience
Nuclear and Particle Science
Nutrition
Pharmacology and Toxicology
Physical Chemistry
Physiology
Phytopathology
Plant Physiology
Psychology
Public Health
Sociology

SPECIAL PUBLICATIONS

Annual Reviews Reprints:
Cell Membranes, 1975–1977
Immunology, 1977–1979

Excitement and Fascination of Science, Vols. 1 and 2

Intelligence and Affectivity, by Jean Piaget

Telescopes for the 1980s

A detachable order form/envelope is bound into the back of this volume.

Ann. Rev. Neurosci. 1988. 11 : 1–12

LOOKING BACK, LOOKING FORWARD

T. P. Feng

Shanghai Institute of Physiology, Academia Sinica, Shanghai, China

INTRODUCTION

Recently my friends both at home and abroad have urged me to write down some reminiscences of my life. While I appreciate such amiable reminders of my getting old, circumstances generally dictate that something more urgent must be done first. The invitation from the Editor to write a prefatory chapter for the *Annual Review of Neuroscience* provides the necessary additional stimulus to sway me finally to write a short summary of my scientific life. My narration chiefly deals with the circumstances and background relating to my development as a neurophysiologist.

MY INITIATION INTO NEUROMUSCULAR PHYSIOLOGY

I started my neurophysiological research in 1929 in the Department of Physiology, University of Chicago, under Ralph Gerard. Before that, immediately after my graduation from the School of Biology, Fudan University in Shanghai, China, I had been a teaching assistant in Physiology to Professor C. Tsai at Fudan University for one year (1926–1927) and then a research fellow in the Department of Physiology, Peking Union Medical College (PUMC) for two years (1927–1929) working under Professor R. K. S. Lim. The two years in PUMC were my first apprenticeship in physiological research. The work in which I took part concerned the nervous and humoral control of gastric secretion. Lim had a strong personality and was an impressive teacher. His operative skill was quite exceptional. Working with him I learned not only elaborate operative techniques for preparing different kinds of gastric pouches, but more importantly, I gained my first practical appreciation of how experimental

1

0147–006X/88/0301–0001$02.00

physiological research is done. During those two years, besides taking part in research on gastric secretion, I read widely in almost every field of physiology. There were two books that especially took my fancy: *Principles of General Physiology* by William Bayliss and *Protoplasmic Action and Nervous Action* by Ralph Lillie. This background prepared me for the next step in my development. In the summer of 1929, I succeeded in winning a competitive examination for the Tsing-Hua University Fellowship to study in the United States. The questions as to where in the States I should go, and what kind of study I should undertake were readily answered: I wanted to go to the University of Chicago to study general physiology under Ralph Lillie. So in the fall of 1929 I went to the University of Chicago, registered as a PhD student in the Department of Physiology, and began my studies by taking the course in General Physiology given by Ralph Lillie. At the same time I tried to learn something about the research going on in Lillie's laboratory and in the laboratories of several other professors, including Ralph Gerard. At that time Lillie was occupied with the study of the iron-wire model of nerve conduction, and Gerard with the study of nerve metabolism. When it came time to decide what research I should take up, the choice between working with a *model* of nerve or working with *real* nerves was quickly made: I wanted to work on real nerves. So Gerard became my first mentor in neurophysiological research.

The problem Gerard put me to work on concerned the mechanism of nerve asphyxiation. The specific question was whether an asphyxiated nerve could be made to recover by soaking it in an oxygen-free solution of certain oxidizing dyes like methylene blue, instead of giving it oxygen. The answer turned out to be no, and so the work was not very interesting. But the process of arriving at that answer was far from straightforward. I quickly discovered that the connective tissue sheath of the nerve was an effective diffusion barrier and prevented the methylene blue from reaching the nerve fibers. This finding invalidated the original experimental design. It took me several tense days before I could alter the design so that my experiments could continue. However, a way was soon found and the answer was obtained. This experience was exciting to me as a beginner in neurophysiology.

When I registered at the University of Chicago as a PhD student I had originally planned to stay there for three years, but an unexpected turn of events radically changed this plan. Shortly after I started working with Gerard, Robert Lim, my old teacher in PUMC made arrangements for me to go to University College London to work with A. V. Hill. I was unaware what Lim had written to Hill about me, but in the spring of 1930 I received a brief letter from A. V. Hill saying: "If you are as good as Lim says you are, come along." That settled the matter: I was to go to London to work with A. V. Hill in the fall of 1930. Before leaving I was able to finish sufficient work for a MS degree from the University of Chicago and I also took the summer course in General Physiology at Woods Hole, according to a plan approved by Lim.

I arrived in London in September 1930. I saw A. V. Hill for the first time while he was doing some work with his assistant, J. L. Parkinson, in his laboratory. My first impression of Hill was that he was rather austere, but I soon felt more at ease when I saw a galvanometer in his laboratory labeled conspicuously, "DANGER 1000 OHMS"! Hill made me start working the day after my arrival. I worked mainly on the heat production first in muscle and later in nerve. In the course of about two and a half years I either did or participated in sufficient work for nine papers, five of which I wrote up myself.

The way Hill dealt with the first paper I wrote, entitled "The Heat-Tension Ratio in Prolonged Tetanic Contractions," is worth mentioning. The problem had been suggested by him and was carried out under his direction with much assistance from Mr. J. L. Parkinson. I naturally put Hill's name on the paper as a co-author. He promptly took his name off the paper, saying: "If this is the only paper you write while you are here, it will not make much difference whether my name is on it or not, and it will not mean much to you." Another remark he made to me in a similar vein toward the end of my stay in London should also be retold. "You have done good work here and you have done most of the work quite independently. But people will still think you are under my direction. You must go back and continue to do good work all by yourself, then you will be recognized as a fully independent worker." I don't know whether A. V. Hill talked to his other students like this, but his words left a deep impression on me.

Although in Hill's laboratory most of my work was concerned with the heat production of muscle and nerve, there was a short period when I was otherwise occupied. I had finished the study suggested by Hill on the thermoelastic properties of muscle and had completed a related study on the effect of length on the resting metabolism of muscle, which described a discovery that I called the "stretch response" (Feng 1932a) but that Hill later dubbed the "Feng effect". Hill then asked me to look into Lapicque's controversial theory of isochronism to see whether I could make something of it. I spent about a month measuring the chronaxies of nerve and muscle under various conditions. I was soon convinced that my results did not support Lapicque but felt that they were not otherwise of much interest, so I was inclined to discontinue the work. I told Hill how my experiments on isochronism had gone and indicated my intention to move on to something else. I had in mind using iodoacetic acid as a tool to address the question of whether lactic acid might play some role in nerve activity. After inquiring as to how I arrived at this idea, Hill encouraged me to go ahead with my proposed experiment. The result was a paper entitled, "The Role of Lactic Acid in Nerve Activity" (Feng 1932b), showing that frog's nerve is capable of utilizing lactic acid by oxidation and that the formation of lactic acid, though not essential to nerve conduction, enables normal nerves to perform long hours of continuous function.

Altogether I stayed with Hill for three years, obtaining my PhD degree in 1933. During those three years, I spent one summer in Plymouth working on the heat production of crustacean nerve, and, for about two months in each place, Hill sent me to E. D. Adrian's lab in Cambridge and then to C. S. Sherrington's lab in Oxford to broaden my research experience. Upon my arrival in Cambridge in 1932, Adrian gave me a problem to solve by myself. A little earlier, he, Cattel, and Hoagland had noticed that if the surface layers of a frog's skin were scraped away, the tactile responses of the cutaneous nerves ceased for a time but eventually returned. How was this reversible inexcitability of the tactile endings in skin to be explained? After about a week's work on the phenomenon, I convinced myself that it was simply due to the release of potassium from the skin by the injury, and that recovery was due to the subsequent removal of potassium by washing. When I told Adrian about this simple solution to the problem he was initially skeptical, saying that it sounded too simple to be true, and adding if I had told him it was due to some complex organic substance he could believe it more readily. He took the trouble to repeat my observations on frog's skin and then extended them to cat's skin before finally accepting my conclusion (Feng 1933). From that experience in Adrian's lab I learned two things. One was the seriousness and carefulness that one should cultivate in scientific work. Adrian set an admirable example of this. The other was the advantage of bringing people with different backgrounds together in scientific research. Coming from A. V. Hill's laboratory, I was already familiar with the reversible inexcitability of muscle due to potassium, and it was a simple matter for me to extend this knowledge to the problem of injured frog's skin.

I should mention that during my stay in Adrian's laboratory, I had the benefit of contact with Bryan Matthews, Rushton, Roughton, Adair, and Willmer. Barcroft was then the head of the Cambridge Physiological Laboratory, and I had frequently met him at the regular afternoon teas.

Hill sent me to Oxford with the remark: "Go to Sherrington and learn how to keep a cat alive." Unfortunately, shortly before I arrived in Oxford, late in 1932, Lady Sherrington died. Sherrington therefore made arrangements for me to work with J. C. Eccles, and I had only occasional contacts with Sherrington himself. Eccles was then working on spinal reflexes in cats, using mainly myographic recordings. I learned various surgical and technical procedures from him and assisted him in minor ways with his experiments. Eccles impressed me as a most energetic worker. During my short stay in Oxford I also had the pleasure of getting acquainted with Ragnar Granit.

My scientific association with A. V. Hill did not end with my stay in London. In 1936, about three years after I had left him, Hill was asked by *Ergebnisse der Physiologie* to write a review on the heat production of nerve. Instead of writing the review himself he recommended that I do it (Feng 1936). Last summer, Professor R. O. Keynes told me that he is planning to write a new review on the heat production of nerve, saying: "You wrote the first review, and I am going to write the last review on the subject."

On the recommendation of A. V. Hill, after leaving London in the summer of 1933, I returned to the United States to spend a year in the newly established Johnson Foundation for Medical Physics in Philadelphia, directed by Detlev Bronk, before returning to China. That new organization with a research staff composed entirely of active young scientists provided a lively and congenial atmosphere, quite different from that prevailing in the old university laboratories in England. During my stay there I was able to see at first hand a rather wide range of biophysical research relating especially to nerve and muscle (Robert Hodes), sympathetic ganglia (Detlev Bronk, M. G. Larrabee), and vision (Keffer Hartline). I did not concentrate on a specific research project, but spent my time mainly learning to make electronic apparatus under the guidance of John Hervey, in preparation for setting up my own laboratory in China.

THE NEUROMUSCULAR JUNCTION
50 YEARS AGO

For my initiation into neuromuscular physiology, I had the good fortune of having eminent men as my teachers. Their example and their generous encouragement were formative influences in my development. In the summer of 1934, I returned to China to take up a teaching position in the Department of Physiology of the PUMC, where I had previously worked as a research student. As no other space was available, Lim, the head of the department, provided me with a long, windowless basement room, isolated from the rest of the department, for my laboratory. Since the teaching duties assigned to me occupied only about six weeks a year, I could devote the rest of the time to research. In planning and doing my research, Lim left me entirely alone; and because of my location in the faraway basement, I was effectively left alone also by everybody else. This solitude turned out to be a good thing, minimizing distractions and interference. Under these circumstances I proceeded to build my own laboratory, partly with equipment purchased and partly with equipment that I had made myself and which I brought back from abroad.

It was clear that I was to do research in neuromuscular physiology. But what, specifically? I expected that I would have to grope about for some time before I could settle down to what I would regard as a serious program of research. I had at my disposal a couple of stimulators capable of delivering stimuli over a wide range of frequencies. The first exploratory experiment I undertook was to give the frog sartorius muscle a one-second tetanic stimulus at increasing frequencies up to about 2000/sec (the stimulating electrodes being placed on the tibial half of the muscle) to see how the tension response recorded on the kymograph varied. I had expected the tension response to decrease progressively and smoothly after a certain maximum, with increasing frequency. To my surprise, the response varied in a periodic manner, i.e. it showed periodic decreases and increases with the progressive increase in stimulus frequency. I then ascertained that this periodic pattern of response was only obtained when the stimulating electrodes were on the nerve-containing tibial half of the sartorius muscle, and not when the electrodes were on the nerve-free pelvic end of the muscle. With nerve stimulation the periodic pattern appeared even more strikingly. Further experiments, with combined nerve and pelvic end muscle stimulation, finally established the important fact that the contraction elicited by direct muscle stimulation could be inhibited by additional nerve stimulation at certain frequencies. This immediately impressed me as something quite new about the neuromuscular junction— something for which I had no ready explanation, but something whose study might disclose other new features of the neuromuscular junction. This intuition led to a quick decision to make the neuromuscular junction the subject of a continuing program of research, and a period of con-centrated work followed. In the course of the next six years (1936–1941) after the first exploratory experiments, in collaboration with a series of students, I published no fewer than 26 papers on the neuromuscular junction in the *Chinese Journal of Physiology* (in English). This work was terminated only by the outbreak of the Pacific war and the closing of the PUMC.

At the time I started my research, the theory of chemical transmission at the neuromuscular junction was still in the formative stage. I was led to my own views by the outcome of my own experiments, without any theoretical preconception. It soon became clear that most of my results fitted well with the new chemical theory of neuromuscular transmission and, indeed, provided support for it. When I was in London working in Hill's laboratory, I had visited the National Institute for Medical Research at Hampstead on several occasions and had the opportunity of getting acquainted with H. H. Dale (then Director of the National Institute for Medical Research), J. H. Gaddum, W. Feldberg, G. L. Brown, and others. I often heard Hill speak of Dale with great respect, much as he did of Sherrington and Adrian. But I could not then imagine that I would later work in an area of research so closely related to that of Dale's group. Some of the more significant findings of my studies on the neuromuscular junction may be briefly summarized here.

1. Accompanying the inhibition produced at the neuromuscular junction by high-frequency indirect stimulation which was mentioned above (and is now generally called junctional inhibition), we found a local contraction surrounding the nerve endings. This local contraction could be greatly exaggerated and could become a prolonged contracture if the muscles were treated with various agents that either inhibited AChE (eserine or prostigmine) or enhanced the sensitivity of the muscle to ACh (barium, methyl alcohol, ethyl alcohol, acetone) or both. The occurrence of this local contraction (or contracture) provided a ready explanation for the inhibiton of directly elicited contraction by high-frequency nerve stimu-lation—it would interfere with the propagation of the muscle action poten-tial. At the same time, it gave a direct demonstration of the stimulated release of ACh by the motor nerve endings, in keeping with the chemical theory of neuromuscular transmission. It may be noted here that Brown, Dale & Feldberg (1936) had shown that eserine greatly potentiates the twitch response of mammalian muscle to maximal single stimuli applied to the nerve, but stated that Z. M. Bacq in their laboratory was unable to find similar twitch potentiation in frog muscle. When I found the proper experimental conditions for demonstrating eserine potentiation of twitches with the amphibian nerve-muscle preparation, I wrote to inform Dale of my experience, and this resulted in a pleasant correspondence with him.

2. Calcium was shown to have various striking effects on the neuro-muscular junction. Raising the calcium concentration in Ringer's solution was first found to greatly intensify the junctional inhibition. Then calcium was found to be a universal "decurarizing" agent, removing or diminishing the neuromuscular block produced by such diverse agents as the following: (a) drugs: curare, eserine, veratrine, nicotine, atropine, ergotoxin, strych-nine, pilocarpine, and novocaine; (b) fatigue and long survival; (c) acid, strontium, magnesium, and barium; and (d) extreme temperatures. Cal-cium was also found to increase the local contraction or contracture produced by high-frequency nerve stimulation. In an attempt to give a unifying explanaton for all of these various effects, I considered several

possibilities and came to the conclusion that the best hypothesis is that calcium causes each individual nerve impulse to liberate a larger or more concentrated amount of ACh from the nerve terminal.

3. Much effort was spent on the study of the facilitation of neuro-muscular transmission during and after prolonged nerve stimulation at various frequencies, and the post-tetanic facilitation or potentiation of the endplate potential, which could last many minutes, was described for the first time.

4. In mammalian muscles the spontaneous twitchings as well as the potentiated twitches due to stimulation, in the presence of eserine, could be shown to be accompanied by repetitive discharges from the motor nerve endings. Likewise the post-tetanically facilitated or potentiated twitches could be shown to be also accompanied by such repetitive activity. In both cases the repetitive activity was readily suppressed by curare. A new prejunctional aspect of the eserine and curare effects and the post-tetanic effects was thus brought to light.

Throughout my research on the neuromuscular junction during the period 1935–1941, I kept an open mind as to the mechanism of neuro-muscular transmission. Although there was no doubt that the weight of my experimental evidence was on the side of chemical transmission, I regarded all my interpretations as tentative. This seemed appropriate because the neuromuscular junction was then still a "black box." Micro-electrophysiology and electron microscopy, to say nothing of molecular biology, had not yet been applied to its study, and we still knew nothing about how the postulated neurotransmitter was released by the pre-junctional motor nerve endings and how it acted on the post-junctional membrane. New, rapid development in the study of the neuromuscular junction soon followed after the end of the Second World War, which served to make the neuromuscular junction a prototype of chemical syn-apses. The enormous advances achieved during the last three to four decades through the efforts of many people has largely opened up this black box, and analysis is now becoming more and more molecular. There is now a new basis for understanding each of my earlier observations. Yet I feel there is still room for further elucidation, and, if circumstances permit, I would like to make a renewed study of some of these in light of our present-day knowledge.

THE EPISODE OF THE NERVE SHEATH CONTROVERSY

After the interruption of my research on the neuromuscular junction by the War and after my departure from the PUMC at the end of 1941, I eventually made my way to Zhongqing, the War-time capital in the interior of China. I was appointed first as Professor of Physiology in the Shanghai Medical College, which had migrated to Zhongqing, and subsequently as the Acting Director of the Medical Research Institute (Preparatory) of the Academia Sinica. In this latter capacity I visited the United States in 1946 to purchase equipment and books for the new institute and also to look into a number of new scientific developments that were of interest to me. On my arrival in New York I went first to see Dr. Herbert Gasser, then Director of the Rockefeller Institute, to seek his advice about possible arrangements for my stay in the US. He told me that since the War had just ended, most laboratories in the States were not yet fully operational. However, Lorente de Nó's laboratory at the Rockefeller Institute was an exception, as Lorente de Nó had worked uninteruptedly throughout the War. Gasser suggested that I might work for a while with Lorente de Nó, at the same time using the Rockfeller institute as my base while I gathered equipment and books for my own institute. I did as Gasser advised, and I should add that Lorente de Nó was very kind and helpful to me. I ended up spending about a year with him.

Lorente de Nó was then intensively engaged in the study of nerve. His monograph, *A Study of Nerve Physiology*, in two large volumes, was then in the process of publication. He expressed interest in my earlier work on the effect of barium on the neuromuscular junction, and suggested that we do some further work together on the effects of barium on nerve. Lorente de Nó was a very energetic worker, who had very strong opinions on most scientific questions. During my stay at the Rockefeller Institute, I was witness to an especially pointed encounter with Kenneth Cole at a small scientific conference on excitable membranes held at the Institute. But on the whole he and I got along together well, probably because I was a good listener, and avoided getting drawn into arguments where I knew that argument would serve no useful purpose.

There was, however, one question in which I was personally involved. I referred above to the discovery I made in Gerard's laboratory of the connective tissue sheath of the nerve as effective diffusion barrier. Lorente de Nó had written in his book, and also told me pointedly that "it is utterly impossible to believe that the connective tissue sheath of frog or bull-frog nerve could act as a diffusion barrier" (Lorente de Nó 1947). It was evidently useless to be drawn into a verbal argument about a scientific opinion so forcibly expressed. On my return to China in 1947, I and my assistant, Dr. Y. M. Liu reexamined the issue experimentally. We managed to finish two papers (Feng & Liu 1949a,b) reporting our new experiments that supported the original conclusion of Feng & Gerard (1930) about the nerve sheath as an effective diffusion barrier, just before the Liberation of

Shanghai in 1949 as a result of the Civil War, which closely followed the Sino-Japanese War. The papers were sent to the *Journal of Cellular and Comparative Physiology*, and copies were sent to Lorente de Nó at the same time. These papers elicited a long reply from him, published in the same journal (Lorente de Nó 1950). During the following two to three years we published several more papers on the question in a Chinese journal. Space does not permit a full account of the controversy that ensued, but I should put on record the fact that throughout it all Lorente de Nó remained friendly toward me.

When later I was a Regent's Professor in the Department of Physiology, at the University of California, Los Angeles in 1981, Lorente de Nó was in the Department of Anatomy. At an official lunch party given in my honor by the Dean of the Medical School, I had the pleasure of meeting Lorente de Nó again. He told me quietly that it was he who had recommended me for the Regent's Professorship—a piece of news that really made me feel pleased.

FROM THE NEUROMUSCULAR JUNCTION TO NERVE-MUSCLE TROPHIC RELATIONS

After the founding of the Peoples Republic of China (PRC) and the establishment of the Chinese Academy of Sciences, my adminstrative and organizational duties and social activities multiplied and kept me distracted from continuous application to scientific work. During the first 30 years of the PRC there were frequent political upheavals, culminating in the so-called "Cultural Revolution," which upset all normal life throughout the country. This "revolution" was bad in every sense, but it had the dialectical virtue of preparing the country for a radical change. This change came in 1979, ushering in a new age in China characterized by far-reaching reforms in all spheres of our national life and opening the country once more to the outside world.

In 1961, after the disturbances of the preceding period—during the so-called "Great Leap Forward Movement" had subsided and the Institute of Physiology of the Chinese Academy of Sciences where I worked seemed to be ready to resume normal scientific life, I planned to start a new research program. The neuromuscular junction which I had studied before had by that time become an active field of research in which many people were engaged. The neuromuscular junction is the locus of the brief, fast events associated with neuromuscular transmission and it was with these that most researchers in the field were concerned. However, it is also the site at which the slower events or trophic transactions between nerve and muscle take place, and at that time these events were receiving much less attention. Partly because of the limitation of my equipment and technical resources, and partly because I desired to do something quite different from what I had done before, I decided to study these slow events.

In trying to initiate a new program of research, there is always the question: How to make an effective start? In my experience, an effective start means discovering something new that opens your eyes to the possibility of many more experiments. This was how my earlier research on the neuromuscular junction had got started. Now how should I begin the study of nerve-muscle trophic relations? I decided to begin by looking into the old phenomenon of muscle atrophy following denervation.

It was known at the time that slow and fast muscles atrophy at somewhat different rates after denervation. Hoping to see this difference in its most striking form, I chose for my first experiments two muscles with the most strikingly different contraction speeds. I happened to know that the anterior and posterior latissimi dorsi (ALD and PLD) muscles on the back of the chick, have very different contraction speeds, ALD being a slow tonic muscle and PLD a fast twitch muscle. Furthermore, these two muscles receive their innervation from the same nerve trunk, thus making their simultaneous denervation a very simple operation. I accordingly asked one of my assistants to do the experiment of cutting the nerve to the ALD and PLD, and to compare their degree of atrophy one month later. The result of this first exploratory experiment was a great surprise: while the PLD showed striking atrophy as I had expected, the ALD not only showed no atrophy but was strikingly hypertrophied! When my assistant, Dr. W. Y. Wu, reported this result to me, I was so excited that I could not help yelling at him: "Either you have made a very bad mistake (meaning he might have inadvertently exchanged the control for the experimental ALD) or you have made a discovery!" Indeed it was a discovery (Feng, Hung & Wu 1962). This most unusual phenomenon of post-denervation hypertrophy was quickly found to be a general phenomenon and not peculiar to the ALD. In fact it occurs in all multiply innervated slow tonic fibers that are widely distributed in many mixed muscles in the chick. It is long-lasting; the post-denervation hypertrophy of the slow fibers in the chick sartorius muscle lasted for at least six months. It is a specifically post-denervation phenomenon and is not due to some extraneous factor such as stretch incidentally accompanying denervation. These two characteristics set it apart from the transient hypertrophy of the rat hemidiaphragm following denervation, previously described by Sola & Martin (1953). This so-called denervation hypertrophy of the rat hemidiaphragm, is transient and was shown (Feng & Lu 1965) to be due to stretch by the opposite intact and rhythmically contracting hemidiaphragm and not simply due to denervation. Unfortunately this work suffered a fate similar to my earlier research on the neuromuscular junction, being again cut short by "war"—this time a very strange kind of internal "war" called "the Cultural Revolution," which lasted for 10 years!

LOOKING FORWARD

Since the dawn of the new age of socialist reconstruction in China in 1979, the whole country has made a surprisingly rapid recovery from the devastation of the preceding decade. Science and technology are now more highly esteemed in China than ever before. And the Shanghai Institute of Physiology, in which my own laboratory is located, has also undergone a phase of regeneration and rennovation. We have resumed and extended our research on nerve-muscle trophic relations (Feng 1986) and are now preparing to launch new experimental ventures. During the past eight years China has had a record of steady progress, which we believe will continue for a long time. I feel happy to have lived long enough to see this new period of constructive development in my country. Though already 80, I look forward with renewed excitement to carrying out new experiments.

ACKNOWLEDGMENTS

The author wishes to thank Y. Cheng and W. D. Li for help in the preparation of this manuscript.

Literature Cited

Feng, T. P., Gerard, R. W. 1930. Mechanism of nerve asphyxiation: With a note on the nerve sheath as a diffusion barrier. *Proc. Exp. Biol. Med.* 27: 1073–76

Feng, T. P. 1932a. The effect of length on the resting metabolism of muscle. *J. Physiol.* 74: 441–54

Feng, T. P. 1932b. The role of lactic acid in nerve activity. *J. Physiol.* 76: 477–86

Feng, T. P. 1933. Reversible inexcitability of tactile endings in skin injury. *J. Physiol.* 79: 103–8

Feng, T. P. 1936. The heat production of nerve. *Ergebn. d. Physiol.* 38: 73–132

Feng, T. P., Liu, Y. M. 1949a. The connective tissue sheath of the nerve as effective diffusion barrier. *J. Cell. Comp. Physiol.* 34: 1–16

Feng, T. P., Liu, Y. M. 1949b. The concentration-effect relationship in the depolarization of amphibian nerve by potassium and other agents. *J. Cell Comp. Physiol.* 34: 33–42

Feng, T. P., Jung, H. W., Wu, W. Y. 1963. The contrasting trophic changes of the anterior and posterior latissimus dorsi of the chick following denervation. In *The Effect of Use and Disuse on Neuromuscular Functions*, ed. E. Gutmann, P. Hnik, pp. 431–41. Prague: Czechoslovak Acad. Sci.

Feng, T. P., Lu, D. X. 1965. New lights on the phenomenon of transient hypertrophy in the denervated hemidiaphragm of the rat. *Sci. Sin.* 14: 1772–84

Feng, T. P. 1986. *The neural determination of skeletal muscle fiber characteristics.* Invited Lecture, 30th Int. Congr. Physiol. Sci., Vancouver, Canada

Lorente de Nó, R. 1947. A study of nerve physiology. *Studies Rockefeller Inst. Med. Res.* 131: 23–24, 58, 132

Lorente de Nó, R. 1950. The ineffectiveness of the connective sheath of nerve as a diffusion barrier. *J. Cell. Comp. Physiol.* 35: 195–240

Sola, O. M., Martin, A. W. 1953. Denervation hypertrophy and atrophy of hemidiaphragm of rat. *Am. J. Physiol.* 172: 324–32

Ann. Rev. Neurosci. 1988. 11 : 13–28

TACHYKININS

J. E. Maggio

Department of Biological Chemistry and Molecular Pharmacology,
Harvard Medical School, Boston, Massachusetts 02115

Introduction

Among the many active principles discovered during the first half of the twentieth century, substance P was the first found in neural tissue that proved to be a peptide. In that sense, substance P is the first neuropeptide, and decades of research and thousands of publications since its discovery have made it arguably the best characterized of the brain-gut peptides. Only recently has it been generally recognized that substance P is but one member of a structurally related family of bioactive peptides, the tachykinins. The literature on substance P has been extensively reviewed elsewhere (Jessell 1983, Jordan & Oehme 1985, Nicoll et al 1980, Pernow 1983, Aronin et al 1983, Barthó & Holzer 1985, Skrabanek & Powell 1977, 1980, 1983a,b); the present article concentrates instead on the tachykinins as a peptide family.

The tachykinins are an active field of research; several hundred new papers enter the literature each year. Much of past work has focused only on substance P, but it is now clear that the exclusion of the other tachykinins from consideration can lead to a picture that is incomplete at best (Erspamer 1981, Maggio 1985). Tachykinins are widely distributed and active in both the CNS and periphery, evoking a variety of biological responses in a variety of tissues. The members of this peptide family excite neurons, evoke behavioral responses, are potent vasodilators and secretagogues, and contract (directly or indirectly via release of classical transmitters) many smooth muscles. Of particular interest is the apparent role of tachykinins in the striatonigral system and in primary sensory afferent transmission and nociception. Of recent and more speculative interest is the possibility that tachykinins may act as growth factors or as messengers between the nervous and immune systems.

13

0147–006X/88/0301–0013$02.00

History

Almost 60 years ago, while mapping the tissue distribution of the newly characterized transmitter acetylcholine, von Euler & Gaddum (1931) discovered in acid ethanol extracts of equine intestine and brain a hypotensive and spasmogenic activity that was clearly different from acetylcholine in that it contracted the rabbit jejunum in the presence of atropine. The active component acquired the name substance P (for preparation), and early work suggested it was proteinaceous in nature. Although it was the object of intense investigation, substance P eluded complete purification and sequencing for four decades. We know now that the tissue extracts known as substance P during this period often contained other bioactive peptides (notably other tachykinins) as impurities; nevertheless many of the major findings of that era proved to be substantially correct.

Two decades later, while screening for biogenic amines, Erspamer (1949) discovered in extracts of the posterior salivary glands of the Mediterranean octopus *Eledone moschata* a hypotensive, sialogogic, and spasmogenic activity that could not be attributed to known substances. The active component, eledoisin, was recognized as a polypeptide with certain similarities to substance P; but eledoisin was present in the cephalopod salivary gland in much higher concentrations than is substance P in mammalian brain. These high concentrations contributed greatly toward making eledoisin the first tachykinin to be sequenced (Erspamer & Anastasi 1962); nevertheless, the scale of purification required (starting material 1450 kg of live octopus) was heroic.

The structure of eledoisin (Table 1) had the notable feature of blocked termini (it lacks free amino and carboxyl ends), which would later be found in many bioactive peptides. It was recognized that eledoisin, although similar to substance P in pharmacological activities and protease sensitivity, could readily be distinguished from the mammalian peptide by parallel bioassays. That is, in spite of the qualitative similarity of their activities, there were significant quantitative differences in the relative potencies of the two peptides in various mammalian systems (Erspamer 1971, Erspamer & Anastasi 1962).

The same group later purified and sequenced the amphibian peptides physalaemin and phyllomedusin (Erspamer et al 1964, Anastasi & Falconieri Erspamer 1970), which shared with eledoisin a similar spectrum of biological activities and a similar chemical structure. Because the active peptides showed a characteristic fast onset of action on tissues of the gut (as compared to the slower acting bradykinins), the term "tachykinin" was coined for peptides of the group. Systematic pharmacological study of synthetic analogues of the peptides (Erspamer & Melchiorri 1973) showed that most of their biological activities depended on the conserved carboxyl-terminal sequence. This pattern led Erspamer (1971) to correctly predict before substance P was sequenced that the mammalian peptide would share with the nonmammalian tachykinins the carboxyl-terminal sequence -Phe-X-Gly-Leu-Met-NH$_2$. Other nonmammalian tachykinins have been subsequently isolated, and the process may be expected to continue; tachykinins probably exist in all phyla from coelenterates to vertebrates. The sequences of the tachykinins described to date are given in Table 1. As discussed below, studies of the nonmammalian tachykinins have been essential to understanding the roles of endogenous tachykinins in mammals.

In the late 1960s, while attempting to isolate a corticotropin-releasing factor from bovine hypothalamus, Leeman & Hammerschlag (1967) discovered a peptide sialogogue that was not blocked by cholinergic or adrenergic antagonists. The sialogogic peptide was purified to homogeneity, recognized as substance P, and sequenced a few years later (Chang & Leeman 1970, Chang et al 1971). Substance P was rediscovered yet again in the early 1970s by Otsuka et al (1972) during a search for sensory neurotransmitters in extracts of bovine dorsal roots.

The availability of synthetic substance P predictably led to an explosion of interest in substance P research. The idea that peptides might transmit information between cells in the nervous system was fairly new in the mid-1970s, and there were few other neuropeptides known at that time; thus substance P was the subject of many of the early attempts to understand the roles of these novel candidate transmitters. In the midst of the great activity in substance P research, interest in the other tachykinins waned on the basis of phylogenetic discrimination, and the assumption was generally made that substance P was the only tachykinin in mammals.

That assumption was not proven invalid until 1983, when four independent groups using three entirely different approaches reported that two more tachykinins exist in mammalian tissue. Maggio et al (1983) adopted the working hypothesis that tachykinin pharmacology could be more economically explained by the existence of a mammalian tachykinin bearing an aliphatic residue at position X in the canonical sequence, and developed an immunochemical assay to search for this hypothetical peptide. They found in extracts of bovine spinal cord a novel cationic tachykinin with a valine residue at this position and named it substance K to reflect its structural homology with the amphibian tachykinin kassinin. The peptide named substance K by these workers was first sequenced by Kimura et al (1983). The latter group had been searching for novel peptides in extracts of porcine spinal cord, using spasmogenic activity on guinea-pig ileum as the assay. They reported the sequence of a novel cationic tachykinin they named neurokinin α, which later proved to be identical to bovine substance K, and the sequence of a novel anionic tachykinin named neurokinin β.

At about the same time, Kangawa et al (1983) were employing a similar strategy in searching for novel mammalian neuropeptides. Among the new peptides sequenced by this group were a novel anionic tachykinin named neuromedin K, which proved identical to neurokinin β, and a cationic tachykinin named neuromedin L (Minamino et al 1984a), which proved identical to neurokinin α or substance K. During the same period, Nawa et al (1983) were at work on the structure of the substance P gene; they revealed the sequence of a mRNA from bovine striatum which encoded not only substance P but also a novel tachykinin which they named substance K. This novel tachykinin sequence proved identical to neurokinin α, neuromedin L, and substance K previously described.

Nomenclature

Reluctant to wait for convergent evolution to settle on a single name for each of the novel peptides, nomenclature committees (Jordan & Oehme 1985, Henry 1987) have attempted to reduce confusion by changing the names chosen by the groups who discovered the peptides. Thus neurokinin A was proposed to replace substance K, neurokinin α, and neuromedin L; neurokinin B was proposed to replace neurokinin β and neuromedin K. To date, the large majority of the published literature refers to the novel mammalian tachykinins as substance K and neuromedin K, but neurokinin A and neurokinin B have recently been gaining in popularity. Thus far, attempts to rename substance P have met with little success.

There are in use at present several systems for classifying and naming tachykinin receptors (Erspamer 1981, Lee et al 1982, Piercey et al 1985, Laufer et al 1985, Buck et al 1984, Henry 1987, Melchiorri & Negri 1984). Most are based on comparisons of agonist potency and do not distinguish between central and peripheral tissues. Specific names for such broad and imprecise classifications tend to mislead rather than enlighten, and thus they are avoided here as much as possible.

Caveats

Peptides are the most recently discovered major class of intercellular messengers and the least completely understood. The temptation to have peptides conform to models of action based on the much better defined classical neurotransmitters has proven difficult to resist, but it is becoming increasingly clear that acetylcholine at the neuromuscular junction is a poor model for the action of many neuropeptides. In many fundamental ways, such as pathways of biosynthesis and receptor coupling, neuro-peptides more closely resemble protein hormones than classical neuro-transmitters.

The very low concentrations of tachykinins and their receptors in tissue (orders of magnitude less than small molecule transmitters) have meant that much of what is known about them has been learned through indirect methods. Recent work has shown that some of the tools used in past studies (e.g. antisera, antagonists, radioligands) lack the speci-ficity to distinguish rigorously between the various tachykinins and tachykinin receptors found in tissue. With hindsight, it is clear that the interpretation of some past experiments could be profitably reexamined (vide infra).

Structure-Activity Relationships

The empirical method of structure-activity studies has been the principal technique for examining the interactions between tachykinins and their receptors. Enough analogues have been examined that some clear patterns have emerged, although the interpretation of these patterns is not so straightforward.

The tachykinins (Table 1) share both the spectrum of biological activities that initially defined the family and the common structural feature that now defines it: the conserved carboxyl-terminal sequence -Phe-X-Gly-Leu-Met-NH_2, where X is an aromatic (Phe, Tyr) or branched aliphatic (Val, Ile) amino acid. Most of the biological activities of the tachykinins depend on the integrity of this sequence (Couture et al 1979, Erspamer & Mel-chiorri 1973, Lee et al 1982, Pernow 1983, Sandberg 1985, Watson et al 1983). It is not true that the amino-terminal portion of the peptides is either without activity or without effect on the activity of the C-terminal sequence, but in general carboxyl fragments of six amino acids or longer retain tachykinin bioactivity. N-terminal fragments do not; removal of the methionineamide or even the C-terminal amide group abolishes activity, although the amide may be replaced with other uncharged functional groups without such a loss. In some systems, the partial C-terminal sequences of tachykinins have been reported to be more potent than the full length peptides. Oxidation of the methionine residue to the sulfoxide form results in a partial loss of activity.

While the tachykinins share a similar spectrum of biological activities, they are distinguished by quantitative differences in their relative potencies in various bioassays. A major determinant of that specificity (Erspamer 1981, Lee et al 1982) is the nature of the amino acid at position X. That is, the tachykinins can be divided into two broad categories according to whether this position is occupied by an aromatic or aliphatic amino acid.

Table 1 Tachykinin sequences[a]

Mammalian	
Substance P[b]	H-Arg -Pro -Lys -Pro-Gln -Gln -PHE -Phe-GLY-LEU -MET-NH$_2$
Substance K/neurokinin A[c,p]	H-His -Lys -Thr-Asp -Ser -PHE -Val-GLY-LEU -MET-NH$_2$
Neuromedin K/neurokinin B[d,p]	H-Asp -Met -His-Asp -Phe -PHE -Val-GLY-LEU -MET-NH$_2$
Nonmammalian	
Scyliorhinin I[e]	H-Ala -Lys -Phe-Asp -Lys -PHE -Tyr-GLY-LEU -MET-NH$_2$
Physalaemin[f]	pGlu-Ala -Asp -Pro-Asn -Lys -PHE -Tyr-GLY-LEU -MET-NH$_2$
Lys^{5}Thr^6-physalaemin[g]	pGlu-Ala -Asp -Pro-Lys -Thr -PHE -Tyr-GLY-LEU -MET-NH$_2$
Uperolein[h]	pGlu-Pro -Asp -Pro-Asn -Ala -PHE -Tyr-GLY-LEU -MET-NH$_2$
SP-like peptide [i,k]	[H-Asp -Ile -Pro -Lys -Pro-Asp -Gln -PHE -Phe-GLY-LEU -MET-NH$_2$]
Hylambatin[q]	H-Asp -Pro -Pro -Asp -Pro-Asp -Arg -PHE -Tyr-GLY-Met -MET-NH$_2$
Entero-hylambatin[k]	H-Asp -Pro -Pro -Asn -Pro-Asp -Arg -PHE -Tyr-GLY-Met -MET-NH$_2$
Phyllomedusin[l]	pGlu-Asn -Pro-Asn -Arg -PHE -Ile -GLY-LEU -MET-NH$_2$

Scyliorhinin II[e]

```
                                  S─────────────────S
                                  |                 |
H-Ser-Pro-Ser-Asn-Ser -Lys-Cys -Pro -Asp -Gly -Pro-Asp -Cys -PHE -Val-GLY-LEU -MET-NH2
```

Glu^{2}Pro^5-kassinin[j,q]	H-Asp -Glu -Pro -Lys -Pro-Asp -Gln -PHE -Val-GLY-LEU -MET-NH$_2$
Kassinin[m]	H-Asp -Val -Pro -Lys -Ser-Asp -Gln -PHE -Val-GLY-LEU -MET-NH$_2$
Entero-kassinin[k]	H-Asp -Glu -Pro -Asn -Ser-Asp -Gln -PHE -Ile -GLY-LEU -MET-NH$_2$
Eledoisin[n]	pGlu-Pro -Ser -Lys-Asp -Ala -PHE -Ile -GLY-LEU -MET-NH$_2$

Canonical tachykinin sequence[o]   PHE -X -GLY-LEU -MET-NH$_2$

---

[a] H- denotes a free N-terminal α-amino group, -N$_2$ a C-terminal α-carboxyamide, pGlu- an N-terminal pyroglutamic acid moiety; references are to sequence determination.
[b] Chang et al (1971).
[c] Alternate names: neurokinin α, α-neurokinin, neuromedin L.
[d] Alternate names: neurokinin β, β-neurokinin.
[e] Conlon et al (1986).
[f] Erspamer et al (1964).
[g] Alternate name: rugosauperolein II. Nakajima et al (1980).
[h] Anastasi et al (1975).
[i] Presumed sequence.
[j] Alternate name: Hylambates-kassinin.
[k] Melchiorri & Negri (1984).
[l] Anastasi & Falconieri Erspamer (1970).
[m] Anastasi et al (1977).
[n] Erspamer & Anastasi (1962).
[o] X is an aromatic or branched aliphatic amino acid.
[p] Kimura et al (1983).
[q] Yasuhara et al (1981).

Thus eledoisin, kassinin, substance K/neurokinin A and neuromedin K/neurokinin B (X aliphatic) are somewhat similar to each other in terms of their potencies in various bioassays (and their affinities for various tachykinin receptors), but quite different from substance P and physalaemin (X aromatic). The variable amino termini of the peptides determines more subtle differences between the various individual tachykinins.

Biosynthesis

Unlike classical transmitters, which are synthesized by enzymes in nerve terminals, neuropeptides are synthesized ribosomally as larger protein precursors in the neuronal cell body and shipped to the terminals as enzymatic processing converts them to mature forms. Thus the sites of biosynthesis and release can be a considerable distance apart, and peptides depleted from a particular terminal may not be quickly replenished.

Recent work in molecular genetics (Kotani et al 1986, Nawa et al 1983, 1984) has contributed greatly to our understanding of tachykinin biosynthesis in mammals. Two remarkably similar genes encode the three known mammalian tachykinins (Figure 1). The preprotachykinin A (PPT-A) gene encodes both substance P (exon 3) and substance K/neurokinin A (exon 6), while the preprotachykinin B gene encodes neuromedin K/neurokinin B (exon 5). Alternative splicing of the transcripts of the former gene gives rise to three mRNAs encoding distinct tachykinin precursor proteins (α-, β-, γ-preprotachykinins); the first of these encodes only substance P and is a minor component, whereas the latter two encode both substance P and substance K/neurokinin A. In the precursor proteins, the C-terminal methionines of the substance P, substance K/neurokinin A, and neuromedin K/neurokinin B sequences are followed immediately by glycine, which is processed to give the essential C-terminal carboxyamide moiety of the mature tachykinins. Furthermore, the sequences are flanked

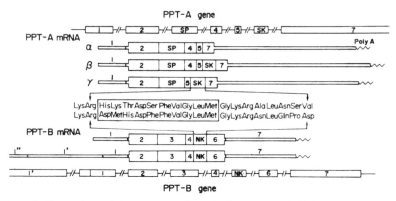

Figure 1 Comparison of the structure of two bovine preprotachykinin genes and their mRNAs. *Larger boxes* represent protein-coding exons; boxes labeled SP, SK, NK represent exons encoding substance P (exon 3), substance K (exon 6), neuromedin K (exon 5), respectively. Substance K and neuromedin K are enclosed in the preprotachykinin partial amino acid sequences displayed in the center of the diagram. Reproduced with permission from Kotani et al (1986).

on each end by paired basic residues, which are targets of processing proteases. Since exon 3 of the PPT-A gene encodes substance P, it is interesting that exon 3 of the PPT-B gene encodes a dodecapeptide sequence flanked by paired basic residues, but it is not a tachykinin sequence.

Intermediates in proteolytic processing of tachykinin precursors can be detected (Kream et al 1985). It is difficult to exclude the possibility that one or more of these intermediates is an active peptide in its own right. Such a claim has been advanced for neuropeptide K, a 36-amino acid peptide with the substance K/neurokinin A sequence at its carboxyl terminus (Tatemoto et al 1985). Since this peptide, which could be a product of β-preprotachykinin, has an intact tachykinin sequence, it would be expected to have some biological activity; whether neuropeptide K is more than an intermediate in substance K/neurokinin A biosynthesis has not been established.

Distribution and Location

Many conclusions about the function of substance P in the mammalian nervous system have been largely based on the presence of immunoreactive substance P within specific microscopic structures (immunohistochemistry) and in extracts of various neural tissues (radioimmunoassay, RIA). That is, if a particular brain region contains a high concentration of substance P-like immunoreactivity, then substance P is likely to be involved in the activities of that region. There are two major potential pitfalls in such reasoning. The first is that the substance P-like immunoreactivity may or may not be substance P; the second is that the presence of substance P need not imply the presence of a receptor for substance P at any particular site, and both the ligand and the receptor are required for activity.

Nearly all antisera employed in research on substance P have been polyclonals directed against the C-terminal portion of the tachykinin. This regional specificity results from the structure of substance P; the N-terminal part of the peptide contains the functional groups through which it is coupled to a protein in the immunogenic complex, leaving the C-terminal portion presented for antibody formation. The regional specificity is unfortunate for discriminating between the various tachykinins, since the C-terminal part of the peptides is most similar across the family (Table 1).

For RIAs, it seems likely that most measurements of substance P-like immunoreactivity will not be significantly compromised by crossreactivity with the aliphatic mammalian tachykinins substance K/neurokinin A and neuromedin K/neurokinin B. The levels of these latter peptides are generally lower than those of substance P, and in many cases crossreactivities have been determined to be within experimental uncertainty. Nevertheless, a rigorous approach to measurement of tachykinins in tissue extracts (Kanazawa et al 1984, Maggio & Hunter 1984, Minamino et al 1984b, Ogawa et al 1985) requires characterization of both the radioimmunoassays (crossreactivity) and the immunoreactive components (chromatography).

The specificity requirements for antisera used in immunohistochemistry are much more stringent than those for antisera used in RIAs. The radioactive tracer contributes a great deal of specificity in RIA by selecting from the whole polyclonal antiserum a particular small subpopulation of antibodies with the highest affinity for that tracer; the other antibodies are silent. No equivalent selection takes place in immunohistochemistry, and RIA specificity is in no sense a guarantee of specificity in staining sections (Maggio 1985). Tachykinin immunohistochemistry requires both adsorption and cross-adsorption controls (Dalsgaard et al 1985, Deutch et al 1985) if the members of the peptide family are to be distinguished from one another. Staining by a C-terminal antiserum against one tachykinin is not, by itself, acceptable evidence for the presence of that particular tachykinin in the region of interest. That caveat applies to monoclonal as well as polyclonal antibodies; monoclonals are necessarily homogeneous, but not necessarily specific.

The structure of the preprotachykinin genes (Figure 1) suggests but does not require a similar regional distribution for substance P and substance K/neurokinin A, and to a first approximation this appears to be the case. In several areas, these two tachykinins are colocalized, cosynthesized, and/or coreleased (Deutch et al 1985, Dalsgaard et al 1985, Lindefors et al 1985, J.-M. Lee et al 1986, Maggio & Hunter 1984). Comparatively little is known about the regional distribution of neuromedin K/neurokinin B, but it appears to be quite different from that of substance P and substance K/neurokinin A (Minamino et al 1984b, Ogawa et al 1985, Kanazawa et al 1984).

Tachykinin Receptors

It has been known for many years that there were significant differences between various tachykinins and their fragments in bioassays. That observation is consistent with the existence of multiple types of tachykinin receptors, but cannot exclude many other possible interpretations. Receptor classification for classical transmitters has been largely based on the use of specific high affinity antagonists; for the tachykinins, and for nearly all active peptides, no such antagonists have yet been discovered. The best tachykinin antagonists reported to date are neither specific nor high affinity

(Hakånson & Sundler 1985, Zachary & Rozengurt 1986). Tachykinin receptors have been classified mainly through the use of agonists in pharmacological experiments; more recently, radioligands have also been employed.

Before the discovery of substance K/neurokinin A and neuromedin K/neurokinin B, the general interpretation of the potency differences among the tachykinins and their fragments in bioassays on mammalian tissues was that multiple subtypes of substance P receptors existed in those tissues. The most popular classification divided mammalian peripheral tissues into two categories based on tachykinin agonist potency. In one group of tissues, the tachykinins tested were roughly equipotent; in the other group of tissues, tachykinins bearing an aliphatic residue at position X were considerably more potent than tachykinins with an aromatic residue at this position (Lee et al 1982). The former group of tissues was said to contain substance P receptors of the SP-P type and the latter group substance P receptors of the SP-E type, *P*hysalaemin and *E*ledoisin being the key discriminant compounds. The division of putative substance P receptors according to the nature of the amino acid at X, position 4 from the C-terminus, was also consistent with the central activities of tachykinins and certain other pharmacological data (Maggio 1985). Recently it has become clear that the SP-E receptor is not a receptor for substance P, but instead a receptor for an aliphatic mammalian tachykinin. Furthermore, it is now recognized that more than two types of tachykinin receptors are found in mammalian tissue (C.-M. Lee et al 1986, Henry 1987, Buck et al 1984, Laufer et al 1985). Nevertheless, the division of mammalian tachykinins and tachykinin receptors into two broad categories based on the nature (aliphatic or aromatic) of the amino acid at position X remains valid. No standard nomenclature for identifying specific subtypes of tachykinin receptors within these broad classifications has yet become convention.

A major advance in tachykinin research came with the development of techniques and high specific activity ligands for studies of substance P binding sites (Too & Hanley 1984), and the binding characteristics and regional distribution substance P receptors are now fairly well established (Hanley et al 1980, Mantyh et al 1984a, Shults et al 1984). As noted, pharmacological data (Hunter & Maggio 1984, Kimura et al 1984, Deutch et al 1985) clearly suggest that substance K/neurokinin A and neuromedin K/neurokinin B receptors are distinct from substance P receptors, and the results of homogenate binding and autoradiographic studies with radiolabeled aliphatic tachykinins confirm this hypothesis (Burcher et al 1984, Buck et al 1984, Cascieri et al 1985, Mantyh et al 1984b, Henry 1987, Beaujouan et al 1986). The fact that binding sites are not necessarily receptors must always be kept in mind when comparing data on radio-ligand binding and pharmacological activity; so must considerations of the particular radioligands and conditions used and the possibility of variation between species.

There is no apparent correlation across all CNS regions between the density of substance P innervation and the density of substance P binding sites (Shults et al 1984, Mantyh et al 1984a). This mismatch was unexpected and took some time to find general acceptance, but there is no *a priori* reason for the regional distribution of a transmitter and its receptor to be identical. In recent years, several other examples of mismatch between the distributions of a peptide and its binding sites have appeared. As time goes on, more and more examples of transmitter coexistence within single neurons are discovered (tachykinins have been found colocalized with several other active peptides and small molecules, notably calcitonin gene-related peptide and serotonin), and it is clear that the presence or absence of postsynaptic receptors is critical in determining the response to released transmitter.

The highest density of substance P (and substance K/neurokinin A) in the brain is found in the substantia nigra, where the peptide is present in fibers and terminals, and can be released by depolarization in a calcium-dependent manner (Nicoll et al 1980, Aronin et al 1983). At one time these findings were interpreted as evidence that substance P must be active at substance P receptors in the nigra, and it was shown that iontophoretic application of substance P could, under some circumstances, excite cells in the area. It now seems quite clear that the substantia nigra contains very few substance P binding sites (Mantyh et al 1984a, Shults et al 1984) and has one of the lowest densities of substance P receptors in the brain. The electrophysiological activity of the exogenous peptide in the area would thus appear to be the result of crosstalk at another receptor or experimental artifact. There is a high density of aliphatic tachykinin binding sites in the pars compacta (Mantyh et al 1984b). The precise role of the tachykinins in the nigrostriatal and other catecholamine pathways has not yet been elucidated (Deutch et al 1985, Bannon et al 1986, Innis et al 1985).

The cerebral cortex shows the inverse situation. Here the concentration of substance P is one of the lowest found in the brain, about 100-fold lower than in the nigra, but the density of substance P binding sites is quite high, particularly in the outer laminae. (The inner laminae of the cortex show a high density of aliphatic tachykinin binding sites.) Other parts of the nervous system (e.g. hypothalamus) show high densities of both substance P and substance P binding sites.

A large fraction of the binding and autoradiographic studies on tachykinin receptors have employed peptides labeled with ^{125}I, because they are easier to synthesize and have higher specific activities than tritiated radioligands. Often the radioiodine is conjugated to the tachykinin as a Bolton-Hunter group, that is a 3-[3-iodo-4-hydroxyphenyl]propionyl group on what was a free amino group in the native peptide. For a peptide the size of a tachykinin, that modification makes a considerable difference in charge, size, and hydrophobicity, and the radioligand may not have the same binding specificity as the native peptide. Recently it has become clear that the Bolton-Hunter modification (C.-M. Lee et al 1986, Henry 1987) is a large enough perturbation that one cannot assume that the specificity of the labeled tachykinin is the same as that of the unlabeled in any system. This caveat is particularly important in experiments on the receptors for aliphatic tachykinins, where addition of the Bolton-Hunter group to a tachykinin may yield a peptide with considerably greater affinity for the receptor than that of the native peptide. Thus while it is clear that aromatic and aliphatic tachykinin receptors are different, it is much more difficult at present to classify the latter. The Bolton-Hunter conjugates of kassinin, eledoisin, substance K/neurokinin A and neuromedin K/neurokinin B all give similar patterns (which are very distinct from the pattern of substance P binding sites) in rat brain autoradiograms, but it is not clear whether the receptors so labeled are heterogeneous, or unselective, or both. Neuromedin K/neurokinin B is presently the leading candidate for physiological ligand at most central aliphatic tachykinin binding sites, while substance K/neurokinin A is the leading candidate at most peripheral aliphatic tachykinin binding sites (Beaujouan et al 1986, Henry 1987, Burcher & Buck 1986, C.-M. Lee et al 1986, Buck et al 1984). But pharmacological experiments have made it quite clear that in vitro, one exogenously applied tachykinin can act as an agonist at another's receptor (crosstalk); tachykinin receptors are not fastidiously specific.

A direct adenyl cyclase response to a tachykinin has not yet been convincingly demonstrated. On the other hand, tachykinins have been shown to evoke a phosphatidylinositol hydrolysis response in several systems (Hanley et al 1980, Mantyh et al 1984c, Watson & Downes 1983, Watson 1984, Hunter et al 1985). The details of the post-receptor intracellular transduction events evoked by the various tachykinins are not known, but it seems likely that a phospholipase C is involved.

Release and Inactivation

If peptides are to fit into the classical model of neurotransmission, then mechanisms for their release and for ending their action must be present. Several studies have shown that a calcium-dependent release of tachykinin-like immunoreactivity can be evoked from nerve terminals (Lindefors et al 1985). Proteolytic degradation is the most popular proposed inactivation mechanism, but no neuropeptide-degrading enzyme analogous to acetylcholinesterase in location and specificity has yet been demonstrated; the fact that central and peripheral tissues contain peptidases capable of degrading neuropeptides does not establish that these enzymes play the physiological role of inactivating released peptides. There are candidate enzymes for inactivation of tachykinins (Krause 1985), but convincing evidence that any of these is the physiological inactivator of released tachykinins is not yet available.

Conclusion

At present three tachykinins and at least that many tachykinin receptors are known to exist in mammals. These numbers should be regarded as lower limits; there is no compelling evidence excluding either more ligands or more receptors, and perhaps the tachykinin family will eventually challenge the opiate family in complexity. The recent discovery that substance P is not the only mammalian tachykinin has led to the realization that some of the biological functions once thought to be mediated by substance P may in fact be mediated by another tachykinin, since some of the tools in past substance P research are simply not specific enough to discriminate between the tachykinins. An important goal for the near future is the development of antisera, antagonists, and radioligands that are specific enough to discriminate between the mammalian tachykinins and tachykinin receptors.

The conclusion that substance P mediates a particular biological activity has often been based on some of the following criteria: (a) exogenous substance P elicits the subject effect (mimicry), (b) substance P antagonists attenuate the subject activity, (c) immunohistochemistry shows immunoreactive substance P at the appropriate location (presence), (d) substance P is released by stimulation of the appropriate tissue, (e) binding and/or autoradiographic studies show that substance P receptors are present at the appropriate target site. In light of recent findings, it is clear that all of the above criteria taken together do not provide compelling evidence for the conclusion. Specifically: (a) because of ligand/receptor crosstalk, exogenous substance P can evoke a response that is physiologically mediated by another tachykinin, (b) the substance P antagonists currently available are not specific and antagonize the actions of all tachykinins, (c) most substance P antisera are C-terminal directed polyclonals and could be staining other tachykinins as well; even if cross-adsorption controls established complete specificity for substance P, another tachykinin may be present within the same neurons, (d) substance P and substance K/neurokinin A are read from the same gene and colocalized within cells; at least in some cases the two are coreleased, (e) although different tachykinin receptors have different regional distributions, many regions have a high density of more than one: just as a cell may express more than one tachykinin, it may express more than one tachykinin receptor.

While some conclusions about the role of substance P may no longer strictly apply to that one member of the tachykinin family, most still apply to the tachykinins as a group. The growing recognition that substance P is not only an interesting neuropeptide but also a member of an interesting peptide family leaves a solid foundation for future research.

Literature Cited

Anastasi, A., Erspamer, V., Endean, R. 1975. Structure of uperolein, a phy-salaemin-like endecapeptide occurring in the skin of *Uperoleia rugosa* and *Uperoleia marmorata. Experientia* 31: 394–95

Anastasi, A., Falconieri Erspamer, G. 1970. Occurrence of phyllomedusin, a phy-salaemin-like decapeptide in the skin of *Phyllomedusa bicolor. Experientia* 26: 866–67

Anastasi, A., Montecucchi, P., Erspamer, V., Visser, J. 1977. Amino acid com-position and sequence of kassinin, a tachy-kinin dodecapeptide from the skin of the African frog *Kassinia senegalensis. Experientia* 33: 857–58

Aronin, N., Difiglia, M., Leeman, S. E. 1983. Substance P. In *Brain Peptides*, ed. D. T. Krieger, M. J. Brownstein, J. B. Martin, pp. 783–803. New York: Wiley. 1032 pp.

Bannon, M. J., Lee, J.-M., Giraud, P., Young, A., Affolter, H.-U., et al. 1986. Dopamine antagonist haloperidol de-creases substance P, substance K and preprotachykinin mRNAs in rat stri-atonigral neurons. *J. Biol. Chem.* 261: 6640–42

Barthó, L., Holzer, P. 1985. Search for a physiological role of substance P in gas-trointestinal motility. *Neuroscience* 16: 1–32

Beaujouan, J. C., Torrens, Y., Saffroy, M., Glowinski, J. 1986. Quantitative auto-radiographic analysis of the distribution of binding sites for [^{125}I]Bolton-Hunter derivatives of eledoisin and substance P in the rat brain. *Neuroscience* 18: 857–76

Buck, S. H., Burcher, E., Shults, C. W., Lovenberg, W., O'Donohue, T. L. 1984. Novel pharmacology of substance K bind-ing sites: A third type of tachykinin recep-tor. *Science* 226: 987–89

Burcher, E., Buck, S. H. 1986. Multiple tachykinin binding sites in hamster, rat

and guinea pig urinary bladder. *Eur. J. Pharmacol.* 128: 165–77

Burcher, E., Shults, C. W., Buck, S. H., Chase, T. N., O'Donohue, T. L. 1984. Autoradiographic distribution of sub-stance K binding sites in rat gas-trointestinal tract: A comparison with substance P. *Eur. J. Pharmacol.* 102: 561–62

Cascieri, M. A., Chicchi, G. G., Liang, T. 1985. Demonstration of two distinct tachykinin receptors in rat brain cortex. *J. Biol. Chem.* 260: 1501–7

Chang, M. M., Leeman, S. E. 1970. Isolation of a sialogogic peptide from bovine hypo-thalamic tissue and its characterization as substance P. *J. Biol. Chem.* 245: 4784–90

Chang, M. M., Leeman, S. E., Niall, H. D. 1971. Amino acid sequence of substance P. *Nature New Biol.* 232: 86–87

Conlon, J. M., Deacon, C. F., O'Toole, L., Thim, L. 1986. Scyliorhinin I and II: Two novel tachykinins from dogfish gut. *FEBS Lett.* 200: 111–16

Couture, R., Fournier, A., Magnan, J., St-Pierre, S., Regoli, D. 1979. Structure-activity studies on substance P. *Can. J. Physiol. Pharmacol.* 57: 1427–36

Dalsgaard, C.-J., Haegerstrand, A., Theo-dorsson-Norheim, E., Brodin, E., Hökfelt, T. 1985. Neurokinin A-like immunoreac-tivity in rat primary sensory neurons: Coexistence with substance P. *Histo-chemistry* 83: 37–40

Deutch, A. Y., Maggio, J. E., Bannon, M. J., Kalivas, P. W., Tam, S.-Y., et al. 1985. Substance K and substance P differentially modulate mesolimbic and mesocortical systems. *Peptides* 6 (Suppl. 2): 113–22

Erspamer, V. 1949. Ricerche preliminaria sulla moschatina. *Experientia* 5: 79

Erspamer, V. 1971. Biogenic amines and active polypeptides of the amphibian skin. *Ann. Rev. Pharmacol.* 11: 327–50

Erspamer, V. 1981. The tachykinin peptide family. *Trends Neurosci.* 4: 267–69

Erspamer, V., Anastasi, A. 1962. Structure and pharmacological actions of eledoisin, the active endecapeptide of the posterior salivary glands of *Eledone. Experientia* 18: 58–59

Erspamer, V., Anastasi, A., Bertaccini, G., Cei, J. M. 1964. Structure and pharmacological actions of physalaemin, the main active polypeptide of the skin of *Physalaemus fuscumaculatus. Experientia* 20: 489–90

Erspamer, V., Melchiorri, P. 1973. Active peptides of the amphibian skin and their synthetic analogs. *Pure Appl. Chem.* 35: 463–94

Håkanson, R., Sundler, F., eds. 1985. *Tachykinin Antagonists.* Amsterdam: Elsevier. 442 pp.

Hanley, M. R., Lee, C.-M., Jones, L. M., Michell, R. H. 1980. Similar effects of substance P and related peptides on salivation and on phosphatidylinositol turnover in rat salivary glands. *Molec. Pharmacol.* 18: 78–83

Henry, J. L., ed. 1987. *Substance P and Neurokinins.* New York: Springer-Verlag. In press

Hunter, J. C., Goedert, M., Pinnock, R. D. 1985. Mammalian tachykinin-induced hydrolysis of inositol phospholipids in rat brain slices. *Biochem. Biophys. Res. Commun.* 127: 616–22

Hunter, J. C., Maggio, J. E. 1984. A pharmacological study with substance K: Evidence for multiple types of tachykinin receptors. *Eur. J. Pharmacol.* 105: 149–53

Innis, R. B., Andrade, R., Aghajanian, G. K. 1985. Substance K excites dopaminergic and non-dopaminergic neurons in rat substantia nigra. *Brain Res.* 335: 381–83

Jessell, T. M. 1983. Substance P in the nervous system. In *Handb. Psychopharmacol.* 16: 1–105

Jordan, C. C., Oehme, P., eds. 1985. *Substance P: Metabolism and Biological Actions.* London: Taylor & Francis. 260 pp.

Kanazawa, I., Ogawa, T., Kimura, S., Munekata, E. 1984. Regional distribution of substance P, neurokinin α and neurokinin β in rat central nervous system. *Neurosci. Res.* 2: 111–20

Kangawa, K., Minamino, N., Fukuda, A., Matsuo, H. 1983. Neuromedin K: A novel mammalian tachykinin identified in porcine spinal cord. *Biochem. Biophys. Res. Commun.* 114: 533–40

Kimura, S., Goto, K., Ogawa, T., Sugita, Y., Kanazawa, I. 1984. Pharmacological characterization of novel mammalian tachykinins, neurokinin α and neurokinin β. *Neurosci. Res.* 2: 97–104

Kimura, S., Okada, M., Sugita, Y., Kanazawa, I., Munekata, E. 1983. Novel neuropeptides, neurokinins α and β, isolated from porcine spinal cord. *Proc. Jpn. Acad. Ser. B* 59: 101–4

Kotani, H., Hoshimaru, M., Nawa, H., Nakanishi, S. 1986. Structure and gene organization of bovine neuromedin K precursor. *Proc. Natl. Acad. Sci. USA* 83: 7074–78

Krause, J. E. 1985. On the physiological metabolism of substance P. See Jordan & Oehme 1985, pp. 13–31

Kream, R. M., Schoenfeld, T. A., Mancuso, R., Clancy, A. N., El-Bermani, W., et al. 1985. Precursor forms of substance P (SP) in nervous tissue: Detection with antisera to SP, SP-Gly, and SP-Gly-Lys. *Proc. Natl. Acad. Sci. USA* 82: 4832–36

Laufer, R., Wormser, U., Friedman, Z. Y., Gilon, C., Chorev, M., et al. 1985. Neurokinin B is a preferred agonist for a neuronal substance P receptor and its action is antagonized by enkephalin. *Proc. Natl. Acad. Sci. USA* 82: 7444–48

Lee, C.-M., Campbell, N. J., Williams, B. J., Iversen, L. L. 1986. Multiple tachykinin receptors in peripheral tissues and in brain. *Eur. J. Pharmacol.* 130: 209–17

Lee, C.-M., Iversen, L. L., Hanley, M. R., Sandberg, B. E. B. 1982. The possible existence of multiple receptors for substance P. *Naunyn-Schmiedeberg's Arch. Pharmacol.* 318: 281–87

Lee, J.-M., McLean, S., Maggio, J. E., Zamir, N., Roth, R. H., et al. 1986. The localization and characterization of substance P and substance K in striatonigral neurons. *Brain Res.* 370: 152–54

Leeman, S. E., Hammerschlag, R. 1967. Stimulation of salivary secretion by a factor from hypothalamic tissue. *Endocrinology* 81: 803–10

Lindefors, N., Brodin, E., Theodorsson-Norheim, E., Ungerstedt, U. 1985. Calcium-dependent potassium-stimulated release of neurokinin A and neurokinin B from rat brain regions *in vitro. Neuropeptides* 6: 453–62

Maggio, J. E. 1985. Kassinin in mammals: The newest tachykinins. *Peptides* 6 (Suppl. 3): 237–43

Maggio, J. E., Hunter, J. C. 1984. Regional distribution of kassinin-like immunoreactivity in rat central and peripheral tissues and the effect of capsaicin. *Brain Res.* 307: 370–73

Maggio, J. E., Sandberg, B. E. B., Bradley, C. V., Iversen, L. L., Santikarn, S., et al. 1983. Substance K: A novel tachykinin in mammalian spinal cord. See Skrabanek & Powell 1983b, pp. 20–21

Mantyh, P. W., Hunt, S. P., Maggio, J. E. 1984a. Substance P receptors: Localization by light microscopic autoradiography using ^{3}H-SP as the radioligand. *Brain Res.* 307: 147–66

Mantyh, P. W., Maggio, J. E., Hunt, S. P. 1984b. The autoradiographic distribution of kassinin and substance K binding sites is different from the distribution of substance P binding sites in rat brain. *Eur. J. Pharmacol.* 102: 361–64

Mantyh, P. W., Pinnock, R. D., Downes, C. P., Goedert, M., Hunt, S. P. 1984c. Substance P receptors: Correlation with substance P-induced inositol phospholipid hydrolysis in the rat central nervous system. *Nature* 309: 795–97

Melchiorri, P., Negri, L. 1984. Evolutionary aspects of amphibian peptides. In *Evolution and Tumour Pathology of the Neuroendocrine System*, ed. S. Falkmer, R. Hakånson, F. Sundler, pp. 231–44. Amsterdam: Elsevier. 278 pp.

Minamino, N., Kangawa, K., Fukuda, A., Matsuo, H. 1984a. Neuromedin L: A novel mammalian tachykinin identified in porcine spinal cord. *Neuropeptides* 4: 157–66

Minamino, N., Masuda, H., Kangawa, K., Matsuo, H. 1984b. Regional distribution of neuromedin K and neuromedin L in rat brain and spinal cord. *Biochem. Biophys. Res. Commun.* 124: 731–38

Nakajima, T., Yasuhara, T., Erspamer, V., Falconieri Erspamer, G., Negri, L., et al. 1980. Physalaemin- and bombesin-like peptides in the skin of the Australian leptodactylid frog *Uperoleia rugosa*. *Chem. Pharm. Bull.* 28: 689–95

Nawa, H., Hirose, T., Takashima, H., Inayama, S., Nakanishi, S. 1983. Nucleotide sequences of cloned cDNAs for two types of bovine substance P precursor. *Nature* 306: 32–36

Nawa, H., Kotani, H., Nakanishi, S. 1984. Tissue-specific generation of two preprotachykinin mRNAs from one gene by alternative RNA splicing. *Nature* 312: 729–34

Nicoll, R. A., Schenker, C., Leeman, S. E. 1980. Substance P as a neurotransmitter candidate. *Ann. Rev. Neurosci.* 3: 227–68

Ogawa, T., Kanazawa, I., Kimura, S. 1985. Regional distribution of substance P, neurokinin α and neurokinin β in rat spinal cord, nerve roots and dorsal root ganglia and the effects of dorsal root section or spinal transection. *Brain Res.* 359: 152–57

Otsuka, M., Konishi, S., Takahashi, T. 1972. The presence of a motoneuron-depolarizing peptide in bovine dorsal roots of spinal nerves. *Proc. Jpn. Acad.* 48: 342–46

Pernow, B. 1983. Substance P. *Pharmacol. Rev.* 35: 85–141

Piercey, M. F., Dobry-Schreur, P. J. K., Masiques, N., Schroeder, L. A. 1985. Stereospecificity of SP_1 and SP_2 substance P receptors. *Life Sci.* 36: 777–80

Sandberg, B. E. B. 1985. Structure-activity relationships for substance P: A review. See Jordan & Oehme 1985, pp. 65–81

Shults, C. W., Quirion, R., Chronwall, B., Chase, T. N., O'Donohue, T. L. 1984. A comparison of the anatomical distribution of substance P and substance P receptors in the rat central nervous system. *Peptides* 5: 1097–1128

Skrabanek, P., Powell, D., eds. 1977, 1980, 1983a. *Substance P*, Vols. 1, 2, 3. Montreal: Eden Press. 181 pp., 175 pp., 184 pp.

Skrabanek, P., Powell, D. 1983b. *Substance P—Dublin 1983*. Dublin: Boole Press. 275 pp.

Tatemoto, K., Lundberg, J. M., Jörnvall, H., Mutt, V. 1985. Neuropeptide K: Isolation, structure and biological activities of a novel brain tachykinin. *Biochem. Biophys. Res. Commun.* 128: 947–53

Too, H. P., Hanley, M. R. 1984. Peripheral and central substance P binding sites. In *Brain Receptor Methodologies, Part B. Amino Acids, Peptides, Psychoactive Drugs*, ed. P. J. Marangos, I. C. Campbell, R. M. Cohen, pp. 151–76. Orlando: Academic. 336 pp.

von Euler, U. S., Gaddum, J. H. 1931. An unidentified depressor substance in certain tissue extracts. *J. Physiol.* 72: 74–87

Watson, S. P. 1984. The action of substance P on contraction, inositol phospholipids and adenylate cyclase in rat small intestine. *Biochem. Pharmacol.* 33: 3733–37

Watson, S. P., Downes, C. P. 1983. Substance P induces hydrolysis of inositol phospholipids in guinea-pig ileum and rat hypothalamus. *Eur. J. Pharmacol.* 93: 245–53

Watson, S. P., Sandberg, B. E. B., Hanley, M. R., Iversen, L. L. 1983. Tissue selectivity of substance P alkyl esters: Suggesting multiple receptors. *Eur. J. Pharmacol.* 87: 77–84

Yasuhara, T., Nakajima, T., Falconieri Erspamer, G., Erspamer, V. 1981. New tachykinins, Glu^2, Pro^5-kassinin (Hylambates-kassinin) and hylambatin, in the skin of the African rhacophorid frog *Hylambates maculatus*. *Biomed. Res.* 2: 613–17

Zachary, I., Rozengurt, E. 1986. A substance P antagonist also inhibits specific binding and mitogenic effects of vasopressin and bombesin-related peptides in Swiss 3T3 cells. *Biochem. Biophys. Res. Commun.* 137: 135–41

Ann. Rev. Neurosci. 1988. 11 : 29–44

MICROTUBULE-ASSOCIATED PROTEINS:
Their Potential Role in Determining Neuronal Morphology

Andrew Matus

Friedrich Miescher Institute, 4002 Basel, Switzerland

INTRODUCTION

The microtubule-associated proteins (MAPs) are a heterogeneous group of molecules that co-purify with tubulin when microtubules are repolymerized from brain homogenates. A variety of data implicate them in regulating the stability of microtubules in axons and dendrites and hence suggest that they are involved in determining neuronal shape and, more importantly, in regulating the balance between rigidity and plasticity in neuronal processes. In this review I consider some properties of the MAPs that bear on this question and attempt to assess critically their interpretation. Necessarily, much detail is omitted, and interested readers are referred to several recent reviews: Olmsted (1986) for a comprehensive general survey of MAPs, and Nunez (1986) and Matus (1987) for other aspects of brain development and microtubules.

MAPs, Microtubule Stability, and Neurite Formation

Experiments with tubulin-depolymerizing drugs have shown that the assembly of microtubules is essential for both the growth of neurites and for the maintenance of neurites that have already formed (Seeds et al 1970, Yamada et al 1970). By observing the fate of labeled tubulin microinjected into living cells, Schulze & Kirschner (1986) have shown that microtubules in nonneuronal cells are unstable and grow rapidly from centrally located organizing centers toward the cell periphery while at the same time being susceptible to sudden catastrophic collapse. It is hardly surprising that such cells do not extend stable neurite-like processes. By the same token

0147–006X/88/0301–0029$02.00

it seems necessary that in order for axons and dendrites to grow, the neuronal cytoplasm must contain molecules that stabilize microtubules and play a cardinal role in neuronal morphogenesis. In this respect brain MAPs acquire particular significance, for various MAPs have been shown to promote the polymerization of tubulin in vitro (MAP1: Kuznetsov et al 1981; MAP2: Sloboda et al 1976; tau: Weingarten et al 1975; MAP3: Huber et al 1986; MAP5: Riederer et al 1986).

MAPs are also believed to be important in organizing the structure of the neuritic cytoplasm. A commonplace but highly significant observation supporting this notion is that microtubules in axons and dendrites are arranged in parallel, not only to one another as fascicles but also to the plasma membrane. It thus appears that the microtubules somehow communicate their position to one another and, either directly or via other cytoskeletal elements, to the surface membrane. The probable means of this communication can be seen in the form of abundant fine fibers that cross-link the neuronal cytoskeleton and that include microtubules, neurofilaments and surface membrane, which impart a structural unity best demonstrated by deep-etched, freeze-fractured techniques (Hirokawa 1982). These observations once again suggest the involvement of MAPs, several of which are long, filamentous molecules (Voter & Erickson 1982) that form filamentous side projections on microtubules (Kim et al 1979, Zingsheim et al 1979).

CHARACTERISTICS AND CELLULAR DISTRIBUTION OF NEURONAL MAPs

Much of the recent interest in MAPs stems from the discovery that certain MAP species are differentially distributed within neurons occurring, for example, only in dendrites (Matus et al 1981). These findings indicate that populations of microtubules of different chemical composition occur in axons and dendrites and that microtubules in growing processes are different from their counterparts in mature brain.

Properties of Individual MAP Species

The more abundant and better studied MAPs can be roughly divided into two groups: a set of very high molecular weight polypeptides, of which MAP1 and MAP2 are the most conspicuous representatives in adult brain, and a group of intermediate-sized proteins (around 60,000 daltons) of which the tau MAPs are the most abundant (Figure 1). There are in addition several dozen minor components, few of which have yet been studied individually.

MAP1a is the largest brain MAP with an apparent molecular weight of 350,000 (Figure 1). Its original title was MAP1, but recently a variety of similarly sized, less abundant proteins have been given names derived from MAP1 even though they may not be chemically related to MAP1 itself. Proteins of this class include MAP1b and MAP1c (Bloom et al 1984) and MAP1x (Calvert & Anderton 1985). As a result of this subdivision of the terminology the original MAP1 is now increasingly referred to as MAP1a (Bloom et al 1984).

Immunocytological studies of MAP1a localization in brain have given conflicting results. Huber & Matus (1984a) found an exclusively neuronal

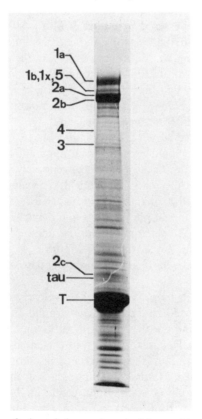

Figure 1 The proteins of microtubules repolymerized from rat brain homogenate as displayed by polyacrylamide gel electrophoresis. The positions of the various microtubule-associated proteins are indicated by their numerical designation or name (see text). T indicates the position of tubulin. The position of MAP4 is calculated from the apparent molecular weight given by Parysek et al (1984). Tau consists of several bands around the position indicated. Other tau bands and the chartins occur between this positon and that of tubulin. The gel was a linear 4 to 14% gradient and was stained with Coomassie Brilliant Blue.

staining, whereas the antibodies used by Bloom et al (1984) stained both neurons and glia. There are several possible reasons why discrepancies in staining patterns with different antibodies against the same molecule may occur. There may be distinct forms of the molecule that differ in cellular expression but have not yet been resolved by chemical analysis. Post-translational modifications of the molecule may produce antigenic variety, as is known to occur with the phosphorylation of neurofilament proteins (Sternberger & Sternberger 1983). Finally, antibodies may be directed against different sites (epitopes) on the molecule that may be accessible in some cell compartments but not in others. In neurons MAP1a is found in the cell body, axons, and dendrites, although dendritic spines are not stained. The antibodies used by Huber & Matus (1984a) stain dendrites more strongly than they stain axons. In the cerebellum anti-MAP1a staining is markedly stronger in Purkinje cells than in the adjacent granule cell neurons (Huber & Matus 1984a). The opposite pattern is observed for anti-MAP2 staining, which is stronger in granule cells than in Purkinje cells (Huber & Matus 1984b). Thus both these molecules may be expressed at various levels in different types of neurons.

The proteins named MAP1x, MAP5, and MAP1b migrate in the same position on polyacrylamide gels with an apparent molecular weight of around 320,000 (Figure 1). Anti-MAP1x exclusively stains axons in brain sections (Calvert & Anderton 1985). Anti-MAP5 staining is neuron specific and is found in both axons and dendrites (Riederer et al 1986). Anti-MAP1b is reported to stain both axons and dendrites and also glial cells (Bloom et al 1985). All three molecules are under strong developmental regulation (Calvert & Anderton 1985, Bloom et al 1985, Reiderer et al 1986) and appear to be more abundant during development than in adult tissue.

MAP2 in the adult brain consists of two closely related polypeptides (MAP2a the larger, MAP2b the smaller) of around 280,000 mol wt. Immunohistochemical staining with a variety of antibodies, both polyclonal and monoclonal, in several laboratories, has shown that aside from a few minor exceptions MAP2 in brain is expressed only in neurons and is limited to dendrites (Bernhardt & Matus 1984, Burgoyne & Cumming 1984, Caceres et al 1984a, De Camilli et al 1984). In cerebellar basket cells MAP2 is present in the initial part of the axon (but not beyond), so its distribution inside the neuron cannot be simply explained by selective transport into dendrites (Bernhardt & Matus 1984). Hippocampal pyramidal cells growing in dispersed cell culture have MAP2 in their dendrites but not in their axons; therefore the partitioning of MAP2 within the cytoplasm of these cells does not depend on extrinsic factors (Caceres et al 1984b, Matus et al 1986). Anti-MAP2 staining has also been reported at a limited number of extra-dendritic sites including motor neuron axons (Papasozomenos et al 1985) and isolated groups of glial cells (Papasozomenos & Binder 1986).

In repolymerized adult brain microtubules MAP3 is present as a pair of polypeptides (MAP3a the larger, MAP3b the smaller) with an apparent molecular weight of 180,000 (Huber et al 1985). In the adult brain MAP3 is prominent in astroglia, and its occurrence in neurons is limited to neurofilament-rich axons (Huber et al 1985). MAP3 promotes tubulin polymerization in vitro and appears to contribute to the overall poly-merization performance of brain microtubules in vitro (Huber et al 1986). MAP4 also consists of a group of polypeptides with an apparent molecular weight around 210,000 and is expressed in glial cells but not neurons (Parysek et al 1984, Olmsted et al 1986).

The tau MAPs are a group of proteins (six components in adult rat brain), in the molecular weight range 55,000 to 65,000, that were first discovered as promoters of tubulin polymerization in vitro (Weingarten et al 1975). Initially, immunohistochemistry with a monoclonal antibody indicated that tau is present only in axons (Binder et al 1985). However, it has since been proved that the epitope recognized by this antibody is present in cell bodies and dendrites where it is masked by phosphorylation. After treatment of sections with phosphatase, tau can be visualized in these sites as well as in axons (Papasozomenos & Binder 1987). The chartins are a group of related proteins in the same size range as tau but distinct from them and less abundant (Magendantz & Solomon 1985). In pheo-chromocytoma PC12 cells they become phosphorylated during nerve growth factor-induced neurite outgrowth (Black et al 1986).

DEVELOPMENTAL REGULATION OF MAP EXPRESSION

Each MAP that has so far been examined undergoes changes in the level and in some cases the form in which it is expressed during brain development. In addition, expression of some MAPs occurs in different cells or different cell compartments in the developing brain than in mature tissue. The changes that occur in rat brain are summarized in Figure 2. Together they suggest that microtubules in developing axons and dendrites differ in chemical composition from their counterparts in mature tissue.

Developmental Changes in Individual MAPs

In the rat embryo and at birth MAP1a is barely detectable. Subsequently, its level increases steadily up to around postnatal day 20, after which there is no further change (Riederer & Matus 1985). In both the neocortex

(Riederer & Matus 1985) and hippocampus (Matus & Riederer 1986) anti-MAP1a dendritic staining appears within 48 hours of birth and increases steadily up to the end of the third week. In the cerebellum the pattern of events is similar but delayed, in accordance with the later development of the cerebellum as compared to the telencephalon, and dendritic staining first appears around day 7 (Bernhardt et al 1985). Axons in the cerebellar white matter show an opposite pattern; they are strongly anti-MAP1a reactive from day 3 until day 10 after which there is a rapid decline to adult levels by day 20 (Bernhardt et al 1985). Granule cell axons (parallel

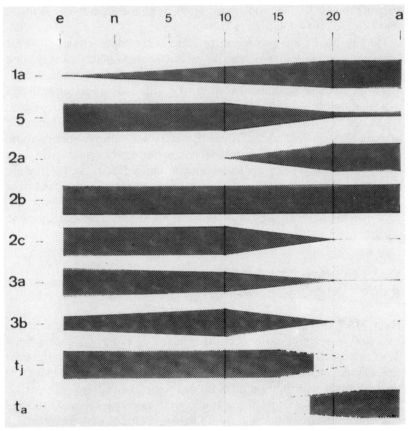

Figure 2 Diagrammatic representation of the changes in abundance of MAPs during postnatal development of the rat brain. The individual MAPs are identified on the left by number except for juvenile tau (t_j) and adult tau (t_a). The ages indicated above are late embryo (e), newborn (n), days after birth indicated by numbers, and adult (a). The greatest change occurs in the period indicated by the vertical lines, between postnatal days 10 and 20.

fibers) are also more strongly reactive, and more axons are stained, in developing cerebellum than in the adult tissue. At all stages of development only a subset of parallel fibers is stained.

MAP5 is some tenfold more abundant in brain tissue of newborn rats than that of adults (Riederer et al 1986). The change from juvenile to adult levels occurs abruptly, beginning around postnatal day 10 and being completed by day 20. The decline is also visible by immunohistochemical staining, which shows a marked fall in anti-MAP5 staining of neurons. Throughout development MAP5 is apparently evenly distributed in the cell bodies, axons, and dendrites of neurons. To date, MAP5 is the only MAP found in neurons prior to terminal differentiation (Reiderer et al 1986). MAP1b is also described as being more abundant in juvenile than in adult brain (Bloom et al 1985), and anti-MAP1x stains axons of juvenile brain more strongly than those of adult (Calvert & Anderton 1985).

Of the two high molecular weight forms of MAP2, MAP2b is present at the same apparent abundance from late embryonic life onward, but MAP2a is absent from the newborn rat brain. MAP2a first appears around postnatal day 10, and its expression rapidly increases to reach its maximum, adult brain, level by postnatal day 20 (Burgoyne & Cumming 1984, Binder et al 1984). A recently discovered low molecular weight form of MAP2, MAP2c, with an apparent molecular weight of 70,000 (Garner et al 1986), shows a pattern of developmental regulation that differs from the pattern found in MAP2a. It is a major MAP component of newborn rat brain, but its levels drop more than tenfold between postnatal days 10 and 20 (Riederer & Matus 1985).

In the late embryo the larger member of the MAP3 doublet, MAP3a, predominates and MAP3b is barely detectable. Immediately after birth, MAP3b levels begin to increase, until by day 10 its abundance is greater than that of MAP3a. Then, between day 10 and day 20, there is a drop of more than tenfold in the level of both forms. Complex changes occur in the sites at which MAP3 is expressed during development. In the neocortex anti-MAP3 stains all neurons until postnatal day 10, after which staining declines rapidly so that only traces are detectable in neurons by day 20. At the same time anti-MAP3 staining appears in proliferating astroglial cells so that in the adult neocortex all the anti-MAP3 staining is of astrocytes and none remains in neurons (Riederer & Matus 1985). In the cerebellum there is strong anti-MAP3 staining of the developing parallel fiber axons until day 20. However, in adult tissue it has completely vanished and the radial fibers of the Bergmann astroglial cells are the predominant anti-MAP3 feature (Bernhardt et al 1985).

MAP4 also undergoes changes in complexity during mouse brain development (Olmsted et al 1986). In the embryo it first appears predominantly as the upper MAP4a band. By birth it has been joined by two slightly smaller bands, MAP4b and MAP4c. Between days 21 and 35 there is a loss of MAP4c component so that in the adult MAP4 is represented by a doublet.

Changes in the composition of tau proteins during brain maturation were the first developmental changes in MAPs to be documented (Mareck et al 1980). It was observed that in newborn rat brain the tau complex consisted of two major components, whereas in adult tissue six or more components were present. Couchie & Nunez (1985) have since found that in neonatal rat brain the protein formerly designated "slow" tau reacts with anti-MAP2 and not with anti-tau. This is presumably the same species we have observed and designated MAP2c (Riederer & Matus 1985, Garner et al 1986). Juvenile tau thus appears to consist of a single major component, molecular weight around 48,000 (Couchie & Nunez 1985). The transition from this relatively simple juvenile state to the adult pattern of more than six bands occurs rather late; the transition begins around postnatal day 15 and is completed by day 35 (Couchie & Nunez 1985). Since tau is primarily localized in axons and since the major phase of axon growth throughout the brain has been completed by day 15, the change in tau composition does not appear to be required for axonal growth (Nunez 1986).

Timing of Developmental Changes and Neurite Growth

One of the most striking features of the changes described above is that they all occur during the short postnatal period of neuronal differentiation in the rat brain (Figure 2). The relatively few MAPs for which such developmental profiles have been determined appear to fall into distinct categories. MAP1a accumulates steadily as neuronal processes are growing and reaches a plateau at the time when this major phase of axon and dendrite formation is coming to a close. The changes in MAPs 2, 3, 5, and 1x all begin and are completed between postnatal days 10 and 20, coinciding with the time when the transition from the growth to the maintenance phase is taking place. Finally, the changes in tau begin with entry into the maintenance phase that marks the achievement of a stable configuration of the major axonal and dendritic branching patterns.

These changes in MAP expression correlate closely in timing with changes in the efficiency of polymerization of brain microtubule proteins. The initial rate of assembly of tubulin from newborn rat brain is some tenfold lower than that of the adult (Fellous et al 1979). This does not appear to be the result of an inhibitor among juvenile brain MAPs, because adding these MAPs in small quantities to adult brain microtubules does not retard polymerization. On the other hand, the polymerization of juvenile brain microtubules is markedly stimulated by the addition of adult brain MAPs (Francon et al 1982). Thus the major and probable sole reason for the differences in polymerization performance in vitro between newborn and adult brain microtubules is that the MAPs of juvenile brain are less effective on a molar basis in driving tubulin to polymerize.

If the microtubules of developing rat brain polymerize less readily in vitro, then one might expect juvenile microtubules to be less stable inside developing neuronal processes. This conjecture is supported by results of Faivre et al (1985). They tested the stability of microtubules in Purkinje cell dendrites by fixing rat brain with a cold fixative solution. In adult rats cold fixation made no difference and both warm and cold fixatives yielded dendrites containing well-preserved microtubules. In juvenile animals, however, microtubule numbers were greatly reduced in the dendrites of tissue fixed with cold solutions compared to those fixed with warm solutions. The proportion of cold-stable microtubules changes from only 25% at postnatal day 10 to more than 90% by day 35. Thus the stability of microtubules changes markedly over exactly the same period that MAP composition changes.

INTERACTION OF MAPs WITH CYTOSKELETAL ELEMENTS OTHER THAN MICROTUBULES

Interactions of MAPs with other cytoplasmic structures must exist if the MAPs are to discharge their putative role as mediators between tubulin polymers and the other structural components of axons and dendrites. One difficulty of investigating such interactions is that in vitro the high affinity of MAPs for tubulin overwhelms their binding to other cytoskeletal elements. However, once tubulin has been removed, such interactions can be observed. By this means both MAP2 and tau have been shown to bind to actin filaments (Nishida et al 1981, Selden & Pollard 1983, Sattilaro 1986) and neurofilaments (Letterier et al 1982, Heimann et al 1985). The binding of MAP2 to neurofilaments appears to be mediated via the 70-kDa protein of neurofilaments at a site that is distinct from its binding site for tubulin (Heimann et al 1985). This suggests that different sequences within the MAP2 molecule may mediate its binding to various structural elements within the cytoplasm. By contrast, the site at which MAP2 binds to actin is on the same 39-kDa fragment of the MAP2 molecule that contains the tubulin-binding site (Sattilaro 1986).

Further evidence for a connection between tubulin and actin in the neuritic cytoplasm comes from a study by Fach et al (1985). They prepared microtubules in vitro without allowing them to depolymerize by using density gradient centrifugation in glycerol-containing stabilization buffers. The composition of these directly isolated microtubules was radically different from that of the conventional variety, which are produced by repolymerizing microtubules from brain homogenates (see Olmsted 1986, for further details). MAP1 and MAP2 were only minor components and, instead, actin and fodrin (brain spectrin) were prominent. When these directly isolated microtubules were subsequently taken through cycles of depolymerization-repolymerization, the actin and fodrin were lost but the MAPs were retained so that their composition came to resemble that of conventionally prepared in vitro microtubules. This suggests that the composition of the much studied, conventionally prepared microtubules may be misleading, because it is skewed by the high affinity of MAPs for tubulin when they are removed from the regulatory influence of organized cytoplasm and therefore does not reflect the true interactions of MAPs with other molecules inside the living cell.

There is also cytological evidence that within living cells MAPs are not always primarily associated with microtubules. For example, MAP3 is present at substantial levels in the axons of cerebellar basket cells, which are rich in neurofilaments but contain very few microtubules. However, all the MAP3 present in brain homogenates is isolated with microtubules and none is present in isolated neurofilaments (Huber et al 1985). Even the paradigm MAP, MAP2, does not co-localize exclusively with micro-tubules in neurons. In cultured cerebellar granule cells MAP2 is localized to a short initial portion of neurites even though tubulin filaments are present throughout their length (Alaimo-Beuret & Matus 1985). In cultured hippocampal neurons all the anti-tubulin staining was present as recognizable filaments and fascicles, but anti-MAP2 staining was spread throughout the cytoplasm (Matus et al 1986).

Modulation of Interactions by Phosphorylation

Phosphorylation is probably of widespread significance in determining how MAPs interact with microtubules and other cytoskeletal elements. Just how important is difficult to judge because the state of phosphorylation of MAPs in vivo is uncertain. MAP2 isolated without microtubule recycling contains much more covalently bound phosphate than that recovered by conventional repolymerization (Tsuyama et al 1986). This could be very significant because phosphorylation diminishes the affinity of MAPs for tubulin polymers (Murthy & Flavin 1983) and promotes their interaction with both neurofilaments (Letterier et al 1982) and actin (Nishida et al 1981). The interaction between MAP2 and actin even appears to change character depending on phosphorylation state; unphosphorylated MAP2 bundles actin filaments, whereas the phosphorylated molecule still binds the filaments, but they are no longer bundled (Sattilaro 1986).

At least three different kinases are known to phosphorylate MAP2 (see Olmsted 1986). The best studied is the brain cyclic AMP-dependent type II protein kinase. This binds via its regulatory subunit to the portion of MAP2 that projects from the microtubule wall (Vallee et al 1981). Thus when cyclic AMP levels inside the neuron are raised, for example as a consequence of dopaminergic transmission, the active catalytic subunit of the enzyme will be released in the immediate vicinity of the MAP2 substrate molecule. Immunohistochemistry has shown that the regulatory subunit (RII) is located in the cell body and dendrites of brain neurons (Miller et al 1982). Thus the distribution of RII mirrors that of MAP2 to which it binds. Recently, De Camilli et al (1986) have found that within the peri-karyon and dendrites of many brain neurons RII co-localizes with Golgi complex membranes. They suggest that RII binds to the *trans*-face of the Golgi membranes and links the membranes to a subset of microtubules via MAP2. This whole multimolecular complex, linking several subcellular elements of the dendritic cytoplasm, would thus be subject to functional modulation by phosphorylation via the localized presence of cAMP-depen-dent protein kinase.

The potential importance of MAP phosphorylation state during brain development is well illustrated in a study by Aoki & Siekevitz (1985). They found that after delaying the onset of the critical period of development in the visual cortex of cats by dark-rearing, a short period of light exposure was sufficient to produce a large increase in the in vitro phosphorylation of MAP2.

PLASTICITY VERSUS RIGIDITY: PHASE CHANGES IN THE NEURITIC CYTOPLASM

When neuroblastoma cells are exposed to microtubule-depolymerizing drugs, there is an intermediate stage in the retraction of neurites when they loose their cylindrical form. Fine motile processes, of similar morphology to the filopodial extensions of growth cones, arise laterally all along the neurite (Bray et al 1978, Anglister et al 1982, Matus et al 1986). In dendrites of hippocampal neurons the loss of characteristic cylindrical uniformity is very striking; not only are lateral filopodia produced but the cylindrical form of the dendrites becomes completely lost and they become quite irregular in outline (Matus 1987).

These observations support the view that actin is present in neurites but that it is constrained from driving cytoplasmic motility by the influence of the microtubules (Bray et al 1978). Complementary evidence has been obtained by Letorneau & Ressler (1984) who applied the drug taxol, which forces tubulin to polymerize, to neurons growing in culture. When this is done, microtubules are formed in the growth cones and they subsequently lose their motility. The conversion from the plasticity of the growth cone to the relative rigidity of the neurite thus seems to depend on a phase change in cytoplasm: from the functional dominance of actin filaments, which promote mobility and constant morphological variation, to the dominance of tubulin filaments, which inhibit cytoplasmic mobility and impart morphological fixity.

What mediates this conversion? A variety of data implicate the MAPs. In the developing brain the levels and forms in which they are expressed change dramatically at exactly the time when growth cones are disappearing and axonal and dendritic form is becoming fixed. Growth phase MAPs do not drive tubulin polymerization as well as mature brain MAPs (Francon et al 1982), and this correlates with the observation that growth phase microtubules are less stable than those of adult brain (Faivre et al 1985). In the PC12 pheochromocytoma cell line, which extends neurites in response to nerve growth factor, the microtubules become more stable as neurite outgrowth proceeds (Black & Greene 1982) and as correlative changes in the levels of tau and a MAP1 species (which MAP1 species is not yet clear) take place (Drubin et al 1985). Changes also take place in the phosphorylation state of chartins (Black et al 1986) during NGF-induced PC12 differentiation.

Distributional differences in MAPs correlated to axon and dendrite formation in cultured neurons have also been described. MAP2 in cerebellar granule cells is spread throughout the entire length of the neurites as they first grow out of the cell body and only later becomes restricted to their initial portion (Alaimo-Beuret & Matus 1985). A similar phenomenon has been described by Caceres et al (1986) in cultured hippocampal pyramidal cells where MAP2 was initially present in axons and only later became restricted to dendrites.

PUZZLES AND PROSPECTS

The changes in MAP distribution in the developing brain present a tantalizing but as yet poorly understood process. What is the meaning of the high levels of MAP1a, MAP1x, MAP3, and MAP5 in growing cerebellar granule cell axons and their much lower levels (or even absence in the case of MAP3) in the same processes in the mature brain? What is the significance of the accumulation of MAP2 that occurs in the secondary branches of Purkinje cell dendrites in the developing brain (Bernhardt & Matus 1982, Matus 1987)?

There are other puzzles. Neuroblastoma cells of several varieties examined in our laboratory make stable neurites without expressing either MAP1, 2, 3, or 5 (B. Brugg & A. Matus, unpublished observations). What, then, do these molecules contribute to axonal and dendritic structure in the primary neurons of the mammalian brain where they are so abundant? And what stabilizes microtubules in these neuroblastoma cell processes? Another puzzle surrounding the role of neuronal MAPs is their apparent dispensability for brain tubulin polymerization in poikilothermic species such as fish. At the temperatures at which these animals live, their brain tubulin polymerizes readily without either needing or binding the MAPs that are present in the surrounding cytoplasm (Langford et al 1986, Williams & Detrich 1986).

Probably the greatest need for future research is to explore MAP function in living cells. A recent paper by Drubin & Kirschner (1986), in which they demonstrate that tau microinjected into nonneuronal cells stabilizes microtubules, provides a welcome entry into this underdeveloped field. A second need is to more accurately define the ligand-binding relationships of MAPs with other cytoplasmic proteins. We need to know not just which other molecules a particular MAP can interact with but the exact amino acid sequences involved in this interaction. The model here is the Arg-Gly-Asp core sequence found in various cell-to-extracellular matrix recognition systems. The application of synthetic peptides containing this sequence has demonstrated its functional involvement in a variety of developmental systems [for example, gastrulation in *Drosophila* (Naidet et al 1987)]. A similar approach may eventually allow the unravelling of functionally important interactions of specific MAP sequences inside neurons. For this reason it seems probable that one of the most illuminating sources of data in the near future will be the use of recombinant DNA techniques. cDNA clones have been isolated for tau (Drubin et al 1984) and MAP2 (Lewis et al 1986a). A cDNA clone for a MAP1-sized species (Lewis et al 1986b) probably encodes MAP5 because antibodies raised against protein expressed from this clone react with MAP5 and stain brain sections with a MAP5 pattern (C. Garner, A. Matus, S. Lewis, and N. Cowan, unpublished observations). The coding sequences of these MAPs will be of great interest. Homologies between different MAPs may help to identify conserved domains associated with particular functions. Comparisons of the tubulin-binding domains should reveal how much sequence specificity is required for MAP-tubulin interaction. Even more significant will be any indications of other conserved modules within MAP sequences that may underlie their interaction with other cytoplasmic structures.

Literature Cited

Alaimo-Beuret, D., Matus, A. 1985. Changes in the cytoplasmic distribution of microtubule-associated protein 2 during the differentiation of cultured cerebellar granule cells. *Neuroscience* 14: 1103–16

Anglister, L., Farber, I. C., Shahar, A., Grinvald, A. 1982. Localization of voltage-sensitive calcium channels and developing neurites: Their possible role in regulating neurite elongation. *Dev. Biol.* 94: 351–65

Aoki, C., Siekevitz, P. 1985. Ontogenic changes in the cyclic adenosine 3′,5′-monophosphate-stimulable phosphorylation of cat visual cortex proteins, particularly of microtubule-associated protein 2 (MAP2): Effects of normal and dark rearing and of exposure to light. *J. Neurosci.* 5: 2465–83.

Bernhardt, R., Huber, G., Matus, A. 1985. Differences in the developmental patterns of three microtubule-associated proteins in brain. *J. Neurosci.* 5: 977–91.

Bernhardt, R., Matus, A. 1982. Initial phase of dendrite growth: Evidence for the involvement of high molecular weight microtubule-associated proteins (HMWPs) before the appearance of tubulin. *J. Cell Biol.* 92: 589–93.

Bernhardt, R., Matus, A. 1984. Light and electron microscopic studies of the distribution of microtubule-associated protein 2 in rat brain: A difference between the dendritic and axonal cytoskeletons. *J. Comp. Neurol.* 226: 203–21.

Binder, L. I., Frankfurter, A., Kim, H., Carceres, A., Payne, M. R., Rebhun, L. I. 1984. Heterogeneity of microtubule-associated protein 2 during rat brain development. *Proc. Natl. Acad. Sci. USA* 81: 5613–17

Binder, L. I., Frankfurter, A., Rebhun, L. I. 1985. The distribution of tau polypeptides in the mammalian central nervous system. *J. Cell Biol.* 101: 1371–78

Black, M. M., Aletta, J. M., Greene, L. A. 1986. Regulation of microtubule composition and stability during nerve growth factor-promoted neurite outgrowth. *J. Cell Biol.* 103: 545–57

Black, M. M., Greene, L. A. 1982. Changes in the colchicine susceptibility of microtubules associated with neurite outgrowth: Studies with nerve growth factor responsive PC12 pheochromocytoma cells. *J. Cell Biol.* 95: 379–86.

Bloom, G. S., Luca, F. C., Vallee, R. B. 1985. Microtubule-associated protein 1B: Identification of a major component of the neuronal cytoskeleton. *Proc. Natl. Acad. Sci. USA* 82: 5404–8

Bloom, G. S., Schoenfield, T. A., Vallee, R.

B. 1984. Widespread distribution of the major polypeptide component of MAP1 (Microtubule-associated protein 1) in the nervous system. *J. Cell Biol.* 98: 320–30

Bray, D., Thomas, C., Shaw, G. 1978. Growth cone formation in cultures of sensory neurons. *Proc. Natl. Acad. Si. USA* 81: 5626–29

Burgoyne, R. D., Cumming, R. 1984. Ontogeny of microtubule-associated protein 2 in rat cerebellum: Differential expression of the doublet polypeptides. *Neuroscience* 11: 156–67

Caceres, A., Banker, G., Binder, L. 1986. Immunocytochemical localization of tubulin and microtubule-associated protein 2 during development of hippocampal neurons in culture. *J. Neurosci.* 6: 714–22.

Caceres, A., Banker, G. A., Steward, O., Binder, L., Payne, M. 1984a. MAP2 is localized to the dendrites of hippocampal neurons which develop in culture. *Dev. Brain Res.* 13: 314–18

Caceres, A., Binder, L. I., Payne, M. R., Bender, P., Rebhun, L., Steward, O., 1984b. Differential subcellular localization of tubulin and the microtubule-associated protein MAP2 in brain tissue as revealed by immunohistochemistry with monoclonal hybridoma antibodies. *J. Neurosci.* 4: 394–410

Calvert, R., Anderton, B. H. 1985. A microtubule-associated protein MAP1 which is expressed at elevated levels during development of rat cerebellum. *EMBO J.* 4: 1171–76

Couchie, D., Nunez, J. 1985. Immunological characterization of microtubule-associated proteins specific for the immature brain. *FEBS Lett.* 188: 331–35

De Camilli, P., Miller, P. E., Navone, F., Theurkauf, W. E., Vallee, R. B. 1984. Distribution of microtubule-associated protein 2 in the nervous system of the rat studied by immunofluorescence. *Neuroscience* 11: 817–46

De Camilli, P., Moretti, M., Donini, S. D., Walter, U., Lohmann, S. M. 1986. Heterogeneous distribution of cAMP receptor protein RII in the nervous system: Evidence for its intracellular accumulation on microtubules, microtubule-organizing centers, and in the area of the Golgi Complex. *J. Cell Biol.* 103: 189–203

Drubin, D. G., Caput, D., Kirschner, M. 1984. Studies in the expression of the microtubule-associated protein, tau, during mouse brain development, with newly isolated complimentary DNA probes. *J. Cell Biol.* 98: 1090–97

Drubin, D. G., Feinstein, S., Shooter, E.,

Kirschner, M. 1985. Nerve growth factor induced outgrowth in PC12 cells involves the coordinate induction of microtubule assembly and assembly promoting factors. *J. Cell Biol.* 101: 1799–1807

Drubin, D. G., Kirschner, M. W. 1986. Tau protein function in living cells. *J. Cell Biol.* 103: 2739–46

Fach, B. L., Graham, S. F., Keates, R. A. B. 1985. Association of fodrin with brain microtubules. *Can. J. Biochem. Cell Biol.* 63: 372–81

Faivre, C., Legrand, Ch., Rabie, A. 1985. The microtubular apparatus of cerebellar Purkinje cells during postnatal development of the rat: The density and cold-stability of microtubules increase with age and are sensitive to thyroid hormone. *Int. J. Dev. Neurosci.* 3: 559–65

Fellous, A., Lennon, A. M., Francon, J., Nunez, J. 1979. Thyroid hormones and neurotubule assembly in vitro during brain development. *Eur. J. Biochem.* 101: 365–76

Francon, J., Lennon, A. M., Fellous, A., Mareck, A., Pierre, M., Nunez, J. 1982. Heterogeneity of microtubule-associated proteins and brain development. *Eur. J. Biochem.* 129: 465–72

Garner, C. C., Brugg, B., Matus, A. 1986. A 70 kDa microtubule-associated protein is a distinct MAP2 species. *J. Cell Biol.* 103: 403a

Heimann, R., Shelanski, M. L., Liem, R. K. H. 1985. Microtubule-associated proteins bind specifically to the 70-kDa neurofilament protein. *J. Biol. Chem.* 260: 12160–66

Hirokawa, N. 1982. Cross-linker system between neurofilaments, microtubules and membranes organelles in frog axons revealed by quick-freeze deep-etching methods. *J. Cell Biol.* 94: 129–42

Huber, G., Alaimo-Beuret, D., Matus, A. 1985. MAP3: Characterization of a novel microtubule-associated protein. *J. Cell Biol.* 100: 496–507

Huber, G., Matus, A. 1984a. Immunohistochemical localization of microtubule-associated protein 1 in rat cerebellum using monoclonal antibodies. *J. Cell Biol.* 98: 777–81

Huber, G., Matus, A. 1984b. Differences in the cellular distributions of two microtubule-associated proteins, MAP1 and MAP2, in rat brain. *J. Neurosci.* 4: 151–60

Huber, G., Pehling, G., Matus, A. 1986. The novel microtubule-associated protein MAP3 contributes to the in vitro assembly of brain microtubules. *J. Biol. Chem.* 261: 2270–73

Kim, H., Binder, L., Rosenbaum, J. L. 1979.

The periodic association of MAP2 with brain microtubules in vitro. *J. Cell Biol.* 80: 266–76

Kuznetsov, S. A., Rodinov, V. I., Gelfand, V. I., Rosenblat, V. A. 1981. Microtubule-associated protein MAP1 promotes microtubule assembly in vitro. *FEBS Lett.* 135: 241–44

Langford, G. M., Williams, E., Peterkin, D. 1986. Microtubule-associated proteins (MAPs) of dogfish brain and squid optic ganglia. *Ann. NY Acad. Sci.* 466: 440–43

Letorneau, P. C., Ressler, A. H. 1984. Inhibition of neurite initiation and growth by taxol. *J. Cell Biol.* 98: 1355–62

Letterier, J. F., Liem, R., Shelanski, M. 1982. Interactions between neurofilaments and microtubule-associated proteins: A possible mechanism for intra-organellar bridging. *J. Cell Biol.* 95: 982–86

Lewis, S. A., Sherline, P., Cowan, N. 1986a. A cloned cDNA encoding MAP1 detects a single copy gene in mouse, and a brain abundant RNA whose level decreases during development. *J. Cell Biol.* 102: 2106–14

Lewis, S. A., Villasante, A., Sherline, P., Cowan, N. 1986b. Brain specific expression of MAP2 detected using a cloned cDNA probe. *J. Cell Biol.* 102: 2098–2105

Magendantz, M., Solomon, F. 1985. Analyzing components of microtubules: Antibodies against chartins, associated proteins from cultured cells. *Proc. Natl. Acad. Sci. USA* 82: 6581–85

Mareck, A., Fellous, A., Francon, A., Nunez, J. 1980. Changes in composition and activity of microtubule-associated proteins during brain development. *Nature* 284: 353–55

Matus, A. 1987. Microtubule-associated proteins and neuronal morphogenesis. In *The Making of the Nervous System*, ed. J. Parnevalas, C. D. Stern, R. V. Stirling. Oxford: Oxford Univ. Press. In press

Matus, A., Bernhardt, R., Bodmer, R., Alaimo, D. 1986. Microtubule-associated protein 2 and tubulin are differently distributed in the dendrites of developing neurons. *Neuroscience* 17: 371–89

Matus, A., Bernhardt, R., Hugh-Jones, T. 1981. High molecular weight microtubule-associated proteins are preferentially associated with dendritic microtubules in brain. *Proc. Natl. Acad. Sci. USA* 78: 3010–14

Matus, A., Riederer, B. 1986. Microtubule-associated proteins in the developing brain. *Ann NY Acad. Sci.* 466: 167–79

Miller, P., Walter, U., Theurkauf, W. E., Vallee, R. B., De Camilli, P. 1982. Frozen tissue sections as an experimental system to reveal specific binding sites for the regulatory subunit of type II cAMP-dependent protein kinase in neurons. *Proc. Natl. Acad. Sci. USA* 79: 5562–66

Murthy, A. S., Flavin, M. 1983. Microtubule assembly using microtubule-associated protein MAP-2 prepared in defined states of phosphorylation with protein kinase and phosphatase. *Eur. J. Biochem.* 137: 37–46

Naidet, C., Semeriva, M., Yamada, K., Thiery, J. P. 1987. Peptides containing the cell-attachment recognition signal Arg-Gly-Asp prevent gastrulation in *Drosophila* embryos. *Nature* 325: 348–50

Nishida, E., Kuwaki, T., Sakai, H. 1981. Phosphorylation of microtubule-associated proteins (MAPs) and pH of the medium control interaction between MAPs and actin filaments. *J. Biochem. Tokyo* 90: 575–78

Nunez, J. 1986. Differential expression of microtubule components during brain development. *Dev. Neurosci.* 8: 125–41

Olmsted, J. B. 1986. Microtubule-associated proteins. *Ann. Rev. Cell Biol.* 2: 421–57

Olmsted, J. B., Asnes, C. F., Parysek, L. M., Lyon, H. D., Kidder, G. M. 1986. Distribution of MAP-4 in cells and in adult and developing mouse tissues. *Ann. NY Acad. Sci.* 466: 292–305

Papasozomenos, S. C., Binder, L. I. 1986. Microtubule-associated protein 2 (MAP2) is present in astrocytes of the optic nerve but absent from astrocytes of the optic tract. *J. Neurosci.* 6: 1748–56

Papasozomenos, C., Binder, L. I. 1987. Phosphorylation determines two distinct species of tau in the central nervous system. *Cell Motil. Cytoskel.* In press

Papasozomenos, S. C., Binder, L. I., Bender, P., Payne, M. R. 1985. Microtubule-associated protein 2 within axons of spinal motor neurons: Associations with microtubules and neurofilaments in normal and b,b'-iminoproprionitrile (IDPN)-treated axons. *J. Cell Biol.* 100: 74–85

Parysek, L. M., Wolosewick, J. J., Olmsted, J. 1984. MAP4: A novel microtubule-associated protein specific for a subset of tissue microtubules. *J. Cell Biol.* 99: 2287–96

Riederer, B., Cohen, R., Matus, A. 1986. MAP5: A novel brain microtubule-associated protein under strong developmental regulation. *J. Neurocytol.* 15: 763–75

Riederer, B., Matus, A. 1985. Differential expression of distinct microtubule-associated proteins during brain development. *Proc. Natl. Acad. Sci. USA* 82: 6006–9

Sattilaro, W. 1986. Interaction of microtubule-associated protein 2 with actin filaments. *Biochemistry* 25: 2003–9

Schulze, E., Kirschner, M. 1986. Microtubule dynamics in interphase cells. *J. Cell Biol.* 102: 1020–31

Seeds, N. W., Gilman, A. G., Amano, T., Nirenberg, M. W. 1970. Regulation of axon formation by clonal lines of a neuronal tumor. *Proc. Natl. Acad. Sci. USA* 60: 160–67

Selden, S. C., Pollard, T. D. 1983. Phosphorylation of microtubule-associated proteins regulates their interaction with actin filaments. *J. Biol. Chem.* 258: 7064–71

Sloboda, R. D., Dentler, W. L., Rosenbaum, J. L. 1976. Microtubule-associated proteins and the stimulation of tubulin assembly in vitro. *Biochemistry* 15: 4497–4505.

Sternberger, L. A., Sternberger, N. H. 1983. Monoclonal antibodies distinguish phosphorylated and non-phosphorylated forms of neurofilaments in situ. *Proc. Natl. Acad. Sci. USA* 80: 6126–30

Tsuyama, S., Bramblett, G. T., Huang, K.-P., Flavin, M. 1986. Calcium/phospholipid-dependent protein kinase recognizes sites in microtubule-associated protein 2 which are phosphorylated in living brain and are not accessible to other kinases. *J. Biol. Chem.* 261: 4110–116

Vallee, R. B., Dibartolomeis, M. J., Theurkauf, W. E. 1981. A protein kinase bound to the projection portion of MAP2 (microtubule-associated protein 2). *J. Cell Biol.* 90: 568–76

Voter, W. A., Erikson, H. P. 1982. Electron microscopy of MAP2 (microtubule-associated protein 2). *J. Ultrastruct. Res.* 80: 374–82

Weingarten, M., Lockwood, A., Hwo, S., Kirschner, M. 1975. A protein factor essential for microtubule assembly. *Proc. Natl. Acad. Sci. USA* 72: 1858–62

Williams, R. C., Detrich, H. W. 1986. Presumptive MAPs and "cold-stable" microtubules from antarctic marine poikilotherms. *Ann. NY Acad. Sci.* 466: 436–39

Yamada, K. M., Spooner, B. S., Wessels, N. K. 1970. Axon growth: Roles of microfilaments and microtubules. *Proc. Natl. Acad. Sci. USA* 66: 1206–12

Zingsheim, H. P., Herzog, W., Weber, K. 1979. Differences in surface morphology of microtubules reconstituted from pure brain tubulin using two different microtubule-associated proteins. *Eur. J. Cell Biol.* 19: 175–83

Ann. Rev. Neurosci. 1988. 11:45–60

5-HYDROXYTRYPTAMINE RECEPTOR SUBTYPES

Stephen J. Peroutka

Departments of Neurology and Pharmacology, Stanford University
Medical Center, Stanford, California 94305

INTRODUCTION

In the four decades since its discovery in the central nervous system
(Twarog & Page 1953), 5-hydroxytryptamine (5-HT) has been implicated
in a number of neuropsychiatric diseases. However, a specific role for 5-
HT in either normal or abnormal brain function has not been identified.
To a significant degree, the inability to determine the pathophysiology of
5-HT in the brain has been due to the lack of specific pharmacological
agents affecting the 5-HT system.

Receptor site analysis has been one of the most productive approaches
to the understanding of the actions of 5-HT in the central and peripheral
nervous systems. In the 30 years since the differentiation of M and D
receptors (Gaddum & Picarelli 1957), it has become clear that multiple 5-
HT receptors exist. The heterogeneity of 5-HT receptors has become
even more apparent in the past decade following the development and
application of radioligand binding techniques. At the present time, at
least five distinct 5-HT binding site subtypes have been differentiated by
radioligand techniques in brain homogenates (Peroutka & Snyder 1979,
Pedigo et al 1981, Hoyer et al 1985b, Heuring & Peroutka 1987). In
addition, 5-HT$_3$ receptors have been identified in the peripheral nervous
system.

This review focuses on current attempts to correlate the pharmacologic
data derived from radioligand binding studies with biochemical and
physiological actions of 5-HT. Although radioligand techniques provide
a rapid and sensitive measure of drug potencies at specific membrane
recognition sites, they cannot differentiate the agonist versus antagonist
properties of drugs, the relative efficacy of agents or drug interactions with

45

0147–006X/88/0301–0045$02.00

other receptor sites involved in a specific physiological response. This type of pharmacologic information can only be derived from physiological experiments. In the following section, a number of physiological models are analyzed for possible relationships to $5-HT_{1A}$, $5-HT_{1B}$, $5-HT_{1C}$, $5-HT_{1D}$, $5-HT_2$, and $5-HT_3$ receptors. As described below, the suggested relationships have been based on the similarities between drug potencies in various physiological systems and their affinities for radioligand binding sites.

Preliminary Binding Studies

The first successful radioligand analysis of 5-HT receptors was reported by Bennett & Aghajanian (1974) and was soon extended and confirmed by other laboratories (Bennett & Snyder 1975, Lovell & Freedman 1976). The second radioligand used to label 5-HT receptors was ^3H-5-HT (Bennett & Snyder 1976, Fillion et al 1978, Nelson et al 1978). Radioligand analysis of 5-HT receptors was considerably advanced by the finding that ^3H-spiperone also labeled 5-HT recognition sites (Leysen et al 1978) in addition to dopamine D_2 receptors.

However, marked differences were noted between the binding characteristics of ^3H-5-HT, ^3H-LSD, and ^3H-spiperone (Peroutka & Snyder 1979). Based on an analysis of drug interactions with each of the three available radioligands, Peroutka & Snyder (1979) concluded that at least two distinct 5-HT membrane recognition sites are present in the central nervous system. The sites labeled by ^3H-5-HT were designated "$5-HT_1$ receptors" and those labeled by ^3H-spiperone were designed "$5-HT_2$ receptors." Since ^3H-LSD had equal affinity for both sites, it was proposed that this ligand could be used to label both $5-HT_1$ and $5-HT_2$ receptors.

Characterization of $5-HT_1$ Binding Site Subtypes

However, $5-HT_1$ binding sites labeled by ^3H-5-HT were soon shown to be heterogeneous. Non-sigmoidal displacement of ^3H-5-HT by spiperone led to the suggestion that sites with high affinity for spiperone should be designated $5-HT_{1A}$ sites, while sites with relatively low affinity for spiperone should be designated $5-HT_{1B}$ sites (Pedigo et al 1981, Schnellmann 1984). A third subtype of total $5-HT_1$ recognition sites (the $5-HT_{1C}$ site) has been identified in the choroid plexus and cortex of various species (Pazos et al 1984a,b, Yagaloff & Hartig 1985, Peroutka 1986). Most recently, a fourth $5-HT_1$ binding site subtype labeled by ^3H-5-HT, the $5-HT_{1D}$ site, has been identified in bovine brain (Heuring & Peroutka 1987). In the past three years, the availability of selective and novel agents has greatly facilitated the analysis of $5-HT_1$ binding site subtypes. A summary of the currently accepted classification system for central 5-HT receptors is presented in

Table 1. A summary of drug potencies at each of the five known 5-HT binding site subtypes is provided in Table 2. These data are discussed in greater detail in the following section.

Table 1 Characteristics of 5-HT$_{1A}$, 5-HT$_{1B}$, 5-HT$_{1C}$, 5-HT$_{1D}$, and 5-HT$_2$ binding sites

	5-HT$_{1A}$	5-HT$_{1B}$	5-HT$_{1C}$	5-HT$_{1D}$	5-HT$_2$
Radiolabeled by	^{3}H-5-HT ^{3}H-8-OH-DPAT ^{3}H-Ipsapirone ^{3}H-WB 4101 ^{3}H-Buspirone ^{3}H-PAPP	^{3}H-5-HT ^{125}I-CYP (rat and mouse only)	^{3}H-5-HT ^{3}H-Mesulergine ^{125}I-LSD	^{3}H-5-HT	^{3}H-Spiperone ^{3}H-Mesulergine ^{125}I-LSD ^{3}H-Ketanserin ^{3}Mianserin ^{3}H-N-methylspiperone ^{125}I-Methyl-LSD ^{3}H-DOB
High density	Raphe nuclei Hippocampus	Substantia nigra Globus pallidus	Choroid plexus	Basal ganglia	Layer IV cortex

Table 2 Drug affinities for 5-HT$_{1A}$, 5-HT$_{1B}$, 5-HT$_{1C}$, 5-HT$_{1D}$, and 5-HT$_2$ receptors[a]

Drug potencies (K_i, nM)	5-HT$_{1A}$	5-HT$_{1B}$	5-HT$_{1C}$	5-HT$_{1D}$	5-HT$_2$
<10	5-CT 8-OH-DPAT 5-HT RU 24969 d-LSD	RU 24969 5-CT 5-HT	Mesulergine Metergoline Methysergide	5-CT 5-HT Metergoline	Spiperone Mesulergine Methysergide Metergoline Mianserin
10–1000	Metergoline Methysergide Spiperone Mesulergine	Metergoline Methysergide d-LSD	Mianserin 5-HT RU 24969 5-CT d-LSD	Methysergide Mianserin 8-OH-DPAT d-LSD RU 24969	d-LSD
>1000	Mianserin	Mianserin Spiperone Mesulergine 8-OH-DPAT	Spiperone 8-OH-DPAT	Mesulergine Spiperone	RU 24969 5-HT 8-OH-DPAT

[a] Data given are derived from Peroutka & Snyder (1979), Peroutka (1986), Hoyer et al (1985b), Heuring & Peroutka (1986) and unpublished observations.

5-HT_{1A} RECEPTORS

Pharmacological Characterization

A number of radioligands have been shown to label the 5-HT_{1A} receptor. Although the 5-HT_{1A} site can be labeled with $^3\text{H-5-HT}$, it can be more directly labeled with $^3\text{H-8-hydroxy-2(di-}n\text{-propyl-amino)tetralin}$ (8-OH-DPAT) (Gozlan et al 1983, Hall et al 1985, Peroutka 1986, Hoyer et al 1985b), $^3\text{H-ipsapirone}$ (formerly called TVX Q 7821) (Dompert et al 1985), $^3\text{H-buspirone}$ (Moon & Taylor 1985) and $^3\text{H-1-(2-(4-aminophenyl)ethyl)-4-(3-trifluro-methylphenyl)piperazine}$ (PAPP) (Asarch et al 1985). In addition $^3\text{H-WB 4101}$, previously considered a selective alpha$_1$-adrenergic radioligand, has been demonstrated to label the 5-HT_{1A} site (Norman et al 1985).

Regardless of the ^3H-ligand used to label the site, it displays high and selective affinity for 8-OH-DPAT, 5-carboxyamido-tryptamine (5-CT), 5-methoxydimethyltryptamine, ipsapirone, and buspirone. The 5-HT_{1A} site is densely present in the CA1 region and dentate gyrus of the hippocampus and in the raphe nuclei (Pazos & Palacios 1985, Hoyer et al 1986a). In addition, $^3\text{H-8-OH-DPAT}$ has been reported to label a presynaptic 5-HT autoreceptor (Gozlan et al 1983, Hall et al 1985) and the "5-HT transporter" (Schoemaker & Langer 1986). Importantly, in the absence of ascorbate, $^3\text{H-8-OH-DPAT}$ also labels glass fiber filter paper (Peroutka & Demopulos 1986, Demopulos & Peroutka 1987). The 5-HT_{1A} binding site is the only site that has been labeled by $^3\text{H-8-OH-DPAT}$ in the presence of ascorbate (Peroutka 1986, Hoyer et al 1985b).

5-HT-Sensitive Adenylate Cyclase

Radioligand binding data support an association between 5-HT_{1A} receptors and a 5-HT-sensitive adenylate cyclase. Regulation of neurotransmitter binding by guanine nucleotides often reflects an association of the agonist binding site with an adenylate cyclase. Thus, GTP and GDP, but not GMP, inhibit the binding of $^3\text{H-8-OH-DPAT}$ to brain homogenates (Hall et al 1985, Schlegel & Peroutka 1986). In addition, guanine nucleotides significantly reduce agonist potencies for $^3\text{H-8-OH-DPAT}$ binding sites whereas antagonist potencies are not affected by nucleotides.

In studies using both rat (Markstein et al 1986) and guinea pig (Shenker et al 1985) hippocampal membranes, a 5-HT-sensitive adenylate cyclase could be stimulated by nanomolar concentrations of 5-HT_{1A} selective agents such as 5-CT and 8-OH-DPAT. In addition, 5-HT can inhibit forskolin-stimulated adenylate cyclase in rat and guinea pig hippocampal membranes (De Vivo & Maayani 1986). The pharmacologic data derived from this system appear to be consistent with a single, homogeneous population of receptors. 8-OH-DPAT, *d*-LSD, and buspirone are similar to 5-HT in their ability to inhibit cyclase activity. By contrast, spiperone is a competitive antagonist at this receptor whereas ketanserin has no effect on 5-HT-induced inhibition of the cyclase activity. The 5-HT_{1A} receptor also appears to mediate inhibition of VIP-stimulated cyclic AMP formation in purified striatal and cortical cultured neurons (Weiss et al 1986). Thus, the 5-HT_{1A} receptor appears to modulate adenylate cyclase activity in certain brain regions.

Neurophysiological Studies

The recent development and characterization of 5-HT_{1A}-selective agents has greatly facilitated the ability to correlate radioligand data with the physiological effects of 5-HT. For example, neurophysiological studies have clearly demonstrated that the 5-HT_{1A} receptor mediates inhibition of raphe nuclei. VanderMaelen & Wildmeran (1984) first showed that buspirone, a 5-HT_{1A}-selective agent, caused complete inhibition of dorsal raphe neuronal firing in the rat. Buspirone also has identical effects in mouse brain slices (Trulson & Arasteh 1985). The ability of other 5-HT_{1A}-selective agents to produce this effect was soon demonstrated by multiple laboratories. For example, ipsapirone, 8-OH-DPAT, and 5-CT were also found to mimic the effect of 5-HT on raphe cell firing (Sinton & Fallon 1986, VanderMaelen et al 1986). By contrast, $(-)$ propranolol, β-adrenergic agent that also displays high affinity for the 5-HT_{1A} receptor (Hiner et al 1986), was shown to reversibly block the inhibitory effects of ipsapirone and 8-OH-DPAT on raphe cell inhibition (Sprouse & Aghajanian 1986).

The hippocampus may be another ideal region in which to study 5-HT_{1A} receptor function since the structure has a high density of 5-HT_{1A} sites (Pazos & Palacios 1985) and is particularly amenable to neurophysiological analysis. Preliminary studies suggested that 5-HT_{1A} drugs may directly inhibit CA1 pyramidal cells (Beck et al 1985, Andrade & Nicoll 1985). However, more detailed analysis suggests that 5-HT_{1A}-selective agents may have effects on hippocampal activity that are not mediated by direct inhibition of CA1 pyramidal cells (Peroutka et al 1987).

Other Systems

5-HT-induced contractions of the canine basilar artery have been proposed to be a functional correlate of the 5-HT_{1A} receptor (Taylor et al 1986, Peroutka et al 1986). Specific components of the 5-HT behavioral syndrome have also been linked to activation of 5-HT_{1A} receptors (Tricklebank 1985, Smith & Peroutka 1986). In addition, studies in male rats have shown that $5\text{-}HT_{1A}$-selective agonists facilitate seminal emissions and/or ejaculations (Kwong et al 1986). The hypotensive potencies in pentobarbitone-anesthesized rats of 8-OH-DPAT and RU 24969 suggest that the $5\text{-}HT_{1A}$ site may mediate these effects (Doods et al 1985). Finally, the thermoregulatory effects of 8-OH-DPAT and RU 24969 also appear to be mediated by the $5\text{-}HT_{1A}$ receptor (Middlemiss et al 1985, Tricklebank 1985, Gudelsky et al 1985).

$5\text{-}HT_{1B}$ RECEPTORS

The putative $5\text{-}HT_{1B}$ site has been more difficult to characterize. Sills et al (1984) defined $5\text{-}HT_{1B}$ binding as specific ^{3}H-5-HT binding in the presence of 1 mM GTP and 2000 nM spiperone. They concluded that RU 24969 and TFMPP were selective $5\text{-}HT_{1B}$ agents. However, these initial selectivity studies were not confirmed when the $5\text{-}HT_{1B}$ site was more directly labeled in rat brain with ^{125}I-cyanopindolol (Pazos et al 1985b, Hoyer et al 1985a,b). The ^{125}I-cyanopindolol site has high affinity for 5-HT and RU 24969 and relatively low affinity for d-LSD and 8-OH-DPAT. The highest densities of $5\text{-}HT_{1B}$ sites in rat brain are found in the globus pallidus, dorsal subiculum, and substantia nigra (Pazos & Palacios 1985). Recent data has demonstrated that this site can also be labeled with ^{3}H-5-HT in rat frontal cortex (Peroutka 1986, Blurton & Wood 1986). Moreover, the $5\text{-}HT_{1B}$ site appears to be species-specific in that it is present in rat and mouse brain but not in guinea pig, cow, chicken, turtle, frog, or human brain membranes (Heuring et al 1986).

The $5\text{-}HT_{1B}$ site has been associated with 5-HT, acetylcholine, and glutamate "autoreceptors" in rat brain. Depolarization of the neurons by potassium or electrical stimulation causes release of the pre-stored neurotransmitter. The release can be inhibited by 5-HT and related agonists, presumably through a "presynaptic autoreceptor." Engel and colleagues (1986) demonstrated that no significant correlation is observed between drug potencies at the 5-HT "autoreceptor" and drug affinities for $5\text{-}HT_{1C}$ or $5\text{-}HT_2$ binding sites. Significant correlations were obtained with drug affinities for both $5\text{-}HT_{1A}$ and $5\text{-}HT_{1B}$ sites. However, selective $5\text{-}HT_{1A}$ drugs such as 8-OH-DPAT and ipsapirone had no effect on the autoreceptor and therefore were not used in the calculation of the correlation coefficient. As a result, the receptor mediating release of 5-HT from nerve terminals is mediated by the $5\text{-}HT_{1B}$ binding site in rat brain. A similar conclusion was reached from an analysis of 5-HT and related drug effects on the release of ^{3}H-acetylcholine from rat hippocampal nerve endings (Maura & Raiteri 1986) and on the release of endogenous glutamate induced by depolarization of rat cerebellum synaptosomes (Raiteri et al 1986).

5-HT$_{1C}$ RECEPTORS

The 5-HT$_{1C}$ site was first characterized in membranes from pig choroid plexus and cortex (Pazos et al 1984a,b, Hoyer et al 1985b). The site was labeled by both ^{3}H-5-HT and ^{3}H-mesulergine. Independently, Yagaloff & Hartig (1985) labeled the site with ^{125}I-LSD in the rat choroid plexus. The site has high affinity for 5-HT, methysergide, and mianserin and relatively low affinity for RU 24969. These pharmacological characteristics are also shared with the putative 5-HT$_{1C}$ site identified in rat cortex (Hoyer et al 1985b, Peroutka 1986).

Biochemical studies have shown that the 5-HT$_{1C}$ site mediates 5-HT-induced phosphoinositide turnover in the rat choroid plexus (Conn et al 1986). The pharmacology of 5-HT-stimulated phosphatidylinositol hydrolysis in choroid plexus was compared to the potency of the drugs versus ^{125}I-LSD binding to 5-HT$_{1C}$ receptors. Mianserin and ketanserin were potent antagonists of 5-HT-induced changes whereas spiperone was more than an order of magnitude less potent in this system. The authors concluded that the 5-HT$_{1C}$ site in choroid plexus is functionally linked to phosphatidylinositol turnover.

5-HT$_{1D}$ RECEPTORS

Recently, a fourth subtype of 5-HT$_1$ site labeled by ^{3}H-5-HT has been identified in bovine brain membranes (Heuring & Peroutka 1987). The addition of either the 5-HT$_{1A}$-selective drug 8-OH-DPAT (100 nM) or the 5-HT$_{1C}$-selective drug mesulergine (100 nM) to the radioligand assay results in a 5–10% decrease in specific ^{3}H-5-HT binding. Competition studies using a series of pharmacologic agents reveal that the sites labeled by ^{3}H-5-HT in bovine caudate in the presence of 100 nM 8-OH-DPAT and 100 nM mesulergine appear to be homogeneous. 5-HT$_{1A}$-selective agents such as 8-OH-DPAT, ipsapirone, and buspirone display micromolar affinities for these sites. RU 24969 and (−)pindolol are approximately two orders of magnitude less potent at these sites than at 5-HT$_{1B}$ sites that have been identified in rat brain. Agents that display nanomolar potencies for 5-HT$_{1C}$ sites such as mianserin and mesulergine are two to three orders of magnitude less potent at the ^{3}H-5-HT binding sites in bovine caudate. In addition, both 5-HT$_2$ and 5-HT$_3$-selective agents are essentially inactive at these binding sites. These ^{3}H-5-HT sites display nanomolar affinity for 5-CT, 5-methoxytryptamine, metergoline, and 5-

HT. Apparent K_i values of 10–100 nM are obtained for d-LSD, RU 24969, methiothepin, tryptamine, methysergide, and yohimbine whereas l-LSD and corynanthine are significantly less potent.

Regional studies demonstrate that this class of sites is most dense in the basal ganglia but exists in all regions of bovine brain. These data identify a homogeneous class of 5-HT$_1$ binding sites in bovine caudate that is pharmacologically distinct from previously defined 5-HT$_{1A}$, 5-HT$_{1B}$, 5-HT$_{1C}$, and 5-HT$_2$ binding site subtypes. The site has been designated the 5-HT$_{1D}$ subtype of binding sites labeled by ^{3}H-5-HT (Heuring & Peroutka 1987). To date, a specific physiological effect of 5-HT has not been correlated with the 5-HT$_{1D}$ binding site. However, the site shares many pharmacological similarities to the inhibitory prejunctional "5-HT$_1$-like" receptor in the isolated perfused rat kidney (Charlton et al 1986) and the receptor mediating 5-HT-induced contractions of the rat stomach fundus (Leysen & Tollenaere 1982, Clineschmidt et al 1985, Cohen & Wittenauer 1986).

5-HT$_2$ RECEPTORS

Pharmacological Characterization

In contrast to the subtypes of the 5-HT$_1$ sites, 5-HT$_2$ sites appear to be homogeneous. ^{3}H-Spiperone, ^{3}H-LSD, ^{3}H-mianserin, ^{3}H-ketanserin, ^{3}H-mesulergine, ^{3}H-N-methylspiperone, ^{125}I-LSD, and N_1-methyl-2-^{125}I-LSD can be used to label the 5-HT$_2$ binding site (Leysen 1981, Leysen et al 1978, 1982, Peroutka & Snyder 1983, Pazos et al 1984a, Engel et al 1984, Kadan et al 1984, Hoffman et al 1985, Lyon et al 1986). Serotonergic antagonists have high affinity for this site while 5-HT and related tryptamines are markedly less potent. The highest level of 5-HT$_2$ binding is in layer IV of the cerebral cortex and caudate, with all other brain regions having substantially fewer binding sites (Pazos et al 1985a, Hoyer et al 1986b). Recently, the hallucinogen ^{3}H-DOB has been shown to label an apparent 5-HT$_2$ recognition site (Titeler & Lyon 1986). Studies are in progress to determine the relevance of this site to 5-HT$_2$ sites labeled by ^{3}H-antagonists.

Phosphatidylinositol Turnover

Besides adenylate cyclase, phosphoinositide turnover is believed to be a common "second messenger" in the transduction of neurotransmitter signals to the cell interior. The hydrolysis of phosphoinositides may modulate a number of intracellular processes including calcium flux, increased arachidonate metabolism, increased cyclic GMP production, and protein kinase C activation. 5-HT has been shown to increase phosphoinositide turnover in the mammalian central nervous system (Brown et al 1984, Conn & Sanders-Bush 1985, Kendall & Nahorski 1985).

5-HT-induced phosphoinositide hydrolysis in rat cerebral cortex (Kendall & Nahorski 1985, Conn & Sanders-Bush 1985) appears to occur as a result of 5-HT_2 receptor activation. 5-HT stimulates phosphoinositide turnover with an EC_{50} of 1 uM. The response to 5-HT is blocked by nanomolar concentrations of ketanserin, and phosphoinositide turnover is not affected by 8-OH-DPAT. Furthermore, tricyclic antidepressants decrease both 5-HT_2 binding sites and 5-HT-induced phosphoinositide turnover (Conn & Sanders-Bush 1986). The rat thoracic aorta (Roth et al 1984), cultured bovine aortic smooth muscle cells (Coughlin et al 1984), and platelets (de Courcelles et al 1985) are three additional systems in which 5-HT appears to modulate phosphoinositide turnover via a receptor that is similar to the 5-HT_2 binding site.

Neurophysiological Studies

To date, two specific neurophysiological effects of 5-HT have been attributed to activation of the 5-HT_2 receptor. Extracellular studies have indicated that 5-HT facilitates excitation of facial motor neurons. The effect of 5-HT could be antagonized by selective 5-HT_2 antagonists such as methysergide, cyproheptadine, and cinanserin (McCall & Aghajanian 1980). Similar data were derived from intracellular studies of this brainstem nucleus. Thus, 5-HT caused a slow depolarization of facial motor neuron membranes that remained subthreshold. The membrane effects of 5-HT could be blocked by methysergide (VanderMaelen & Aghajanian 1980). These data suggest that 5-HT_2 receptors mediate 5-HT-induced excitation of facial motor neurons.

More recently, Davies et al (1987) demonstrated that 5-HT caused a slow depolarization of 68% of cortical neurons, which was associated with a decreased conductance. The response displayed some voltage dependency and was desensitized by repeated 5-HT applications. The selective 5-HT_2 receptor antagonists, ritanserin and cinanserin, blocked the depolarizing effects of 5-HT. Therefore, the effects of 5-HT on cortical pyramidal neurons share many similarities to the depolarizing effects of 5-HT in the facial motor nucleus and also appear to be mediated by 5-HT_2 receptors.

Other Systems

In contrast to 5-HT-induced contractions of the canine basilar artery, 5-HT-induced contractions of many other vascular tissues are mediated by 5-HT_2 receptors (Peroutka 1984). A number of behavioral effects of 5-HT have also been attributed to 5-HT_2 receptors. For example, drug antagonism of the "head-shake" or "head-twitch component" of the 5-HT behavioral syndrome has clearly been related to a blockade of 5-HT_2 receptors (Peroutka et al 1981, Leysen et al 1982, 1984). Likewise, tryptamine-induced seizure activity can be prevented by selective 5-HT_2 antagonists (Leysen et al 1978). Behavioral studies of the discriminative cue properties of 5-HT agonists have concluded that this behavioral response may be mediated by 5-HT_2 receptors (Glennon et al 1983).

Antagonism of 5-HT induced forepaw edema in the rat by 22 antagonists correlates with drug affinity for 5-HT_2 sites labeled by ^{3}H-spiperone (Ortmann et al 1982). Similarly, drug antagonism of tracheal smooth muscle contraction, in vivo bronchoconstriction and contraction of guinea pig ileum is consistent with mediation by 5-HT_2 sites (Leysen et al 1984). 5-HT_2 receptors have also been implicated in the regulation of aldosterone production (Matsuoka et al 1985). 5-HT induction of platelet shape changes and aggregation may also be mediated by 5-HT_2 receptors (Leysen et al 1984).

5-HT_3 RECEPTORS

The pharmacological characteristics of peripheral 5-HT receptors are quite distinct from the 5-HT binding sites identified in brain. In general, 5-HT effects in the peripheral nervous system are not affected by 5-HT_1 and/or 5-HT_2-selective drugs such as 8-OH-DPAT and ketanserin. Nevertheless, 5-HT has multiple effects in the periphery that can often be blocked with drugs such as MDL 72222, (−)cocaine, metoclopramide, and ICS 205-930 (Fozard 1984, Richardson et al 1985, Richardson & Engel 1986). Recently, it has been suggested that this class of peripheral 5-HT receptors be designated 5-HT_3 receptors (Bradley et al 1986).

However, it is clear that heterogeneity exists within the 5-HT_3 class of receptors. Significant differences exist between the potency of selective 5-HT_3 antagonists in various physiological systems. For example, MDL 72222 has a pA_2 of 9.1 against 5-HT-induced effects on postganglionic sympathetic and parasympathetic neurons in the rabbit heart but is inactive against putative 5-HT_3 receptors in the guinea pig ileum (Richardson & Engel 1986). The development of more selective and varied pharmacological agents should further define the pharmacological characteristics and functional correlates of 5-HT_3 receptors and their probable subtypes.

FUTURE TRENDS

In the present report, evidence is provided from a variety of biochemical, physiological, and behavioral studies that multiple 5-HT receptors exist in the central nervous system. The differentiation of 5-HT receptors into 5-HT_{1A}, 5-HT_{1B}, 5-HT_{1C}, 5-HT_{1D}, 5-HT_2, and 5-HT_3 subtypes still appears to be the most relevant classification system. Indeed, a number of functional correlates of these "5-HT binding sites" have been proposed (Table 3). In many of these systems, significant correlations have been documented between physiological drug potencies and drug affinities for specific binding site subtypes.

The development of selective 5-HT_1 subtype agonists such as 8-OH-DPAT and RU 24969 has greatly facilitated research into the analysis of 5-HT_1 binding site subtypes. To a significant degree, the recent advancements in the analysis of 5-HT receptor subtypes are a direct result of radioligand binding techniques. Although binding techniques have been criticized in terms of their relevance to functional receptor sites, the current data strongly support an association between the majority of serotonergic binding sites defined in radioligand studies and distinct physiological responses.

Further advancements in the understanding of the physiological effects of 5-HT receptors would have important implications. To the basic scien-

Table 3   Proposed functional correlates of 5-HT receptor subtypes

5-HT_{1A}	Adenylate cyclase modulation
	Raphe cell and CA1 hippocampal cell inhibition
	Canine basilar artery contractions
	Forepaw treading, tremor, head-weaving
	Facilitation of ejaculation and/or seminal emissions
	Thermoregulation
	Hyportensive effects
5-HT_{1B}	"Autoreceptor"
5-HT_{1C}	Phosphatidylinositol turnover
5-HT_{1D}	? Rat kidney perfusion
	? Rat stomach fundus
5-HT_2	Phosphatidylinositol turnover
	Contraction of vascular smooth muscle
	5-Hydroxytryptophan or mescaline induced head twitches
	Tryptamine induced seizures
	Rat forepaw edema
	Discriminative cue properties
	Contraction of bronchial smooth muscle
	Platelet shape changes and aggregation
	Smooth muscle prostacyclin synthesis
5-HT_3	Depolarization of postgaglionic autonomic neurons
	Contraction of ileal smooth muscle
	Pain, wheal, and flare reaction

tist, characterization of all specific 5-HT receptors would greatly clarify the role of 5-HT in the central nervous system. Clinically, 5-HT has been implicated in a number of human disorders such as anxiety, depression, migraine, vasospasm, and epilepsy, although the exact role of 5-HT in these disorders remains speculative. Analysis of 5-HT receptor subtypes and their functional role in the central nervous system should greatly elucidate the pathophysiological basis of many of these human diseases.

By defining the pharmacological characteristics of each 5-HT receptor subtype, an opportunity is created to develop potent and selective agents to interact with each site. In the recent past, this approach to drug design has proven tremendously successful in the development of β-adrenergic and histamine$_2$ antagonists. Similar progress in the 5-HT field is likely to yield important information about 5-HT function in normal and abnormal brain and may lead to the development of unique therapeutic agents in neuropsychiatry.

ACKNOWLEDGMENTS

I thank Faith H. Smith for assistance with manuscript preparation. This work was supported in part by the John A. and George L. Hartford Foundation, Alfred P. Sloan Foundation, the McKnight Foundation, and National Institute of Health Grants NS12151-12 and NS23560-01.

Literature Cited

Andrade, R., Nicoll, R. A. 1985. The novel anxiolytic buspirone elicits a small hyperpolarization and reduces serotonin responses at putative 5-HT$_1$ receptors on hippocampal CA1 pyramidal cells. *Soc. Neurosci. Abstr.* 11: 597

Asarch, K. B., Ransom, R. W., Shih, J. C. 1985. [^{3}H]-1-(2-(4-aminophenyl)ethyl)-4-(3-trifluoromethylphenyl) Piperazine: A selective radioligand for 5-HT$_{1A}$ receptors in rat brain. *Soc. Neurosci.* 11: 1257

Beck, S. G., Clarke, W. P., Goldfarb, J. 1985. Spiperone differentiates multiple 5-hydroxytryptamine responses in rat hippocampal slices in vitro. *Eur. J. Pharmacol.* 116: 195–97

Bennett, J. L., Aghajanian, G. K. 1974. *d*-LSD binding to brain homogenates: Possible relationship to serotonin receptors. *Life Sci.* 15: 1935–44

Bennett, J. P. Jr., Snyder, S. H. 1975. Stereospecific binding of *d*-lysergic acid diethylamide (LSD) to brain membranes: Relationship to serotonin receptors. *Brain Res.* 94: 523–44

Bennett, J. P. Jr., Snyder, S. H. 1976. Serotonin and lysergic acid diethylamide binding in rat brain membranes: Relationship to postsynaptic serotonin receptors. *Mol. Pharmacol.* 12: 373–89

Blurton, P. A., Wood, M. D. 1986. Identification of multiple binding sites for [^{3}H]5-hydroxytryptamine in the rat CNS. *J. Neurochem.* 46: 1392–98

Bradley, P. B., Engel, G., Feniuk, W., Fozard, J. R., Humphrey, P. P. A., Middlemiss, D. N., Mylecharane, E. J., Richardson, B. P., Saxena, P. R. 1986. Proposals for the classification and nomenclature of functional receptors for 5-hydroxytryptamine. *Neuropharmacology* 25: 563–76

Brown, E., Kendall, D. A., Nahorski, S. R. 1984. Inositol phospholipid hydrolysis in rat cerebral cortical slices: I. Receptor characterization. *J. Neurochem.* 42: 1379–87

Charlton, K. G., Bond, R. A., Clarke, D. E. 1986. An inhibitory prejunctional 5-HT$_1$-like receptor in the isolated perfused rat kidney. *Naunyn Schmiedeberg's Arch. Pharmacol.* 332: 8–15

Clineschmidt, B. V., Reiss, D. R., Pettibone, J., Robinson, J. L. 1985. Characterization of 5-hydroxytryptamine receptors in rat stomach fundus. *J. Pharmacol. Exp. Ther.* 235: 696–708

Cohen, M. L., Wittennauer, L. A. 1986. Further evidence that the serotonin receptor in the rat stomach fundus is not 5-HT$_{1A}$ or 5-HT$_{1B}$. *Life Sci.* 38: 1–5

Conn, P. J., Sanders-Bush, E. 1985. Serotonin-stimulated phosphoinositide turnover: Mediation by the S$_2$ binding site in rat cerebral cortex but not in subcortical regions. *J. Pharmacol. Exp. Ther.* 234: 195–203

Conn, P. J., Sanders-Bush, E. 1986. Regulation of serotonin-stimulated phosphoinositide hydrolysis: Relation to the 5-HT-2 binding site. *J. Neurosci.* 6: 3369–79

Conn, P. J., Sanders-Bush, E., Hoffman, B. J., Hartig, P. R. 1986. A unique serotonin receptor in choroid plexus is linked to phosphatidylinositol turnover. *Proc. Natl. Acad. Sci. USA* 83: 4086–88

Coughlin, S. R., Moskowitz, M., Levine, L. 1984. Identification of a serotonin type 2 receptor linked to prostacyclin synthesis in vascular smooth muscle cells. *Biochem. Pharmacol.* 33: 692–95

Davies, M. F., Deisz, R. A., Prince, D. A., Peroutka, S. J. 1987. Two distinct effects of 5-hydroxytryptamine on cortical neurons. *Brain Res.* In press

de Courcelles, D., Leysen, J. E., De Clerck, F., Van Belle, H., Janssen, P. A. 1985. Evidence that phospholipid turnover is the signal transducing system coupled to serotonin-S$_2$ receptor sites. *J. Biol. Chem.* 260: 7603–8

De Vivo, M., Maayani, S. 1986. Characterization of the 5-hydroxytryptamine$_{1A}$ receptor-mediated inhibition of forskolin-stimulated adenylate cyclase activity in guinea pig and rat hippocampal membranes. *J. Pharmacol. Exp. Ther.* 238: 248–53

Demopulos, C. M., Peroutka, S. J. 1987. "Specific" ^{3}H-8-OH-DPAT binding to glass fiber filter paper: Implications for the analysis of serotonin binding site subtypes. *Neurochem. Int.* 10: 371–76

Dompert, W. U., Glaser, T., Traber, J. 1985. ^{3}H-TVX Q 7821: Identification of 5-HT$_1$ binding sites as target for a novel putative anxiolytic. *Naunyn-Schmiedeberg's Arch. Pharmacol.* 328: 467–70

Doods, H. N., Kalkman, H. O., De Jonge, A., Thoolen, M., Wilffert, B., Timmermans, P., Van Zwieten, P. A. 1985. Differential selectivities of RU 24969 and 8-OH-DPAT for the purported 5-HT$_{1A}$ and 5-HT$_{1B}$ binding sites. Correlation between 5-HT$_{1A}$ affinity and hypotensive activity. *Eur. J. Pharmacol.* 112: 363–70

Engel, G., Gothert, M. K., Hoyer, D., Schlicker, E., Hillenbrand, K. 1986. Identity of inhibitory presynaptic 5-hydroxytryptamine (5-HT) autoreceptors in the rat brain cortex with 5-HT$_{1B}$ binding sites. *Naunyn-Schmiedeberg's Arch. Pharmacol.* 357: 1–7

Engel, G., Muller-Schweinitzer, E., Palacios, J. M. 1984. 2-[^{125}Iodo] LSD, a new ligand for the characterization and localization of 5HT$_2$ receptors. *Naunyn-Schmiedeberg's Arch. Pharmacol.* 325: 328–36

Fillion, G. M. B., Rousselle, J., Fillion, M., Beaudoin, D. M., Goiny, M. R., Deniau, J., Jacob, J. J. 1978. High-affinity binding of [^{3}H]5-Hydroxytryptamine to brain synaptosomal membranes: Comparison with [^{3}H]lysergic acid diethylamide binding. *Mol. Pharmacol.* 14: 50–59

Fozard, J. R. 1984. Neuronal 5-HT receptors in the periphery. *Neuropharmacology* 23: 1473–86

Gaddum, J. H., Picarelli, Z. P. 1957. Two kinds of tryptamine receptor. *Br. J. Pharmacol. Chemother.* 12: 323–28

Glennon, R. A., Young, R., Rosecrans, J. A. 1983. Antagonism of the effects of the hallucinogen DOM and the purported 5-HT agonist quipazine by 5-HT$_2$ antagonists. *Eur. J. Pharmacol.* 91: 189–96

Gozlan, H., El Mestikawy, S., Pichat, L., Glowinski, J., Hamon, M. 1983. Identification of presynaptic serotonin autoreceptors using a new ligand: ^{3}H-PAT. *Nature* 305: 140–42

Gudelsky, G. A., Koenig, J. I., Meltzer, H. Y. 1985. Serotonin receptor subtypes and thermoregulation. *Abstr. Am. Soc. Neuropsychopharmacol.* 24: 64

Hall, M. D., El Mestikawy, S., Emerit, M. B., Pichat, L., Hamon, M., Gozlan, H. 1985. [^{3}H]8-hydroxy-2-(Di-n-propylamino)tetralin binding to pre- and postsynaptic 5-hydroxytryptamine sites in various regions of the rat brain. *J. Neurochem.* 44: 1685–96

Heuring, R. E., Schlegel, J. R., Peroutka, S. J. 1986. Species variations in 5-HT$_{1B}$ and 5-HT$_{1C}$ binding sites defined by RU 24969 competition studies. *Eur. J. Pharmacol.* 122: 279–82

Heuring, R. E., Peroutka, S. J. 1987. Characterization of ^{3}H-5-HT binding in bovine caudate. *J. Neurosci.* 7: 894–903

Hiner, B. C., Roth, H. L., Peroutka, S. J. 1986. Antimigraine drug interactions with 5-hydroxytryptamine$_{1A}$ receptors. *Ann. Neurol.* 19: 511–13

Hoffman, B. J., Karpa, M. D., Lever, J. R., Hartig, P. R. 1985. N_1-methyl-2-I^{125} LSD (I^{125} MIL), a preferred ligand for serotonin

58 PEROUTKA

5-HT$_2$ receptors. *Eur. J. Pharmacol.* 110: 147–48

Hoyer, D., Engel, G., Kalkman, H. O. 1985a. Characterization of the 5-HT$_{1B}$ recognition site in rat brain: Binding studies with (−) [^{125}I] iodocyanopindolol. *Eur. J. Pharmacol.* 118: 1–12

Hoyer, D., Engel, G., Kalkman, H. O. 1985b. Molecular pharmacology of 5-HT$_1$ and 5-HT$_2$ recognition sites in rat and pig brain membranes: Radioligand binding studies with [^{3}H]5-HT, [^{3}H]8-OH-DPAT, (−) [^{125}I]iodocyanopindolol, [^{3}H]mesulergine and [^{3}H]ketanserin. *Eur. J. Pharmacol.* 118: 13–23

Hoyer, D., Pazos, A., Probst, A., Palacios, J. M. 1986a. Serotonin receptors in the human brain: I. Characterization and autoradiographic localization of 5-HT$_{1A}$ recognition sites. Apparent absence of 5-HT$_{1B}$ recognition sites. *Brain Res.* 376: 85–96

Hoyer, D., Pazos, A., Probst, A., Palacios, J. M. 1986b. Serotonin receptors in the human brain: II. Characterization and autoradiographic localization of 5-HT$_{1C}$ and 5-HT$_2$ recognition sites. *Brain Res.* 376: 97–107

Kadan, M. J., Krohn, A. M., Evans, M. J., Waltz, R. L., Hartig, P. R. 1984. Characterization of ^{125}I-lysergic acid diethylamide binding to serotonin receptors in rat frontal cortex. *J. Neurochem.* 43: 601–6

Kendall, D. A., Nahorski, S. R. 1985. 5-hydroxytryptamine-stimulated inositol phospholipid hydrolysis in rat cerebral cortex slices: Pharmacological characterization and effects of antidepressants. *J. Pharmacol. Exp. Ther.* 233: 473–79

Kwong, L. L., Smith, E. R., Davidson, J. M., Peroutka, S. J. 1986. Differential interactions of "prosexual" drugs with 5-hydroxytryptamine$_{1A}$ and alpha$_2$-adrenergic receptors. *Behav. Neurosci.* 100: 664–68

Leysen, J. E. 1981. Serotonergic receptors in brain tissue: Properties and identification of various ^{3}H-ligand binding studies in vitro. *J. Physiol. Paris* 77: 351–62

Leysen, J. E., de Courcelles, D. C., De Clerck, F., Niemegeers, J. E. Van Nueten, J. M. 1984. Serotonin-S$_2$ receptor binding sites and functional correlates. *Neuropharmacology* 23: 1493–1501

Leysen, J. E., Niemegeers, C. J. E., Tollenaere, J. P., Laduron, P. M. 1978. Serotonergic component of neuroleptic receptors. *Nature* 272: 163–66

Leysen, J. E., Niemegeers, C. J. E., Van Nueten, J. M., Laduron, P. M. 1982. ^{3}H-Ketanserin (R 41 468), a selective ^{3}H-ligand for receptor binding sites. *Mol.*

Pharmacol. 21: 301–14

Leysen, J. E., Tollenaere, J. P. 1982. Biochemical models for serotonin receptors. *Ann. Rev. Med. Chem.* 17: 1–10

Lovell, R. A., Freedman, D. X. 1976. Stereospecific receptor sites for *d*-lysergic diethylamide in rat brain: Effects of neurotransmitters, amine antagonists, and other psychotropic drugs. *Mol. Pharmacol.* 12: 620–30

Lyon, R. A., Titeler, M., Frost, J. J., Whitehouse, P. J., Wong, D. F., Wagner, H. N., Dannals, R. F., Links, J. M., Kuhar, M. J. 1986. ^{3}H-3-*N*-Methylspiperone labels dopamine D$_2$ receptors in basal ganglia and S$_2$ serotonin receptors in cerebral cortex. *J. Neurosci.* 6: 2941–49

Markstein, R., Hoyer, D., Engel, G. 1986. 5-HT$_{1A}$-receptors mediate stimulation of adenylate cyclase in rat hippocampus. *Naunyn-Schmiedeberg's Arch. Pharmacol.* 333: 335–41

Maura, G., Raiteri, M. 1986. Cholinergic terminals in rat hippocampus possess 5-HT$_{1B}$ receptors mediating inhibition of acetylcholine release. *Eur. J. Pharmacol.* 129: 333–37

Matsuoka, H., Ishii, M., Goto, A., Sugimoto, T. 1985. Role of serotonin type 2 receptors in regulation of aldosterone production. *Am. J. Physiol.* 234: 234–38

McCall, R. B., Aghajanian, G. K. 1980. Pharmacological characterization of serotonin receptors in the facial motor nucleus: A microiontophoretic study. *Eur. J. Pharmacol.* 65: 175–83

Middlemiss, D. N., Neill, J., Tricklebank, M. D. 1985. Subtypes of the 5-HT receptor involved in hypothermia and forepaw treading induced by 8-OH-DPAT. *Br. J. Pharmacol.* 85: 251P

Moon, S. L., Taylor, D. P. 1985. *In vitro* autoradiography of ^{3}H-buspirone and ^{3}H-2-deoxyglucose after buspirone administration. *Soc. Neurosci. Abstr.* 11: 114

Nelson, D. L., Herbet, A., Bourgoin, S., Glowinski, J., Hamon, M. 1978. Characteristics of central 5-HT receptors and their adaptive changes following intracerebral 5,7-dihydroxytryptamine administration in the rat. *Mol. Pharmacol.* 14: 983–95

Norman, A. B., Battaglia, G., Morrow, A. L., Creese, I. 1985. [^{3}H]WB4101 labels S$_1$ serotonin receptors in rat cerebral cortex. *Eur. J. Pharmacol.* 106: 461–62

Ortmann, R., Bischoff, S., Radeke, E., Buech, O., Delini-Stula, A. 1982. Correlations between different measures of antiserotonin activity of drugs. *Naunyn-Schmiedeberg's Arch. Pharmacol.* 321: 265–70

Pazos, A., Palacios, J. M. 1985. Quantitative

autoradiographic mapping of serotonin receptors in the rat brain. I. Serotonin-1 receptors. *Brain Res.* 346: 205–30

Pazos, A., Cortes, R., Palacios, J. M. 1985a. Quantitative autoradiographic mapping of serotonin receptors in the rat brain. II. Serotonin-2 receptors. *Brain Res.* 346: 231–49

Pazos, A., Engel, G., Palacios, J. M. 1985b. Beta-adrenoceptor blocking agents recognize a subpopulation of serotonin receptors in brain. *Brain Res.* 343: 403–8

Pazos, A., Hoyer, D., Palacios, J. M. 1984a. Mesulergine, a selective serotonin-2 ligand in the rat cortex, does not label these receptors in porcine and human cortex: Evidence for species differences in brain serotonin-2 receptors. *Eur. J. Pharmacol.* 106: 531–38

Pazos, A., Hoyer, D., Palacios, J. M. 1984b. The binding of serotonergic ligands to the porcine choroid plexus: Characterization of a new type of serotonin recognition site. *Eur. J. Pharmacol.* 106: 539–46

Pedigo, N. W., Yamamura, H. I., Nelson, D. L. 1981. Discrimination of multiple [^{3}H]5-hydroxytryptamine binding sites by the neuroleptic spiperone in rat brain. *J. Neurochem.* 36: 220–26

Peroutka, S. J. 1984. Vascular serotonin receptors: Correlation with 5-HT$_1$ and 5-HT$_2$ binding sites. *Biochem. Pharmacol.* 33: 2349–53

Peroutka, S. J. 1986. Pharmacological differentiation and characterization of 5-HT$_{1A}$, 5-HT$_{1B}$ and 5-HT$_{1C}$ binding sites in rat frontal cortex. *J. Neurochem.* 47: 529–40

Peroutka, S. J., Demopulos, C. M. 1986. ^{3}H-8-OH-DPAT "specifically" labels glass fiber filter paper. *Eur. J. Pharmacol.* 129: 199–200

Peroutka, S. J., Lebovitz, R. M., Snyder, S. H. 1981. Two distinct central serotonin receptors with different physiological functions. *Science* 212: 827–29

Peroutka, S. J., Huang, S., Allen, G. S. 1986. Canine basilar artery contractions mediated by 5-hydroxytryptamine$_{1A}$ receptors. *J. Pharmacol. Exp. Ther.* 237: 901–6

Peroutka, S. J., Mauk, M. D., Kocsis, J. D. 1987. Modulation of hippocampal neuronal activity by 5-hydroxytryptamine and 5-hydroxytryptamine$_{1A}$ selective drugs. *Neuropharmacology* 26: 139–46

Peroutka, S. J., Snyder, S. H. 1979. Multiple serotonin receptors: Differential binding of ^{3}H-serotonin, ^{3}H-lysergic acid diethylamide and ^{3}H-spiroperidol. *Mol. Pharmacol.* 16: 687–99

Peroutka, S. J., Snyder, S. H. 1983. Multiple serotonin receptors and their physiological significance. *Fed. Proc.* 42: 213–17

Raiteri, M., Maura, G., Bonanno, G., Pittaluga, A. 1986. Differential pharmacology and function of two 5-HT$_1$ receptors modulating transmitter release in cerebellum. *J. Pharmacol. Exp. Ther.* 237: 644–48

Richardson, B. P., Engel, G. 1986. The pharmacology and function of 5-HT$_3$ receptors. *Trends Neurosci.* 7: 424–28

Richardson, B. P., Engel, G., Donatsch, P., Stadler, P. A. 1985. Identification of serotonin M-receptor subtypes and their specific blockade by a new class of drugs. *Nature* 336: 126–31

Roth, B. L., Nakaki, T., Chuang, D. M., Costa, E. 1984. Aortic recognition sites for serotonin (5-HT) are coupled to phospholipase C and modulate phosphatidylinositol turnover. *Neuropharmacology* 23: 1223–25

Schlegel, J. S., Peroutka, S. J. 1986. Nucleotide interactions with 5-HT$_{1A}$ binding sites directly labeled by [^{3}H]-8-hydroxy-2-(di-n-propylamino)tetralin ([^{3}H]8-OH-DPAT). *Biochem. Pharmacol.* 35: 1943–49

Schnellmann, R. G., Waters, S. J., Nelson, D. L. 1984. [^{3}H]5-hydroxytryptamine binding sites: Species and tissues variation. *J. Neurochem.* 42: 65–70

Schoemaker, H., Langer, S. Z. 1986. [^{3}H]8-OH-DPAT labels the serotonin transporter in the rat striatum. *Eur. J. Pharmacol.* 124: 371–73

Shenker, A., Maayani, S., Weinstein, H., Green, J. P. 1985. Two 5-HT receptors linked to adenylate cyclase in guinea pig hippocampus are discriminated by 5-carboxamidotryptamine and spiperone. *Eur. J. Pharmacol.* 109: 427–29

Sills, M. A., Wolfe, B. B., Frazer, A. 1984. Determination of selective and nonselective compounds for the 5-HT$_{1A}$ and 5-HT$_{1B}$ receptor subtypes in rat frontal cortex. *J. Pharmacol. Exp. Ther.* 231: 480–87

Sinton, C. M., Fallon, S. L. 1986. Differences in response of dorsal and median raphe serotonergic neurons to 5-HT1 receptor ligands. *Soc. Neurosci. Abstr.* 12: 1239

Smith, L. M., Peroutka, S. J. 1986. Differential effects of 5-hydroxytryptamine$_{1A}$ selective drugs on the 5-HT behavioral syndrome. *Pharmacol. Biochem. Behav.* 24: 1513–19

Sprouse, J. S., Aghajanian, G. K. 1986. Inhibition of serotonergic dorsal raphe cell firing by 5-HT$_{1A}$ agonists: Stereoselective blockade by propranolol. *Soc. Neurosci. Abstr.* 12: 1240

Taylor, E. W., Duckles, S. P., Nelson, D. L. 1986. Dissociation constants of serotonin agonists in the canine basilar artery correlate to K_i values at the 5-HT$_{1A}$ binding site. *J. Pharmacol. Exp. Ther.* 236: 118–25

Titeler, M., Lyon, R. A. 1986. ³H-DOB, a radioactive hallucinogen, labels a guanine nucleotide-sensitive state (5-HT$_{2H}$) of the 5-HT$_2$ serotonin receptor. *Soc. Neurosci. Abstr.* 12: 311

Tricklebank, M. D. 1985. The behavioral response to 5-HT receptor agonists and subtypes of the central 5-HT receptor. *Trends Pharmacol. Sci.* 6: 403–7

Trulson, M. E., Arasteh, K. 1985. Buspirone decreases the activity of 5-hydroxy-tryptamine-containing dorsal raphe neurons in-vitro. *J. Pharm. Pharmacol.* 38: 380–82

Twarog, B. M., Page, I. 1953. Serotonin content of some mammalian tissues and urine and a method for its determination. *Am. J. Psychiat.* 175: 157–61

VanderMaelen, C. P., Aghajanian, G. K. 1980. Intracellular studies showing modulation of facial motoneurone excitability by serotonin. *Nature* 287: 346–47

VanderMaelen, C. P., Gehlbach, G., Yocca, F. D., Mattson, R. J. 1986. Inhibition of serotonergic dorsal raphe neurons in rat brain slice by the 5-HT$_1$ agonist 5-carboxyamidotryptamine. *Soc. Neurosci. Abstr.* 12: 1239

VanderMaelen, C. P., Wilderman, R. C. 1984. Buspirone, a non-benzodiazepine anxiolytic drug, causes inhibition of serotonergic dorsal raphe neurons in the rat. *Soc. Neurosci. Abstr.* 10: 259

Weiss, S., Sebben, M., Kemp, D., Bockaert, J. 1986. Serotonin 5-HT$_1$ receptors mediate inhibition of cyclic AMP production in neurons. *Eur. J. Pharmacol.* 120: 227–30

Yagaloff, K. A., Hartig, P. R. 1985. ¹²⁵I-LSD binds to a novel serotonergic site on rat choroid plexus epithelial cells. *J. Neurosci.* 5: 3178–83

Ann. Rev. Neurosci. 1988. 11:61–80

EXCITATORY AMINO ACID NEUROTRANSMISSION: NMDA Receptors and Hebb-Type Synaptic Plasticity

Carl W. Cotman and Daniel T. Monaghan

Department of Psychobiology, University of California, Irvine, California 92717

Alan H. Ganong

Division of Neuroscience, Beckman Research Institute of the City of Hope, Duarte, California 91010

Introduction

The study of excitatory amino acids as neurotransmitters has recently resulted in many new and fundamental concepts in neuroscience. Much of this progress centers upon the role of N-methyl-D-aspartate (NMDA) receptors in CNS synaptic transmission and plasticity. In this review, we describe the properties of the excitatory amino acid receptor classes, particularly those of the NMDA class, and indicate how properties of this receptor permit a unifying concept for understanding plasticity associated with development and learning.

Excitatory amino acids, primarily L-glutamate and related derivatives, are the major excitatory neurotransmitters in the vertebrate CNS. Extensive biochemical, anatomical and electrophysiological analysis has shown that in corticofugal, corticocortical, and other pathways, the excitatory amino acids have the essential properties indicative of neurotransmitters (for reviews see Fagg & Foster 1983, Fonnum 1984, Cotman & Monaghan 1987). The excitatory amino acids act through multiple receptor classes, of which the NMDA receptor class is the most well understood. NMDA

0147–006X/88/0301–0061$02.00

receptor activation may initiate developmental plasticity or long-term potentiation, a synaptic analogue of memory, or, if excessively activated, result in seizures and/or injury to the cell. Thus, NMDA receptors may be a final common pathway for plasticity but also, ironically, pathology. What molecular and synaptic mechanisms make NMDA receptors pivotal?

We begin the review with a definition of NMDA receptors relative to other subtypes, and proceed to their role in the mature nervous system as signal amplifiers and their properties that allow such functions. We outline their role in learning, CNS pathology, and developmental plasticity. Many of the present data converge to indicate that NMDA receptors may serve as a molecular mechanism for Hebb-type synapses, thereby providing a molecular approach to understanding learning.

Pharmacological Definition of NMDA Receptors

The clarification of the role of NMDA receptors in a variety of neural systems has followed the development of specific antagonists that distinguish NMDA receptors from other subtypes of excitatory amino acid receptors. The first demonstration of selective NMDA receptor antagonist activity by a series of monoamino dicarboxylic acids (Hall et al 1977, Biscoe et al 1977) was followed by the development of a series of more potent phosphonic acid derivatives. As in the original series, the phosphonic acid derivatives with four or six carbons separating the terminal acidic groups (2-amino-5-phosphonopentanoate, or AP5, and 2-amino-7-phosphonoheptanoate, or AP7, respectively) are the most potent NMDA receptor antagonists, with the activity residing in the D-isomer (Davies et al 1981, Evans et al 1982, Perkins et al 1981). The antagonists AP5 and AP7 have been the key compounds in identifying NMDA receptor action in a variety of CNS regions in both electrophysiological and biochemical assays. The most potent and selective NMDA antagonist currently available is a cyclic analogue of AP7, 3-((+)-2-carboxypiperazin-4-yl)-propyl-1-phosphonate (CPP) (Davies et al 1986a, Harris et al 1986).

NMDA receptors represent one of at least four classes of excitatory amino acid receptor subtypes (for review see; Foster & Fagg 1984, Watkins 1984, Cotman & Monaghan 1987). Of these, the NMDA, kainate, and quisqualate classes are defined by their selective agonists. Both kainate and quisqualate receptors are unaffected by the selective NMDA antagonists and are distinguished from one another by the preferential blockade of quisqualate responses by glutamate diethyl ester. A fourth class is defined by the antagonist action of L-2-amino-4-phosphonobutyrate. Structure-activity studies have shown that the general requirements necessary for a compound to mimic or inhibit the activity of L-glutamate include (*a*) at least one acidic group (usually carboxyl), (*b*) an amino group on the same carbon as the acidic function, and (*c*) a second acidic (carboxylic, phosphonic, phosphoric, or sulfonic) or at least a polar group at a relative distance of 2–4 carbon units (after folding) from the other acidic group. The characteristics that differentiate the classes, however, are unclear (Watkins 1984).

NMDA is not present endogenously in the brain. For all of the agonist-defined receptors, glutamate is the highest affinity endogenous ligand and is the strongest candidate for the transmitter of excitatory amino acid pathways in the CNS. Several other acidic amino acid molecules, particularly L-aspartate (Collins et al 1981) and L-homocysteate (Do et al 1986), have been proposed as transmitters for some pathways in the CNS that may employ NMDA receptors. However, whether glutamate or another molecule is the endogenous substance acting on NMDA receptors at each of these synapses is not yet known with certainty.

NMDA Receptors Mediate Induction of Long-term Potentiation

A major concept in neurosciences has been that activity-induced changes in the efficacy of existing synapses are responsible for changes at the behavioral level during learning. Recent studies have implicated NMDA receptors as an essential part of the cellular mechanisms that underlie certain forms of long-term potentiation (LTP), and suggest that NMDA receptors may be an important component of the processes involved in learning. LTP is a long-lasting increase (potentiation) of synaptic efficacy that is induced by a train of high-frequency afferent stimulation. This use-dependent increase in synaptic strength can be rapidly induced (by less than a second of high frequency afferent activity) and persists for hours or days. LTP of a monosynaptic excitatory pathway was first described and studied in the rabbit hippocampus (Bliss & Lømo 1973); similar phenomena have since been described in a number of other vertebrate CNS and PNS pathways (Teyler & Discenna 1987). In the hippocampus, where LTP has been most rigorously studied, different afferent inputs to the same postsynaptic cells exhibit associative requirements for the induction of LTP analogous to the stimulus requirements necessary for associative learning at a behavioral level (Levy & Steward 1979, Barrionuevo & Brown 1983).

Pharmacological studies have demonstrated that activation of NMDA receptors is required for the induction of LTP in several hippocampal pathways. Focal applications of DL-AP5 in hippocampal slices prevent the induction of LTP of the Schaffer collateral-commissural pathway, as measured by potentiation of the pyramidal cell population spike

(Collingridge et al 1983) or by extracellularly recorded synaptic potentials (Wigström & Gustafsson 1984, Harris et al 1984). Furthermore, the antagonism of LTP of the Schaffer-CA1 pathway by NMDA antagonists parallels the efficacy with which different antagonists block NMDA receptor activation (Harris et al 1984, 1986). The conclusion that can be drawn from these studies is that the presence of an NMDA receptor antagonist precludes the induction of LTP of the Schaffer-CA1 pathway and that the NMDA receptors are therefore normally activated during LTP-inducing stimulation. A considerable amount of evidence indicates that LTP is probably a critical component of learning (Thompson 1986). In support of this hypothesis, intraventricular administration of DL-AP5 (but not the inactive L-isomer of AP5) can both prevent LTP of the perforant path input to the dentate gyrus and selectively impair learning of a spatial discrimination task (Morris et al 1986).

The NMDA Receptor as a "Conditional" Receptor

Kainate and/or quisqualate receptors appear to work in concert with NMDA receptors at many CNS synapses. Kainate and quisqualate receptors mediate voltage-independent fast-acting conductances whereas NMDA receptor responses are voltage-dependent and slower acting. Since NMDA responses are preferentially activated at depolarized potentials, NMDA receptors will amplify the synaptic currents only after a particular level of depolarization is achieved. Thus, the conductance response following transmitter binding to the NMDA receptor is "conditional" upon concurrent postsynaptic depolarization.

MULTIPLE RECEPTOR TYPES MEDIATE EXCITATORY POSTSYNAPTIC RESPONSES
Upon a single activation of a presynaptic input, the postsynaptic response is dominated by the conductance activated by the non-NMDA receptors. Although NMDA receptor antagonists block LTP of the Schaffer-CA1 pathways, these compounds do not block synaptic responses in this pathway elicited by single-pulse afferent stimulation (Koerner & Cotman 1982, Collingridge et al 1983). Detailed pharmacological studies using the most potent antagonists available indicate a parallel antagonism of epsps and kainate or quisqualate receptors (Ganong et al 1986). Similar pharmacologic comparisons have been made between synaptic and excitatory amino acid receptors in other CNS regions such as the spinal cord, caudate nucleus, olfactory cortex, and lateral septum (for a review see Mayer & Westbrook 1987). Consistent with the voltage-independent Na^+/K^+ conductances associated with kainate and quisqualate receptors (Mayer & Westbrook 1984, Jahr & Stevens 1987), synaptic responses at several excitatory CNS neuronal synapses are relatively voltage-inde-

pendent (Finkel & Redman 1983, Brown & Johnston 1983, Nelson et al 1986). Thus, the prevailing view is that epsps at many CNS excitatory amino acid synapses are mediated by kainate or quisqualate receptors.

Under certain conditions, it is possible to demonstrate directly an NMDA receptor-mediated component to the epsps. An AP5-sensitive long-latency component of the Schaffer-CA1 synaptic response can be revealed by decreasing extracellular Mg^{2+}, blocking Cl^--mediated inhibition, or by repetitive presynaptic stimulation (Coan & Collingridge 1985, Herron et al 1986, King & Dingledine 1986, Wigström et al 1986a). Thus, it appears that certain CNS synapses (such as the Schaffer-CA1 synapse) have postsynaptic receptors of both the NMDA and non-NMDA type. Some synaptic responses in neocortex (Thomson 1986), spinal cord (Dale & Roberts 1985), and red nucleus (Davies et al 1986b) appear to be primarily mediated by NMDA receptors. However, at most excitatory amino acid pathways that have been rigorously studied, it appears that both NMDA and non-NMDA receptors are present at the same postsynaptic site. The presence of an NMDA receptor component in the epsp depends upon the membrane potential of the postsynaptic cell.

NMDA RECEPTOR-MEDIATED CONDUCTANCE IS VOLTAGE-DEPENDENT   A voltage-dependence of responses induced by NMDA receptor agonists were first noted in cat spinal motor neurons (Engberg et al 1979). In voltage-clamp experiments, which are necessary to distinguish between agonist-induced and other voltage-dependent currents, NMDA-induced inward currents are maximal at membrane potentials from about -30 to -20 mV and are reduced both at more hyperpolarizing and depolarized levels (MacDonald et al 1982, Flatman et al 1983). This response is therefore voltage-dependent. The reversal potential for NMDA agonist-induced responses of around 0 mV indicates that these channels are permeable to both Na^+ and K^+ (Nowak et al 1984, Mayer & Westbrook 1985a). At normal resting potentials of -70 to -80 mV, NMDA receptor-activated currents are suppressed as compared to currents elicited by the same agonist application at more depolarized potentials.

The voltage-dependence of the NMDA receptor-activated currents is due to a block by extracellular Mg^{2+} (Nowak et al 1984, Mayer et al 1984). The Mg^{2+} block is voltage-dependent, only occurring at potentials more negative than about -20 mV (in physiological concentrations of Mg^{2+} of about 1 mM). The discovery of the voltage-dependent block of the NMDA receptor explains earlier pharmacological experiments on spinal neurons that showed the sensitivity of NMDA-induced depolarizations to extracellular Mg^{2+} (Ault et al 1980). The blocking action of Mg^{2+} has been proposed to occur by the binding of Mg^{2+} to a site within the channel complex that is sensitive to the potential across the membrane; the affinity of the binding site for Mg^{2+} (and the block of current flow through the channel by Mg^{2+}) increases with hyperpolarization (Mayer & Westbrook 1987). The binding of divalent cations to the voltage-dependent site is likely to be complex, and probably involves interactions between various cations (Mayer & Westbrook 1987). However, it is clear that current flow through the NMDA receptor-associated channel is conditional upon sufficient depolarization of the membrane to remove the Mg^{2+} block of the channel. Thus, at the level of the synapse, NMDA receptor-induced responses depend upon (a) presynaptic release of transmitter and (b) concurrent postsynaptic depolarization.

Anatomical Organization of NMDA Receptors

NMDA RECEPTOR DISTRIBUTION   Which brain systems use NMDA receptors, and how does the distribution of NMDA receptors compare to that of kainate and/or quisqualate receptors? NMDA receptors may be localized in the brain by receptor autoradiography with the ligands L-[^{3}H]glutamate (NMDA-displacable portion), D-[^{3}H]AP5, or [^{3}H]CPP (Monaghan et al 1983, 1984b, Olverman et al 1986). Subcellular fractionation studies indicate that NMDA-sensitive L-[^{3}H]glutamate binding sites are concentrated in synapse-enriched fractions and are even associated with the postsynaptic density protein complex (Monaghan & Cotman 1986, Fagg & Matus 1984).

NMDA sites in rat brain are predominately located in cortical structures, basal ganglia, and sensory-associated systems (Monaghan & Cotman 1985). Highest levels are found in the hippocampal CA1 termination zone of the Schaffer collateral pathway. High levels are also found in the lateral septum where hippocampal fibers terminate. In neocortex, binding is densest in the superficial layers (Layers I–III and Va in parietal cortex of rat). Among cortical regions, frontal, anterior cingulate, and perirhinal corticies show higher levels of binding than parietal, posterior cingulate, and entorhinal regions. In general, regions receiving specific thalamic innervation show lower binding, particularly in layer IV. In the basal ganglia, nucleus accumbens has highest levels, caudate/putamen moderate levels, and the globus pallidus quite low levels.

NMDA receptors are associated at various levels within sensory systems. Within the olfactory system, significant binding levels are found in the external plexiform layer of the olfactory bulb, the anterior olfactory nuclei, the pyriform cortex, and the nucleus of the lateral olfactory tract. In the auditory and vestibular systems, higher levels of NMDA sites are found in the granule cell layer of the dorsal cochlear nucleus and the medial vestibular nucleus. NMDA sites are also associated with the visual system (dorsal lateral geniculate and superficial gray of the superior colliculus) and with somatosensory input (spinal trigeminal nucleus, substantia gelatinosa, thalamic ventrobasal complex, and nucleus of the solitary tract). It is not yet known how NMDA sites modify sensory input, but experiments indicate that NMDA sites are a major component of the thalamic response to natural somatosensory stimulation (Salt 1986).

COMPARISON TO NON-NMDA EXCITATORY AMINO ACID RECEPTOR DISTRIBUTIONS   As described above, kainate and/or quisqualate receptors are thought to mediate the fast epsp at NMDA receptor-using synapses. In most regions of the brain, quisqualate receptors are colocalized with NMDA sites (Monaghan et al 1984a, Rainbow et al 1984). Kainate sites, on the other hand, have a more dissimilar distribution (Monaghan & Cotman 1982, Unnerstall & Wamsley 1983).

Of those pathways which use excitatory amino acids, only select ones possess NMDA receptors. Glutamate is thought to be the transmitter at many synapses within the cerebellar molecular layer, the pontine nucleus, the mammillary bodies, and the hippocampal mossy fiber pathway, and these regions are relatively poor in NMDA-sensitive L-[^{3}H]glutamate binding sites (Monaghan & Cotman 1985). As might be expected, parallel fiber synapses in the cerebellar molecular layer do not exhibit the type of long-term potentiation found in the hippocampal Schaffer collateral system (Kano & Kato 1987). Likewise, the hippocampal mossy fibers also exhibit a form of LTP plasticity that differs from that found in the Schaffer system, and this form of LTP is not blocked by NMDA antagonists (Harris & Cotman 1986). Regions that use excitatory amino acids and are low in NMDA receptors generally contain significant levels of kainate receptors. Examples of such regions include hypothalamus, deep cortical layers, reticular nucleus of thalamus, mammillary bodies, pontine nucleus, and hippocampal mossy fibers.

NMDA Receptor-ionophore Complex

Recent evidence indicates that the NMDA receptor consists of multiple components that interact allosterically. The NMDA receptor complex has binding sites for (*a*) its transmitter, glutamate; (*b*) an allosteric potentiator, glycine; and (*c*) a site closely associated with the channel, phencyclidine. The NMDA receptor regulated channel is permeable to Ca^{2+}, and other cations and the same channel may be coupled to kainate and quisqualate receptors.

CALCIUM ENTRY THROUGH NMDA RECEPTOR ASSOCIATED CHANNELS   Channels regulated by NMDA receptors are permeable to Ca^{2+} as well as Na^+ and K^+. An increase in intracellular Ca^{2+} following activation of NMDA receptors has been suggested on the basis of extracellular (Pumain & Heinemann 1985) and intracellular (Bürhle & Sönnhof 1983) measurements with ion-sensitive electrodes. Voltage-clamp studies of cultured neurons injected with a calcium-sensitive dye directly indicate that Ca^{2+} can enter into these cells following application of NMDA agonists (MacDermott et al 1986). In addition, measurements of reversal potentials of NMDA agonist-induced currents indicate that the reversal potentials of the currents are sensitive to changes of extracellular Ca^{2+} (Mayer & Westbrook 1985b, Ascher & Nowak 1986, Jahr & Stevens 1987). Under normal physiological conditions, the activation of NMDA receptors results in significant local Ca^{2+} entry. Intracellular Ca^{2+} could then lead to subsequent biochemical/structural processes necessary for long-term changes in synaptic strength (Lynch & Baudry 1984, Miller & Kennedy 1986). Thus, Ca^{2+} entry through voltage-dependent Ca^{2+} channels could account for LTP that is not mediated by NMDA receptors (Harris & Cotman 1986). On the other hand, whether or not LTP occurs at an NMDA receptor-using synapse would depend upon the presence of these Ca^{2+}-activated processes that express the LTP.

ALLOSTERIC REGULATION OF NMDA RECEPTORS BY PCP   As Lodge and colleagues first reported (Anis et al 1983), intravenous or iontophoretically applied ketamine and phencyclidine (PCP) potently block NMDA responses in rat and cat spinal cords. Responses to kainate and quisqualate are unaffected, while acetylcholine responses are moderately reduced. The selectivity of PCP and σ opiate compounds as NMDA antagonists has been shown in a variety of preparations, including rat cortical slices (Harrison & Simmonds 1985, Thomson 1986), striatal slices (Snell & Johnson 1986), and mouse hippocampal slices (Duchen et al 1985).

PCP antagonism of NMDA action does not occur at the NMDA recognition site. NMDA dose-response curves in the presence of ketamine indicate a noncompetive interaction (Schild plot slope < 1; Martin & Lodge 1986). Analysis of antagonist interactions among AP5, ketamine, and Mg^{2+} suggests a separate site for each of these antagonists (Harrison & Simmonds 1985, Martin & Lodge 1986). Furthermore, ligand binding to the NMDA recognition site is not displaced by ketamine and other PCP-like compounds (Monaghan & Cotman 1986).

[³H]PCP binds to two distinct sites in rat brain, a PCP-preferring site and a σ opiate/halperidol-sensitive site (Largent et al 1986). The PCP-preferring site corresponds to the site of NMDA antagonism, and this site is labeled by [³H]1-[1-(2-thienyl)cyclohexyl]piperidine ([³H]TCP; Vignon et al 1983) and by [³H]MK-801 (Wong et al 1986). The activity of PCP-like compounds as NMDA antagonists corresponds to their activity as

[^{3}H]MK-801 displacers ($r = 0.99$, Wong et al 1986). Binding at this PCP site is enhanced by NMDA agonists (Loo et al 1986, Fagg 1987) whereas NMDA antagonists decrease the binding levels. Thus NMDA and PCP sites appear to be positioned close to each other and interact allosterically.

The PCP binding site may be within the NMDA receptor's ion channel. PCP blockade of NMDA-induced responses is voltage dependent (Honey et al 1985). Furthermore, PCP block is enhanced by agonist presence, i.e. it is use-dependent (Honey et al 1985, Wong et al 1986), suggesting that PCP binds to the open channel conformation. Such observations are strikingly similar to that found for blockade of nicotinic acetylcholine receptor ion channels by PCP (Albuquerque et al 1980). The model of PCP blockade of the channel is also consistent with the observation that NMDA agonists increase the binding of [^{3}H]TCP. As expected, the distribution of NMDA sites (labeled by L-[^{3}H]glutamate) is very similar to (Maragos et al 1986) but not identical (Cotman et al 1987b) to that found for PCP sites.

GLYCINE ENHANCEMENT OF NMDA RESPONSES   Glycine at concentrations below 1 μM greatly enhances NMDA-induced responses in mouse cultured neurons (Johnson & Ascher 1987). This effect is not blocked by strychnine and is therefore distinct from the inhibitory strychnine-sensitive glycine receptor. Patch clamp analysis of single channels indicates that glycine increases the frequency of channel opening and not the current amplitude. Since this effect is present in outside-out patches, the effect probably does not involve an intracellular signal.

Radioligand binding experiments directly demonstrate a glycine interaction with NMDA receptors. [^{3}H]Glycine-binding sites have an NMDA receptor-like distribution (Monaghan & Cotman 1985, Bristow et al 1986). Glycine stimulates L-[^{3}H]glutamate binding to NMDA sites; glutamate (and other NMDA agonists, but not NMDA antagonists) stimulates [^{3}H]glycine binding (Nguyen et al 1987). The action of glycine in both electrophysiological and radioligand binding systems is of similar potency (EC$_{50}$ between 0.1 and 1 μM), is insensitive to strychnine, and may be mimicked by serine and alanine (Johnson & Ascher 1987, Nguyen et al 1987). Further experiments are necessary to determine the in vivo role of this allosteric interaction and to determine what other endogenous compounds may also be acting at this site.

NMDA RECEPTOR COUPLING TO SECOND MESSENGERS   NMDA and other excitatory amino acids can activate the phosphoinositol and cyclic nucleotide second messenger systems. NMDA stimulates inositol phospholipid (IP) metabolism in primary cultures of striatal neurons (Sladeczek et al 1985) and cerebellar granule cells (Nicoletti et al 1986b). In the adult rat hippocampus, NMDA will inhibit muscarinic receptor–activated IP metabolism (Baudry et al 1986) and L-glutamate will inhibit noradrenergic activation of IP metabolism (Nicoletti et al 1986a). In cerebellar slices of neonatal rats (Roberts et al 1982) and primary cultures of cerebellar granule cells (Novelli et al 1987), NMDA application results in large increases in cyclic GMP production. The magnitudes of L-glutamate–induced IP metabolism (Nicoletti et al 1986a) and L-glutamate–induced cyclic GMP formation (Garthwaite & Balazs 1978) exhibit large declines during development, but the specific role of NMDA receptors during this transition is not yet known. In summary, though excitatory amino acid receptors appear to interact with the second messenger systems, their precise role in excitatory amino acid receptor response requires further clarification and will likely be of considerable significance.

A SINGLE CHANNEL FOR NMDA, KAINATE, AND QUISQUALATE RECEPTORS   The NMDA receptor complex has multiple regulatory sites and is sensitive to noncompetitive agonists and antagonists. In some way this complex may also be coupled to kainate and/or quisqualate receptors. Recent work using single channel recording techniques has revealed that excitatory amino acid agonists activate several different conductance states (Jahr & Stevens 1987, Cull-Candy & Usowicz 1987). The relative proportion of channel openings to the different substates varies between agonists. NMDA tends to activate a relatively large conductance, and only infrequently activates several smaller conductance states. Kainate and quisqualate tend to preferentially (although not exclusively) activate the smaller conductance states. General agonists such as glutamate activate the larger and smaller conductance states. The analysis of transitions between the multiple conductances, and the lack of apparent simultaneous openings of more than one channel, suggest to these investigators that the activity may be due to multiple conductance states of a single channel that is coupled to both NMDA and non-NMDA receptors.

NMDA RECEPTOR SUBTYPES   The distribution of NMDA-sensitive L-[^{3}H]glutamate binding sites and [^{3}H]CPP/D-[^{3}H]AP5 binding sites are not identical (Olverman et al 1986). Although the patterns are quite similar, NMDA-sensitive L-[^{3}H]glutamate binding sites show relatively greater binding in outer cerebral cortical layers, cerebellar granule cell layer, medial striatum, and lateral septum. In contrast, [^{3}H]CPP shows relatively greater levels of binding in the lateral portions of thalamus. Comparison of the reported values for displacement potencies of various glutamate analogues indicates that agonists are better displacers of NMDA-sensitive L-[^{3}H]glutamate binding whereas antagonists are better displacers of radiolabeled antagonist binding (Olverman et al 1984, 1986, Fagg & Matus

1984, Monaghan et al 1983, 1985, Monaghan & Cotman 1986). Hill plot analysis also suggests that NMDA interacts with two sites (D. T. Monaghan and C. W. Cotman, unpublished observations).

The two patterns of NMDA receptor distribution are similar to the variations seen with the PCP and glycine binding sites. The [³H]TCP binding pattern is more similar to that of [³H]CPP (e.g. relatively high binding in lateral thalamus); [³H]glycine binding is more similar to NMDA-sensitive L-[³H]glutamate binding sites (with relatively high binding in cerebellar granule cell layer, medial striatum, and superficial cerebral cortical layers). Although the reason for these variations has not been determined (differences in association rates, dissociation rates, or maximal binding site density), these results further suggest a heterogeneity of NMDA receptors and also suggest that certain regions may have differential sensitivities to drugs that act at the different sites on the NMDA receptor complex. These results might, for example, explain why the NMDA agonist quinolinate displays greater potency in the rostral striatum than in caudal striatum (Perkins & Stone 1983, Lehmann et al 1983).

Neuropathology and the Protective Action of Excitatory Amino Acid Antagonists

NMDA receptors provide a unique mechanism for the CNS to enhance the synaptic response and to increase Ca^{2+} influx. Both of these properties also lead to excitotoxicity. Several recent investigations have provided evidence that damage to the brain produced by anoxia/ischemia and hypoglycemia are mediated by the excitatory amino acid receptors, particularly of the NMDA type (Meldrum 1985, Rothman & Olney 1986, Wieloch 1985a,b). As a result of such trauma, excessive amounts of glutamate (up to eightfold) are released into the extracellular space (Benveniste et al 1984) and are toxic to neurons if the levels remain elevated for periods beyond several minutes (Rothman & Olney 1986). Several glutamate antagonists, particularly those in the phosphonic acid series, can be protective, at least partially, if applied either immediately before or shortly after insult (Simon et al 1984, Meldrum 1985). Neuronal death results when cultures are maintained in a partially ischemic environment and exposed to glutamate levels of approximately 100 μM (Rothman 1984). Glutamate antagonists, particularly those of the NMDA class, protect against cell death in these systems (Rothman 1984, Rothman & Olney 1986).

It is well known that epileptic attacks also cause neuronal damage and eventual cell death (Meldrum & Brierly 1973). Recently it has become clear that antagonists against acidic amino acid receptors have strong and selective anticonvulsant action in rodent and primate models and are active at very low concentrations, with minimal side effects (Coutinho-Netto et al 1981, Czuczwar & Meldrum 1982, Meldrum et al 1983). Excessive synaptic activation of pathways that use excitatory amino acids can result in neuronal loss, with a pattern of cell death identical to that obtained following direct injection of an excitatory amino acid (Slovitor 1983). Furthermore, antagonists to excitatory amino acid receptors can protect against epileptic seizures and cell loss. Thus it appears that damage due to either excessive excitation or ischemic conditions share a common feature, the excitotoxic action of excitatory amino acids. A specific involvement of NMDA receptors in Huntington's disease is suggested by the observation that the pattern of cell death in the striatum following administration of the NMDA agonist quinolinate, but not following quisqualate or kainate, is similar to Huntington's disease (Beal et al 1986, Koh et al 1986).

NMDA and Other Excitatory Amino Acid Receptors in Alzheimer's Disease

Whereas overactivation of NMDA receptors would lead to neuro-pathology, the loss of NMDA receptor function in various neurological diseases could lead to losses in cognition. Changes in [^{3}H]glutamate binding in Alzheimer's disease have been reported that could indicate a loss of NMDA receptors in hippocampus and cortex in Alzheimer's disease (Greenamyre et al 1985, 1987). Direct determinations of NMDA receptor density in the hippocampus indicate little or no change in area CA1, the dentate gyrus, or the subiculum except in cases in which there was severe cell loss (Geddes et al 1986). These observations were further supported by the analysis of [^{3}H]TCP binding (Monaghan et al 1987). The relative preservation of NMDA sites under conditions of partial cell loss, and the expansion of the kainic acid receptor field associated with hippocampal afferent reorganization in Alzheimer's disease (Geddes et al 1985), may lead to a partial preservation of circuitry function but also to a greater vulnerability to excitotoxicity (Cotman et al 1987a).

NMDA Receptors Can Regulate Developmental Plasticity

It appears that NMDA receptors can play a critical role in early learning and synaptic specificity. The development of the visual system, and prob-ably other systems as well, depends on staged events: first, fibers are guided to their proper targets, and second, their synaptic position is refined in an activity-dependent process. This fine-tuning appears to involve NMDA receptor activation. In surgically produced three-eyed *Rana pipien* tadpoles, retinal ganglion cell fibers from normal and supranumery eyes segregate in the optic tectum into stereotyped eye-specific stripes. If action potentials of retinal ganglion cells are blocked by TTX, segregation is prevented (Reh & Constantine-Paton 1985). Thus neural activity is a necessary step in establishing the retinal-tectal map. Chronic application of AP5 to the optic tectum of three-eyed tadpoles also desegregates retinal ganglion terminals reversibly. Conversely, NMDA application appears to promote the development of sharper borders, and the stripes have fewer forks and turns (Cline et al 1987). Neural activity and NMDA receptors also appear to participate in the development of the mammalian visual cortex (see Singer 1987, Rauschecker & Hahn 1987). In the kitten visual cortex, blocking impulse traffic prevents formation of ocular dominance columns (Stryker & Harris 1986), and chronic infusion of AP5 prevents the ocular dominance shift normally seen in response to monocular experience (Singer et al 1986).

In the developing olfactory system, NMDA receptors appear to participate in the mechanisms of early learning. Neonatal rats learn to prefer either the odor of their mother or an artificial odor experienced with appropriate tactile stimulation, apparently due to a large increase in the size of the glomeruli that are coding for the learned odor. The increased size of select glomeruli is apparently due to the survival of additional tufted cells. The NMDA antagonist AP5 blocks growth (monitored by 2-deoxyglucose uptake) in focal areas of the glomerular layer, as well as the specific neurobehavioral response to early olfactory learning. NMDA receptors and therefore neural activity appear to mediate this process (Lincoln et al 1986, 1987). Thus the involvement of NMDA receptors in developmental plasticity may be a general mechanism in developmental plasticity and learning.

NMDA Receptors as the Molecular Basis for Hebbian Synapses

Synapses with Hebbian properties have been proposed to be involved with neural plasticity in many different contexts, including various aspects of associative learning (Viana Di Prisco 1984, Tesauro 1986) and the formation and maintenance of appropriate synaptic connections during development (Stent 1973, Linkser 1986). Hebb (1949) postulated that "when an axon of cell A is near enough to excite a cell B and repeatedly or persistently takes part in firing it, some growth process or metabolic change takes place in one or both cells such that A's efficiency as one of the cells firing B, is increased." The essence of Hebb's postulate is that strengthening the synapse is contingent upon concurrent pre- and post-synaptic activity. Thus, some mechanism of the synapse must respond only to concurrent pre- and post-synaptic activity. The properties of the NMDA receptor fulfill this requirement. The NMDA receptor channel complex allows current flow only when two conditions are met: (a) presynaptic activity releases transmitter that binds to the receptor, and (b) sufficient post-synaptic depolarization relieves the voltage-dependent Mg^{2+}-block of the channel.

A Hebb-like mechanism can account for the stabilization of appropriate synaptic connections during development, such as in retinal-tectal topography. Coincident activity of neighboring retinal ganglion cells (with overlapping receptor fields) will result in summated postsynaptic depolarization in the target tectal neurons. A Hebb-like mechanism would strengthen these synapses and therefore conserve a retino-topic organization of the projections. The desegregation of eye-specific innervation in three-eyed tadpoles by AP5 may indicate that NMDA receptors endow the retinal tectal synapses with Hebbian properties (Cline et al 1987). Similarly, ocular dominance plasticity in kitten visual cortex has been explained by invoking a Hebb-like mechanism that is provided by post-synaptic NMDA receptors (Singer 1987).

The Hebb-like properties of the NMDA receptor complex can also explain the requirements for the induction of LTP in Schaffer-CA1 synapses. These requirements are simultaneous presynaptic stimulation and postsynaptic depolarization. Hyperpolarizing the postsynaptic neuron during high-frequency afferent stimulation blocks the induction of LTP of the synaptic response that would otherwise occur (Malinow & Miller 1986). Postsynaptic depolarization via an intracellular electrode enables the induction of LTP when specifically paired with a stimulation otherwise subthreshold for the induction of LTP (Wigström et al 1986b, Kelso et al 1986). These results can be viewed as consistent with the notion of synaptic strengthening proposed by Hebb (1949); a synapse will be stabilized when there is simultaneous pre- and post-synaptic activity.

The two requirements for the activation of NMDA receptor-mediated currents are met during stimulation that is sufficient to induce LTP. Transmitter necessary for the activation of the receptor is released from the presynaptic nerve terminals, and postsynaptic depolarization relieves the Mg^{2+} blockade of the channel. Thus, when these conditions are met, current, carried in part by Ca^{2+}, can flow through the channel associated with NMDA receptors, consistent with the previous suggestion of the critical importance of postsynaptic Ca^{2+} for the induction of LTP (Lynch et al 1983). The elevation of Ca^{2+} via NMDA receptor activation could show a high degree of spatial specificity within the postsynaptic neuron if the NMDA receptors were localized to the subsynaptic region of a spine head. Intracellular Ca^{2+} buffering and the thin spine neck present in many neurons would provide the isolation of Ca^{2+} increases between the spine head and the main dendritic shaft. The spine apparatus may be a Ca^{2+}-sequestering structure that restricts the distribution of LTP-associated changes to local dendritic domains. In developing neurons, or those without spines, postsynaptic depolarization and Ca^{2+} may spread over a greater extent of the neuron and therefore involve a greater number of connections, as required for mechanisms such as retinal-tectal topography.

Conclusion

Excitatory synaptic transmission along pathways with NMDA receptors appears distinct from conventional fast-acting synaptic transmission such as that at the neuromuscular junction. NMDA synapses operate on a dual receptor system: kainate or quisqualate receptors serve fast-acting responses whereas NMDA receptors amplify the synaptic response, and permit Ca^{2+} entry, if the condition of concurrent postsynaptic depolarization is met. Responses can be enhanced (via the glycine receptor) or deactivated (via the σ/PCP receptor). There is evidence that there will be receptor heterogeneity and variations within this central framework.

Studies on the NMDA receptor are the focus of intense research activity because they appear to impact on so many areas in the neurosciences. There are examples to illustrate involvement in developmental plasticity, learning, neurodegenerative diseases, epilepsy, and brain damage. The coupling to the PCP receptor may indicate relationships to other as yet unexplored mental functions.

The conjunctive requirement of both pre- and post-synaptic activation for NMDA receptor response embodies the NMDA synapse with the essential properties of Hebb-type synapses. NMDA receptors may thus represent a unitary concept underlying both developmental plasticity and activity-induced neural changes in the mature CNS. Thus, the study of the NMDA receptor provides a new set of concepts, pharmacological approaches, and techniques to probe and define higher order brain function at an unprecedented level of detail.

Literature Cited

Albuquerque, E. X., Tsai, M.-C., Aronstam, R. S., Witkop, B., Eldefrawi, A. T., Eldefrawi, M. E. 1980. Phencyclidine interactions with the ionic channel of the acetylcholine receptor and electrogenic membrane. *Proc. Natl. Acad. Sci. USA* 77: 1224–28

Anis, N. A., Berry, S. C., Burton, N. R., Lodge, D. 1983. The dissociative anaesthetics ketamine and phencyclidine, selectively reduce excitation of central mammalian neurons by *N*-methyl-aspartate. *Br. J. Pharmac.* 79: 565–75

Ascher, P., Nowak, L. 1986. Calcium permeability of the channels activated by *N*-methyl-D-aspartate (NMDA) in mouse central neurons. *J. Physiol.* 377: 35p

Ault, B., Evans, R. H., Francis, A. A., Oakes, D. J., Watkins, J. C. 1980. Selective depression of excitatory amino acid induced depolarizations by magnesium ions in isolated spinal cord preparations. *J. Physiol.* 307: 413–28

Barrionuevo, G., Brown, T. H. 1983. Associative long-term potentiation in hippocampal slices. *Proc. Natl. Acad. Sci. USA* 80: 7347–51

Baudry, M., Evans, J., Lynch, G. 1986. Excitatory amino acids inhibit stimulation of phosphatidylinositol metabolism by aminergic agonists in hippocampus. *Nature* 319: 329–31

Beal, M. F., Kowall, N. W., Ellison, D. W., Mazurek, M. F., Swartz, K. J., Martin, J. B. 1986. Replication of the neurochemical characteristics of Huntington's disease by quinolinic acid. *Nature* 321: 168–71

Benveniste, H., Drejer, J., Schousboe, A., Diemer, N. H. 1984. Elevation of the extracellular concentrations of glutamate and aspartate in rat hippocampus during transient cerebral ischemia monitored by intracerebral microdialysis. *J. Neurochem.* 43: 1369–74

Biscoe, T. J., Evans, R. H., Francis, A. A., Martin, M. R., Watkins, J. C., Davies, J. Dray, A. 1977. D-α-aminoadipate as a selective antagonist of amino acid-induced and synaptic excitation of mammalian spinal neurones. *Nature* 270: 743–45

Bliss, T. V. P., Lømo, T. 1973. Long-lasting potentiation of synaptic transmission in the dentate area of the anaesthetized rabbit following stimulation of the perforant path. *J. Physiol.* 232: 331–56

Bristow, D. R., Bowery, N. G., Woodruff, G. N. 1986. Light microscopic autoradiographic localisation of [³H]strychnine binding sites in rat brain. *Eur. J. Pharmacol.* 126: 303–7

Brown, T. H., Johnston, D. 1983. Voltage clamp analysis of mossy fiber synaptic input to hippocampal neurons. *J. Neurophysiol.* 50: 487–507

Bürhle, C. P., Sönnhof, U. 1983. The ionic mechanism of the excitatory action of L-glutamate upon the membranes of motoneurones of the frog. *Pflugers Archiv.* 396: 154–62

Cline, H. T., Debski, E., Constantine-Paton, M. 1987. NMDA receptor antagonist desegregates eye specific stripes. *Proc. Natl. Acad. Sci. USA* 84: 4342–45

Coan, E. J., Collingridge, G. L: 1985. Magnesium ions block an *N*-methyl-D-aspartate receptor-mediated component of synaptic transmission in rat hippocampus. *Neurosci. Lett.* 53: 21–26

Collingridge, G. L., Kehl, S. J., McLennan, H. 1983. Excitatory amino acids in synaptic transmission in the Schaffer collateral-commissural pathway of the rat hippocampus. *J. Physiol. London* 334: 33–46

Collins, G. G. S., Anson, J., Probett, G. A. 1981. Patterns of endogenous amino acid release from slices of rat and guinea-pig olfactory cortex. *Brain Res.* 204: 103–20

Cotman, C. W., Geddes, J. W., Monaghan, D. T., Anderson, K. J. 1987a. Excitatory amino acid receptors in Alzheimer's disease. *Banbury Rep.* In press

Cotman, C. W., Monaghan, D. T. 1987.

Chemistry and anatomy of excitatory amino acid systems. In *Psychopharmacology: Next Generation of Progress*, ed. H. Y. Meltzer et al. New York: Raven. In press

Cotman, C. W., Monaghan, D. T., Ottersen, O. P., Storm-Mathisen, J. 1987b. Anatomical organization of excitatory amino acid receptors and their pathways. *Trends Neurosci.* 10: 273–80

Coutinho-Netto, J., Abul-Ghani, A. S., Collins, J. F., Bradford, H. F. 1981. Is glutamate a trigger factor in epileptic hyperactivity? *Epilepsia* 22: 289–96

Cull-Candy, S. G., Usowicz, M. M. 1987. Multiple-conductance channels activated by excitatory amino acids in cerebellar neurons. *Nature* 325: 525–28

Czuczwar, S. J., Meldrum, B. 1982. Protection against chemically induced seizures by 2-amino-7-phosphonoheptanoic acid. *Eur. J. Pharmacol.* 83: 335–38

Dale, N., Roberts, A. 1985. Dual-component amino acid-mediated synaptic potentials: Excitatory drive for swimming in *Xenopus* embryos. *J. Physiol.* 363: 35–59

Davies, J., Evans, R. H., Herrling, P. L., Jones, A. W., Olverman, H. J., Pook, P., Watkins, J. C. 1986a. CPP, a new potent and selective NMDA antagonist. Depression of central neuron responses, affinity for D-[³H]AP5 binding sites on brain membranes and anticonvulsant activity. *Brain Res.* 382: 169–73

Davies, J., Miller, A. J., Sheardown, M. J. 1986b. Amino acid receptor mediated excitatory synaptic transmission in the cat red nucleus. *J. Physiol.* 376: 13–29

Davies, J., Francis, A. A., Jones, A. W., Watkins, J. C. 1981. 2-amino-5-phosphonovalerate (2-APV), a potent and selective antagonist of amino acid-induced and synaptic excitation. *Neurosci. Lett.* 21: 77–81

Do, K. Q., Mattenberger, M., Streit, P., Cuenod, M. 1986. In vitro release of endogenous excitatory sulfur-containing amino acids from various rat brain regions. *J. Neurochem.* 46: 779–86

Duchen, M. R., Burton, N. R., Biscoe, T. J. 1985. An intracellular study of the interactions of *N*-methyl-DL-aspartate with ketamine in the mouse hippocampal slice. *Brain Res.* 342: 149–53

Engberg, I., Flatman, J. A., Lambert, J. D. C. 1979. The actions of excitatory amino acids on motoneurones in the feline spinal cord. *J. Physiol.* 288: 227–61

Evans, R. H., Francis, A. A., Jones, A. W., Smith, D. A. S., Watkins, J. C. 1982. The effects of a series of ω-phosphonic α-carboxylic amino acids on electrically evoked and amino acid induced responses in isolated spinal cord preparations. *Br. J. Pharmacol.* 75: 65–75

Fagg, G. E. 1987. Phencyclidine and related drugs bind to the activated NMDA receptor-channel complex in rat brain membranes. *Neurosci. Lett.* 76: 221–27

Fagg, G. E., Foster, A. C. 1983. Amino acid neurotransmitters and their pathways in the mammalian central nervous system. *Neuroscience* 9: 701–19

Fagg, G. E., Matus, A. 1984. Selective association of N-methyl aspartate and quisqualate types of L-glutamate receptor with brain postsynaptic densities. *Proc. Natl. Acad. Sci. USA* 81: 6876–80

Finkel, A. S., Redman, S. J. 1983. The synaptic current evoked in cat spinal motoneurones by impulses in a single group 1a axons. *J. Physiol.* 342: 615–32

Flatman, J. A., Schwindt, P. C., Crill, W. E., Stafstrom, C. E. 1983. Multiple actions of N-methyl-D-aspartate on cat neocortical neurons *in vitro. Brain Res.* 266: 169–73

Fonnum, F. 1984. Glutamate: A neurotransmitter in mammalian brain. *J. Neurochem.* 42: 1

Foster, A. C., Fagg, G. E. 1984. Acidic amino acid binding sites in mammalian neuronal membranes: Their characteristics and relationship to synaptic receptors. *Brain Res. Rev.* 7: 103–64

Ganong, A. H., Jones, A. W., Watkins, J. C., Cotman, C. W. 1986. Parallel antagonism of synaptic transmission and kainate/quisqualate responses in the hippocampus by piperazine-2,3-dicarboxylic acid analogs. *J. Neurosci.* 6: 930–37

Garthwaite, J., Balazs, R. 1978. Supersensitivity to the cyclic GMP response to glutamate during cerebellar maturation. *Nature* 275: 328–29

Geddes, J. W., Chang-Chui, H., Cooper, S. M., Lott, I. T., Cotman, C. W. 1986. Density and distribution on NMDA receptors in the human hippocampus in Alzheimer's disease. *Brain Res.* 399: 156–61

Geddes, J. W., Monaghan, D. T., Cotman, C. W., Lott, I. T., Kim, R. C., Chang-Chui, H. 1985. Plasticity of hippocampal circuitry in Alzheimer's disease. *Science* 230: 1179–81

Greenamyre, J. T., Penney, J. B., D'Amato, C. J., Young, A. B. 1987. Dementia of the Alzheimer's type: Changes in hippocampal L-[³H]glutamate binding. *J. Neurochem.* 48: 543–51

Greenamyre, J. T., Penney, J. B., Young, A. B., D'Amato, C. J., Hicks, S. P., Shoulson, I. 1985. Alterations in L-glutamate binding in Alzheimer's and Huntington's disease. *Science* 227: 1496

Hall, J. G., McLennan, H., Wheal, H. V.

1977. The actions of certain amino acids as neuronal excitants. *J. Physiol.* 272: 52–53p

Harris, E. W., Cotman, C. W. 1986. Long-term potentiation of guinea pig mossy fiber responses is not blocked by N-methyl-D-aspartate antagonists. *Neurosci. Lett.* 70: 132–37

Harris, E. W., Ganong, A. H., Cotman, C. W. 1984. Long-term potentiation in the hippocampus involves activation in N-methyl-D-aspartate receptors. *Brain Res.* 323: 132–37

Harris, E. W., Ganong, A. H., Monaghan, D. T., Watkins, J. C., Cotman, C. W. 1986. Action of 3-((+)-2-carboxypiperazin-4-yl)-propyl-1-phosphonic acid (CCP): A new and highly potent antagonist of N-methyl-D-aspartate receptors in the hippocampus. *Brain Res.* 382: 174–77

Harrison, N. L., Simmonds, M. A. 1985. Quantative studies on some antagonists of N-methyl-D-aspartate in slices of rat cerebral cortex. *Br. J. Pharmacol.* 84: 381–91

Hebb, D. O. 1949. *The Organization of Behavior.* New York: Wiley

Herron, C. E., Lester, R. A. J., Coan, E. J., Collingridge, G. L. 1986. Frequency-dependent involvement in NMDA receptors in the hippocampus: A novel synaptic mechanism. *Nature* 322: 265–68

Honey, C. R., Miljkovic, Z., MacDonald, J. F. 1985. Ketamine and phencyclidine cause a voltage-dependent block of responses to L-aspartic acid. *Neurosci. Lett.* 61: 135–39

Jahr, C. E., Stevens, C. F. 1987. Glutamate activates multiple single channel conductances in hippocampal neurons. *Nature* 325: 522–25

Johnson, J. W., Ascher, P. 1987. Glycine potentiates the NMDA response in cultured mouse brain neurons. *Nature* 325: 529–31

Kano, M., Kato, M. 1987. Quisqualate receptors are specifically involved in cerebellar synaptic plasticity. *Nature* 325: 276–79

Kelso, S. R., Ganong, A. H., Brown, T. H. 1986. Hebbian synapses in hippocampus. *Proc. Natl. Acad. Sci. USA* 83: 5326–30

King, G. L., Dingledine, R. 1986. Evidence for activation of the N-methyl-D-aspartate receptor during epileptic discharge. In *Excitatory Amino Acids and Epilepsy,* ed. R. Schwarcz, Y. Ben-Ari, pp. 465–74. New York/London: Plenum. 735 pp.

Koerner, J. F., Cotman, C. W. 1982. Response of Shaffer collateral-CAI pyramidal cell synapses of the hippocampus to analogues of acidic amino acids. *Brain Res.* 251: 105–15

Koh, J.-Y., Peters, S., Choi, D. W. 1986.

Neurons containing NADPH-diaphorase are selectively resistant to quinolinate toxicity. *Science* 234: 73–76

Largent, B. L., Gundlach, A. L., Snyder, S. H. 1986. Pharmacological and autoradiographic discrimination of sigma and phencyclidine receptor binding sites in brain with (+)-[³H] SKF 10,047, (+)-[³H] - 3 - [3 - hydroxyphenyl] - N - (1-propyl) piperidine and [³H]-1-(2-thienyl) cyclohexyl]piperidine. *J. Pharmac. Exper. Ther.* 238: 739–48

Lehmann, J., Schaefer, P., Ferkany, J. W., Coyle, J. T. 1983. Quinolinic acid evoked [³H]acetylcholine release in striatal slices: Mediation by NMDA-type excitatory amino acid receptors. *Eur. J. Pharmacol.* 96: 111–15

Levy, W. B., Steward, O. 1979. Synapses as associative memory elements in the hippocampal formation. *Brain Res.* 175: 233–45

Lincoln, J., Coopersmith, R., Harris, E. W., Cotman, C. W., Leon, M. 1987. NMDA receptor activation is required for the neural basis of early learning. Submitted

Lincoln, J., Coopersmith, R., Harris, E. W., Monaghan, D., Cotman, C. W., Leon, M. 1986. NMDA receptor blockade prevents the neural and behavioral consequences of early olfactory experience. *Soc. Neurosci.* 12: 124

Linsker, R. 1986. From basic network principles to neural architecture: Emergence of spatial-opponent cells. *Proc. Natl. Acad. Sci. USA* 83: 7508–12

Loo, P., Braunwalder, A., Lehmann, J., Williams, M. 1986. Radioligand binding to central phencyclidine recognition sites is dependent on excitatory amino acid receptor antagonists. *Eur. J. Pharmacol.* 123: 467–68

Lynch, G., Baudry, M. 1984. The biochemistry of memory: A new and specific hypothesis. *Science* 224: 1057–63

Lynch, G., Larson, J., Kelso, S., Barrionuevo, G., Schottler, F. 1983. Intracellular injections of EGTA block induction of hippocampal long-term potentiation. *Nature* 305: 719–21

MacDermott, A. B., Mayer, M. L., Westbrook, G. L., Smith, S. J., Barker, J. L. 1986. NMDA-receptor activation increases cytoplasmic calcium concentration in cultured spinal cord neurones. *Nature* 321: 519–22

MacDonald, J. F., Porietis, A. V., Wojtowicz, J. M. 1982. L-Aspartic acid induces a region of negative slope conductance in the current voltage relationship of cultured spinal cord neurons. *Brain Res.* 237: 248–53

Malinow, R., Miller, J. P. 1986. Postsynaptic

hyperpolarization during conditioning reversibly blocks induction of long-term potentiation. *Nature* 320: 529–30

Maragos, W. F., Chu, D. C. M., Greenamyre, J. T., Penney, J. B., Young, A. B. 1986. High correlation between localization of [³H]TCP binding and NMDA receptors. *Eur. J. Pharmacol.* 123: 173–74

Martin, D., Lodge, D. 1986. Ketamine acts as a non-competitive N-methyl-D-aspartate antagonist on frog spinal cord *in vitro*. *Neuropharmacology* 24: 999–1003

Mayer, M. L., Westbrook, G. L. 1984. Mixed-agonist action of excitatory amino acids on mouse spinal cord neurons under voltage clamp. *J. Physiol.* 354: 29–53

Mayer, M. L., Westbrook, G. L. 1985a. The action of N-methyl-D-aspartic acid on mouse spinal cord neurones in culture. *J. Physiol.* 361: 65–90

Mayer, M. L., Westbrook, G. L. 1985b. Divalent cation permeability of N-methyl-D-aspartate channels. *Soc. Neurosci. Abstr.* 11: 785

Mayer, M. L., Westbrook, G. L. 1987. The physiology of excitatory amino acids in the vertebrate central nervous system. *Prog. Neurobiol.* 28: 197–276

Mayer, M. L., Westbrook, G. L., Guthrie, P. B. 1984. Voltage-dependent block by Mg^{2+} of NMDA responses in spinal cord neurones. *Nature* 309: 261–63

Meldrum, B. 1985. Possible therapeutic applications of antagonists of excitatory amino acid neurotransmitters. *Clin. Sci.* 68: 113–22

Meldrum, B., Brierley, J. B. 1973. Prolonged epileptic seizures in primates: Ischemic cell change and its relation to ictal physiologic effects. *Arch. Neurol.* 28: 10–17

Meldrum, B. S., Croucher, M. J., Badman, G., Collins, J. F. 1983. Antiepileptic action of excitatory amino acid antagonists in the photosensitive baboon, *Papio papio*. *Neurosci. Lett.* 39: 101–4

Miller, S. G., Kennedy, M. B. 1986. Regulation of brain type II Ca^{2+}/calmodulin-dependent protein kinase by autophosphorylation: A Ca^{2+}-triggered molecular switch. *Cell* 44: 861–70

Monaghan, D. T., Cotman, C. W. 1982. Distribution of ³H-kainic acid binding sites in rat CNS as determined by autoradiography. *Brain Res.* 252: 91–100

Monaghan, D. T., Cotman, C. W. 1985. Distribution of NMDA-sensitive L-³H-glutamate binding sites in rat brain as determined by quantitative autoradiography. *J. Neurosci.* 5: 2909–19

Monaghan, D. T., Cotman, C. W. 1986. Identification and properties of NMDA receptors in rat brain synaptic plasma membranes. *Proc. Natl. Acad. Sci. USA* 83: 7532–36

Monaghan, D. T., Geddes, J. W., Yao, D., Chung, C., Cotman, C. W. 1987. [³H]TCP binding sites in Alzheimer's disease. *Neurosci. Lett.* 83: 197–200

Monaghan, D. T., Holets, V. R., Toy, D. W., Cotman, C. W. 1983. Anatomical distributions of four pharmacologically distinct ³H-L-glutamate binding sites. *Nature* 306: 176–79

Monaghan, D. T., Yao, D., Cotman, C. W. 1984a. Distribution of ³H-AMPA binding sites in rat brain as determined by quantitative autoradiography. *Brain Res.* 324: 160–64

Monaghan, D. T., Yao, D., Cotman, C. W. 1985. L-³H-glutamate binds to kainate-, NMDA-, and AMPA-sensitive binding sites: An autoradiographic analysis. *Brain Res.* 340: 378–83

Monaghan, D. T., Yao, D., Olverman, H. J., Watkins, J. C., Cotman, C. W. 1984b. Autoradiography of D-[³H]2-amino-5-phosphonopentanoate binding sites in rat brain. *Neurosci. Lett.* 52: 253–58

Morris, R. G. M., Anderson, E., Lynch, G. S., Baudry, M. 1986. Selective impairment of learning and blockade of long-term potentiation by an N-methyl-D-aspartate receptor antagonist, AP5. *Nature* 319: 774–76

Nelson, P. G., Pun, R. Y. K., Westbrook, G. L. 1986. Synaptic excitation in culture of mouse spinal cord neurones: Receptor pharmacology and behaviour of synaptic currents. *J. Physiol.* 372: 169–90

Nguyen, L., Monaghan, D. T., Cotman, C. W. 1987. Glycine binding sites reciprocally interact with glutamate binding sites at the NMDA receptor complex. *Soc. Neurosci. Abstr.* 13: In press

Nicoletti, F., Iadarola, M. J., Wroblewski, J. T., Costa, E. 1986a. Excitatory amino acid recognition sites coupled with inositol phospholipid metabolism: Developmental changes and interaction with alpha₁-adrenoceptors. *Proc. Natl. Acad. Sci. USA* 83: 1931–35

Nicoletti, F., Wroblewski, J. T., Novelli, A., Alho, H., Guidotti, A., Costa, E. 1986b. The activation of inositol phospholipid metabolism as a signal-transducing system for excitatory amino acids in primary cultures of cerebellar granule cells. *J. Neurosci.* 6: 1905–11

Novelli, A., Nicoletti, F., Wroblewski, J. T., Alho, H., Costa, E., Guidotti, A. 1987. Excitatory amino acid receptors coupled with guanylate cyclase in primary cultures of cerebellar granule cells. *J. Neurosci.* 7(1): 40–47

Nowak, L., Bregestovski, P., Ascher, P.,

Herbet, A., Prochiantz, A. 1984. Magnesium gates glutamate-activated channels in mouse central neurones. *Nature* 307: 462–65

Olverman, H. J., Jones, A. W., Watkins, J. C. 1984. L-Glutamate has higher affinity than other amino acids for [³H]D-AP5 binding sites in rat brain membranes. *Nature* 307: 460–62

Olverman, H. J., Monaghan, D. T., Cotman, C. W., Watkins, J. C. 1986. [³H]CPP, a new competitive ligand for NMDA receptors. *Eur. J. Pharmacol.* 131: 161–62

Perkins, M. N., Stone, T. W. 1983. Pharmacology and regional variations of quinolinic acid-evoked excitations in the rat central nervous system. *J. Pharmacol. Exp. Ther.* 226: 551–57

Perkins, M. N., Stone, T. W., Collins, J. F., Curry, K. 1981. Phosphonate analogues of carboxylic acids as amino acid antagonists on rat cortical neurones. *Neurosci. Lett.* 23: 333–36

Pumain, R., Heinemann, U. 1985. Stimulus- and amino acid-induced calcium and potassium changes in the rat neocortex. *J. Neurophysiol.* 53: 1–16

Rainbow, T. C., Wieczorek, C. M., Halpain, S. 1984. Quantitative autoradiography of binding sites for ³H-AMPA, a structural analogue of glutamic acid. *Brain Res.* 309: 173–77

Rauschecker, J. P., Hahn, S. 1987. Ketamine-zyloazine anesthesia blocks consolidation of ocular dominance changes in kitten visual cortex. *Nature* 326: 183–85

Reh, T. A., Constantine-Paton, M. 1985. Eye specific segregation requires neural activity in three-eyed *Rana pipiens. J. Neurosci.* 5: 1132–43

Roberts, P. J., Foster, G. A., Sharif, N. A., Collins, J. F. 1982. Phosphonate analogues of acidic amino acids: inhibition of excitatory amino acid transmitter binding to cerebellar membranes and of the stimulation of cerebellar cyclic GMP levels. *Brain Res.* 238: 475–79

Rothman, S. 1984. Synaptic release of excitatory amino acid neurotransmitter mediates anoxic neuronal death. *J. Neurosci.* 4: 1884–91

Rothman, S. M., Olney, J. W. 1986. Glutamate and the pathology of hypoxic/ischemic brain damage. *Ann. Neurol.* 19: 105

Salt, T. E. 1986. Mediation of thalamic sensory input by both NMDA receptors and non-NMDA receptors. *Nature* 322: 263–65

Simon, R. P., Swan, J. H., Griffiths, T., Meldrum, B. S. 1984. Blockage of N-methyl-D-aspartate receptors may protect against ischemic damage in the brain. *Science* 226: 850–52

Singer, W. 1987. Activity-dependent self-organization of synaptic connections as a substrate of learning. *Life Sci. Res. Rep. Neural Mol. Mech. Learning* 38: 301–35

Singer, W., Kleinschmidt, A., Bear, M. F. 1986. Infusion of an NMDA receptor antagonist disrupts ocular dominance plasticity in kitten striate cortex. *Abstr. Soc. Neurosci.* 12: 786

Sloviter, R. S. 1983. "Epileptic" brain damage in rats induced by sustained electrical stimulation of the perforant path. I. Acute electrophysiological and light microscopic studies. *Brain Res. Bull.* 10: 675–97

Sladeczek, F., Pin, J.-P., Recasens, M., Bockaert, J., Weiss, S. 1985. Glutamate stimulates inositol phosphate formation in striatal neurones. *Nature* 317: 717–19

Snell, L. D., Johnson, K. M. 1986. Characterization of the inhibition of excitatory amino acid-induced neurotransmitter release in the rat striatum by phencyclidine-like drugs. *J. Pharmacol. Exper. Ther.* 238: 938–46

Stent, G. S. 1973. A physiological mechanism for Hebb's postulate of learning. *Proc. Natl. Acad. Sci. USA* 70: 997–1001

Stryker, M. P., Harris, W. A. 1986. Binocular impulse blockade prevents the formation of ocular dominance columns in the cat visual cortex. *J. Neurosci.* 6: 2117–33

Tesauro, G. 1986. Simple neural models of classical conditioning. *Biol. Cybernet.* 55: 187–200

Teyler, T. J., Discenna, P. 1987. Long-term potentiation. *Ann. Rev. Neurosci.* 10: 131–61

Thomson, A. M. 1986. A magnesium-sensitive post-synaptic potential in rat cerebral cortex resembles neuronal responses to N-methylaspartate. *J. Physiol. London* 370: 531–49

✓ Thompson, R. F. 1986. The neurobiology of learning and memory. *Science* 233: 941–47

Unnerstall, J. R., Wamsley, J. K. 1983. Autoradiographic localization of high-affinity ^{3}H-kainic acid binding sites in the rat forebrain. *Eur. J. Pharmacol.* 86: 361–71

Viana Di Prisco, G. 1984. Hebb synaptic plasticity. *Prog. Neurobiology* 22: 89–102

Vignon, J., Chicheportiche, R., Chicheportiche, M., Kamenka, J.-M., Geneste, P., Lazdunski, M. 1983. [^{3}H]TCP: A new tool with high affinity for the PCP receptor in rat brain. *Brain Res.* 280: 194–97

Watkins, J. C. 1984. Excitatory amino acids and central synaptic transmission. *TIPS* 5: 373–76

Wieloch, T. 1985a. Hypoglycemia-induced neuronal damage prevented by an N-methyl-D-aspartate antagonist. *Science* 230: 681–83

Wieloch, T. 1985b. Neurochemical correlates to selective neuronal vulnerability. *Prog. Brain Res.* 63: 69–85

Wigström, H., Gustafsson, B. 1984. A possible correlate of the post-synaptic condition for long-lasting potentiation in the guinea pig hippocampus *in vitro. Neurosci. Lett.* 44: 327–32

Wigström, H., Gustafsson, B., Huang, Y.-Y. 1986a. Mode of action of excitatory amino acid receptor antagonists on hippocampal long-lasting potentiation. *Neuroscience* 17: 1105–15

Wigström, H., Gustafsson, B., Huang, Y.-Y., Abraham, W. C. 1986b. Hippocampal long-term potentiation is induced by pairing single afferent volleys with intracellularly injected depolarizing current pulses. *Acta Physiol. Scand.* 126: 317–19

Wong, E. H. F., Kemp, J. A., Priestley, T., Knight, A. R., Woodruff, G. N., Iverson, L. L. 1986. The anti-convulsant MK-801 is a potent N-methyl-D-aspartate antagonist. *Proc. Natl. Acad. Sci. USA* 83: 7104–8

Ann. Rev. Neurosci. 1988. 11:81–96

MPTP TOXICITY:
Implications for Research
in Parkinson's Disease[1]

I. J. Kopin

National Institutes of Health, National Institute of Neurological and
Communicative Disorders and Stroke, Bethesda, Maryland 20892

S. P. Markey

National Institutes of Health, National Institute of Mental Health,
Bethesda, Maryland 20892

INTRODUCTION

Most of the major clinical features of Parkinson's disease were described
by James Parkinson in 1817. Resting tremor, slowness of movement (bradykinesia), and rigidity are the three hallmark symptoms. In addition there
may be flat facial expression, flexed posture, loss of postural reflexes,
shuffling gait, loss of finger dexterity, oily skin, and difficulty in swallowing.
Recently it has been discovered that drug abusers who self-administer
1-methyl-4-phenyl-1,2,3,6-tetrahydropyridine (MPTP) rapidly develop a
syndrome clinically undistinguishable from advanced Parkinson's disease
(Davis et al 1979, Langston et al 1983). Subsequent studies in animals have
shown that the brain lesions, biochemical changes, and motor abnormalities produced by MPTP closely resemble those found in spontaneous
Parkinson's disease (Burns et al 1983). Since Parkinson's disease does not
occur more frequently in identical twins than among fraternal twins, and
epidemiological studies do not shown familial concentrations of the
disorder, it appears unlikely that heredity is an important determinant in

---

[1] The US Government has the right to retain a nonexclusive, royalty-free license in and to
any copyright covering this paper.

the development of the disorder (Duvoisin 1986). The epidemiological evidence that environmental factors are important in the pathogenesis of Parkinson's disease and the discovery that exposure to a chemical substance may result in a highly specific form of brain damage and cause neurologic deficits almost identical to those of Parkinson's disease has excited new efforts (see Markey et al 1986) directed at understanding the actions of MPTP and identifying the factors that might be involved in the pathogenesis of the disease.

SIMILARITIES BETWEEN MPTP TOXICITY AND PARKINSON'S DISEASE

The availability of suitable animal models for human diseases greatly facilitates research and accelerates progress by providing investigators with access to tissue, with the capacity to control extraneous factors, and the opportunity to attempt new therapies that are not available when only humans are available for study. The more faithfully the models resemble the spontaneously occurring disorder, the more likely that valid conclusions can be obtained. It is important, therefore, to compare closely the model with the characteristic features of the disease in humans.

MPTP Toxicity in Humans and Parkinson's Disease

The parkinsonian syndrome induced by MPTP in man is clinically indistinguishable from idiopathic Parkinson's disease except for one major feature. Idiopathic Parkinson's disease develops slowly over a period of years, so gradually that patients cannot recall the onset of the first symptoms. In contrast, MPTP-induced parkinsonism develops so suddenly (5–15 days) that the first patient encountered was diagnosed initially as having catatonic schizophrenia (Davis et al 1979). The pathological changes in Parkinson's disease are well defined, but so far only the brain of this single patient with MPTP toxicity has been examined.

Brains from patients who have died from Parkinson's disease exhibit a profound loss of the large pigmented neurons of the substantia nigra that normally project into the striatum. This neuronal cell loss has been associated with eosinophilic intracytoplasmic inclusions, or Lewy bodies, in the substantia nigra. Other areas showing degenerative changes include the locus coeruleus, the dorsal motor nucleus of the vagus, and the basal nucleus of Meynert; these changes are also accompanied by the presence of Lewy bodies.

In the single patient, age 23, who committed suicide 18 months after MPTP exposure, there was a significant cell loss of pigmented neurons in the substantia nigra and a single eosinophilic inclusion that did not meet the usual criteria for a Lewy body. No other notable pathological changes were found (Davis et al 1979, Langston et al 1983).

Biochemically, the loss of substantia nigra neurons in Parkinson's disease is attended by a decrease in the dopamine content as well as a decrease in the levels of the catecholamine biosynthetic enzymes (tyrosine hydroxylase and dopa-decarboxylase) in the affected areas of the brain (see Hornykiewicz 1982). In untreated patients, lowered cerebrospinal fluid (CSF) levels of homovanillic acid (HVA), the major dopamine metabolite, indicate the relative importance of nigral neurons to whole brain dopamine production. At least 60% of HVA in the CSF derives from these neurons, but there are similarly large decrements in HVA levels in the plasma or urine of Parkinsonian patients, since dopamine is also produced and metabolized in peripheral sympathetic neurons (see Kopin 1985). Decrements in CSF levels of HVA have been reported in both Parkinson's disease and in MPTP-exposed patients (Burns et al 1984a), with mean levels of 15 and 16 ng/ml, respectively, compared to the mean figure for controls of 47 ng/ml. Levels of 5-hydroxyindoleacetic acid (5HIAA), the major serotonin metabolite in the CSF, are not altered in either patient group, but there are consistently higher levels of the norepinephrine metabolite, 4-hydroxy-3-methoxyphenylethylenglycol (MHPG) in MPTP-exposed patients (7.7 ± 0.5 ng/ml) compared to those with Parkinson's disease (4.3 ± 0.4) or controls (5.4 ± 0.4). MPTP-exposed patients with damage limited to dopaminergic neurons may have compensatory central noradrenergic activity; this would not be possible in patients with Parkinson's disease who also have damage to the locus coeruleus.

The loss of dopaminergic neurons has been demonstrated using positron emission tomography with ^{18}F-6-fluorodopa. Given peripherally, this synthetic DOPA analogue is transported into the brain, taken up by neurons, and metabolized to ^{18}F-6-fluorodopamine, which is stored in functional dopaminergic nerve terminals. After an interval of nonspecific distribution, the images of positron emission from ^{18}F reveal relatively high activity in the caudate/putamen of control subjects, but relatively low activity in these areas in Parkinsonian or MPTP-exposed patients. Most interestingly, four subjects known to have been exposed to MPTP but free from Parkinsonian symptoms had striatal ^{18}F-dopamine activity intermediate between normal and Parkinson's disease patient groups (Calne et al 1985).

MPTP-induced Parkinsonism in Monkeys

Since the clinical and biochemical profiles of MPTP-induced and idiopathic Parkinsonism in man are nearly identical, MPTP is a valuable tool with which to create a Parkinsonian condition in subhuman primates as an experimental model to study the pathogenesis and treatment of

Parkinson's disease (Burns et al 1983, Langston et al 1984a). All species of primates tested so far (rhesus, squirrel, and cynomologus monkeys, marmosets, and baboons) begin to develop Parkinsonian symptoms within several days after MPTP injections (ranging in dose from 1 to 10 mg/kg); the disorder then gradually increases in severity for several weeks (see reviews in Markey et al 1986). The degree of motor impairment is highly variable and difficult to predict, but the age and sex of the animal are important (older animals and females are more vulnerable). Some animals recover from a condition of severe impairment at one month to nearly normal locomotion without medication six to nine months later. Neither progression nor improvement were evident in a chemist who had accidently been exposed to MPTP and had been treated with L-DOPA for nine years (Burns et al 1984b). Consequently there are important differences between MPTP-induced parkinsonism and Parkinson's disease. Idiopathic Parkinson's disease is progressive, at individually varying rates. Neurotoxin-induced parkinsonism has at least three phases—acute, chronic, and compensatory. The acute effects of MPTP last less than 24 hours, but the long brain half-life (approximately 10 days in the rhesus monkey) of the major metabolite MPP^+ (Markey et al 1984) suggests that direct drug effects may persist for several months. The stabilization of locomotor disability and the compensatory mechanisms that produce some functional recovery in monkeys suggest that MPTP-treated animals may be significantly different from victims of the human disease; stabilization of the motor disorder in animals may require several months prior to their use for assessment of therapeutic regimens. Demonstration of the toxic response vs gradual recovery cycle was shown in baboons by positron emission tomography (PET) studies using ^{76}Br-spiperone to visualize D_2-dopamine receptors (Hantraye et al 1986). In the first few days following a series of daily MPTP treatments, D_2-receptors in the striatum were diminished relative to those in the cerebellum. Behavioral symptoms (akinesia, rigidity, etc) were maximal when receptor levels were lowest, and disappeared when receptor levels returned to near normal. Repetition of the treatment cycle reproduced the same D_2-receptor reduction with attendant motor deficits, followed by recovery over several weeks until the syndrome was established at a stable level after a third treatment series.

The pattern of substantia nigral cell death in MPTP-treated monkeys is similar to that following knife cuts into the nigro-striatal pathway; catecholamine fluorescence diminishes first in the cell bodies, with enlarged, distorted axons (Jacobowitz et al 1984). One month after treatment, the remaining cells contain increased pigment granules. In areas where Lewy bodies are found in humans, eosinophilic inclusion bodies have been reported in squirrel monkeys (substantia nigra, dorsal motor nucleus of the vagus, nucleus basalis of Meynert, and dorsal raphe nucleus; Forno et al 1986). Areas other than the substantia nigra, and particularly the locus coeruleus, have been the subject of several investigations because of their involvement in idiopathic Parkinson's disease (Mitchell et al 1985). Those studies which have demonstrated involvement of locus coeruleus histologically have used monkeys killed during the acute phases of MPTP treatment, but in studies conducted after stabilization for six weeks, locus coeruleus involvement was not found (Jacobowitz et al 1984, Burns et al 1983).

The neurochemical sequellae of MPTP treatment in primates supports the notion that there are several phases of MPTP toxicity. During the acute phase, CSF metabolites of serotonin, norepinephrine, and dopamine are markedly affected, being 50, 75, and 50% of their respective baseline values (Burns et al 1983). After three months, the norepinephrine metabolite returns to 75% of its previous value, the serotonin metabolite to 86%, but the dopamine metabolite remains at only 50% of the baseline value. These measures suggest that MPTP produces a dopaminergic lesion, but that all neurotransmitter systems are involved, possibly because of inhibition of monoamine oxidase (see below) during the early phase. In fact, monkeys killed during the first ten days after MPTP treatment actually have increased dopamine levels in the caudate and putamen, but these levels drop markedly by the eighth week, coincident with the disappearance of cell bodies in the substantia nigra. The specificity of the dopaminergic lesion caused by MPTP in the primate has been confirmed by extensive biochemical measures of both neurotransmitters and their synthetic enzymes (see reviews in Markey et al 1986). Unlike 6-hydroxydopamine, which destroys all dopaminergic cells, MPTP's toxicity appears confined mainly to those substantia nigra neurons that project to the striatum. Levels of dopamine are not decreased in the nucleus accumbens; in the other parts of the striatum, dopamine depletion appears proportional to the degree of locomotor deficit. In severely affected monkeys, there may be 95% depletion of dopamine in caudate and putamen, whereas the dopamine content of the nucleus accumbens remains within 10% of the control value. Tyrosine hydroxylase and dopa-decarboxylase activities mirror the changes in dopamine levels, being significantly depressed (5–10% of control) in the caudate and putamen but near control values (85–90%) in the nucleus accumbens.

Species Differences in Vulnerability to MPTP Toxicity

While the question of dopaminergic tissue vulnerability to MPTP in primates remains unresolved, a related question regarding species vulnerability is equally puzzling. In several strains of laboratory rats that were administered nearly lethal doses of MPTP (5–10 mg/kg) by a variety of routes (i.v., i.p., i.c.v., intranigral), striatal dopamine levels remained relatively unaffected when measured after a few days (Chiueh et al 1984, Enz et al 1984, Sahgal et al 1984, Boyce et al 1984). In contrast, significant prolonged striatal dopamine depletion results from multiple high dose administration (10–30 mg/kg) of MPTP to mice, particularly those of the C57Bl/6 strain (Hallman et al 1984, 1985, Heikkila 1984a,b, 1985). Although the extent and specificity of neuronal damage in mice appears to parallel that seen in primates, over a six-month period there is a recovery of dopamine levels, demonstrating a neuronal plasticity in the mouse that is not evident in primates. In mice, locomotor disability produced by MPTP disappears within hours after treatment, whereas biochemical and histological measures demonstrate loss of dopamine synthesis in the substantia nigra that persist for weeks. This gradual reappearance of dopamine-containing nigral neurons is presumably due to the recovery of damaged cells rather than the genesis of new cells. Whatever the mechanism, the process of recovery is different in primates where histological and biochemical measures have clearly demonstrated nigral cell death. In primates that have regained normal locomotor function, compensatory mechanisms (higher tonic levels of brain dopamine, supersensitive D_2-receptors) are probably responsible for the apparent functional recovery.

The effects of MPTP in other animal species have been explored, partly to find alternatives to primates for research purposes and partly to probe mechanisms and hypotheses for MPTP toxicity. For example, because neuromelanin is present in man and subhuman primates, it has been postulated that species which have neuromelanin-containing cells in the substantia nigra might be more susceptible to MPTP toxicity (Burns et al 1984b). Beagle dogs have neuromelanin and exhibit the same extensive and specific loss of nigral neurons following MPTP as primates, and at similar dose levels (Parisi & Burns 1986). In contrast, cats and guinea pigs are significantly less affected by MPTP (Schneider et al 1986, Chiueh et al 1984). Guinea pigs are of particular interest because, like man, they require dietary ascorbic acid, a protective antioxidant (Perry et al 1985b). Frogs have large pigmented areas on their skin, and thus the effects of MPTP were studied to determine whether the skin pigment correlated with brain neuromelanin and nigral neuron toxicity (Barbeau et al 1985a,b). Frogs are not particularly sensitive to MPTP, and are probably less useful than the mouse for study. The effects of MPTP on the leech nervous system have also been studied because the simple nervous system of the invertebrate affords an opportunity to study specific and accessible neurons (Lent 1986). The observations that pigmented leeches were both more susceptible and relatively rich in monoamine oxidase-B (MAO-B) than nonpigmented leeches are consistent with observations on other species.

MECHANISM OF MPTP TOXICITY AND IMPLICATIONS FOR TREATMENT OF PARKINSON'S DISEASE

The factors that determine the effects of a toxin include the processes responsible for its bioactivation or inactivation, the accessibility of the toxin to sites at which it can interact with cellular components, and the comparative rates at which the toxin damages and the cells repair their vital constituents. Interference with these processes may diminish or enhance toxicity, and if similar processes are involved in naturally occurring disease states, new therapies might be devised based on measures that prevent the toxic effects. This has been the basis for the great interest in determining the factors that contribute to the toxic effects of MPTP.

Bioactivation of MPTP by Monoamine Oxidase

MPTP is a lipid soluble molecule that readily penetrates into the brain. The metabolism of administered MPTP to its pyridinium derivative, MPP^+, was demonstrated by Markey et al (1984) in monkeys and in mice. The oxidation of MPTP in the brain to a positively charged quarternary compound, MPP^+, results in intracellular trapping of the metabolite. Accumulation of MPP^+ in either the striatal dopaminergic nerve terminals (Markey et al 1984) or in substantia nigral cell bodies has been proposed to underlie its neurotoxic specificity (Irwin & Langston 1985).

Crude mitochondrial preparations from rat brain transform MPTP to MPP^+ (Chiba et al 1984). Since pargyline (a nonselective MAO-inhibitor) and deprenyl (a selective inhibitor of MAO-B) but not clorgyline (selective for MAO-A) blocked the formation of MPP^+, MAO-B was implicated as the enzyme involved in the metabolism of MPTP. The initial oxidation of MPTP results in a dihydropyridine intermediate, $MPDP^+$, which may autoxidize by disproportionation to MPTP and MPP^+, or be converted enzymatically to MPP^+.

The importance of MAO-B mediated bioactivation in the toxicity of MPTP has been established by the demonstration that the inhibition of MAO-B by pretreatment with pargyline, deprenyl, or other inhibitors of MAO-B (but not clorgyline) not only blocks formation of MPP^+ but prevents the depletion of dopamine in the brains of mice (Heikkila et al 1984b, Markey et al 1984, Fuller & Hemrick-Luecke 1985) and protects against the development of MPTP-induced parkinsonism in monkeys (Langston et al 1984b, Cohen et al 1984).

MAO Inhibitors in the Treatment of Parkinson's Disease

The discoveries that dopamine is deficient in the brains of patients with Parkinson's disease (Hornykiewicz 1982) and that MAO inhibitors increase brain dopamine levels in experimental animals led to attempts to use MAO inhibitors for treating Parkinson's disease, alone (Rosen 1969), or in combination with DOPA (Birkmayer & Hornykiewicz 1961, McGeer et al 1961). The use of MAO inhibitors alone was marginally effective, whereas the combination of drugs, while effective, frequently caused unacceptable fluctuating and uncontrolled increases in blood pressures.

The introduction of L-deprenyl (Knoll et al 1965), a selective MAO-B inhibitor that does not potentiate the pressor effects of sympathomimetic agents, made possible safe MAO inhibition as an adjunct to L-DOPA in the treatment of Parkinson's disease. In parkinsonian patients who died while being treated with deprenyl, the concentrations of dopamine in the substantia nigra and its dopaminergic projection regions were found to be about 75% of normal, whereas in untreated patients the dopamine levels were only about 25% of normal (Riederer & Youdim 1986). In treated patients, the MAO-B activity was markedly inhibited in all regions of the brain examined. Several studies have confirmed the DOPA-potentiating effect of L-deprenyl (Lees et al 1977, Yahr 1978, Rinne 1983) but in a recent study (Birkmayer et al 1985) the combination of L-deprenyl with DOPA appeared to have also increased the life expectancy of the patients over a nine year interval, compared to that of patients being treated with DOPA alone. While the mechanism by which deprenyl might increase the life expectancy of parkinsonian patients is unknown, the fact the MAO-B inhibition prevents MPTP-induced parkinsonism in experimental animals has led to the speculation that MAO-B substrates may also be involved in the mechanisms responsible for the degenerative process of Parkinson's disease. The origins of such substrates have not been determined, but analogues of MPTP that might be converted to MPP^+-like compounds (Ansher et al 1986, Johannessen 1987) are currently being sought.

MPTP and the Dopamine Uptake System

The toxic effects of MPP^+ can be demonstrated in many tissues, in vivo and in vitro. Although highly polar MPP^+ does not enter the brain, it is formed there from MPTP, which, because of its lipid solubility, readily penetrates the blood brain barrier. MPTP is converted to MPP^+ by the action of MAO-B, which is localized within glia and serotonergic neurons (Westlund et al 1985). MPP^+ is therefore formed outside the neurons that it will ultimately damage. Because MPP^+ has a high affinity for both dopamine and norepinephrine uptake sites (Javitch et al 1985, Javitch &

Snyder 1984), it can be concentrated to high levels in catecholaminergic neurons, and drugs that inhibit the uptake of dopamine prevent MPTP-induced damage to dopamine neurons in mice (Javitch et al 1985, Melamed et al 1985, Mayer et al 1986). The specific uptake of MPP^+ adequately explains the sites of toxicity in mice, since both noradrenergic and dopaminergic neurons are affected, but it does not explain the resistance of rats, guinea pigs, or other species to MPTP toxicity, or the selectivity of MPTP toxicity in primates where high concentrations of MPP^+ accumulate without producing damage in hypothalamus, nucleus accumbens, and adrenal (Markey et al 1984). Inhibition of uptake is effective in blocking MPTP-induced nigrostriatal damage in primates only if the inhibitor is given for prolonged intervals after the administration of MPTP (Schultz et al 1986).

Mechanisms of Cellular Toxicity of MPP^+

OXIDATIVE STRESS AND FREE RADICALS   Formation of free radicals is a well-known and an important mechanism mediating the toxic effects of a variety of chemicals (Trush et al 1982). The similarity in structure of MPP^+ and paraquat suggest that the mechanisms of their toxicity might be similar (Kopin et al 1985). Bus et al (1976) proposed that the toxic effects of paraquat were mediated by redox cycling, in the presence of oxygen, to generate superoxide free radicals (O_2). After depletion of protective antioxidants (e.g. glutathione), O_2^-, H_2O_2, and OH are formed and cause membrane lipid peroxidation, oxidative damage to DNA or RNA, or inactivation of vital enzymes. Normally, superoxide dismutase, catalase, and glutathione peroxidase, together with soluble reducing substances, prevent the accumulation of excess free radicals. When free radical formation is accelerated, however, the protective mechanisms may be inadequate.

Support for the role of redox cycling and free radical involvement in MPP^+ toxicity has been obtained largely by indirect evidence. Sinha et al (1986) could not demonstrate by electron spin resonance free radical formation from MPP^+ under conditions in which paraquat-free radicals were apparent. By using a spin-trapping technique, they did find, however, significant stimulation by MPP^+ of superoxide and hydroxyl radical formation during incubation of NADPH cytochrome P-450 reductase with NADPH in the presence of oxygen. Formation of these radicals was less rapid than from paraquat.

Pretreatment of mice with diethyl dithiocarbamate (DDC), which chelates copper and inhibits superoxide dismutase, potentiates the toxicity of both paraquat (Goldstein et al 1979) and MPTP (Corsini et al 1985). DDC, however, inhibits other copper-dependent enzymes, including aldehyde dehydrogenase. Although ethanol, a good hydroxyl radical scavenger, inhibits MPP^+ potentiation of free radical formation in vitro (Sinha et al 1986), it and its metabolite, acetaldehyde, potentiate the toxicity of MPTP in mice (Corsini et al 1987).

Like paraquat, MPP^+, when administered systematically, causes several toxic damage in the lungs and increases the plasma levels of glutathione disulfide, indicating an oxidative stress (Johannessen et al 1986). In isolated hepatocytes, both MPP^+ and paraquat cause rapid depletion of ATP and, in the presence of BCNU, an inhibitor of glutathione reductase of glutathione (Di Monte et al 1986), the toxicity of paraquat is enhanced. MPP^+ cytotoxicity, however, is not enhanced by BCNU nor diminished by an antoxidant (N,N-diphenyl-p-phenylenediamine). Similarly, although MPTP causes glutathione depletion from the brain stem (Yong et al 1986) and substantia nigra (Ferraro et al 1986) of mice, MPP^+ toxicity does not appear to be diminished by agents that elevate tissue levels of glutathione (Perry et al 1986).

Attempts have been made to prevent or diminish the toxicity of MPTP by treatment with one or more antoxidants administered to mice before, during and after the toxin. Perry et al (1985a) partially prevented the depletion of striatal dopamine one month after the administration of a single dose of MPTP (40 mg/kg) by treatment with α-tocopherol (2.5 g/kg), β-carotene (100 mg/kg), L-ascorbic acid (100 mg/kg) or N-acetylcysteine (50 mg/kg) daily for five days beginning two days before administration of the toxin. Using similar methods, however, Baldessarini et al (1986) and Martinovits et al (1986) could not demonstrate protection from the toxicity at seven or 30 days.

The potential role of free radicals in degenerative processes of several neurological disorders has been the subject of speculation for a number of years (see e.g. Clausen 1984). Free radicals are believed to be involved in the cross-linking reactions leading to the formation of lipofuscin, which is deposited alone or as a matrix for neuromelanin. It has been suggested that antoxidant therapy, e.g., with α-tocopherol, by reducing lipid peroxidation, may diminish the rate of lipofuscin deposition and prolong neuronal life.

INHIBITION BY MPP^+ OF MITOCHONDRIAL ENZYMES   Since interference with essential mitochondrial functions could lead to cell death, the effects of MPTP and its metabolites on mitochondrial function are of interest.

In vitro, MPP^+ interferes wth NADH-linked oxidation of pyruvate or glutamate by purified rat or mouse brain mitochondria, without affecting the oxidation of succinate. This suggests that the site of interaction with the respiratory process is at complex I (Nicklas et al 1985). In slices of mouse neostriatum, MPP^+ or MPTP increase lactate formation and enhance the accumulation of glutamate and aspartate, consistent with inhibition of mitochondrial oxidation. The effects of MPTP are prevented by pretreatment with pargyline, indicating that its toxicity is dependent on its oxidation, presumably to MPP^+ (Vyas et al 1986). This is consistent with reversible inhibition by MPP^+, but not by MPTP, of NADH dehydrogenase cytochrome C reductase (Poirier & Barbeau 1985).

Ramsay et al (1986a,b, Ramsay & Singer 1986) and Frei & Richter (1986) showed that MPP^+ is accumulated by an energy-dependent uptake process into intact mitochondria. The uptake process appears to be dependent on the electrochemical gradient of the mitochondrial membrane, since valinomycin and K^+, which abolish the gradient, abolish MPP^+ uptake. Mitochondrial uptake of MPP^+ is not inhibited by dopamine uptake inhibitors (e.g. mazindol) but is blocked by respiratory enzyme inhibitors and uncouplers. The efficacy of MPP^+ in blocking mitochondrial respiration is far less in inverted mitochondria or isolated mitochondrial inner membranes than in intact mitochondria (Ramsay & Singer 1986), consistent with the view that high concentrations of $MPP+$ are attained as a result of its energy-dependent uptake by intact mitochondria (Ramsay et al 1986b). Calcium ions interfere with the uptake of MPP^+ by isolated rat mitochondria (Frei & Richter 1986); they may compete for the same electrochemical gradient. Although neither 6-OH-DA nor MPP^+ separately have much effect on the rate of release of calcium sequestered in mitochondria, together they greatly increase the rate of calcium efflux. MPTP inhibits calcium efflux and almost completely prevents the enhanced calcium release attending combined MPP^+-6-OH-DA treatment. Since excess calcium release has been implicated as a cause of cell death, it was suggested that the toxic effects of MPTP might be mediated through release of calcium from mitochondria.

With glutamate or malate as substrates, the respiratory rate of mitochondria from brains of 24-month-old rats was lower than that of mitochondria from brains of 2-month-old animals and glutamate uptake was markedly lower (Victorica et al 1985, Victorica & Satrüstegui 1986). There was no difference, however, when succinate was used as substrate. The rate of calcium entry in mitochondria from aged rats is also lower than in mitochondria from young rats. It is perhaps not fortuitous that aging produces changes in mitochondrial respiration and calcium distribution that parallel the effects of MPP^+ and increased toxicity of MPTP with age.

Neuromelanin and MPP^+

As indicated above, the occurrence of neuromelanin in primates but not in mice or rats suggested that this pigment might be related to the vulnerability of primates to MPTP toxicity (Burns et al 1984b). Support for this speculation was obtained when it was shown that dogs, a species in which neuromelanin accumulates, are also vulnerable to MPTP. Furthermore, both brain neuromelanin content and the susceptibility of monkeys to MPTP toxicity increase with age (Burns et al 1984b). However, MPTP toxicity also increases with age in mice (Jarvis & Wagner 1985). MPP^+ binds with relatively high affinity to neuromelanin (D'Amato et al 1986), as do a variety of drugs; this property has been suggested to account for MPP^+ retention and vulnerability of neuromelanin-containing neurons.

Neuromelanin, itself a redox polymer containing many free radicals, may, if free radical mechanisms are important for MPP^+ toxicity, make the cell more vulnerable to toxic damage. The possibility that neuromelanin accumulation occurs only under conditions that also make the cells vulnerable to MPP^+ toxicity must be considered. If this should prove to be the case, neuromelanin would become an indicator of vulnerability rather than a participant in the toxic mechanism.

Summary

In summary, the parkinsonism induced by MPTP in man closely resembles time-telescoped Parkinson's disease. Parkinsonian symptoms can be duplicated in all aspects under controlled conditions in subhuman primates; the biochemical changes are replicated in mice, dogs, and to varying degrees in other species. Mechanisms of bioactivation by MAO-B of MPTP to MPP^+, concentration of MPP^+ in neurons with a catecholamine uptake system, and vulnerability to cellular toxic effects of MPP^+ are the basis for the specificity of MPTP targeting of nigrostriatal dopaminergic neurons. It is hoped that an understanding of the mechanism of species specificity and cellular toxicity will in time explain the pathogenesis of idiopathic Parkinson's disease and suggest new opportunities for effective therapy.

Literature Cited

Ansher, S. S., Cadet, J. L., Jakoby, W. B., Baker, J. K. 1986. Role of N-methyltransferases in the neurotoxicity associated with the metabolites of L-methyl-4-phenyl-1,2,3,6-tetrahydropyridine (MPTP) and other 4-substituted pyridines present in the environment. *Biochem. Pharmacol.* 35: 3359–63

Baldessarini, R. J., Kula, N. S., Francoeur, D., Finklestein, S. P. 1986. Antioxidants fail to inhibit depletion of striatal dopamine by MPTP [letter]. *Neurology* 36: 735

Barbeau, A., Dallaire, L., Buu, N. T., Poirier, J., Rucinska, E. 1985a. Comparative behavioral, biochemical and pigmentary effects of MPTP, MPP^+ and paraquat in *rana pipiens. Life Sci.* 37: 1529–38

Barbeau, A., Dallaire, L., Buu, N. T., Veielleux, F., Boyer, H., deLanney, L. E., Irwin, I., Langston, E. G., Langston, J. W. 1985b. New amphibian models for the study of 1-methyl-4-phenyl-1,2,3,6-tetrahydropyridine (MPTP). *Life Sci.* 36: 1125–34

Birkmayer, W., Hornykiewicz, O. 1961. Der

L-3,4-dioxyphenylalanin (= DOPA)-Effekt bei der Parkinson-Akinese. *Wein. Klin. Wschr.* 73: 787–88

Birkmayer, W., Knoll, J., Riederer, P., Youdim, M. B., Hars, V., Marton, J. 1985. Increased life expectancy resulting from addition of L-deprenyl to Madopar treatment in Parkinson's disease: A long-term study. *J. Neural Transm.* 64(2): 113–27

Boyce, S., Kelly, E., Reavill, C., Jenner, P., Marsden, C. D. 1984. Repeated administration of N-methyl-4-phenyl-1,2,3,6-tetrahydropyridine to rats is not toxic to striatal dopamine neurons. *Biochem. Pharmacol.* 33: 1747–52

Burns, R. S., Chiueh, C. C., Markey, S. P., Ebert, M. H., Jacobowitz, D. M., Kopin, I. J. 1983. A primate model of parkinsonism: Selective destruction of dopaminergic neurons in the pars compacta of the substantia nigra by N-methyl-4-phenyl-1,2,3,6-tetrahydropyridine. *Proc. Natl. Acad. Sci. USA* 80: 4546–50

Burns, R. S., LeWitt, P. A., Ebert, M. H., Pakkenberg, H., Kopin, I. J. 1984a. The clinical syndrome of striatal dopamine deficiency: Parkinsonism induced by 1-methyl - 4 - phenyl - 1,2,3,6 - tetrahydropyridine (MPTP). *New Eng. J. Med.* 312: 1418–21

Burns, R. S., Markey, S. P., Phillips, J. M., Chiueh, C. C. 1984b. The neurotoxicity of 1 - methyl - 4 - phenyl - 1,2,3,6 - tetrahydropyridine in the monkey and man. *Can. J. Neurol. Sci.* 11(1 Suppl): 166–68

Bus, J. S., Cogen, S., Olgaard, M., Gibson, J. E. 1976. A mechanism of paraquat toxicity in mice and rats. *Toxicol. Appl. Pharmacol.* 35: 501–13

Calne, D. B., Langston, J. S., Martin, W. R., Stoessl, A. J., Ruth, T. J., Adam, M. J., Pate, B. D., Schulzer, M. 1985. Positron emission tomography after MPTP: Observations relating to the cause of Parkinson's disease. *Nature* 317: 246–48

Chiba, K., Trevor, A., Castagnoli, N. Jr. 1984. Metabolism of the neurotoxic tertiary amine, MPTP, by brain monoamine oxidase. *Biochem. Res. Commun.* 120: 574–78

Chiueh, C. C., Markey, S. P., Burns, R. S., Johannessen, J. N., Jacobowitz, D. M., Kopin, I. J. 1984. Neurochemical and behavioral effects of 1-methyl-4-phenyl-1,2,3,6-tetrahydropyridine (MPTP) in rat, guinea pig, and monkey. *Psychopharmacol. Bull.* 20: 548–53

Clausen, J. 1984. Demential syndromes and the lipid metabolism. L-DOPA treatment of Parkinson's disease. *Acta Neurol. Scand.* 70: 345–55

Cohen, G., Pasik, P., Cohen, B., Leist, A.,

Mytilineou, C., Yahr, M. D. 1984. Pargyline and deprenyl prevent the neurotoxicity of 1-methyl-4-phenyl-1,2,3,6-tetrahydropyridine (MPTP) in monkeys. *Eur. J. Pharmacol.* 106: 209–10

Corsini, G. U., Pintus, S., Chiueh, C. C., Weiss, J. F., Kopin, I. J. 1985. 1-Methyl - 4 - phenyl - 1,2,3,6 - tetrahydropyridine (MPTP) neurotoxicity in mice is enhanced by pretreatment with diethyldithiocarbamate. *Eur. J. Pharmacol.* 119: 127–28

Corsini, G. U., Zuddas, A., Bonuccelli, U., Schinelli, S., Kopin, I. J. 1987. 1-Methyl - 4 - phenyl - 1,2,3,6 - tetrahydropyridine (MPTP) neurotoxicity in mice is enhanced by ethanol or acetaldehyde. *Life Sci.* 40: 827–32

D'Amato, R. J., Lipman, Z. P., Snyder, S. H. 1986. Selectivity of the parkinsonian neurotoxin MPTP: Toxic metabolite MPP$^+$ binds to neuromelanin. *Science* 231: 987–89

Davis, G. C., Williams, A. C., Markey, S. P., Ebert, M. H., Caine, E. D., Reichert, C. M., Kopin, I. J. 1979. Chronic parkinsonism secondary to intravenous injection of meperidine analogues. *Psychiatry Res.* 1: 249–54

DiMonte, D., Sandy, M. S., Ekström, G., Smith, M. T. 1986. Comparative studies on the mechanisms of paraquat and 1-methyl-4-phenylpyridine (MPP$^+$) cytotoxicity. *Biochem. Biophys. Res. Commun.* 137: 303–9

Duvoisin, R. C. 1986. Etiology of Parkinson's disease: Current concepts. *Clin. Neuropharm.* 9: S3–S11

Enz, A., Hefti, F., Frick, W. 1984. Acute administration of 1-methyl-4-phenyl-1,2,3,6-tetrahydropyridine (MPTP) reduces dopamine and serotonin but accelerates norepinephrine metabolism in the rat brain. Effect of chronic pretreatment with MPTP. *Eur. J. Pharmacol.* 101: 37–44

Ferraro, T. N., Golden, G. T., DeMattei, M., Hare, T. A., Fariello, R. G. 1986. Effect of 1-methyl-4-phenyl-1,2,3,6-tetrahydropyridine (MPTP) on levels of glutathione in the extrapyramidal system of the mouse. *Neuropharmacology* 25(9): 1071–74

Forno, L. S., Langston, J. W., DeLanney, L. E., Irwin, I., Ricaurte, G. A. 1986. Locus ceruleus lesions and eosinophilic inclusions in MPTP-treated monkeys. *Ann. Neurol.* 20: 449–55

Frei, B., Richter, C. 1986. 1-Methyl-4-phenylpyridine (MPP$^+$) together with 6-hydroxydopamine or dopamine stimulates Ca^{2+} release from mitochondria. *FEBS Lett.* 198: 99–102

Fuller, R. W., Hemrick-Leucke, S. K. 1985. Influence of selective, reversible inhibitors of monoamine oxidase on the prolonged depletion of striatal dopamine by 1-methyl - 4 - phenyl - 1,2,3,6 - tetrahydropyridine in mice. *Life Sci.* 37: 1098–96

Goldstein, B. D., Rozen, M. G., Quintavalla, J. C., Amoruso, M. A. 1979. Decrease in mouse lung and liver glutathione peroxidase activity and potentiation of the lethal effects of ozone and paraquat by the superoxide dismutase inhibitor diethyldithiocarbamate. *Biochem. Pharmacol.* 29: 27–30

Hallman, H., Lange, J., Olson, L., Stromberg, I., Jonsson, G. 1985. Neurochemical and histochemical characterization of neurotoxic effects of 1-methyl-4-phenyl-1,2,3,6-tetrahydropyridine on brain catecholamine neurons in the mouse. *J. Neurochem.* 44: 117–20

Hallman, H., Olson, L., Jonsson, G. 1984. Neurotoxicity of the meperidine analogue *N*-methyl-4-phenyl-1,2,3,6-tetrahydropyridine on brain catecholamine neurons in the mouse. *Eur. J. Pharmacol.* 197: 133–36

Hantraye, P., Loc'h, C., Tacke, U., Riche, D., Stulzaft, O., Doudet, D., Guibert, B., Maquet, R., Mazïere, B., Mazïere, M. 1986. In vivo: Visualization by positron emission tomography of the progressive striatal dopamine receptor damage occurring in MPTP-intoxicated non-human primates. *Life Sci.* 39: 1375–82

Heikkila, R. E., Hess, A., Duvoisin, R. C. 1984a. Dopaminergic neurotoxicity of 1-methyl - 4 - phenyl - 1,2,3,6 - tetrahydropyridine in mice. *Science* 244: 1451–53

Heikkila, R. E., Hess, A., Duvoisin, R. C. 1985. VI. Dopaminergic neurotoxicity of 1 - methyl - 4 - phenyl - 1,2,3,6 - tetrahydropyridine (MPTP) in the mouse: Relationships between monoamine oxidase, MPTP metabolism and neurotoxicity. *Life Sci.* 36: 231–36

Heikkila, R. E., Mazino, L., Cabbat, F. S., Duvoisin, R. C. 1984b. Protection against the dopaminergic neurotoxicity of 1-methyl - 4 - phenyl - 1,2,3,6 - tetrahydropyridine by monoamine oxidase inhibitors. *Nature* 311: 467–69

Hornykiewicz, O. 1982. Imbalance of brain monoamines and clinical disorders. *Prog. Brain Res.* 55: 419–29

Irwin, I., Langston, J. W. 1985. Selective accumulation of MPP$^+$ in the substantia nigra: A key to neurotoxicity? *Life Sci.* 36: 213–18

Jacobowitz, D. M., Burns, R. S., Chiueh, C. C., Kopin, I. J. 1984. *N*-methyl-4-phenyl-1,2,3,6-tetrahydropyridine (MPTP) causes destruction of the nigrostriatal but not the mesolimbic dopamine system in the monkey. *Psychopharmacol. Bull.* 20: 416

Jarvis, M. F., Wagner, G. C. 1985. Age-dependent effects of 1-methyl-4-phenyl-1,2,3,6-tetrahydropyridine (MPTP). *Neuropharmacology* 24: 581–83

Javitch, J. A., D'Amato, R. J., Strittmatter, S. M., Snyder, S. H. 1985. Parkinsonism-inducing neurotoxin, *N*-methyl-4-phenyl-1,2,3,6-tetrahydropyridine: Uptake of the metabolite *N*-methyl-4-phenylpyridine by dopamine neurons explains selective toxicity. *Proc. Natl. Acad. Sci. USA* 82: 2173–77

Javitch, J. A., Snyder, S. H. 1984. Uptake of MPP (+) by dopamine neurons explains selectivity of parkinsonism-inducing neurotoxin, MPTP. *Eur. J. Pharmacol.* 106: 455–56

Johannessen, J. N., Adams, J. D., Schuller, H. M., Bacon, J. P., Markey, S. P. 1986. 1-Methyl-4-phenylpyridine (MPP$^+$) induces oxidative stress in the rodent. *Life Sci.* 38: 743–49

Johannessen, J. N., Savitt, J. M., Markey, C. J., Bacon, J. P., Weisz, A., Hanselman, D. S., Markey, S. P. 1987. *Life Sci.* 40: 697–704

Knoll, J., Ecseri, Z., Kelemen, K., Nievele, J., Knoll, B. 1965. Phenylisopropylmethylpropinylamine (E-250), a new spectrum psychic energizer. *Arch. Int. Pharmacodyn. Ther.* 155: 154–64

Kopin, I. J. 1985. Catecholamine metabolism: Basic aspects and clinical significance. *Pharm. Rev.* 37: 333–64

Kopin, I. J., Markey, S. P., Burns, R. S., Johannessen, J. N., Chiueh, C. C. 1985. Mechanisms of neurotoxicity of MPTP. In *Recent Developments in Parkinson's Disease*, ed. S. Fahn, C. D. Marsden, P. Jenner, P. Teychenne, pp. 165–73. New York: Raven

Lent, C. M. 1986. MPTP depletes neuronal monoamines and impairs the behavior of the medicinal leech. See Markey et al 1986, pp. 105–18

Langston, J. W., Ballard, P., Tetrud, J. W., Irwin, I. 1983. Chronic parkinsonism in humans due to a product of meperidine-analog synthesis. *Science* 219: 979–80

Langston, J. W., Forno, L. S., Rebert, C. S., Irwin, I. 1984a. Selective nigral toxicity after systemic administration of 1-methyl - 4 - phenyl - 1,2,3,6 - tetrahydropyridine (MPTP) in the squirrel monkey. *Brain Res.* 292: 390–94

Langston, J. W., Irwin, I., Langston, E. B., Forno, L. S. 1984b. Pargyline prevents MPTP-induced parkinsonism in primates. *Science* 225: 1480–82

Lees, A. J., Shaw, K. M., Kohout, L. J., Stern, G. M., Elsworth, J. D., Sandler, M., Youdim, M. B. 1977. Deprenyl in Parkinson's disease. *Lancet* 2: 791–95

Markey, S. P., Castagnoli, N. Jr., Trevor, A., Kopin, I. J., eds. 1986. *MPTP: A Neurotoxin Producing a Parkinsonian Syndrome.* Orland, FL: Academic

Markey, S. P., Johannessen, J. N., Chiueh, C. C., Burns, R. S., Herkenham, M. A. 1984. Intraneuronal generation of a pyridinium metabolite may cause drug-induced parkinsonism. *Nature* 311: 464–67

Martinovits, G., Melamed, E., Cohen, O., Rosenthal, J., Uzzan, A. 1986. Systemic administration of antioxidants does not protect mice against the dopaminergic neurotoxicity of 1-methyl-4-phenyl-1,2,3,6-tetrahydropyridine (MPTP). *Neurosci. Lett.* 69: 192–97

Mayer, R. A., Kindt, M. V., Heikkila, R. E. 1986. Prevention of the nigrostriatal toxicity of 1-methyl-4-phenyl-1,2,3,6-tetrahydropyridine by inhibitors of 3,4-dihydroxyphenylethylamine transport. *J. Neurochem.* 47: 1073–79

McGeer, P. L., Boulding, J. E., Gibson, W. C., Foulkes, R. G. 1961. Drug-induced extrapyramidal reactions. *J. Am. Med. Assoc.* 177: 665–70

Melamed, E., Rosenthal, J., Cohen, O., Globus, M., Uzzan, A. 1985. Dopamine but not norepinephrine or serotonin uptake inhibitors protect mice against neurotoxicity of MPTP. *Eur. J. Pharmacol.* 116: 179–81

Mitchell, I. J., Cross, A. J., Sambrook, M. A., Crossman, A. R. 1985. Sites of the neurotoxic action of 1-methyl-4-phenyl-1,2,3,6-tetrahydropyridine in the macque monkey include the ventral tegmental area and the locus coeruleus. *Neurosci. Lett.* 61: 195–200

Nicklas, W. J., Vyas, I., Heikkila, R. E. 1985. Inhibition of NADH-linked oxidation in brain mitochondria by 1-methyl-4-phenyl-1,2,3,6-tetrahydropyridine. *Life Sci.* 36: 2503–8

Parisi, J. E., Burns, R. S. 1986. The neuropathology of MPTP-induced parkinsonism in man and experimental animals. See Markey et al 1986, pp. 141–48

Perry, T. L., Yong, V. W., Clavier, R. M., Jones, K., Wright, J. M., Foulks, J. G., Wall, R. A. 1985a. Partial protection from the dopaminergic neurotoxin 1-methyl-4-phenyl-1,2,3,6-tetrahydropyridine by four different antioxidants in the mouse. *Neurosci. Lett.* 60: 109–14

Perry, T. L., Yong, V. W., Ito, M., Jones, K., Wall, R. A., Foulks, J. G., Wright, J. M., Kish, S. J. 1985b. 1-Methyl-4-phenyl-1,2,3,6-tetrahydropyridine (MPTP) does not destroy in nigrostriatal neurons in the scorbutic guinea pig. *Life Sci.* 36: 1233–38

Perry, T. L., Yong, V. W., Jones, K., Wright, J. M. 1986. Manipulation of glutathione contents fails to alter dopaminergic nigrostriatal neurotoxicity of 1-methyl-4-phenyl-1,2,3,6-tetrahydropyridine (MPTP) in the mouse. *Neuro. Sci. Lett.* 70: 261–65

Poirier, J., Barbeau, A. 1985. 1-Methyl-4-phenylpyridinium-induced inhibition of nicotinamide adenosine dinucleotide cytochrome C reductase. *Neurosci. Lett.* 62: 7–11

Ramsey, R. R., Dadgar, J., Trevor, A., Singer, T. P. 1986a. Energy-driven uptake of N-methyl-4-phenylpyridine by brain mitochondria mediates the neurotoxicity of MPTP. *Life Sci.* 39: 581–88

Ramsey, R. R., Salach, J. I., Singer, T. P. 1986b. Uptake of the neurotoxin 1-methyl-4-phenylpyridine (MPP^+) by mitochondria and its relation to the inhibition of the mitochondrial oxidation of NAD^+-linked substrates by MPP^+. *Biochem. Biophys. Res. Commun.* 134: 743–48

Ramsey, R. R., Singer, T. P. 1986. Energy-dependent uptake of N-methyl-4-phenylpyridinium, the neurotoxic metabolite of 1-methyl-4-phenyl-1,2,3,6-tetrahydropyridine, by mitochondria. *J. Biol. Chem.* 261: 7585–87

Riederer, P., Youdim, M. B. H. 1986. Monoamine oxidase activity and monoamine metabolism in brains of parkinsonian patients treated with *l*-deprenyl. *J. Neurochem.* 44: 1359–65

Rinne, U. K. 1983. Deprenyl (selegiline) in the treatment of Parkinson's disease. *Acta Neurol. Scand. Suppl.* 95: 107–11

Rosen, J. A. 1969. The effect of monoamine oxidase inhibitor on the bradykinesia of human parkinsonism. In *Progress in Neurogenetics,* ed. A. Barbeau, J. R. Brunette 1: 346–48. *Int. Congr. Ser.* No. 175: 346–48. Amsterdam: Excerpta Medica Found.

Sahgal, A., Andrews, J. S., Biggins, J. A., Candy, J. M., Edwardson, J. A., Keith, A. B., Turner, J. D., Wright, C. 1984. N-methyl-4-phenyl-1,2,3,6-tetrahydropyridine (MPTP) affects locomotor activity without producing a nigrostriatal lesion in the rat. *Neurosci. Lett.* 48: 179–84

Schneider, J. S., Yuwiler, A., Markham, C. H. 1986. Production of a Parkinson-like syndrome in the cat with N-methyl-4-phenyl-1,2,3,6-tetrahydropyridine (MPTP): Behavior, histology, and biochemistry. *Exp. Neurol.* 91: 293–307

Schultz, W., Scarnati, E., Sundström, E., Tsutsumi, T., Jonsson, G. 1986. The catecholamine uptake blocker nomifensine protects against MPTP-induced parkinsonism in monkeys. *Exp. Brain Res.* 63: 216–20

Sinha, B. K., Singh, Y., Krishna, G. 1986. Formation of superoxide and hydroxyl radicals from 1-methyl-4-phenlypyridinium ion MPP$^+$: Reductive activation by NADPH cytochrome P-450 reductase. *Biochem. Biophys. Res. Commun.* 135: 283–88

Trush, M. A., Mimnaugh, E. G., Gram, T. E. 1982. Activation of pharmacologic agents to radical intermediates. Implications for the role of free radicals in drug action and toxicity. *Biochem. Pharmacol.* 31: 3335–46

Vitorica, J., Clark, A., Machado, A., Satrüstegui, J. 1985. Impairment of glutamate uptake and absence of alterations in the energy-transducing ability of old rat brain mitochondria. *Mech. Ageing Dev.* 29: 255–66

Vitorica, J., Satrüstegui, J. 1986. Involvement of mitochondria in the age-dependent decrease in calcium uptake of rat brain synaptosomes. *Brain Res.* 16(378): 36–48

Vyas, I., Heikkila, R. E., Nicklas, W. J. 1986. Studies on the neurotoxicity of 1-methyl-4-phenyl-1,2,3,6-tetrahydropyridine: Inhibition of NAD-linked substrate oxidation by its metabolite, 1-methyl-4-phenylpyridinium. *J. Neurochem.* 46: 1501–7

Yahr, M. D. 1978. Overview of present day treatment of Parkinson's disease. *J. Neural Transm.* 43: 227–38

Yong, V. W., Perry, T. L., Krisman, A. A. 1986. Depletion of glutathione in brainstem of mice caused by N-methyl-4-phenyl-1,2,3,6-tetrahydropyridine is prevented by antioxidant pretreatment. *Neurosci. Lett.* 639: 56–60

Ann. Rev. Neurosci. 1988. 11:97–118

ADENOSINE 5'-TRIPHOSPHATE-SENSITIVE POTASSIUM CHANNELS

Frances M. Ashcroft

University Laboratory of Physiology, Parks Road, Oxford OX1 3PT, England

INTRODUCTION

In 1983 Akinori Noma described a new type of potassium channel in single channel recordings from cardiac cell membranes. This channel was characterized by a pronounced inhibition of channel openings when the adenosine-5'-triphosphate (ATP) concentration at the intracellular membrane surface was increased to millimolar levels. The ATP-sensitive K-channel has now been described in the membranes of cardiac muscle (Noma 1983, Kakei & Noma 1984, Trube & Hescheler 1984), skeletal muscle (Spruce et al 1985, 1987), and pancreatic β-cells (Cook & Hales 1984, Findlay et al 1985a, Rorsman & Trube 1985). It has a unitary conductance of between 50 and 80 pS when exposed to symmetrical concentrations of 150 mM K^+ and is highly potassium selective. A new class of channel blockers, the sulphonylureas, selectively inhibit the ATP-sensitive K-channel (Sturgess et al 1985, Trube et al 1986). When studied in cell-attached patches on intact cells, the channel can be activated by manipulations that decrease intracellular ATP such as hypoxia or inhibitors of cell metabolism (Kakei & Noma 1984, Trube & Hescheler 1984, Ashcroft et al 1985). In pancreatic β-cells, where channel activity can be recorded under resting conditions, the channel can also be induced to close by substances like glucose that act as substrates for β-cell metabolism (Ashcroft et al 1984, 1987a, Rorsman & Trube 1985, Misler et al 1986).

My aim in this review is to summarize what is known about the properties of this ATP-sensitive K-channel and its regulation by cell metabolism. Other types of channels permeable to K^+ and inhibited by intra-

97

0147–006X/88/0201–0097$02.00

cellular ATP have been reported briefly (Ashford et al 1966, Findlay et al 1985a, Dunne et al 1986) but are not considered here.

Almost all information on the properties of the ATP-sensitive K-channel has been obtained using the standard variants of the patch-clamp method (Hamill et al 1981) and an additional configuration, the open-cell patch (Kakei et al 1985). In the latter method the pipette remains in the cell-attached mode but the cell is permeabilized (usually with saponin) to allow access to the intracellular membrane surface from the bath solution; the idea is that the cytoskeleton remains intact and cytoplasmic constituents that may be required for channel function will be lost less rapidly than in the inside-out patch. With a few interesting exceptions these methods yield comparable results.

PERMEABILITY AND KINETIC PROPERTIES

Table 1 gives values obtained for the unitary conductance (γ) of the ATP-sensitive K-channel in different tissues, measured as the slope conductance

Table 1   Properties of ATP-sensitive K-channels

Tissue	γ(pS)	Configuration	K_i ATP (μM)	Hill coefficient	Ref.
Cardiac muscle	80	i/o	100		Noma (1983)
					Trube & Hescheler (1984), Kakei & Noma (1984)
		o/c	500	3–4	Kakei & Noma (1984)
		w/c	500		Noma & Shibasaki (1985)
Skeletal muscle	42	i/o	135	1	Sturgess et al (1985)
Pancreatic β-cell					
human	64	i/o			Ashcroft et al (1987d)
	56	c/a			Ashcroft et al (1987d)
rat (neonatal)	54	i/o	15	1.2	Cook & Hales (1984)
rat (adult)	51	c/a			Ashcroft et al (1985)
	60–65	i/o	20–200	1	Misler et al (1986)
	75	i/o			Findlay et al (1985a)
mouse (adult)	51	c/a			Rorsman & Trube (1985)
	56	i/o			Rorsman & Trube (1985)
			18	1.8	Ohno-Shosaku et al (1987)
mouse (obese)	66	c/a			Arkhammar et al (1987)
RINm5F	90	c/a			Dunne et al (1986)
	50	i/o, c/a	78	1.8	Ribalet & Ciani (1987)
CR1-G1	55	i/o	13	1.2	Sturgess et al (1986)

Single channel conductance measured with an external K-concentration of approximately 140 mM except in the case of skeletal muscle where $[K]_o$ was 60 mM. $[K]_i$ was between 100–140 mM. The recording configuration abbreviations refer to cell-attached (c/a), open-cell (o/c), whole-cell (w/c), or inside-out (i/o). K_i is the ATP concentration (μM) that produces half-maximal inhibition of channel activity.

over the linear region of the current-voltage relation. For external K concentrations ($[K]_o$) around 140 mM, γ lies between 50–80 pS, the lower values being for β-cells. A conductance of around 20 pS is obtained for physiological $[K]_o$ of 5 mM (Noma 1983, Kakei et al 1985, Findlay et al 1985a, Trube et al 1986).

The current-voltage relation measured for the single ATP-sensitive K-channel shows inward rectification when the channel is exposed to roughly symmetrical K-concentrations (Noma 1983, Cook & Hales 1984, Spruce et al 1985) with outward currents being significantly smaller than inward. A similar inward rectification is also found in cell-attached patch clamp recordings (Ashcroft et al 1984, Kakei & Noma 1984). It is now clear that the primary cause of the rectification is a voltage-dependent block of outward K-currents by intracellular cations, principally Na^+ and Mg^{2+} (Horie et al 1987).

A small degree of rectification persists, however, even when all internal cations (other than K^+) are absent (Spruce et al 1987, Ashcroft et al 1987c), and this is greater than that predicted by the Goldman-Hodgkin-Katz (GHK) equation (Goldman 1943, Hodgkin & Katz 1949). It is still too early to say whether the remaining rectification is due to an inherent property of the open channel (for example, a binding site for K^+ within the channel to which K ions bind and so block their own movement) or due to channel kinetics that at positive potentials become too fast to be resolved by the patch-clamp amplifier and so produce a time-averaged current of reduced amplitude.

Potassium Permeability

The unitary conductance of the ATP-sensitive K-channel depends on the external K-concentration and increases as this is raised. In the β-cell this increase is approximately that expected if K^+ ions move through the pore independently of one another, that is if the single channel *permeability* is independent of external potassium (Ashcroft et al 1987b). By contrast, in both cardiac (Kakei et al 1985, Horie et al 1987) and skeletal (Spruce et al 1987) muscle the unitary conductance shows significantly more saturation than predicted by the GHK equation. This difference may be explained by assuming a binding site for K^+ within the channel conduction pathway that has a much higher affinity for K^+ ions in muscle than in the β-cell.

Single-channel Kinetics

The kinetics of the ATP-sensitive K-channel are complex, consisting of bursts of channel openings separated by relatively long closed periods. There is general agreement that at least one open state and two closed states are required to account for the ATP-sensitive K-channel kinetics (Kakei & Noma 1984, Trube & Hescheler 1984, Kakei et al 1985, Rorsman

& Trube 1985, Spruce et al 1987, Ashcroft et al 1987b), thus suggesting a reaction scheme of the following form (Colquhoun & Hawkes 1981, 1982):

$$----C_2 \underset{k_{-2}}{\overset{k_2}{\rightleftharpoons}} C_1 \underset{k_{-1}}{\overset{k_1}{\rightleftharpoons}} O --- \quad \text{or} \quad -- C_2 \underset{k_{-2}}{\overset{k_2}{\rightleftharpoons}} O \underset{k_{-1}}{\overset{k_1}{\rightleftharpoons}} C_1,$$

where the rate constant k_1 is fast in relation to k_{-2}. Closed state 1 corresponds to the short closings that occur within a burst of openings and closed state 2 to the longer interburst intervals. However, further open and closed states have also been postulated. For example, very long recordings suggest that the bursts of openings are clustered, indicating there may be three, or even more, closed states in skeletal muscle (Spruce et al 1987), and evidence has also been found for two open states (Spruce et al 1985, 1987, Kakei et al 1985).

Voltage-dependence of single channel kinetics has been reported by several investigators, depolarization increasing the mean open time(s) (cardiac muscle, Trube & Hescheler 1984; skeletal muscle, Spruce et al 1985, 1987). Our own studies on β-cells (F. M. Ashcroft and M. Kakei, unpublished) support this observation but, additionally, we find the kinetics depend on the direction of current flow in such a way that outward currents are associated with long open times whereas inward currents show burst kinetics. Although this result is reminiscent of the presence of a blocking ion in the external solution it has not been possible to identify such an ion (Horie et al 1987).

The principal effect of ATP on the channel kinetics is to reduce both the mean open time of the channel and the number of openings per burst (Kakei et al 1985, Spruce et al 1987). This effect is similar to that produced by glucose in cell-attached recordings (Ashcroft et al 1987b) and provides support for the idea that ATP is an intracellular regulator of channel activity. No effect of ATP on the amplitude of the single channel current is seen.

A complete kinetic analysis aims to describe the ATP-sensitive K-channel's kinetic behavior in terms of the length of time spent in a particular state (open or closed) and, ultimately, how the dwell times in these states are influenced by binding ATP. Unfortunately we are still a long way from achieving this ideal. This is at least partly due to the problems inherent in the analysis of the long closed times found for the ATP-sensitive K-channel. These cannot be interpreted easily because long periods of recording are required in order to obtain sufficient data to analyze; for cell-attached patches, it is unlikely that metabolism will remain stable during this time, as is assumed by the analysis. Isolated patches seem to be plagued either by the presence of too many channels or by the problem that after excising the patch the channel activity rapidly declines (a phenomenon known as "rundown"). A final caution is that as earlier studies were unaware of the existence of ionic block, channel lifetimes obtained for outward currents might include contributions from both this ionic block and channel gating.

The difficulties encountered in analyzing the channel kinetics also apply to measurements of open probability. With these provisos, depolarization has been reported to increase the open probability in muscle (Kakei & Noma 1984, Spruce et al 1985) or to have little effect on the fraction of time spent in the open state (Kakei et al 1985, Ashcroft et al 1984, Rorsman & Trube 1985).

Ionic Selectivity of the Channel

The ATP-sensitive K-channel is permeable to K^+ and Rb^+ but discriminates strongly against external (but not internal) Na^+ when permeability is assessed from measurements of reversal potential. The relative permeability to rubidium (P_{Rb}/P_K) measured under bi-ionic conditions (Rb_o, K_i) is 0.76 in skeletal muscle (Spruce et al 1987), and similar in the β-cell (Ashcroft et al 1987c) suggesting that Rb^+ can enter the channel from the inside almost as easily as K^+. But, as described below, Rb^+ may bind more tightly to sites within the pore than K^+ because Rb currents are very small and Rb^+ ions can also block K-currents.

Sodium is much less permeant than potassium. ATP-sensitive K-channels in all tissues discriminate strongly against external Na, as judged by the fact that the reversal potential approximates the K-equilibrium potential when $[K]_o$ is replaced by $[Na]_o$ (Kakei & Noma 1984, Trube & Hescheler 1984, Kakei et al 1985, Ashcroft et al 1987b, Spruce et al 1987). Permeability ratios (P_{Na}/P_K) of 0.015 and of 0.07 have been measured for skeletal muscle (Spruce et al 1987) and β-cells (Ashcroft et al 1987c, Misler et al 1986), respectively. Interestingly, internal Na^+ ions are relatively more permeant ($P_{Na}/P_K = 0.33$; Ashcroft et al 1987c).

DISTRIBUTION

ATP-sensitive K-channels have been identified by single channel recording in three types of tissue (see Table 1 for references): cardiac muscle, adult frog skeletal muscle, and pancreatic β-cells. The channel is absent in mammalian myotubes (Spruce et al 1987). Circumstantial evidence suggests that ATP-sensitive K-channels may also be present in other cell membranes since electrical activity in a number of cells is influenced by glucose and by metabolic poisons. For example, glucose reduces membrane K-permeability and depolarizes the glucoreceptor neurons of the nucleus tractus solitarii (Mizuno & Oomura 1984) and type C neurons of the ventro-medial hypothalamic neurons (Minami et al 1966). The resting K-permeability of hippocampal neurons increases and the membrane hyperpolarizes in response to hypoxia (Fujiwara et al 1987). There is also evidence that a lowering of intracellular ATP links hypoxia to excitation of nerve endings in the carotid body (Joels & Neil 1968); this raises the interesting possibility that ATP-sensitive K-channels may play a role in mediating the response of the carotid body chemoreceptors to hypoxia. The selective inhibition of ion flux through ATP-sensitive K-channels produced by sulphonylureas (Sturgess et al 1985, Trube et al 1986) and the demonstration of saturable, high-affinity, membrane sites for these drugs (Kaubisch et al 1982, Geisen et al 1985) suggests they may be used to localize ATP-sensitive K-channels. High affinity binding sites for sulphonylureas have been described in rat cerebral cortex (Kaubisch et al 1982, Geisen et al 1985).

Density of ATP-sensitive K-channels

It is evident from the very large number of channels (10–20) that can be observed in excised patches by using standard size patch electrodes that the ATP-sensitive K-channel must occur at high density in the membranes of muscle and pancreatic β-cells. Estimates of channel density range from $0.5/\mu m^2$ in ventricular cells (Noma & Shibasaki 1965, Belles et al 1987) to $1.5–8/\mu m^2$ in β-cells (Ohno-Shosaku et al 1987, Misler et al 1986) and may be more than $10/\mu m^2$ in skeletal muscle (Spruce et al 1985). These values are representative of the total number of channels that can be activated in a membrane patch. The number of channels functional at any given time in the intact cell is considerably lower; less than $0.3/\mu m^2$ is estimated for resting β-cells (Ashcroft et al 1984) and even lower values are indicated for muscle.

MECHANISMS OF CHANNEL INHIBITION

Inhibition of Channel Activity by Nucleotides

ADENOSINE 5′-TRIPHOSPHATE   As their name indicates, ATP-sensitive K-channels are characteristically closed by adenosine 5′-triphosphate. The nucleotide is only effective when applied to the intracellular membrane surface, and inhibition is independent of membrane potential (Sturgess et al 1986). Thus although ATP carries a strong negative charge, its ability to block the channel does not appear to result from the molecule entering the pore and blocking K-current flow. Rather, the evidence favors the idea that there is a specific binding site for ATP associated with the channel. In skeletal muscle a 1 : 1 binding of ATP to this site is indicated, although in cardiac cells, and in some β-cells, there is a cooperativity between binding sites, as the Hill coefficient is greater than 1 (for references see Table 1). The sensitivity of the binding site for ATP varies between different tissues (Table 1), with values for half maximal inhibition ranging from 10 to 200 μM when measured in the inside-out patch. About 1 mM ATP is sufficient to inhibit more than 95% of channel activity in the inside-out patch. In general, the β-cell channels seem somewhat more sensitive to ATP (Table 1).

A major problem associated with the construction of a dose-response curve for ATP is the rundown of channel activity that follows patch excision. Despite this, a similar ATP sensitivity is found when precautions are taken to avoid rundown (Ohno-Shosaku et al 1987; and see below).

Estimates of the threshold concentration of ATP required for complete inhibition, and of the K_i, obtained using the open-cell or whole-cell configurations, invariably yield higher values (Kakei et al 1985, Noma & Shibasaki 1985, Dunne et al 1986, Belles et al 1987). The reason for this is not clear. Noma & Shibasaki (1985), having blocked both contraction and the Na-K-ATPase, assumed that the ATP consumption of ventricular myocytes would be low. They therefore concluded that an ATP concentration gradient through the cell was unlikely to account for the reduced sensitivity in the open-cell patch. Instead they postulated that patch excision altered the channel protein(s) in some way. Others have reached a different conclusion. The observation that ATP-sensitive K-currents were only transiently suppressed in myocytes dialyzed for long periods with 20 mM ATP led Belles et al (1987) to suggest that the slow diffusion of ATP may not be sufficient to compensate for its catabolism in dialyzed cells. This idea receives support from the finding that the addition of creatine phosphate to the internal solution increases the blocking efficacy of ATP (Noma & Shibasaki 1985). Against their hypothesis, however, is the fact that nonhydrolyzable ATP analogues are unable to permanently inhibit whole-cell K-currents (Belles et al 1987).

Mechanism of ATP inhibition   It is well established that phosphorylation of the channels or associated control sites by membrane bound protein kinases is not the mechanism by which ATP inhibits the channel, since the nonhydrolyzable analogues of ATP, adenylylimidodiphosphate (AMP-PNP) and adenylylmethylenediphosphonate (AMP-PCP) are also effective inhibitors and ATP inhibits in the absence of Mg^{2+} (Cook & Hales 1984, Kakei et al 1985, Spruce et al 1985, 1987, Ashcroft & Kakei 1987). The ATP^{4-} ion appears to be more potent than MgATP at producing channel inhibition in β-cells, as the block is potentiated when the free ion is increased by Mg^{2+} removal (Ashcroft & Kakei 1987). Unfortunately, it is very difficult to test whether the MgATP complex also inhibits because

Mg^{2+} itself reduces channel activity (see below). The relative potency of the ATP^{4-} and MgATP is of physiological significance, however, since intracellular ATP^{4-} is low.

A variety of analogues of ATP exist that may prove useful in elucidating the interaction of ATP with its binding site (Yount 1975). In particular, the relative efficacy of AMP-PNP and AMP-PCP is of interest since the structure of these nucleotides differs from ATP only in that the terminal phosphate group is linked by a nitrogen (AMP-PNP) or a carbon (AMP-PCP) rather than an oxygen (ATP) atom (Yount 1975). Substitution of a nitrogen atom has little effect on the bond angle of the P–X–P linkage; this may explain why AMP-PNP and ATP are equipotent (Spruce et al 1987) whereas AMP-PCP, with a smaller bond angle, is a less effective substitute for ATP (F. M. Ashcroft and M. Kakei, unpublished observations).

Specificity for ATP   Alterations of every part of the ATP molecule reduce its ability to close the channel. Although other adenine nucleotides are able to block the channel, they are considerably less effective than ATP, the relative potency being ATP > ADP > AMP > adenosine (Cook & Hales 1984, Kakei et al 1985, Spruce et al 1987, Ribalet & Ciani 1987). Likewise, modifying the ribose sugar either by reduction to 2-deoxy-ATP or by oxidation to ATP-2,3-dialdehyde substantially decreases activity (Spruce et al 1987). Replacement of the adenosine moiety by other purine or by pyrimidine nucleotides also reduces the degree of inhibition. At a concentration of 1 mM, GTP, ITP, XTP, CTP, and UTP were about ten times less effective than ATP in inhibiting the channel in skeletal muscle, producing between 40–80% inhibition (Spruce et al 1987). GTP and UTP (2 mM) also produced partial channel block in ventricular myocytes (Kakei et al 1985). A controversy currently exists with regard to the effects of GTP on ATP-sensitive K-channels in pancreatic β-cells. Cook & Hales (1984) initially reported that 1 mM GTP did not inhibit the channel, but more recently others have found that GTP (0.01–1 mM) is actually able to increase channel activity (Dunne & Petersen 1986b). At higher concentration (3 mM), however, GTP has a clear inhibitory action (G. Trube, personal communication).

Agents capable of modifying the ATP-sensitivity of the channel at physiological concentration are likely to be of importance in regulating channel activity in the intact cell. As yet, only ADP (1–2 mM) (Kakei et al 1986, Dunne & Petersen 1986a, Ribalet & Ciani 1987) has been shown to have such an effect. Millimolar concentrations of AMP, adenosine, inorganic phosphate, and GTP did not substantially alter ATP inhibition (Kakei et al 1986). The observation that ADP both blocks the channel itself and shifts the ATP dose-response curve to higher concentrations suggests that ADP relieves ATP-induced inhibition by binding competitively to the same site as ATP (Kakei et al 1986). The fact that no marked decrease in channel activity is found when patches are excised into solutions containing ADP and ATP at their measured (total, not free) cytosolic concentrations is consistent with the idea that ADP might also contribute to channel regulation in β-cells (Kakei et al 1986).

Some Pharmacological Channel Blockers

The discovery that the ATP-sensitive K-channel is the major determinant of the β-cell resting potential immediately suggested that those pharmacological agents producing a decrease in β-cell resting K-permeability might act as channel blockers. This idea has been supported by patch clamp experiments and has led to the identification of a new group of specific channel blockers, the sulphonylureas. These drugs are of especial interest as they are used clinically as hypoglycemic agents in the treatment of non-insulin-dependent diabetes mellitus; their therapeutic effect is thought to result from their ability to decrease β-cell resting K-permeability and so, ultimately, to increase insulin release. Single channel recordings have confirmed that, as predicted, the ATP-sensitive K-channel is blocked by the sulphonylureas tolbutamide and glibenclamide (Sturgess et al 1985, Trube et al 1986, Belles et al 1987, Ashcroft et al 1987d). Inhibition appears to be specific, as tolbutamide has no effect on any of the following: voltage-activated K currents (β-cell delayed rectifier; Rorsman & Trube 1986), Ca currents (β-cell, Rorsman & Trube 1986; ventricular muscle, Belles et al 1987), Ca-activated K-channels (β-cell, Trube et al 1986), or inwardly-rectifying K-channels (ventricular cells, Belles et al 1987). Tolbutamide is able to block the channel both from the inside or the outside of the membrane and, in cell-attached patches, is effective if it is added only to the bath solution and not to that in the pipette. It has therefore been proposed that the sulphonylurea, a lipophilic molecule, may reach its target site by dissolving in the lipid phase of the membrane (Trube et al 1986) a concept supported by the observation that the potency of sulphonylureas increases with their hydrophobicity (Gylfe et al 1984).

In both β-cells and cardiac muscle the dose-response curve for inhibition can be fitted with a Hill coefficient of around 1, thus suggesting a 1:1 relationship between the drug and the channel. However, the β-cell is relatively more sensitive to tolbutamide: 50% inhibition of the whole-cell current activated by ATP removal is produced by a drug concentration of 7 μM (Trube et al 1986) as compared to 400 μM for ventricular myocytes (Belles et al 1987). A further point of difference is that whereas the sulphonylurea had immediate effects on the β-cell K-currents, in ventricular myocytes inhibition of whole cell-currents slowly developed over the course of 20–30 min and 30% of isolated patches were unaffected by the drug (Belles et al 1987). Thus in cardiac muscle, extracellular tolbutamide may not block the channel directly. Glibenclamide is at least 100 times more effective than tolbutamide in β-cells (Sturgess et al 1985, Trube et al 1986) and blocks irreversibly in cell-attached patches. Interestingly, diazoxide, a drug structurally related to the sulphonylureas, activates ATP-sensitive K-channels (Trube et al 1986).

The isolation and purification of ion channel proteins is facilitated if a specific, high-affinity ligand for the channel is available. How far do the sulphonylureas constitute such a ligand for the ATP-sensitive K-channel? Equilibrium binding data to crude membrane fractions from islet-cells demonstrate saturable binding sites for sulphonylureas with a binding constant of 6–10 μM (Geisen et al 1985), similar to that found for the ATP-sensitive K-channel in the β-cell (Trube et al 1986). However, binding of sulphonylureas to artificial phospholipid bilayers meets the same criteria for specificity (Deleers & Malaisse 1984), so raising doubts about the specificity of sulphonylurea binding. Furthermore, in cardiac muscle, there is evidence for an indirect action of tolbutamide on the channel. It therefore seems premature to conclude that sulphonylureas bind solely, or even directly, to the ATP-sensitive K-channel.

The well-known potassium channel blockers, quinine, 4-aminopyridine (4-AP), 9-aminoacridine (9-AA), and tetraethylammonium ions (TEA$^+$) are also effective inhibitors of the ATP-sensitive K-channel (Cook & Hales 1984, Findlay et al 1985b, Kakei et al 1985, Spruce et al 1987). Dose-response curves for these agents have not been reported, but a consideration of the literature suggests that quinine is the most potent blocker whether applied to the external or internal membrane surface. ATP-sensitive K-channels are considerably less sensitive to block by TEA$^+$ than the large conductance Ca-activated K-channel (Findlay et al 1985b). Barium ions (1 mM) block outward K-currents when added to either the external or internal solution (Kakei & Noma 1984). The block appears as a decrease in the frequency of channel openings without a change in unitary current amplitude. The mechanism has not been established.

Voltage-dependent Ionic Block

A number of cations block outward K-currents through the ATP-sensitive K-channel when added to the internal solution (Cook & Hales 1984, Horie et al 1987). The block is voltage dependent and increases with depolarization, consistent with the idea that these ions enter the channel conduction pathway but are unable to pass through the selectivity filter. They thereby block K$^+$ flux. The rapidity of the blocking kinetics gives rise to single channel currents of reduced amplitude and, together with the voltage-dependence of the block, results in a strong negative slope region in the outward current-voltage relation.

Horie et al (1987) have carried out an extensive investigation of the Na^+ and Mg^{2+} block of K-currents in cardiac muscle. Their studies show that the blocking effects of these ions can be explained by assuming a binding site for one Mg^{2+} or two Na^+ located approximately 30% of the way through the membrane voltage field from the inner mouth of the channel. Magnesium ions bind more tightly to the site than Na^+; half maximal block is produced by 0.5 mM Mg^+ as compared with 15 mM for Na^+ (at +40 mV). As intracellular free Mg^{2+} lies between 0.5–3 mM and Na^+ between 5–10 mM (Gupta et al 1984), ionic block can account for the inward rectification of the current-voltage relation observed in cell-attached patches (Horie et al 1987).

External Na^+, Ca^{2+}, and Mg^{2+} do not produce a measurable block of the ATP-sensitive K-channel and are not responsible for the rapid burst kinetics of the channel at negative potentials (Horie et al 1987). External rubidium ions, however, block inward K-currents in a way consistent with Rb^+ acting as a permeant voltage-dependent blocking ion (Ashcroft et al 1987c). In other words, the block initially increases and is subsequently relieved by increasing hyperpolarization, an effect that can be attributed to Rb^+ being driven through the channel by the strong voltage field.

Inhibition of Channel Activity by Ions

In addition to the voltage-dependent block of outward currents described above, there is evidence that divalent cations reduce channel activity by another mechanism. Internal Mg^{2+} ions (in β-cells, Ashcroft & Kakei 1987) or Ca^{2+} ions (in cardiac muscle, Kakei & Noma 1984) decrease the frequency of channel openings at all potentials without affecting the amplitude of the single channel current. Significant channel inhibition is produced at $[Ca]_i$ of 10^{-4} M and by $[Mg]_i$ greater than 10^{-3} M. Regulation of channel activity by Mg^{2+} ions may be of physiological relevance, since cytosolic Mg^{2+} is between 0.5–3 mM (Gupta et al 1984, Corkey et al 1986).

MECHANISMS OF CHANNEL ACTIVATION

Is Phosphorylation Required to Maintain Channel Availability?

Evidence is accumulating that ATP has a dual effect on the ATP-sensitive K-channel of β-cells and that, in addition to its inhibitory action, the nucleotide is required to maintain channel activity. Several studies have shown that the rundown in channel activity characteristic of isolated patches can be prevented or reversed by exposure to MgATP (Findlay & Dunne 1986, Misler et al 1986, Ohno-Shosaku et al 1987). Non-hydrolyzable ATP analogues such as AMP–PNP and AMP–PCP are unable to substitute for ATP in channel reactivation (Ohno-Shosaku et al 1987). Together with the requirement for Mg^+, this finding favors the idea that hydrolysis of ATP is required to maintain availability of the ATP-sensitive K-channel. Like certain other types of channel (Levitan 1985) phosphorylation of the ATP-sensitive K-channel itself, or of associated control proteins, may bias the channel toward the activatable state. If this is the case, the protein kinase responsible for phosphorylation must be closely associated with the membrane as ATP alone is sufficient to reactivate channels in the majority of isolated patches.

The question as to whether dephosphorylation is responsible for the channel rundown found in excised patches remains open. The ATP-analogue, adenosine thiophosphate (ATPγS), which donates a thiophosphate group resistant to enzymic hydrolysis (Yount 1975), produces sustained activation of molluscan Ca channels (Chad & Eckert 1986) but was unable to prevent rundown of ATP-sensitive K-currents in β-cells (Ohno-Shosaku et al 1987). This does not argue against dephosphorylation as the origin of channel rundown, however, because thiophosphorylation of the channel may not be an adequate substitute for phosphorylation in maintaining activity. Further experiments, for example using phosphatases or their inhibitors, are required to resolve this issue.

The possible activating effect of ATP on cardiac and skeletal muscle channels requires investigation. Although preliminary experiments suggest MgATP does not reactivate channels in patches excised from ventricular myocytes (Trube & Hescheler 1984; G. Trube, personal communication), this may simply be because the requisite kinase is not present in the isolated patch.

Effects of Other Nucleotides

Activating effects of other nucleotides have also been described for ATP-sensitive K-channels in β-cells. In particular, channel activity may be increased in both inside-out and open-cell patches by adding low concentrations (10–100 μM) of ADP, GTP, or GDP (Dunne & Petersen 1986a,b) to the internal solution. No clear activation was seen when AMP, cyclic AMP, and adenosine were tested (Dunne & Petersen 1986a).

The effect of ADP on channel activity is rather confusing, since both the nucleotide concentration and the extent to which rundown of channel activity has occurred appear to influence whether ADP increases or decreases channel activity. The inhibitory effect of ADP dominates in the fully activated patch whereas activation is favored when channel activity has substantially declined (Dunne & Petersen 1986a). In relatively stable patches, the dose-response curve for ADP resembles a parabola with low concentrations increasing, and high concentrations decreasing, channel activity (F. M. Ashcroft and M. Kakei, unpublished observations). As described above, the guanine nucleotides have been variously reported to potentiate (Dunne & Petersen 1986b) or to reduce (Cook & Hales 1984) channel activity in β-cells. One way of reconciling these conflicting results would be if GTP and GDP were found to act in a way similar to ADP.

How these other nucleotides activate the ATP-sensitive K-channel remains a mystery but the mechanism apparently differs from that of ATP. At least in the case of the guanine nucleotides, activation does not involve phosphorylation, because nonhydrolyzable analogues are also effective (Dunne & Petersen 1986b). Somewhat surprisingly, perhaps, ADPβS did not produce channel activation (Dunne & Petersen 1986a).

DOES ATP SERVE AS THE PHYSIOLOGICAL SECOND MESSENGER?

Channel Activity in the Intact Cell Is Regulated by Metabolism

The cell-attached patch clamp configuration is well suited for investigating the physiological regulation of single channel activity because the cell remains intact (Hamill et al 1981). Metabolic and second messenger systems thus remain undisturbed. Cell-attached recordings have established that the ATP-sensitive K-channel in the intact cell is regulated by metabolism. This is perhaps most clearly evident in the case of the β-cell, where both up- and down-regulation of channel activity can be observed. Unlike cardiac and skeletal muscle, many openings of ATP-sensitive K-channels can be recorded from intact β-cells lacking an exogenous fuel supply (Ashcroft et al 1984, Rorsman & Trube 1985). Substrates for β-cell metabolism produce a rapid, reversible and dose-dependent decrease of channel activity that can be reversed by the subsequent addition of metabolic inhibitors (Ashcroft et al 1984, 1985, 1987a,b, Rorsman & Trube 1985, Misler et al 1986, Ribalet & Ciani 1987). The ability of inhibitors of aerobic metabolism [rotenone, azide, dinitrophenol (DNP)] to rapidly increase channel activity blocked by glucose suggests that mitochondrial metabolism may be more important than glycolysis in generating intracellular mediators of channel inhibition.

Although channel activity is rarely observed in cell-attached patches on resting cardiac muscle, inhibition of metabolism by cyanide or DNP does produce marked channel activation (Kakei & Noma 1984, Trube & Herscheler 1984).

ATP as an Intracellular Second Messenger

A question of fundamental importance is how the ATP-sensitive K-channel is regulated in the intact cell. There is growing agreement that the primary determinant of channel activity is intracellular ATP and that local changes in ATP concentration constitute the mechanism by which cell metabolism is linked to channel activity. The properties required for an intracellular second messenger linking metabolic and membrane events have been enumerated previously (Ashcroft et al 1986), and several lines of evidence support such a role for ATP. As discussed above, channel activity in cell-attached patches is potentiated by metabolic inhibitors that lower $[ATP]_i$. In β-cells, nutrient secretogogues both increase the average cytosolic ATP level (Ashcroft et al 1973, Malaisse et al 1979) and inhibit the channel (Ashcroft et al 1987a,b). Indeed there is a striking correlation between the glucose dependence of channel inhibition and that of the increase in ATP (Ashcroft et al 1986). Furthermore, prior to channel inhibition, glucose induces a transient activation of ATP-sensitive K-channels, accompanied by membrane hyperpolarization (Arkhammar et al 1987), which parallels the transient fall in ATP measured by Malaisse et al (1979). Finally, the effect of ATP on channel kinetics in isolated patches resembles that found for glucose in cell-attached recordings in that both decrease the burst duration.

The main argument against ATP as the principal physiological regulator of channel activity is the clear discrepancy between the ATP-sensitivity measured for the inside-out patch (Table 1) and the observation that channel activity can be recorded from intact cells. Numerous studies have shown that the average cytosolic ATP concentration lies in the millimolar range in cardiac (see Panten & Lenzen 1987 for references) and skeletal (Dawson et al 1980) muscle and in β-cells (Ashcroft et al 1973, 1987c). Additionally, ATP is well buffered by creatine phosphate, so that changes in its concentration produced by glucose or anoxia are small (Gudbjarnason et al 1970, Ashcroft et al 1973). It would therefore at first seem unlikely that the channel would ever be open!

Several hypotheses have been put forward to account for this discrepancy: (a) the channel may have a low open probability in the intact cell; (b) the ATP-sensitivity of the channel may be altered in the isolated patch due to the loss of additional channel modulators; (c) by removing cytoskeletal connections, patch excision may alter channel properties; (d) ATP^{4-} may be more important at regulating channel activity than MgATP; (e) ATP may be compartmentalized within the cell such that the local concentration close to the channel binding site is very low. All these mechanisms probably contribute to regulate channel activity in the intact cell.

First, there is clear evidence in favor of the idea that even in the absence of exogenous nutrients only a fraction of the ATP-sensitive K-channels available are active. Channel openings are very infrequent in resting skeletal and cardiac muscle (Kakei & Noma 1984, Trube & Herscheler 1984, Spruce et al 1987). Even in the β-cell, considerable resting inhibition of channel activity is indicated because metabolic inhibitors produce substantial channel activation in glucose-free solutions, and channel activity increases on patch excision (Ashcroft et al 1985, Findlay et al 1985a, Ribalet & Ciani 1987).

Several other modulators of ATP-sensitive K-channels in β-cell membranes have been discussed in this review, some of which may be of physiological relevance. For example, changes in submembrane ADP levels may well be implicated in the response of the β-cell to glucose. Although in most cells free $[ADP]_i$ is very low ($< 100 \ \mu M$) because of substantial binding to cytosolic proteins (Hebisch et al 1984, Gamkema et al 1983), submembrane concentrations may well be significantly higher as a result of the activity of membrane ATP-ases. By altering the ATP-sensitivity of the channel, reciprocal changes in ADP might serve to amplify the effects of increased ATP (Kakei et al 1986, Misler et al 1986, Ribalet & Ciani 1987). Estimates of cytosolic GTP and GDP levels in the β-cell (Zunckler et al 1987) are also consistent with a role for these nucleotides in determining the resting level of channel activity. Finally, Horie et al (1987) also point out that an elevation of $[Mg]_i$ may be expected in anoxia due to release of Mg^{2+} bound to ATP, an event that might reduce both outward K-currents and, in β-cells, channel activity. The precise role of each of these other channel modulators needs to be elucidated.

A further possibility is that the submembrane ATP concentration is significantly lower than the cytosolic ATP level. Processes that might contribute to such compartmentalization of ATP have been reviewed in detail elsewhere (Panten & Lenzen 1987, Jones 1986) and include: sequestration within organelles; binding to nondiffusible cytosolic macromolecules; cytosolic gradients of ATP concentration resulting from regional differences in the rates of ATP supply and utilization; cytosolic Donnan equilibria that exclude the ATP polyanion.

In the β-cell, the total cellular ATP concentration provides a reasonable approximation to the average cytosolic level, as mitochondrial and granular pools of ATP are relatively small. Although mitochondrial volume is larger in cardiac and skeletal muscle, the average cytosolic ATP levels still lie around 8 mM (Panten & Lenzen 1987).

Evidence for cellular diffusion gradients for ATP has been reviewed in some detail by Jones (1986). As he argues, intracellular gradients for ATP can arise without membrane compartmentation if the rate of ATP consumption at cytosolic and/or membrane sites is greater than its rate of supply by the mitochondria. This hypothesis receives support from recent studies on hepatocytes that indicate Na-K ATP-ase activity in the plasma membrane is more sensitive to cellular $[ATP]_i$ than is that of cytosolic ATP-ases, the former being totally inhibited when $[ATP]_i$ falls by 40% to 2 mM (Aw & Jones 1985). Like the ATP-sensitive K-channel, the affinity of the Na-K-ATPase for ATP lies in the micromolar range (Schwarz et al 1975). A similar microheterogeneity of cytosolic ATP concentration therefore, if it exists in muscle and β-cells, may account for part of the discrepancy between measured cellular ATP levels and the activity of the ATP-sensitive K-channel. If this idea turns out to be correct, then the ATP-sensitive K-channel may prove a useful tool for assessing sub-membrane ATP levels, in the same way that the Ca-activated K-channel has been used to estimate submembrane Ca-concentrations.

The importance of the creatine phosphate/creatine/creatine kinase system in buffering $[ATP]_i$ is well established (Gudbjarnason et al 1970). Creatine kinase is present in both cardiac and skeletal muscle, although apparently not in isolated islets (Panten & Lenzen 1987). It has therefore been argued that the creatine kinase system will buffer $[ATP]_i$ and maintain $[ADP]_i$ at a low level. This argument, however, rests on the assumption that creatine kinase will be localized at the plasma membrane and that the activity of membrane ATP-ases will not alter the submembrane ATP/ADP ratio, points that have not been established.

Much of the cytosolic ATP exists as the MgATP complex. Since ATP^{4-} turns out to be a potent channel inhibitor (Ashcroft & Kakei 1987), cytosolic ATP^{4-}, rather than MgATP, might regulate channel activity. Therefore it will be important to establish the cytosolic concentration of this ion, and its modulation by metabolism.

PHYSIOLOGICAL FUNCTION OF THE ATP-SENSITIVE K-CHANNEL

Pancreatic β-cell

The ATP-sensitive K-channel plays a key role in initiating insulin release from pancreatic β-cells. Stimulatory concentrations of glucose have long been known to reduce the resting membrane K-permeability, thereby depolarizing the β-cell and initiating a characteristic pattern of electrical activity (see review by Henquin & Meissner 1984). Single channel recordings have established that in the absence of glucose the ATP-sensitive K-channel regulates the β-cell resting K-permeability and that glucose causes a rapid, reversible, and dose-dependent inhibition of channel activity (Ashcroft et al 1984, Rorsman & Trube 1985, Misler et al 1986). It is clear that glucose exerts its effect on the ATP-sensitive K-channel via an intracellular second messenger generated as a consequence of its metabolism. As is the case for insulin secretion (Sener & Malaisse 1984), the ability of the sugar to close the channel is prevented by metabolic inhibitors (Ashcroft et al 1984, 1985, Misler et al 1986). Studies using other nutrient secretogogues, or inhibitors of the electron transport chain, have shown that glycolysis is not essential for channel inhibition and that mitochondrial metabolism is capable also of regulating channel activity (Ashcroft et al 1985, 1987a,b, Misler et al 1986). The close relationship between β-cell metabolism and channel inhibition and insulin secretion is further emphasized by studies using insulin-secreting cell lines. Glucose itself neither stimulates insulin release nor inhibits the ATP-sensitive K-channel in these cell lines (for an exception see Ribalet & Ciani 1987), but both insulin secretion and channel inhibition do occur in response to other nutrients: leucine (Cr1-G1; Sturgess et al 1986) or glyceraldehyde (RINm5F; Dunne et al 1986). Therefore the lack of glucose sensitivity in these cell lines may reflect a defect in their metabolism rather than in the ATP-sensitive K-channel.

There is still debate regarding the identity of the coupling factor that links metabolism to closing of the ATP-sensitive K-channel. Clearly, any substance undergoing a change in concentration or activity during metabolism must be considered a candidate: the various possibilities are the subject of a recent review (Malaisse et al 1984). Perhaps the most attractive idea is that the increased intracellular ATP resulting from metabolism of glucose and other nutrient secretogogues (Ashcroft et al 1973, Malaisse et al 1979) mediates channel inhibition. The other possibilities, however, should not be forgotten.

How does the closing of ATP-sensitive K-channels lead to insulin secretion? In the absence of glucose, the β-cell is electrically silent. Although the resting potential is mainly determined by the ATP-sensitive K-channel, a small inward current must also be present as the resting potential is less negative than the equilibrium potential for K^+. Because ATP-sensitive K-channels close in response to glucose metabolism, membrane K-permeability falls and the contribution of the inward current to the membrane permeability therefore will increase. This produces depolarization. If depolarization is sufficient to exceed the threshold for the opening of the voltage-dependent Ca-channels, electrical activity is initiated (Rorsman & Trube 1985). The increased influx of Ca^{2+} into the cell (Arkhammar et al 1987) raises intracellular Ca^{2+} to levels required for exocytosis of insulin granules. Closure of ATP-sensitive K-channels, by depolarizing the β-cell, therefore initiates insulin release.

Channel inhibition occurs in response to glucose concentrations as low as 2 mM, but membrane depolarization is not significant until the glucose level exceeds 5 mM, as evidenced by the lack of a decrease in the single channel current amplitude in cell-attached patches (Ashcroft et al 1987b, Arkhammar et al 1987, Ribalet & Ciani 1987) and by measurements of resting membrane potential (Henquin & Meissner 1984). This suggests that almost all of the ATP-sensitive K-channels in the cell have to close to produce depolarization and that the resting inward leak must be small.

Cardiac and Skeletal Muscle

Noma & Shibasaki (1985) have considered the role of the ATP-sensitive K-channel in the heart. They concluded that activation of the ATP-sensitive K-channel underlies the outward current induced by anoxia and is primarily responsible for the shortening of the plateau of the ventricular action potential. Because a decrease in action potential duration is accompanied by a reduced contraction, Noma (1983) also suggested this might provide a mechanism for preventing $[ATP]_i$ from falling to damagingly low levels. In addition, the outward shunt conductance that results from channel activation will increase the threshold for electrical excitation and so slow pacemaker activity.

Discussion of the functional role of the ATP-sensitive K-channel in skeletal muscle is speculative at present, as it does not appear to be activated significantly under resting conditions (J. M. Quayle, referred to in Spruce et al 1987). The possibility of channel activation in cell-attached patches subjected to anoxia has not yet been investigated. However, it is worth noting that the resting K-permeability is increased in metabolically exhausted frog skeletal muscle (Fink & Luttgau 1976) and that the associated increase in ^{86}Rb efflux shows a pharmacological sensitivity characteristic of ATP-sensitive K-channels (Castle & Haylett 1987).

Activation of ATP-sensitive channels may also be expected to increase potassium efflux from the cell, an event that may have important physiological effects. It is well known that K^+ is a potent vasodilator, and the increase in interstitial potassium that occurs during exercise is thought to provide a local mechanism for regulating muscle blood flow (Ganong 1985). An additional role suggested for ATP-sensitive K-channels in muscle is that the increased K-efflux associated with their activation contributes to the vasodilation of muscle arterioles (Spruce et al 1987). Potassium efflux through ATP-sensitive K-channels may also influence the electrical activity of adjacent cells if the extracellular volume is small enough to allow K^+ ions to accumulate. In addition, by modulating K^+ fluxes, the ATP-sensitive K-channel might contribute to the regulation of cell volume.

SUMMARY

This review has focused on the properties of the ATP-sensitive K-channel found in cardiac and skeletal muscle, and in pancreatic β-cells. It is conceivable that this channel will be found in other cell types. In particular, it would be worthwhile looking for its presence in those cells in which electrical activity is linked to metabolism, glucose concentration, or oxygen levels. Obvious examples are the glucoreceptor neurons of mammalian brain and chemoreceptors such as those of the carotid body.

While ATP-sensitive K-channels in cardiac and skeletal muscle membranes are rather similar, there are a few significant differences between these channels and that found in the β-cell. Most notably, the latter is more sensitive to inhibition by ATP and sulphonylureas. It remains to be seen whether they also differ in the ability of nucleotides to activate the channel. Considerable confusion also still surrounds the physiological regulation of the ATP-sensitive K-channel in intact cells. Although the general consensus seems to be that $[ATP]_i$ modulates channel activity, the role of other nucleotides and ions as well as the way in which their concentrations alter with metabolism requires further elucidation. A combined electrophysiological and biochemical approach is likely to prove most successful in establishing which second messenger systems contribute to the physiological regulation of the ATP-sensitive K-channel.

Finally, the close correlation between cell metabolism and the activity of the ATP-sensitive K-channel raises the intriguing possibility that disorders of cell metabolism might produce alterations in channel activity and consequent changes in cell function.

ACKNOWLEDGMENTS

I am grateful to Drs. Stephen Ashcroft, Masafumi Kakei, Uwe Panten, and Gerhardt Trube for valuable discussion and to many colleagues for allowing me to read their manuscripts in press. I thank the Royal Society, the British Diabetic Association, the Wellcome Trust, and the British Heart Foundation for support.

Literature Cited

Arkhammar, P., Nilsson, T., Rorsman, P., Berggren, P.-O. 1987. Inhibition of ATP-regulated K⁺ channels precedes depolarisation induced increase in cytoplasmic free Ca²⁺ concentration in pancreatic β-cells. *J. Biol. Chem.* 262: 5448–54

Ashford, M. J., Hales, C. N., Sturgess, N. C. 1986. Adenosine triphosphate inhibits the activity of the calcium-activated non-selective cation channel in a rat insulinoma cell line. *J. Physiol.* 381: 116P

Ashcroft, F. M., Harrison, D. E., Ashcroft, S. J. H. 1984. Glucose induces closure of single potassium channels in isolated rat pancreatic β-cells. *Nature* 312: 446–48

Ashcroft, F. M., Ashcroft, S. J. H., Harrison,

D. E. 1985. The glucose-sensitive potassium channel in rat pancreatic beta-cells is inhibited by intracellular ATP. *J. Physiol.* 369: 101P

Ashcroft, F. M., Harrison, D. E., Ashcroft, S. J. H. 1986. A potassium channel modulated by glucose metabolism in rat pancreatic β-cells. In *Biophysics of the Pancreatic β-cell*, ed. I. Atwater, pp. 53–63. New York: Plenum

Ashcroft, F. M., Kakei, M. 1987. Effects of internal Mg^{2+} on ATP-sensitive K-channels in isolated rat pancreatic β-cells. *J. Physiol.* In press

Ashcroft, F. M., Ashcroft, S. J. H., Harrison, D. E. 1987a. Effects of 2-ketoisocaproate on insulin release and single potassium channel activity in dispersed rat pancreatic β-cells. *J. Physiol.* 385: 517–31

Ashcroft, F. M., Ashcroft, S. J. H., Harrison, D. E. 1987b. Properties of single potassium channels modulated by glucose in rat pancreatic β-cells. *J. Physiol.* In press

Ashcroft, F. M., Kakei, M., Kelly, R. P. 1987c. Rubidium permeability of ATP-sensitive K-channels in isolated pancreatic β-cells. *J. Physiol.* Submitted

Ashcroft, F. M., Kakei, M., Kelly, R. P., Sutton, R. 1987d. ATP-sensitive K-channels in isolated human pancreatic β-cells. *FEBS Lett.* 215: 9–12

Ashcroft, S. J. H., Weerasinghe, L. C. C., Randle, P. J. 1973. Interrelationship of islet metabolism, adenosine triphosphate content, and insulin release. *Biochem. J.* 132: 223–31

Aw, T. Y., Jones, D. P. 1985. ATP concentration gradients in cytosol of liver during hypoxia. *Am. J. Physiol.* 249: C385–92

Belles, B., Hescheler, J., Trube, G. 1987. Changes of membrane current in cardiac cells induced by long whole-cell recordings and tolbutamide. *Pflügers Arch.* 409: 582–88

Castle, N. A., Haylett, D. G. 1987. Effect of channel blockers on potassium efflux from metabolically exhausted frog skeletal muscle. *J. Physiol.* 383: 31–45

Chad, J. E., Eckert, R. 1986. An enzymatic mechanism for calcium current inactivation in dialysed *Helix* neurones. *J. Physiol.* 378: 31–53

Colquhoun, D., Hawkes, A. G. 1981. On the stochastic properties of single ion channels. *Proc. R. Soc. London Ser. B* 199: 231–62

Colquhoun, D., Hawkes, A. G. 1982. On the stochastic properties of bursts of single ion channel openings and clusters of bursts. *Philos. Trans. R. Soc. London Ser. B* 300: 1–59

Cook, D. L., Hales, C. N. 1984. Intracellular ATP directly blocks K^+ channels in pancreatic β-cells. *Nature* 311: 271–73

Corkey, B. E., Duszynski, J., Rich, T. L., Matchinsky, B., Williamson, J. R. 1986. Regulation of free and bound magnesium in rat hepatocytes and isolated mitochondria. *J. Biol. Chem.* 261: 2567–74

Dawson, M. J., Gadian, D. G., Wilkie, D. R. 1980. Mechanical relaxation rate and metabolism studies in fatiguing muscle by phosphorus nuclear magnetic resonance. *J. Physiol.* 299: 465–84

Deleers, M., Malaisse, W. J. 1984. Binding of hypoglycaemic sulphonylureas to an artificial phospholipid bilayer. *Diabetologia* 26: 55–59

Dunne, M. J., Petersen, O. 1986a. Intercellular ADP activates K^+ channels that are inhibited by ATP in an insulin-secreting cell line. *FEBS Lett.* 208: 58–62

Dunne, M. J., Petersen, O. H. 1986b. GTP and GDP activation of K^+ channels that can be inhibited by ATP. *Pflugers Arch.* 407: 564–65

Dunne, M. J., Findlay, I., Petersen, O. H., Wollheim, C. B. 1986. ATP-sensitive K^+ channels in an insulin-secreting cell line are inhibited by D-glyceraldehyde and activated by membrane permeabilization. *J. Memb. Biol.* 93: 271–79

Findlay, I., Dunne, M. J. 1986. ATP maintains ATP-inhibited K^+ channels in an operational state. *Pflugers Arch.* 407: 238–40

Findlay, I., Dunne, M. J., Petersen, O. H. 1985a. ATP-sensitive inward rectifier and voltage- and calcium activated K^+ channels in cultured pancreatic islet cells. *J. Memb. Biol.* 88: 165–72

Findlay, I., Dunne, M. J., Ullrich, S., Wollheim, C. B., Petersen, O. H. 1985b. Quinine inhibits Ca^{2+}-independent K^+ channels whereas tetraethylammonium inhibits Ca^{2+}-activated K^+ channels in insulin secreting cells. *FEBS Lett.* 185: 4–6

Fink, R., Luttgau, H. Ch. 1976. An evaluation of the membrane constants and the potassium conductance in metabolically exhausted muscle fibres. *J. Physiol.* 263: 215–38

Fujiwara, N., Higashi, H., Shimoji, K., Yoshimura, M. 1987. Effects of hypoxia on rat hippocampal neurons *in vitro*. *J. Physiol.* 384: 131–53

Gamkema, H. S., Groen, A. K., Wanders, R. J. A., Tager, J. M. 1983. Measurement of binding of adenine nucleotides and phosphate to cytosolic proteins in permeabilised rat liver cells. *Eur. J. Biochem.* 131: 447

Ganong, W. F. 1985. *Review of Medical Physiology.* Los Altos, Calif: Lange

Geisen, K., Hitzel, V., Okomonpoulous, R.,

Punter, J., Weyer, R., Summ, H. D. 1985. Inhibition of ^{3}H-glibenclamide binding to sulphonylurea receptors by oral antidiabetics. *Arznemittelforsch/Drug. Res.* 35: 707–12

Goldman, D. E. 1943. Potential, impedance and rectification in membranes. *J. Gen. Physiol.* 27: 37–60

Gudbjarnason, S., Mathes, P., Ravens, K. G. 1970. Functional compartmentation of ATP and creatine phosphate in heart muscle. *J. Molec. Cell. Cardiol.* 1: 325–39

Gupta, R. K., Gupta, P., Moore, R. D. 1984. NMR studies of intracellular metal ions in intact cells and tissues. *Ann. Rev. Biophys. Bioeng.* 13: 221–46

Gylfe, E., Hellman, B., Sehlin, J., Taljedal, I. B. 1984. Interaction of the sulphonylurea with the pancreatic β-cell. *Experientia* 40: 1126–34

Hamill, O. P., Matry, A., Neher, E., Sakmann, B., Sigworth, F. 1981. Improved patch-clamp techniques for high-resolution current recordings from cells and cell-free membrane patches. *Pflugers Archiv.* 391: 85–100

Hebisch, S., Soboll, S., Schwenen, M., Sies, H. 1984. Compartmentation of high energy phosphates in resting and working rat skeletal muscle. *Biochem. Biophys. Acta* 764: 117

Henquin, J. C., Meissner, H. P. 1984. Significance of ionic fluxes and changes in membrane potential for stimulus-secretion coupling in pancreatic β-cells. *Experientia* 40: 1043–52

Hodgkin, A. L., Katz, B. 1949. The effects of sodium ions on the electrical activity of the giant axon of the squid. *J. Physiol.* 108: 37–77

Horie, M., Irisawa, H., Noma, A. 1987. Voltage-dependent magnesium block of adenosine-triphosphate-sensitive potassium channel in guinea-pig ventricular cells. *J. Physiol.* 387: 251–72

Joels, N., Neil, E. 1968. In *Arterial Chemoreceptors*, ed. R. W. Torrance, pp. 153–78. Blackwells: Oxford

Jones, D. P. 1986. Intracellular diffusion gradients of O_2 and ATP. *Am. J. Physiol.* 250: C663–75

Kakei, M., Noma, A. 1984. Adenosine 5'-triphosphate-sensitive single potassium channel in the atrioventricular node cell of the rabbit heart. *J. Physiol.* 352: 265–84

Kakei, M., Noma, A., Shibasaki, T. 1985. Properties of adenosine-triphosphate-regulated potassium channels in guinea-pig ventricular cells. *J. Physiol.* 363: 441–62

Kakei, M., Kelly, R. P., Ashcroft, S. J. H., Ashcroft, F. M. 1986. The ATP-sensitivity of K^+ channels in rat pancreatic β-cells is modulated by ADP. *FEBS Lett.* 208: 63–66

Kaubisch, N., Hammer, R., Wolheim, C., Renold, A. E., Offord, R. E. 1982. Specific receptors for sulphonylureas in brain and in a β-cell tumor of the rat. *Biochem. Pharmacol.* 31: 1171–74

Levitan, I. B. 1985. Phosphorylation of ion channels. *J. Memb. Biol.* 87: 177–90

Malaisse, W. J., Malaisse-Lagae, F., Sener, A. 1984. Coupling factors involved in insulin release. *Experientia* 40: 1035–43

Malaisse, W. J., Hutton, J. C., Kawazu, S., Herchuelz, A., Valverde, I., Sener, A. 1979. The stimulus secretion coupling of glucose induced insulin release. XXXV. The links between metabolic and cationic events. *Diabetologia* 16: 331–41

Misler, D. S., Falke, L. C., Gillis, K., McDaniel, M. L. 1986. A metabolite regulated potassium channel in rat pancreatic β cells. *Proc. Natl. Acad. Sci. USA* 83: 7119–23

Minami, T., Oomura, Y., Sugimori, M. 1986. Electrophysiological responsiveness of guinea-pig ventromedial hypothalamic neurones *in vitro*. *J. Physiol.* 380: 127–43

Mizuno, Y., Oomura, Y. 1984. Glucose-responding neurones in the nucleus tractus solitarius of the rat: *In vitro* study. *Brain Res.* 307: 109–16

Noma, A. 1983. ATP-regulated K^+ channels in cardiac muscle. *Nature* 305: 147–48

Noma, A., Shibasaki, T. 1985. Membrane current through adenosine-triphosphate-regulated potassium channels in guinea-pig ventricular cells. *J. Physiol.* 363: 463–80

Ohno-Shosaku, T., Zunckler, B., Trube, G. 1987. Dual effects of ATP on K^+ currents of mouse pancreatic β-cells. *Pflugers Arch.* 408: 133–38

Panten, U., Lenzen, S. 1988. Alterations in energy metabolism of secretory cells. In *The Energetics of Secretion Responses*, ed. J. W. N. Akkerman. Boca Raton, Fla: CRC. In press

Ribalet, B., Ciani, S. 1987. Regulation and function of a metabolite-dependent K channel in insulin-secreting β-cells (RINm5F). *Proc. Natl. Acad. Sci. USA.* In press

Rorsman, P., Trube, G. 1985. Glucose dependent K^+ channels in pancreatic β-cells are regulated by intracellular ATP. *Pflugers Arch.* 405: 305–9

Rorsman, P., Trube, G. 1986. Calcium and delayed potassium currents in mouse pancreatic β-cells under voltage clamp conditions. *J. Physiol.* 374: 531–50

Schwarz, A., Lindenmayer, G. E., Allen, J. C. 1975. The sodium potassium adenosine triphosphatase: Pharmacological, physiological and biochemical aspects. *Pharmacol. Rev.* 27: 3–134

Sener, A., Malaisse, W. J. 1984. Nutrient metabolism in islet cells. *Experientia* 40: 1026–35

Spruce, A. E., Standen, N. B., Stanfield, P. R. 1985. Voltage-dependent ATP-sensitive potassium channels of skeletal muscle membrane. *Nature* 316: 736–38

Spruce, A. E., Standen, N. B., Stanfield, P. R. 1987. Studies on the unitary properties of adenosine-5′-triphosphate-regulated potassium channels of frog skeletal muscle. *J. Physiol.* 382: 213–37

Sturgess, N. C., Ashford, M. L. J., Cook, D. L., Hales, C. N. 1985. The sulphonylurea receptor may be an ATP-sensitive potassium channel. *Lancet* 8453: 474–75

Sturgess, N. C., Ashford, M. L. J., Carrington, C. A., Hales, C. N. 1986. Single channel recordings of potassium currents in an insulin-secreting cell line. *J. Endocrinol.* 109

Trube, G., Hescheler, J. 1984. Inward-rectifying channels in isolated patches of the heart cell membrane: ATP-dependence and comparison with cell-attached patches. *Pflügers Arch.* 401: 178–84

Trube, G., Rorsman, P., Ohno-Shosaku, T. 1986. Opposite effects of tolbutamide and diazoxide on the ATP-dependent K^+ channel in mouse pancreatic β-cells. *Pflugers Arch.* 407: 493–99

Yount, R. G. 1975. ATP analogs. *Adv. Enzymol.* 43: 1–56

Zunckler, B. J., Lenzen, S., Panten, U. 1987. D-glucose enhances GTP content in mouse pancreatic islets. *IRCS Med. Sci.* 354: 354–55

Ann. Rev. Neurosci. 1988. 11 : 119–36

MODULATION OF ION CHANNELS IN NEURONS AND OTHER CELLS

Irwin B. Levitan

Graduate Department of Biochemistry, Brandeis University, Waltham, Massachusetts 02254

INTRODUCTION

The electrical activity of nerve cells, and indeed of all cells, is governed by the flow of ion current across the plasma membrane. Although some of the current is contributed by ion transport systems or carriers, these are intrinsically slow, and the major source of membrane current is the passive flow of ions down their concentration gradients through ion channels in the membrane. Ion channels allow the movement of ions at enormous rates, of the order of 10^7 to 10^9 ions per second (some four orders of magnitude faster than most active transport systems), and are responsible for most electrical excitation phenomena in nerve and muscle (Hille 1984, Miller 1987). Although it was thought for many years that ion channels were restricted to this special class of "excitable" cells, the recent application of patch clamp and reconstitution techniques has made it clear that these intrinsic membrane proteins are ubiquitous, being found in the plasma membranes of such "nonexcitable" cell types as fibroblasts (Chen & Hess 1987), lymphocytes (Cahalan et al 1985), and even plant cells and prokaryotes (Hille 1984). Indeed ion channels are not even restricted to the plasma membrane, but are found in intracellular membranes (e.g. Smith et al 1985), where they probably function to regulate the distribution of ions between intracellular compartments.

If ion channels were simply membrane pores that are always open to the passive movement of ions, they would be relatively boring (at least to the investigator interested in cellular regulatory mechanisms). But most channels are far from passive pores. Rather they can be opened or closed

119

0147–006X/88/0201–0119$02.00

(gated) by a variety of stimuli, including the transmembrane voltage, and the binding of small ligands such as neurotransmitters or ions. Furthermore, these gating transitions, which tend to occur on a time scale of milliseconds, are subject to much longer term modulation by mechanisms such as the interaction of channels with other membrane proteins (see below), or covalent modification of the ion channel proteins themselves (Levitan 1985 and below). Such longer term modulation of ion channels, which may profoundly influence the activity of the cells in which they reside, is the subject of this review. I do not restrict myself to nerve cells, since principles of channel modulation are emerging that clearly are applicable to a wide variety of cell types. These principles are illustrated with selected examples, as the expansion of this field during the last several years makes it impossible to present a comprehensive summary in a review of this length.

ION CHANNEL MODULATION BY PROTEIN PHOSPHORYLATION

The ion channel modulatory mechanism that has been most thoroughly investigated is protein phosphorylation. This does not mean that phosphorylation is the only or even necessarily the most important way of modulating ion channel properties. However, the availability of specific biochemical probes such as pure protein kinases and protein kinase inhibitors, together with the conceptual framework provided by extensive investigations of the role of protein phosphorylation in metabolic regulation (Cohen 1982), have made it relatively easy to think about and study this mode of ion channel modulation. Because this area has been reviewed recently (Levitan 1985), the present discussion is confined to a brief summary of earlier work and a description of some new examples of channel modulation by phosphorylation that have arisen during the last two years.

Cyclic AMP–Dependent Protein Kinase

Work from a number of different laboratories over the last decade has provided convincing evidence that cyclic AMP (cAMP)–dependent protein phosphorylation can modulate ion channel properties, and that this form of channel regulation occurs in cells under physiological conditions. The active catalytic subunit of the cAMP-dependent protein kinase is easily purified from mammalian heart or skeletal muscle (Peters et al 1977), as is the 10,000 dalton specific protein inhibitor of this kinase (Walsh et al 1971). Because their active sites appear to have been very well conserved during evolution, these purified proteins from mammalian sources can be used as specific biochemical probes in such diverse groups as mammals and molluscs (e.g. Kaczmarek et al 1980, Castellucci et al 1980, Adams & Levitan 1982, DePeyer et al 1982, Osterrieder et al 1982, Alkon et al 1983, Doroshenko et al 1984, Eckert & Chad 1984, Ewald et al 1985, Shuster et al 1985, Huganir et al 1986).

POTASSIUM AND CALCIUM CHANNELS   The intracellular injection of these probes has demonstrated that calcium currents and several different types of potassium currents are regulated by cAMP-dependent phosphorylation in intact cells (reviewed by Levitan 1985). Furthermore, measurements of single channel activity in isolated membrane patches (Shuster et al 1985, Ewald et al 1985) or in artificial phospholipid bilayers (Ewald et al 1985) have provided evidence that the phosphorylation target is either the ion channel protein itself or some intimately associated regulatory component. In an elegant recent extension of the bilayer approach, Flockerzi et al (1986) have shown that a purified dihydropyridine binding protein from skeletal muscle (Borsotto et al 1984, Curtis & Catterall 1984) can be reconstituted to form a functional calcium channel, and that the activity of this purified reconstituted channel can be modulated by cAMP-dependent phosphorylation. Biochemical studies (Curtis & Catterall 1985, Hosey et al 1986) also indicate that the dihydropyridine binding protein is a substrate for protein kinases. Such co-purification of the phosphorylatable regulatory site with a functional calcium channel provides evidence that modulatability is an inherent property of this ion channel, and possibly of others as well. This purified reconstituted calcium channel appears to lack voltage-dependent gating (Flockerzi et al 1986), in marked contrast to its behavior in the native muscle plasma membrane. Although the reasons for this are not clear, it is striking that at least in this case regulatability by phosphorylation may be more an intrinsic property of the channel than voltage-dependent gating.

THE ACETYLCHOLINE RECEPTOR/CHANNEL   It has been known for a number of years that the nicotinic acetylcholine receptor/channel complex is a good substrate for cAMP-dependent protein kinase (Gordon et al 1977, Teichberg et al 1977, Saitoh & Changeux 1981, Huganir & Greengard 1983). In addition several other protein kinases can phosphorylate the purified acetylcholine receptor/channel at specific sites (Huganir 1987). Recently, Huganir et al (1986) investigated the functional significance of the cAMP-dependent phosphorylation, by phosphorylating the acetylcholine receptor/channel in *Torpedo californica* membrane fractions with the catalytic subunit of cAMP-dependent protein kinase, then purifying the phosphorylated receptor/channel and reconstituting it into phospholipid vesicles. Rapid flux measurements were used to examine acetylcholine-dependent ion transport by phosphorylated and nonphosphorylated preparations. Although phosphorylation has no effect on the initial rate of ion transport, the rate of desensitization of the acetylcholine receptor/channel is much higher following phosphorylation. Furthermore, the extent of the increase in the desensitization rate is correlated with the stoichiometry of the phosphorylation (Huganir et al 1986). These results are consistent with the finding that in intact muscle the adenylate cyclase activator forskolin can increase dramatically the rate of desensitization of the response to acetylcholine (Albuquerque et al 1986, Middleton et al 1986). In addition, this elegant study, with a well-defined receptor/channel complex in which the probable phosphorylation sites have been identified (Huganir 1987), provides perhaps the strongest evidence that regulation by phosphorylation is an intrinsic property of some ion channels.

CHLORIDE CHANNELS   The modulation of chloride channels has been much less thoroughly studied than that of potassium and calcium channels. Although there are examples of neuronal chloride currents that are subject to modulation (Lotshaw et al 1986, Madison et al 1986), the explosion in single channel studies during the last few years largely passed chloride channels by. The reasons for this are not at all clear, but this state of affairs now appears to be changing. One particularly interesting example of chloride channel regulation is in the apical membrane of many epithelial cells. Chloride transport across the epithelial cell apical membrane, which can be stimulated by β-adrenergic agonists and cAMP analogs, is essential for normal water and electrolyte secretion. Quinton (1983) was the first to demonstrate that chloride permeability is abnormally low in sweat gland epithelia from patients suffering from cystic fibrosis. Chloride transport and its regulation by cAMP are also defective in airway epithelia in cystic fibrosis patients (Knowles et al 1983, Widdicombe et al 1985), and this probably contributes to the accumulation of mucus that compromises pulmonary function in these patients. The defect in chloride transport has now been localized to the regulation of a chloride channel in the apical membranes of airway epithelial cells. Welsh (1986) described patch clamp studies of a voltage- and calcium-independent 26 pS chloride channel, which is responsive to the β-adrenergic agonist isoproterenol, in primary cultures of normal human tracheal epithelium. This channel is also present in cells from cystic fibrosis patients (Welsh & Liedtke 1986), but it is not activatable by isoproterenol and can be observed only in excised membrane patches (either in the presence or absence of calcium). Frizzell et al (1986) independently reported the presence of two types of chloride channel, with single channel conductances of about 20 pS and 50 pS, in upper airway cells from both normal individuals and cystic fibrosis patients. In normal cells these channels are evident in the cell-attached recording mode, and can be activated by β-adrenergic agonists or a cAMP analog. However, the channels from diseased cells are not responsive to β-adrenergic stimulation and can be observed only when the membrane patches are excised from the cell into a calcium-containing solution. Although some differences between the studies of Welsh & Liedtke (1986) and Frizzell et al (1986) with respect to the single channel conductances and calcium sensitivity remain to be resolved, it is clear from these experiments that apical chloride channels are present in airway epithelia from cystic fibrosis patients, but their regulation by β-adrenergic stimulation is defective. In view of the fact that isoproterenol-induced accumulation of cAMP is normal in cystic fibrosis cells (Welsh & Liedtke 1986), the defect in cystic fibrosis must be somewhere downstream from the production of cAMP. One obvious possibility is that cAMP-dependent phosphorylation of chloride channels is defective, resulting from either a defective protein kinase or a missing or altered regulatory site on the channel itself. It will be of great interest to examine the effects of the purified catalytic subunit of cAMP-dependent protein kinase on the activity of epithelial chloride channels in detached membrane patches.

Cyclic GMP–Dependent Protein Kinase

It is becoming clear that cyclic GMP (cGMP) plays an essential role in the regulation of the light-dependent conductance in both vertebrate and invertebrate photoreceptors (Fesenko et al 1985, Yau & Nakatani 1985, Zimmerman & Baylor 1986, Johnson et al 1986). This appears to be due to a direct interaction of cGMP with the light-dependent ion channel (Haynes et al 1986) and does not involve cGMP-dependent protein phosphorylation. Aside from photoreceptors there are few examples of ion currents that are regulated by cGMP. In an early study, Krnjevic et al (1976) demonstrated that intracellular injection of cGMP can alter the membrane properties of cat spinal motoneurons. Connor & Hockberger (1984) found that cGMP modulates a sodium current in some molluscan neurons; however, this same current is modulated in the same way by cAMP, and thus it is possible that cGMP is simply interacting with the cAMP regulatory system to produce this effect. A cGMP analog alters the pattern of endogenous bursting activity in the identified *Aplysia* neuron R15 (Levitan & Norman 1980), and this has recently been shown to be due to inhibition of a voltage-dependent calcium current that contributes to generation of the bursting pattern (Levitan & Levitan 1988). The possible role of cGMP-dependent protein phosphorylation in this phenomenon has not yet been investigated.

In a group of identified ventral neurons of the snail *Helix aspersa*, cGMP appears to produce an opposite effect, an increase in the calcium current (Paupardin-Tritsch et al 1986a, Gerschenfeld et al 1986). Serotonin causes an increase in the duration of the action potential in these neurons, as it does in other *Helix* (Paupardin-Tritsch et al 1981) and some *Aplysia* (Klein & Kandel 1978) neurons. In these latter cases the actions of serotonin are mediated by cAMP and involve a decrease in a specific potassium current, the S current (Deterre et al 1981, Siegelbaum et al 1982). However, in the *Helix* ventral neurons, potassium currents are not affected by serotonin, and the change in action potential duration can be attributed entirely to an increase in the calcium current (Paupardin-Tritsch et al 1986a). Furthermore, the action potential duration and the calcium current are not altered by the injection of cAMP or application of forskolin, but they are enhanced by the injection of cGMP or by the application of zaprinast, an inhibitor of cGMP-phosphodiesterase (Paupardin-Tritsch et al 1986a, Gerschenfeld et al 1986). Since serotonin and zaprinast also appear to increase cGMP levels in these ventral neurons (unpublished results quoted in Paupardin-Tritsch et al 1986b), the results are consistent with the possibility that cGMP mediates this serotonin-induced increase in the calcium current. In two examples in which cGMP effects have been examined under voltage clamp in molluscan neurons, different changes in ion currents have been observed (Paupardin-Tritsch et al 1986a, Levitan & Levitan 1988). This is reminiscent of the far more thoroughly studied case of cAMP, where effects on different kinds of ion channels are observed routinely in different cells (Siegelbaum & Tsien 1983, Levitan et al 1983, Levitan 1985) and even within the same cell (Lotshaw et al 1986, Strong & Kaczmarek 1987).

Paupardin-Tritsch et al (1986b) have recently extended their studies of the *Helix* ventral neurons to examine the role of cGMP-dependent protein phosphorylation. When the purified cGMP-dependent protein kinase from bovine lung (Walter et al 1981) is injected intracellularly, it has little effect by itself on either the action potential duration or the calcium current, but it enhances the changes evoked by low concentrations of serotonin or zaprinast. The interpretation of these results is that the cGMP concentration in the cell is sufficient to activate the injected kinase only after the application of serotonin or zaprinast. In support of this is the finding that, if the kinase is activated by preincubation with cGMP prior to its injection, the activated enzyme can increase the calcium current in the absence of serotonin or zaprinast. These results (Paupardin-Tritsch et al 1986b) clearly implicate cGMP-dependent protein phosphorylation in the modulation of neuronal ion currents, and indicate that the molecular mechanism of cGMP action is very different in photoreceptors and in these *Helix* neurons. Investigations of the mechanism of other examples of cGMP modulation will be of interest with this distinction in mind.

Protein Kinase C

A number of different laboratories have now examined the effects of protein kinase C activators such as phorbol esters or diacylglycerols (Nishizuka 1984) on ion currents in a variety of neurons and other cells. The many changes that have been reported include decreases in potassium (Baraban et al 1985, Malenka et al 1986, Farley & Auerbach 1986, Higashida & Brown 1986, Grega et al 1987) and chloride (Madison et al 1986) currents, and both increases (DeRiemer et al 1985, Farley & Auerbach 1986) and decreases (Rane & Dunlap 1986, Hammond et al 1986, 1987) in calcium currents. Thus, here again, as in the cases of cAMP and cGMP, multiple ion currents even within the same neuron seem to be modulated by phorbol esters and diacylglycerols. Although these agents are strongly hydrophobic and might generate nonspecific effects on membranes, in most of these studies the effective concentrations are consistent with a role for protein kinase C in ion current modulation.

In several cases the role of protein kinase C has been studied more directly by injecting it intracellularly. DeRiemer et al (1985) found that intracellular pressure injection of protein kinase C purified from bovine brain (Kikkawa et al 1982) into *Aplysia* neurosecretory bag cell neurons enhances the amplitude of the action potential. This effect, which is mimicked by application of phorbol ester, is due to selective enhancement of the calcium current with no changes in potassium currents (DeRiemer et al 1985). In a recent extension of this work to the single channel level, Strong et al (1987) have demonstrated that the enhancement of calcium current by phorbol ester involves the recruitment of a novel calcium channel in bag cell neuron membranes. In control neurons the calcium current appears to be carried by a class of voltage-dependent calcium channels with a single channel conductance of about 12 pS. After exposure of the neurons to phorbol ester or diacylglycerol these channels are still present, but in addition there appears a new class of calcium channel, with single channel conductance of approximately 24 pS, which is never seen in control cells. Since the smaller conductance channels are observed with about the same frequency in control and phorbol ester-treated neurons, the large conductance channels probably do not arise from conversion of the small ones. One intriguing feature of this study is that the appearance of the new channel is dependent on the pretreatment of the cells with phorbol ester, and is not observed when the protein kinase C activator is added to the bathing medium after formation of the gigaseal membrane patch (Strong et al 1987). Although the reasons for this remain to be elucidated, an interesting possibility is that patch formation may interfere with the translocation of protein kinase C between cytoplasmic and membrane sites following activation (Nishizuka 1984).

A similar intracellular injection approach has been used to implicate protein kinase C as the mediator of the cholecystokinin (CCK)-induced modulation of certain identified *Helix* neurons (Hammond et al 1987). CCK application decreases the duration of the action potential in these neurons, by decreasing the calcium current. These changes are mimicked by the extracellular application of diacylglycerol or phorbol ester, and by the intracellular pressure injection of purified protein kinase C. In addition, the effects of low concentrations of CCK are enhanced by the injection of protein kinase C (Hammond et al 1987). These results provide strong evidence for the mediation of the effects of a peptide neurotransmitter by protein kinase C. The opposite actions of protein kinase C on calcium currents in these *Helix* neurons and in the *Aplysia* bag cell neurons (DeRiemer et al 1985, Strong et al 1987) further reinforce the notion of diversity in ion channel regulatory mechanisms.

Protein kinase C has also been injected intracellularly into *Hermissenda* photoreceptors, where it produces multiple changes in ion currents (Farley & Auerbach 1986). Among the modulatory effects observed are decreases in the calcium-dependent and the transient voltage-dependent potassium currents, and enhancement of the calcium current. These changes are also produced by associative training of the animal or by application of serotonin, and serotonin occludes the effects of protein kinase C injection, leading Farley & Auerbach (1986) to suggest that the serotonin-induced activation of protein kinase C might contribute to the cellular changes that underlie associative learning. There are several curious features about these experiments. First the protein kinase C was injected iontophoretically. Since the injection electrode also contained 20 mM Tris-HCl and 350 mM NaCl, and the small ions would be expected to have much higher transport numbers than a large protein, it is surprising that any enzyme could be ejected from the electrode at all; however, the authors do report that protein kinase C activity could be detected outside the electrode following the application of iontophoretic current (Farley & Auerbach 1986). Second, a partially purified protein kinase C preparation was used for this study, and in recent experiments with what is described as a purified enzyme, very different effects on ion currents have been reported in these same *Hermissenda* photoreceptors (Kubota et al 1986). Since these latter experiments were also done by iontophoretic injection, and in view of the conflicting results between the two studies, the precise role (if any) of protein kinase C in this system remains to be demonstrated.

Phosphorylation—A Fundamental Modulatory Mechanism

There no longer is any question that ion channel properties can be modulated by protein phosphorylation. In a number of cases the phosphorylation target has been demonstrated to be closely associated with the ion channel protein itself (Shuster et al 1985, Ewald et al 1985, Huganir et al 1986, Flockerzi et al 1986), and the modulation of channel properties by phosphorylation often has profound physiological consequences. As the field has developed it has become evident that no unique type of ion channel is regulated in this way, and indeed it is possible that under appropriate conditions *all* ion channels may be subject to modulation by protein phosphorylation. Similarly, phosphorylation produces its effects by several different molecular mechanisms. The membrane current I carried by a given population of ion channels can be described by the equation

$$I = N * p * i$$

where N is the number of functional channels in the membrane, p is the probability that an individual channel will be open, and i is the unitary current carried by a single open channel. A change in I measured at the whole cell level could be due in principle to a change in any one or several of these parameters N, p, and i. In fact, changes in both p (Cachelin et al 1983, Ewald et al 1985, Flockerzi et al 1986, Belardetti et al 1987) and N (Bean et al 1984, Shuster et al 1985, Gunning 1987) have been reported [the appearance of a novel calcium channel in bag cell neurons (Strong et al 1987) can also be thought of as a change in N]. Although channels seem to be reluctant to alter their unitary conductances, it would not be surprising in the future to find situations where i is modulated as well. The bottom line seems to be that protein phosphorylation is a fundamental regulatory mechanism in the control of cellular membrane properties, and that a number of different kinds of membrane ion channels can be modulated by phosphorylation via several different molecular mechanisms.

ION CHANNEL MODULATION BY GUANYL NUCLEOTIDE–BINDING PROTEINS

One method of modulating ion channel properties that has come to the fore recently involves the interaction of channels with the ubiquitous class of membrane proteins that bind guanyl nucleotides (the so-called G proteins). This can be described as the current growth area in the general field of channel modulation, with some half dozen major publications on this topic appearing in print during the month prior to the preparation of this review. Thus, although this account can be no more than a summary of the rapidly changing state of the art as of February 1987, the clear importance of this area justifies its inclusion here.

G proteins are membrane proteins that mediate signal transduction in a wide variety of cell types by interacting with receptor proteins (Rodbell 1980). Binding of an agonist to a receptor allows the receptor to interact with a G protein, causing the latter to release GDP and bind GTP. In its GTP-bound form the G protein is active and can regulate the activity of enzymes in the membrane; the cycle is terminated by the hydrolysis of the bound GTP to GDP by a GTPase activity intrinsic to the G protein. Although G proteins were originally described as regulators of adenylate cyclase activity (Rodbell 1980), it is now clear that they can modulate the activity of a wide variety of membrane proteins, including a number of different enzymes and ion channels. Furthermore, G proteins are themselves structurally heterogeneous. They are heterotrimers, containing α chains that bind GTP and are thought to determine the specificity for particular receptors and effectors (but see below) and β and γ chains that form a $\beta\gamma$ complex that is thought to anchor the G protein to the plasma membrane (Bourne 1986). At least eight different α-subunit genes have been identified to date, and the protein products of at least four of these genes have been isolated (Bourne 1987).

Molecular Probes for Studying the Role of G Proteins

Several different molecular probes that regulate G protein activity are available and have been used to examine the modulation of ion channels. The GDP analog guanosine 5'-O-(2-thiodiphosphate) (GDPβS) competitively inhibits the binding of GTP and thus inhibits G protein activity (Eckstein et al 1979). In contrast, nonhydrolyzable GTP analogs such as guanylylimidodiphosphate [Gpp(NH)p] or guanosine 5'(β-thio) triphosphate (GTPγS) cause persistent activation of G proteins (Rodbell 1980). Furthermore, at least two bacterial toxins modulate the activity of G proteins by ADP-ribosylating the α subunits (Rodbell 1980). Cholera toxin causes activation of one type of G protein called G_s (so named because it stimulates the activity of adenylate cyclase), whereas pertussis toxin (also known as islet-activating protein or IAP) inhibits the actions of the G proteins G_i (which inhibits adenylate cyclase) and G_o (some other function) (Bourne 1986). These probes have been used to confirm a role for adenylate cyclase activation in the regulation of neuronal ion currents (Treistman & Levitan 1976, Treistman 1981, Lemos & Levitan 1984) and more recently to demonstrate interaction of G proteins with ion channels via mechanisms that do not involve adenylate cyclase and cAMP.

Modulation of Ion Currents by G Protein Probes

Pfaffinger et al (1985) and independently Breitwieser & Szabo (1985) examined the muscarinic receptor–mediated activation of inwardly rectifying potassium channels by acetylcholine in the heart. Intracellular application of Gpp(NH)p causes a persistent agonist-dependent activation of the potassium current (Breitwieser & Szabo 1985). Furthermore, intracellular GTP is required for the normal activation of the current by acetylcholine, and this activation is blocked by pretreatment of the cardiac cells with pertussis toxin (Pfaffinger et al 1985). These findings have recently been extended to the single channel level by the demonstration (Kurachi et al 1986, Yatani et al 1987) that single potassium channels in detached "inside-out" (Hamill et al 1981) membrane patches from these cells can be activated by GTPγS. Since cAMP is not involved in the muscarinic regulation of the inwardly rectifying potassium current in the heart (Trautwein et al 1982, Nargeot et al 1983), these studies suggest the mediation by some pertussis toxin-sensitive G protein (such as G_i or G_o), which in this case is not coupled to adenylate cyclase.

A similar conclusion has been drawn by Holz et al (1986), who investigated the inhibition of calcium current in dorsal root ganglion neurons by noradrenaline or γ-amino butyric acid (GABA). This effect of the neurotransmitters, which is not dependent on changes in cAMP levels (Forscher & Oxford 1985), is blocked by intracellular GDPβS or pretreatment with pertussis toxin. Andrade et al (1986) demonstrated that pertussis toxin pretreatment or intracellular application of GDPβS block the increase in potassium current evoked by serotonin or baclofen (a selective GABA receptor agonist) in hippocampal pyramidal cells. Furthermore, intracellular GTPγS causes an enhancement of potassium conductance and occludes the effects of serotonin and baclofen. Again cAMP is not involved in these effects (Andrade et al 1986). Identical findings were reported by Sasaki & Sato (1987) in *Aplysia* neurons in which potassium conductance is increased by dopamine, histamine, and acetylcholine. These authors were able to inject pertussis toxin intracellularly into the large *Aplysia* neurons and monitor the slow time course of the decrease in response to the neurotransmitters; this presumably reflects the time course of ADP-ribosylation of the relevant G protein. In this latter study, as in several other examples of ion current modulation by pertussis toxin (e.g. Aghajanian & Wang 1986), a role for adenylate cyclase inhibition has not yet been excluded. An interesting feature of all of these experiments is that within each cell type, several different neurotransmitter receptors are coupled to a single population of ion channels via G proteins.

Direct Modulation of Membrane Ion Channels by Purified G Proteins

If adenylate cyclase is not involved, what is the mechanism by which ion channels are coupled to G proteins in these cells? One interesting possibility raised by Holz et al (1986) is that the coupling is provided by activation of a phospholipase C with the consequent release of diacylglycerol and activation of protein kinase C. There is evidence for the receptor-mediated activation of phospholipases via G proteins (e.g. Cockcroft & Gomperts 1985), and in dorsal root ganglion neurons the G protein–mediated decrease in calcium current (Holz et al 1986) is mimicked by diacylglycerol and phorbol ester (see above and Rane & Dunlap 1986). However, this cannot be the explanation for all of the above results, since in the hippocampal pyramidal neurons, for example, phorbol ester blocks the G protein-mediated increase in potassium current evoked by serotonin (Andrade et al 1986). Another intriguing possibility, for which experimental support is beginning to emerge, is that G proteins can interact directly with ion channels just as they do with enzymes such as adenylate cyclase. Yatani et al (1987) examined single potassium channels in detached membrane patches from guinea pig atrial cells. These channels appear to be responsible for the inwardly rectifying potassium current that is activated by acetylcholine in these cells (Pfaffinger et al 1985, Breitwieser & Szabo 1985, Kurachi et al 1986). In detached "inside-out" patches, in which the cytoplasmic membrane face is exposed to the bathing medium, potassium channel activity is virtually absent. However, when a purified activated G_i from human erythrocytes is added to the medium, robust gating of potassium channels is observed (Yatani et al 1987). This effect is observed with picomolar concentrations of G_i if it has been activated by prior incubation with GTPγS. In contrast, nonactivated G_i, and both activated and nonactivated G_s, do not modulate potassium channel gating even at much higher concentrations. In view of the potency of G_i in activating potassium channels, Yatani et al (1987) suggest that it be renamed G_k.

An extension of this approach to purified G protein subunits has been reported from several laboratories. Since a number of distinct α subunits have been identified, it has been suggested that these subunits are responsible for the specificity of transduction to different effector systems (Bourne 1986). However, Logothetis et al (1987) reported that two different kinds of α subunit (those from G_i and G_o), purified from bovine cerebral cortex, do not alter potassium channel activity in detached "inside-out" patches from chick atrial cells. On the other hand, the purified $\beta\gamma$ complex, which displays little heterogeneity and is thought to play a structural role in anchoring the G protein to the plasma membrane (Bourne 1986), surprisingly is a potent activator of potassium channels in these patches at nanomolar concentrations. Furthermore, this effect of $\beta\gamma$ can be blocked by preincubating the latter with an excess of α subunits. These authors (Logothetis et al 1987) point out that it is possible that the α subunits for some reason are not able to interact with the membrane and that they cannot exclude entirely a role for these subunits in potassium channel gating. Nevertheless, the activation of the potassium channels by $\beta\gamma$ may necessitate a reevaluation (Bourne 1987) of the view (Bourne 1986) that these subunits play only a structural role. It is important to point out that as of this writing these results are highly controversial. Codina et al (1987), working with the same cardiac potassium channel, have found that α but not $\beta\gamma$ subunits increase channel opening probability. Since the experiments appear to have been done in a very similar manner in the two laboratories, the reasons for the different results are not at all clear at this time.

Another case in which α subunits appear to be effective is in neuroblastoma/glioma hybrid cells, in which an opioid peptide inhibits the calcium current (Hescheler et al 1987). This effect is mimicked by intracellular GTPγS, and is inhibited by intracellular GDPβS and by pertussis toxin pretreatment, reminiscent of the effects of noradrenaline and GABA on calcium current in dorsal root ganglion neurons (Holz et al 1986). Hescheler et al (1987) report that the opioid peptide effect can be restored in pertussis toxin–treated cells by intracellular perfusion of the purified α subunit from pig brain G_o. In contrast to the findings of Logothetis et al (1987) in the heart, $\beta\gamma$ is without effect by itself on the neuroblastoma/glioma calcium current, and it does not influence the effectiveness of the α subunit (Hescheler et al 1987).

Whither Modulation by G Proteins?

What can we conclude at this time about the physiological importance of ion channel modulation via G proteins? This field is in a stage of rapid expansion, with novel results appearing in print almost weekly, and thus prognostication is not only difficult but dangerous. Nevertheless it seems likely that G protein/channel interactions will turn out to be a major mechanism of channel modulation, at least in heart and nerve cells, and may even displace protein phosphorylation from its present position of supremacy. Certainly some of the results to date have already necessitated a rethinking of current dogma, both about G proteins and ion channels, and new and exciting developments can be anticipated in this area during the next few years.

SUMMARY

The field of ion channel modulation has entered a stage of maturity. The development of powerful new molecular biological and biophysical approaches has provided important new insights into the structure and function of ion channels and has revealed them as dynamic entities whose activity can be regulated. The physiological consequences of such regulation are obvious for so-called excitable cells like nerve and muscle cells, but it is also evident that modulation occurs in many other cell types, where its effects on the lifestyle of the cell are less clear. Furthermore, the number of ion channels that have been shown to be subject to modulation continues to increase, and the end is not yet in sight. For example, exciting information is beginning to emerge about gating and conduction properties of the large class of channels coupled to excitatory amino acid receptors (Mayer 1987), but their modulation has not yet been studied in any detail. In any event, there no longer is any doubt that modulatability is an intrinsic property of many and perhaps all membrane ion channels.

The mechanisms of channel modulation are also turning out to be more diverse than was first suspected. One possible explanation for this diversity is to provide a broad temporal spectrum for the regulation of channel activity. One temporal extreme may be a directly coupled system such as the nicotinic acetylcholine receptor/channel, where interactions between the several subunits of a single macromolecular complex determine rapid channel gating. The other extreme may be covalent modification (by protein phosphorylation or other covalent change), which results in a functional change that can long outlast the initial stimulus. G-protein modulation, which involves the (presumably) noncovalent interaction between distinct membrane proteins, may provide for intermediate duration changes in channel properties. Whether these mechanisms will be joined by others that provide for an even more subtle temporal discrimination is an exciting question for the future.

ACKNOWLEDGMENTS

Supported by grants from the National Institute of Health (NS17910) and National Science Foundation (BNS84-00875). I am grateful to Drs. R. Frizzell, L. Kaczmarek, R. Kramer, C. Miller, and P. Reinhart for providing me with unpublished data and critical comments on an earlier version of the manuscript.

Literature Cited

Adams, W. B., Levitan, I. B. 1982. Intracellular injection of protein kinase inhibitor blocks the serotonin-induced increase in K^+ conductance in *Aplysia* neuron R15. *Proc. Natl. Acad. Sci. USA* 79: 3877–80

Aghajanian, G. K., Wang, Y.-Y. 1986. Pertussis toxin blocks the outward currents evoked by opiate and α_2-agonists in locus coeruleus neurons. *Brain Res.* 371: 390–94

Albuquerque, E. X., Deshpande, S. S., Aracava, Y., Alkondon, M., Daly, J. W. 1986. A possible involvement of cyclic AMP in the expression of desensitization of the nicotinic acetylcholine receptor: A study with forskolin and its analogs. *FEBS Lett.* 199: 113–20

Alkon, D. L., Acosta-Urquidi, J., Olds, J., Kuzma, G., Neary, J. T. 1983. Protein kinase injection reduces voltage-dependent potassium current. *Science* 219: 303–6

Andrade, R., Malenka, R. C., Nicoll, R. A. 1986. A G protein couples serotonin and $GABA_B$ receptors to the same channels in hippocampus. *Science* 234: 1261–65

Baraban, J. M., Snyder, S. H., Alger, B. E. 1985. Protein kinase C regulates ionic conductance in hippocampal pyramidal neurons: Electrophysiological effects of phorbol esters. *Proc. Natl. Acad. Sci. USA* 82: 2538–42

Bean, B. P., Nowycky, M. C., Tsien, R. W. 1984. β-Adrenergic modulation of calcium channels in frog ventricular heart cells. *Nature* 307: 371–75

Belardetti, F., Kandel, E. R., Siegelbaum, S. A. 1987. Neuronal inhibition by the peptide FMRFamide involves opening of S K^+ channels. *Nature* 325: 153–56

Borsotto, M., Barhanin, J., Norman, R. I., Lazdunski, M. 1984. Purification of the dihydropyridine receptor of the voltage-dependent Ca^{2+} channel from skeletal muscle transverse tubules using $(+)$ [³H]PN 200-110. *Biochem. Biophys. Res. Commun.* 122: 1357–66

Bourne, H. R. 1986. One molecular machine can transduce diverse signals. *Nature* 321: 814–16

Bourne, H. R. 1987. "Wrong" subunit regulates cardiac potassium channels. *Nature* 325: 296–97

Breitwieser, G. E., Szabo, G. 1985. Uncoupling of cardiac muscarinic and β-adrenergic receptors from ion channels by a guanine nucleotide analogue. *Nature* 317: 538–40

Cachelin, A. B., DePeyer, J. E., Kokubun, S., Reuter, H. 1983. Calcium channel modulation by 8-bromocyclic AMP in heart cells. *Nature* 304: 462–64

Cahalan, M. D., Chandy, K. G., DeCoursey, T. E., Gupta, S. 1985. A voltage-gated potassium channel in human T lymphocytes. *J. Physiol.* 358: 197–237

Castellucci, V. F., Kandel, E. R., Schwartz, J. H., Wilson, F. D., Nairn, A. C., Greengard, P. 1980. Intracellular injection of the catalytic subunit of cyclic AMP-dependent protein kinase simulates facilitation of transmitter release underlying behavioral sensitization in *Aplysia*. *Proc. Natl. Acad. Sci. USA* 77: 7492–96

Chen, C., Hess, P. 1987. Calcium channels in mouse 3T3 and human fibroblasts. *Biophys. J.* 51: 226a

Cockcroft, S., Gomperts, B. D. 1985. Role of guanine nucleotide binding protein in the activation of polyphosphoinositide phosphodiesterase. *Nature* 314: 534–36

Codina, J., Yatani, A., Grenet, D., Brown, A. M., Birnbaumer, L. 1987. The α subunit of the GTP binding protein G_k opens atrial potassium channels. *Science* 236: 442–45

Cohen, P. 1982. The role of protein phosphorylation in neural and hormonal control of cellular activity. *Nature* 296: 613–20

Connor, J. A., Hockberger, P. 1984. A novel membrane sodium current induced by injection of cyclic nucleotides into gastropod neurones. *J. Physiol.* 354: 139–62

Curtis, B. M., Catterall, W. A. 1984. Purification of the calcium antagonist receptor of the voltage-sensitive calcium channel from skeletal muscle transverse tubules. *Biochemistry* 23: 2113–18

Curtis, B. M., Catterall, W. A. 1985. Phosphorylation of the calcium antagonist receptor of the voltage-sensitive calcium channel by cAMP-dependent protein kinase. *Proc. Natl. Acad. Sci. USA* 82: 2528–32

DePeyer, J. E., Cachelin, A. B., Levitan, I. B., Reuter, H. 1982. Ca^{2+}-activated K^+ conductance in internally perfused snail neurons is enhanced by protein phosphorylation. *Proc. Natl. Acad. Sci. USA* 79: 4207–11

DeRiemer, S. A., Strong, J. A., Albert, K. A., Greengard, P., Kaczmarek, L. K. 1985. Enhancement of calcium current in *Aplysia* neurones by phorbol ester and protein kinase C. *Nature* 313: 313–16

Deterre, P., Paupardin-Tritsch, D., Bockaert, J., Gerschenfeld, H. M. 1981. Role of cyclic AMP in a serotonin-evoked slow inward current in snail neurones. *Nature* 290: 783–85

Doroshenko, P. A., Kostyuk, P. G., Martynyuk, A. E., Kursky, M. D., Vorobetz, Z. D. 1984. Intracellular protein kinase and calcium inward currents in perfused neurones of the snail helix pomatia. *Neuroscience* 11: 263–67

Eckert, R., Chad, J. E. 1984. Inactivation of calcium channels. *Prog. Biophys. Molec. Biol.* 44: 215–67

Eckstein, F., Cassel, D., Levkowitz, H., Lowe, M., Selinger, Z. 1979. Guanosine 5'-O-(2-thiodiphosphate): An inhibitor of adenylate cyclase stimulation by guanine nucleotides and fluoride ions. *J. Biol. Chem.* 254: 9829–34

Ewald, D., Williams, A., Levitan, I. B. 1985. Modulation of single Ca^{2+}-dependent K^+-channel activity by protein phosphorylation. *Nature* 315: 503–6

Farley, J., Auerbach, S. 1986. Protein kinase C activation induces conductance changes in *Hermissenda* photoreceptors like those seen in associative learning. *Nature* 319: 220–23

Fesenko, E. F., Kolesnikov, S. S., Lyubarsky, A. L. 1985. Induction by cyclic GMP of cationic conductance in plasma membrane of retinal rod outer segment. *Nature* 313: 310–13

Flockerzi, V., Oeken, H.-J., Hofmann, F., Pelzer, D., Cavalie, A., Trautwein, W. 1986. Purified dihydropyridine-binding site from skeletal muscle t-tubules is a functional calcium channel. *Nature* 323: 66–68

Forscher, P., Oxford, G. S. 1985. Modulation of calcium channels by norepinephrine in internally dialyzed avian sensory neurons. *J. Gen. Physiol.* 85: 743–63

Frizzell, R. A:, Rechkemmer, G., Shoemaker, R. L. 1986. Altered regulation of airway epithelial cell chloride channels in cystic fibrosis. *Science* 233: 558–60

Gerschenfeld, H. M., Hammond, C., Paupardin-Tritsch, D. 1986. Modulation of the calcium current of molluscan neurones by neurotransmitters. *J. Exp. Biol.* 124: 73–91

Gordon, A. S., Davis, C. G., Milfay, D., Diamond, I. 1977. Phosphorylation of acetylcholine receptor by endogenous membrane protein kinase in receptor-enriched membranes of *Torpedo californica*. *Nature* 267: 539–40

Grega, D. S., Werz, M. A., Macdonald, R. L. 1987. Forskolin and phorbol esters reduce the same potassium conductance of mouse neurons in culture. *Science* 235: 345–48

Gunning, R. 1987. Increased numbers of ion channels promoted by an intracellular second messenger. *Science* 235: 80–82

Hamill, P. P., Marty, A., Neher, E., Sakmann, B., Sigworth, F. J. 1981. Improved patch-clamp techniques for high-resolution current recording from cells and cell-free membrane patches. *Pflugers Arch.* 391: 85–100

Hammond, C., Paupardin-Tritsch, D., Gerschenfeld, H. M. 1986. Cholecystokinin 8 (CCK 8) induces in an identified snail neuron an "irreversible" decrease of calcium current probably involving the activation of protein kinase C. *Neurosci. Abstr.* 12: 16

Hammond, C., Paupardin-Tritsch, D., Nairn, A. C., Greengard, P., Gerschenfeld, H. M. 1987. Cholecystokinin induces a decrease in Ca^{2+} current in snail neurons that appears to be mediated by protein kinase C. *Nature* 325: 809–11

Haynes, L. W., Kay, A. R., Yau, K.-W. 1986. Single cyclic GMP-activated channel activity in excised patches of rod outer segment membrane. *Nature* 321: 66–70

Hescheler, J., Rosenthal, W., Trautwein, W., Schultz, G. 1987. The GTP-binding protein, G_0, regulates neuronal calcium channels. *Nature* 325: 445–46

Higashida, H., Brown, D. A. 1986. Two polyphosphatidylinositide metabolites control two K^+ currents in a neuronal cell. *Nature* 323: 333–35

Hille, B. 1984. *Ionic Channels of Excitable Membranes.* Sunderland, Mass: Sinauer Assoc. 426 pp.

Holz, G. G. IV, Rane, S. G., Dunlap, K. 1986. GTP-binding proteins mediate transmitter inhibition of voltage-dependent calcium channels. *Nature* 319: 670–72

Hosey, M. M., Borsotto, M., Lazdunski, M. 1986. Phosphorylation and dephosphorylation of dihydropyridine-sensitive voltage-dependent Ca^{2+} channel in skeletal muscle membranes by cAMP- and Ca^{2+}-dependent processes. *Proc. Natl. Acad. Sci. USA* 83: 3733–37

Huganir, R. L. 1987. Phosphorylation of purified ion channel proteins. In *Neuromodulation: The Biochemical Control of Neuronal Excitability*, ed. L. K. Kaczmarek, I. B. Levitan, Chapt. 5, pp. 86–99. New York: Oxford University Press

Huganir, R. L., Delcour, A. H., Greengard, P., Hess, G. P. 1986. Phosphorylation of the nicotinic acetylcholine receptor regulates its rate of desensitization. *Nature* 321: 774–76

Huganir, R. L., Greengard, P. 1983. cAMP-dependent protein kinase phosphorylates the nicotinic acetylcholine receptor. *Proc. Natl. Acad. Sci. USA* 80: 1130–34

Johnson, E. C., Robinson, P. R., Lisman, J. E. 1986. Cyclic GMP is involved in the excitation of invertebrate photoreceptors. *Nature* 324: 468–70

Kaczmarek, L. K., Jennings, K. R., Strumwasser, F., Nairn, A. C., Walter, U.,

Wilson, F. D., Greengard, P. 1980. Microinjection of catalytic subunit of cyclic AMP-dependent protein kinase enhances calcium action potentials of bag cell neurons in cell culture. *Proc. Natl. Acad. Sci. USA* 77: 7487–91

Kikkawa, U., Takai, Y., Minakuchi, R., Inohara, S., Nishizuka, Y. 1982. Calcium-activated, phospholipid-dependent protein kinase from rat brain. Subcellular distribution, purification, and properties. *J. Biol. Chem.* 257: 13341–48

Klein, M., Kandel, E. 1978. Presynaptic modulation of voltage dependent Ca^{2+} current: Mechanism for behavioral sensitization in *Aplysia californica*. *Proc. Natl. Acad. Sci. USA* 75: 3512–16

Knowles, M. R., Stutts, M. J., Spock, A., Fischer, N., Gatzy, J. T., Boucher, R. C. 1983. Abnormal ion permeation through cystic fibrosis respiratory epithelium. *Science* 221: 1067–70

Krnjevic, K., Puil, E., Werman, R. 1976. Is cyclic guanosine monophosphate the internal "second messenger" for cholinergic actions on central neurons? *Can. J. Physiol. Pharmacol.* 54: 172–76

Kubota, M., Alkon, D. L., Naito, S., Rasmussen, H. 1986. Regulation of membrane currents by C-kinase before and after activation with phorbol ester. *Neurosci. Abstr.* 12: 559

Kurachi, Y., Nakajima, T., Sugimoto, T. 1986. On the mechanism of action of muscarinic K^+ channels by adenosine in isolated atrial cells: Involvement of GTP-binding proteins. *Pflugers Arch.* 407: 264

Lemos, J. R., Levitan, I. B. 1984. Intracellular injection of guanyl nucleotides alters the serotonin-induced increase in potassium conductance in *Aplysia* neuron R15. *J. Gen. Physiol.* 83: 269–85

Levitan, E. S., Levitan, I. B. 1988. A cyclic GMP analog decreases the currents underlying bursting activity in the *Aplysia* neuron R15. *J. Neurosci.* In press

Levitan, I. B. 1985. Phosphorylation of ion channels. *J. Membr. Biol.* 87: 177–90

Levitan, I. B., Lemos, J. R., Novak-Hofer, I. 1983. Protein phosphorylation and the regulation of ion channels. *Trends Neurosci.* 6: 496–99

Levitan, I. B., Norman, J. 1980. Different effects of cAMP and cGMP derivatives on the activity of an identified neuron: Biochemical and electrophysiological analysis. *Brain Res.* 187: 415–29

Logothetis, D. E., Kurachi, Y., Galper, J., Neer, E. J., Clapham, D. E. 1987. The $\beta\gamma$ subunits of GTP-binding proteins activate the muscarinic K^+ channel in heart. *Nature* 325: 321–26

Lotshaw, D. P., Levitan, E. S., Levitan, I. B.

1986. Fine tuning of neuronal electrical activity: Modulation of several ion channels by intracellular messengers in a single identified nerve cell. *J. Exp. Biol.* 124: 307–22

Madison, D. V., Malenka, R. C., Nicoll, R. A. 1986. Phorbol esters block a voltage-sensitive chloride current in hippocampal pyramidal cells. *Nature* 321: 695–97

Malenka, R. C., Madison, D. V., Andrade, R., Nicoll, R. A. 1986. Phorbol esters mimic some chlonergic actions in hippocampal pyramidal neurons. *J. Neurosci.* 6: 475–80

Mayer, M. 1987. Two channels reduced to one. *Nature* 325: 480–81

Middleton, P., Jaramillo, F., Schuetze, S. M. 1986. Forskolin increases the rate of acetylcholine receptor desensitization at rat soleus endplates. *Proc. Natl. Acad. Sci. USA* 83: 4967–71

Miller, C. M. 1987. How ion channel proteins work. In *Neuromodulation: The Biochemical Control of Neuronal Excitability*, ed. L. K. Kaczmarek, I. B. Levitan, Chapt. 3, pp. 39–63. New York: Oxford Univ. Press

Nargeot, J., Nerbonne, J. M., Engels, J., Lester, H. A. 1983. Time course of the increase in the myocardial slow inward current after a photochemically generated concentration jump of intracellular cAMP. *Proc. Natl. Acad. Sci. USA* 80: 2395–99

Nishizuka, Y. 1984. Turnover of inositol phospholipids and signal transduction. *Science* 225: 1365–70

Osterrieder, W., Brum, G., Hescheler, J., Trautwein, W., Flockerzi, V., Hofmann, F. 1982. Injection of subunit of cyclic AMP-dependent protein kinase into cardiac myocytes modulates Ca^{2+} current. *Nature* 298: 576–78

Paupardin-Tritsch, D., Deterre, P., Gerschenfeld, H. M. 1981. Relationship between two voltage-dependent serotonin responses of molluscan neurones. *Brain Res.* 217: 201–6

Paupardin-Tritsch, D., Hammond, C., Gerschenfeld, H. M. 1986a. Serotonin and cyclic GMP both induce an increase of the calcium current in the same identified molluscan neurons. *J. Neurosci.* 6: 2715–23

Paupardin-Tritsch, D., Hammond, C., Gerschenfeld, H. M., Nairn, A. C., Greengard, P. 1986b. cGMP-dependent protein kinase enhances Ca^{2+} current and potentiates the serotonin-induced Ca^{2+} current increase in snail neurones. *Nature* 323: 812–14

Peters, K., Demaille, J., Fischer, E. 1977. Adenosine $3':5'$-monophosphate dependent protein kinase from bovine heart.

Characterization of the catalytic subunit. *Biochemistry* 16: 5691–97

Pfaffinger, P. J., Martin, J. M., Hunter, D. D., Nathanson, N. M., Hille, B. 1985. GTP-binding proteins couple cardiac muscarinic receptors to a K channel. *Nature* 317: 536–38

Quinton, P. M. 1983. Chloride impermeability in cystic fibrosis. *Nature* 301: 421–22

Rane, S. G., Dunlap, K. 1986. Kinase C activator 1,2-oleoylacetylglycerol attenuates voltage-dependent calcium current in sensory neurons. *Proc. Natl. Acad. Sci. USA* 83: 184–88

Rodbell, M. 1980. The role of hormone receptors and GTP-regulatory proteins in membrane transduction. *Nature* 284: 17–22

Saitoh, T., Changeux, J.-P. 1981. Change in the state of phosphorylation of acetylcholine receptor during maturation of the electromotor synapse in *Torpedo marmorata* electric organ. *Proc. Natl. Acad. Sci. USA* 78: 4430–34

Sasaki, K., Sato, M. 1987. A single GTP-binding protein regulates K^+-channels coupled with dopamine, histamine and acetylcholine receptors. *Nature* 325: 259–62

Shuster, M., Camardo, J., Siegelbaum, S., Kandel, E. R. 1985. Cyclic AMP-dependent protein kinase closes the serotonin-sensitive K^+ channels of *Aplysia* sensory neurones in cell-free membrane patches. *Nature* 313: 392–95

Siegelbaum, S. A., Camardo, J. S., Kandel, E. R. 1982. Serotonin and cyclic AMP close single K^+ channels in *Aplysia* sensory neurones. *Nature* 299: 413–17

Siegelbaum, S. A., Tsien, R. W. 1983. Modulation of gated ion channels as a mode of transmitter action. *Trends Neurosci.* 6: 307–13

Smith, J. S., Coronado, R., Meissner, G. 1985. Sarcoplasmic reticulum contains adenine nucleotide-activated calcium channels. *Nature* 316: 446–49

Strong, J. A., Fox, A. P., Tsien, R. W., Kaczmarek, L. K. 1987. Stimulation of protein kinase C recruits covert calcium channels in *Aplysia* bag cell neurones. *Nature* 325: 714–17

Strong, J. A., Kaczmarek, L. K. 1987. Potassium currents that regulate action potentials and repetitive firing. In *Neuromodulation: The Biochemical Control of Neuronal Excitability*, ed. L. K. Kaczmarek, I. B. Levitan, Chapt. 7, pp. 119–

37. New York: Oxford Univ. Press

Teichberg, V. I., Sobel, A., Changeaux, J.-P. 1977. In vitro phosphorylation of the acetylcholine receptor. *Nature* 267: 540–42

Trautwein, W., Taneguchi, J., Noma, A. 1982. The effect of intracellular cyclic nucleotides and calcium on the action potential and acetylcholine response of isolated cardiac cells. *Pflugers Arch.* 392: 307–14

Treistman, S. 1981. Effect of adenosine $3',5'$-monophosphate on neuronal pacemaker activity: A voltage clamp analysis. *Science* 211: 59–61

Treistman, S., Levitan, I. B. 1976. Intraneuronal guanylylimidodiphosphate injection mimics long-term synaptic hyperpolarization in *Aplysia*. *Proc. Natl. Acad. Sci. USA* 73: 4689–92

Walsh, D. A., Ashby, C. D., Gonzalez, C., Calkins, D., Fischer, E. H., Krebs, E. G. 1971. Purification and characterization of a protein inhibitor of adenosine $3',5'$-monophosphate-dependent protein kinases. *J. Biol. Chem.* 246: 1977–85

Walter, U., Miller, P., Wilson, F., Menkes, D., Greengard, P. 1981. Immunological distinction between guanosine $3':5'$-monophosphate-dependent and adenosine $3':5'$-monophosphate-dependent protein kinases. *J. Biol. Chem.* 255: 3757–62

Welsh, M. J. 1986. An apical-membrane chloride channel in human tracheal epithelium. *Science* 323: 1648–50

Welsh, M. J., Liedtke, C. M. 1986. Chloride and potassium channels in cystic fibrosis airway epithelia. *Nature* 322: 467–70

Widdicombe, J. H., Welsh, M. J., Finkbeiner, W. E. 1985. Cystic fibrosis decreases the apical membrane chloride permeability of monolayers cultured from cells of tracheal epithelium. *Proc. Natl. Acad. Sci. USA* 82: 6167–71

Yatani, A., Codina, J., Brown, A. M., Birnbaumer, L. 1987. Direct activation of mammalian atrial muscarinic potassium channels by GTP regulatory protein G_k. *Science* 235: 207–11

Yau, K.-W., Nakatani, K. 1985. Light-suppressible, cyclic GMP-sensitive conductance in the plasma membrane of a truncated rod outer segment. *Nature* 317: 252–55

Zimmerman, A. L., Baylor, D. A. 1986. Cyclic GMP-sensitive conductance of retinal rods consists of aqueous pores. *Nature* 321: 70–72

Ann. Rev. Neurosci. 1988. 11 : 137–56

TOPOGRAPHY OF COGNITION:
Parallel Distributed Networks
in Primate Association Cortex

Patricia S. Goldman-Rakic

Section of Neuroanatomy, Yale University School of Medicine,
New Haven, Connecticut 06510

Introduction

The association cortices, those regions interposed between the primary
sensory and primary motor areas, are thought to mediate a variety of
cognitive functions. Among these, the prefrontal cortex has traditionally
been considered the least understood and most complex. The only point
that seems to marshall agreement among basic neuroscientists is that
prefrontal functions and operations, acknowledged to be important in
human behavior, are less accessible to analysis than are many other areas
of the cortex. One reason for this perception may be the lack of an agreed
upon anatomical blueprint that firmly establishes the relationship between
prefrontal cortex and other, better understood, parts of the brain.
However, this circumstance has changed radically in the last decade, and
the structure and functions of the frontal lobes, particularly the prefrontal
"silent" portion, which have intrigued investigators since the last century
(Ferrier 1886, Bianchi 1895, Franz 1907), have recently again become the
subject of intense interest (for reviews, see Damasio 1979, Fuster 1980,
Goldman-Rakic 1984, 1987, Milner et al 1985, Mesulam 1986, Ingvar
1983, Stuss & Benson 1984). The availability of solid new findings in
experimental animals and human patients and the promise of further
discoveries are undoubtedly the basis of this renewed interest.

Several extraordinarily influential papers paved the way for the modern
era of frontal lobe research. Without a doubt, anatomical studies of prefrontal cortex were stimulated by Nauta's 1971 paper on "The Problem
of the Frontal Lobe: A Reinterpretation." This review made an indelible
impression both for its eloquence and fundamental validity, suggesting as

137

0147–006X/88/0301–0137$02.00

it did that the prefrontal cortex was unique among cortical areas in its relationships with interoceptive as well as exteroceptive sensory domains and hence the foremost structure that could synthesize the inner and outer sensory worlds (Nauta 1971). Additional impetus for frontal lobe research was clearly provided with the analysis of cortico-cortical connections by workers using the Nauta method. Using this technique, Pandya & Kuypers (1969) and Jones & Powell (1970) established the anatomical under-pinnings for viewing the prefrontal cortex as the ultimate target of sen-sory cascades. These studies gave credence to what can be considered the dominant theme in almost all treatises on prefrontal cortex from the middle of the eighteenth century—that the prefrontal cortex stands at the "common end-point" for diverse sensory afferent channels and is "privy" to all incoming information. This view may be considered the prefronto-centric theory of cortical function, which is both appealing and yet prob-lematic in its simplicity.

Apart from the philisophical issue of whether cognitive function can be localized to the same degree as "sensory" and "motor" representations, the prefronto-centric view of higher cortical function has the problem that there is no single place in the prefrontal cortex upon which the output of the sensory centers converge (Jones & Powell 1970, Pandya & Kuypers 1969). Indeed, as this chapter will reveal, the best description of projections of the sensory association areas is that they project in parallel upon differ-ent subareas of prefrontal cortex. Thus, one major issue for analysis of prefrontal function is how and where integration across modalities occurs. Commonly, various synthetic functions like "spatial memory," "response inhibition," "short-term memory," "polymodal integration," "planning," "the temporal structuring of behavior" were allocated to different regions of prefrontal cortex (e.g. Bignall & Imbert 1969, Fuster 1980, Jacobsen 1936, Mishkin 1964, Rosenkilde 1979, Stuss & Benson 1984). However, the fact that such functions may be supramodal or integrative or localized in prefrontal cortex actually told us very little about how they might be related to functions carried out in other parts of the cortex and subcortical areas. In the present chapter, I discuss anatomical and physiological evi-dence that places subdivisions of the prefrontal cortex within parallel systems of distributed neural networks and emphasizes their cooperative rather than preeminent role in cognitive operations. I have argued else-where that prefrontal cortex is necessary for regulating behavior guided by representations or internalized models of reality and is not required for behavior guided by external stimuli in the outside world (Goldman-Rakic 1987). According to our neurobiological model, each subdivision of pre-frontal cortex performs a similar operation and differences between areas lie mainly in the nature of information upon which the operation is performed. It should be emphasized at the outset that neither the functions nor organization of prefrontal cortex can be understood without reference to its connections with other structures, whose functions may be better known.

Parcellation of Posterior Parietal Cortex by Cortico-Cortical Connections

The posterior parietal cortex is a convenient point of departure for considering the organization of cortical connections. For most of this century, parietal association cortex has been treated as a more or less homogeneous region, while the temporal neocortex, by contrast, was accorded a modest degree of heterogeneity, with evidence that the superior and inferior temporal gyri contained auditory and visual association areas, respectively. Upon closer scrutiny by modern anatomical and physiological techniques, the compartmentalization of the posterior parietal cortex has likewise become recognized. In nonhuman primates, it can be divided into several specialized information centers characterized by distinctive connections with sensory and limbic systems. For example, area 7a, the portion of the posterior parietal cortex that is situated most caudally and medially on the lateral surface, is interconnected with the anterior and posterior cingulate cortex, the dorsal parahippocampal gyrus and the medial portion of Brodmann's area 19, including the anterior part of a newly defined visual area, PO (Colby et al 1987), and with the upper bank of the superior temporal sulcus (STS) in the temporal lobe (Stanton et al 1977, Mesulam et al 1977, Cavada & Goldman-Rakic 1986) (Figure 1).

Another subdivision of the posterior parietal cortex that has been recognized on cytoarchitectonic and hodological grounds lies in the caudal bank of the intraparietal sulcus and is variously named POa (Pandya & Seltzer 1982a), 7ip (Cavada & Goldman-Rakic 1986) and LIP (Asanuma et al 1985) (see Figure 2). Area 7ip (as this area is designated here) is also connected with the anterior and posterior cingulate cortex, the parahippocampal gyrus, area 19, and the STS. Thus, the connections of 7ip follow a plan remarkably similar to those of area 7a. Direct comparisons between the connections of these two areas, however, reveal distinct topographic separation of their connections (Cavada & Goldman-Rakic 1986). For example, area 7ip is innervated by posterior portions of area PO in the preoccipital sulcus (Seltzer & Pandya 1980, Cavada & Goldman-Rakic 1987) whereas 7a is connected with a more anterior region of 19 (Cavada & Goldman-Rakic 1986). Another example is that area 7ip is innervated by afferents from the ventral bank of the superior temporal sulcus, including area MT (Maunsell & Van Essen 1983, Ungerleider & Desimone 1986), while 7a receives input from along most of the dorsal bank of this sulcus

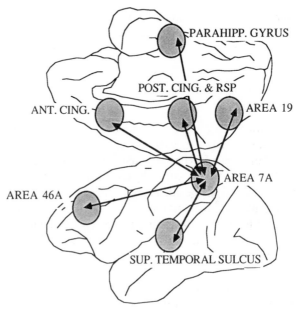

Figure 1 Simplified diagram of the sensory, limbic, premotor, and prefrontal connections
with area 7a of the posterior parietal cortex. For purposes of illustration, many parts of the
circuit are left out—these include orbital, dorsomedial, and prearcuate and premotor sites.

(Cavada & Goldman-Rakic 1986). Although the connections of 7ip and
7a with STS and dorsomedial 19 suggest that both areas are involved in
processing information concerned with visuo-spatial or directional information (Bruce et al 1981, Colby et al 1987), the differential topography of
their connections with STS and PO implies some functional specializations
that have not yet been revealed.

   Other recognized subdivisions of posterior parietal cortex include area
7b, the more rostral and lateral part of posterior parietal cortex, and an
area on the medial surface of the hemisphere (referred to here as 7m) (see
Figure 2). Each of these subareas also receives a distinctive set of inputs.
For instance, the sensory innervation of area 7m originates in parts of
medial area 19 different from those that project to 7b and 7ip and has few
if any afferents from STS (Cavada & Goldman-Rakic 1986). Area 7b, on
the other hand receives no input from area 19 but instead receives a heavy
projection from somatosensory cortices (SI and SII), and is thus more
related to somatic than to visual sensation.

   Based on the anatomical connections with sensory association cortices,
it would appear that separate subdivisions of posterior parietal cortex, by

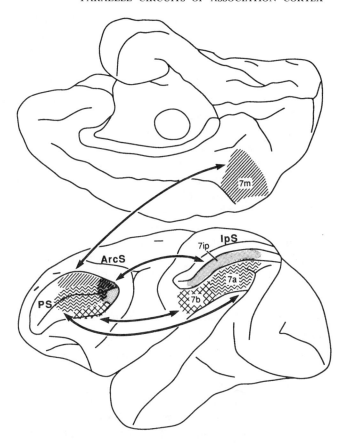

Figure 2 Diagrammatic representation of reciprocal, parallel circuits linking the posterior parietal cortex and caudal principal sulcus (PS). Area 7ip is connected with the caudal end of the principal sulcus, 7m with the dorsal rim, 7a with the fundus, and 7b with the ventral rim of the principal sulcus [from Goldman-Rakic (1987)].

virtue of their distinctive connections, are specialized for different, though possibly related, information processing functions. This is consistent with available electrophysiological data summarized by Hyvarinen (1982, pp. 126–39). In essence, cellular activity in area 7b is related more to somatic stimuli, while that in area 7a is more related to visual and oculomotor mechanisms. Within 7b there is a weak somatotopy with cells related to mouth located more rostrally than those related to head and arm (Robinson & Burton 1980, Hyvarinen 1982). The predominant type of neuron so far recorded from 7a and 7ip has been a visually responsive cell, e.g. visual fixation neurons or neurons that fire during reaching movements toward desired visually presented objects (Mountcastle et al 1984). Recently, cells in 7ip have been shown to discharge in relation to impending eye movements (Gnadt et al 1986). So far, neural activity related to auditory stimuli has not been recorded in any of these parietal subareas (Hyvarinen 1982).

Among the posterior parietal subdivisions, areas 7a and 7ip appear to be particularly strongly connected to dorsal and ventral parts of the parahippocampal gyrus, respectively, as well as to separate territories of the posterior cingulate cortex (Mesulam et al 1977, Cavada & Goldman-Rakic 1986). In contrast, 7b has little or no parahippocampal interactions and 7m has only minor connections with these regions. The strong limbic connections of areas 7a and 7ip may account in part for the remarkable characteristic that many visually triggered neurons recorded from these areas fire more intensely when the stimulus has behavioral significance for the animal (Hyvarinen 1982, Mountcastle et al 1984, Lynch et al 1977, Robinson et al 1978). Whatever their function, the limbic associations, like those of the sensory association areas, exhibit strong compartmentalization within the various subdivisions of the posterior parietal cortex.

Parcellation of Prefrontal Cortex: Parieto-Prefrontal Projections

A parcellation of prefrontal association cortex is suggested by its connections with the posterior parietal cortical subdivisions (Figure 2). Thus, area 7m is interconnected with the dorsal rim and bank of the principal sulcus (Brodmann's area 46); area 7a is connected with the fundus of the principal sulcus, and area 7b is connected with its ventral rim; area 7ip is interconnected with the caudal tip of the principal sulcus and anterior arcuate (frontal eyefield) cortex (Figure 3, Cavada & Goldman-Rakic 1987). Thus, each sector of parietal cortex is connected in parallel with a particular sector of the principal sulcus and presumably transposes some of its sensory-limbic specializations to these areas accordingly. Further, the parieto-prefrontal projections terminate in a "feedforward" pattern, i.e. parietal axons terminate in layers I, IV, and VI of prefrontal cortex, whereas prefrontal axons in parietal cortex avoid layer IV but terminate in layers I and VI (Selemon & Goldman-Rakic 1985, 1988; M. L. Schwartz and P. S. Goldman-Rakic, unpublished), i.e. in a "feedback" pattern.

Although similarly detailed studies do not yet exist for the forward projections of the superior and inferior posterior temporal gyri, common principles of connectivity may apply to these regions as well. Thus, a number of HRP or autoradiographic tracing studies have shown that areas of the temporal neocortex project topographically to the inferior convexity and orbital prefrontal cortex (Jones & Powell 1970, Pandya & Kuypers

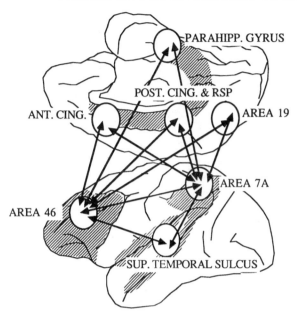

Figure 3 Summary of findings in a double-label experiment of Selemon and Goldman-Rakic showing only some of the multiple areas with which the parietal and prefrontal cortex are interconnected. Cross-hatched areas are those that are innervated by the medial pulvinar nucleus of the thalamus, as discussed in the text. [Medial pulvinar projections based on Baleydier & Maugiere (1985), Trojanoski & Jacobson (1974), Goldman-Rakic & Porrino (1985), Giguere & Goldman-Rakic (1985). Cortico-cortical connections based on Cavada & Goldman-Rakic (1986), Mesulam et al (1977).]

1969, Markowitsch et al 1985, Moran et al 1987). Therefore, parallel pathways not only connect sectors of posterior parietal with dorsolateral prefrontal sectors, but temporal lobe projections to ventral prefrontal regions may be organized in the same manner as the parieto-prefrontal projections.

Distributed Neural Networks: Revelations of a Double Labeling Paradigm

In the study of connections, as in other areas of research, the particular results and the general overview of brain organization that emerge are greatly dependent on the methods and strategies used. Accordingly, for more than a century, ideas about cortical circuitry have relied on reconstructions of circuits from the study of individual cases with, by current standards, large lesions (Pandya & Kuypers 1969, Jones & Powell 1970). By necessity, the connections of each area had to be examined in one animal at a time. It is from such analyses that our knowledge of brain circuitry has been forced into "source and sink" conceptions, i.e. parieto-prefrontal, occipito-temporal, prefronto-cingulate links. Of course, the limitations of technique have never inhibited anatomists from constructing complex circuit diagrams and flow charts, but these are no substitute for direct determination of how many specific populations of cells are directly connected with one another. However, this issue can now be addressed more directly with the strategy of double labeling two cortical areas that are connected to each other in the same hemisphere of the same animal (Selemon & Goldman-Rakic 1985). The question can be asked whether two areas that project to each other also project to other cortical areas in the same hemisphere. If they do, are the connections mutually exclusive or convergent? If convergent, do they totally or partially overlap and with what relationship to the columnar architecture of the target structures? Recent studies in rhesus monkeys that were given simultaneous injections of two distinguishable anterograde tracers into the principal sulcus and posterior parietal cortex, respectively, reveal the remarkable finding that these two areas are mutually interconnected with as many as 15 other cortical areas (Figure 3). Thus, posterior parietal and the principal sulcal cortex project in common to the anterior cingulate, posterior cingulate, and supplementary motor cortex, the ventral and dorsal premotor areas, the orbital prefrontal, prearcuate, frontal opercular, insula, and superior temporal cortex, the parahippocampal cortex, the presubiculum and caudomedial lobule, and, finally, the medial prestriate cortex (Selemon & Goldman-Rakic 1985, 1988). In this double label experiment, no attempt was made to confine the injection sites to one or another parietal or prefrontal subdivision. However, analysis of single label experiments involving subareas 7a and 7ip, reviewed above, suggests the type of parallel circuitry shown in Figure 4.

The prefrontal and parietal axons within these "third party" targets terminate in one of two characteristic modes: either as interdigitated spatially distinct fiber columns or in complementary layers within a single column or a set of columns (Goldman-Rakic & Selemon 1988). In cingulate, retrosplenial, and parahippocampal cortices (medial areas), the mode of termination is the interdigitated pattern, i.e. prefrontal-afferent fiber columns alternate with parietal afferent columns. In the lateral convexity cortex, i.e. the parietal operculum and superior temporal sulcus, the pattern is the complementary intralaminar mode (Figure 5, Goldman-Rakic & Selemon 1988). In the latter, parietal axons terminate in layer IV and VI, whereas prefrontal projections are highly concentrated in layer I and are much less dense in layers III and V/VI (Figure 5). The two different patterns of cortical termination suggest that the integration of prefrontal

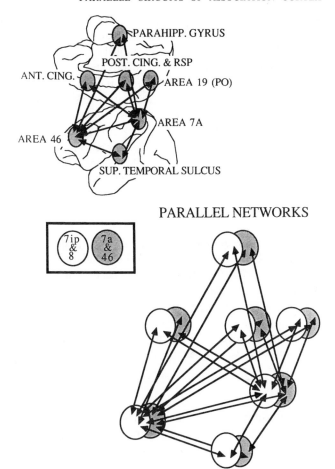

Figure 4   Schematic model of parallel circuits linking areas 7ip and 7a with the caudal tip and the fundus of the principal sulcus, respectively. Area 7ip projections based on the literature cited in the text.

and parietal information in these areas differs accordingly. In the cingulate cortices, for example, parietal and prefrontal terminals presumably terminate on different sets of cells, whereas in temporal lobe and parietal operculum, prefrontal and parietal afferents could terminate upon different parts of the dendritic arbor of the very same cells, much as hippocampal inputs are distributed on proximal and distal segments of pyramidal neurons in Ammon's horn. These anatomical findings open up new issues and possibilities for physiological and electromicroscopic analysis of cortical networks.

Neurons in over a dozen target areas that receive afferents from both the prefrontal and parietal cortex reciprocate these projections (M. L. Schwartz and P. S. Goldman-Rakic, unpublished). For the most part, separate populations of cells within these targets project back to each association area. For example, although a few cingulate neurons have been shown to send one collateral to prefrontal cortex in one hemisphere and another to parietal cortex in the other, most cingulate neurons project either to the prefrontal or to the parietal cortex (Schwartz & Goldman-Rakic 1982). Furthermore, the prefrontal, parietal, limbic, and temporal areas that are linked by cortico-cortical connections are unified also by their thalamic input from the medial pulvinar nucleus of the thalamus (Baleydier & Mauguiere 1980, 1985, Trojanowski & Jacobson 1974, Kievit & Kuypers 1977, Giguere & Goldman-Rakic 1985, Goldman-Rakic & Porrino 1985, Moran et al 1987, Vogt et al 1971). Thus, the medial pulvinar projects to the posterior cingulate, and parahippocampal cortex and also to the superior temporal and fronto-parietal operculum, as well as to the principal sulcus and posterior parietal cortex (see Figure 3). This thalamic nucleus, which is prominent only in primates (Hartung et al 1972), is thereby in position to recruit an entire neural system defined by cortico-cortical connectivity and possibly by common dedication to the complex function of being oriented in time and space (for a discussion of this function see Goldman-Rakic 1987; also Fuster 1980).

Distributed Networks vs Hierarchical Models of Cortical Organization

The conclusion traditionally reached in virtually all comprehensive studies of cortical connections is that they are organized in a step-wise hierarchical sequence proceeding from relatively raw sensory input at the primary sensory cortices through "successive stages of intramodality elaboration allowing progressively more complex discriminations of the features of a particular stimulus. Then, by a series of further connections, this sensory information, now in a highly complex form, is conveyed to polymodal zones for cross-modal interchange of information, to paralimbic and limbic areas for investment with emotional tone and placement in memory, and to the frontal association areas where both sensory and limbic data are integrated in preparation for the organism to respond to sensory stimuli by an appropriate response" (Pandya & Seltzer 1982b). According to hierarchical models of cortical organization, sensory signals are elaborated at successive stages in sensory association cortices, and information flow is mainly unidirectional, i.e. from sensory through associational to motor centers. Further, some sort of convergence occurs at each stage along the hierarchy such that integration of the different sensory inputs takes place in key polymodal areas like the posterior parietal (Mesulam et al 1977), the superior temporal polysensory area (Bruce et al 1981), and/or prefrontal (Bignall & Imbert 1969, Nauta 1971, Pandya & Kuypers 1969) cortex. Without denying that some convergence occurs in association cortical regions, considerations discussed in the present chapter lead toward a different emphasis—one that focuses on the distributed functions in several parallel systems.

Parallel Circuits for Guidance of Eye and Hand Movements

The posterior parietal cortex of the parietal lobe and the caudal principal sulcus of frontal lobe, which have been shown to be reciprocally connected with each other and also interconnected with at least 15 other cortical areas as well as with the medial pulvinar of the thalamus, are functionally related as well. Lesion studies demonstrate the involvement of both areas in spatial abilities, and physiological studies show that their neurons share certain properties. For example, neurons in the frontal eyefield (Walker's area 8A) increase their activity when a monkey withholds eye movements while remembering the location of a target (Bruce & Goldberg 1984). According to a recent preliminary report, neurons in the lateral bank of the intraparietal sulcus (area 7ip) may behave exactly the same way (Gnadt et al 1986). At present, it seems that the apparent similarities in task-related activity in the two areas can be interpreted as evidence that portions of parietal and prefrontal cortex are part of a common circuit and that both have a role in the guidance of eye movements by visuo-spatial information. Presumably, the parietal component (area 7ip) is important for sensorimotor integration and formation of spatial representations, while the prefrontal component (Walker's areas 46 and 8a) is necessary under certain circumstances for the utilization of these representations to generate eye movement commands (see Goldman-Rakic 1987 for further discussion).

Given that areas 7ip and posterior prefrontal (Walker's areas 46 and 8a) constitute part of a cortical system for regulating eye movements, it is possible that guidance of the responses of the hand might similarly be a "distributed" function of another "in parallel" neural circuit that reciprocally links areas 7a and the depths of the principal sulcus (part of Walker's area 46) (Figure 3). The neuronal activity that has been recorded from tissue in or near the principal sulcus in relation to performance on delayed-response tasks has mainly been in studies of manual performance (for review, see Goldman-Rakic 1987). Principal sulcal neurons exhibit spatially coded firing during the delay period, i.e. discharging more to a target on the left (or right) when that target is no longer in view (e.g. Niki &

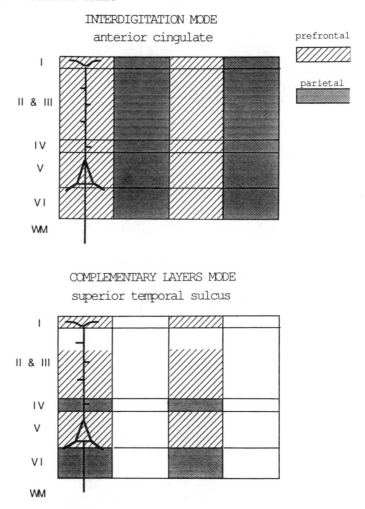

Figure 5   The modes of "third party" interactions are illustrated in this schematic diagram. Prefrontal and parietal terminals terminate in complementary layers in the parietal operculum and in alternating columns in the anterior cingulate cortex.

Watanabe 1976, Kojima & Goldman-Rakic 1982). Very little data as yet exist regarding the role of area 7a neurons in spatial memory tasks. However, many neurons in this area are strongly related to intentional movements of the hand (Mountcastle et al 1984) and, Batuev et al (1985) have recorded from this area during delayed-response performance and reported a small percentage of neurons that exhibited "spatioselective" discharge patterns during the delay period and in response to cue displays. A preliminary study conducted in my laboratory a number of years ago likewise uncovered delay-related activity in recordings from the posterior parietal cortex of immature (one year old) and adult rhesus monkeys performing a conventional delayed-response task (see Figure 6, Alexander 1982). The percentage of delay-related neurons in parietal cortex was lower than in other areas recorded, but even their low incidence in parietal cortex was surprising. We did not know then what to make of the fact that our "control" data resembled our prefrontal findings. Today,

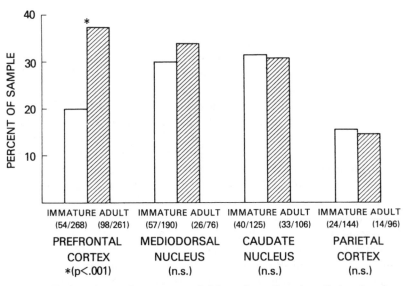

Figure 6   Various classes of neurons recorded from the prefrontal, mediodorsal nucleus, caudate nucleus, and posterior parietal cortex in monkeys performing the spatial delayed-response task. All structures are connected with the principal sulcus. The unpublished results of this study indicated that there was a difference in the incidence of delay-related neurons in one-year-old monkeys compared to adults. I present these data here as evidence that neurons activated during the delay phase of the task can be found in posterior parietal cortex, albeit at apparently lower incidence than in other structures. The low proportion of such neurons could also be related to the fact that the recordings were obtained from the most rostral part of area 7a at the border with area 7b. Perhaps other recording sites would reveal a larger fraction of delayed-related cells. [From Alexander (1982).]

however, in light of Batuev's results and new anatomical data, we might speculate that the similar profiles of activation in parietal and prefrontal cortex indicate a commonality of function and shared circuitry. Again, the parietal contribution is presumably to form and maintain the spatial coordinates representing the location of an object in space, and the prefrontal contribution is to use that knowledge to guide a response, in this case, a hand movement.

However, a strict segregation of hand and eye movement control should not be expected in the prefrontal regions. Recently Funahashi et al (1986) recorded spatially discriminative activity from delay-related neurons in the middle of the principal sulcus, indicating that this region can regulate motor commands to both the hand and the eye, presumably via its multiple connections with the relevant motor centers. Future studies need to work out the precise nature of motor control exerted by different prefrontal regions.

The contribution of other components of the neural networks defined in anatomical studies also need to be examined. Neuronal recording in behaving monkeys has not been attempted in many of the cortical areas connected to prefrontal and parietal cortex, e.g. the posterior cingulate or retrosplenial or parahippocampal cortex. Further, in cortical regions like the superior temporal sulcus that have been studied physiologically, delay tasks have not generally been employed (e.g. Bruce et al 1981). However, it must be noted that delay-enhanced discharge during delayed-response tasks has been reported in several key structures with which posterior parietal areas, the principal sulcus, and frontal eye fields are connected, e.g. the hippocampus (Watanabe & Niki 1985), the head of the caudate nucleus (Niki et al 1972, Hikosaka & Sakamoto 1986) and the mediodorsal nucleus of the thalamus (Alexander & Fuster 1973), though not from the cholinergic system of basal forebrain nuclei (Richardson & DeLong 1986). Also, results from 2-deoxyglucose studies of monkeys performing delayed-response tasks show elevated metabolic activity in prefrontal cortex (Bugbee & Goldman-Rakic 1981), in the caudomedial lobule and hippocampus proper (Friedman & Goldman-Rakic 1985) and mediodorsal nucleus of the thalamus (Friedman et al 1987), compared with animals performing other cognitive tasks. Thus, while not necessarily revealing the functional specialization of each structure in the network, these physiological and metabolic studies are consistent with the supposition of a distributed richly interconnected system of neural structures engaged in spatial information processing. In such systems, integrative functions may emerge from the dynamics of the entire network rather than from linear computations performed at each nodal point in the circuit (e.g. Edelman 1979).

Parallel Circuits For Spatial (Where) and Object (What) Memory

It is well established that monkeys with bilateral principal sulcus lesions exhibit profound and selective deficits on spatial delayed-response tasks that require memory for the location of objects in space while leaving memory for the features of stimuli or objects unaffected (e.g. Blum 1952, Gross & Weiskrantz 1964, Butters et al 1972, Mishkin & Manning 1978, Goldman & Rosvold 1970, Goldman et al 1971). In contrast, monkeys with lesions of the inferior convexity and/or orbital prefrontal cortex exhibit deficits on tasks requiring memory for visual features (e.g. color, shape) of objects but not for their location (Goldman 1971, Passingham 1972, 1975, Mishkin & Manning 1978, Bachavalier & Mishkin 1986). Electrophysiological studies support this division of labor in prefrontal areas. Analysis of unit activity in prefrontal cortex of awake behaving monkeys reveals delay-activated neuronal activity in spatial delay tasks from the region of the principal sulcus and in nonspatial memory tasks (e.g. matching-to-sample) from the inferior convexity and orbital prefrontal cortex (Rosenkilde et al 1981). The conclusion from these studies is that the dorsal (principal sulcus, Walker's area 46) and ventral (inferior convexity and orbital cortex, Walker's areas 11–13) subdivisions of prefrontal cortex may be specialized for working memory of spatial and nonspatial knowledge, respectively, i.e. each area accesses a different class of information. Given that each prefrontal area represents one node in a different distributed circuit, it appears possible that parallel cortical networks subserve spatial and nonspatial information processing.

Parallel Distributed Functions in Human Cortex

Blood flow studies in normal subjects performing psychological tasks support a parallel distributed processing model. The act of thinking increases blood flow in multiple cortical fields in homotypical cortical zones outside the immediate sensory association areas and, predictably, the constellation of cortical areas activated differs with different "types" of thinking or internal operations. Spatial thinking in the form of a mental route-finding task activates the superior occipital, the posterior parietal, the posterior inferotemporal cortex, and several zones within prefrontal cortex; mathematical thinking activates other sets of cortical areas of posterior and anterior association cortex, and linguistic (jingo) thinking activates still other sets of areas (Roland & Friberg 1985). One cannot escape the conclusion that the constellation of areas activated by spatial thought processes in these studies represents the same type of circuitry that is interconnected by cortico-cortical and thalamo-cortical connections described in the nonhuman primate and that mathematical and linguistic thinking engage similarly organized parallel circuits. If this is so, then traditional ideas of hierarchical processing that may apply within some systems is not the dominant mode of functional organization of the associative cortex. Rather, higher cortical functions seem to be carried out by a finite number of dedicated networks of reciprocally interconnected areas. Further, since we already know that different features of the visual world are processed in parallel in visual cortical areas (e.g. Hubel & Livingston 1985, Shipp & Zeki 1985), it seems possible that this segregation of input is "respected" and maintained in the association networks. For example, area MT projects mainly to area 7ip whereas area 7b receives its input primarily from somatosensory association cortex. Since these parietal subareas project to distinct subareas of prefrontal cortex, the opportunity would seem to be at hand for bridging the sensory and executive processes of the cerebral cortex.

Integration Across Systems

If parallel systems of circuits subserve various distinct information-processing tasks as the foregoing analysis suggests, it is appropriate to raise the issue of integration across cortical networks. The field of cortical systems research will have to address the mechanisms by which knowledge of the color or form of an object is integrated with knowledge of its position in space, as such knowledge would appear to involve crosstalk between two different functional systems. If there is but one central executive structure in the prefrontal cortex as cognitive studies in humans have suggested (Shallice 1982), then its essence may be discovered by the nature of the interconnections between neural networks. This could take the form of local cortico-cortical connections between, for example, subdivisions of posterior parietal (Seltzer & Pandya 1986) or prefrontal (Barbas & Mesulam 1981, 1985) cortex or possibly the multiple innervation of all components of a network by a thalamic nucleus. As previously mentioned, in primates the medial pulvinar nucleus projects to the posterior parietal, prefrontal, anterior cingulate, superior temporal sulcus, and other areas of the cortex, i.e. it projects to a system of cortical areas that are interconnected. The question of how the brain organizes its subsystems to produce integrated behavior is perhaps the most challenging that can be posed; hopefully, new approaches to this issue can be informed by anatomical findings and insights.

The picture that emerges from the new anatomy is that of a highly integrated but distributed machinery whose resources are allocated to several basic parallel functional systems that bridge all major subdivisions of the cerebrum. Importantly, this view is supported by recent developmental findings: synaptogenesis proceeds at the same rate and reaches peak values at the same age in areas of sensory, motor, limbic, and association cortex, indicating an unexpected degree of integration in maturational sequence (Rakic et al 1986). If subdivisions of limbic, motor, sensory, and associative cortex exist in developmentally linked and functionally unified networks, as the anatomical, physiological, and behavioral evidence reviewed here suggests, it may in the future be more useful to study the cortex in terms of information processing functions and systems rather than traditional but artificially segregated sensory, motor, or limbic components and individual neurons within only one of these components. Furthermore, in light of the detailed knowledge about specific interconnections at the cortical level, it is to be expected that more and more of this information will be used to guide physiological analysis of higher cortical function.

Literature Cited

Alexander, G. E., Fuster, J. M. 1973. Effects of cooling prefrontal cortex on cell firing in the nucleus medialis dorsalis. *Brain Res.* 61: 93–105

Asanuma, C., Andersen, R. A., Cowan, W. M. 1985. The thalamic relations of the caudal inferior parietal lobule and the lateral prefrontal cortex in monkeys: divergent cortical projections from cell clusters in the medial pulvinar nucleus. *J. Comp. Neurol.* 241: 357–81

Bachevalier, J., Mishkin, M. 1986. Visual recognition impairment follows ventromedial but not dorsolateral prefrontal lesions in monkeys. *Behav. Brain Res.* 20: 249–61

Baleydier, C., Mauguiere, F. 1980. The duality of the cingulate gyrus in monkey: Neuroanatomical study and functional hypothesis. *Brain* 103: 525–54

Baleydier, C., Mauguiere, F. 1985. Anatomical evidence for medial pulvinar connections with the posterior cingulate cortex, the retrosplenial area, and the posterior parahippocampal gyrus in monkeys. *J. Comp. Neurol.* 232: 219–28

Barbas, H., Mesulam, M.-M. 1981. Organization of afferent input to subdivisions of area 8 in the rhesus monkey. *J. Comp. Neurol.* 200: 407–31

Barbas, H., Mesulam, M.-M. 1985. Cortical afferent input to the principalis region of the rhesus monkey. *Neuroscience* 15: 619–37

Batuev, A. S., Shaefer, V. I., Orlov, A. A. 1985. Comparative characteristics of unit activity in the prefrontal and parietal areas during delayed performance in monkeys. *Behav. Brain Res.* 16: 57–70

Bianci, L. 1895. The functions of the frontal lobes. *Brain* 18: 497–530

Bignall, K. E., Imbert, M. 1969. Polysensory and cortico-cortical projections to frontal lobe of squirrel and rhesus monkey. *Electroenceph. Clin. Neurophysiol.* 26: 206–15

Blum, R. A. 1952. Effects of subtotal lesions of frontal granular cortex on delayed reaction in monkeys. *AMA Arch. Neurol. Psychiatry* 67: 375–86

Bruce, C. J., Desimone, R., Gross, C. G. 1981. Visual properties of neurons in a polysensory area in superior temporal sulcus of the macaque. *J. Neurophysiol.* 46: 369–84

Bruce, C. J., Goldberg, M. E. 1984. Physiology of the frontal eye fields. *Trends Neurosci.* 7: 436–41

Bugbee, N. M., Goldman-Rakic, P. S. 1981. Functional 2-deoxyglucose mapping in association cortex: Prefrontal activation in monkeys performing a cognitive task. *Soc. Neurosci. Abstr.* 7: 416

Butters, N., Pandya, D., Stein, D., Rosen, J. 1972. A search for the spatial engram within the frontal lobes of monkeys. *Acta Neurobiol. Exp.* 32: 305–29

Cavada, C., Goldman-Rakic, P. S. 1986. Subdivisions of area 7 in the rhesus monkey exhibit selective patterns of connectivity with limbic, visual and somatosensory cortical areas. *Soc. Neurosci. Abstr.* 12: 262

Colby, C. L., Gattass, R., Olson, C. R., Gross, C. G. 1987. Topographic organization of cortical afferents to extrastriate area PO in macaque: A dual tracer study. *J. Comp. Neurol.* In press

Damasio, A. R. 1979. The frontal lobes. In

154   GOLDMAN-RAKIC

Clinical Neuropsychology, ed. K. M. Heilman, E. Valenstein, pp. 360–411. New York: Oxford Univ. Press

Edelman, G. M. 1979. Group selection and phasic reentrant signaling: A theory of higher brain functions. In The Mindful Brain, ed. G. M. Edelman, V. B. Mountcastle, pp. 51–100. Cambridge, MA: MIT Press

Ferrier, D. 1886. The Functions of the Brain. New York: Putnam. 2nd ed.

Franz, S. I. 1907. On the function of the cerebrum: The frontal lobes. Arch. Psychol. 2: 1–64

Friedman, H., Goldman-Rakic, P. S. 1985. Enhancement of trisynaptic pathway in the hippocampus during performance of spatial working memory tasks: A 2-DG behavioral study in rhesus monkeys. Soc. Neurosci. Abstr. 11: 460

Funahashi, S., Bruce, C. J., Goldman-Rakic, P. S. 1986. Perimetry of spatial memory representation in primate prefrontal cortex: Evidence for a mnemonic hemianopia. Soc. Neurosci. Abstr. 12: 554

Fuster, J. M. 1973. Unit activity in prefrontal cortex during delayed-response performance: Neuronal correlates of transient memory. J. Neurophysiol. 36: 61–78

Fuster, J. M. 1980. The Prefrontal Cortex. New York: Raven

Fuster, J. M., Alexander, G. E. 1971. Neuron activity related to short-term memory. Science 173: 652–54

Giguere, M., Goldman-Rakic, P. S. 1985. Disjunctive distribution of mediodorsal thalamic afferents in the prefrontal cortex of rhesus monkey. Soc. Neurosci. Abstr. 11: 677

Gnadt, J. W., Anderson, R. A., Blatt, G. J. 1986. Spatial memory and motor planning properties of saccade related activity in the lateral intraparietal area of macaque. Soc. Neurosci. Abstr. 12: 458

Goldman, P. S. 1971. Functional development of the prefrontal cortex in early life and the problem of neuronal plasticity. Exp. Neurol. 32: 366–87

Goldman, P. S., Rosvold, H. E. 1970. Localization of function within the dorsolateral prefrontal cortex of the rhesus monkey. Exp. Neurol. 27: 291–304

Goldman, P. S., Rosvold, H. E., Vest, B., Galkin, T. W. 1971. Analysis of the delayed alternation deficit produced by dorsolateral prefrontal lesions in the rhesus monkey. J. Comp. Physiol. Psychol. 77: 212–20

Goldman-Rakic, P. S. 1984. The frontal lobes: Uncharted provinces of the brain. Trends Neurosci. 7: 425–29

Goldman-Rakic, P. S. 1987. Circuitry of the prefrontal cortex and the regulation of

behavior by representational memory. Handb. Physiol. 5(Part 1, Ch. 9): 373–417

Goldman-Rakic, P. S., Porrino, L. J. 1985. The primate mediodorsal (MD) nucleus and its projections to the frontal lobe. J. Comp. Neurol. 242: 535–60

Goldman-Rakic, P. S., Selemon, L. D. 1988. Evidence for distributed cortical networks in nonhuman primates. Submitted for publication

Gross, C. G., Weiskrantz, L. 1964. Some changes in behavior produced by lateral frontal lesions in the macaque. In The Frontal Granular Cortex and Behavior, ed. J. M. Warren, K. Akert, pp. 74–101. New York: McGraw-Hill

Hartung, J. K., Hall, W. C., Diamond, I. T. 1972. Evolution of the pulvinar. Brain Behav. Evol. 6: 424–52

Hikosaka, O., Sakamoto, M. 1986. Cell activity in monkey caudate nucleus preceding saccadic eye movements. Exp. Brain Res. 63: 659–62

Hubel, D. H., Livingstone, M. S. 1985. Complex-unoriented cells in a subregion of primate area 18. Nature 315: 325–27

Hyvarinen, J. 1982. The Parietal Cortex of Monkey and Man. Berlin/Heidelberg: Springer-Verlag

Ingvar, D. H. 1983. Several aspects of language and speech related to prefrontal cortical activity. Human Neurobiol. 2: 177–89

Jacobsen, C. F. 1936. Studies of cerebral function in primates. Comp. Psychol. Monogr. 13: 1–68

Jones, E. G., Powell, T. P. S. 1970. An anatomical study of converging sensory pathways within the cerebral cortex of the monkey. Brain 93: 793–820

Kievit, J., Kuypers, H. G. J. M. 1977. Organization of the thalamo-cortical connexions to the frontal lobe in the rhesus monkey. Exp. Brain Res. 29: 299–322

Kojima, S., Goldman-Rakic, P. S. 1982. Delay-related activity of prefrontal cortical neurons in rhesus monkeys performing delayed response. Brain Res. 248: 43–49

Kubota, K., Niki, H. 1971. Prefrontal cortical unit activity and delayed cortical unit activity and delayed alternation performance in monkeys. J. Neurophysiol. 34: 337–47

Lynch, J. C., Mountcastle, V. B., Talbot, W. H., Yin, T. C. T. 1977. Parietal lobe mechanisms of directed visual attention. J. Neurophysiol. 40: 362–89

Markowitsch, H. J., Emmans, D., Irle, E., Streicher, M., Preilowski, B. 1985. Cortical and subcortical afferent connections of the primate's temporal pole: A study of rhesus monkeys, squirrel monkeys, and marmosets. *J. Comp. Neurol.* 242: 425–58

Maunsell, J. H. R., Van Essen, D. C. 1983. The connections of the middle temporal visual area (MT) and their relationship to a cortical hierarchy in the macaque monkey. *J. Neurosci.* 3: 2563–86

Mesulam, M.-M. 1986. Frontal cortex and behavior. *Ann. Neurol.* 19: 320–24

Mesulam, M.-M., Van Hoesen, G. W., Pandya, D. N., Geschwind, N. 1977. Limbic and sensory connections of the inferior parietal lobule (area PG) in the rhesus monkey: A study with a new method for horseradish peroxidase histochemistry. *Brain Res.* 136: 393–414

Milner, B., Petrides, M., Smith, M. L. 1985. Frontal lobes and the temporal organization of memory. *Human Neurobiol.* 4: 137–42

Mishkin, M. 1964. Perseveration of central sets after frontal lesions in monkeys. In *The Frontal Granular Cortex and Behavior*, ed. J. M. Warren, K. Akert, pp. 219–41. New York: McGraw-Hill

Mishkin, M., Manning, F. J. 1978. Nonspatial memory after selective prefrontal lesions in monkeys. *Brain Res.* 143: 313–23

Moran, M. A., Mufson, E. J., Mesulam, M.-M. 1987. Neural inputs into the temporopolar cortex of the rhesus monkey. *J. Comp. Neurol.* 256: 88–103

Mountcastle, V. B., Motter, B. C., Steinmetz, M. A., Duffy, C. J. 1984. Looking and seeing: The visual functions on the parietal lobe. In *Dynamic Aspects of Neocortical Function*, ed. G. M. Edelman, W. E. Gall, W. M. Cowan, pp. 159–93. New York: Wiley

Nauta, W. J. H. 1971. The problem of the frontal lobe: A reinterpretation. *J. Psychiat. Res.* 8: 167–87

Niki, H., Sakai, M., Kubota, K. 1972. Delayed alternation performance and unit activity of the caudate head and medial orbito-frontal gyrus in the monkey. *Brain Res.* 38: 343–53

Niki, H., Watanabe, M. 1976. Prefrontal unit activity and delayed response: Relation to cue location versus direction of response. *Brain Res.* 105: 78–88

Pandya, D. N., Kuypers, H. G. J. M. 1969. Cortico-cortical connections in the rhesus monkey. *Brain Res.* 13: 13–36

Pandya, D. N., Seltzer, B. 1982a. Intrinsic connections and architectonics of posterior parietal cortex in the rhesus monkey. *J. Comp. Neurol.* 204: 196–210

Pandya, D. N., Seltzer, B. 1982b. Association areas of the cerebral cortex. *Trends Neurosci.* 5: 386–90

Passingham, R. 1975. Delayed matching after selective prefrontal lesions in monkeys (*Macaca mulatta*). *Brain Res.* 92: 89–102

Passingham, R. E. 1972. Visual discrimination learning after selective prefrontal ablations in monkeys (*Macaca mulatta*). *Neuropsychologia* 10: 27–39

Rakic, P., Bourgeois, J.-P., Zecevic, N., Eckenhoff, M. F., Goldman-Rakic, P. G. 1986. Concurrent overproduction of synapses in diverse regions of the primate cerebral cortex. *Science* 232: 232–35

Richardson, R. T., DeLong, M. R. 1986. Nucleus basalis of Meynert neuronal activity during a delayed response task in monkey. *Brain Res.* 399: 364–68

Robinson, C. J., Burton, H. 1980. The organization of somatosensory receptive fields in cortical areas 7b, retroinsular, postauditory and granular insula of M. fascicularis. *J. Comp. Neurol.* 192: 69–92

Robinson, D. L., Goldberg, M. E., Stanton, G. B. 1978. Parietal association cortex in the primate: Sensory mechanisms and behavioral modulations. *J. Neurophysiol.* 41: 910–32

Roland, P. E., Friberg, L. 1985. Localization of cortical areas activated by thinking. *J. Neurophysiol.* 53: 1219–43

Rosenkilde, C. E., Bauer, R. H., Fuster, J. M. 1981. Single cell activity in ventral prefrontal cortex of behaving monkeys. *Brain Res.* 209: 275–94

Rosenkilde, K. E. 1979. Functional heterogeneity of the prefrontal cortex in the monkey: A review. *Behav. Neurol. Biol.* 25: 301–45

Schwartz, M. L., Goldman-Rakic, P. S. 1982. Single cortical neurones have axon collaterals to ipsilateral and contralateral cortex in fetal and adult primates. *Nature* 299: 154–56

Selemon, L. D., Goldman-Rakic, P. S. 1985. Common cortical and subcortical target areas of the dorsolateral prefrontal and posterior parietal cortices in the rhesus monkey. *Soc. Neurosci. Abstr.* 11: 323

Selemon, L. D., Goldman-Rakic, P. S. 1988. Common cortical and subcortical target areas of the prefrontal posterior cortices in the Rhesus monkey: A double anterograde label study of distributed neural networks. Submitted for publication

Seltzer, B., Pandya, D. N. 1980. Converging visual and somatic sensory cortical input to the intraparietal sulcus of the rhesus monkey. *Brain Res.* 192: 339–51

Seltzer, B., Pandya, D. N. 1986. Posterior parietal projections to the intraparietal sulcus of the rhesus monkey. *Exp. Brain Res.* 62: 459–69

Shallice, T. 1982. Specific impairments in

planning. *Philos. Trans. R. Soc. London Ser. B* 298: 199–209

Shipp, S., Zeki, S. 1985. Segregation of pathways leading from area V2 to areas V4 and V5 of macaque monkey visual cortex. *Nature* 315: 322–25

Stanton, G. B., Cruce, W. L. R., Goldberg, M. E., Robinson, D. L. 1977. Some ipsilateral projections to area PF and PG of the inferior parietal lobule in monkeys. *Neurosci. Lett.* 6: 243–50

Stuss, D. T., Benson, D. F. 1984. Neuropsychological studies of the frontal lobes. *Psych. Bull.* 95: 3–28

Trojanowski, J. Q., Jacobson, S. 1974. Medial pulvinar afferents to frontal eye fields in rhesus monkey demonstrated by horseradish peroxidase. *Brain Res.* 80: 395–411

Ungerleider, L., Desimone, R. 1986. Cortical connections of visual area MT in the macaque. *J. Comp. Neurol.* 248: 190–222

Vogt, B. A., Rosene, D. L., Pandya, D. N. 1971. Thalamic and cortical afferents differentiate anterior from posterior cingulate cortex in the monkey. *Science* 204: 205–7

Watanabe, T., Niki, H. 1985. Hippocampal unit activity and delayed response in the monkey. *Brain Res.* 325: 241–54

References added in proof:

Alexander, G. E. 1982. Functional development of frontal association cortex in monkeys: Behavioral and electrophysiological studies. *Neurosci. Res. Program Bull.* 20: 471–78

Friedman, H., Janas, J., Goldman-Rakic, P. S. 1987. Metabolic activity in the thalamus and mammillary bodies of the monkey during spatial memory performance. *Soc. Neurosci. Abstr.* 13: 207

Ann. Rev. Neurosci. 1988. 11 : 157–98

mRNA IN THE MAMMALIAN CENTRAL NERVOUS SYSTEM

J. Gregor Sutcliffe

Department of Molecular Biology, Research Institute of Scripps Clinic, La Jolla, California 92037

INTRODUCTION

Scientists have studied brain molecules for decades, but it is only in the last few years that a consensus field of molecular neurobiology has emerged (cf *Cold Spring Harbor Symposium* 1983). A significant aspect of molecular neurobiology is tracking the flow of information from genes into proteins, and that is the topic of this article.

Tremendous progress has been made in understanding the structures and biological properties of macromolecules. This is largely due to technical advances brought about by molecular biologists studying bacterial phage and animal viruses. These advances, which include the following, have made it possible to investigate the structures of genes, RNA transcripts, and proteins with a rapidity and precision far beyond previous technical know-how: the ability to isolate and amplify fragments of interesting DNA by fusing them with bacterial replicons such as plasmids or phage (DNA cloning); the availability of restriction endonucleases for sequence-specific DNA cleavage and hence controlled handling and analysis of cloned DNA fragments; the use of the retroviral enzyme reverse transcriptase for copying RNA molecules into complementary DNA (cDNA),[1] which is extremely stable compared to chemically labile RNA,

---

[1] Abbreviations used: AChR, acetylcholine receptor; CAMs, cell adhesion molecules; cDNA, complementary DNA; CNS, central nervous system; GAD, glutamic acid decarboxylase; GAP, GnRH-associated peptide; GFAP, glial fibrillary acidic protein; GnRH, gonadotropin-releasing hormone; hnRNA, heterogeneous nuclear RNA; MAG, myelin-associated glycoprotein; MAPs, microtubule-associated proteins; MBP, myelin basic protein; mRNA, messenger RNA; N-CAM, neural cell adhesion molecule; NF, neurofilament; NGF, nerve growth factor; NSE, neuron-specific enolase; ORF, open reading frame, protein-coding region of an mRNA; PIF, prolactin-inhibiting factor; PLP, proteolipid protein; PNS, peripheral nervous system; PPT, preprotachykinin; TH, tyrosine hydroxylase; TRH, thyrotropin-releasing hormone; UT, untranslated region.

157

0147–006X/88/0301–0157$02.00

hence simplifying the analysis of RNA structures; the development of reproducible gel electrophoresis systems that have single nucleotide-resolving power over chain lengths of from a few to several hundred nucleotides and other gel systems that (with lower resolution) can separate molecules several thousands of nucleotides in length; the invention of rapid techniques for DNA sequence determination such that extremely accurate sequences of many thousands of nucleotides can now be obtained in a few months; the refinement of nucleic acid hybridization techniques such that particular nucleotide sequences can be located in vitro within DNA molecules (Southern blotting) or RNA molecules (Northern blotting) or in situ on chromosomes or in whole tissue slices; the automation of chemical synthetic methods for producing oligonucleotides and oligopeptides; the use of chemically and bacterially synthesized peptide fragments of proteins to raise polyclonal and monoclonal antibody reagents for protein identification and isolation; the promise of methods still in their technical infancy such as site-directed mutagenesis of cloned nucleotide sequences and the reintroduction of cloned genes into (transgenic) animals. These modern tools have also, to some extent, freed the molecular biologist from the previous absolute necessity of working on simple model systems: Hybridization and antibody probes allow single nucleic acid or protein species to be examined in complex mixtures or in situ. Thus reductionistic strategies for examining tissues even as complicated as vertebrate brain are feasible.

To be sure, much successful molecular brain research preceded the advent of these modern techniques. In particular, in the closely allied field of endocrinology, molecular studies have been dramatically successful in characterizing hormone and receptor molecules. Molecular endocrinologists have made many of the most important contributions to the development of recombinant DNA-based analysis of vertebrate physiology: The cloning and structural characterization of mRNAs for insulin (Ullrich et al 1977, Villa-Komaroff et al 1978), growth hormone (Seeburg et al 1977), somatostatin (Hobart et al 1980, Goodman et al 1980), and proopiomelanocortin (Nakanishi et al 1979) captured the scientific and public attention they deserved and proved the value of the modern molecular tools for structural analysis.

THE BRAIN IS MADE OF PROTEINS

Moore & McGregor (1965) recognized the importance of knowing the constituent molecules—the hardware—of the brain in order to understand its functioning. They set out by protein fractionation to identify molecules unique to the brain and were able to isolate a few abundant species, including a glial protein, called S100 because of its solubility in 100% ammonium sulphate (Moore 1965), and a neuronal protein, originally termed 14-3-2 because of its electrophoretic and chromatographic properties (Moore & Perez 1966) and later identified as a neuron-specific enolase isozyme. We discuss recent molecular studies on S100 and neuron-specific enolase below. This general strategy for identifying brain-specific molecules has never been widely applied for several reasons. Most abundant brain proteins are also shared with other tissues. Thus selecting individual, low abundance proteins for study may require several serial fractionations of brain and control tissues to identify novel candidate protein species, usually as bands or dots on a gel, or peaks from a gradient. Further studies require purifying each protein essentially to homogeneity and raising an antiserum in order to establish brain specificity rigorously, since that would initially be the most salient characteristic of the novel species. In such studies, little can be learned until a pure protein fraction is in hand, and further characterizations can be extremely slow and tedious, partly because of the technical difficulties of protein chemistry and partly because one starts out with neither functional nor structural information about each candidate protein.

With the discovery and application of monoclonal antibody technology, a way to circumvent some of the shortcomings of the tedious protein purification approach was found. As reviewed in this series (McKay 1983, Valentino et al 1985), panels of monoclonal antibodies raised against crude neural protein preparations can be rapidly screened to find antibodies that show the desired neural specificity. Because each monoclonal antibody in a panel is elicited in response to a different single component in the initially complex immunogen, the process shortcuts the necessity for purification of individual proteins to raise antisera: Cloning the hybridoma lines accomplishes the same result. This approach has resulted in many dramatic successes in revealing molecules unique to single cells or small subsets of cells within the nervous systems of primitive and complex animals (reviewed by McKay 1983, Valentino et al 1985). Such studies have led to important conclusions about the diversity of protein expression in neurons, the existence of previously unsuspected relationships among neurons of different systems within the CNS, the discovery of molecules with interesting distributions, and the subcellular localization of brain-specific antigens.

Some monoclonal antisera have been excellent reagents for early characterizations of neural proteins. They can be used to demonstrate the anatomic and subcellular distribution of the molecule and also may be used to identify it in biochemical assays. Eventually the structure of the protein must be elucidated. The most extensive pursuit to date of a neural protein discovered via monoclonal antibodies to a crude starting antigen is that of Zipursky, Benzer and colleagues (1984, 1985). A monoclonal antibody that reacted exclusively with structures within photoreceptor cells of the retina of the fruit fly *Drosophila melanogaster* was used to describe the anatomical and subcellular localization of the protein Ag24B10 during ontogeny, then was further used to purify this 160 kDa glycoprotein by immunoaffinity (Zipursky et al 1984). The purified retinal protein was subjected to partial amino acid sequence analysis and the resulting sequence data were used by reverse translation of the genetic code to design an oligonucleotide putatively complementary to the messenger RNA (mRNA) for protein Ag24B10. The complementary oligonucleotide was used as a probe to isolate a clone of the gene for Ag24B10, and nucleotide sequence analysis provided the structure of the 5' end of the gene (Zipursky et al 1985). The genomic clone was used to identify, by blot hybridization, a 4300 nucleotide mRNA found in the head but not the body of the fly, and the clone was also used to map the Ag24B10 gene to chromosome band 100B at the distal end of chromosome 3. Further experiments are necessary to obtain cDNA clones of the mRNA for Ag24B10 and to learn its structure, and hence to learn the protein sequence of the 160 kDa glycoprotein Ag24B10. That sequence can then be compared to the sequences of other proteins to obtain functional insights. The information about the location and structure of the gene will allow the performance of genetic experiments that should lead to an understanding of the function of this retina-specific protein.

PROTEINS ARE ENCODED BY mRNAS

In the present era of technology, the structure of an mRNA is the currency of experiment. With a small number of known exceptions (such as the genes for immunoglobulin in B-cells, and T-cell receptor in T-cells), the primary structures of genes are thought to be the same in cells of all types. In higher eukaryotic organisms, genes, located on chromosomes in the cell nucleus, encode proteins in discontinuous segments (called exons), which are interrupted by noncoding regions (called introns) (reviewed by Breathnach & Chambon 1981). In each appropriate cell, a gene (Figure 1) is transcribed by the enzyme RNA polymerase II, generating a single stranded RNA molecule called a heterogeneous nuclear RNA (hnRNA). The hnRNA molecule is a complete copy of the gene sequence and still carries noncoding introns that interrupt coding exons. The hnRNA molecule has a 5'-cap structure that is added soon after transcription is initiated, and ends with a 3' run of polyadenylate 100–200 residues in length (reviewed by Darnell 1982). This 3' poly(A) tail is not encoded in

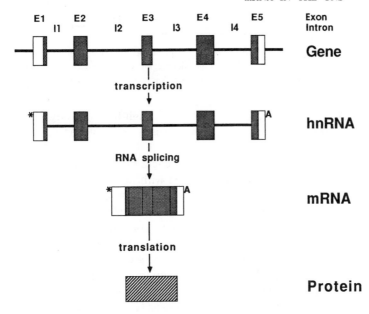

Figure 1   The central dogma of molecular biology. Genes, made up of exons (E, *boxes*) and introns (I, *solid lines*), are transcribed by RNA polymerase II to produce a single stranded heterogeneous nuclear RNA (hnRNA) copy, to which a 5′ cap (*) and a 3′ tract of poly-adenylate (A) are added. RNA splicing removes the introns, fusing the exons to produce messenger RNA (mRNA), which still carries the cap and poly(A) tract and is translated cytoplasmically into protein. The protein-coding region (also known as the ORF) is *darkly shaded* in the boxes of the gene, hnRNA, and mRNA.

the genome, but rather is added post-transcriptionally. The process of RNA splicing removes the noncoding intron regions and fuses the exons to generate mRNA, which is translocated from the nucleus to the cytoplasm. In the cytoplasm, the mRNA molecule with its 5′ cap and 3′ poly(A) tail (shown schematically in Figure 2) is translated on ribosomes to generate a protein whose amino acid sequence is informationally equivalent to the coding region of the mRNA. The protein may be post-translationally modified by proteolytic cleavage, glycosylation, phosphorylation, sulfation, acylation, or other reactions. It then goes about its functional business in the cell (or outside the cell if it is secreted) until it is inactivated, usually by proteolytic degradation. Proteins constitute, or produce via catalysis, all the molecules of the cell and hence determine its structural properties and ultimately allow its function. The generation of mRNA is the critical intermediate step in producing the cellular protein hardware from the original DNA blueprint.

With present techniques, the isolation and structural characterization

Figure 2 Anatomy of an mRNA molecule. The protein-coding open reading frame (ORF) is preceded by a 5′ untranslated (UT), or noncoding region, and followed by a 3′ UT to which a poly(A) tract is attached. The 5′ end of the mRNA is capped with 7-methyl guanine (mG).

of cDNA copies of mRNAs is generally straightforward, although non-trivial. Thus mRNAs are now the most easily definable macromolecular entities. They are also the most easily decoded informational form of the organism's hardware. For example, the edges of genes are hard to define and their coding regions are interrupted with introns; hence a gene may be undecipherable to the scientist, even if its entire nucleotide sequence is known. In contrast to genes, mRNAs have definable 5′ and 3′ ends, and the sequences of their coding regions exactly correspond to the amino acid sequences of their primary translation products. Also, because of the relative ease and high accuracy of nucleotide sequence analysis compared to protein sequence analysis, most new amino acid sequences come from deciphering nucleotide sequences. Because they are easily definable, mRNAs may represent the best indicators as to whether a particular gene is expressed.

The characterization of the structures of mRNA molecules is presently one of the major endeavors of molecular biology. This is partly because mRNA structures can tell us about protein structures and hence the cells that contain the proteins, and partly because very little is known about the precise nature of the genetic control mechanisms that govern cell type–specific transcription. Thus by characterizing the products of transcription exclusive to a particular tissue, much may be learned about the tissue-specific processes that generate them. In this article I review some of what has been learned about the structures of mRNA molecules expressed in the central nervous system (CNS) and the processes that generate them. The review's scope is limited to CNS mRNAs from mammals with few exceptions. A more complete overview of modern molecular neurobiology would also incorporate molecular studies of nematodes, *Aplysia* (recently reviewed in this series by Kaldany et al 1985), leeches, and *Drosophila*, and would overflow more into endocrinology.

GENERAL STRUCTURE OF AN mRNA

Let us consider in general terms the structure of an individual brain mRNA molecule (Figure 2). It has a 5′ end at which it is capped, and extends for hundreds or thousands of nucleotides until a 3′-terminal region of poly(A) is reached. The 5′ and 3′ ends of the mRNAs correspond to regions on a chromosomal gene. Thus, there must be sequences within the genes that dictate in brain cells exactly where RNA polymerase II initiates and terminates transcription, as well as sequence details that determine at what rate initiation occurs and in which particular brain cells. The structure of the mRNA molecule resembles the structure of the gene except that the mRNA has precisely definable ends whereas the gene does not, and the intronic regions of the gene are absent from the mRNA. Within the nucleotide sequence of a brain mRNA molecule is a contiguous region that encodes a brain protein. At initial stages of analysis the relationship between an mRNA and its protein product is a putative one. The protein-coding region of an mRNA is initially recognized as a so-called open reading frame or ORF: a region that contains a suitable translational initiator codon (AUG for methionine), followed by an uninterrupted series of triplet codons whose cognate amino acids can be translated via the genetic code into a protein sequence, and ending with one of the three triplets (UGA, UAG, UAA) that specify translational termination. An mRNA has more "personality" than simply encoding a protein. It contains both 5′ and 3′ noncoding regions, referred to here as untranslated regions. These regions, either because of primary sequence or because of the way they affect the solution structure of the mRNA, may influence the rate at which it is translated, its stability and, perhaps, in what subcellular compartment translation occurs.

Thus, in the analysis of an mRNA structure, one can learn about the 5′ end, the 5′ untranslated region, the ORF (or coding region), the 3′ untranslated region, and the 3′ end. In the abstract, such information, although not trivial to establish, seems interminably dull. Its realization has been far from that. Scrutiny of mRNA structures has revealed that the 5′ ends may vary in mRNAs transcribed from single genes; that 5′ untranslated regions, which are thought to direct translational initiation, vary greatly in sequence characteristics and length and thus might carry different translational regulatory information; that coding regions mature in alternative ways from their hnRNA precursors, thus leading to ORFs with different protein-coding capacity coming from a single gene; that 3′ untranslated regions can vary greatly in length; that 3′ end formation can vary in different transcripts from the same gene (Leff et al 1986). Not least, mRNA sequences provide protein sequences, and these by their very nature have led to insights about protein function.

A set of questions about brain mRNAs is thus easy to formulate. How many mRNAs are expressed in the brain? How many in other tissues? How many in common? What is known about 5′ ends, 3′ ends, mRNA length? Is alternative splicing of hnRNA precursors a rare or frequent strategy for brain gene expression? Is the generation of mRNA the primary controlling step in brain gene expression? These points are considered in the discussions that follow.

MANY mRNAS ARE EXPRESSED IN THE BRAIN

It seems obvious that the brain is a highly complex organ in terms of function and anatomy, and therefore that there must be many genes dedicated to giving this organ its unique properties. Analysis backs up this natural assertion. The brain is molecularly a very complex entity with about one third of the mammalian genome exclusively dedicated to its function (Bantle & Hahn 1976, Chikaraishi et al 1978). However, among mammals, brain molecular complexity does not vary greatly (Kaplan & Finch 1982).

How many genes does an organ or tissue express? This question classically has been addressed by isolating the mRNA from that tissue and then measuring its hybridization parameters, either the extent of hybridization to an excess of genomic DNA or the kinetics of hybridization to reverse-transcribed copies of the mRNA. (Since hybridization is a bimolecular reaction, the rate at which a known concentration of nucleic acids forms hybrids is a function of the number of distinct molecular species in the reaction mixture.) The answers from either method come out in numbers of nucleotides expressed in a particular tissue. For brain mRNA, the answer varies from study to study (reviewed by Milner et al 1987), the more recent measurements ranging from 10^8 to 2×10^8 nucleotides of mature polyadenylated cytoplasmic RNA (Bantle & Hahn 1976, Grouse et al 1978, Chikaraishi 1979, Van Ness et al 1979, Colman et al 1980, Beckmann et al 1981, Chaudhari & Hahn 1983) transcribed from 32–42% of the rodent genome (Bantle & Hahn 1976, Chikaraishi et al 1978). Within a study, the amount of RNA sequences expressed in non-neuronal tissues such as liver or kidney is always two to three fold lower than the amount of RNA expressed in brain. Most of the mRNA exclusive to brain is present at low individual abundance.

In order to calculate from the amount of RNA expressed in a tissue the number of genes expressed there, the lengths of the mRNA molecules must be determined. Milner & Sutcliffe (1983) estimated brain mRNA length by isolating almost 200 random cDNA clones of individual brain mRNAs and measuring the size, abundance, and tissue distribution of each mRNA by Northern blot analysis. It was found that brain-specific mRNAs of the low-abundance class, the class that makes up most of the mRNA mass, have average lengths of 5000 nucleotides or more. Thus it could be calculated that the $1–2 \times 10^8$ nucleotides of brain mRNA are accounted for by approximately 30,000 distinct mRNA species of an average length of

5000 nucleotides. It was found by kinetic analyses that about 65% of brain mRNA molecules were not shared with other tissues (Chaudhari & Hahn 1983) and by clonal analysis that between 40–56% of the mass of brain mRNA was brain specific (Milner & Sutcliffe 1983). Thus, of the estimated 30,000 brain mRNAs, roughly 20,000 are brain specific, at least in the sense that they are not expressed in detectable amounts in liver and kidney, the two tissues routinely used as nonneural controls.

The above discussion is based on mRNA molecules with 3′ poly(A) tails. It was observed by Chikaraishi (1979) and verified by Van Ness and colleagues (1979) that, as isolated, a large fraction of the nucleotide complexity of brain cytoplasmic ribosome-associated RNA lacks a poly(A) tail. The cytoplasmic RNAs of other tissues do not contain much so-called poly(A)$^-$ RNA. Chaudhari & Hahn (1983) showed that most of the brain poly(A)$^-$ RNA has postnatal developmental onset. These RNAs, which are individually rare but comprise up to 1.7×10^8 nucleotides, have little overlap with the poly(A)$^+$ RNA population. While there are precedents for some mRNAs, such as those for histones (Adesnik & Darnell 1972), lacking a poly(A) tail, the brain may have many more such species. Alternatively, the extreme length of rarer brain mRNAs (Milner & Sutcliffe 1983) may make it technically difficult to isolate intact copies of many rare brain mRNAs, hence their 5′ fragments might appear to be poly(A)$^-$ and behave in kinetic analyses as though there was little overlap with the poly(A)$^+$ population. Such 5′ fragments of long poly(A)$^+$ mRNAs might also be ribosome-associated and serve as templates for translation in in vitro protein synthesis systems. The possibility of poly(A)$^-$ RNAs encoding a substantial portion of the postnatal onset brain mRNAs remains a controversial one, and its resolution requires the demonstration that several individual bonafide poly(A)$^-$ RNA species encode postnatal onset proteins. To date, a few clones of short, randomly selected genomic fragments that hybridize to rare, apparently brain-specific RNAs found predominantly in the poly(A)$^-$ fraction have been isolated, but analysis has not yet established that these RNAs encode proteins or that they are representative of the bulk of the poly(A)$^-$ species (Brilliant et al 1984).

Even if a large class of truly poly(A)$^-$ RNAs were eventually shown not to exist, but rather were largely explained by the difficulty of obtaining unbroken, very long mRNAs, it has been shown that there is a significant amount (1.1 to 1.7×10^8 nucleotides) of brain RNA (Chikaraishi 1979, Chaudhari & Hahn 1983) that has postnatal onset. A limited clonal analysis has shown that 30–40% of the brain-specific poly(A)$^+$ mRNAs first appear postnatally (Sutcliffe et al 1986). It was previously argued (Milner & Sutcliffe 1983) that long mRNAs might be products of recently evolved genes that had not yet become streamlined. Thus if poly(A)$^-$ RNAs represent 5′ ends of long broken poly(A)⁺ molecules, it may well be that ontogeny (the postnatal developmental onset of these long species) is again recapitulating phylogeny, with the most recently evolved genes becoming active in late brain development.

The structure of the brain is obviously complex, as any reader of this series is aware. Complexity is evident anatonomically, cellularly, functionally, and chemically: Virtually every technique that has been used to examine the brain, from the unaided eye to the electron microscope, has revealed regional heterogeneity. Are there regionally specific mRNAs to direct the synthesis of region-specific proteins? This question has only recently been addressed. Previous RNA complexity analyses have shown there were modest (5–10%) RNA complexity differences between various brain regions (Kaplan & Finch 1982), but with one exception these differences were at the limits of experimental error. The exception was that the cerebellum consistently exhibited considerably less complexity than other regions. Clonal analyses in human (Wood et al 1986), monkey, and rat (Travis et al 1987) have revealed that there are very few if any relatively abundant (0.1% or greater) region-specific mRNAs, but that several lower abundance class mRNAs (cumulatively totalling less than 1% of the regional RNA) do show profound regional specialization (Travis et al 1987). Thus some region-specific proteins can account for some of the functional differences in brain regions, but these are only a small subset of the total brain proteins. The conceptual property of "neuron" seems dominant at the molecular level to that of "region." Present hybridization analyses of cloned molecules, however, do not detect the rarest mRNA species, those which contribute most to the molecular diversity. Hence, as some of the conclusions drawn here are based on studies of small numbers of the relatively more abundant mRNAs, and this should be kept in mind when generalizing to the total RNA population.

KNOWN MAMMALIAN mRNA STRUCTURES

The analysis of the mRNA for a particular protein begins with the isolation of a cDNA clone. The clone may then be used in blot hybridization experiments to learn about the size and abundance of the mRNA as well as its anatomical and developmental distribution. Ultimately, the nucleotide sequence of a cDNA clone that is a full-length copy of the mRNA, or of a series of clones that cumulatively cover the entire mRNA, is required to solve the mRNA structure.

Let us consider a number of mammalian brain mRNAs for which most or all of the structure is known. The list is not exhaustive because of time and space limitations. Any omissions are unintentional as no selection criteria other than mammalian and CNS were imposed, and I apologize to workers whose contributions I may have inadvertently neglected in my literature search. Our present interest is mostly structural, so often some of the important biological discoveries in the studies reviewed are overlooked. The structural data are reviewed tersely (and compiled in Table 1); a discussion of general principles that might be inferred from the collection follows. Within this review I use the convention that the ORF excludes the translational termination triplet. Although the terminator is certainly part of the coding information, its exclusion allows the simple inter-conversion between ORF length and amino acid chain length.

mRNAs for Myelin Proteins

Myelin, the multilamellar membrane that wraps around axons and acts as an insulator facilitating transmission of nerve impulses, is one of the most obvious specific structural features of vertebrate nervous tissues. Purified myelin contains only a few abundant proteins. In the myelin of the CNS, myelin basic protein (MBP) and myelin proteolipid protein (PLP) are the major protein components. MBP is associated with the cytoplasmic face of the myelin membrane and is thought to mediate myelin compaction, especially the interactions between the adjacent cytoplasmic membrane surfaces. PLP is an integral membrane protein thought to mediate inter-actions between opposing extracellular membrane surfaces. In the myelin of Schwann cells in the peripheral nervous system (PNS), P_0 protein is the major constituent; MBP is also expressed, although at lower levels than in the CNS. P_0 is thought to be involved in linking adjacent extracellular membrane surfaces during compaction.

The structures of the mRNAs for rat and mouse MBP (see Figure 3) have been worked out in detail by analysis of cDNA and genomic clones (Roach et al 1983, Zeller et al 1984, Takahashi et al 1985, de Ferra et al 1985, Kimura et al 1985). The genetic locus for mouse MBP is *shiverer* (*shi*) located on chromosome 18, as the MBP gene is partially deleted in the *shi* haplotype (Roach et al 1983, Kimura et al 1985, Roach et al 1985, Molineaux et al 1986). The single 32 kb MBP gene is made up of seven exons (Takahashi et al 1985, de Ferra et al 1985). Transcription begins during the first week of postnatal development (Zeller et al 1984) from a discrete promoter, as shown both by S1-nuclease protection and by primer extension studies (Takahashi et al 1985). There are four distinct MBP mRNAs produced by alternative RNA splicing, which range in size from 1920 to 2121 nucleotides, plus a 3′ poly (A) tail. These four related MBP mRNAs share exons 1, 3, 4, 5, and 7 differ from each other by the inclusion of the second or sixth exon, or neither or both, in the mature spliced mRNA (de Ferra et al 1985, Takahashi et al 1985). The four mRNAs

MBP

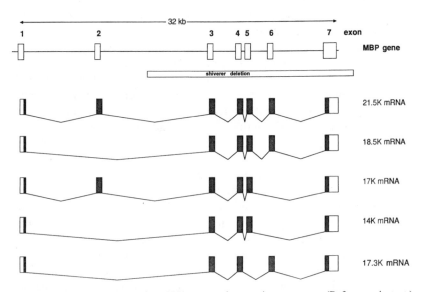

Figure 3 Myelin basic protein (MBP) gene and transcript structures. (References in text.) The sizes of the exons are exaggerated and are not necessarily proportional. The 32 kb MBP gene has 7 exons; exons 3–7 and part of the second intron are deleted in the *shiverer* haplotype. The protein-coding regions of the spliced mRNAs are *shaded*. The poly(A) tail is not indicated in the drawing. The mRNA designations refer to the sizes of the encoded proteins.

(which have 47 nucleotide 5′ untranslated regions and 1486 nucleotide 3′ untranslated regions) have ORFs of 128, 154, 169, and 195 triplets. These ORFs account for the four forms of MBP (14 kDa, 17 kDa, 18.5 kDa, 21.5 kDa) found in rodent myelin (Barbarese et al 1977). Recently, cDNA clones of a fifth MBP mRNA, which excludes exons 2 and 5 but includes exon 6 (as well as 1, 3, 4, and 7) and thus encodes a 17.3 kDa MBP, have been described for both humans and rodents (Roth et al 1986, Kamholz et al 1986, Newman et al 1987). The removal of the N-terminal methionine is the only apparent post-translational modification of MBP other than phosphorylation. Thus the single MBP gene gives rise to at least five distinct proteins that share a common amino acid sequence, including common N- and C-termini, but that differ by the inclusion or exclusion of one or more of three exons. This exon shuffling is possible because all splice sites are located between codons, hence each splice variant maintains the same ORF (Takahashi et al 1985, de Ferra et al 1985). It is thus possible that more combinations will be found.

The single gene for PLP (Milner et al 1985) is located on the X-chromosome (Willard & Riordan 1985, Dautigny et al 1986) and is the rodent gene *jimpy*, since *Jp*/y mice produce an mRNA with a 74 nucleotide deletion in the PLP coding region (Nave et al 1986), apparently because of defective RNA splicing. The PLP gene spans 15 kb and is made up (Figure 4) of seven exons (Diehl et al 1986). Transcription of the rat PLP gene begins postnatally and is coincident with the onset of myelination. Initiation occurs at either of three sites located within 75 nucleotides; polyadenylation of PLP transcripts occurs at either of three sites (Milner et al 1985). The PLP mRNAs have 5′ untranslated regions of 89, 124, or 154 nucleotides, share a common protein-encoding ORF of 277 triplets, and differ in their 3′ untranslated regions, which are either 2066, 1340, or

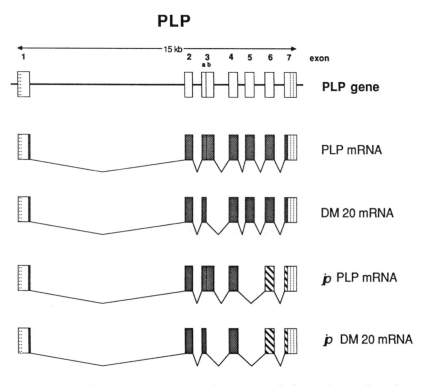

Figure 4 Proteolipid (PLP) gene and transcript structures. (References in text.) Exon sizes are exaggerated and are not proportional. The 15 kb PLP gene contains 7 exons. Heterogeneity in transcriptional initiation sites and in polyadenylation sites is indicated with *dotted lines*. Protein-coding regions of spliced exons are *shaded*. Note that exon 3 is spliced differently to produce PLP and DM20. In *jimpy* mice, exon 5 is present in the genome but not in either PLP or DM20 mRNAs. The C-terminus of the *jimpy* proteins is *shaded differently* to indicate a frame-shifted amino acid sequence.

450 nucleotides (Milner et al 1985). Thus the rat PLP mRNAs occur as two abundant families of approximately 3200 and 1600 nucleotides and a less abundant family of approximately 2400 nucleotides. Other species such as mouse and monkey have much lower levels of the 1600 nucleotide RNAs. Whether there are functional differences in the three families of RNA is unclear. A comparison of the protein-sequencing data (Lees et al 1983, Stoffel et al 1983) with the amino acid sequence deduced from the cDNA clones (Dautigny et al 1985, Naismith et al 1985, Milner et al 1985) suggests that the N-terminal methionine of PLP (but no additional residues) is removed posttranslationally, and that residue 198 (and other yet unidentified residues) is acylated to produce the 30 kDa mature protein. Thus, this intrinsic membrane protein does not utilize a signal peptide for insertion in the myelin membrane. A second myelin proteolipid called DM20 (which has faster gel mobility) is translated from an alternatively spliced PLP mRNA lacking the coding region for residues 116–151 (Nave et al 1987). Interestingly, the intron-exon structure of the human and mouse PLP genes (Diehl et al 1986, Nave et al 1987) suggests that the rarely encountered alternate selection of 5′ splice donor sites rather than an independent exon shuffling accounts for the alternate splicing. Thus, both MBP and PLP transcripts utilize alternative splicing to produce internal variants with altered coding capacities.

Transcription of the single gene for the peripheral myelin protein P_0 increases rapidly in rat sciatic nerve after birth, giving rise to a 1900–2000 nucleotide mRNA (Lemke & Axel 1985). The sequence of a nearly full-length cDNA clone of the P_0 mRNA shows it to contain at least 32 5′ untranslated nucleotides, an ORF of 247 triplets, and a 3′ untranslated region of at least 252 nucleotides (Table 1). The first 28 amino acids encoded by the ORF do not appear in the mature P_0 protein, thus they probably represent a signal peptide to direct translation to the endoplasmic reticulum (Lemke & Axel 1985). P_0 is generated by removal of the N-terminal signal to produce a 219 residue polypeptide which is glycosylated at a single site to account for the 31.5 kDa mature species. The sequence of the mature product shows that P_0 is a member of the immunoglobulin superfamily of proteins (Lai et al 1987), thus accounting for the postulated homotypic binding activity of P_0 in peripheral myelin compaction (Lemke & Axel 1985).

Components of the Cytoskeleton

The cytoskeletons of brain cells, as well as all other cell types, are composed of several major structural features: actin filaments, microtubules, and intermediate filaments. Tubulin, the major protein of microtubules, exists as two peptides, α and β tubulin, each of which exhibits brain-specific isoelectric point heterogeneity (Gozes & Littauer 1978, Denoulet et al 1982). The heterogeneity is accounted for partly by post-translational modifications and in partly by the expression of members of α and β tubulin multigene families (reviewed by Cleveland & Sullivan 1985). Generally, α and β tubulin expression is high in early brain development but decreases at later stages (Bond & Farmer 1983). Some members of the α and β tubulin gene families appear to be expressed by rats and mice in a brain-specific or highly brain-enriched fashion (Bond et al 1984, Lewis et al 1985, Miller et al 1987a,b). The mouse brain-enriched $M\alpha 1$ tubulin, expressed in brain, lung, and testes, is predominantly detected in brain during early stages of brain development (Lewis et al 1985). The apparent rat analogue $T\alpha 1$, has been shown by in situ hybridization to be expressed at elevated levels in neurons involved in process extension (Miller et al 1987b). A rat β tubulin, RBT1 (Bond et al 1984), has similar developmental expression. Several mouse β tubulin genes have enriched expression in brain, especially during early development. However, one form, $\beta 4$, appears exclusive to brain but with late developmental accumulation (Lewis et al 1985). The cDNA clones used to identify the various α and β tubulins have been sequenced and the parameters of the important brain forms are compiled in Table 1.

Polymerized microtubules contain, at lower levels than tubulin, other proteins called microtubule-associated proteins (MAPs), several of which are brain-specific or highly enriched in brain (reviewed by Olmsted 1986). MAPs are thought to mediate microtubule assembly and, since they project from polymerized microtubules, interact with cellular elements. Each of the brain MAPs is a heterogeneous collection of proteins. cDNA clones for three MAPs, MAP1, MAP2, and tau, have been isolated and their genes and mRNAs analyzed. No sequence information is yet available. The single mouse MAP1 gene gives rise to an mRNA of greater than 10 kb, whose expression is at least 500-fold higher in brain than other tissues (Lewis et al 1986a). Three forms of MAP1, each about 350 kDa, are products of the mRNA; this suggesting either alternative splicing of MAP1 hnRNA or posttranslation modifications of MAP1 protein. Even though MAP1 protein accumulates developmentally, steady state MAP1 mRNA concentration decreases, thus suggesting the possibility of translational control (Lewis et al 1986a). The single mouse gene for MAP2 gives rise with similar developmental parameters (with the same implications) to a 9 kb brain-specific mRNA (Lewis et al 1986a,b). Two 250 kDa forms of MAP2, which are found in neuronal soma and dendrites, are products of the single gene. The human MAP2 gene has been localized on chromosome 2 (Neve et al 1986). The complex tau family of 55–70 kDa proteins, found largely in axons, are products of one or more brain-specific 6 kb mRNAs transcribed from a single mouse gene (Drubin et al 1984). The mouse cDNA clone was used as a probe to isolate and map the human tau gene to chromosome 17 (Neve et al 1986).

The intermediate filament proteins are cell type–specific in vertebrates and exist in four forms in brain: glial fibrillary acidic protein (GFAP) in astrocytic glial cells, and three related neurofilament (NF) proteins in neurons of 68 kDa, 150 kDa, 200 kDa (reviewed by Lazarides 1982, Osborn & Weber 1983). The intermediate filament proteins show considerable sequence similarity and share structural features within their central rod regions, which are thought to be involved in filament formation, but each has its unique characteristics at the N- and C-termini, which are thought to provide different functionality in the various cells of specific expression (Geisler & Weber 1982, Steinert et al 1985). Partial structures of the mRNAs for GFAP, NF68, and NF150 have been described.

A single 10 kb gene for mouse GFAP gives rise to an approximately 2660 nucleotide mRNA with a 50 nucleotide 5' untranslated region and 1353 nucleotide 3' untranslated region (Lewis et al 1984, Balcarek & Cowan 1985). The protein sequence deduced from the 418 triplet ORF shows a high degree of sequence similarity (65 and 67%) with two other intermediate filament proteins, desmin and vimentin (Lewis et al 1984). The GFAP gene contains 9 exons and their placement is the same as in the genes for other intermediate filaments, a finding that suggests a common evolutionary ancestor (Balcarek & Cowan 1985).

The single mouse gene for NF68 is transcribed to generate two mRNA species of approximately 2500 and 4000 nucleotides (Lewis & Cowan 1985) whose relationship to each other has not been established. The two NF68 mRNAs are detectable in 11-day-old mouse embryos and their abundance increases until P5 (Lewis & Cowan 1985, Julien et al 1986). Sequence analysis of partial cDNA clones reveals a 5' untranslated region of at least 33 nucleotides and a 3' untranslated region of at least 343 nucleotides (Lewis & Cowan 1985, Julien et al 1986). The deduced protein sequence matches well with that of the porcine NF68 protein determined by direct amino acid sequence analysis (Geisler et al 1983). The gene for NF150 is located within 30 kb of the NF68 gene, and a partial cDNA sequence has been generated (Julien et al 1986). Transcription of the gene is developmentally coordinated with that of NF68 and gives rise to a 3 kb mRNA. A genomic clone for NF200 has been analyzed and used to show that appearance of the 4.5 kb neurofilament 200 mRNA lags developmentally compared to the other neurofilaments. The available sequences show that sequence similarities between the three neurofilaments extend beyond the central rod region, highly conserved among all intermediate filaments, into the normally hypervariable tail region; this suggests that the three neurofilament genes share a recent evolutionary antecedent (Julien et al 1986).

Molecules of Development

One of the primary processes of organogenesis and the maintenance of histological integrity is the mechanical interaction between cells mediated by cell adhesion molecules (CAMs). During early development of neural and some nonneural tissues, neural cell adhesion molecule (N-CAM) is the major CAM (Edelman 1986). At the structural level, both the chicken and mouse N-CAMs seem to have largely equivalent properties and are discussed herein together. The single gene for N-CAM (Murray et al 1984, Goridis et al 1985) has been mapped to mouse chromosome 9 (near but distinct from the *staggerer* and Thy-1 loci) and the syntenic region of human chromosome 11 (D'Eustachio et al 1985, Nguyen et al 1986). The chicken gene is larger than 50 kb and contains (Figure 5) at least 19 coding exons (Owens et al 1987). The first 14 exons are shared between the three known mRNAs. Alternative splicing produces three forms of mRNA (see Figure 5), with ORFs encoding proteins of 120 kDa, 140 kDa, and 170 kDa (Murray et al 1984, Goridis et al 1985, Hemperly et al 1986a,b, Owens et al 1987). Although complete 5′ sequences have not yet been reported, the three ORF products are known to share N-terminal regions (encoded by exons 1–14) that have five repeated domains resembling immunoglobulin family members and differ in their respective C-termini. The Ig-like domains are likely to be involved in the ligand interactions of cell adhesion. Each N-CAM form has a large amount of carbohydrate added post-translationally. The shortest N-CAM protein, whose mRNA appears late in development, ends in a 25 amino acid region, distinct from the larger forms, through which it is attached via the lipid phosphatidylinositol to the membrane (Hemperly et al 1986b). The two larger forms, present from early embryonic stages, share a membrane-spanning region but differ in their C-terminal tails, which are thought to modulate the activities of the N-termini and, in this regard, may be phosphorylation substrates (Hemperly et al 1986a). The largest of the three alternatively spliced RNAs is exclusive to neural tissues (Murray et al 1986). N-CAM cDNA clones also hybridize with smaller RNAs whose structures are not yet known (Goridis et al 1985, Hemperly et al 1986b, Owens et al 1987).

A second CAM, L1/Ng-CAM, is involved in early neural interactions. During later neural cell interactions, N-CAM and L1/Ng-CAM appear to be superceded by yet a third CAM, myelin-associated glycoprotein (MAG) (Quarles 1984, Martini & Schachner 1986). MAG and brain protein 1B236 (Sutcliffe et al 1983) have recently been shown to be indistinguishable

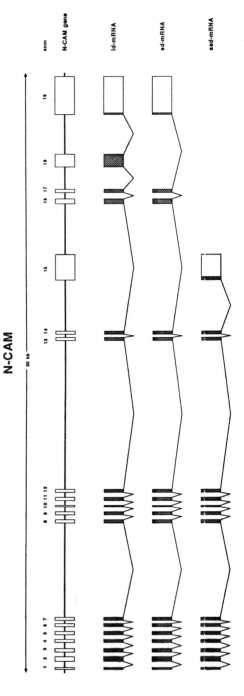

Figure 5   The neural cell adhesion molecule (N-CAM) gene and transcript structures. (References in text.) The 50 kb N-CAM gene is composed of 19 exons, which are shown approximately proportionally but are exaggerated in size relative to the introns. Protein-coding regions of the spliced exons are *shaded*. The 5'-end of the gene has not yet been mapped. The nomenclature for the three alternate splice forms are those of Owens et al (1987).

products of the same single 15 kb rat gene (Lai et al 1987, Arquint et al 1987). Transcription of 1B236/MAG (see Figure 6) begins postnatally in the rat brain from a GC-rich promoter at a series of at least seven sites spaced across 50 nucleotides. Alternative RNA splicing produces from the 13-exon gene several distinct mRNAs of about 2500 nucleotides that differ from one another in both coding and 5′ untranslated regions (Lai et al 1987). At least two sites are used for polyadenylation. During early postnatal rat brain development, 1B236/MAG RNAs are detected by in situ hybridization primarily in oligodendrocytes within white matter areas; in adult rats 1B236/MAG is expressed at lower levels but is more prevalent in neurons in grey matter regions than in white matter (Higgins et al 1987). The 626 triplet ORF found in an early developmental form shows that the 1B236/MAG protein has an N-terminal signal peptide followed by a large, glycosylated N-terminal region, a membrane-spanning domain, and a C-terminal tail of 92 residues: the mature protein is 100 kDa (Lai et al 1987, Arquint et al 1987). The N-terminal region is composed of five domains related in sequence to each other and to immunoglobulin-like molecules, and is especially related to N-CAM. This suggests that 1B236/MAG and N-CAM make similar sorts of interactions during cell adhesion processes (Lai et al 1987). A developmentally later form is an alternatively spliced 1B236/MAG mRNA encoding a molecule of 582 residues with a shorter C-terminal tail. The two C-terminal tails, which are probably cytoplasmic, undergo proteolytic cleavage (Malfroy et al 1985) and are possibly phosphorylation substrates (Arquint et al 1987), either of which might modulate the activity of the protein. As 1B236/MAG is present in both neurons and oligodendrocytes (Bloom et al 1985, Higgins et al 1987), it may homotypically mediate interactions between these cell types as well as interactions between glial cells. The 1B236/MAG gene has been shown by restriction fragment length polymorphism analysis to be within 1 centimorgan of *Gpi-1* on mouse chromosome 7, and thus is tightly linked to the neurological defect *quivering* (Blatt et al 1987).

Trophic factors are required for neuronal growth. Nerve growth factor (NGF) is a molecule made of three polypeptide chains that is necessary in the PNS both for neuron survival and for growth and maintenance of fibers. The β chain is the active constituent. NGF is secreted by the target tissue, interacts with a neuronal receptor, and is retrogradely transported to the nerve cell bodies where it serves in an unknown manner, possibly through a cascade involving the products of the cellular homologues of the *myc*, *fos*, *src*, and *ras* oncogenes, to induce the production of enzymes for catecholamine and polyamine synthesis. NGF is also present in the CNS and may act there similarly to how it acts trophically in the PNS to induce synthesis of neuronal products. β-NGF cDNA clones have been

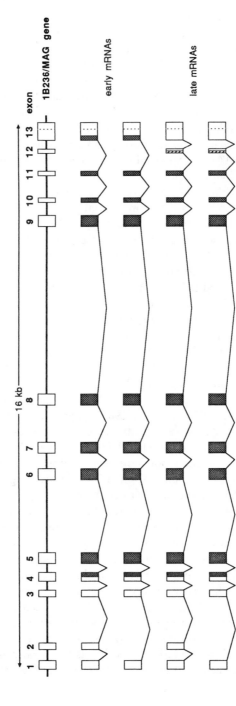

Figure 6 The 1B236/MAG (myelin-associated glycoprotein) gene and transcript structures. (References in text.) The 16 kb 1B236/MAG gene is composed of 13 exons, which are shown approximately proportionally but are exaggerated in size relative to the introns. Protein-coding regions of the spliced exons are *shaded*; the alternate C-terminus of the late developmental mRNAs is *shaded differently*. *Dots* in the 3' UT indicate different sites of polyadenylation. The three sites of proteolytic cleavage in the C-terminal cytoplasmic tail encoded by exons 11 and 13 are too closely spaced to be indicated, as are the 7 transcriptional initiation sites.

isolated from mouse submaxillary gland, where the NGF accounts for about 0.1% of the protein (Scott et al 1983, Ullrich et al 1983). There is a single gene for β-NGF (Figure 7), which gives rise to an approximately 1300 nucleotide mRNA with 160 nucleotides of 3′ untranslated region. Alternative RNA splicing produces two RNAs, differing by the inclusion or exclusion of exon 2, with 5′ untranslated regions of either 100 or 172 nucleotides (Edwards et al 1986). Two different methionines act as initiators for the ORFs, which are either 307 or 241 triplets (Scott et al 1983, Ullrich et al 1983). The shorter pre-pro-β-NGF has an N-terminal signal; the longer ORF may utilize the signal differently, perhaps as a membrane anchor (Edwards et al 1986). The pro-β-NGF molecules (which contain three possible sites for N-glycosylation) are cleaved at a series of dibasic residues to release the 118 residue β-NGF molecule, which is at the C-termini of both ORFs, and possibly as many as three other peptides, ranging from 27 to 89 residues. The longest of these is truncated in the product of the shorter splice form, thus suggesting another possible functional explanation for the splicing difference. No functions for these putative cryptic peptides are yet known.

For NGF to activate a neuron, that neuron must express an NGF

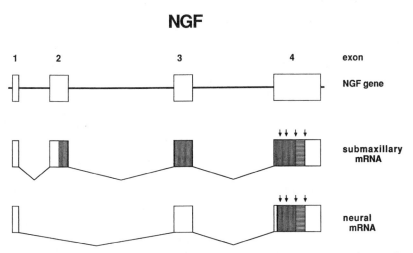

Figure 7 The nerve growth factor (NGF) gene and transcript structures. (References in text.) The length of the first intron is not known. The sizes of the exons are approximately proportional, but are exaggerated relative to the sizes of the introns. The protein-coding regions of the spliced exons are *striped*. The *horizontal stripes* represent the β-NGF coding region; the *vertical stripes* represent cryptic regions. The *arrows* are the putative sites of proteolytic cleavage in the protein product. Note that the initiator methionine codon utilized in the submaxillary gland mRNA is absent from the neural mRNA form because of alternative splicing, hence a later initiator is utilized in the neural molecule.

receptor. NGF receptors have been characterized by binding and cross-linking studies, and exist as two forms: high affinity and lower affinity. Genomic and cDNA clones for the human and rat lower affinity form have been isolated and characterized (Chao et al 1986, Johnson et al 1986, Radeke et al 1987). The single human NGF receptor gene is greater than 18 kb in length and is located on chromosome 17 (Huebner et al 1986). The cDNA clones hybridize to an approximately 3800 nucleotide mRNA in cells that express predominantly either high or low affinity forms, a finding that suggests both forms are products of the same gene. The human clone has a 427 triplet ORF encoding a protein with a 28 residue signal peptide followed by a 160 residue region composed of 4 repeated cysteine-rich domains, a 51 residue serine-threonine-rich region, a 28 residue membrane-spanning region, and a 150 residue C-terminal (presumably cytoplasmic) tail (Johnson et al 1986). Glycosylation accounts for the difference between the size of the ORF and the observed 83 kDa protein. The rat sequence is nearly identical in structure and highly conserved in sequence except that the signal region is 29 residues and the C-terminal tail is only 147 residues (Radeke et al 1987). Despite the high degree of sequence conservation in the ORF, the 3′ untranslated regions differ considerably both in length (human 1992, rat 1881) and sequence.

Enzymes for Transmitter Biosynthesis

The transfer of electrical signals at synapses is mediated in part by small molecules such as acetylcholine, dopamine, norephinephrine, serotonin, GABA, glycine, and glutamic acid. The biosynthesis, degradation, storage, and uptake of these small molecules requires enzymes and receptors, most of which are expected to be restricted to neural tissues. Many of the enzymes in the synthetic pathways are known, and cDNA clones of mRNAs for tyrosine hydroxylase (TH) (Lamouroux et al 1982, Chikaraishi et al 1983) and glutamic acid decarboxylase (GAD) have been isolated.

The sequence of a 1770 nucleotide clone of the 1900 nucleotide rat TH mRNA has been reported (Grima et al 1985). The clone is incomplete at its 5′ end and is thought to correspond to 11 nucleotides of putative 5′ untranslated region, an ORF of 498 triplets, and a complete 3′ untranslated region of 265 nucleotides. It has not yet been firmly established that the 5′ end of the ORF is entirely complete. The present ORF encodes a 56 kDa protein that corresponds to the 62 kDa TH species observed on SDS gels.

GAD is the 60–66 kDa synthetic enzyme for the inhibitory neurotransmitter GABA. A feline cDNA clone has been isolated that, in bacteria, directs the synthesis of a fusion protein with GAD catalytic activity (Kaufman et al 1986). The clone hybridizes to a 3.7 kb brain-specific mRNA (Wood et al 1986), thus indicating that the RNA contains more noncoding than coding sequence (assuming that the primary translation product is not extensively processed).

Peptide Hormones

A large number of peptides are secreted by both endocrine cells and neurons and appear to have dual functions as blood-borne hormones for regulating peripheral organs and in brain as transmitter or modulatory substances. The mRNAs for precursors of many such peptides have been closed and sequenced, and their gene structures have been worked out. Excellent reviews by Douglass and colleagues (1984) and Lynch & Snyder (1986) cover the structural analysis of proopiomelanocortin, preproenkephalin, preprodynorphin, preprovasopressin, preprooxytocin, preproCRF, preproGRF, preprosomatostatin, preproCGRP, preprocholcystokinin, and preproVIP and the processing pathways that generate them. Hence, I do not discuss these here. I discuss a few more recent examples as representatives of this apparently large class of proteins with important, although not always exclusive, function in the brain.

Thyrotropin-releasing hormone (TRH) is a C-terminally amidated tripeptide synthesized in subregions of the hypothalamus and involved in regulating the hypothalamic-pituitary-thyroid axis. It is generated by proteolytic cleavage from a precursor (prohormone) encoded by a 1700 nucleotide hypothalamic mRNA (Lechan et al 1986). A partial cDNA clone of the rat mRNA reveals that there are at least 102 5' untranslated region nucleotides, an ORF of 255 triplets, and a 3' untranslated region of at least 454 nucleotides. The sequence of the preprohormone translated from the ORF reveals a putative N-terminal signal and five copies of the TRH sequence flanked by tandem basic amino acids, thus indicating that each prohormone molecule can give rise via protolysis to five molecules of TRH. The fate of the seven surrounding proteolytic fragments has not been determined, but some of these may be functional (Lechan et al 1986).

The tachykinins are a family of structurally related peptides with similar functional properties, the best known of which is substance P, a C-terminally amidated undecapeptide with both CNS and peripheral distribution. Although strictly speaking not a brain-specific molecule, substance P and other tachykinins have neurotransmitter or neuromodulatory properties. Transcription of the single bovine gene (Figure 8) for the precursor for substance P (preprotachykinin, PPT) initiates at five major sites clustered within five nucleotides of each other and also at four minor sites; this results in mRNAs with 5' untranslated regions of 115–120 (major) or 141–155 (minor) nucleotides (Nawa et al 1984a). Alternative RNA splicing

PPT

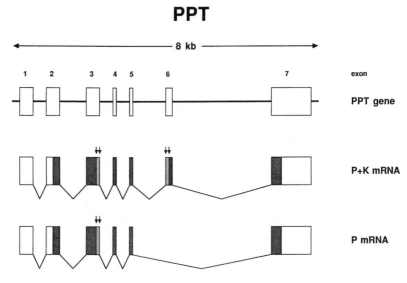

Figure 8 The preprotachykinin (PPT) gene and transcript structures. (References in text.) The 8 kb PPT gene covers 7 exons. The sizes of the exons are approximately proportional but are exaggerated relative to the sizes of the introns. *Shaded regions* in the spliced products indicate protein-coding regions. The *arrows* represent sites for proteolytic processing in the protein products. Processing releases substance P from the products of both mRNAs and substance K from the product of the P+K mRNA; the coding regions for these two neuropeptides are more *lightly shaded* than the rest of the coding regions.

of the bovine PPT hnRNA, which contains seven exons, produces ORFs of 112 and 130 triplets, called α and β respectively, which differ by the inclusion or exclusion of exon 6. The ratio of α and β forms of PPT mRNA vary; β is more abundant than α in peripheral tissues. Thus RNA splicing is regulated in a tissue-specific fashion, either because of factors with different splicing specificities or variations in the concentrations of common splicing factors. Translation of the α and β ORFs generates the preprohormones, which are processed by removal of their signal sequence and cleavage at paired basic amino acids coupled with the formation of the C-terminal amides. The α ORF contains a single substance P unit, but the β ORF encodes both one unit of substance P and a novel tachykinin-like unit called substance K. Substance K has been identified in vivo and synthetic substance K has tachykinin-like activity that differs from substance P (Nawa et al 1984b). Thus splicing alters the ratios of substance P to substance K, with P much more abundant than K in CNS but approximately equimolar to K in peripheral tissues. A possible third alternatively spliced bovine form was detected, but its structure is unknown

(Nawa et al 1983, 1984a). Recently the structure of three mRNAs that are the products of a single rat PPT gene have been described (Krause et al 1987). The rat α and β forms are equivalent to the two bovine PPT mRNAs; however, the most abundant rat form, γ, lacks exon 4 but includes exon 6; hence the γ PPT mRNA encodes both substance P and K.

The single human and rat genes for gonadotropin-releasing hormone (GnRH), each composed of at least four exons spanning 4 kb (Adelman et al 1986), are expressed in hypothalamus and other regions of the CNS, as well as in gonads, mammary glands, and placenta. Because of its low abundance, the GnRH mRNA has not been visualized by blotting, but much of its structure has been deduced from cDNA and genomic clones (Seeburg & Adelman 1984, Adelman et al 1986). cDNA clones of hypo-thalamic mRNAs reveal 5′ untranslated regions of at least 70 (rat) or 32 (human) nucleotides preceding an ORF of 276 nucleotides and 3′ untranslated regions of 148 (rat) or 162 (human) nucleotides. GnRH cDNA clones from placenta show a 5′ untranslated region in excess of 1000 nucleotides, including a region that is clearly an intron in the hypo-thalamus, thus suggesting that alternative RNA splicing (and possibly also alternative transcriptional initiation sites) might generate a placental GnRH mRNA with an extremely large 5′ untranslated region. The hypo-thalamic and placental mRNAs have identical ORFs encoding a 92 residue precursor in which a signal sequence (23 residues) is followed by the GnRH decapeptide, Gly-Lys-Arg cleavage-transamination signal, and a 56 residue C-terminal peptide called GAP (GnRH-associated protein). Pulsatile release of the GnRH decapeptide from the hypothalamus triggers release of luteinizing and follicle-stimulating hormones from the anterior pituitary. The co-released GAP peptide (also called PIF: prolactin-inhibiting factor) acts to inhibit prolactin release (Nikolics et al 1985).

Signal Transduction and Channels

In the retina, light is absorbed by visual pigment molecules that consist of an opsin protein covalently linked to the chromophore retinal. Excitation by a photon causes an isomerization of the chromophore, resulting in activating conformational changes in the opsin. The activated opsin func-tions catalytically to activate transducin, a retinal-specific G-protein; this in turn leads to decreases in cyclic GMP levels. cGMP acts as a second messenger, regulating a cascade of reactions that lead to a signal being produced by a particular neuron in this CNS sensory organ. The neurons of the retina are among the most specialized and stereotypical in the CNS; one opsin type is produced in lower mammals, four in higher primates. Rhodopsin, the pigment of rod cells, is the product of a single 5 exon gene of about 6.5 kb in cows and humans; the human gene is on chromosome

3 (Nathans & Hogness 1983, 1984, Nathans et al 1986a). There are two bovine rhodopsin mRNAs of about 1800 and 2600 nucleotides (Nathans & Hogness 1983, Koike et al 1983, Kuo et al 1986), which carry a 96 nucleotide 5′ untranslated region and a 1044 nucleotide ORF and differ in their sites of polyadenylation, generating 3′ untranslated regions of either 365 or 1391 nucleotides. The human mRNA has a similar structure, except that initiation occurs at either of two sites generating RNAs with 5′ untranslated regions of either 95 or 93 nucleotides (Nathans & Hogness 1984). The encoded 348 residue protein contains seven apparent membrane-crossing regions and thus is structurally analogous to bacterial rhodopsin and the muscarinic acetylcholine receptor (AChR).

Higher primates have three color-detecting cone pigments. The human 3308 nucleotide, 5-exon, blue pigment gene, located on chromosome 7, gives rise to an 1108 nucleotide mRNA (Nathans et al 1986a,b). The encoded 348 residue blue opsin is 42% identical in sequence to rhodopsin, and shares its major structural properties. The genes for the red and green cone pigments are located on the X-chromosome (Nathans et al 1986a). In males with normal color vision, there are one red and one or more (up to at least four) copies of the green pigment gene (Nathans et al 1986b). These genes have identical 6-exon structures and very similar sequences and produce 1242 nucleotide mRNAs with 41 nucleotide 5′ untranslated regions, 1092 nucleotide ORFs, and 109 nucleotide 3′ untranslated regions. The 364 residue red and green opsins again have 7 membrane crossing regions and are very highly similar to each other and also similar to blue opsin and rhodopsin. The number of green genes in an individual is hypothesized to be determined by unequal recombination events between tandem homologous red and green genes: Green gene number may increase or decrease in the various meiotic reciprocal segregants and thus gene number variation appears to be a dynamic process in the population (Nathans et al 1986b). Errors in this process that result in interchanges between the red and green genes, producing hybrid red-green opsins or deleting one type, account for most common types of color blindness (Nathans et al 1986a).

The process of light transduction in the retina involves the activation of a G-protein, transducin, a three-subunit enzyme with α, β, and γ chains of 39 kDa, 36 kDa, and 8 kDa, that couples the photoactivation of opsins to changes in cGMP levels in the photoreceptor cells. Upon activation, the α subunit exchanges GDP for GTP, in which state it stimulates a cGMP phosphodiesterase until its intrinsic GTPase activity self-deactivates. Two similar but not identical structures for the mRNA for the bovine rod outer segment transducin α subunit have been elucidated. One form, the product of a single gene, has at least 174 5′ untranslated region nucleotides, an

ORF of 1062 nucleotides, and a 3' untranslated region in excess of 426 nucleotides (Lochrie et al 1985). A second form, the product of an 8 kb gene expressed in retina but not brain, liver, or heart, has a 5' untranslated region of at least 121 nucleotides, an ORF of 1050 nucleotides, and a 3' untranslated region of 1259 nucleotides (Tanabe et al 1985, Medynski et al 1985, Yatsunami & Khorana 1985). The two ORFs encode very similar proteins of 354 and 350 residues that, like other G protein α subunits, are related in sequence to the *ras* protein and to protein synthesis elongation factors. The level of α mRNA is regulated during the light-dark cycle (Brann & Cohen 1987).

Whereas the α subunits of transducin are specific to retina, the β subunit is thought to be the same for all G-proteins. The β mRNA in bovine retina has a 5' untranslated region of 121 nucleotides, an ORF of 1020 nucleotides, and has 3' untranslated regions of either 306 or 1664 nucleotides as a result of differential poly(A) site selection (Sugimoto et al 1985). The ORF and body of these 1447 and 2805 nucleotide retinal mRNAs are shared with brain and other tissues; however, the 5' untranslated regions differ between the β mRNAs in retina and brain, thus indicating that different sites of transcriptional initiation on the same gene are used in different cell types, presumably for regulatory reasons (Sugimoto et al 1985). Alternative RNA splicing must connect the different 5' ends to the same mRNA body.

The γ subunit of bovine transducin, which is required for opsin stimulation of the α subunit, is encoded by an mRNA of about 500 nucleotides (Hurley et al 1984, Yatsunami et al 1985). A 448 nucleotide cDNA sequence contains a 5' untranslated region of 67 nucleotides (not known to be complete), an ORF of 222 nucleotides, and a complete 3' untranslated region of 145 nucleotides. Curiously, there are significant differences in the 5' untranslated regions in the two studies (Hurley et al 1984, Yatsunami et al 1985). Maturation of the γ protein involves removal of the N-terminal methionine residue.

Three G-proteins other than transducin modulate second-messenger levels in response to transmitters and hormones in the brain and in peripheral tissues. In each case the G-protein is a heterotrimer and the β and γ subunits are the same as found in transducin. The α subunit of G_s, which has at least two forms of 52 kDa and 45 kDa, dissociates from $\beta\gamma$ after activation and directly stimulates adenylate cyclase. The single bovine αG_s gene gives rise to 1900 nucleotide mRNA products whose abundance in brain and adrenal is much higher than in liver (Harris et al 1985). The 52 kDa bovine protein is the product of an mRNA with at least 52 5' untranslated region nucleotides, an ORF of 1182 nucleotides, and known 3' untranslated regions ranging from 309 to 365 nucleotides (Robishaw et al

1986a, Nukada 1986). The mRNA for the rat 52 kDa species has a similar structure (Itoh et al 1986). Alternative RNA splicing produces an mRNA with a 42 nucleotide internal deletion, generating a shorter ORF of 1140 nucleotides and accounting for the 45 kDa α subunit (Robishaw et al 1986b). The ratios of these two RNAs differ between tissues. Studies on the human α subunits (which are 95% identical to the bovine and rat subunits) suggest that ambiguity in choosing the alternate splice junctions may actually lead to four splice variants with encoded proteins of 378, 379, 393, and 394 residues (Bray et al 1986). The functional consequences of these internal variations in α subunit structures have not yet been established. A second brain G protein, G_i, with a 41 kDa α subunit, acts to inhibit G_s by providing β, which binds to the activated α_s subunit. The G_i α subunit is the product of an mRNA with a 5′ untranslated region of at least 48 nucleotides, 1065 nucleotide ORF, and 3′ untranslated region of at least 562 nucleotides (Itoh et al 1986). The partial sequence of the α subunit of a third G protein G_0, which may act similarly to G_i, is also known (Itoh et al 1986). The two forms of α subunit from transducin and the α subunits from G_s, G_i, and G_0 show substantial sequence relationships (Itoh et al 1986).

Receptors for transmitters such as acetylcholine transduce the external chemical signal to an intracellular chemical message. Clones of the mRNAs for the multisubunit nicotinic AChR from nonmammalian and muscle sources have been described by several groups (reviewed by McCarthy et al 1986) and are not reviewed here. The single-subunit porcine cerebrum muscarinic AChR, in conjunction with G-proteins, modulates potassium channels in response to acetylcholine. Its 2300 nucleotide mRNA, which shows differential localization within the brain, has a 443 nucleotide 5′ untranslated region and 1061 nucleotide 3′ untranslated region (Kubo et al 1986). Its 460 triplet ORF encodes a protein with no apparent signal sequence but with seven putative membrane spanning domains, thus suggesting a membrane topography similar to that for rhodopsin and the β-adrenergic receptor (Kubo et al 1986).

Two mRNAs of 9000–9100 nucleotides encode distinct rat brain voltage-gated sodium channels (I and II), and there is evidence for a third distinct channel (Noda et al 1986a). The two mRNAs, which are expressed in brain and in cardiac muscle, appear to be products of distinct genes based on the presence of differences throughout the two sequences. The channel I and II mRNAs have at least 251 and 106 5′ untranslated region nucleotides, ORFs of 6027 and 6015 nucleotides, and 2120 and 2328 nucleotide 3′ untranslated regions, respectively. The two encoded proteins (2009 and 2005 residues) are 87% identical, and each contains four repeats of an approximately 300 residue domain. Each domain contains six putative membrane-spanning regions, hence suggesting a model for the structure and function of the channels (Noda et al 1986a). *Xenopus* oocyte micro-injection of either of the two rat channel mRNAs directs the formation of functional sodium channels, a finding that indicates that the products of these mRNAs are the only specific channel components (Noda et al 1986b).

Other Brain-specific Molecules

Some molecules first became known because they were found to have the property of brain-specific distribution. Further studies of such molecules have provided progressively more thorough descriptions of their bio-chemical, anatomical, and superstructural properties. Interpretations based upon these cumulative descriptions have led to models for possible functions for some brain-specific proteins.

S100 was one of the early "brain-specific" proteins described by Moore (1965). This astrocytic protein is related in structure to the calcium-binding species calmodulin and intestinal vitamin D-dependent Ca-binding protein. In rat, S100 is a homodimer with two 11 kDa β chains. The structure of the S100 β mRNA, which has gel mobility of 1500 nucleotides, was deduced from a series of overlapping partial cDNA clones (Kuwano et al 1984). The 1488 nucleotide composite sequence reveals a 5′ untranslated region of at least 120 nucleotides and an ORF of 92 triplets. Poly(A) is added at either of two sites; this results in 3′ untranslated regions of 1076 or 1092 nucleotides. The initiator methionine is removed post-translationally to produce the mature 91 residue S100 protein. Transcription of S100 is at low levels in fetal brain and increases gradually during early postnatal development, then steeply between days 10 and 30. Recently the structure of bovine S100 α mRNA, which is expressed in bovine but not rat brain, was described (Kuwano et al 1986). The approximately 700–800 nucleotide α mRNA has a 5′ untranslated region of at least 89 nucleotides, an ORF of 282 nucleotides, and 161 3′ untranslated region nucleotides.

Neuron-specific enolase (NSE), originally described by Moore & Perez (1966) as an abundant brain-specific protein (called 14-3-2), is a dimer of identical 46 kDa subunits. The rat NSE mRNA is expressed from a single gene (Sakimura et al 1985), and transcription initiates at any of 8 major 5′ sites (Forss-Petter et al 1986), generating an mRNA with variable length 5′ untranslated regions of 56–112 nucleotides, an ORF of 434 triplets, and a 3′ untranslated region of 852 nucleotides. NSE mRNA is detected at low levels in midgestational rat fetuses. It achieves an intermediate plateau of 30% of the adult level at embryonic day 20, which persists until postnatal day 5 and then increases to near adult levels by day 25 (Forss-Petter et al

1986). The NSE protein accumulates with a time course much delayed compared to NSE mRNA, thus indicating that some sort of translational control may regulate NSE expression.

Synapsin I is a vesicle-associated phosphoprotein found presynaptically in both PNS and CNS neurons, whose developmental appearance correlates with synapse formation. There are two rat synapsin I mRNAs of 4500 and 5800 nucleotides (Kilimann & deGennaro 1985) which are products of a single gene (McCaffery & deGennaro 1986). The human gene has been localized on the X chromosome by using the rat clone (Yang-Feng et al 1986). A partial sequence reveals a probably complete 691 triplet ORF encoding a 73 kDa polypeptide with some sequence similarity to actin-binding proteins (McCaffery & deGennaro 1986). There are two forms of synapsin I of 74 and 78 kDa that differ at their C-termini and are likely to be alternative splice products from the single gene.

The O-44 mRNA family is known from the sequence of a rat brain cDNA clone (Tsou et al 1986). The single O-44 gene is transcribed in both neural and nonneural tissues, but its 5' end formation differs between brain and other tissues. In brain, transcription initiates at any of 22 sites, producing RNAs with 5' untranslated regions of 65–328 nucleotides. In nonneural tissues only 11 of the shorter 5' ends are found. There are two ORFs of 116 and 127 triplets generated by alternative RNA splicing, and two different polyadenylation sites producing 3' untranslated regions of 91 and 179 nucleotides. This complex family of mRNAs encodes two proteins of unknown function that differ from each other at their C-termini. Sequence analysis of the O-44 gene in the region upstream from the 22 transcription initiation sites reveals neither of the commonly found transcriptional initiation signals, CAAT or TATA.

There is a single gene for the retina-specific 950 nucleotide pCR18 mRNA (Miki et al 1986). In situ hybridization demonstrated that this mRNA is expressed by a subset of ganglion cells. The mRNA has a 5' untranslated region of 137 nucleotides, an ORF of 441 nucleotides, and a 3' untranslated region of 246 nucleotides (Miki et al 1986). The sequence of the encoded 147 residue protein is novel.

STRUCTURAL GENERALIZATIONS

The structural information about 39 mRNAs whose expression is specific to or highly enriched in the CNS are collected in Table 1. Most of these would have been classified as Class III "brain-specific" by the criteria used in the study of Milner & Sutcliffe (1983), with the exceptions of the mRNAs for P_0, tubulins, PPT, the G proteins (except α transducin), and O-44. This set of 39 mRNAs and the subset of 30 "brain-specific" mRNAs will be used to derive the structure of the "average" brain-specific mRNA. A further subset of 15 "rarer" abundance mRNAs has been culled from the brain-specific group by eliminating the abundant mRNAs for major proteins of myelin, major cytoskeletal proteins, major retinal proteins, NSE, and S100.

The average length of mRNAs listed in Table 1 is 3243 nucleotides; the brain-specific mRNAs average 3690 nucleotides and the rarer members of the group 4945 nucleotides (Table 2). The nonbrain-specific mRNAs have the considerably shorter average length of 1623 nucleotides. Previously, Milner & Sutcliffe (1983) estimated from a randomly collected set of brain mRNAs that the average length of brain-specific molecules was 3660, and for the rarer class 4960, nucleotides. The values tabulated here based on sequence data are obviously in good agreement. Thus, as previously predicted, brain-specific mRNAs are on average considerably larger than molecules not specifically expressed in the brain and 5000 nucleotides appears to be a valid estimate of average chain length for the rarer species that make up most of the complexity of brain mRNA.

How is the increased number of nucleotides distributed in the mRNA molecule? For the molecules whose 5′ ends have been mapped (a limited subset), 5′ untranslated regions range from 7 to 443 nucleotides, with an average of 139 (Table 2). This is generally a small fraction of the mRNA total length, and no compelling brain-specific effect is apparent in the small sample.

The ORFs range from 222 to 6027 nucleotides, with an average of 1385 (Table 2). Brain-specific, and especially rarer brain-specific, ORFs are considerably larger, and this trend will increase when the sequences for MAPs and neurofilaments become known. Thus, one reason brain-specific mRNAs are abnormally long is that on average they encode large proteins, perhaps larger than 76 kDa.

The 3′ untranslated regions whose structures are known average 1113 nucleotides in length (Table 2) and 1669 nucleotides for the rarer brain-specific class. Hence, these mRNAs also have anomalously long 3′ untranslated regions. Of course, there are mRNAs with higher and lower values than the average for each of these observations and for those observations discussed below, but the average brain-specific mRNA has both larger coding and larger 3′ untranslated regions than the average nonbrain-specific mRNA.

How prevalent a phenomena are multiple mRNAs from a single gene? Of the 14 mRNAs whose 5′ ends have been mapped, 8 show 5′-end heterogeneity; this indicates that transcription by RNA polymerase II initiates at anywhere from a few to several sites on the gene (Table 3). Both brain-specific and nonspecific mRNAs show this feature. It has been suggested that heterogeneous initiation sites may be characteristic of transcripts of so-called housekeeping genes that exhibit little tissue specificity (Dynan 1986). The present examples show that the phenomenon is not an exclusive property of housekeeping genes and can be expected to be encountered in a large percentage of both tissue-specific and common transcripts. No functional role for 5′ end heterogeneity is known. It seems

Table 1 Properties of brain mRNAs

RNA[a]	Length[b]	5′ Ends[c]	5′ UT[d]	ORF[e]	3′ UT[f]	3′ Ends[g]	Alt. splice[h] forms
MBP	1920–2121	1	47	384–585	1486	1	5
PLP/DM20	1600–3200	3	89–154	723–831	450–2066	3	2
P$_0$	1900–2100	nd	≥32	741	≥252	nd	1
Tubulin Mα1	1800	nd	≥48	1350	140	1	1
Tubulin Mβ4	1900	nd	nd	1335	≥317	nd	1
MAP 1	>10,000	nd	nd	nd	nd	nd	nd
MAP 2	9000	nd	nd	nd	nd	nd	nd
tau	6000	nd	nd	(1300)	(4600)	nd	nd
GFAP	2657	1	50	1254	1353	1	1
NF68	2500–4000	nd	≥33	1623	≥343	nd	1
NF150	3000	nd	nd	nd	nd	nd	1
NF200	4500	nd	nd	nd	nd	nd	1
N-CAM	4200–7000	nd	nd	~1797–~2892	>664	2	3
1B236/MAG	2450–2500	7	137–187	1746–1878	349–454	2	4
NGF	1055–1181	1	100–172	723–921	160	1	2
NGF-R	3800	nd	≥113	1281	1992	1	1
TH	1900	nd	≥11	1494	265	1	1
GAD	3700	nd	nd	1800	(1800)	nd	1
TRH	1700	nd	≥102	765	>454	nd	nd
PPT	1000–1094	9	115–155	336–390	549	1	3
GnRH	>500	2	≥70	276	162	1	2
Rhodopsin	1800–2600	1, 2	95	1044	365–1391	2	1
Blue opsin	1108	1	7	1044	57	1	1
Red opsin	1242	1	41	1092	109	1	1
Green opsin	1242	1	41	1092	109	1	1
TαI	nd	nd	≥174	1062	>426	1	1
TαII	2446	nd	≥121	1050	1259	1	1
Tβ	1447–2805	2	≥121	1020	306–1664	2	2
Tγ	500	nd	≥67	222	145	1	1
G$_s$α	1900	nd	≥52	1134–1182	309–365	2	4
G$_i$α	nd	nd	≥48	1065	562	1	1
mAChR	3000	nd	443	1380	1061	1	1
NaChan I	9000	nd	≥251	6027	2120	1	1
NaChan II	9100	nd	≥106	6015	2328	1	1
NSE	2210–2266	8	56–112	1302	852	1	1
S100	1472–1488	nd	120	276	1076–1092	2	1
Synapsin	4500–5800	nd	nd	2073	(2400–3700)	1	nd
O-44	504–888	22	65–328	348–381	91–179	2	2
CR18	950		137	441	246	1	1

Table 2   The "average" brain mRNA[a]

	Total set[b]	Brain-specific[c]	Others[d]	Rarer specifics[e]
Length[f]	3243	3690	1623	4945
5' UT[g]	139 (115)	123 (117)	241 (106)	224 (155)
ORF[h]	1385	1569	854	2082
3' UT[i]	1113 (1197)	1303 (1430)	515 (547)	1669 (1774)

[a] All lengths in nucleotides, calculated from Table 1.
[b] From entire set of 39 listed mRNAs.
[c] Excludes P_0, tubulins, PPT, $T\beta$, $T\gamma$, $G_s\alpha$, $G_i\alpha$, O-44.
[d] Those excluded in C.
[e] Excludes MBP, PLP/DM20, GFAP, NF68, NF150, NF200, rhodopsin, blue, green, red opsin, $T\alpha$I, $T\alpha$II, NSE, S100, CR18.
[f] Average length from Table 1; for multiple RNAs, longest was chosen.
[g] Average length from Table 1 of untranslated region (UT) from molecules with defined 5' ends; parenthetical numbers include lengths of molecules with undefined 5' ends.
[h] Open reading frame (ORF) average length calculated from data in Table 1; for multiple ORFs, longest was chosen.
[i] Average length of 3' untranslated region (UT) calculated from data in Table 1 for available sequence; parenthetical numbers include estimated lengths of 3' untranslated regions when required coding length is considered and assuming typical 5' UT.

more likely that it signifies a different mode for initiating transcription than a means for generating structural diversity in the transcripts. Nonetheless, it has not been ruled out that such heterogeneity could influence the translational efficiency of the mRNA, its stability, or the subcellular compartment in which it is translated.

---

[a] The properties of mRNAs are listed for myelin basic protein (MBP); proteolipid protein and its smaller form DM20 (PLP/DM20); peripheral myelin protein P_0; mouse tubulin isotypes $M\alpha$1 and $M\beta$4; microtubule-associated proteins MAP-1, MAP-2, and tau; glial fibrillary acidic protein (GFAP); neurofilament proteins NF68, 150, and 200; neural cell adhesion molecule (N-CAM); myelin-associated glycoprotein (1B236/MAG); nerve growth factor (NGF) and its receptor (NGF-R); tyrosine hydroxylase (TH); glutamic acid decarboxylase (GAD); thyrotropin-releasing hormone (TRH); preprotachykinin (PPT); gonadotropin-releasing hormone (GnRH); rhodopsin and the three color pigment opsins; transducin α ($T\alpha$I, $T\alpha$II), β ($T\beta$), and ($T\gamma$) subunits; G protein α subunits ($G_s\alpha$, $G_i\alpha$); muscarinic acetylcholine receptor (mAChR); sodium channel (Na Chan I, II); neuron specific enolase (NSE); S100; synapsin; O-44; and CR18 (all references in text).
[b] Length in nucleotides as determined by gel mobility except when the entire mRNA sequence is known and gives a unique length; nd, length not determined.
[c] Number of 5' ends determined by S1-nuclease protection, primer extension, or both; nd, not determined.
[d] Length or range of lengths in nucleotides of 5' untranslated regions; nd, not determined; $\geq$, sequence from cDNA clone has not been shown to be complete.
[e] Length or range of lengths in nucleotides of protein-coding open reading frame (ORF); nd, not determined; parenthetical entry for tau is based on the size of the protein; > for N-CAM reflects unpublished N-terminal sequence.
[f] Length or range of lengths in nucleotides of 3' untranslated regions; nd, not determined; $\geq$ sequence from cDNA clone has not been shown to be complete; > sequence is clearly not complete; parenthetical entries estimated by difference between ORF and mRNA size, assuming average size 5' untranslated region.
[g] Number of identified sites of polyadenylation; nd, not enough data to determine.
[h] Demonstrated alternate splice forms; nd, not enough data to determine: value is assumed to be 1 unless existing data prove alternative splicing or suggest it may account for observed protein heterogeneity.

Of the 34 molecules about which enough structural data are known, ten exhibit alternative splicing (Table 3). This mechanism of generating molecular diversity from a single gene has been discussed extensively in the excellent review by Leff and colleagues (1986). Alternative splices allow a regulatable diversity of the final gene product whereby proteins with similar general properties can differ in their fine specificities. As the examples show, a substantial portion of genes utilize this mechanism.

Of the 29 mRNAs whose 3′ untranslated regions are characterized, eight show alternative poly(A) site selection (Table 3). No function is known for this variation, but again it could affect the properties of the mature mRNA.

Of the mRNA molecules characterized, almost half show one or more kinds of diversity: heterogeneous 5′ ends, alternative splices, or different polyadenylation sites. The functional significance of these variations, and also the reason that brain-specific mRNAs are so large, have not yet been assimilated by either the molecular or neurobiological scientific communities. Indeed, much of the data reviewed here was published while the review was being prepared, and more data are needed to form a solid basis for interpretation and extrapolation from mRNA structures to functions. To add to the relevant database, I request that reprints of any articles I have inadvertently missed in this review, as well as those on topics explicitly excluded from the review, and all new articles relating to brain mRNAs be mailed to me to aid in the beginning of a brain molecular database. Comments on what would be appropriate to include in the database are welcome.

Table 3   Diversity in mRNA products of single genes

	5′ Ends[a]	Alt splice forms[b]	3′ Ends[c]
1	6 (6)	24 (19)	21 (17)
2	3 (2)	5 (3)	7 (4)
more	5 (3)	5 (3)	1 (1)
nd	25 (19)	5 (5)	10 (8)
	— —	— —	— —
	39 (30)	39 (30)	39 (30)

[a] Classification of the mRNAs from Table 1 by the number of distinct 5′ transcriptional initiation sites as determined by S1-nuclease protection, primer extension, or both: 1 site, 2 sites, more than 2 sites, or nd (data not available). Parenthetical numbers apply to subset of 30 brain-specific mRNAs identified in Table 2.
[b] From Table 1, number of known alternate splice forms of total set or (parenthetically) of brain-specific set.
[c] From Table 1, number of known differential poly(A) addition sites of total set or (parenthetically) of brain specific set.

THE ROLE OF MOLECULAR BIOLOGY IN NEUROSCIENCE

As the RNA complexity studies discussed at the beginning of this article have shown, the vast majority of brain proteins have brain-specific distributions and as yet unknown functions. The structural studies reviewed here show that modern molecular analytic techniques offer powerful tools for working out the molecular hardware of the brain. The value of solid structural information should not be underestimated; proven molecular structures provide insights into molecular function and also lead to explicit, testable hypotheses and molecular assays. Coupled with studies on anatomical distributions of molecules, molecular techniques can be used to address some of the most fundamental neurobiological questions—those concerning the molecular identities of individual neurons and the process of neural development. They also provide a sound basis for working out the functions of particular brain proteins.

Other strategies now allow many brain-specific molecules to be identified and characterized. These include strategies using monoclonal antibodies to individual antigenic components of complex brain protein mixtures (reviewed by McKay 1983, Valentino et al 1985) and strategies for isolating cDNA clones of brain specific mRNAs, determining DNA sequence, identifying and translating ORFs, and producing antibody either to chemically synthesized regions of the putative amino acid sequence or to bacterial fusion proteins in order to elicit reagents to describe the encoded brain-specific protein (Sutcliffe et al 1983). Such studies are crucial to progress in neurobiology for two reasons. First, such a small portion of brain proteins are presently known that it is difficult to assess the generality of what has thus far been learned about these proteins without first determining how representative they are of the many thousands of brain proteins. Second, studies that can provide detailed information about proteins *not* selected because of their particular functional properties, but randomly selected based only on distributional properties, are much more likely to turn up new types of molecules than are studies based only on either step-by-step extensions of present knowledge or sudden leaps of insight into new areas.

What would one learn by knowing all of the genes? Certainly this information alone would not solve the two current major issues in the study of biology—how genes are regulated by the developmental program and how the developed brain functions to interpret the environment and initiate behavioral responses. Thus it would be foolish to advocate total commitment of resources to determining the sequences of all genes. Modern analysis can be deeply illuminating, however, and has the potential to provide unambiguous biochemical assays for physiological phenomena. As the molecular basis for the control of gene expression is determined, the factors that stimulate (or repress) transcription of particular sets of genes will be described. Of particular interest will be elucidating the number of sets of coregulated genes, since each might be correlated to functional units within the brain. Learning the identities of the members of each set and also learning how the regulatory molecules themselves are regulated will reveal the underlying physiology that determines how various sets of neurons contribute to behavior. Thus these analytic approaches should be considered as partners with other methods of neuroscientific inquiry.

ACKNOWLEDGMENTS

I thank Rob Milner, Sonja Forss-Petter, Patria Danielson, Ute Hochgeschwender, Miles Brennan, Ian Brown, Frank Burton, and Floyd Bloom for helpful criticisms, many of the scientists whose work I have reviewed for sending copies of published and unpublished work, and Linda Elder for preparing the manuscript. I thank the National Institutes of Health (NS22111, GM32355, NS22347) for partial support. This is publication no. 4727-MB from the Research Institute of Scripps Clinic.

Literature Cited

Adelman, J. P., Mason, A. J., Hayflick, J. S., Seeburg, P. H. 1986. Isolation of the gene and hypothalamic cDNA for the common precursor of gonadotropin-releasing hormone and prolactin release-inhibiting factor in human and rat. *Proc. Natl. Acad. Sci. USA* 83: 179–83

Adesnik, M., Darnell, J. E. 1972. Biogenesis and characterization of histone messenger RNA in HeLa cells. *J. Mol. Biol.* 67: 397–406

Arquint, M., Roder, J., Chia, L.-S., Down, J., Wilkinson, D., et al. 1987. Molecular cloning and primary structure of myelin-associated glycoprotein. *Proc. Natl. Acad. Sci. USA* 84: 600–4

Balcarek, J. M., Cowan, N. J. 1985. Structure of the mouse glial fibrillary acidic protein gene: Implications for the evolution of the intermediate filament multigene family. *Nucl. Acids Res.* 13: 5527–43

Bantle, J. A., Hahn, W. E. 1976. Complexity and characterization of polyadenylated RNA in mouse brain. *Cell* 8: 139–50

Barbarese, E., Braun, P. E., Carson, J. H. 1977. Identification of prelarge and presmall basic proteins in mouse myelin and their structural relationship to large and small basic proteins. *Proc. Natl. Acad. Sci. USA* 74: 3360–64

Beckmann, S. L., Chikaraishi, D. M., Deeb, S. S., Sueoka, N. 1981. Sequence complexity of nuclear and cytoplasmic RNAs from clonal neurotumor cell lines and brain sections of the rat. *Biochemistry* 20: 2684–92

Blatt, C., Weiner, L., Sutcliffe, J. G., Nesbitt, M. N., Simon, M. I. 1987. Mapping of brain-specific genes: MAG and the QV locus. Submitted for publication

Bloom, F. E., Battenberg, E. L. F., Milner, R. J., Sutcliffe, J. G. 1985. Immunocytochemical mapping of 1B236: A brain specific neuronal protein deduced from the sequence of its mRNA. *J. Neurosci.* 5: 1781–1802

Bond, J. F., Farmer, S. R. 1983. Regulation of tubulin and actin mRNA production in rat brain: Expression of a new β-tubulin mRNA with development. *Mol. Cell. Biol.* 3: 1333–42

Bond, J. F., Robinson, G. S., Farmer, S. R. 1984. Differential expression of two neural cell-specific β-tubulin mRNAs during rat brain development. *Mol. Cell. Biol.* 4: 1313–19

Brann, M. R., Cohen, L. V. 1987. Diurnal expression of transducin mRNA and translocation of transducin in rods of rat retina. *Science* 235: 585–87

Bray, P., Carter, A., Simons, C., Guo, V., Puckett, C., et al. 1986. Human cDNA clones for four species of Gα, signal transduction protein. *Proc. Natl. Acad. Sci. USA* 83: 8893–97

Breathnach, R., Chambon, P. 1981. Organization and expression of eukaryotic split genes coding for proteins. *Ann. Rev. Biochem.* 50: 349–83

Brilliant, M. H., Sueoka, N., Chikaraishi, D. M. 1984. Cloning of DNA corresponding to rare transcripts of rat brain: Evidence of transcriptional and post-transcriptional control and of the existence of non-polyadenylated transcripts. *Mol. Cell. Biol.* 4: 2187–97

Chao, M. V., Bothwell, M. A., Ross, A. H., Koprowski, H., Lanahan, A. A., Buck, C. R., Sehgal, A. 1986. Gene transfer and molecular cloning of the human NGF receptor. *Science* 232: 518–21

Chaudhari, N., Hahn, W. E. 1983. Genetic expression in the developing brain. *Science* 220: 924–28

Chikaraishi, D. M. 1979. Complexity of cytoplasmic polyadenylated and non-adenylated rat brain ribonucleic acids. *Biochemistry* 18: 3250–56

Chikaraishi, D. M, Deeb, S. S., Sueoka, N. 1978. Sequence complexity of nuclear RNAs in adult rat tissues. *Cell* 13: 111–20

Chikaraishi, D. M, Brilliant, M. H., Lewis, E. J. 1983. Cloning and characterization of rat-brain-specific transcripts: Rare, brain-specific transcripts and tyrosine hydroxylase. *Cold Spring Harbor Symp. Quant. Biol.* 48: 309–18

Cleveland, D. W., Sullivan, K. F. 1985. Molecular biology and genetics of tubulin. *Ann. Rev. Biochem.* 54: 331–65

Cold Spring Harbor Symp. on Quant. Biol. 1983. *Mol. Neurobiol.* Vol. 48

Colman, P. D., Kaplan, B. B., Osterburg, H. H., Finch, C. E. 1980. Brain poly(A)+ RNA during aging: Stability of yield and sequence complexity in two rat strains. *J. Neurochem.* 34: 335–45

Darnell, J. E. Jr. 1982. Variety in the level of gene control in eukaryotic cells. *Nature* 297: 365–71

Dautigny, A., Alliel, P. M., d'Auriol, L., Pham Dinh, D., Nussbaum, J.-L., et al. 1985. Molecular cloning and nucleotide sequence of a cDNA clone coding for rat brain myelin proteolipid. *FEBS Lett.* 188: 33–36

Dautigny, A., Mattei, M.-G., Morello, D., Alliel, P. M., Pham-Dinh, D., et al. 1986. The structural gene coding for myelin-associated proteolipid protein is mutated in jimpy mice. *Nature* 321: 867–69

de Ferra, F., Engh, H., Hudson, L., Kamholz, J., Puckett, C., et al. 1985. Alternative splicing accounts for the four forms of myelin basic protein. *Cell* 43: 721–27

Denoulet, P., Jeantet, C., Gros, F. 1982. Tubulin microheterogeneity during mouse liver development. *Biochem. Biophys. Res. Commun.* 105: 806–13

D'Eustachio, P., Owens, G. C., Edelman, G. M., Cunningham, B. A. 1985. Chromosomal location of the gene encoding the neural cell adhesion molecule (N-CAM) in the mouse. *Proc. Natl. Acad. Sci. USA* 82: 7631–35

Diehl, H.-J., Schaich, M., Budzinski, R.-M., Stoffel, W. 1986. Individual exons encode the integral membrane domains of myelin proteolipid protein. *Proc. Natl. Acad. Sci. USA* 83: 9807–11

Douglass, J., Civelli, O., Herbert, E. 1984. Polyprotein gene expression: Generation of diversity of neuroendocrine peptides. *Ann. Rev. Biochem.* 53: 665–715

Drubin, D. G., Caput, D., Kirschner, M. 1984. Studies on the expression of the microtubule-associated protein, tau, during mouse brain development, with newly isolated complementary DNA probes. *J. Cell Biol.* 98: 1090–97

Dynan, W. S. 1986. Promoters for housekeeping genes. *Trends in Genetics* 2: 196–97

Edelman, G. M. 1986. Cell adhesion molecules in the regulation of animal form and tissue pattern. *Ann. Rev. Cell Biol.* 2: 81–116

Edwards, R. H., Selby, M. J., Rutter, W. J. 1986. Differential RNA splicing predicts two distinct nerve growth factor precursors. *Nature* 319: 784–87

Forss-Petters, S., Danielson, P., Sutcliffe, J. G. 1986. Neuron-specific enolase: Complete structure of rat mRNA, multiple transcriptional start sites, and evidence suggesting post-transcriptional control. *J. Neurosci. Res.* 16: 141–56

Geisler, N., Kaufman, E., Fischer, S., Plessmann, U., Weber, K. 1983. Neurofilament architecture combines structural principles of intermediate filaments with carboxy-terminal extensions increasing in size between triplet proteins. *EMBO J.* 2: 1295–1302

Geisler, N., Weber, K. 1982. The amino acid sequence of chicken muscle desmin provides a common structural model for intermediate filament proteins. *EMBO J.* 1: 1649–56

Goodman, R. H., Jacobs, J. W., Chin, W. W., Lund, P. K., Dee, P. C., Habener, J.

F. 1980. Nucleotide sequence of a cloned structural gene coding for a precursor of pancreatic somatostatin. *Proc. Natl. Acad. Sci. USA* 77: 5869–73

Goridis, C., Hirn, M., Santoni, M.-J., Gennarini, G., Deagostini-Bazin, H., et al. 1985. Isolation of mouse N-CAM-related cDNA: Detection and cloning using monoclonal antibodies. *EMBO J.* 4: 631–35

Gozes, I., Littauer, U. Z. 1978. Tubulin microheterogeneity increases with rat brain maturation. *Nature* 276: 411–13

Grima, B., Lamouroux, A., Blanot, F., Faucon Biguet, F., Mallet, J. 1985. Complete coding sequence of rat tyrosine hydroxylase mRNA. *Proc. Natl. Acad. Sci. USA* 82: 617–21

Grouse, L. D., Schrier, B. K., Bennett, E. L., Rosenzweig, M. R., Nelson, P. G. 1978. Sequence diversity studies of rat brain RNA: Effects of environmental complexity on rat brain RNA diversity. *J. Neurochem.* 30: 191–203

Harris, B. A., Robishaw, J. D., Mumby, S. M., Gilman, A. G. 1985. Molecular cloning of complementary DNA for the alpha subunit of the G protein that stimulates adenylate cyclase. *Science* 229: 1274–77

Hemperly, J. J., Murray, B. A., Edelman, G. M., Cunningham, B. A. 1986a. Sequence of a cDNA clone encoding the polysialic acid-rich and cytoplasmic domains of the neural cell adhesion molecule N-CAM. *Proc. Natl. Acad. Sci. USA* 83: 3037–41

Hemperly, J. J., Edelman, G. M., Cunningham, B. A. 1986b. cDNA clones of the neural cell adhesion molecule (N-CAM) lacking a membrane-spanning region consistent with evidence for membrane attachment via a phosphatidylinositol intermediate. *Proc. Natl. Acad. Sci. USA* 83: 9822–26

Higgins, G. A., Schmale, H., Bloom, F. E., Wilson, M. C., Milner, R. J. 1987. Development shift in the cellular expression of the brain-specific gene 1B236 revealed by in situ hybridization. Submitted for publication

Hobart, P., Crawford, R., Shen, L., Pictet, R., Rutter, R. J. 1980. Cloning and sequence analysis of cDNAs encoding two distinct somatostatin precursors found in the endocrine pancreas of angler fish. *Nature* 288: 137–41

Huebner, K., Isobe, M., Chao, M., Bothwell, M., Ross, A. H., et al. 1986. The nerve growth factor receptor gene is at human chromosome region 17q12-17q22, distal to the chromosome 17 breakpoint in acute leukemias. *Proc. Natl. Acad. Sci. USA* 83: 1403–7

Hurley, J. B., Fond, H. K. W., Teplow, D. B., Dreyer, W. J., Simon, M. I. 1984. Isolation and characterization of a cDNA clone for the γ subunit of bovine retinal transducin. *Proc. Natl. Acad. Sci. USA* 81: 6948–52

Itoh, H., Kozasa, T., Nagata, S., Nakamura, S., Katada, T., et al. 1986. Molecular cloning and sequence determination of cDNAs for α subunits of the guanine nucleotide-binding proteins G_s, G_i, and G_0 from rat brain. *Proc. Natl. Acad. Sci. USA* 83: 3776–80

Johnson, D., Lanahan, A., Buck, C. R., Sehgal, A., Morgan, C., et al. 1986. Expression and structure of the human NGF receptor. *Cell* 47: 545–54

Julien, J.-P., Dies, M., Flavell, D., Hurst, J., Grosveld, F. 1986. Cloning and developmental expression of the murine neurofilament gene family. *Mol. Brain Res.* 1: 243–50

Kaldany, R. J., Nambu, J. R., Scheller, R. H. 1985. Neuropeptides in identified *Aplysia* neurons. *Ann. Rev. Neurosci.* 8: 431–55

Kamholz, J., deFerra, F., Puckett, C., Lazzarini, R. 1986. Identification of three forms of human myelin basic protein by cDNA cloning. *Proc. Natl. Acad. Sci. USA* 83: 4962–66

Kaplan, B. B., Finch, C. E. 1982. The sequence complexity of brain ribonucleic acids. In *Molecular Approaches to Neurobiology*, ed. I. R. Brown, pp. 71–98. New York: Academic

Kaufman, D. L., McGinnis, J. F., Krieger, N. R., Tobin, A. J. 1986. Brain glutamate decarboxylase cloned in λgt-11: Fusion protein produces γ-aminobutyric acid. *Science* 232: 1138–40

Kilimann, M. W., deGennaro, L. J. 1985. Molecular cloning of cDNAs for the nerve-cell specific phosphoprotein, synapsin I. *EMBO J.* 4: 1997–2002

Kimura, M., Inoko, H., Katsuki, M., Ando, A., Sato, T., et al. 1985. Molecular genetic analysis of myelin-deficient mice: *Shiverer* mutant mice show deletion in gene(s) coding for myelin basic protein. *J. Neurochem.* 44(3): 692–96

Koike, S., Nabeshima, Y., Ogata, K., Fukui, T., Ohtsuka, E., Ikehara, M., Tokunaga, F. 1983. Isolation and nucleotide sequence of a partial cDNA clone for bovine opsin. *Biochem. Biophys. Res. Commun.* 116: 563–67

Krause, J. E., Chirgwin, J. M., Carter, M. S., Xu, Z. S., Hershey, A. D. 1987. Three rat preprotachykinin mRNAs encode the neuropeptides substance P and neurokinin A. *Proc. Natl. Acad. Sci. USA* 84: 881–85

Kubo, T., Fukuda, K., Mikami, A., Maeda, A., Takahashi, H., et al. 1986. Cloning, sequencing and expression of complementary DNA encoding the muscarinic acetylcholine receptor. *Nature* 323: 411–16

Kuo, C.-H., Yamagata, K., Moyzis, R. K., Bitensky, M. W., Miki, N. 1986. Multiple opsin mRNA species in bovine retina. *Mol. Brain Res.* 1: 251–60

Kuwano, R., Usui, H., Maeda, T., Fukui, T., Yamanari, N., et al. 1984. Molecular cloning and the complete nucleotide sequence of cDNA to mRNA for S-100 protein of rat brain. *Nucl. Acids Res.* 12: 7455–65

Kuwano, R., Maeda, T., Usui, H., Araki, K., Yamakuni, T., et al. 1986. Molecular cloning of cDNA of S100α subunit mRNA. *FEBS Lett.* 202: 97–101

Lai, C., Brow, M. A., Nave, K.-A., Noronha, A. B., Quarles, R., Bloom, F. E., Milner, R. J., Sutcliffe, J. G. 1987. Two forms of 1B236/myelin-associated glycoprotein (MAG), a cell adhesion molecule for postnatal neural development, are produced by alternative splicing. *Proc. Natl. Acad. Sci. USA* 84: 4337–41

Lamouroux, A., Faucon Biguet, N., Samolyk, D., Privat, A., Salomon, J. C., et al. 1982. Identification of cDNA clones coding for rat tyrosine hydroxylase antigen. *Proc. Natl. Acad. Sci. USA* 79: 3881–85

Lazarides, E. 1982. Intermediate filaments: A chemically heterogeneous, developmentally regulated class of proteins. *Ann. Rev. Biochem.* 51: 219–50

Lechan, R. M., Wu, P., Jackson, I. M. D., Wolf, H., Cooperman, S., Mandel, G., Goodman, R. H. 1986. Thyrotropin-releasing hormone precursor: Characterization in rat brain. *Science* 231: 159–61

Lees, M. B., Chao, B. H., Lin, L.-F. H., Samiullah, M., Laursen, R. A. 1983. Amino acid sequence of bovine white matter proteolipid. *Arch. Biochem. Biophys.* 226: 643–56

Leff, S. E., Rosenfeld, M. G., Evans, R. M. 1986. Complex transcriptional units: Diversity in gene expression by alternative RNA processing. *Ann. Rev. Biochem.* 55: 1091–2117

Lemke, G., Axel, R. 1985. Isolation and sequence of a cDNA encoding the major structural protein of peripheral myelin. *Cell* 40: 501–8

Lewis, S. A., Balcarek, J. M., Krek, V., Shelanski, M., Cowan, N. J. 1984. Sequence of a cDNA clone encoding mouse glial fibrillary acidic protein: Structural conversation of intermediate filaments. *Proc. Natl. Acad. Sci. USA* 81: 2743–46

Lewis, S. A., Cowan, N. J. 1985. Genetics, evolution and expression of the 68Kd neurofilament protein: Isolation of a cloned cDNA probe. *J. Cell Biol.* 100: 843–50

Lewis, S. A., Lee, M. G. S., Cowan, N. J. 1985. Five mouse tubulin isotypes and their regulated expression during development. *J. Cell Biol.* 101: 852–61

Lewis, S. A., Sherline, P., Cowan, N. J. 1986a. A cloned cDNA encoding MAP 1 detects a single copy gene in mouse, and a brain-abundant RNA whose level decreases during development. *J. Cell Biol.* 102: 2106–14

Lewis, S. A., Villasante, A., Sherline, P., Cowan, N. J. 1986b. Brain specific expression of MAP 2 detected using a cloned cDNA probe. *J. Cell Biol.* 102: 2098–2105

Lochrie, M. A., Hurley, J. B., Simon, M. I. 1985. Sequence of the alpha subunit of photoreceptor G protein: Homologies between transducin, *ras*, and elongation factors. *Science* 228: 96–99

Lynch, D. R., Snyder, S. H. 1986. Neuropeptides: Multiple molecular forms, metabolic pathways, and receptors. *Ann. Rev. Biochem.* 55: 773–99

Malfroy, B., Bakhit, C., Bloom, F. E., Sutcliffe, J. G., Milner, R. J. 1985. Brain-specific polypeptide 1B236 exists in multiple molecular forms. *Proc. Natl. Acad. Sci. USA* 82: 2009–13

Martini, R., Schachner, M. 1986. Immunoelectron microscopic localization of neural cell adhesion molecules (L1, N-CAM, and MAG) and their shared carbohydrate epitope and myelin basic protein in developing sciatic nerve. *J. Cell Biol.* 103: 2439–48

McCaffery, C. A., deGennaro, L. J. 1986. Determination and analysis of the primary structure of the nerve terminal specific phosphoprotein, Synapsin I. *EMBO J.* 5: 3167–73

McCarthy, M. P., Earnest, J. P., Young, E. F., Choe, S., Stroud, R. M. 1986. The molecular neurobiology of the acetylcholine receptor. *Ann. Rev. Neurosci.* 9: 383–413

McKay, R. D. G. 1983. Molecular approaches to the nervous system. *Ann. Rev. Neurosci.* 6: 527–46

Medynski, D. C., Sullivan, K., Smith, D., Van Dop, C., Chang, F.-H., et al. 1985. Amino acid sequence of the α subunit of transducin deduced from the cDNA sequence. *Proc. Natl. Acad. Sci. USA* 82: 4311–15

Miki, N., Kuo, C.-H., Nakagawa, Y., Bitensky, M. W., Ishii, K. 1986. Cloning and characterization of retina-specific cDNAs. In *Molecular Genetics in Developmental Neurobiology*, ed. Y. Tsukada, pp. 167–76. Tokyo: Japan Sci. Soc. Press

Miller, F. D., Naus, C. C. G., Higgins, G. A., Bloom, F. E., Milner, R. J. 1987a. Developmentally regulated rat brain mRNAs: Molecular and anatomical characterization. *J. Neurosci.* 7: 2433–44

Miller, F. D., Naus, C. C. G., Durand, M., Bloom, F. E., Milner, R. J. 1987b. Isotypes of α-tubulin are differentially regulated during neuronal maturation. *J. Cell Biol.* In press

Milner, R. J., Lai, C., Nave, K.-A., Lenoir, D., Ogata, J., Sutcliffe, J. G. 1985. Nucleotide sequences of two mRNAs for rat brain myelin proteolipid protein. *Cell* 42: 931–39

Milner, R. J., Bloom, F. E., Sutcliffe, J. G. 1987. Brain specific genes: Strategies and issues. *Curr. Top. Dev. Biol.* 21: 117–50

Milner, R. J., Sutcliffe, J. G. 1983. Gene expression in rat brain. *Nucl. Acids Res.* 11: 5497–5520

Molineaux, S. M., Engh, H., deFerra, F., Hudson, L., Lazzarini, R. A. 1986. Recombination within the myelin basic protein gene created the dysmyelinating shiverer mouse mutation. *Proc. Natl. Acad. Sci. USA* 83: 7542–46

Moore, B. W. 1965. A soluble protein characteristic of the nervous system. *Biochem. Biophys. Res. Commun.* 19: 739–44

Moore, B. W., McGregor, D. 1965. Chromatographic and electrophoretic fractionation of soluble proteins of brain and liver. *J. Biol. Chem.* 240: 1647–53

Moore, B. W., Perez, V. J. 1966. Specific acidic proteins of the nervous system. In *Physiological and Biochemical Aspects of Nervous Integration*, ed. F. D. Carlson, pp. 343–59. Engelwood Cliffs, NJ: Prentice Hall

Murray, B. A., Hemperly, J. J., Gallin, W. J., MacGregor, J. S., Edelman, G. M., Cunningham, B. A. 1984. Isolation of cDNA clones for the chicken neural cell adhesion molecule (N-CAM). *Proc. Natl. Acad. Sci. USA* 81: 5584–88

Murray, B. A., Owens, G. C., Prediger, E. A., Crossin, K. L., Cunningham, B. A., Edelman, G. M. 1986. Cell surface modulation of the neural cell adhesion molecule resulting from alternative mRNA splicing in a tissue-specific developmental sequence. *J. Cell Biol.* 103: 1431–39

Naismith, A. L., Hoffman-Chudzik, E., Tsui, L.-C., Riordan, J. R. 1985. Study of the expression of myelin proteolipid protein (lipophilin) using a cloned complementary DNA. *Nucl. Acids Res.* 13: 7413–25

Nakanishi, S., Inoue, A., Kita, T., Nakamura, M., Chang, A. C. Y., Cohen, S. N., Numa, S. 1979. Nucleotide sequence of cloned cDNA for bovine corticotropin-β-lipotropin precursor. *Nature* 278: 423–27

Nathans, J., Hogness, D. S. 1983. Isolation, sequence analysis, and intron-exon arrangement of the gene encoding bovine rhodopsin. *Cell* 34: 807–14

Nathans, J., Hogness, D. S. 1984. Isolation and nucleotide sequence of the gene encoding human rhodopsin. *Proc. Natl. Acad. Sci. USA* 81: 4851–55

Nathans, J., Piatanida, T. P., Eddy, R. L., Shows, T. B., Hogness, D. S. 1986a. Molecular genetics of inherited variation in human color vision. *Science* 232: 203–10

Nathans, J., Thomas, D., Hogness, D. S. 1986b. Molecular genetics of human color vision: The genes encoding blue, green, and red pigments. *Science* 232: 193–202

Nave, K.-A., Lai, C., Bloom, F. E., Milner, R. J. 1986. Jimpy mutant mouse: A 74 base deletion in the mRNA for myelin proteolipid protein (PLP) and evidence for a primary defect in RNA splicing. *Proc. Natl. Acad. Sci. USA* 83: 9264–68

Nave, K.-A., Lai, C., Bloom, F. E., Milner, R. J. 1987. Splice site selection in the proteolipid protein (PLP) gene transcript and primary structure of the DM-20 protein of CNS myelin. *Proc. Natl. Acad. Sci. USA* 84: 5665–69

Nawa, H., Hirose, T., Takashima, H., Inayama, S., Nakanishi, S. 1983. Nucleotide sequences of cloned cDNAs for two types of bovine brain substance P precursor. *Nature* 306: 32–36

Nawa, H., Kotani, H., Nakanishi, S. 1984a. Tissue-specific generation of two preprotachykinin mRNAs from one gene by alternative RNA splicing. *Nature* 312: 729–34

Nawa, H., Doteuchi, M., Igano, K., Inouye, K., Nakanishi, S. 1984b. Substance K: A novel mammalian tachykinin that differs from substance P in its pharmacological profile. *Life Sci.* 34: 1153–60

Neve, R. L., Harris, P., Kosik, K. S., Kurnit, D. M., Donlon, T. A. 1986. Identification of cDNA clones for the human microtubule-associated protein tau and chromosomal localization of the genes for tau and microtubule-associated protein 2. *Mol. Brain Res.* 1: 271–80

Newman, S., Kitamura, K., Campagnoni, A. T. 1987. Identification of a cDNA coding for a fifth form of myelin basic protein in mouse. *Proc. Natl. Acad. Sci. USA* 84: 886–90

Nguyen, C., Mattei, M.-G., Mattei, J.-F., Santoni, M.-J., Goridis, C., Jordan, B. R. 1986. Localization of the human NCAM gene to band q23 of chromosome 11: The third gene coding for a cell interaction molecule mapped to the distal portion of the long arm of chromosome 11. *J. Cell Biol.* 102: 711–15

Nikolics, K., Mason, A. J., Szonyi, E., Ramachandran, J., Seeburg, P. H. 1985. A prolactin-inhibiting factor within the precursor for human gonadotropin-releasing hormone. *Nature* 316: 511–17

Noda, M., Ikeda, T., Kayano, T., Suzuki, H., Takeshima, H., et al. 1986a. Existence of distinct sodium channel messenger RNAs in rat brain. *Nature* 320: 188–92

Noda, M., Ikeda, T., Suzuki, H., Takeshima, H., Takahashi, T., et al. 1986b. Expression of functional sodium channels from cloned cDNA. *Nature* 322: 826–28

Nukada, T., Tanabe, T., Takahashi, H., Noda, M., Hirose, T., et al. 1986. Primary structure of the α-subunit of bovine adenylate cyclase-stimulating G-protein deduced from the cDNA sequence. *FEBS Lett.* 195: 220–24

Olmsted, J. B. 1986. Microtubule-associated proteins. *Ann. Rev. Cell Biol.* 2: 421–57

Osborn, M., Weber, K. 1983. Biology of disease. Tumor diagnosis by intermediate filament typing: A novel tool for surgical pathology. *Lab. Invest.* 48: 372–94

Owens, G. C., Edelman, G. M., Cunningham, B. A. 1987. Organization of the neural cell adhesion molecule (N-CAM) gene: Alternative exon usage as the basis for different membrane-associated domains. *Proc. Natl. Acad. Sci. USA* 84: 294–98

Quarles, R. H. 1984. Myelin-associated glycoprotein in development and disease. *Dev. Neurosci.* 6: 285–303

Radeke, M. J., Misko, T. P., Hsu, C., Herzenberg, L. A., Shooter, E. M. 1987. Gene transfer and molecular cloning of the rat nerve growth factor receptor. *Nature* 325: 593–97

Roach, A., Boylan, K., Horvath, S., Prusiner, S. B., Hood, L. E. 1983. Characterization of cloned cDNA representing rat myelin basic protein: Absence of expression in brain of shiverer mutant mice. *Cell* 34: 799–806

Roach, A., Takahashi, N., Pravtcheva, D., Ruddle, F., Hood, L. 1985. Chromosomal mapping of mouse myelin basic protein gene and structure and transcription of the partially deleted gene in shiverer mutant mice. *Cell* 42: 149–55

Robishaw, J. D., Russell, D. W., Harris, B. A., Smigel, M. D., Gilman, A. G. 1986a. Deduced primary structure of the α subunit of the GTP-binding stimulatory protein of adenylate cyclase. *Proc. Natl. Acad. Sci. USA* 83: 1251–55

Robishaw, J. D., Smigel, M. D., Gilman, A. G. 1986b. Molecular basis for two forms of the G protein that stimulates adenylate

cyclase. *J. Biol. Chem.* 261: 9587–90

Roth, H. J., Kronquist, K., Pretorius, P. J., Crandall, B. F., Campagnoni, A. T. 1986. Isolation and characterization of a cDNA coding for a novel human 17.3K myelin basic protein (MBP) variant. *J. Neurosci. Res.* 16: 227–38

Sakimura, K., Kushiya, E., Obinata, M., Takahashi, Y. 1985. Molecular cloning and the nucleotide sequence of cDNA for neuron-specific enolase messenger RNA of rat brain. *Proc. Natl. Acad. Sci. USA* 82: 7453–57

Scott, J., Selby, M., Urdea, M., Quiroga, M., Bell, G. I., Rutter, W. J. 1983. Isolation and nucleotide sequence of a cDNA encoding the precursor of mouse nerve growth factor. *Nature* 302: 538–40

Seeburg, P. H., Adelman, J. P. 1984. Characterization of cDNA for precursor of human luteinizing hormone releasing hormone. *Nature* 311: 666–68

Seeburg, P. H., Shine, J., Martial, J. A., Baxter, J. D., Goodman, H. M. 1977. Nucleotide sequence and amplification in bacteria of the structural gene for rat growth hormone. *Nature* 270: 486–94

Steinert, P. M., Steven, A. C., Roop, D. R. 1985. The molecular biology of intermediate filaments. *Cell* 42: 411–19

Stoffel, W., Hillen, H., Schroeder, W., Deutzmann, R. 1983. The primary structure of bovine brain lipophilin (proteolipid apoprotein). *Hoppe-Seyler's Z. Physiol. Chem.* 364: 1455–66

Sugimoto, K., Nukada, T., Tanabe, T., Takahashi, H., Noda, M., et al. 1985. Primary structure of the β-subunit of bovine transducin deduced from the cDNA sequence. *FEBS Lett.* 191: 235–40

Sutcliffe, J. G., Milner, R. J., Shinnick, T. M., Bloom, F. E. 1983. Identifying the protein products of brain-specific genes with antibodies to chemically synthesized peptides. *Cell* 33: 671–82

Sutcliffe, J. G., McKinnon, R. D., Tsou, A.-P. 1986. Gene expression in mammalian brain. In *Role of RNA and DNA in Brain Function*, ed. A. Giuditta, B. B. Kaplan, C. Zomzely-Neurath, pp. 23–31. Boston: Martinus Nijhoff

Takahashi, N., Roach, A., Teplow, D. B., Prusiner, S. B., Hood, L. 1985. Cloning and characterization of the myelin basic protein gene from mouse: One gene can encode both 14 kd and 18.5 kd MBPs by alternate use of exons. *Cell* 42: 139–48

Tanabe, T., Nukada, T., Nishikawa, Y., Sugimoto, K., Suzuki, H., et al. 1985. Primary structure of the α-subunit of transducin and its relationship to *ras* proteins. *Nature* 315: 242–45

Travis, G. H., Naus, C. G., Morrison, J.

H., Bloom, F. E., Sutcliffe, J. G. 1987. Subtractive cDNA cloning and analysis of primate neocortex mRNAs with regionally-heterogeneous distributions. *Neuropharmacology* 26: 845–54

Tsou, A.-P., Lai, C., Danielson, P., Noonan, D. J., Sutcliffe, J. G. 1986. Structural characterization of a heterogeneous family of rat brain mRNAs. *Mol. Cell. Biol.* 6: 768–78

Ullrich, A., Shine, J., Chirgwin, J., Pictet, R., Tischer, E., Rutter, W. J., Goodman, H. M. 1977. Rat insulin genes: Construction of plasmids containing the coding sequences. *Science* 196: 1313–19

Ullrich, A., Gray, A., Berman, C., Dull, T. J. 1983. Human β-nerve growth factor gene sequence highly homologous to that of mouse. *Nature* 303: 821–25

Valentino, K. L., Winter, J., Reichardt, L. F. 1985. Applications of monoclonal antibodies to neuroscience research. *Ann. Rev. Neurosci.* 8: 199–232

Van Ness, J., Maxwell, I. H., Hahn, W. E. 1979. Complex population of nonpolyadenylated messenger RNA in mouse brain. *Cell* 18: 1341–49

Villa-Komaroff, L., Efstratiadis, A., Broome, S., Lomedico, P., Tizard, R., Naber, S., Chick, W., Gilbert, W. 1978. A bacterial clone synthesizing proinsulin. *Proc. Natl. Acad. Sci. USA* 75: 3727–31

Willard, H. F., Riordan, J. R. 1985. Assignment of the gene for myelin proteolipid protein to the X chromosome: Implications for X-linked myelin disorders. *Science* 230: 940–42

Wood, T. L., Frantz, G. D., Menkes, J. H., Tobin, A. J. 1986. Regional distribution of messenger RNAs in postmortem human brain. *J. Neurosci. Res.* 16: 311–24

Yang-Feng, T. L., deGennaro, L. J., Francke, U. 1986. Genes for synapsin I, a neuronal phosphoprotein, map to conserved regions of human and murine X chromosomes. *Proc. Natl. Acad. Sci. USA* 83: 8679–83

Yatsunami, K., Khorana, H. G. 1985. GTPase of bovine rod outer segments: The amino acid sequence of the α subunit as derived from the cDNA sequence. *Proc. Natl. Acad. Sci. USA* 82: 4316–20

Yatsunami, K., Pandya, B. V., Oprian, D. D., Khorana, H. G. 1985. cDNA-derived amino acid sequence of the subunit of GTPase from bovine rod outer segments. *Proc. Natl. Acad. Sci. USA* 82: 1936–40

Zeller, N. K., Hunkeler, M. J., Campagnoni, A. T., Sprague, J., Lazzarini, R. A. 1984. Characterization of mouse myelin basic protein messenger RNAs with a myelin basic protein cDNA clone. *Proc. Natl. Acad. Sci. USA* 81: 18–22

Zipursky, S. L., Venkatesh, T. R., Teplow, D. B., Benzer, S. 1984. Neuronal development in the *Drosophila* retina: Monoclonal antibodies as molecular probes. *Cell* 36: 15–26

Zipursky, S. L., Venkatesh, T. R., Benzer, S. 1985. From monoclonal antibody to gene for a neuron-specific glycoprotein in *Drosophila. Proc. Natl. Acad. Sci. USA* 82: 1855–59

Ann. Rev. Neurosci. 1988. 11 : 199–223

ANIMAL SOLUTIONS TO PROBLEMS OF MOVEMENT CONTROL: The Role of Proprioceptors

Z. Hasan and D. G. Stuart

Department of Physiology, University of Arizona, Health Sciences Center, Tucson, Arizona 85724

INTRODUCTION

In this review we address some of the problems encountered by the central nervous system (CNS) for the control of limb movement, and the strategies employed for their resolution by the utilization of proprioceptive information.[1] The problems to be discussed arise partly from the mechanics of the musculoskeletal system and partly from the interactions of this system with the physical environment; they are subsequently considered under these two headings.

That the CNS is able to make the available mechanical structures do its bidding is self evident. The feasibility of such proficient control in the field of robotics, however, remains only a future possibility. In fact, the motor performance of biological organisms is regarded as an existence proof of this feasibility (Raibert 1986). Although in this review we do not discuss the strategies adopted by roboticists (e.g. Shin & Malin 1984), we recognize that the elucidation of these strategies and the study of biological motor

---

[1] We use the term *proprioceptors* in its conventional sense. These are the receptors that respond to the mechanical variables associated with muscles and joints. The proprioceptors, for which adequate stimuli arise from the actions of the organism itself (Evarts 1981), include the muscle spindles, Golgi tendon organs, and joint receptors in vertebrates, and a variety of muscle-receptor and chordotonal organs in invertebrates. In contrast, exteroceptors (such as tactile receptors) respond primarily to external stimuli. The force-sensitive receptors in the exoskeleton of many invertebrates, however, can function as proprioceptors (Libersat et al 1987).

199

0147–006X/88/0301–0199$02.00

control are likely to be of mutual benefit (Hinton 1984). Accordingly, the approach we have adopted is intended to emphasize the issues common to the control of any articulated limb, be it that of a vertebrate, an arthropod, or a robot.

We have, however, limited the scope of this review to the consideration of some issues whose resolution seems to necessitate afferent input to the CNS. This emphasis is in keeping with the current swing of the pendulum of opinion in the field of motor control. It has become increasingly clear since the late 1970s that afferents play a crucial role in modulating the broad features of motor output, even though the latter can be generated autonomously (Grillner 1985). The motor patterns generated in the absence of afferent input are quite labile (Grillner & Zangger 1974, Pearson 1985). In the 1960s, afferents were considered to provide a tonic bias to the central networks for the generation of movement programs (e.g. Wilson & Wyman 1965), but this view seems no longer sufficient. (See Altman 1982, Horsmann et al 1983, Lennard & Hermanson 1985, Pearson 1985, Delcomyn 1985.)

The new information concerning afferent effects in the course of movement shows that these effects are far more complex and interesting than was envisaged when the traditional concepts of servomechanistic control systems held sway. For instance, the central pattern generator has been found to gate the afferent inflow so that the effect of the latter varies during the movement cycle (Bayev & Kostyuk 1982, Reichert & Rowell 1986, Sillar & Skorupski 1986). In addition, the afferent inflow can be crucial in selecting among the available motor patterns; these patterns encompass not only different gaits of locomotion (Wetzel & Stuart 1976), but entirely different modes such as flight (Ritzman et al 1980) or swimming (Bévengut et al 1986). Analogously, the design of an artificial walking machine has employed peripheral sensors for roles that go beyond the role of servo control (Hirose 1984). In the case of biological organisms, however, the diverse roles of proprioceptors have not all received their due share of attention from experimenters. We are therefore unable to treat the different roles evenly.

Whatever may be the rules whereby afferent signals sculpt the CNS output to the muscles for generating a movement, there can be no question that the efferent output signals must be tailored so as to take into account the complex mechanical interactions among the joints and between the limb and the external world. That the mechanical interactions can have profound effects is illustrated by the fact that some of the kinematic features of the locomotor cycle can be elaborated without the necessity of precisely timed efferent signals (McMahon 1984). Thus, the observation that the knee flexes at a certain time in the cat's step cycle does not imply that knee flexor muscles must be active at that time (Hoy & Zernicke 1985). Similarly, the transition from flexion to extension in the knee can be entirely due to a whipping motion initiated from the pelvis (Loeb et al 1985; see also Williams 1985). Evidently, the CNS can exploit the physics of the limb to simplify the generation of the efferent pattern (Greene 1982). These and other observations underscore the timeliness of an "outside-to-inside" approach to the study of motor control, in which the starting point is the consideration of the problems posed by the mechanical system, which are solved by the CNS (Loeb 1987). This is the approach we have adopted in this review.

PROPRIOCEPTORS AND STABILITY

According to a recent reinterpretation of Sherrington's ideas (Evarts 1981, 1985), the classification of peripheral sensors into exteroceptors and proprioceptors is of crucial importance in regard to the types of modulation of a motor program that result from their activity (cf Lennard 1985, Möhl 1985). Exteroceptors respond to conditions in the external environment, and their activity is responsible for significant, even drastic, alterations in the motor output (Forssberg 1979). Proprioceptors, according to Evarts' interpretation, sense primarily the effects of self-generated actions, and are only secondarily responsive to external perturbations. Moreover, their effects are "mild" inasmuch as their activity results in small alterations in the motor output.

The mild effects of proprioceptive afferents, however, should not be considered inconsequential simply because of their small magnitude. For example, the seemingly simple task of maintaining a constant position or force, even in the absence of external perturbations, cannot be performed adequately without input from muscle afferents (Marsden et al 1984). Experiments, in particular, on the monkey's jaw (Goodwin et al 1978), the human leg (Mauritz & Dietz 1980, Diener et al 1984), the human thumb (Rothwell et al 1982), and the human wrist (Sanes et al 1985), have revealed pronounced tremor and/or drift resulting from the absence of normal proprioceptive feedback. Also, Cussons et al (1980) have shown that elbow tremor is enhanced when tendon vibration is used to occlude the responsiveness of muscle spindles to changes in length. Thus, even without externally applied perturbations, proprioceptors are important for stabilization of self-generated motor output in postural states. This point of view differs sharply from the one usually encountered in the design of servo-mechanisms, namely, that feedback is likely to create instability whereas the open-loop system is stable.

With regard to the control of centrally generated movement, the servomechanistic point of view has failed to admit the possibility that the role played by proprioceptive activity could be different in movement control than in postural regulation (Hasan et al 1985). The observed efficacy ("loop gain") of the proprioceptive feedback during centrally generated movement, however, is too small to support this point of view (Bizzi et al 1978, Stein 1982). Nevertheless, feedback can, in a sense, stabilize the overall pattern of movement, even if the feedback is not efficacious at every moment of time. For instance, feedback is not essential for the generation of the wing-flapping rhythm of the locust (Wilson 1961). But a locust does not exhibit stable flight after extirpation of the muscle receptors because the appropriate fine adjustments of the wing beats are not possible without muscle receptors (Wendler 1974, Altman 1982, Möhl 1985). Evidently, what is being stabilized is not a servomechanism that controls the displacement of the wing according to a prescribed template on a moment-to-moment basis, but rather the tendency (which is slow in comparison with the wing beat frequency) for the mechanical structure to topple due to slight morphological asymmetries and whiffs of air currents.

Even this type of stability, however, may not be desirable during intended changes in the direction of flight. As suggested by Brown (1963) in the context of bird flight, "While an engineer designs an aircraft with the maximum stability consistent with the manoeuvrability required, the animal evolves a system with the maximum manoeuvrability, and consequent instability, that the nervous system can control." There is, therefore, no reason to expect that the stabilizing role of proprioceptive activity would remain unchanging throughout the course of a movement. (Indeed, as is discussed below, destabilizing effects of afferent activity have been observed during movements.) Centrally mediated stabilization based on afferent feedback has the advantage that it can be turned off temporarily in the interests of maneuverability, unlike the stabilization conferred by peripheral structures. A bicycle, for instance, which achieves its stability via the nervous system of the rider, is far more maneuverable than a tricycle controlled by the same rider.

MUSCULOSKELETAL CONSIDERATIONS

For the present we focus attention on the mechanics of the musculoskeletal system comprising a limb, and leave out of consideration any changes in the conditions external to the limb. Ranging from the mechanics of a single muscle to multijoint coordination, we attempt to identify some of the problems the CNS must solve. For each of these problems we provide a rationale as to how the CNS might utilize input from proprioceptors to effect an appropriate solution, and summarize the experimental evidence that bears on the issue.

Control of a Muscle

Consider the task of activating a muscle, which is free to change its length, so that it maintains a constant length in the presence of a constant opposing force. In view of the well-known length-tension relationship, a certain level of muscle activation (i.e. a certain number of recruited motor units discharging at certain rates) should enable the muscle to oppose the applied force exactly at a certain length, but, in fact, this state of equilibrium may or may not be a stable one. A small mismatch between the muscular force and the applied force, which allows the muscle to be stretched beyond its short-range elastic limit, would cause the muscle force to decline substantially (Rack & Westbury 1974), an event that would further accentuate the mismatch and lead to further elongation of the muscle. This "yielding" behavior of muscle is clearly a potential source of instability. Other complications of muscle mechanics include the decline in steady-state force below the isometric value during constant velocity stretch (Joyce et al 1969), history-dependent effects such as "sag" (Burke et al 1976), "catch" (Burke et al 1971), "rest-shortening," and "rest lengthening" (Houk et al 1971), not to speak of the phenomena collectively known as muscle fatigue (Porter & Whelan 1981, Bigland-Ritchie et al 1986). It is, therefore, a problem just to keep the muscle at a certain length against a constant force (Marsden et al 1984). The problem can be solved if muscle activation, instead of being kept constant, were adjusted in response to changes in muscle length in such a way that even if the increase in length exceeded the short-range elastic limit, the muscle force would not decline. Adjustment of muscle activation in response to change in length does not necessarily imply the regulation of muscle length via a set-point servomechanism; the adjustment need only assure stability around whatever the muscle length happens to be.

While the evidence for servoregulation of muscle length is weak (Stein 1982), there is clear experimental support for the idea that the yielding behavior of muscle is prevented when autogenetic reflexes are present. Nichols & Houk (1976) demonstrated in decerebrate cats that the variation of soleus muscle force was very different in response to the same applied stretch, depending upon whether or not the appropriate dorsal roots were intact. These authors made sure that the pre-stretch state of the muscle was the same in the two cases, and showed that whereas muscle force fell during continuing muscle stretch in the deafferented preparation, such behavior was not exhibited when reflex modulation of muscle activity was allowed to take place. (See also Nichols 1985.)

In mammals, the receptors responsible for yield compensation are most likely to be muscle spindle afferents (Houk & Rymer 1981). The spindle afferents, which are particularly well suited to the detection of small stretches of the muscle, may therefore be seen as signaling the impending decline in muscle force; the decline, however, is preempted by reflex action. As to the reason the reflexly recruited motor units themselves do not exhibit a force decline during stretch, it is apparently a property of newly recruited muscle fibers (Cordo & Rymer 1982). The peculiar properties of muscle fibers, which may have evolved for metabolic efficiency rather than for mechanical stability, are used to advantage for stabilization via the stretch reflex.

The preceding discussion has been focused on the role of muscle receptors in stabilization when the aim is to maintain equilibrium in a postural state. Relatively little information is available concerning this role during movements. It is clear, however, that yield compensation would be important during lengthening contractions. During shortening contractions, though, there seems to be little need for stabilization mediated by muscle receptors, since in this case the behavior of a muscle is similar in the presence or in the absence of reflexes (Nichols & Houk 1976). In fact, for relatively fast movements in man, the reflex responsiveness of an actively shortening muscle is depressed (Gottlieb & Agarwal 1980, 1984). The situation with regard to slow movements is less clear, as the reflex effects are far from negligible (Soechting et al 1981) and may play a role in smoothing out slight irregularities of movement. The primary spindle afferents are exquisitely sensitive to such irregularities (Burke 1980, Vallbo 1981). This role of the proprioceptors would be in line with the notion that these receptors serve to sculpt the motor output during self-generated actions.

It is not clear, a priori, how the muscle spindle would respond when muscle length and the efferent drive to the receptor are both changing simultaneously (Hulliger 1984, Prochazka 1986). In the presence of the normal efferent drive, the afferent signal from a spindle in an active muscle may indicate the departure from expected kinematics, rather than the absolute muscle length and movement velocity (Loeb 1984). On the other hand, the signals from a spindle in a passively stretched, antagonist muscle are more likely to provide information in absolute terms about the kinematic state of the joint (Hasan & Stuart 1984). Although less often studied, spindle signals from passive antagonist muscles are likely to be equally important as those from the active movers. Experiments in man that have utilized the technique of increasing spindle discharge by tendon vibration have shown that spindle signals from antagonist muscles are as potent as those from agonists insofar as the reflex response of the agonist muscles to imposed joint movement is concerned (Matthews & Watson 1981). Spindles in antagonist muscles are also potent in generating errors in the final joint position (Capaday & Cooke 1981, Bullen & Brunt 1986). In a different type of experiment, carried out in the spinalized turtle, it has been found that afferent signals from the antagonist muscles are crucial in order for the limb to exhibit resistance against an obstruction placed in its path (Hasan & Sasaki 1986). Perhaps the efferent control of spindles in active, agonist muscles is more pertinent for fine corrections and stabilization than for keeping track of absolute position or velocity.

Control of a Joint

Three types of problems, which arise from the following considerations, are discussed.

1. For the performance of fast movements muscle force cannot be turned on instantaneously.
2. The mechanical advantage (leverage) of a muscle varies with joint angle.
3. Many joints have more than one degree of freedom of movement, and even single-freedom joints have several muscles acting across them.

1. In the case of fast, ballistic movements of a joint, the smoothing and stabilizing roles of proprioceptive signals discussed above would be vitiated by propagation delays, nor could they be very useful for such movements. Other roles have been discovered, however, even in the case of such a quintessentially ballistic movement as the locust's jump. This movement is discussed below in relative detail since it exemplifies the importance of mechanical considerations in understanding proprioceptive effects.

The jumping movement of the locust is comprised of preparatory and triggering phases. The preparatory phase involves cocontraction of the flexor and extensor tibiae muscles, a cocontraction that is maintained by afferents from a variety of peripheral receptors that respond to tension in the extensors (Heitler & Burrows 1977a,b). It is noteworthy that in an aroused locust, at least one such receptor (the subgenual organ) produces progressively increasing excitation in the extensor muscles via positive feedback (Heitler & Burrows 1977b). The usual emphasis on error correction by negative feedback mechanisms is obviously misleading in this instance.

The energy stored in elastic (e.g. cuticular) elements during cocontraction is released subsequently by strong inhibition of the flexor motoneurons, and results in a sudden extension that underlies the jump. An identified interneuron is responsible for the inhibition (Pearson et al 1980). The peripheral input to the inhibitory interneuron is such that the triggering of the jump is contingent on the successful completion of the preparatory phase (Pearson 1981). These effects of proprioceptive input are thus seen to be components of an orderly scheme in which flexor activity is at first promoted in order to store elastic energy and later inhibited in order to release the energy.

This scheme can be seen as an elegant solution to the problem posed by the sluggish responsiveness of muscle to neural commands (Partridge & Benton 1981). If muscles could respond with maximal force immediately upon receiving the command, a jump could perhaps be performed simply by activation of the extensors. Slow rise of muscle force precludes this simple strategy and necessitates a preparatory phase (cf Zajac 1985).

The example of the locust's jump demonstrates that the proprioceptive effects can be understood in the context of the mechanics of the entire, multiphase performance. In the case of mammals this understanding is presently lacking, as insufficient experimental emphasis has been placed on the complex mechanical interactions of muscular effects and the storage and transfer of energy. A further complicating factor is that an important class of proprioceptors, the muscle spindles, are subject to efferent control like the muscle receptors in crustacean limbs (Bush & Cannone 1985) but unlike certain receptors in insect limbs. Although a good deal is known about the wealth of the central actions of mammalian proprioceptive afferents as revealed by electrophysiological techniques (Baldissera et al 1981, McCrea 1986), it is difficult to sort out the relative significance of the various synaptic effects in the performance of different actions. These observed effects, however, provide important constraints for any theory of motor control.

2. Another problem in the control of a joint arises from the circumstance that the mechanical advantage for any muscle spanning the joint depends upon the angle of the joint (with the possible exception of muscles whose tendons wrap around the joint as if on a pulley). In other words, the same muscle force translates into a small amount of joint torque when the muscle's lever arm is short, compared to the torque contribution of the muscle when the lever arm is longer. Since the geometry of the origin and insertion of many muscles is such that the lever arm is short when the joint is fully extended or flexed, and longest at an intermediate angle of the joint (e.g. An et al 1981), it follows that the mechanical advantage is greatest at the intermediate position. This phenomenon in conjunction with the length-tension relationship for muscle implies that, for a given level of muscle activation, the torque contribution of the muscle would not vary uniformly with joint angle, but would rather exhibit a peak at a certain angle. This peak is indeed observed, and as a consequence the same torque is exerted at two different joint angles for the same degree of muscle activation (Hasan & Enoka 1985).

The nonuniqueness of joint angle for the same torque presents a problem for the control of movement. For instance, if a flexion movement is desired, a transient increase in flexor muscle activity would certainly initiate the movement, but the steady-state flexor activity at the final, more flexed angle may need to be smaller than that at the initial angle, depending upon the two angles (Hasan & Enoka 1985). In other words, the generation of the initial and the steady-state components of flexor muscle activity would have to be based on quite different rules.

Conceptually, there is a simple way of solving the problem posed by the nonuniform change in lever arm with joint angle. Afferent signals from the muscle can be utilized to augment muscle activation in such a way that the decrease in the lever arm with increase in muscle length, which occurs over a certain range, is offset by increase in muscle activation over this range. The torque, then, would be a uniformly increasing function of joint angle, for a given amount of descending excitation impinging on the motoneurons. There is no direct evidence for this scheme, but the experiments of Feldman (1980) have demonstrated that the torque-angle relationship does not exhibit a peak, but is rather a monotonic one over the range of motion, when the joint angle and the torque change as a result of unexpected perturbation. The reflex (revealed by the unexpected perturbation) seems to compensate automatically for lever-arm variations.

3. The control of a set of muscles acting on a single joint presents several problems. One arises from the apparent redundancy of the muscles, i.e. the number of muscles is usually so large that it is possible, in principle, to effect the same movement by using more than one combination of muscles and activation patterns. The problem of how the CNS decides among the options has received scant experimental attention, perhaps because of the wide variation in muscle use among individuals (cf Howard et al 1986). It is clear, however, that during the acquisition of a skill, the muscular involvement is altered progressively (Cooke 1980b), thus implicating the contribution of muscle receptors, albeit indirectly (cf Marsden et al 1984). Similarly indirect are the psychophysical studies that have revealed the importance of muscle receptors for the subjective sense of the angular position of a joint. (See McCloskey 1981, Matthews 1982, Clark et al 1985.)

Another problem associated with the control of a joint arises from the fact that many joints exhibit more than one degree of freedom of movement, and the muscles spanning the joint cannot be partitioned into sets, one for each degree of freedom (Buchanan et al 1986). For example, the effect of the activation of the biceps brachii on the elbow is a combination of flexion and supination; the relative strengths of these actions, moreover, depend on the current angles of flexion and supination. In order to perform either of these movements, or a combination thereof, the CNS must engage the biceps, along with other muscles, to varying degrees depending upon the *initial configuration* of the joint (Gielen & van Zuylen 1986). It is difficult to imagine how muscle activity could be apportioned correctly in the absence of afferent information about the initial configuration (Lacquaniti & Soechting 1982).

The idea has been advanced that neural commands specify only the final equilibrium position (Polit & Bizzi 1979), a notion that renders irrelevant the information about the initial position. This notion cannot be generally applicable in view of the following observations. (*a*) The equilibrium point, as specified by the neural commands, does not jump to the final position, but rather makes a gradual transition from the initial to the final position (Bizzi et al 1982, 1984). (*b*) When a viscous load is present the EMG pattern is modified suitably (Day & Marsden 1982). (*c*) As discussed above, the initial and steady-state changes in EMG follow quite different rules. One-degree-of-freedom, unloaded movements to a target, however, can be performed after deafferentation of the limb (Polit & Bizzi 1979). The afferent information about the initial configuration may be of greater importance in movements with multiple degrees of freedom.

Multijoint Coordination

In the experiments of Polit & Bizzi (1979), the animal's task was to perform elbow flexion-extension movements directed toward visual targets, while the shoulder joint was stabilized by the apparatus. As these authors report, when the shoulder was placed in a position different from the one in which the task had originally been learned, the animal after deafferentation could not perform the elbow task. This shows that an important role is played by afferents in coordinating the motions about the various joints in a limb.

Even in spinal-sectioned animals, provided the afferents are intact, the initial configuration of a limb influences subsequent, reflexly elicited motor activity in that and other limbs. For example, the classic crossed-extension reflex has been reported to convert to a crossed-flexion reflex when the limb was initially in an extended configuration (Rossignol & Gauthier 1980). A more "global" interaction has been described by Fukson et al (1980) in spinalized frogs, concerning the wiping reflex of the hindlimb elicited by irritation of a spot of skin on the forelimb. The trajectory of movement of the wiping hindlimb was altered when the initial configuration of the forelimb was changed, even though the spot of skin that was stimulated remained the same. Clearly, therefore, the proprioceptive information about the initial configuration has far reaching consequences for the spinal generation of subsequent movement.

We consider now the dynamical interactions among the joints *during* the course of a multijoint movement. When more than one joint in a limb is free to move, a torque due to a muscle spanning only one joint can result in rotations about several joints. The apparent torque at these other joints is a purely mechanical phenomenon, which arises from the inertias of the linked segments of a limb. We refer to this phenomenon as the *inertial interaction* among the segments, though other authors have subdivided this interaction into inertial and other components (Hollerbach & Flash 1982, Hoy & Zernicke 1985). As shown by Hollerbach & Flash (1982), this interaction has significant consequences even in the case of relatively low-speed movements of the human arm. An extreme example of the importance of inertial interactions is provided by the transient antagonism of the effects of soleus and gastrocnemius muscles during recovery from a disturbance of upright posture (Gordon et al 1986), despite the fact that these muscles are considered synergists. Dynamics, it would appear, can be counterintuitive.

Inertial interactions among the joints obviously present a problem for the CNS in controlling a multijoint movement, especially since the inter-action terms, as calculated from classical dynamics, depend in complicated ways on the joint angles and angular velocities. The idea that afferent information may play an important part in solving this problem is sup-ported by certain observations on a rather specialized behavior: a cat's paw-shaking movements in response to mild stimuli (Smith & Zernicke 1987). During this behavior, fast, rhythmic movements are exhibited about the ankle joint, whereas the knee joint shows little movement even though it is not locked. Hoy et al (1985) have shown that the knee joint is actively stabilized during the paw-shake response, i.e. it would have moved far more as a result of the inertial interaction with the shaking paw, were it not for the muscular torques about the knee that resist the inertial effect. (Cf Green 1982 for the arm.) Interestingly, when a plaster cast was used to immobilize the knee and ankle joints, there was little effect on the activity of ankle muscles, but knee extensor activity was reduced sub-stantially (Sabin & Smith 1984). (In the normal, unimpeded behavior there is little movement about the knee.) Similarly, the addition of weights to the paw affected the activation of knee muscles but not of ankle muscles (Hart et al 1987).

One can conclude from these observations that the fast movements about the ankle are generated without taking into account the afferent feedback, but the muscular torques that are necessary about the knee joint (in order to counteract the pronounced inertial effects due to the shaking paw) are generated in the light of the peripheral mechanical events (Smith & Zernicke 1987). Thus, in the case of the paw-shake response, afferent signals play an important role in stabilizing the knee, which entails solving the dynamics problem presented by inertial interactions. Whether the afferents are needed for solving the dynamics problem in other contexts remains to be investigated.

Most movements exhibited by animals involve the coordination of rotations about several joints in a manner which is recognized immediately as "graceful," as contrasted with the fractionated movements one associates with present-day robots. This qualitative difference, one may surmise, stems from the common practice in robotics to ignore inertial interactions in order to avoid the enormous computational load that would otherwise be entailed (Craig 1986). It is not known, in general, how this problem is solved in living systems, but kinematic studies have revealed certain rules that seem to govern the solution (Atkeson & Hollerbach 1985).

For example, when the human arm reaches out to grasp an object that is so placed as to require shoulder flexion and elbow extension, the angular velocities at the shoulder and the elbow reach their peaks at the same time (Soechting & Lacquaniti 1981, Morasso 1981). Note that because of inertial effects the elbow would tend to extend anyway at the initiation of shoulder flexion. Soechting & Lacquaniti (1981) have suggested that the observed kinematic coupling, which serves to take advantage of the inertial interaction, can result from feedback regulation based on peripheral force receptors. This suggestion has been supported by recent simulation studies (Lacquaniti & Soechting 1986a). However, when the target is so placed that the inertial interaction between the shoulder and the elbow must be resisted by muscular activity, the peak velocities at the two joints do not coincide in time (Morasso 1981, Lacquaniti & Soechting 1982). Perhaps this indicates that a different strategy is employed when inertial interactions are of no avail.

If the utilization of inertial interactions involves an afferent-based mechanism, as seems to be suggested by the examples given above, there is pressing need to reconcile this purported mechanism with what is known about the synaptic connections of proprioceptive afferents to motoneurons that have their effects on neighboring joints as well as their "own" joint (Jankowska 1984).

The generation of appropriate efferent signals for a multijoint movement is complicated not only by the inertial interactions among the joints but also by the fact that most muscles span more than one joint. Simultaneous flexion of the hip and the knee, as for example occurs in the phylogenetically old withdrawal reflex, involves the activation of some hip flexor muscles that also tend to extend the knee, and of some knee flexor muscles that tend to extend the hip (Sherrington 1910). It is remarkable that in the cat's hindlimb there is only one, relatively small muscle (the medial portion of sartorius) that has the anatomical connections to cause flexion around both the knee and hip joints. Therefore the coordination of even a primitive behavior such as the withdrawal reflex entails the activation of biarticular muscles whose "wrong" action on one of the joints must be overcome.

Many biarticular muscles, however, exhibit some degree of compartmentalization (reviewed in Stuart et al 1988), in that a part of the muscle exerts its action primarily on one joint while another part has the greater effect on the other joint. Electromyographic recordings during locomotion have revealed differences in the activation patterns of the different compartments, in some though not all cases. The parcellation of activity among the compartments is most likely a feature of the central pattern generator, yet it is noteworthy that the primary spindle endings located in one compartment show a stronger monosynaptic influence on motoneurons innervating the same compartment as compared to other motoneurons that innervate other compartments. The roles of such synaptic influence (including the roles discussed above in the context of the control of a muscle) may therefore be compartment-specific rather than specific to each anatomically defined muscle. If this is the case, the biarticular nature of the muscle as a whole may not present a special problem.

The preceding argument is buttressed by a negative finding on a rather specialized biarticular muscle. The cat's semitendinosus is comprised of two muscular portions in series, each with its own nerve branch. These two "compartments" obviously cannot exert different effects on the two joints spanned by the semitendinosus. In consonance with this lack of difference, EMG recordings during locomotion have revealed simultaneous activation of the two portions (Murphy et al 1981, English & Letbetter 1981), and it has been found that the primary spindle afferents originating in one portion do not have a preferential synaptic effect on motoneurons innervating that portion (Botterman et al 1983).

Although the experimental findings concerning compartmentalization and reflex partitioning are few in number (Stuart et al 1988), the view is emerging that, from the standpoint of the CNS, each compartment of a biarticular muscle may be separately controllable (cf Loeb 1987). The separate controllability seems to go hand in hand with some difference in the synaptic projection of the respective afferents.

ENVIRONMENTAL CONSIDERATIONS

The ability of animals to adjust their movement sequences according to environmental circumstances is so commonplace that one seldom stops to marvel at it. In the words of Gallistel (1980), "The flexible and rapid adoption of new action sequences in response to changes in circumstance is . . . closely allied to our intuitions about intelligence itself." Of course, as argued above, proprioceptors are not the primary source of information about the environment. Nevertheless, their signals are the first to reach the CNS in many instances, and there is evidence that they do modify the motor sequence. In what follows, we attempt to describe some of these modifications from the point of view of functional appropriateness.

Unexpected Perturbations

The study of muscle responses to unexpected perturbations in postural states has permeated most of the developments, both conceptual and technical, in the field of motor control. The functional significance of the well-known stretch reflex, however, remains elusive, especially in light of the changeability of the reflex (Diener et al 1983, Nichols & Steeves 1986). Its possible role in stabilizing the mechanical variables against the destabilizing tendencies inherent in muscles and in some joints is discussed above. (See also, Akazawa et al 1983.) Here we emphasize the proprioceptively mediated responses to sudden perturbation that cannot be subsumed readily under the rubric of the stretch reflex.

In an animal maintaining a postural state, a sudden displacement of the center of mass by some external forces can result in toppling unless corrective action is taken. Traditionally, the corrective action has been thought of in terms of the resistance of each limb muscle to its own stretch, via the negative-feedback pathways of the stretch reflex (Sherrington 1924). If the displacement is in the anteroposterior direction ("sway"), this notion appears to be substantiated by experiments on standing humans in which the support platform was displaced in the anterior or posterior direction (Nashner & Woollacott 1979). In these experiments, however, the monosynaptic response was relatively insignificant. Moreover, the significant EMG responses occurred earlier in the more distal muscles compared to those in proximal muscles of the leg. This distal to proximal progression appears to be of biomechanical importance, since its reversal, in certain pathological states, is associated with collapse of the knees consequent to platform displacement (Nashner et al 1983).

Other types of platform perturbation, however, elicit responses that cannot be described as stretch reflex effects (Nashner et al 1979). For instance, if the support surface for the legs is raised suddenly, an EMG response in the gastrocnemius is observed even if the stretch of this muscle is prevented by simultaneous rotation of the ankle. More remarkably, if the platform supporting one foot is raised suddenly while the other foot is lowered simultaneously, the corrective action is seen in the form of EMG increases in the shortening muscles. The alternative strategy of greater activation of the stretched muscles, which is not the strategy adopted, would have led to a one-legged stance and a possible lateral fall. All these various observed strategies are exhibited even on the first trial, a finding that suggests that they are not the products of learning. (See also Rushmer et al 1983.) Consequently the view has emerged that a sudden perturbation elicits one of a discrete set of strategies (or "functional synergies") depending on the type of perturbation (Nashner & McCollum 1985); the synergies are not comprised simply of the stretch reflexes for all the muscles. It is possible, however, that the synergies are not discrete, but that each muscle responds in a particular manner to a given set of inputs (Pellionisz & Llinás 1982).

In the case of anteroposterior displacements, when such displacements are applied via a manipulandum attached to the hand of a standing human, there still seems to be a biomechanical necessity for the corrective actions of the leg muscles to occur in a distal-to-proximal sequence, which is what is exhibited (Cordo & Nashner 1982). It would appear that these stereotyped synergies are well suited to the requirements of mechanically appropriate corrective action, and they represent the first line of defence since the effects of vestibular and visual inputs come into play at longer latencies (Allum 1983, Nashner & McCollum 1985).

As in the example of reciprocal vertical platform displacements, muscle activation can increase in response to imposed shortening (Angel 1983). Lacquaniti & Soechting (1986b) have found consistent examples of such behavior in the case of the human biceps brachii muscle (which they distinguish from the classical "shortening reaction"). They report such reflex reversal when both the elbow and the shoulder are free to move, and thus the inertial interaction between them can come into play. Specifically, a posterior displacement of the upper arm leads initially to elbow flexion because of the inertia of the forearm; although this entails a net decrease in biceps length, the muscle consistently exhibits increased EMG. They advance the argument that corrective action is taken in response to changes in some globally defined mechanical variable rather than local, muscle variables. It may be recalled that in the context of self-generated multijoint movements we had emphasized the possible stabilizing role within submuscular compartments. The data concerning responses to imposed perturbations, on the other hand, point toward the importance of global effects involving muscles across several joints. The contradiction between these two views may only be a superficial one, stemming from classical servomechanistic ideas in which no distinction is made between errors arising from internal or external disturbances.

Turning now to the effect of environmental perturbations during self-generated movement, corrective actions in the nature of resistance to the perturbation have been described (Cooke 1980a, Gottlieb & Agarwal 1980, Mortimer et al 1981, Soechting et al 1981, Evoy & Ayers 1982, Marsden et al 1983, Lee et al 1986). The efficacy of the reflex as judged by the EMG response seems, however, to be reduced considerably during voluntary movement (Gottlieb & Agarwal 1980, 1984). Whereas the modulation, and even reversal, of reflexes elicited by exteroceptive stimuli has been studied extensively during movement (Forrsberg 1979, Abraham et al 1985), it is more difficult experimentally to separate the effects of position or force perturbations from the records of ongoing movement. Andersson & Grillner (1981) have studied the effects, recorded as electroneurographic responses, of mechanical perturbations during "fictive" locomotion in the cat with low-spinal transection. According to these authors, imposed hip flexion (as well as extension), if it occurred early in the flexor part of the cycle, reinforced the flexor activity and prolonged the cycle duration. Toward the end of the normally occurring flexor burst, however, the effect was sensitive to the direction of the imposed perturbation, but the responses exhibited "positive feedback" rather than the negative feedback one associates with resistance to the perturbation. In this phase the perturbation also caused a marked shortening of the cycle duration. These complicated reflex effects do not support the idea that the nervous system is poised to resist imposed perturbations whenever they may occur in the locomotor cycle. Rather, the reflex effects may play a role in synchronizing the internal, neural oscillator to the external, mechanical rhythm, as is discussed below.

Positive feedback effects of proprioceptive input have been known for some time in the invertebrate literature, where the term "assistance reflex" (contrasted with "resistance reflex") is commonly employed (Evoy & Ayers 1982). What is often observed is that a resting animal responds to an externally applied perturbation by resisting it, whereas an aroused animal responds in the opposite manner, via an assistance reflex. (See, for example, DiCaprio & Clarac 1981, Vedel 1982, Skorupski & Sillar 1986.) One may surmise that during ongoing, self-generated movements it would be counterproductive for passively stretched reflex muscles to exhibit resistance reflexes. Keeping in mind that a stretch can only result in absorption of energy from the agency that stretches the muscle, it may be appropriate to shut down the reflex or even to pump energy into the mechanical system by reversing the sign of the reflex. Whether or not assistance reflexes occur during self-generated movement (Grigg et al 1978, Akazawa et al 1982), however, the evidence strongly suggests that resistance reflexes come into play when the afferent input does not correspond to what might have been "expected" from the motor output (Barnes 1977).

Locomotor Adjustments

In the case of locomotion on a moving treadmill belt, it is well known that even spinal-sectioned animals adjust their gait as well as the cadence of their steps to the speed of the moving belt by shortening the stance phase duration when the belt speed is higher (Wetzel & Stuart 1976). This attests to the ability of the afferent signals to modify the timing of the central pattern generator. Lennard (1985) has found in the swimming turtle that the artificial injection of a proprioceptive signal can permanently reset the timing generator whereas cutaneous signals cause a temporary phase shift. The experiments of Andersson & Grillner (1981) on fictive locomotion of the spinal cat described above, also point to the importance of proprioceptors in altering the timing of the central pattern generator. Moreover, Grillner & Rossignol (1978) have shown that in spinal cats walking on a treadmill the swing phase can be initiated in one limb when the hip is extended by hand to a certain threshold, while the contralateral limb continues to walk. The transition, however, is inhibited if the ankle extensor muscle force is sufficiently large, at least in premammillary cats (Duysens & Pearson 1980) and in cockroaches (Pearson & Duysens 1976).

The transition in phase from stance to swing seen in the experiments of Grillner & Rossignol (1978) in response to extension of the hip is reminiscent of the observations of Gray & Lissmann (1940) on the spinal toad. These latter authors had interpreted their results as supporting the reflex-chain theory against the central program theory of locomotor rhythm generation. However, as pointed out by Bässler (1985, 1986) in the context of stick insect walking, the transition from stance to swing can be performed by central as well as peripheral influences. The principle that the phase transition can occur in response to proprioceptive signaling of a sufficiently retracted position of the limb, but can also occur without afferent activity if the central excitability is sufficiently high, seems to be a general principle that holds across phyla. The result of the operation of this principle is easily interpreted as allowing the animal to adjust its locomotion to moving belts. Surely, though, the survival value of this functional adjustment could not have been predicated on the existence of treadmills.

In a locomoting animal with several legs, however, the motions of any one leg are influenced not only by the muscular activity in that leg but also by the progression of the body with respect to the substrate, which makes the situation somewhat analogous to that on a treadmill. Thus, the external environment (the ground in this case) can influence the coordination of the various limbs (English & Lennard 1982; see Shik & Orlovsky 1976 for walking on paired treadmills). The mechanical coupling among the legs via the ground has important effects on the neural output, and this has been studied extensively in arthropods. For instance, in a decapod crustacean, the normal, alternating pattern of interlimb coordination has been shown to be contingent on ground contact (Clarac & Chasserat 1979, Clarac 1985). In stick insects, Dean & Wendler (1982) have found that if the swing phase of a leg is prolonged by blocking it with a rod, the swing phase of the leg in front of it is delayed. When interlimb interactions mediated via the ground are abolished by rendering the surface slippery, the legs on the same side can move at different speeds (Epstein & Graham 1983). One could multiply such examples of reflexly mediated interlimb coordination; indeed, many of these phenomena have been incorporated in a model by Cruse (1983), which may be of interest to designers of multilegged walking machines. (See also Cruse & Graham 1985, Chasserat & Clarac 1986.)

To reiterate: when any one leg of a walking animal hits the ground, it not only extends actively against the ground to generate propulsion, it is also extended passively by the forward momentum of the animal's body. Teleologically, the latter effect must not be allowed to interfere with the propulsive thrust. Indeed, that is the case as revealed by the experiments of Andersson & Grillner (1983) on the spinalized cat, in which the central network was uncoupled from generating any active movements by curarization. These authors imposed sinusoidal movements on the hip, and found that the central rhythm could be entrained by the peripheral rhythm. The phase of the entrainment was such that the efferents to different extensor muscles were active during imposed extension of the hip. This is the phase relationship one would expect from the teleological requirement mentioned above but is contrary to what would be required for resisting imposed movements of the hip.

It goes without saying that the animal's repertoire for responding to unexpected perturbations and to the substrate extends far beyond the relatively stereotyped, proprioceptively mediated effects described here. (See, for example, Pearson & Franklin 1984, Forssberg 1985.) But it is noteworthy that even these limited aspects of the repertoire exhibit a complexity whose neural basis remains to be elucidated (cf Loeb 1987).

Other Issues

It is well known that the addition of inertial or viscous loads to a limb affects the activation pattern of the muscles, both agonist and antagonist, involved in performing subsequent movements (Lestienne 1979, Day & Marsden 1982, Karst & Hasan 1987). The learning of the load must clearly involve afferent information. But, beyond stating this truism, it is difficult to be more specific. Similarly, Pearson (1981) has pointed out the necessity of afferent information in dealing with peripheral changes caused by growth, maturation, and use (cf Loeb 1983). Although Pearson (1981) has provided an example (viz the swing angle of a walking cockroach leg remains constant during growth because hair plate afferents are responsible for terminating the swing), there is little by way of systematic studies of this issue.

How proprioceptive information is utilized by the higher centers is largely unknown except in very broad terms (Brooks 1986). Perhaps it is necessary first to catalog and understand the low-level utilization of this information. Even these phenomena are far from simple.

CONCLUSIONS

The problems of movement control involve many levels of analysis. The role of proprioceptive information too is multifarious, but may be broken down into two categories. One involves the smoothing and stabilization of internally generated motor patterns, by way of rectification of certain peculiar properties of muscle, compensation for lever-arm variations, and correction for interjoint interaction effects. The other category is comprised of the roles of proprioception with respect to the physical environment. Resistance to unexpected perturbations is only one such role. Perhaps more importantly, proprioceptive signals can help select the appropriate "synergy" of response, and can even be used to assist external forces; the assistance may be important in interlimb coordination in the presence of the mechanical coupling among the limbs provided by the environment.

The "outside-to-inside" approach we have adopted in this review for describing proprioceptive effects was intended to foster a collective attack on problems of movement control by neurophysiologists and investigators of biomechanics, including roboticists. This approach is yet to be supplemented by an in-depth account of the neural circuitry that underlies the phenomenal motor performance of living things.

Acknowledgments

For their criticisms of a draft of this manuscript we would like to thank Drs. François Clarac, Keir Pearson, and John Soechting. The work we have cited from our laboratories was supported by USPHS grants HL 07249 (Department of Physiology), NS 19407 and NS 19438 (Z.H.), and NS 07888 (D.G.S.).

218   HASAN & STUART

Literature Cited

Abraham, L. D., Marks, W. B., Loeb, G. E. 1985. The distal hindlimb musculature of the cat: Cutaneous reflexes during locomotion. *Exp. Brain Res.* 58: 594–603

Akazawa, K., Aldridge, J. W., Steeves, J. D., Stein, R. B. 1982. Modulation of stretch reflexes during locomotion in the mesencephalic cat. *J. Physiol. London* 329: 553–67

Akazawa, K., Milner, T. E., Stein, R. B. 1983. Modulation of reflex EMG and stiffness in response to stretch of human finger muscle. *J. Neurophysiol.* 49: 16–27

Allum, J. H. J. 1983. Organization of stabilizing reflex responses in tibialis anterior muscles following ankle flexion perturbations of standing man. *Brain Res.* 264: 297–301

Altman, J. 1982. The role of sensory inputs in insect flight motor pattern generation. *Trends Neurosci.* 5: 257–60

An, K. N., Hui, F. C., Morrey, B. F., Linscheid, R. L., Chao, E. Y. 1981. Muscles across the elbow joint: A biomechanical analysis. *J. Biomech.* 14: 659–69

Andersson, O., Grillner, S. 1981. Peripheral control of the cat's step cycle: I. Phase dependent effects of ramp-movements of the hip during "fictive locomotion." *Acta Physiol. Scand.* 113: 89–101

Andersson, O., Grillner, S. 1983. Peripheral control of the cat's step cycle: II. Entrainment of the central pattern generators for locomotion by sinusoidal hip movements during "fictive locomotion." *Acta Physiol. Scand.* 118: 229–39

Angel, R. W. 1983. Muscular contractions elicited by passive shortening. *Motor Control Mechanisms in Health and Disease. Adv. Neurol.* 39: 555–63

Atkeson, C. G., Hollerbach, J. M. 1985. Kinematic features of unrestrained vertical arm movements. *J. Neurosci.* 5: 2318–30

Baldissera, F., Hultborn, H., Illert, M. 1981. Integration in spinal neuronal systems. *Handb. Physiol. (Sec. 1: The Nervous System), Motor Control* 2(1): 509–95

Barnes, W. J. P. 1977. Proprioceptive influences on motor output during walking in the crayfish. *J. Physiol. Paris* 73: 543–64

Bässler, U. 1985. Proprioceptive control of stick insect walking. In *Insect Locomotion,* ed. M. Gewecke, G. Wendler, pp. 43–48. Berlin/Hamburg: Paul Parey

Bässler, U. 1986. On the definition of central pattern generator and its sensory control. *Biol. Cybern.* 54: 65–69

Bayev, K. V., Kostyuk, P. G. 1982. Polarization of primary afferent terminals of lumbosacral cord elicited by the activity of spinal locomotor generator. *Neuroscience* 7: 1401–9

Bévengut, M., Libersat, F., Clarac, F. 1986. Dual locomotor activity selectively controlled by force- and contact-sensitive mechanoreceptors. *Neurosci. Lett.* 66: 323–27

Bigland-Ritchie, B. R., Dawson, N. J., Johansson, R. S., Lippold, O. C. J. 1986. Reflex origin for the slowing of motoneurone firing rates in fatigue of human voluntary contractions. *J. Physiol. London* 379: 451–59

Bizzi, E., Accornero, N., Chapple, W., Hogan, N. 1982. Arm trajectory formation in monkeys. *Exp. Brain Res.* 46: 139–43

Bizzi, E., Accornero, N., Chapple, W., Hogan, N. 1984. Posture control and trajectory formation during arm movement. *J. Neurosci.* 4: 2738–44

Bizzi, E., Dev, P., Morasso, P., Polit, A. 1978. Effect of load disturbances during centrally initiated movements. *J. Neurophysiol.* 41: 542–46

Botterman, B. R., Hamm, T. M., Reinking, R. M., Stuart, D. G. 1983. Distribution of monosynaptic Ia excitatory post-synaptic potentials in the motor nucleus of the cat semitendinosus muscle. *J. Physiol. London* 338: 379–93

Brooks, V. B. 1986. *The Neural Basis of Motor Control.* New York/Oxford: Oxford Univ. Press

Brown, R. H. J. 1963. The flight of birds. *Biol. Rev.* 38: 460–89

Buchanan, T. S., Almdale, D. P. J., Lewis, J. L., Rymer, W. Z. 1986. Characteristics of synergic relations during isometric contractions of human elbow muscles. *J. Neurophysiol.* 56: 1225–41

Bullen, A. R., Brunt, D. 1986. Effects of tendon vibration on unimanual and bimanual movement accuracy. *Exp. Neurol.* 93: 311–19

Burke, D. 1980. Muscle spindle function during movement. *Trends Neurosci.* 3: 251–53

Burke, R. E., Rudomin, P., Zajac, F. E. 1971. Catch property in single mammalian motor units. *Science* 168: 122–24

Burke, R. E., Rudomin, P., Zajac, F. E. 1976. The effect of activation history on tension production by individual muscle units. *Brain Res.* 109: 515–29

Bush, B. M. H., Cannone, A. J. 1985. How do crabs control their muscle receptors? In *Feedback and Motor Control in Invertebrates and Vertebrates,* ed. W. J. P. Barnes, M. H. Gladden, pp. 145–66. London: Croom Helm

Capaday, C., Cooke, J. D. 1981. The effect of muscle vibration on the attainment of intended final arm position during voluntary human arm movements. *Exp. Brain Res.* 42: 228–30

Chasserat, C., Clarac, F. 1986. Basic processes of locomotor coordination in the rock lobster. II. Simulation of leg coupling. *Biol. Cybern.* 55: 171–85

Clarac, F. 1985. Stepping reflexes and the sensory control of walking in crustacea. In *Feedback and Motor Control in Invertebrates and Vertebrates*, ed. W. J. P. Barnes, M. H. Gladden, pp. 379–400. London: Croom Helm

Clarac, F., Chasserat, C. 1979. Experimental modification of interlimb coordination during locomotion of a crustacean. *Neurosci. Lett.* 12: 271–76

Clark, F. J., Burgess, R. C., Chapin, J. W., Lipscomb, W. T. 1985. Role of intramuscular receptors in the awareness of limb position. *J. Neurophysiol.* 54: 1529–40

Cooke, J. D. 1980a. The role of stretch reflexes during active movements. *Brain Res.* 181: 493–97

Cooke, J. D. 1980b. The organization of simple, skilled movements. In *Tutorials in Motor Behavior*, ed. G. E. Stelmach, J. Requin, pp. 199–212. Amsterdam: North-Holland

Cordo, P. J., Nashner, L. M. 1982. Properties of postural adjustments associated with rapid arm movements. *J. Neurophysiol.* 47: 287–302

Cordo, P. J., Rymer, W. Z. 1982. Contributions of motor-unit recruitment and rate modulation to compensation for muscle yielding. *J. Neurophysiol.* 47: 797–809

Craig, J. J. 1986. *Introduction to Robotics. Mechanics and Control*. Reading, Mass: Addison-Wesley

Cruse, H. 1983. The influence of load and leg amputation upon coordination in walking crustaceans: A model calculation. *Biol. Cybern.* 49: 119–25

Cruse, H., Graham, D. 1985. Models for the analysis of walking in arthropods. In *Coordination of Motor Behaviour*, ed. B. M. H. Bush, F. Clarac, pp. 284–301. Cambridge: Cambridge Univ. Press

Cussons, P. D., Matthews, P. B. C., Muir, R. B. 1980. Enhancement by agonist or antagonist muscle vibration of tremor at the elastically loaded human elbow. *J. Physiol. London* 302: 443–61

Day, B. L., Marsden, C. D. 1982. Accurate repositioning of the human thumb against unpredictable dynamic loads is dependent upon peripheral feedback. *J. Physiol. London* 327: 393–407

Dean, J., Wendler, G. 1982. Stick insects walking on a wheel: Perturbations induced by obstruction of leg protraction. *J. Comp. Physiol.* 148: 195–207

Delcomyn, F. 1985. Insect locomotion: Past, present and future. In *Insect Locomotion*, ed. M. Gewecke, G. Wendler, pp. 1–18. Berlin/Hamburg: Paul Parey

DiCaprio, R. A., Clarac, F. 1981. Reversal of a walking leg reflex elicited by a muscle receptor. *J. Exp. Biol.* 90: 197–203

Diener, H. C., Bootz, F., Dichgans, J., Bruzek, W. 1983. Variability of postural "reflexes" in humans. *Exp. Brain Res.* 52: 423–28

Diener, H. C., Dichgans, J., Guschlbauer, B., Mau, H. 1984. The significance of proprioception on postural stabilization as assessed by ischemia. *Brain Res.* 296: 103–9

Duysens, J., Pearson, K. G. 1980. Inhibition of flexor burst generation by loading ankle extensor muscles in walking cats. *Brain Res.* 187: 321–32

English, A. W., Lennard, P. R. 1982. Interlimb coordination during stepping in the cat: In-phase stepping and gait transitions. *Brain Res.* 245: 353–64

English, A. W., Letbetter, W. D. 1981. Intramuscular "compartmentalization" of the cat biceps femoris and semitendinosus muscles: Anatomy and EMG patterns. *Soc. Neurosci. Abstr.* 7: 557

Epstein, S., Graham, D. 1983. Behaviour and motor output of stick insects walking on a slippery surface. I. Forward walking. *J. Exp. Biol.* 105: 215–29

Evarts, E. V. 1981. Sherrington's concept of proprioception. *Trends Neurosci.* 4: 44–46

Evarts, E. V. 1985. Transcortical reflexes: Their properties and functional significance. *Hand Function and the Neocortex. Exp. Brain Res. Suppl.* 10: 130–54

Evoy, W. H., Ayers, J. 1982. Locomotion and control of limb movements. In *Biology of Crustacea*, ed. D. E. Bliss, 4: 61–105. New York: Academic

Feldman, A. G. 1980. Superposition of motor programs. I. Rhythmic forearm movements in man. *Neuroscience* 5: 81–90

Forssberg, H. 1979. Stumbling corrective reaction: A phase-dependent compensatory reaction during locomotion. *J. Neurophysiol.* 42: 936–53

Forssberg, H. 1985. Phase dependent adaptations during human locomotion. In *Feedback and Motor Control in Invertebrates and Vertebrates*, ed. W. J. P. Barnes, M. H. Gladden, pp. 451–64. London: Croom Helm

Fukson, O. I., Berkinblit, M. B., Feldman, A. G. 1980. The spinal frog takes into account the scheme of its body during the wiping reflex. *Science* 209: 1261–63

Gallistel, C. R. 1980. *The Organization of Action: A New Synthesis.* Hillsdale, NJ: L. Erlbaum

Gielen, C. C. A. M., van Zuylen, E. J. 1986. Coordination of arm muscles during flexion and supination: Application of the tensor analysis approach. *Neuroscience* 17: 527–39

Goodwin, G. M., Hoffman, D., Luschei, E. S. 1978. The strength of the reflex response to sinusoidal stretch of monkey jaw closing muscles during voluntary contraction. *J. Physiol. London* 279: 81–111

Gordon, M. E., Zajac, F. E., Hoy, M. G. 1986. Postural synergies dictated by segmental accelerations from muscles and physical constraints. *Soc. Neurosci. Abstr.* 12: 1425

Gottlieb, G. L., Agarwal, G. C. 1980. Response to sudden torques about ankle in man. III. Suppression of stretch-evoked responses during phasic contraction. *J. Neurophysiol.* 44: 233–46

Gottlieb, G. L., Agarwal, G. C. 1984. Modulation of human spinal reflexes. In *Brainstem Control of Spinal Cord Function,* ed. C. D. Barnes, pp. 1–26. Orlando: Academic

Gray, J., Lissmann, H. W. 1940. Ambulatory reflexes in spinal amphibians. *J. Exp. Biol.* 17: 237–51

Greene, P. H. 1982. Why is it easy to control your arms? *J. Motor Behav.* 14: 260–86

Grigg, P., Harrigan, E. P., Fogarty, K. E. 1978. Segmental reflexes mediated by joint afferent neurons in cat knee. *J. Neurophysiol.* 41: 9–14

Grillner, S. 1985. Neurobiological bases of rhythmic motor acts in vertebrates. *Science* 228: 143–49

Grillner, S., Rossignol, S. 1978. On the initiation of the swing phase of locomotion in chronic spinal cats. *Brain Res.* 146: 269–77

Grillner, S., Zangger, P. 1974. Locomotor movements generated by the deafferented spinal cord. *Acta Physiol. Scand.* 91: 38A–39A

Hart, T. J., Cox, E. M., Hoy, M. G., Smith, J. L., Zernicke, R. F. 1987. Intralimb kinetics of perturbed paw-shake response. In *Biomechanics X,* ed. B. Jonsson, pp. 471–78. Champaign, Ill: Human Kinetics Publ.

Hasan, Z., Enoka, R. M. 1985. Isometric torque-angle relationship and movement-related activity of human elbow flexors: Implications for the equilibrium-point hypothesis. *Exp. Brain Res.* 59: 441–50

Hasan, Z., Enoka, R. M., Stuart, D. G. 1985. The interface between biomechanics and neurophysiology in the study of move-

ment: Some recent approaches. *Exerc. Sport Sci. Rev.* 13: 169–234

Hasan, Z., Sasaki, S.-I. 1986. Afferents from passive antagonist muscles underlie the limb's resistance to an obstruction during spinal locomotion in turtles. *Proc. 30th Congr. IUPS* 16: 513

Hasan, Z., Stuart, D. G. 1984. Mammalian muscle receptors. In *Handbook of the Spinal Cord,* ed. R. A. Davidoff, 3: 559–607. New York: Marcel Dekker

Heitler, W. J., Burrows, M. 1977a. The locust jump. I. The motor programme. *J. Exp. Biol.* 66: 203–20

Heitler, W. J., Burrows, M. 1977b. The locust jump. II. Neural circuits of the motor programme. *J. Exp. Biol.* 66: 221–42

Hinton, G. 1984. Parallel computation for controlling an arm. *J. Motor Behav.* 16: 171–94

Hirose, S. 1984. A study of design and control of a quadruped walking vehicle. *Int. J. Robotics Res.* 3(2): 113–33

Hollerbach, J. M., Flash, T. 1982. Dynamic interactions between limb segments during planar arm movement. *Biol. Cybern.* 44: 67–77

Horsmann, U., Heinzel, H.-G., Wendler, G. 1983. The phasic influence of self-generated air current modulations on the locust flight motor. *J. Comp. Physiol.* 150: 427–38

Houk, J. C., Rymer, W. Z. 1981. Neural control of muscle length and tension. *Handb. Physiol. (Sec. 1: The Nervous System), Motor Control* 2(1): 257–323

Houk, J. C., Singer, J. J., Henneman, E. 1971. Adequate stimulus for tendon organs with observations on mechanics of ankle joint. *J. Neurophysiol.* 34: 1051–65

Howard, J. D., Hoit, J. D., Enoka, R. M., Hasan, Z. 1986. Relative activation of two human elbow flexors under isometric conditions: A cautionary note concerning flexor equivalence. *Exp. Brain Res.* 62: 199–202

Hoy, M. G., Zernicke, R. F. 1985. Modulation of limb dynamics in the swing phase of locomotion. *J. Biomech.* 18: 49–60

Hoy, M. G., Zernicke, R. F., Smith, J. L. 1985. Contrasting roles of inertial and muscle moments at knee and ankle during paw-shake response. *J. Neurophysiol.* 54: 1282–94

Hulliger, M. 1984. The mammalian muscle spindle and its central control. *Rev. Physiol. Biochem. Pharmacol.* 101: 1–110

Jankowska, E. 1984. Interneuronal organization in reflex pathways from proprioceptors. In *Frontiers in Physiological Research,* ed. D. G. Garlick, P. I. Korner, pp. 228–37. Canberra: Austral. Acad. Sci.

Joyce, G. C., Rack, P. M. H., Westbury, D. R. 1969. The mechanical properties of cat soleus muscle during controlled lengthening and shortening movements. *J. Physiol. London* 204: 461–74

Karst, G. M., Hasan, Z. 1987. Antagonist muscle activity during human forearm movements under varying kinematic and loading conditions. *Exp. Brain Res.* 67: 391–401

Lacquaniti, F., Soechting, J. F. 1982. Coordination of arm and wrist motion during a reaching task. *J. Neurosci.* 2: 399–408

Lacquaniti, F., Soechting, J. F. 1986a. Simulation studies on the control of posture and movement in a multi-jointed limb. *Biol. Cybern.* 54: 367–78

Lacquaniti, F., Soechting, J. F. 1986b. EMG responses to load perturbations in the upper limb: Effect of dynamic coupling between shoulder and elbow motion. *Exp. Brain Res.* 61: 482–96

Lee, R. G., Lucier, G. E., Mustard, B. E., White, D. G. 1986. Modification of motor output to compensate for unanticipated load conditions during rapid voluntary movements. *Can. J. Neurol. Sci.* 13: 97–102

Lennard, P. R. 1985. Afferent perturbations during "monopodal" swimming movements in the turtle: Phase-dependent cutaneous modulation and proprioceptive resetting of the locomotor rhythm. *J. Neurosci.* 5: 1434–45

Lennard, P. R., Hermanson, J. W. 1985. Central reflex modulation during locomotion. *Trends Neurosci.* 8: 483–86

Lestienne, F. 1979. Effects of inertial load and velocity on the braking process of voluntary limb movements. *Exp. Brain Res.* 35: 407–18

Libersat, F., Clarac, F., Zill, S. 1987. Force-sensitive mechanoreceptors of the dactyl of the crab: Single-unit responses during walking and evaluation of function. *J. Neurophysiol.* 57: 1618–37

Loeb, G. E. 1983. Finding common ground between robotics and physiology. *Trends Neurosci.* 6: 203–4

Loeb, G. E. 1984. The control and responses of mammalian muscle spindles during normally executed motor tasks. *Exerc. Sport Sci. Rev.* 12: 157–204

Loeb, G. E. 1987. Hard lessons in motor control from the mammalian spinal cord. *Trends Neurosci.* 10: 108–13

Loeb, G. E., Marks, W. B., Rindos, A. J., He, J., Roberts, W. M., Levine, W. S. 1985. Kinematics and dynamics of the cat hindlimb during locomotion. *Soc. Neurosci. Abstr.* 11: 1030

Marsden, C. D., Obeso, J. A., Rothwell, J. C. 1983. The function of the antagonist muscle during fast limb movements in man. *J. Physiol. London* 335: 1–13

Marsden, C. D., Rothwell, J. C., Day, B. L. 1984. The use of peripheral feedback in the control of movement. *Trends Neurosci.* 7: 253–57

Matthews, P. B. C. 1982. Where does Sherrington's "muscular sense" originate? Muscles, joints, corollary discharges? *Ann. Rev. Neurosci.* 5: 189–218

Matthews, P. B. C., Watson, J. D. G. 1981. Effect of vibrating agonist or antagonist muscle on the reflex response to sinusoidal displacement of the human forearm. *J. Physiol. London* 321: 297–316

Mauritz, K.-H., Dietz, V. 1980. Characteristics of postural instability induced by ischemic blocking of leg afferents. *Exp. Brain Res.* 38: 117–19

McCloskey, D. I. 1981. Corollary discharges: Motor commands and perception. *Handb. Physiol. (Sec. 1: The Nervous System), Motor Control* 2(2): 1415–47

McCrea, D. A. 1986. Spinal cord circuitry and motor reflexes. *Exerc. Sport Sci. Rev.* 14: 105–41

McMahon, T. A. 1984. *Muscles, Reflexes, and Locomotion.* Princeton: Princeton Univ. Press

Möhl, B. 1985. Sensory aspects of flight pattern generation in the locust. In *Insect Locomotion,* ed. M. Gewecke, G. Wendler, pp. 139–48. Berlin/Hamburg: Paul Parey

Morasso, P. 1981. Spatial control of arm movements. *Exp. Brain Res.* 42: 223–27

Mortimer, J. A., Webster, D. D., Duckich, T. G. 1981. Changes in short and long-latency stretch responses during the transition from posture to movement. *Brain Res.* 229: 337–51

Murphy, K., Roy, R. R., Bodine, S. C. 1981. Recruitment of the proximal and distal portions of the cat semitendinosus during running and jumping. *Med. Sci. Sport Exerc.* 13: 127–28

Nashner, L. M., McCollum, G. 1985. The organization of human postural movements: A formal basis and experimental synthesis. *Behav. Brain Sci.* 8: 135–72

Nashner, L. M., Shumway-Cook, A., Marin, O. 1983. Stance posture control is select groups of children with cerebral palsy: Deficits in sensory organization and muscular coordination. *Exp. Brain Res.* 49: 393–409

Nashner, L. M., Woollacott, M. 1979. The organization of rapid postural adjustments of standing humans: An experimental-conceptual model. In *Posture and Movement,* ed. R. E. Talbott, D. R. Humphrey, pp. 243–57. New York: Raven

Nashner, L. M., Woollacott, M., Thuma, G. 1979. Organization of rapid responses to

222   HASAN & STUART

postural and locomotor-like perturbations of standing man. *Exp. Brain Res.* 36: 463–76

Nichols, T. R. 1985. Autogenetic reflex action in tibialis anterior compared with that in soleus muscle in the decerebrate cat. *Exp. Brain Res.* 59: 232–41

Nichols, T. R., Houk, J. C. 1976. Improvement in linearity and regulation of stiffness that results from actions of stretch reflex. *J. Neurophysiol.* 39: 119–42

Nichols, T. R., Steeves, J. D. 1986. Resetting of resultant stiffness in ankle flexor and extensor muscles in the decerebrate cat. *Exp. Brain Res.* 62: 401–10

Partridge, L. D., Benton, L. A. 1981. Muscle, the motor. *Handb. Physiol. (Sec. 1: The Nervous System), Motor Control* 2(1): 43–106

Pearson, K. G. 1981. Function of sensory input in insect motor systems. *Can. J. Physiol. Pharmacol.* 59: 660–66

Pearson, K. G. 1985. Are there central pattern generators for walking and flight in insects? In *Feedback and Motor Control in Invertebrates and Vertebrates*, ed. W. J. P. Barnes, M. H. Gladden, pp. 307–15. London: Croom Helm

Pearson, K. G., Duysens, J. 1976. Function of segmental reflexes in the control of stepping in cockroaches and cats. In *Neural Control of Locomotion*, ed. R. E. Herman, S. Grillner, P. S. G. Stein, D. G. Stuart, pp. 519–37. New York: Plenum

Pearson, K. G., Franklin, R. 1984. Characteristics of leg movements and patterns of coordination in locusts walking on rough terrain. *Int. J. Robotics Res.* 3(2): 101–12

Pearson, K. G., Heitler, W. J., Steeves, J. D. 1980. Triggering of locust jump by multimodal inhibitory interneurons. *J. Neurophysiol.* 43: 257–78

Pellionisz, A., Llinás, R. 1982. Space-time representation in the brain. The cerebellum as a predictive space-time metric tensor. *Neuroscience* 7: 2949–70

Polit, A., Bizzi, E. 1979. Characteristics of motor programs underlying arm movements in monkeys. *J. Neurophysiol.* 42: 183–94

Porter, R., Whelan, J., eds. 1981. *Human Muscle Fatigue: Physiological Mechanisms. Ciba Found. Symp. 82.* London: Pitman Medical

Prochazka, A. 1986. Proprioception during voluntary movement. *Can. J. Physiol. Pharmacol.* 64: 499–504

Rack, P. M. H., Westbury, D. R. 1974. The short range stiffness of active mammalian muscle and its effect on mechanical properties. *J. Physiol. London* 240: 331–50

Raibert, M. H. 1986. *Legged Robots that Balance.* Cambridge, Mass: MIT Press

Reichert, H., Rowell, C. H. F. 1986. Neuronal circuits controlling flight in the locust: How sensory information is processed for motor control. *Trends Neurosci.* 9: 281–83

Ritzmann, R. E., Tobias, M. L., Fourtner, C. R. 1980. Flight activity initiated via giant interneurons of the cockroach: Evidence for bifunctional trigger interneurons. *Science* 210: 443–45

Rossignol, S., Gauthier, L. 1980. An analysis of mechanisms controlling the reversal of crossed spinal reflexes. *Brain Res.* 182: 31–45

Rothwell, J. C., Traub, M. M., Day, B. L., Obeso, J. A., Thomas, P. K., Marsden, C. D. 1982. Manual motor performance in a deafferented man. *Brain* 105: 515–42

Rushmer, D. S., Russell, C. J., Macpherson, J., Phillips, J. O., Dunbar, D. C. 1983. Automatic postural responses in the cat: Responses to headward and tailward translation. *Exp. Brain Res.* 50: 45–61

Sabin, C., Smith, J. L. 1984. Recovery and perturbation of paw-shake responses in spinal cats. *J. Neurophysiol.* 51: 680–88

Sanes, J. N., Mauritz, K.-H., Dalakas, M. C., Evarts, E. V. 1985. Motor control in humans with large-fiber sensory neuropathy. *Human Neurobiol.* 4: 101–14

Sherrington, C. S. 1910. Flexion reflex of the limb, crossed extension reflex, and reflex stepping and standing. *J. Physiol. London* 40: 28–121

Sherrington, C. S. 1924. Problems of muscular receptivity. *Nature* 113: 929–32

Shik, M. L., Orlovsky, G. N. 1976. Neurophysiology of locomotor automatism. *Physiol. Rev.* 56: 465–501

Shin, K. G., Malin, S. B. 1984. A hierarchical system structure for coordinated control of industrial manipulators. In *International Conference on Robotics*, pp. 609–19. New York: IEEE Comput. Soc. Press

Sillar, K. T., Skorupski, P. 1986. Central input to primary afferent neurons in crayfish *Pacifastacus leniusculus* is correlated with rhythmic motor output of thoracic ganglia. *J. Neurophysiol.* 55: 678–88

Skorupski, P., Sillar, K. T. 1986. Phase-dependent reversal of reflexes mediated by the thoracocoxal muscle receptor organ in the crayfish *Pacifastacus leniusculus. J. Neurophysiol.* 55: 689–95

Smith, J. L., Zernicke, R. F. 1987. Predictions for neural control based on limb dynamics. *Trends Neurosci.* 10: 123–28

Soechting, J. F., Dufresne, J. R., Lacquaniti, F. 1981. Time-varying properties of myotatic response in man during some simple motor tasks. *J. Neurophysiol.* 46: 1226–43

Soechting, J. F., Lacquaniti, F. 1981.

Invariant characteristics of a pointing movement in man. *J. Neurosci.* 1: 710–20

Stein, R. B. 1982. Which muscle variable(s) does the nervous system control in limb movements? *Behav. Brain Sci.* 5: 535–77

Stuart, D. G., Hamm, T. M., Vanden Noven, S. 1988. Partitioning of monosynaptic Ia EPSP connections with motoneurons according to neuromuscular topography: Generality and functional implications. *Prog. Neurobiol.* In press

Vallbo, A. 1981. Basic patterns of muscle spindle discharge in man. In *Muscle Receptors and Movement*, ed. A. Taylor, A. Prochazka, pp. 263–75. London: Macmillan

Vedel, J.-P. 1982. Reflex reversal resulting from active movements in the antenna of the rock lobster. *J. Exp. Biol.* 101: 121–33

Wendler, G. 1974. The influence of proprioceptive feedback on locust flight coordination. *J. Comp. Physiol.* 88: 173–200

Wetzel, M. C., Stuart, D. G. 1976. Ensemble characteristics of cat locomotion and its neural control. *Prog. Neurobiol.* 7: 1–98

Williams, K. R. 1985. Biomechanics of running. *Exerc. Sport Sci. Rev.* 13: 389–441

Wilson, D. M. 1961. The central nervous control of flight in a locust. *J. Exp. Biol.* 38: 471–90

Wilson, D. M., Wyman, R. J. 1965. Motor output patterns during random and rhythmic stimulation of locust thoracic ganglia. *Biophys. J.* 5: 121–43

Zajac, F. E. 1985. Thigh muscle activity during maximum-height jumps by cats. *J. Neurophysiol.* 53: 979–94

Ann. Rev. Neurosci. 1988. 11 : 225–51

SEXUALLY DIMORPHIC BEHAVIORS

Darcy B. Kelley

Department of Biological Sciences, Columbia University, New York, New York 10027

Sexually Dimorphic Behaviors

Many behaviors exhibited by females differ from those characteristic of males. In extreme instances, one sex displays a behavioral pattern never exhibited by the other. More often, the form or frequency of the behavioral act differs between the sexes. These behaviors are often called "sexually dimorphic" by analogy with vertebrate secondary sex characteristics such as the combs of roosters or mammalian mammary glands. This review examines dimorphic behaviors, with emphasis on proximate neural and endocrine mechanisms governing sex-typical behaviors in vertebrates. A number of more comprehensive reviews have recently dealt with selected aspects of this topic (Clemens & Gladhue 1979, Goy & McEwen 1980, Pfaff 1980, Arnold & Gorski 1984, Konishi 1985, Kelley 1986b, Crews & Moore 1986). My aim here is to focus on a few sexually dimorphic behavioral systems for which extensive neural and endocrine data are available. These systems are exemplars of three sexually dimorphic characters: parental care, mating reflexes, and courtship. For parental care, the most extensive body of neural and endocrine research has been on mammalian maternal care. Relatively less attention has been paid to species with biparental (birds) or paternal (fish, frogs) systems. Our understanding of neural and endocrine control of copulatory postures (e.g. rat lordosis and penile reflexes, amplexus in frogs) has progressed recently due, in part, to the development of certain tractable experimental paradigms (reviewed in Kelley 1986b). Finally, I consider courtship—which includes both mate attraction and intrasexual competition—in several systems, including bird and frog song, duetting in tropical birds, and polyandry in the jacana and phalarope.

225

0147–006X/88/0301–0225$02.00

Evolution of Sexually Dimorphic Behaviors

Why are some behaviors the exclusive province of one sex in a particular phyletic group, yet are monomorphic or even "sex reversed" in closely related species? Sex differences in behavior are the result of evolution (Darwin 1871, Williams 1966, Trivers 1972, Maynard Smith 1978). Both natural and sexual selection contribute to sexual dimorphism (e.g. Lande 1980, Partridge & Halliday 1984). Natural selection affects those characters directly connected with the propagation of species including fertilization, successful insemination or the rearing of offspring (Darwin 1871). Thus, sex differences in reproductive organs, in appendages used in copulation or in structures used in the nourishment of young are attributed to natural selection as are their behavioral counterparts, copulatory reflexes and parental behaviors. In some species, males and females differ markedly in structures not associated with reproduction, e.g. bill size and shape in birds (Selander 1972). Such structural dimorphisms also arise by natural selection and contribute indirectly to the propagation of the species (Darwin 1871).

Certain differences between the sexes are not so easily characterized as contributing to the propagation of species. Darwin (1871) considered the evolution of these characters to be driven by a separate process, sexual selection. These characters impart a competitive advantage in contests for mating opportunities. Either the character is particularly helpful in same sex competitions for mate access (intrasexual selection) or it imparts an advantage in attracting a member of the opposite sex (intersexual selection). In intrasexual selection, the selective driving force is predominantly on the behavior of the sex competing for access to mates whereas intersexual selection also acts on the choice behavior of the courted individual. In both intrasexual and intersexual selection, behaviors of both partners have consequences for reproductive success and for fitness. Current theoretical approaches to evolution of these interactions [notably the models of Lande (1980) and the notions of "evolutionary stable strategies" and of mathematical games theory analyses (Maynard Smith 1982)] appropriately emphasize these interactions among individuals.

Intersexual selection is commonly assumed to drive sexual dimorphism in two behavioral systems reviewed here, bird song and frog mate calling. Songs of male birds are believed to function in attracting females by advertising desirable attributes (Thorpe 1961, Searcy & Andersson 1986). For example, in species that add new song types with each season, repertoire size should be a good guide to age and thus survival ability—a useful attribute in a potential mate. In canaries, more complex songs are known to be more attractive to females and more stimulatory to reproductive activity than their simpler counterparts (Kroodsma 1976). In frogs, advertisement calls attract females to the breeding site; in some species vocal courtship by individuals contributes to mate selection (Wells 1977). Bird song also functions in territorial defense; for example, in acoustic competition between neighboring males in temperate bird species. Whether females can assess overall song complexity (intersexual selection) or whether they are instead merely the prizes in an elaborate acoustic battle between males (intrasexual selection; see Kelley 1986b) is not yet clear. Recent studies suggest that other dimorphic characters such as elaborate male plumage in birds of paradise, function to intimidate rival males rather than attract females (Le Croy 1981).

In frogs, an advertisement call can convey two messages. In the Puerto Rican tree frog, *Eleutherodactylus coqui*, males utter a "coqui" call during the breeding season. The "co" note serves as a territorial or agonistic signal to adjacent males, the "qui" note is attractive to females (Narins & Capranica 1980). In some anuran species, of which *E. coqui* is an example, males breed over a relatively prolonged period and defend sites suitable for calling, oviposition and/or feeding (Wells 1977). In others, the "explosive breeders," conditions suitable for spawning are only briefly available; females do not choose individual males, but instead locate the entire calling chorus (Arak 1983). There is little evidence for ritualized male/male competition in explosive breeders (Wells 1977).

As the above examples suggest, males compete for females in most vertebrate species. Courtship is thus an example of a characteristic male behavior. The predominance of male courtship and competition follows, I believe, from two characteristics of vertebrate mating systems: anisogamy and parental care. One sex (female) puts effort into producing a large gamete (anisogamy), a resource for which the other sex (male) competes. In some species, effort directed at offspring extends beyond gamete production and fertilization. This "parental investment"—defined broadly by Trivers (1972) as all allocation of effort in rearing existing offspring at the expense of producing additional offspring—includes the parental care seen in some species. Maynard Smith (1978, 1984) has classified the most common patterns of parental care as follows: the most common "system" is no parental care; if there is care it is typically uniparental, if there is internal fertilization, care is usually maternal, and if there is external fertilization, care is usually paternal but may be biparental. Parental care, when it exists, is thus usually a sexually dimorphic behavioral characteristic. In frogs and fish, there are numerous examples of paternal care (nests, fanning the eggs, mouth brooding, and the like). In birds, the most common pattern is biparental; the second most common pattern is maternal care. Care includes feeding the young and even extends, in some

Columbiformes, to the production of crop "milk" by both sexes. In mammals, however, the most common pattern is maternal care. Male contributions, though important when present, are largely indirect, e.g. territory defense. Parental care in rats, then, is an example of a characteristic female behavior.

Mate attraction and parental care are closely connected. There are two predominant patterns in birds: monogamy with biparental care and polygyny with maternal care. A rare and fascinating exception is the polyandry with paternal care found in some shore birds, jacanas and phalaropes (Jenni & Collier 1972, Jenni & Betts 1978). What accounts for this role reversal? Mating systems can be considered as evolutionary stable strategies (Maynard Smith 1982). The goal is to maximize the number of surviving offspring. Each sex has two available strategies: guarding or deserting the young (Maynard Smith 1984). Which approach is taken depends both on the probability that the young will survive to reproduce without further care and on the strategy assumed by the mate. Maynard Smith (1984) considers the reversal seen in jacanas to have arisen from the more generalized "double clutching" tendency seen in other shorebirds—females lay one clutch cared for by the male and another by herself. Polyandry in jacanas and phalaropes could have arisen if a female instead obtained a new partner for the second clutch, thus freeing her to lay a third clutch and so on. These examples illustrate the phyletic specificity of the way sexual dimorphisms in mating strategy evolve. While the evolutionary factors shown to operate in birds, for example, are among those candidates for effects in other groups (e.g. primates), they by no means can be assumed to operate in the same way, no matter how appealing the analogy. Polyandry in human societies (Daly & Wilson 1978) usually does not involve role reversal in courtship or parental care.

The elements of competition and response in sexual selection lead to the powerful and rapid evolution of sexually dimorphic characters. Expression, however, is also subject to natural selection, which can either enhance or subdue sexual dimorphism (Lande & Arnold 1985). Sexually dimorphic traits may enable males and females to optimally exploit different resources and minimize competition. Dimorphism in bill size or shape may permit dimorphic foraging strategies (Selander 1972). Natural selection should also serve to check deleterious excesses attendant on sexual selection (e.g. the peacock's tail, Darwin 1871). The additional burden of parental care should attenuate extremes of sexual dimorphism; feeding or defending chicks can interfere with survival of the parent. Conversely, bright coloration of the male could increase the chances of attracting a predator to the nest. Thus one might invoke natural selection to explain why biparental, monogamous species (especially those living under stringent but stable environmental circumstances) are less sexually dimorphic in morphology and behavior than are closely related maternal, polygynous species.

Finally, with regard to the interpretation of proximal control of sexually dimorphic behaviors, we must bear in mind that most physiological studies of animal behavior are conducted on a limited number of species under laboratory conditions that may not match those encountered by the animal in its native habitat (Crews & Moore 1986). For animals living in groups, for example, we cannot neglect the social context in which behaviors are observed (Goldfoot & Neff 1985). When tested alone, male and female Rhesus monkeys respond to distress signals from an isolated infant with equivalent frequencies and latencies (Gibber 1981). However, if the same male and female are housed together, only the female responds to the infant. Thus the unwary observer might conclude from the first testing situation that the sexes share parental care or from the second that care is the exclusive province of the female.

The Design of Sexually Dimorphic Systems

In principle, sex differences in behavior are produced in two ways (Figure 1). One is to provide adults of both sexes with all the requisite sensory, neural, and motor components necessary to produce the behavior but to arrange that the sexes operate in different environments. External environmental duality is achieved only in very rare instances of extreme dimorphism: for example, when the male is a parasite living on the female (Darwin 1871). Within the context of a social group, environmental duality can be achieved if certain social stimuli are only given to one sex. In the example of Rhesus monkeys and the stranded infant above, a male might never be alone with the infant and care-giving behavior would not be exhibited. A sex difference in environment could also result from differences in the internal milieu; for example, sex differences in gonadal and placental hormones. Many courtship and parental behaviors are controlled by hormones circulating in adults (Kelley & Pfaff 1978, Kelley 1978, Moore 1983, Crews & Silver 1985, Rosenblatt et al 1985). One way to produce female-only behaviors would be to arrange that one or more of the sensory, neural, and motor elements that produce sexually dimorphic behaviors function only in the presence of female-specific hormones.

Endocrine Control of Sex Differences

Circulating hormones exert very powerful control over reproductive behaviors in many species (for others, see Crews & Moore 1986). Sex differences in circulating titers of hormones (e.g. gonadal steroids) can suffice to account for sex differences in behavioral frequencies. The evi-

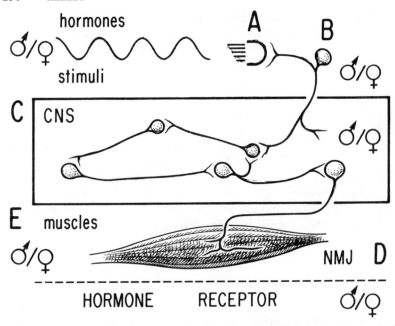

Figure 1 Possible mechanisms for generating sexually dimorphic behavior. In one, both sexes have all the necessary sensory (A), neural (B and C), and neuromuscular (D and E) components required, but one sex lacks access to the requisite stimuli or hormones (*above*). In the other, sex differences are due to sensory receptor (A), neural (B, C, D), or muscular dimorphisms (E), including sex differences in hormone receptor expression (*below*).

dence suggesting that androgens control male reproductive behaviors and that estrogens (and progestins) control female pro- and receptivity in a wide variety of species has been reviewed extensively elsewhere (Kelley & Pfaff 1978, Clemens & Gladhue 1979, Goy & McEwen 1980, Crews & Silver 1985). An example of an androgen-mediated behavior is clasping, a male-typical sex behavior in *Xenopus laevis* (Kelley & Pfaff 1976). A sexually active male initiates amplexus by swimming toward a female and grasping her with his forelegs. Castrated males cease clasping within one to two weeks; the behavior can be reinstated by treatment with exogenous androgens, but not estrogen. Completely male-like behavior can be induced in adult, ovariectomized females by androgen treatment. Brain regions that participate in the control of clasping appear to be capable of behavioral control in both sexes (Hutchison & Poynton 1963). Neurons that contain intracellular receptors for androgen are found in spinal cord motor pools that innervate arm muscles used in clasping in both sexes. Androgen induces increased excitability of these cells (Erulkar et al 1981). Thus, the sensory, neural, and motor elements necessary for clasping are present in both sexes in *X. laevis* and females do not normally exhibit the behavior because of insufficient titers of androgen (Lambdin & Kelley 1986). Still not clear is whether androgen acts at one or more sites (sensory, neural, motor) to permit behavioral expression or whether one particular site (for example, arm muscle motor neurons) is limiting for behavioral expression.

Endocrine Control of Maternal Behavior in Rats

Steroid hormone treatment of adults may not produce an exact phenocopy of behavior in the opposite sex; the behavior seen may be less frequent, or it may differ in form or be less vigorous. One reason may be partial sexual differentiation—underlying sex differences in behavioral effector elements due to hormone action during development (discussed below). Another set of factors, however, are the combined actions of many hormones, levels of which differ between the sexes. One of the most intriguing systems in this regard is the control of maternal behavior in rats.

Under laboratory conditions, male rats usually exhibit little, if any, parental care when first exposed to newborn pups (paternal care is more frequent in other rodents, Elwood 1983). In contrast, parturient females— even those giving birth for the first time—immediately display the full array of parental activities: retrieving, grouping, licking, crouching, nursing, and nest building. Males (and virgin females) are fully capable of displaying parental behaviors and will do so if exposed to pups repeatedly (5–7 days; Rosenblatt 1967). Once parental behavior has been induced, it is always exhibited with short latency. The difference between primiparous parturient females and other adults is in the latency of the full response and the percentage of responsive animals upon *first* exposure to pups. All primiparous females are immediately fully responsive, while very few males or virgin females display "spontaneous" maternal behavior. The difference in latency between parturient females and others has been attributed to olfactory cues; latency to pup retrieval is reduced by anosmia (Fleming & Rosenblatt 1974). If newborn pups produce aversive olfactory cues, the increased "emotionality" of the virgin female may also contribute to pup avoidance (Fleming & Luebke 1981).

What accounts for the rapid onset of maternal behaviors after birth? A variety of evidence in many species implicates endocrine changes associated with pregnancy and birth (see Rosenblatt et al 1985 for a recent review). Four hormones have been implicated: estrogen, progesterone, prolactin and oxytocin (Moltz et al 1970, Zarrow et al 1971, Siegel & Rosenblatt 1975, Pedersen et al 1982). During pregnancy, estradiol rises gradually to maximum levels of approximately 155 pg/ml at day 15 of gestation; between days 20 and 22 (birth) levels fall abruptly to 10 pg/ml. Progesterone in circulation increases throughout pregnancy, reaching maximum maintained levels (approximately 70 pg/ml) two days before birth and then dropping precipitously (Bridges 1984). These endocrine changes result in very high P/E ratios around the time of parturition. Prolactin levels rise abruptly at the end of pregnancy (Morishige et al 1973). Oxytocin levels rise throughout gestation (Boer et al 1979).

Treatment with high doses of estradiol alone can reduce the latency to exhibit maternal behavior in virgin females or males (Bridges 1984). Progesterone by itself is ineffective, but acts synergistically with estrogen to reduce latencies to first display of full maternal responsiveness. Prior treatment with progesterone sensitizes females to the behavioral effects of estradiol (Bridges 1984). The onset of decreasing levels of progesterone appears to control the time of onset of behavioral responsiveness in primigravid or estradiol treated nulliparous rats (Siegel & Rosenblatt 1975, Bridges & Russell 1981). Progesterone and estradiol secretion during pregnancy appear to *prepare* the primigravid female to exhibit relatively short latency maternal behavior; neither hormone need be present when behavior is displayed. Effects on maternal behavior are proportional to the duration of treatment or of pregnancy (Bridges 1984).

Treatment with steroids alone, however, does not reproduce the very rapid, full behavioral responsiveness seen in parturient females (100% of animals tested are fully responsive to young within seconds or minutes of birth). This responsiveness is not necessarily related to parturition itself as full-term primigravidas and females delivered by caesarian section are also responsive to young (reviewed in Rosenblatt et al 1985). Two nonsteroid hormones have been extensively examined as possible mediators of short-latency maternal responsiveness: prolactin and oxytocin, both of which are elevated at parturition. Virgin females induced to show parental behavior by exposure to pups or steroid treatment also show increases in plasma prolactin levels whereas similarly treated males do not (Bridges 1983, Samuels & Bridges 1983, Tate-Ostroff & Bridges 1985). However, latencies to exhibit parental behavior (nest building, retrieval, crouching) are no different in parental females and males, even when steroid treated (Samuels & Bridges 1983). More powerful evidence for prolactin effects comes from studies showing that nulliparous female rats do not show steroid-induced decreases in responsive latency unless an intact pituitary is present (Bridges et al 1985). If an ectopic pituitary graft is given, some steroid-treated females (50%) display maternal responsiveness within 30 min of test initiation; only 10% of control females display such rapid responsiveness. Prolonged treatment with prolactin also significantly decreases maternal response latencies in steroid-treated females. Whether any prolactin-treated females showed the very rapid maternal responsiveness seen in animals with pituitary grafts remains unclear. Comparable results in males have not yet been reported.

Recent evidence, though still somewhat controversial, also implicates the hormone oxytocin in maternal behavior. Intracerebroventricular (ICV) injection of oxytocin into estrogen-primed ovariectomized rats induces immediate increases in the percentage of females displaying maternal behavior (Pedersen & Prange 1979, Pedersen et al 1982). Under optimal test conditions in a responsive strain (Sprague-Dawley rats obtained from Zivic-Miller), approximately 90% of ovariectomized estrogen-primed nulliparous females obtained high maternal behavior scores within one hour of contact with foster-pups (Fahrbach et al 1986). Some additional evidence that endogenous oxytocin participates in maternal responsiveness comes from results of studies in which oxytocin antagonists or antisera were applied ICV to pregnancy-terminated, estrogen-treated nulliparous females (Fahrbach et al 1985). These treatments significantly delay the onset of maternal responsiveness as well as the percentage of females showing maternal behavior within the first hour of testing. Maternal behavior of lactating females five days postpartum is not disrupted by treatment with an oxytocin antagonist, thus suggesting that oxytocin does not maintain maternal responsiveness. Unlike the effects of prolactin, short-term disruption of oxytocin (1 hr) was as effective as long-term treatment.

There are two probable endogenous sources of prolactin and oxytocin: the pituitary or CNS neurons. Studies by Bridges and colleagues (1985) on ectopic pituitaries strongly suggest that the behaviorally important source of prolactin is the hypophysis. The source of oxytocin is less clear since comparable studies on hypophysectomized animals have not been carried out. Oxytocin is synthesized by neurons in the paraventricular (PVN) and supraoptic nuclei of the brain. There are multiple CNS targets of oxytocinergic fibers including fibers that end in the posterior pituitary and supply circulating oxytocin. Based on ICV administration of the hormone, antagonists, and antisera, oxytocin appears to work on the CNS (Fahrbach et al 1985). Knife cuts severing the efferents of the PVN nucleus do not disrupt ongoing maternal behavior (Numan & Corodimas 1985). While it is also assumed that prolactin acts directly on the CNS (Bridges et al 1985), direct proof is not yet available.

Neural Control of Maternal Behavior in Rats

As described above, maternal behavior in the laboratory rat consists of a number of different behaviors (e.g. pup retrieval, licking, crouching), each quite complex and independent, in terms of ongoing motor output, of other components. Many of these activities, however, are eliminated by lesions of the preoptic area (POA) of the hypothalamus (Numan et al 1985). In addition, implants of crystalline estradiol into the POA reduce the latency of primed, nulliparous females to respond to pups (Fahrbach & Pfaff 1986, Numan 1987).

Estrogen plays an important role in the production of short-latency maternal behavior (see above), and the POA contains many estradiol-concentrating neurons (Pfaff & Keiner 1973). The disruptive effects of POA lesions appear to be due to projections of medial POA neurons to the lateral POA (Numan et al 1985). These LPOA cells, in turn, are postulated to affect maternal responses by means of various efferent projections, notably those to the ventral tegmental area (VTA). How the VTA influences motor activity related to maternal behavior is unclear. One possibility is via projections to nucleus accumbens and subsequent pallidal outflow. Another is via efferents back to the POA (Numan 1985). It should be emphasized that results of comparable lesions in other POA targets (e.g. central gray) on maternal behavior have not yet been reported. In addition, there is no reason to believe that the POA to VTA to accumbens outflow is important only for maternal behavior; this pathway may function in a wide variety of "motivated" behaviors.

Sexual Dimorphism in Parental Behavior of Laboratory Rats

As described above, sex differences in parental responsiveness to pups consist primarily of the latency with which behaviors are exhibited on initial exposure: seconds or minutes for parturient females, days for males or nulliparous females. A hormonal regime, mimicking critical features of pregnancy, can produce short-latency parental responsiveness in males or nulliparous females (see above). Thus, I consider sex differences in parental behavior discussed here to result from sex differences in the endocrine environment of males and females (Figure 1). Both sexes can and do display full maternal responsiveness following prolonged exposure to pups, thus suggesting that all the requisite sensory, neural, and motor systems are present. The action of hormones can be seen as lowering a sensory barrier (aversive smell) to pup approach and facilitating, probably by means of hormone action on the POA, the motor components of parental care.

One of the most interesting features of the maternal experience is its permanence. Even in males or nulliparous females, sensitization by pup exposure produces a permanent decrease in latency of the maternal response. Parturition and maternal behavior have been reported (Hatton & Ellisman 1982) to be associated with a change in synaptic ultrastructure in the female PVN. Determining whether such changes also occur in maternal males and nulliparous females as well as in other brain nuclei associated with the parental response would be of great interest.

A few reports indicate that males may be endogenously less responsive than females. First, mammary glands are not present in adult males, so actual suckling, milk ejection, etc does not occur. Second, neonatal administration of androgen to females reduces the percentage retrieving pups in a T-maze in adulthood (Bridges et al 1973). Exposure to pups induces increases in circulating prolactin in parturient females and steroid-primed females, but not in males (Bridges 1983, Samuels & Bridges 1983). Recall, however, that all three groups are equivalently responsive to young under these test conditions. Compared with the major sex differences in behavior associated with certain courtship systems (see below), sex differences in ability to display parental care are quite minor.

Sex-reversed Behaviors—Polyandry

One way to test the generality of exclusive endocrine control of behavioral dimorphism is to examine species that are sex reversed; for example, species in which females perform all courtship and males all parental care. As described above, biparental care is more common in birds than in mammals. In the American jacana (*Jacana semispinosa*), however, males provide almost all parental care; female contributions are limited to territorial defense (Jenni & Collier 1972, Jenni & Betts 1978). Females are polyandrous and defend territories that encompass those of up to four males. Do species exhibiting a reversal of the predominant behavioral pattern (males—courtship, females—parental care) also show a corresponding reversal in type of gonadal hormone secreted (males—estrogen, females—androgen)?

Rissman & Wingfield (1984) report that in another related polyandrous shorebird (*Actitis macularia*), androgen levels are greater in preincubating males than in females or incubating males; estrogen levels in males are lower than those of females. Thus despite some reversal in courtship and parental care, the usual pattern of steroid secretion (male—androgen; female—estrogen) persists. Interestingly, however, male levels of prolactin (associated with parental care in many species) are greater than levels in females (Oring et al 1986). Elevated male prolactin levels are also seen in another polyandrous species, Wilson's phalarope (*Phalaropus tricolor*). In this latter species, again, androgen levels in nonincubating males are greater than those in females or incubating males; estrogen levels are higher in females than in incubating males (Fivizzani et al 1986).

Thus a dichotomy in hormone/behavior associations exists in these sex-role reversed birds. Levels of gonadal steroids are similar to those seen in monogamous or polygynous species. Levels of prolactin—of pituitary origin—are reversed: the incubating male has levels as high as those of females in species in which only females incubate. Gonadal steroids thus appear largely restricted by sex: estrogen to females, androgen to males. Prolactin, however, is not restricted; why? Estrogens and androgens have pleotrophic effects: testosterone is necessary for maturation of the male and contributes to fertility and estradiol plays an analogous role in females. The requirements of sex-specific gamete production and reproductive tract development may limit flexibility in secretion of gonadal steroids. If this is the case, "sex reversed" behaviors can be produced in at least two other ways in species in which the ancestral pattern was male-courtship-testosterone/female-incubation-estrogen. One is to emancipate behaviors from gonadal steroid control entirely (see Crews & Moore 1986); the other is to express receptors for sex-typical steroids in brain regions devoted to the "sex-reversed" behavior.

Duetting in Tropical Song Birds

In temperate birds, song is the exclusive province of the male. In the tropics, however, both sexes sing in many species; often a close temporal relationship exists between songs of a mated pair. Such synchronized songs are termed duets (Thorpe 1972, Farabaugh 1982). In many song birds, song production in adulthood is controlled by androgen secretion (Prove 1983). Brain regions implicated in song control in finches (Nottebohm et al 1976) contain cells that concentrate androgenic steroids (Arnold et al 1976). The absolute number of hormone-concentrating cells, as well as the percentage of such cells, is greater in male than in female zebra finches (Arnold & Saltiel 1979, Nordeen et al 1986). Brenowitz & Arnold (1985) examined androgen accumulation in bay wrens (*Thyrothorus nigricapillus*), a species in which females always sing the leading song in duets (Levin 1983, 1985, Brenowitz et al 1985). After administration of radioactive testosterone, males and females exhibited the same proportion of steroid-accumulating cells in two telencephalic song nuclei, hyperstriatum ventrale, pars caudalis (HVc) and magnocellular nucleus of the anterior neostriatum (MAN) (Brenowitz & Arnold 1985). Whether testosterone itself or one of its major metabolites (estradiol or dihydrotestosterone) is the active form of the hormone accumulated in these brain regions remains unclear. In zebra finches, cells in HVc accumulate testosterone or dihydrotestosterone but very little, if any, estradiol (Arnold et al 1976). If we assume that hormone activation of HVc is essential for song production, it may be that female bay wrens achieve song by androgen secretion. Alternatively, cells in HVc of bay wrens might express estrogen receptors, and be activated by estrogen in females and by testosterone aromatized to estradiol in situ in males. Considering the conservative nature of hormone secretion in birds (see above), the latter scenario seems more likely.

The accumulation of androgens in cells of HVc in male zebra finches is regulated by male-specific secretion of estradiol around the time of hatching (Nordeen et al 1986, Hutchison et al 1984). Thus, in bay wrens, either HVc cells are emancipated from the requirement for neonatal estradiol in both sexes or both sexes secrete estradiol (Brenowitz & Arnold 1985). Unlike jacanas and phalaropes, however, bay wrens are not sex-reversed. Both sexes contribute to parental care; certain aspects of courtship are male specific; females contribute most nest building (R. Levin, personal communication). If early estrogen secretion is responsible for "masculinization" of song nuclei in bay wrens, alternative mechanisms must operate to preserve other behavioral dimorphisms.

Sensory Sexual Dimorphism

Sensory structures are frequently sexually dimorphic in invertebrates. An example is antennal olfactory receptive organs of moths (Hildebrand 1985). In *Manduca sexta*, the female releases pheromones by extrusion of an abdominal gland. Both sexes have antennal receptors that respond to plant odors, but only the male antennae exhibit sensilla with receptors for the mate attraction pheromone (Schneiderman & Hildebrand 1985). In both sexes, antennal olfactory receptors project to glomeruli in the olfactory lobe (Schneiderman et al 1982). Only males have specialized olfactory glomeruli that contain processes of interneurons responsive to female-produced pheromones (Schneiderman et al 1982).

The neural system that drives male approach to the signalling female is governed by sex differences in the sensory periphery. If the imaginal disc containing antennal precursor cells is transplanted from a male to a female larva, a male antenna will develop as will a male-like olfactory glomerulus with pheromonal responsive interneurons (Schneiderman et al 1982). The available data support the hypothesis that male afferents arriving at the antennal lobe are responsible for organizing a masculine neuropil. Still unclear is whether the response characteristics of the glomerular interneurons are changed by exposure to male antennal afferents or whether afferents affect the survival of a preexisting class of responsive cells. In either case, the possession of a male antenna has a profound influence on the behavior of the moth. Schneiderman et al (1986) have recently shown that a female receiving transsexual grafts will display male-like zigzag flight approaches when placed within a pheromone odor plume. In the most extreme case, the implanted female attempted to mate with the odor source. Zigzag flight patterns were shown even by females receiving a unilateral antennal graft. Masculine sensory input could be responsible for neural reorganization, including the production of male-like motor patterns. However, genetic females will sometimes display zigzag flight patterns toward a food source, thus suggesting that input from grafted male antennae may access an existing flight pattern generator. Some additional evidence in support of this interpretation is that the behavior of the transsexually grafted female described above culminated in oviposition.

Sexual Differentiation Within the Nervous System

For some sexually dimorphic behaviors, central nervous system participants have been described. These behaviors include the lordosis reflex of female rats (Pfaff 1980), copulatory reflexes in male rats (Hart 1980, Sachs 1981), song in birds (Nottebohm et al 1976) and calling in frogs (Schmidt 1976, Wetzel et al 1985, Kelley 1986a). Many of these systems show marked sex differences in brain nuclei that control dimorphic behaviors. For example, the volume of the CNS nucleus, number of cells, dendritic extent, synaptic input, axonal arborizations, post-synaptic targets, and number of hormone-accumulating cells have all been shown to be sexually dimorphic (Nottebohm & Arnold 1976, Greenough et al 1977, Gorski et al 1978, Gurney 1981, DeVoogd & Nottebohm 1981, Konishi & Akutagawa 1985, Wetzel et al 1985, Nordeen et al 1986). That these sex differences contribute to differences in behavioral expression seems reasonable (Gurney 1982, Kelley 1986b). For example, sex differences in the volume of telencephalic song nuclei in finches (HVc, hyperstriatum ventrale, pars caudalis; RA, nucleus robustus archistriatalis) correlate closely with behavioral differences. In zebra finches, song in females is exceedingly rare, even under appropriate endocrine conditions (see below). The male/female ratio in volume of telencephalic song control nuclei is approximately 5/1, depending on the region; one nucleus (X) is not even discernable in females (Nottebohm & Arnold 1976). The volume of telencephalic song control regions in adult zebra finches and white crowned sparrows does not vary seasonally nor can the nuclei be induced to grow by hormone treatment in adulthood (Gurney & Konishi 1980, Baker et al 1984). In canaries, however, some androgen-treated females sing. Song frequency and size of vocal control nuclei can be increased by hormone treatment. Brain nuclei, while dimorphic, are normally less so than in zebra finches (Nottebohm & Arnold 1976, Nottebohm 1980, 1981). In some tropical species, both males and females sing (see above). Song control nuclei in several duetting species have been examined (Brenowitz et al 1985, Brenowitz & Arnold 1986). In species such as the bay wren, where females sing as frequently as males, there is no sex difference in volume of song control nuclei (Brenowitz & Arnold 1986). In duetting species with greater male song complexity (e.g. the rufous and white wren), song nuclei are dimorphic though less so than in canaries or zebra finches (Brenowitz & Arnold 1986). Thus, behavioral and brain dimorphisms are well correlated.

As described above, many sex differences in behavior are attributable to sex-specific hormone secretion patterns. Some effects of hormones are reversible. An example is the induction of clasping by androgen in adult clawed frogs (Kelley & Pfaff 1976). A classic example of an irreversible effect of hormones is the masculinization and defeminization of neonatal female rats given testosterone (Goy & McEwen 1980). Androgen administration produces permanent changes in behavioral capabilities, including sensitivity to the behavioral effects of steroids. Permanent effects usually, though not necessarily, occur during a circumscribed period early in development (Goy & McEwen 1980). Reversible effects are usually (though again not necessarily) characteristic of sexually mature adults. Both sorts of effects require steroid hormone receptors; mutants defective in receptor expression are behaviorally insensitive to these hormones (Olsen 1979).

Sex differences in dimorphic brain nuclei are also controlled by secretion of gonadal steroids (Gurney & Konishi 1980, Arnold & Gorski 1984). Some dimorphisms are due to the action of circulating hormones on the adult brain and are reversible. Song control nuclei in canaries wax and wane in volume seasonally in response to changes in circulating androgen (Nottebohm 1981). Other sex differences in the brain appear to result from the irreversible action of steroids during development of the nervous system (Raisman & Field 1973, Gurney 1981, Jacobson et al 1981). The volume of telencephalic song control nuclei in zebra finches is irreversibly established by hormone-stimulated growth at hatching and prior to the first singing season (Gurney & Konishi 1980, Hutchison et al 1984). Early hormone secretion regulates the number of androgen-responsive cells in these areas (Arnold & Saltiel 1979, Nordeen et al 1986). Organizational and activational effects of steroids on behavior and on responsible brain nuclei are thus closely correlated.

Cellular Contributions to CNS Sexual Dimorphism— Cell Number

At the cellular level, there are many kinds of sex differences in behavioral effector brain nuclei: volume, cell number, somal size, dendritic branching, afferent volume, percentage of hormone-target cells (reviewed in Arnold & Gorski 1984, Konishi 1985, Kelley 1986b). Behaviorally relevant brain dimorphisms must ultimately contribute to differences in synaptic connectivity. Thus we might expect to find certain axonal projections in one sex and not in another or to find sex differences in the robustness or efficacy of synaptic connections. This last mechanism has been considered in detail at certain sexually dimorphic neuromuscular synapses (see below). In CNS nuclei that contribute to vocalization in frogs and birds, some connections are reduced or absent in females (Wetzel et al 1985, Konishi & Akutagawa 1985). The number and placement of synapses also can differ in telencephalic and diencephalic nuclei (DeVoogd & Nottebohm 1981, Raisman & Field 1973). Whether differences in synaptic connectivity between the sexes arise from different projections and/or from different frequencies of homologous neurons remains to be clarified.

Many sexually dimorphic brain nuclei exhibit differences in cell number. In song bird nuclei HVc and RA, the rat spinal nucleus of the bulbocavernosus (SNB), and motor nucleus N.IX-X of clawed frogs, the number of neurons is greater in males than in females (Gurney 1981, Breedlove & Arnold 1981, Kelley 1986a). The number and efficacy of contacts between neurons contribute to cell number—afferent input, connection to appropriate targets, and synaptic competition are all important (Hamburger & Oppenheim 1982). Thus manipulation of cell number in a key brain nucleus can have dramatic effects on cells synaptically connected to that nucleus and on cells in competition for afferents and efferents.

The effects of changes in cell number during development on synaptic connectivity are diagrammed in Figures 2 and 3. Consider the consequences of adding additional cells to brain nucleus B (Figure 2). Neurons in nucleus A, which otherwise would have died (x) due to an insufficient target in B, now survive. The additional neurons in B now send axons to C and effectively compete with axons from D; this leads to the death (X) of the latter cells. The end result is a robust connection from A to B to C and no connection from D to C. If, instead of cell addition, nucleus B loses neurons, a very different scenario ensues (Figure 3). Cells in A do die (x) due to an insufficient target in nucleus B. Cells in B die due to lack of afferents. Cells in D that project to C now survive due to lack of competition. In the first case, the A to B to C connection is established and D to C lost. In the second, the A to B to C connection is lost, but D to C survives. These alternative cases illustrate how cell number in a key brain nucleus can make dramatic contributions to synaptic connectivity. If, for example, androgen maintains cell number in nucleus B, its secretion in males (Figure 2) and not in females (Figure 3) would generate sex-specific patterns of CNS circuitry. There is quite good evidence that a series of events, similar to that outlined here, contributes to sexual dimorphism in connectivity in zebra finch brain (Gurney 1981, Arnold & Gorski 1984, Konishi & Akutagawa 1985, Nordeen et al 1987b).

Neuron number in a CNS nucleus is governed by proliferation, migration, and death. Administration of steroid hormones can affect cell

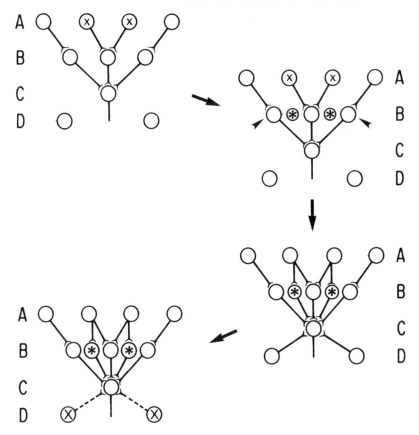

Figure 2 Generation of sex dimorphisms in neuronal circuitry by differences in cell number. *Ab initio* (*upper left*), some neurons (x) in nucleus A are destined to die because of the limited target availability in nucleus B. The addition of neurons (*) to nucleus B rescues these neurons in A, preserving and strengthening the A-to-B connection. However, afferents arriving at nucleus C from D now (*third panel*) face stiff competition and ultimately die (X, *lower left*).

number in dimorphic, behavioral effector nuclei, thus suggesting that steroids control one of the above processes (Gurney 1981, Arnold & Gorski 1984, Arnold 1984, Konishi & Akutagawa 1985, Sengelaub & Arnold 1986). For the most part, neurogenesis is complete before steroids are secreted or affect cell number, and before steroid receptors appear (Jacobson & Gorski 1981, Breedlove et al 1983, Gorlick & Kelley 1986, 1987). However, one song nucleus in the brain of canaries and zebra finches, HVc, continues to add new neurons in adulthood by migration from an overlying stem cell population (Goldman & Nottebohm 1983). Some new

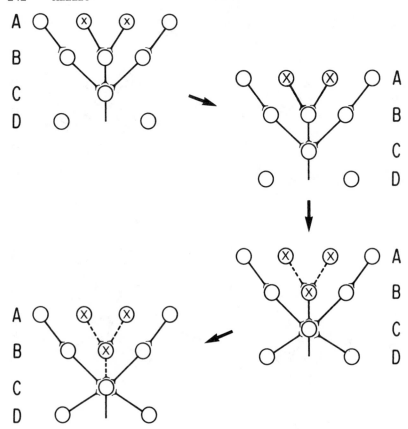

Figure 3 Consider instead the case of cell loss in nucleus B. Consequences include loss of neurons (x) in A, loss of targets in B, and ultimate degeneration of the A-to-B connection. The D-to-C connection, however, is preserved in this case, resulting in a set of connections different from those established by cell addition to B. Sex-specific hormone regulation of cell number in one, strategically placed brain nucleus (by proliferation, migration or death) could, in this fashion, create sex differences in connectivity.

neurons in HVc are responsive to acoustic stimuli, thus indicating their incorporation into functional circuitry (Paton & Nottebohm 1984). New neurons in HVc thus might contribute to sex differences in cell number (Nordeen et al 1987b).

In rat spinal cord, the sexually dimorphic SNB neurons arrive at their characteristic location by secondary migration from an adjacent motor nucleus (Sengelaub & Arnold 1986). Sex differences in neuron number, however, seem to be primarily due to prevention of ontogenetic cell death (Nordeen et al 1985). Cell death has also been postulated to account for male/female differences in the song nucleus, RA, of zebra finches (Gurney 1981, Konishi & Akutagawa 1985). Steroids could prevent ontogenetic cell death by affecting synaptic connectivity (more cells, larger somata and dendritic trees, more numerous axonal projections) or by exerting direct or indirect trophic control of developing targets in a manner analogous to developmental actions of nerve growth factor. One very puzzling finding has been that, although estradiol increases somal size and cell number in HVc and RA of nestling zebra finches, these regions have few, if any, intracellular receptors for estradiol either in adults (Arnold et al 1976) or neonates (Nordeen et al 1987a). Estradiol may be acting via membrane receptors or afferents from estrogen target neurons. Alternatively, estradiol may be promoting the secretion of a growth factor or synthesis of growth factor receptors in dimorphic target neurons.

One contribution of steroid hormones to regulation of sexual dimorphism may be control of expression of cell types. This has been a difficult question to address experimentally because of the paucity of cell type markers. Male and female brain nuclei appear to contain the same morphological cell types (Gurney 1981, DeVoogd & Nottebohm 1981, Kelley 1986b). However, output cells in female zebra finch HVc project only to a rim of tissue surrounding RA and not into RA itself (Konishi & Akutagawa 1985). It is not clear whether male and female RA/HVc nuclei contain different output neuron types or different projections of homologous neurons. Perhaps the best evidence on cell type comes from studies of androgen-concentrating cells in the zebra finch telencephalon; females have many fewer such cells in MAN and HVc (Arnold & Saltiel 1979). During development, estrogen acts to preserve androgen-concentrating MAN cells in the face of massive ontogenetic death of other MAN cell types (Nordeen et al 1987b). Thus some sex differences in song nuclei are likely to be due to differences in cell type. Estrogen may rescue existing androgen-concentrating cells or else increase receptor expression in previously insensitive neurons.

Control of Sexually Dimorphic Behaviors at the Neuromuscular Junction

Finally, sex differences in behavioral expression can also be controlled by differences in motor neurons and muscles (Figure 1). In the extreme, muscles may be completely absent in one sex, as in the levator ani/ bulbocavernosus muscles (LA/BC) in rats. These muscles control copulatory reflexes in males that are essential for successful insemination (Hart 1980). At birth, LA/BC muscles are present in both sexes but involute in females unless supplied with androgen (Breedlove & Arnold 1983b). The motor neurons supplying these muscles (the SNB nucleus) are approximately five times more numerous in adult males (Breedlove & Arnold 1983a,b). SNB motor neurons express androgen receptor in adults (Breedlove & Arnold 1981), as do their targets, the LA/BC muscles (Jung & Baulieu 1972). Administration of androgen during the perinatal period prevents ontogenetic cell death in female SNB cells (Nordeen et al 1985). Androgens rescue male SNB neurons from cell death by acting on the LA/BC target muscles. Autoradiographic studies show that there is no androgen accumulation in SNB cells at developmental times when rescue occurs (Breedlove 1986a). SNB cells that do not express functional androgen receptor can be rescued from ontogenetic cell death provided that both androgen and androgen responsive LA/BC muscles are present (Breedlove 1986b; see also Kelley 1986b). Cells of the SNB in males remain responsive to androgen into adulthood; testosterone can induce extensive dendritic growth in castrates (Kurz et al 1986).

Even when homologous muscles are present in both sexes, sex differences in muscle fibers or at the neuromuscular junction may constrain behavioral expression. The most dramatic example is the larynx of clawed frogs, *Xenopus laevis* (see Kelley 1986a for a more extensive review). Male frogs use mate calls, trills with alternating fast and slow phases, to attract and excite females (Wetzel & Kelley 1983). Females use a slower, monotonous trill-ticking to avert or terminate male clasp attempts (Weintraub et al 1985). While males also tick when clasped, females do not mate call even when given the appropriate male-typical hormones in adulthood (Hannigan & Kelley 1986). Brain regions that participate in vocal production have been identified (Schmidt 1976, Wetzel et al 1985) and some are sexually dimorphic in cell number, dendritic extent, somal size, and connectivity (Kelley 1986a). The male central nervous system generates the mate call temporal pattern and rate of activity in laryngeal motor neurons (Tobias & Kelley 1987).

Laryngeal muscles and neuromuscular junctions are responsible for sex-typical constraints on the rate at which calls can be produced (Sassoon et al 1987, Tobias & Kelley 1987, 1988). The isolated male larynx can be induced to mate call by providing the mate call pattern of stimulation to the laryngeal nerves; the female larynx cannot call at mate call rates under these conditions (Tobias & Kelley 1987). Adult males have eight times the female number of muscle fibers, and cells are of the fast-twitch, fatigue resistant type (Sassoon & Kelley 1986, Sassoon et al 1987). Levels of androgen receptor in adult male muscle are four times that of females (Segil et al 1987). Male muscle fibers require repetitive stimulation to produce action potentials, probably because most synaptic terminals release only small amounts of neurotransmitter in response to a single shock to the laryngeal nerve (Tobias & Kelley 1988). Adult female laryngeal muscles are heterogeneous in fiber type; most are slow twitch (Sassoon et al 1987). All female fibers produce action potentials when the nerve is stimulated. Further, muscle fibers are extensively dye coupled in females, thus suggesting that they function syncytially (Tobias & Kelley 1988).

Characteristics of the muscle fibers themselves and of their synaptic innervation are well matched to the demands of vocal production in each sex. The slow rate of ticking permits relaxation of female, slower twitch fibers between each contraction. The high safety factor at female neuromuscular junctions and coupling of fibers ensures that the small number of fibers will contract together effectively enough to produce a click. The large number of fibers in the male provides sufficient cells for recruitment; fiber characteristics permit fast twitches and fatigue resistance. Low safety factor synapses with attendant requirements for facilitation in males ensure that some fibers will not contract during every nerve volley; this mitigates fatigue and facilitates recruitment.

Certain developmental events contributing to laryngeal sex differences have been identified. As is the case in the central nervous system, control of cell number and type in the neuromuscular periphery is a key component. At metamorphosis, laryngeal muscle fiber number in both sexes is the same as in adult females (Sassoon & Kelley 1986). In response to rising androgen levels, males add new fibers for the next ten months; no net addition of muscle fibers occurs in females (Sassoon et al 1986, Sassoon & Kelley 1986). In addition to controlling fiber number, androgen also influences fiber type expression. In juvenile females, exogenous androgen converts the heterogeneous pattern of female fiber types (predominantly slow) to a homogenous, all-fast profile (Sassoon et al 1987). Androgen treatment does not similarly convert fiber types in adults.

Regulation of synaptic input to laryngeal muscle fibers during development has not yet been examined. In adults, some male muscle fibers are multiply innervated (Tobias & Kelley 1988). One possibility is that the multiple innervation of single fibers seen in males reflects androgen-induced prevention of synapse elimination. This mechanism has been proposed in the androgen-sensitive LA/BC muscle of male rats, which is retarded, relative to other muscles, in withdrawal of supernumerary axon terminals (Jordan et al 1986).

Summary and Conclusions

Sex differences in behavior are the result of natural and sexual selection. The dimorphic classes of behavior described here, courtship, copulatory, and parental behaviors, reflect both kinds of evolutionary selective pressures. We can further distinguish two kinds of mechanisms that produce differences in male and female behaviors. In one, both sexes can perform a behavior but one does not because of sex differences in the external stimuli or the endocrine milieu. Maternal behavior in rodents falls into this category, as do certain other reproductive behaviors. In the other, the sensory, CNS, or motor components that produce behaviors are different in males and females. Many courtship and copulatory behaviors are in this category. I have considered some cellular mechanisms that generate sex differences in behavioral effector neurons, including sensitivity to hormones, cell number, and synaptic connectivity. A common feature of many such systems is a degree of developmental arrest: sexually dimorphic, hormone-sensitive neurons or muscles are immature at stages when other cells have completed differentiation. The cellular and molecular processes whereby hormones harness the developmental programs of behavioral effector cells remain largely unknown and are the focus of active investigation.

ACKNOWLEDGMENTS

Research in the author's laboratory is supported by National Institute of Health grants NS19949 and NS23684. I wish to thank Marianne Hayes for help in preparing the manuscript. Eliot Brenowitz and Michael Scudder provided critical reviews and useful suggestions. The Neurosciences Institute located at The Rockefeller University kindly furnished a room not only of one's own but also with a view. I am most grateful to Dr. G. Edelman and Dr. E. Gall for the respite.

Literature Cited

Arak, A. 1983. Sexual selection by male-male competition in natterjack toads. *Nature* 306: 261–62

Arnold, A. P. 1984. Androgen regulation of motor neuron size and number. *Trends Neurosci.* 7: 239–42

Arnold, A. P., Gorski, R. 1984. Gonadal steroid induction of structural sex differences in the CNS. *Ann. Rev. Neurosci.* 7: 413–42

Arnold, A. P., Saltiel, A. 1979. Sexual difference in pattern of hormone accumulation in the brain of a song bird. *Science* 205: 702–5

Arnold, A. P., Nottebohm, F., Pfaff, D. 1976. Hormone concentrating cells in vocal control and other areas of the brain of the zebra finch (*Poephilia guttata*). *J.*

Comp. Neurol. 165: 487–512

Baker, M. C., Bottjer, S. W., Arnold, A. P. 1984. Sexual dimorphism and lack of seasonal changes in vocal control regions of the White-crowned sparrow brain. *Brain Res.* 295: 85–89

Boer, K., Dogterom, J., Snijdewint, F. 1979. Neurohypophysial hormone release during pregnancy and parturition. *J. Endocrinol.* 80: 41–42

Breedlove, S. M. 1986a. Cellular analysis of hormone influence on motoneuronal development and function. *J. Neurobiol.* 17: 157–76

Breedlove, S. M. 1986b. Absence of androgen accumulation by motoneurons of neonatal rats. *Soc. Neurosci. Abstr.* 12: 1220

Breedlove, S. M., Arnold, A. P. 1981. Sexually dimorphic motor nucleus in the rat lumbar spinal cord: Response to adult hormonal manipulation, absence in androgen-insensitive rats. *Brain Res.* 225: 297–307

Breedlove, S. M., Arnold, A. P. 1983a. Hormonal control of a developing neuromuscular system. I. Complete demasculinization of the male rat spinal nucleus of the bulbocavernosus using the antiandrogen Flutamide. *J. Neurosci.* 3: 417–23

Breedlove, S. M., Arnold, A. P. 1983b. Hormonal control of a developing neuromuscular system. II. Sensitive periods for the androgen-induced masculinization of the rat spinal nucleus of the bulbocavernosus. *J. Neurosci.* 3: 424–32

Breedlove, S. M., Jordan, C. L., Arnold, A. P. 1983. Neurogenesis of motoneurons in the sexually dimorphic spinal nucleus of the bulbocavernosus in rats. *Dev. Brain Res.* 9: 39–43

Brenowitz, E. A., Arnold, A. P. 1985. Lack of sexual dimorphism in steroid accumulation in vocal control brain regions of duetting song birds. *Brain Res.* 344: 172–75

Brenowitz, E. A., Arnold, A. P. 1986. Interspecific comparisons of the size of neural song control regions and song complexity in duetting birds: Evolutionary implications. *J. Neurosci.* 6: 2875–79

Brenowitz, E. A., Arnold, A. P., Levin, R. N. 1985. Neural correlates of female song in tropical duetting birds. *Brain Res.* 343: 104–12

Bridges, R. S. 1983. Sex differences in prolactin secretion in parental male and female rats. *Psychoneuroendocrinology* 8: 109–16

Bridges, R. S. 1984. A quantitative analysis of the role of dosage, sequence and duration of estradiol and progesterone exposure in the regulation of maternal behavior in the rat. *Endocrinology* 114: 930–40

Bridges, R. S., Russell, D. 1981. Steroidal interactions in the regulation of maternal behaviour in virgin female rats: Effects of testosterone, dihydrotestosterone, estradiol, progesterone and the aromatase inhibitor, 1,4,6-androstatriene-3-17-dione. *J. Endocrinol.* 90: 31–40

Bridges, R. S., Zarrow, M. X., Denenberg, V. H. 1973. The role of neonatal androgen in the expression of hormonally induced maternal responsiveness in the adult rat. *Horm. Behav.* 4: 315–22

Bridges, R. S., DiBiase, R., Loundes, D. D., Doherty, P. C. 1985. Prolactin stimulation of maternal behavior in female rats. *Science* 227: 782–84

Clemens, L., Gladhue, B. 1979. Neuroendocrine control of adult sexual behavior. *Rev. Neurosci.* 4: 73–103

Crews, D., Moore, M. C. 1986. Evolution of mechanisms controlling mating behavior. *Science* 231: 121–25

Crews, D., Silver, R. 1985. Reproductive physiology and behavior interactions in nonmammalian vertebrates. In *Reproduction*, ed. N. Adler, R. Goy, D. Pfaff, pp. 101–82. New York: Plenum

Daly, M., Wilson, M. 1978. *Sex, Evolution and Behavior.* Belmont, Calif: Wadsworth

Darwin, C. 1871. *The Descent of Man and Selection in Relation to Sex.* London: Murray

DeVoogd, T. J., Nottebohm, F. 1981. Sex differences in dendritic morphology of a song control nucleus in the canary: A quantitative Golgi study. *J. Comp. Neurol.* 196: 309–16

Elwood, R. W. 1983. Paternal behavior of rodents. In *Parental Behavior of Rodents*, ed. R. W. Elwood, pp. 235–53. New York: Wiley

Erulkar, S., Kelley, D., Jurman, M., Zemlan, F., Schneider, G., Krieger, N. 1981. The modulation of the neural control of the clasp reflex in male *Xenopus laevis* by androgens. *Proc. Natl. Acad. Sci. USA* 78: 5876–80

Fahrbach, S. E., Pfaff, D. W. 1986. Effect of preoptic region implants of dilute estradiol on the maternal behavior of ovariectomized, nulliparous rats. *Horm. Behav.* 20: 354–63

Fahrbach, S. E., Morrell, J. I., Pfaff, D. W. 1985. Possible role for endogenous oxytocin in estrogen-facilitated maternal behavior in rats. *Neuroendocrinology* 40: 526–32

Fahrbach, S. E., Morrell, J. I., Pfaff, D. W. 1986. Effect of varying the duration of pretest cage habituation on oxytocin induction of short-latency maternal behavior. *Physiol. Behav.* 37: 135–39

Farabaugh, S. M. 1982. The ecological and social significance of duetting. In *Acoustic Communication in Birds*, 2: 85–124. New York: Academic Press

Fivizzani, A. J., Colwell, M. A., Oring, L. W. 1986. Plasma steroid hormone levels in free-living Wilson's phalaropes, *Phalaropus tricolor*. *Gen. Comp. Endocrinol.* 62: 137–44

Fleming, A., Rosenblatt, J. 1974. Olfactory regulation of maternal behavior in rats. I. Effects of olfactory bulb removal in experienced and inexperienced lactating and cycling females. *J. Comp. Physiol. Psychol.* 86: 221–32

Fleming, A. S., Luebke, C. 1981. Timidity

prevents the virgin female rat from being a good mother: Emotionality differences between nulliparous and parturient females. *Physiol. Behav.* 27: 863–68

Gibber, J. R. 1981. *Infant-directed behaviors in male and female rhesus monkeys.* PhD thesis, Univ. Wisconsin-Madison, Dept. Psychol.

Goldman, S., Nottebohm, F. 1983. Neuronal production, migration and differentiation in a vocal control nucleus of the adult female canary brain. *Proc. Natl. Acad. Sci. USA* 80: 2390–94

Goldfoot, P., Neff, P. 1985. On measuring behavioral sex differences in social contexts. See Crews & Silver 1985, pp. 767–83

Gorlick, D. L., Kelley, D. B. 1986. The ontogeny of androgen receptors in the CNS of *Xenopus laevis* frogs. *Dev. Brain Res.* 26: 193–200

Gorlick, D. L., Kelley, D. B. 1987. Neurogenesis in the vocalization pathway of *Xenopus laevis. J. Comp. Neurol.* 257: 614–27

Gorski, R., Gordon, J., Shryne, J., Southam, A. 1978. Evidence for a morphological sex difference within the medial preoptic area of the rat brain. *Brain Res.* 148: 333–46

Goy, R., McEwen, B. 1980. *Sexual Differentiation of the Brain.* Cambridge: MIT Press

Greenough, W. T., Carter, C. S., Steerman, C., DeVoogd, T. J. 1977. Sex differences in dendritic patterns in hamster preoptic area. *Brain Res.* 126: 63–72

Gurney, M. E. 1981. Hormonal control of cell form and number in the zebra finch song system. *J. Neurosci.* 1: 658–73

Gurney, M. E. 1982. Behavioral correlates of sexual differentiation in the zebra finch song system. *Brain Res.* 231: 153–72

Gurney, M. E., Konishi, M. 1980. Hormone induced sexual differentiation of brain and behavior in zebra finches. *Science* 208: 1380–82

Hamburger, V., Oppenheim, R. 1982. Naturally occurring neuronal death in vertebrates. *Neurosci. Comment.* 1: 39–55

Hannigan, P., Kelley, D. 1986. Androgen-induced alterations in vocalizations of female *Xenopus laevis*: Modifiability and constraints. *J. Comp. Physiol. A* 158: 517–28

Hart, B. 1980. Neonatal spinal transection in male rats: Differential effects on penile reflexes and other reflexes. *Brain Res.* 185: 423–28

Hatton, J., Ellisman, M. 1982. A restructuring of hypothalamic synapses is associated with motherhood. *J. Neurosci.* 2: 704–7

Hildebrand, J. G. 1985. Metamorphosis of the insect nervous system: Influences of the periphery on the postembryonic development of the antennal sensory pathway in the brain of *Manduca sexta.* In *Model Neural Networks and Behavior*, ed. A. I. Selverston, pp. 129–480. New York: Plenum

Hutchison, J., Poynton, J. 1963. A neurological study of the clasp reflex in *Xenopus laevis* (Daudin). *Behavior* 22: 41–60

Hutchison, J. B., Wingfield, J. C., Hutchison, R. E. 1984. Sex differences in plasma concentrations of steroids during the sensitive period for brain differentiation in the zebra finch. *J. Endocrinol.* 103: 363–69

Jacobson, C., Gorski, R. 1981. Neurogenesis of the sexually dimorphic nucleus of the preoptic area in the rat. *J. Comp. Neurol.* 196: 519–29

Jacobson, C. D., Csernus, V. J., Shryne, J. E., Gorski, R. A. 1981. The influence of gonadectomy, androgen exposure, or a gonadal graft in the neonatal rat on the volume of the sexually dimorphic nucleus of the preoptic area. *J. Neurosci.* 1: 1142–47

Jenni, D. A., Betts, B. J. 1978. Sex differences in nest construction, incubation, and parental behaviour in the polyandrous American jacana (*Jacana semispinosa*). *Animal Behav.* 26: 207–18

Jenni, P., Collier, G. 1972. Polyandry in the American jacana (*Jacana semispinosa*). *Auk* 89: 743–65

Jordan, C., Letinsky, M., Arnold. A. 1986. Androgen prevents synapse elimination in a hormone-sensitive muscle of the rat. *Soc. Neurosci. Abstr.* 12: 1213

Jung, I., Baulieu, E. E. 1972. Testosterone cytosol receptor in the rat levator ani muscle. *Nature New Biol.* 237: 24–26

Kelley, D. 1978. Neuroanatomical correlates of hormone-sensitive sex behaviors in frogs and birds. *Amer. Zool.* 18: 477–88

Kelley, D. 1986a. Neuroeffectors for vocalization in *Xenopus laevis*: Hormonal regulation of sexual dimorphism. *J. Neurobiol.* 17: 231–48

Kelley, D. 1986b. The genesis of male and female brains. *Trends Neurosci.* 9: 499–502

Kelley, D., Pfaff, D. 1976. Hormone effects on male sex behavior in adult South African clawed frogs, *Xenopus laevis. Horm. Behav.* 7: 159–82

Kelley, D., Pfaff, D. 1978. Generalizations from comparative studies on neuroanatomical and endocrine mechanisms of sexual behavior. In *Biological Determinants of Sexual Behavior*, ed. J. Hutchison, pp. 225–54. Chichester: Wiley

Konishi, M. 1985. Birdsong: From behavior to neuron. *Ann. Rev. Neurosci.* 8: 125–70

Konishi, M., Akutagawa, E. 1985. Neuronal growth, atrophy and death in a sexually dimorphic song nucleus in the zebra finch brain. *Nature* 315: 145–47

Kroodsma, D. 1976. Reproductive development in a female songbird: Differential stimulation by quality of male song. *Science* 192: 574–75

Kurz, E. M., Sengelaub, D. R., Arnold, A. P. 1986. Androgens regulate the dendritic length of mammalian motoneurons in adulthood. *Science* 232: 395–98

Lambdin, L., Kelley, D. 1986. Organization and activation of sexually dimorphic vocalizations: Androgen levels in developing and adult *Xenopus laevis*. *Soc. Neurosci. Abstr.* 12: 1213

Lande, R. 1980. Sexual dimorphism, sexual selection, and adaptation in polygenic characters. *Evolution* 34: 292–305

Lande, R., Arnold, S. J. 1985. Evolution of mating preference and sexual dimorphism. *J. Theor. Biol.* 117: 651–64

Le Croy, M. 1981. The genus *Paradisea*— Display and evolution. *Amer. Museum Novitates*, No. 2714

Levin, R. N. 1983. The adaptive significance of vocal duetting in the bay wren (*Thryothorus nigricapillus*). *Abstr. 101st Stated Meet. Amer. Ornithol. Union.* 54: 282

Levin, R. N. 1985. The function of vocal duetting in the bay wren (*Thryothorus nigricapillus*). *Amer. Zool.* 25: 3A

Maynard Smith, J. 1978. *The Evolution of Sex*. Cambridge: Cambridge Univ. Press

Maynard Smith, J. 1982. *Evolution and the Theory of Games*. Cambridge: Cambridge Univ. Press

Maynard Smith, J. 1984. The ecology of sex. In *Behavioral Ecology—An Evolutionary Approach*. Oxford: Blackwell. 2nd ed.

Moltz, H., Lubin, M., Leon, M., Numan, M. 1970. Hormonal induction of maternal behavior in the ovariectomized nulliparous rat. *Physiol. Behav.* 5: 1373–77

Moore, F. L. 1983. Behavioral endocrinology of amphibian reproduction. *Bioscience* 33: 557–61

Morishige, W., Pepe, A., Rothchild, I. 1973. Serum luteinizing hormone, prolactin and progesterone levels during pregnancy in the rat. *Endocrinology* 92: 1527–30

Narins, P. M., Capranica, R. R. 1980. Neural adaptations for processing the two-note call of the Puerto Rican treefrog, *Eleutherodactylus coqui*. *Brain Behav. Evol.* 17: 48–66

Nordeen, E. J., Nordeen, K. W., Sengelaub, D., Arnold, A. P. 1985. Androgens prevent normally occurring cell death in a sexually dimorphic spinal nucleus. *Science* 229: 671–73

Nordeen, K. W., Nordeen, E. J., Arnold, A. P. 1986. Estrogen establishes sex differences in androgen accumulation in zebra finch brain. *J. Neurosci.* 6: 734–38

Nordeen, K. W., Nordeen, E. J., Arnold, A. P. 1987a. Estrogen accumulation in zebra finch song control nuclei: Implications for sexual differentiation and adult activation of song behavior. *J. Neurobiol.* 18: 569–82

Nordeen, K. W., Nordeen, E. J., Arnold A. P. 1987b. Sexual differentiation of androgen accumulation within the zebra finch brain through selective cell loss and addition. *J. Comp. Neurol.* 259: 393–99

Nottebohm, F. 1980. Testosterone triggers growth of brain vocal control nuclei in adult female canaries. *Brain Res.* 189: 429–36

Nottebohm, F. 1981. A brain for all seasons: Cyclical anatomical changes in song control nuclei of the canary brain. *Science* 214: 1368–70

Nottebohm, F., Arnold, A. P. 1976. Sexual dimorphism in vocal control areas of the song bird brain. *Science* 194: 211–13

Nottebohm, F., Stokes, T., Leonard, C. 1976. Central control of song in the canary (*Serinus canarius*). *J. Comp. Neurol.* 207: 344–57

Numan, M. 1985. Brain mechanisms and parental behavior. See Crews & Silver 1985, pp. 537–605

Numan, M. 1986. The role of the medial preoptic area in the regulation of maternal behavior in the rat. *Ann. NY Acad. Sci.* 474: 226–33

Numan, M., Corodimas, K. P. 1985. The effects of paraventricular hypothalamic lesions on maternal behavior in rats. *Physiol. Behav.* 35: 417–25

Numan, M., Morrell, J. I., Pfaff, D. W. 1985. Anatomical identification of neurons in selected brain regions associated with maternal behavior deficits induced by knife cuts of the lateral hypothalamus in rats. *J. Comp. Neurol.* 237: 552–64

Olsen, K. 1979. Induction of male mating behavior in androgen-insensitive (tfm) and normal (King-Holtzman) male rats: Effect of testosterone propionate, estradiol benzoate and dihydrotestosterone. *Horm. Behav.* 61: 105–15

Oring, L. W., Fivizzani, A., ElHalawani, M., Goldsmith, A. 1986. Seasonal changes in prolactin and luteinizing hormone in the polyandrous spotted sandpiper, *Actitis macularia*. *Gen. Comp. Endocrinol.* 62: 394–403

Partridge, L., Halliday, T. 1984. Mating patterns and mate choice. In *Behavioral Ecology—An Evolutionary Approach*, ed. J. Krebs, N. Davies. Sunderland: Sinauer Assoc.

Paton, J., Nottebohm, F. 1984. Neurons generated in the adult brain are recruited into functional circuits. *Science* 225: 1046–48

Pedersen, C., Prange, A. 1979. Induction of maternal behavior in virgin rats after intracerebroventricular administration of oxytocin. *Proc. Natl. Acad. Sci. USA* 76: 6661–65

Pedersen, C., Ascher, J., Monroe, Y., Prange, A. 1982. Oxytocin induces maternal behavior in virgin female rats. *Science* 216: 648–50

Pfaff, D. 1980. *Estrogens and Brain Function.* New York: Springer-Verlag

Pfaff, D., Keiner, M. 1973. Atlas of estradiol-concentrating cells in the central nervous system of the female rat. *J. Comp. Neurol.* 151: 121–58

Prove, E. 1983. Hormonal correlates of behavioral development in zebra finches. In *Hormones and Behavior in Higher Vertebrates,* ed. J. Balthazar, E. Prove, R. Gilles, pp. 368–74. Berlin: Springer-Verlag

Raisman, G., Field, P. M. 1973. Sexual dimorphism in the neuropil of the preoptic area of the rat and its dependence on neonatal androgen. *Brain Res.* 54: 1–29

Rissman, E., Wingfield, J. 1984. Hormonal correlates of polyandry in the spotted sandpiper, *Actitis macularia. Gen. Comp. Endocrinol.* 56: 401–5

Rosenblatt, J. 1967. Nonhormonal basis of maternal behavior in the rat. *Science* 156: 1512–14

Rosenblatt, J., Mayer, A., Siegel, H. 1985. Maternal behavior among the nonprimate mammals. See Crews & Silver 1985, pp. 229–98

Sachs, B. 1981. Role of the rat's striated penile muscles in penile reflexes, copulation and the induction of pregnancy. *J. Reprod. Fertil.* 66: 433–43

Samuels, M. H., Bridges, R. S. 1983. Plasma prolactin concentrations in parental male and female rats: Effects of exposure to rat young. *Endocrinology* 113: 1647–54

Sassoon, D., Kelley, D. 1986. The sexually dimorphic larynx of *Xenopus laevis*: development and androgen regulation. *Am. J. Anat.* 177: 457–72

Sassoon, D., Gray, G., Kelley, D. 1987. Androgen regulation of muscle fiber type in the sexually dimorphic larynx of *Xenopus laevis. J. Neurosci.* 7: 3198–3206

Sassoon, D., Segil, N., Kelley, D. 1986. Androgen-induced myogenesis and chondrogenesis in the larynx of *Xenopus laevis. Dev. Biol.* 113: 135–40

Schmidt, R. 1976. Neural correlates of frog calling. Isolated brain stem. *J. Comp. Physiol.* 108: 99–113

Schneiderman, A. M., Hildebrand, J. G. 1985. Sexually dimorphic development of the insect olfactory pathway. *Tr. Neurosci.* 8: 494–99

Schneiderman, A. M., Hildebrand, J. G., Brennan, M., Tumlinson, H. 1986. Trans-sexually grafted antennae alter pheromone-directed behaviour in a moth. *Nature* 323: 801–3

Schneiderman, A. M., Matsumoto, S. G., Hildebrand, J. G. 1982. Trans-sexually grafted antennae influence development of sexually dimorphic neurones in moth brain. *Nature* 298: 844–46

Searcy, W., Andersson, M. 1986. Sexual selection and the evolution of song. *Ann. Rev. Ecol. Syst.* 17: 507–33

Segil, N., Silverman, L., Kelley, D. B. 1987. Androgen binding levels in the sexually dimorphic muscle of *Xenopus laevis. Gen. Comp. Endocrinol.* 66: 95–101

Selander, R. 1972. Sexual selection and dimorphism in birds. In *Sexual Selection and the Descent of Man,* ed. B. Campbell, pp. 180–230. London: Heinemann

Sengelaub, D., Arnold, A. 1986. Development and loss of early projections in a sexually dimorphic rat spinal nucleus. *J. Neurosci.* 6: 1613–20

Siegel, H., Rosenblatt, J. 1975. Hormonal basis of hysterectomy-induced maternal behavior during pregnancy in the rat. *Horm. Behav.* 6: 211–22

Tate-Ostroff, B., Bridges, R. S. 1985. Plasma prolactin levels in parental male rats: Effects of increased pup stimuli. *Horm. Behav.* 19: 220–26

Thorpe, W. 1961. *Bird Song: The Biology of Vocal Communication and Expression in Birds.* Cambridge: Cambridge Univ. Press

Thorpe, W. 1972. Duetting and antiphonal singing in birds: Its extent and significance. *Behavior* (Suppl. 18)

Tobias, M., Kelley, D. 1987. Vocalizations of a sexually dimorphic isolated larynx: Peripheral constraints on behavioral expression. *J. Neurosci.* 7: 3191–97

Tobias, M., Kelley, D. 1988. Electrophysiology and dye-coupling are sexually dimorphic characteristics of individual laryngeal muscle fibers in *Xenopus laevis. J. Neurosci.* In press

Trivers, R. 1972. Parental investment and sexual selection. In *Sexual Selection and the Descent of Man,* ed. B. Campbell. Chicago: Aldine

Weintraub, A., Bockman, R., Kelley, D. 1985. Prostaglandin E_2 induces sexual receptivity in female *Xenopus laevis. Horm. Behav.* 19: 386–99

Wells, K. 1977. The social behavior of anuran amphibians. *Anim. Behav.* 25: 666–93

Wetzel, D., Kelley, D. 1983. Androgen and gonadotropin control of the mate calls of male South African clawed frogs, *Xenopus laevis. Horm. Behav.* 17: 388–404

Wetzel, D., Haerter, U., Kelley, D. 1985. A proposed efferent pathway for mate calling in South African clawed frogs, *Xenopus laevis*: tracing afferents to laryngeal

motor neurons with HRP-WGA. *J. Comp. Physiol. A* 157: 749–61

Williams, G. 1966. *Sex and Evolution.* Princeton: Princeton Univ. Press

Zarrow, M., Gandelman, R., Denenberg, V. 1971. Prolactin: Is it an essential hormone for maternal behavior in mammals? *Horm. Behav.* 12: 343–54

Ann. Rev. Neurosci. 1988. 11 : 253–88

ANATOMICAL ORGANIZATION OF MACAQUE MONKEY STRIATE VISUAL CORTEX

Jennifer S. Lund

Departments of Psychiatry, Neurology, Ophthalmology and
Neurobiology, Anatomy and Cell Science, University of Pittsburgh
School of Medicine, Pittsburgh, Pennsylvania 15261

Introduction

The internal neural anatomy of the primate primary visual cortex—
variously known as striate cortex, area 17, or V1—is the subject of this
review. In primates, unlike many other mammalian groups, nearly all
visual information enters the other cortical processing areas via area V1.
Visual information is relayed from the retina via the dorsal lateral geniculate nucleus (LGN) of the thalamus to V1. The geniculo-cortical pathway is composed of multiple channels, each of which is believed to have
different functional characteristics. In addition to area V1, about 20 separate regions of visually related cerebral cortex have now been recognized,
each of which is ultimately dependent either directly or indirectly on
information processed in V1. The relays out of V1 to various cortical and
subcortical destinations arise from different laminae and different cell
groups within V1. Each of these efferent relays carries its own particular
type of visual information derived within the neuropil of V1 from one or
more of the inputs from the LGN. The evolution of multiple new abstrac-
tions of the visual information within V1 places a tremendous demand on
the neuronal circuitry of this cortical area.

Early physiological exploration of V1 supported the view that neurons
are arranged in a mosaic of small functionally distinct regions. Because
neurons exhibit some of the same functional properties at any one cortical
point through several layers in cortical depth, this mosaic was thought of
as a set of columns, each with a particular function. We now know,
however, that each lamina in cortical depth has its own unique properties

253

0147–006X/88/0301–0253$02.00

and functional mosaic, its own pattern of dependency on different inputs from the LGN, and its own efferent relay pattern. In this review I consider current knowledge of the anatomical substrates for this laminar organization of area V1 circuitry. An anatomical approach to this complex neuropil may be useful to investigations of other regions of cerebral cortex. The primates used for the majority of the studies described were macaque monkeys. While the description is often applicable to other primates, there appear to be subtle anatomical differences between the visual area of macaques and that of New world monkeys, apes, and man. However, the significance of these differences is not understood. In this review I cover those specific points that seem to be of current interest, rather than providing an exhaustive treatment of all that is known of this area. The extensive literature on the functional organization of the V1 region is not discussed in any detail. Figure 1, A–D, provides a set of landmarks that in part define the laminar architecture of V1.

Thalamic Afferents

A natural starting point for exploring the functional architecture of the striate cortex has been the definition of zones that receive input from the lateral geniculate nucleus (LGN) (Hubel & Wiesel 1972, Lund 1973, Hendrickson et al 1978, Livingstone & Hubel 1982, Blasdel & Lund 1983, Fitzpatrick et al 1983a). In the macaque these regions stand out clearly because they contain an especially rich content of the mitochondrial enzyme cytochrome oxidase (Wong-Riley 1979, Horton 1984), which provides valuable reference points and boundaries against which other anatomical features and functional patterns of the cortex can be compared (Figures 1C and 3). The use of tritiated amino acids or horseradish peroxidase (HRP) conjugated wheat-germ agglutinin (WGA) as well as earlier degeneration techniques has shown a parcellation of different thalamic fiber groups to different laminae in cortical depth; careful study of retrogradely transported HRP has indicated in which layers of the LGN (magnocellular, parvocellular or intercalated) these various afferent fiber groups originate. The diagram of Figure 2 summarizes the information currently available, and shows at least seven separate channels of LGN input that are known to exist. The most abundant inputs are to laminae 4Cα and 4Cβ, but only laminae 4B and 5 fail to receive at least some direct inputs from the LGN.

Intra-axonal filling with HRP by micropipette after recording from individual afferent fibers and filling by bulk injection of HRP has shown that in the adult the various thalamic fiber groups differ markedly from one another in the extent of lateral spread of each in their zones of termination (Blasdel & Lund 1983). It is believed, at least for laminae 4C

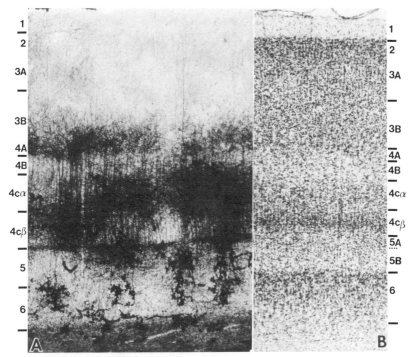

Figure 1(A–B)   A Golgi-Rapid impregnation of cortical area V1 from a *Macaca nemestrina* monkey is shown in *A*. This impregnation reveals a tufted axon plexus in lamina 4A and 3B and a narrow horizontal fiber plexus at the junction of laminae 4Cβ and 5. Both these plexuses arise from the axon projections of spiny stellate neurons lying in 4Cβ. The junctional region, where laminae 4Cβ and 5 meet, is labelled lamina 5A in the Nissl stained section of V1 shown in *B*. Lamina 5A is an important recipient of axon relays from local circuit neurons lying in thalamic recipient layers 4A, 4Cα, 4Cβ, and 6; its projections can be seen in Figure 7. (From Lund 1987.)

and 4A, that the degree of lateral spread of single fiber arbors relates to the size of the receptive field to which each fiber responds. This in turn establishes the overall representation of the retinal surface within the particular cortical layer.

The degree of spatial overlap of terminal arbors for these geniculate inputs is an important issue, because it may determine the field size and properties of the postsynaptic neurons. Attempts to estimate this spatial overlap face the problem that there seem to be several different kinds of afferent fiber emerging from even a single parvocellular or magnocellular layer. If different populations terminate within a single cortical lamina, we must first disentangle the particular numerical balance of fiber per area of cortex for each *single*, functionally distinct population before a meaningful

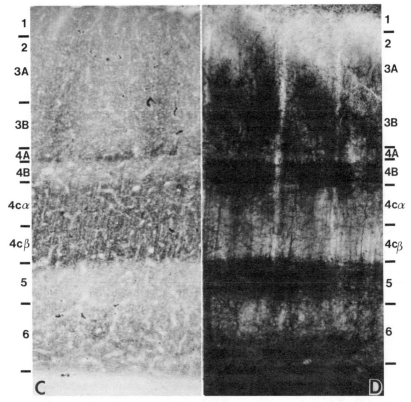

Figure 1(C–D)   A section of V1 reacted for cytochrome oxidase is shown in C. The boundaries of thalamic recipient regions 4C and 4A are clearly seen with two cytochrome rich "blob" regions evident in layer 3. In *D*, a Golgi Rapid preparation of V1, laminae 4B and 5 show heavily impregnated fiber plexuses. Lamina 4A appears as a slightly translucent cleft immediately above the upper boundary of the fiber plexus of 4B. (From Lund 1987.)

statement of internal overlap can be obtained. This presumes of course that each fiber type provides coverage of the entire visual field for a particular quality of visual input [which, for instance, blue cone input does not (deMonasterio et al 1981, Williams et al 1981)] and distributes evenly over the lamina concerned (which, for instance, the input to the cytochrome-rich patches of lamina 2-3 does not).

While the macaque has perhaps the clearest laminar separation of different thalamic fiber groups of any mammalian visual cortex so far described, very probably different fiber groups overlap at least partially in cortical depth. For instance, Blasdel & Lund (1983) showed an example of a magnocellular axon—one of the largest filled with HRP—that had

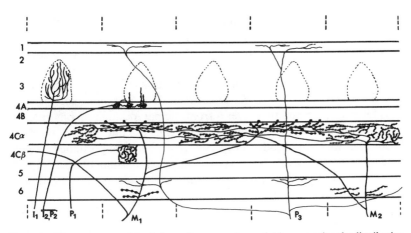

Figure 2   Diagram summarizing information currently available concerning the distribution of thalamic inputs from the LGN to V1. M_1, M_2 = inputs from magnocellular LGN layers; P_1, P_2, P_3 = inputs from parvocellular layers; I_1, I_2 = inputs from intercalated layers. The ocular dominance band dimensions are indicated by dashed lines, 450 μm apart (Hubel & Wiesel 1972, Hendrickson et al 1978, Blasdel & Lund 1983, Fitzpatrick et al 1983a). (Adapted from Fitzpatrick et al 1985.)

its terminal field limited to the upper half of lamina 4Cα, whereas the other magnocellular axons had fields that spanned the entire depth of this lamina. Another example of such an HRP-filled axon is currently under study by K. A. C. Martin and P. Somogyi (personal communication). This suggests that two populations of magnocellular axons with partial overlap in upper lamina 4Cα may exist in the macaque. We shall see later that intrinsic anatomical features distinguish upper lamina 4Cα from its lower part.

There is also some evidence that the input to lamina 4A comes from both the parvocellular layers and the intercalated layers (Fitzpatrick et al 1983a). Blasdel & Lund (1983) also showed that axons to 4A can contribute rising collaterals to the more superficial laminae (see Figure 2). These rising collaterals may converge upon the punctate, cytochrome-rich zones, which receive input from the intercalated layers of the LGN (Livingstone & Hubel 1982, Fitzpatrick et al 1983a). We shall also see that lamina 4Cβ—a primary input zone for parvocellular LGN input—is distinctly subdivided into upper and lower halves, yet the fields of HRP-filled axons from the LGN seem to span the entire depth of the lamina (Figure 2). It may well be that some, as yet unidentified, thalamic axon group may terminate uniquely in the lower of these β subdivisions and thus resemble the partial overlap of input to the α subdivision described above.

Thalamic axons also project to lamina 6 (Hubel & Wiesel 1972,

Hendrickson et al 1978). The magnocellular axons filled with HRP (with major terminal field in lamina 4Cα) have been found to contribute relatively sparse collaterals within lamina 6 where their distribution favors the deeper half of the layer (Blasdel & Lund 1983). Hendrickson et al (1978) show clear label in upper lamina 6 after injections of tritiated amino acids into the macaque LGN parvocellular layers. However, no collaterals have thus far been seen entering layer 6 from the parvocellular axons entering lamina 4Cβ or from those entering 4A (Blasdel & Lund 1983). One HRP-filled axon with color-coded properties (and therefore presumed to emanate from the parvocellular layers) was, however, filled by Blasdel & Lund, it was a widespreading, exceedingly fine-caliber axon that contributed terminals to upper lamina 6 and to lamina 1.

It is important to ascertain whether or not the thalamic inputs to layer 6 are relatively insignificant, in terms of relaying visual information, since their terminal fields in layer 6 are sparse relative to input to layers that are more superficial. However, a strategically placed, but sparse input might be as effective functionally as a numerically larger but less efficiently positioned input, in terms of the region of the postsynaptic cell contacted. Or it might be more likely that these thalamic inputs to lamina 6 target a very specific and, perhaps for the layer, numerically small population of neurons, which may explain the apparently sparse input. We shall see later that layer 6 contains a very diverse population of neurons and that all inputs to the layer appear sparse and are equally diverse in their origins.

The uncertainties concerning the distribution of subpopulations of thalamic axons are compounded by lack of specific information concerning the functional characteristics of these different thalamic inputs. We certainly do not yet know the difference between parvocellular inputs to laminae 6, 4Cβ, 4A, and 1, nor do we have any information as to why there are two types of magnocellular fibers ending in lamina 4Cα (though perhaps they correspond to the X and Y classes described by Blakemore & Vital-Durand 1986). The role of different fiber populations in relation to color-coded information is also obscure. Color response properties are clearly present in the LGN parvocellular layers (Wiesel & Hubel 1966, Schiller & Malpeli 1978, Creutzfeldt et al 1979, Schiller & Colby 1983). Clearly, there is color-coded input to 4Cβ, but it may not be a significant stimulus for all the cells of 4Cβ that lie postsynaptic to the LGN parvocellular inputs (e.g. Dow & Gouras 1973, Gouras 1974, Michael 1978, Livingstone & Hubel 1984a). The distribution of color-coded cells in the superficial layers may be discontinuous (Livingstone & Hubel 1984a), which raises the question as to whether the representation of color-coding in lamina 4Cβ is also patchy. The correlation suggested for color responses and cytochrome patches is somewhat enigmatic since in primates, in general, the small cell, intercalated layers of the LGN, which provide input to the patches, are not noticeably color coded; perhaps the relay from lamina $4C\beta$ to the patches is responsible for the color coding of these regions, as suggested by Michael (1986); this indeed is an area in need of further study.

The physiological information available shows that the parvocellular inputs to lamina $4C\beta$ provide small receptive fields and, in terms of the postsynaptic responses, produce an exceedingly precise point-by-point representation of the visual field in lateral progression across the cortex within lamina $4C\beta$ (Hubel et al 1974, Blasdel & Fitzpatrick 1984, Parker & Hawken 1984). The magnocellular inputs to lamina $4C\alpha$ clearly provide larger receptive fields; but an equally precise progression of these fields can be recorded in the postsynaptic cell responses in lamina $4C\alpha$, in register with the progression in $4C\beta$ (Blasdel & Fitzpatrick 1984). These inputs to laminae $4C\beta$ and α differ physiologically in a number of important characteristics other than field size and are served in part by different ganglion cell populations (Leventhal et al 1981, Shapley & Perry 1986). The magnocellular system projecting to laminae $4C\alpha$ has for some time been known to respond vigorously to motion (eg. see Dow 1974, Movshon et al 1985). Recently this system has been suggested to be an important element in establishing a foundation for depth perception (M. S. Livingstone and D. Hubel, personal communication). The sustained responses, color selectivity, and fine scale of representation in the inputs to laminae $4C\beta$ have suggested their role is a more detailed qualitative analysis of the visual world (see recent reviews by Schiller 1986, Shapley & Perry 1986).

The laminar distribution of different thalamic fiber types is only one dimension of the distribution pattern of these axons. Another important anatomical feature is the laterally distributed patterning of the thalamic axon terminations within their lamina of choice. The study of patterned distribution of inputs to striate cortex began with the demonstration by Hubel & Wiesel (1969, 1972) that right eye and left eye inputs are segregated—the ocular dominance stripe distribution of thalamic fibers to laminae $4C\alpha$ and $4C\beta$. We now know (Hendrickson et al 1978, Livingstone & Hubel 1982, Weber et al 1983, Fitzpatrick et al 1983a) that in squirrel monkey and macaque the neurons of the intercalated LGN layers provide input to patches or blobs in lamina 2-3, aligned in rows above the centers of ocular dominance stripes in layer 4, and that physiologically these patches seem also to be monocular (Figure 3). An even more detailed patterning is seen in the honeycomb distribution pattern of the terminals of thalamic fibers, which also show ocular dominance banding, to lamina 4A (Hendrickson et al 1978, Blasdel & Lund 1983) (see Figure 3).

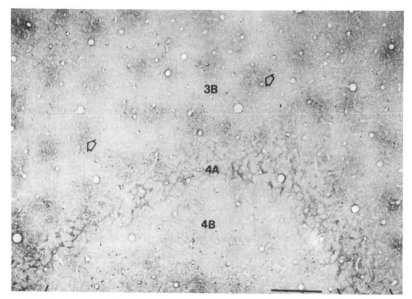

Figure 3   A tangential section through layers 4B, 4A, and 3B of V1 reacted for cytochrome oxidase. The thalamic input regions are seen as a dark reticular network in laminar 4A and as dark patches in lamina 3B. Scale bar = 500 µm. (From Fitzpatrick et al 1985.)

The origin of the patterns of ocular dominant stripes in lamina 4C with aligned lamina 4A inputs and patches in the supragranular layers seems to occur prenatally. Horton (1984) has observed the cytochrome-rich "blob" pattern in layer 2-3 at E144 and points out the correspondence in time with the start of ocular segregation of fibers, which was shown, using densitometry, by Rakic (1976). Whether the framework for this patterning is built into the cortical matrix, and the incoming axons fit into this framework in early growth, or whether competitive interaction between axons in early growth determines the pattern, is one of the central questions in cortical development. Although Wiesel and Hubel (Hubel et al 1977, LeVay et al 1980) showed that the relative width of ocular dominance stripes could be changed by postnatal monocular occlusion, the framework spacing remains intact throughout all manipulations. For the "blobs" this framework changes subtly, over the surface of the visual cortex of the normal monkey, depending on the existence of monocular or binocular inputs and on eccentricity (Horton 1984). In apes and man the honeycomb lattice of input to 4A, as seen in macaques, seems to be lacking (Horton 1984, Tigges & Tigges 1986) and the cytochrome-rich patches extend down through 4B. In prosimians the cytochrome blobs are not present in all species (McGuiness et al 1986). The distribution of efferent neurons in the superficial layers of macaque is clearly part of an intricate pattern that includes the cytochrome patch (blob) regions, and patterns of intrinsic connectivity and physiological patterning also occur in regular repeating sequences that are seemingly related to this same intrinsic framework. Understanding the origins and constraints of these anatomical and physiological patterns and whether or not they are derived from the thalamic inputs will be extremely important to our overall grasp of visual cortex function.

Postsynaptic Targets of Thalamic Inputs

The evidence for channelling of different types of input to different laminae in the depth of the cortex, though as yet far from complete, provides the rationale for a close examination of the cellular constituents of these thalamic recipient regions and for an attempt to trace the further projections of the neurons in these layers. The information about LGN fiber termination presented above suggests the following questions:

1. Do the neurons lying in these laminae restrict their dendritic surface to a particular thalamic fiber termination zone or do the dendrites reach into other laminae?
2. Do the thalamic afferents end on all available dendritic and somal surfaces in the layer or do they selectively establish contacts on particular cell types?
3. What are the distribution patterns of the axons arising from different types of neurons in the layer, both within and outside the lamina?
4. What inputs to the layer exist, other than the thalamic inputs that we have used to define the limits of the layer?
5. Do we believe that we have defined a single functional zone by using the thalamic input distribution to define the boundaries of each of the layers? These questions are discussed below.

The thalamic input zones of laminae 4Cα, 4Cβ, and 4A, are characterized by the presence of spiny stellate neurons (Lund 1973, 1984) and an almost complete lack of pyramidal neurons in 4C (Mates & Lund 1983). Intermingled with them are aspinous or sparsely spined stellate neurons, which comprise at least 15–19% of the total cell population on the basis of counts of neurons positive for GAD and GABA immunoreactivity—which most of these neurons show (Ribak 1978, Somogyi et al 1983a, Fitzpatrick et al 1983b, 1987). In addition, these laminae are transversed by the apical dendrites of pyramidal neurons with somata and basal dendrites in laminae 5 and 6. The apical dendrites of particular populations of layer 6 pyramidal neurons arborize within 4Cα, and this is accompanied by recurrent axonal arbors from these same layer 6 pyramidal cells within α. Another layer 6 pyramidal population provides dendritic arbors off the main apical dendrite at the border of laminae 5 and 4Cβ (lamina 5A), and a terminal tuft of dendrites on the 4A region (Lund & Boothe 1975); it is curious that while these lamina 6 pyramidal neuron dendritic arbors appear reluctant to enter lamina 4Cβ, these same neurons send vigorous axonal arbors into the β division with collateral extensions terminating in 4A. The lamina 5 pyramidal neuron apical dendrites pass through these layers without arborization but with variable spine populations (Lund & Boothe 1975). It is noticeable that spine number on the apical dendrite shafts and on intruding dendritic arbors falls off sharply in layer 4Cβ (Lund 1973).

These various dendritic arbor patterns suggest that the deeper pyramidal neurons, particularly those of layer 6, may receive thalamic input within 4C, particularly 4Cα, and within 4A; the basal dendrites of these same pyramidal neurons may also receive direct thalamic input in layer 6. The work of Katz et al (1984) suggests but does not prove that in primates these pyramidal neurons with dendritic and axonal relationship to the thalamic input zones of layer 4 comprise the neuron populations projecting back to the LGN known to exist in 6 (Lund et al 1975). Now that the cytochrome oxidase method has allowed a clearer definition of the depth and borders of lamina 4B (which is cytochrome poor but sandwiched between the clearly defined margins of cytochrome-rich laminae 4A and 4α), it seems probable, from examination of Golgi impregnations, that upper lamina 4Cα may interrelate to a different set of lamina 6 pyramidal neurons than does lower lamina 4Cα (Lund, unpublished observations). Lamina 4B itself may not have a significant relationship to layer 6 pyramidal neurons because, injection of HRP into lamina 6 produces very little label in lamina 4B (Blasdel et al 1985).

The recurrent axon collaterals of lamina 6 pyramidal neurons that give terminal arbors to divisions of lamina 4 differ in terminal morphology from the thalamic axons entering the same layers. The recurrent pyramidal neuron axons terminals are made from complex side-spine arrays (Fitzpatrick et al 1985) whereas the thalamic axons have en-passant, large diameter, beaded axon terminals; electron microscopy has shown that both are of type 1 morphology and therefore are believed to be excitatory. The physiological implications of this difference in axon morphology may be important, and in cat it seems to be coupled with a difference in the numerical balance of the postsynaptic site. The thalamic terminals end predominantly but not exclusively on spines (Garey & Powell 1971, Winfield & Powell 1983, McGuire et al 1984, LeVay 1986), whereas the layer 6 pyramidal neuron axons end predominantly but not exclusively on dendritic shafts of smooth dendritic neurons within layer 4 (McGuire et al 1984).

The projection of layer 6 upon layer 4 seems to rival the thalamic input in density, but the relative strength of these two important sources of afferents on single postsynaptic neurons—if the thalamic and layer 6 inputs do in fact share common postsynaptic targets—is unknown. It should also be remembered that some of the aspinous neurons of lamina 6 send direct axon projections to at least 4Cα, and so the influence of layer 6 upon layer 4 may be both excitatory and inhibitory (Lund 1987). Only by laborious EM reconstruction of the relationship between neurons (whose morphology is first identified by Golgi or HRP methods) and labelled afferent axons will the pattern of distribution of specific populations of terminals be solved. This approach is already underway in other species (Somogyi 1978, Peters et al 1979, Hornung & Garey 1981, Freund et al 1985), and is beginning in the primate (Kisvarday et al 1987), but its accuracy depends upon a clear identification of different classes of thalamic axon (or other afferents) as well as of the postsynaptic neurons. It is suspected that in single laminae there are different classes of spiny stellate neurons; certainly there are many kinds of smooth dendritic cells (Lund 1987). We have no idea of the degree or kind of patterns of synaptic specificities, and the danger is that small samples may be misleading. However, the importance of obtaining this information is undeniable and the rewards great, since the data obtained may illustrate general features of cortical organization.

Single axon terminal fields of the parvocellular LGN afferents to 4Cβ, identified so far by HRP filling, bridge the depth of the lower half of 4C (Blasdel & Lund 1983); the upper and lower limits of this 4Cβ zone were defined earlier by injecting tritiated amino acids into or making lesions in the parvocellular LGN layers (Hubel & Wiesel 1972, Lund 1973, Hendrickson et al 1978, Fitzpatrick et al 1983a). The dendrites of the neurons whose somata lie in 4Cβ may overlap into lamina 5A and into lamina 4Cα when they lie near the limits of the zone defined by the parvocellular inputs. Heavy cytochrome oxidase staining comes to an abrupt end at the lower border of lamina 4Cβ, but it is not entirely certain that this rich cytochrome zone excludes lamina 5A, which is an important but very narrow (approximately 50 μm in depth) lamina. Therefore, while the thalamic axons to 4Cβ seem to have a reasonably sharply defined upper and lower boundary to their distribution, the dendrites of potential postsynaptic neurons do not seem as sharply constrained to obey these boundaries. The same phenomenon is seen in lamina 4Cα, where the thalamic axons either bridge the depth of 4Cα or run along its upper half, but where the neurons lying at the borders of 4Cα have dendrites that stray into laminae 4Cβ and 4B respectively. There are, moreover, a group of spiny stellate neurons whose dendrites bridge the depth of the 4C division (Mates & Lund 1983).

Although they seem to be infrequent, these neurons may play an important role. These observations rise the question whether we can expect a clear physiological difference between neurons in the α and β halves of lamina 4C or whether a gradual transition of neuronal properties through the depth of 4C is more likely. This question is partially addressed in the papers of Blasdel & Fitzpatrick (1984), Parker & Hawken (1984), and Livingstone & Hubel (1984a), but a clear-cut answer may be difficult to obtain.

Axonal projections from different varieties of aspinous neurons, whose dendrites are restricted to either the α or β division, cross between the α and β thalamic input zones to provide arbors to both divisions and to 4A (see Figure 4). In addition, at least one class of lamina 4A aspinous neuron sends a prolific axon arbor back to 4C (Lund 1986). These axon relays—perhaps GABAergic—should be taken into account when considering the functional role of each channel of input. Equally important, but not yet clearly explored, is the question of whether the spiny stellate neurons of lamina 4Cα and β project between the α and β subdivisions. Although they may provide quite dense axon collaterals to the level at which their dendrite field lies within α or β (L. Katz, K. Martin, personal communications) it is not clear to what extent, if at all, axon collaterals of single, spiny stellate cells establish synaptic contacts in both α and β divisions.

Intrinsic Relays of Thalamic-Recipient Neuron Populations

Because of the clear boundaries established for lamina 4Cβ by the parvo-cellular afferents, it was surprising to find evidence of a sharp subdivision in the layer that split it into an upper and a lower half (Fitzpatrick et al 1985). This subdivision is demonstrated by the retrograde labelling of cell bodies in 4Cβ after injecting HRP into the principal projection region of the 4Cβ spiny stellate neurons, lamina 3B and adjacent 4A (see Figure 5). Following such an injection, the upper half of lamina 4Cβ and the lower part of 4Cα contain a population of well-labelled neurons; these labelled neurons extend laterally at least as far as the margins of the overlying injection site. However, in lower β, only a very narrow column of cells is retrogradely labelled, lying directly under the center of the injection site. The boundary between the upper and lower zones of retrograde labelled cells in 4Cβ is very sharp. Although nothing in the dendritic field organ-ization of the neurons in 4C has so far alerted us to this division (the labeled neurons would be expected to have dendrites passing freely across this boundary between upper and lower β), differences in the way the axons of these cells of upper and lower β target the 4A-3B region sharply define a neuronal population that resides in upper 4Cβ and lower 4Cα but which is absent from the lower half of 4Cβ. Fitzpatrick et al (1985) suggest

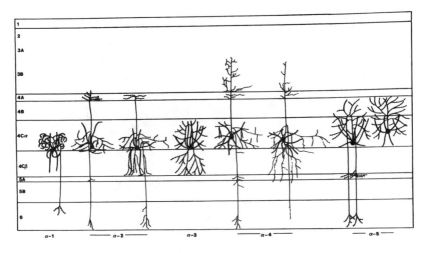

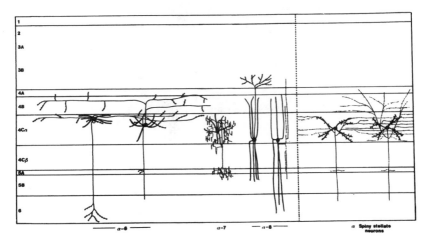

Figure 4 Diagrams summarizing the varieties of local circuit neurons (usually GABAergic) found in lamina 4Cα. Note the patterns of interlaminar axon projections differ for each variety. The axon projection patterns of two varieties of spiny stellate neurons of lamina 4Cα are shown for comparison. (From Lund 1987.)

a difference in the lateral spread of the terminal fields of the 4Cβ neurons in laminae 4A and 3B. One might suppose that the deeper β cells have a more local axon field and that the upper neurons a wider terminal distribution, but this hypothesis needs careful confirmation. So far, a physiological correlate of this split has not been discovered, though the current

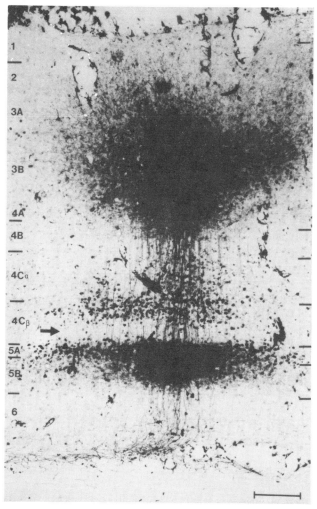

Figure 5 Photomicrograph showing the distribution of labelled neurons following a large injection of HRP into laminae 3B and 4A. Note the presence of a distinct cleft (*arrow*) of unlabelled neurons in lower lamina 4Cβ. Scale bar = 200 μm. (From Fitzpatrick et al 1985.)

source density analysis of Mitzdorf & Singer (1979) may show such a boundary.

Some years ago, using Golgi impregnations, Lund (1973) found that spiny stellate neurons of lamina 4Cα differed in their axonal projection from those of 4Cβ. Because magnocellular LGN projects to 4Cα and the parvocellular to 4Cβ, this observation suggests a continued separation of magno- and parvocellularly derived information within the cortex. While the principal projection of the 4Cβ spiny stellate neurons is to laminae 3B and 4A, the projection of the majority of 4Cα spiny stellate neurons is to lamina 4B (Lund 1973, Fitzpatrick et al 1985) or solely within 4Cα itself (K. Martin, personal communication). It now seems probable that there is a gradual diminution, rather than a sharp cutoff, of neurons projecting to the 3B lamina region, as one moves up from lamina 4Cβ into lower lamina 4Cα, and a gradually increasing number contribute instead to projections to lamina 4B and within lamina 4Cα itself (Fitzpatrick et al 1985). The lateral spread of the 4Cα axon projections can be considerable, and periodic clusters of termination are visible in the upper half of the 4Cα layer after HRP injections within 4Cα (Fitzpatrick et al 1985). Both spiny and aspinous neurons contribute laterally spreading axons, so both may add terminals to these clusters (K. Martin, personal communication, Lund 1987). Although some projections into lamina 4B from the 4Cα neurons are wide spreading, the bulk of the lamina 4Cα to lamina 4B projection forms a more narrowly focused vertical cylinder (Fitzpatrick et al 1985). The lateral spread of axons within lamina 4Cα or passing from lamina 4Cα to 4B is probably 1.0–2.0 mm on each side of a single point. Laterally spreading connectivity is further emphasized by neurons lying within lamina 4B whose axons spread over 4.5 mm to either side of a small injection of HRP tracer—again, with periodic terminal zones occurring every 400–500 μm. Both pyramidal and spiny stellate neurons contribute to these connections (Rockland & Lund 1983).

Within the neuropil of upper 4Cα itself and, with clearest definition, in 4B, the physiological properties of orientation specificity, binocular fusion, and recognition of direction of motion are all generated from the magnocellular thalamic input, an input that at entry is comprised of circularly symmetric receptive fields (Blasdel & Fitzpatrick 1984, Parker & Hawken 1984, Livingstone & Hubel 1984a). In 4B lie neurons with efferent projections to at least areas V2, V3, PO, and MT (Lund et al 1975, Rockland & Pandya 1979, Lund et al 1981, Colby et al 1983 and personal communication, Van Essen 1984), although it is not yet known if the same neurons project to all regions or if separate groups project to individual cortical destinations. The efferent neurons are clearly of different sizes, even when projecting to a single area, and the largest neurons of the layer, the spiny stellate neurons, are included in the efferent neuron pool projecting to MT as well as pyramidal neurons. The periodicity of organization in the layer may also be reflected in clustering of efferent neurons projecting to these different destinations.

The diversity of smooth dendritic neurons within lamina 4Cα is considerable (Figure 4), and this probably reflects the multiple tasks accomplished functionally within the layer (Lund 1987). These neurons have not only local axon projections within the layer but also axon projections to many intralaminar destinations outside the confines of 4Cα. With the exception of one variety of aspinous neuron that contributes laterally widespreading axon arbors within 4Cα and 4B (Figure 6), the axons of these neurons have a columnar trajectory to their interlaminar projections, which target one or more of the following laminae: 6, 5A, 4Cβ, 4B, 4A, and 3B (Lund 1987). The projection of these neurons to 3B must contribute to the rather fine orthograde label in 3B after injections of HRP into 4Cα (Fitzpatrick et al 1985). The organization of the smooth dendritic neurons in 4Cα also suggests a division between upper and lower halves of the layer, with different cell types associated with the two regions (Lund 1987).

The population of aspinous neurons in lamina 4Cβ, while complex, does not include such a large number of varieties as is seen in lamina 4Cα. Indeed, physiologically, the generation of new functional complexity in the 4Cβ layer seems less than that accomplished within 4Cα; the neurons largely retain characteristics of the circular field organization that is seen

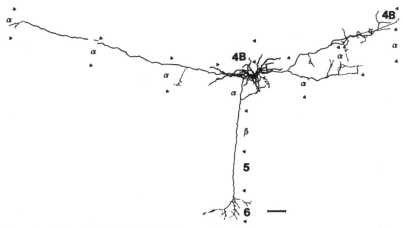

Figure 6 A drawing of a Golgi Rapid impregnation of a variety of interneuron, probably GABAergic, in lamina 4Cα of monkey V1 cortex. This variety is recognized for its large diameter axon trunks that spread away from the neuron for long distances in 4Cα and 4B. Short, beaded collaterals with predominantly vertical orientation branch off from these trunks, suggesting that the neuron is a "basket" neuron with axon contacts onto somata and initial dendritic segments of pyramidal neurons. A descending axon trunk passes to lamina 6, with a terminal arbor of limited lateral extent. Another tier of similar basket neurons is seen at the junction of laminae 5 and 6 (see Figure 11). These two tiers are the regions in monkey V1 where direction-selective neurons are located. Scale bar = 50 μm. (From Lund 1987.)

in the parvocellular afferents entering the layer (Bullier & Henry 1980, Blasdel & Fitzpatrick 1984, Parker & Hawken 1984, Linvingstone & Hubel 1984a). Orientation specificity and binocular fusion do not seem to be generated to any marked extent, at least within the lower β division. The aspinous neurons of lamina $4C\beta$, like those of $4C\alpha$, contribute axon projections to other laminae, in tight columnar trajectories, and the long distance lateral projections of smooth dendrite neurons seen in lamina $4C\alpha$ do not occur in the $4C\beta$ lamina. The axon projections of these local circuit neurons of $4C\beta$ target one or more of laminae 6, $4C\alpha$, 4A, 3B and seem also to include 5A.

The smooth dendritic neurons of $4C\alpha$ and $4C\beta$ have interlaminar projections that share many of the same target regions, namely 6, 5A, 4A, and 3B, plus the other half of the 4C division—α or β; the α division differs from the β division because it has projections to 4B that the β division lacks. It is also evident that neither division projects to 5B or apparently above lamina 3B. The projections to laminae 5A and 6 are accompanied by fine caliber, weak axonal projections from the spiny stellate neurons of the $4C\alpha$ and $4C\beta$ divisions. Again: how does one assess the importance of quantitatively weak projections? The strong axonal projections of the α and β spiny neurons to 4B, and to 4A and 3B respectively suggest that the predominant "forward" flow of information from 4C passes to these two destinations rather than to infragranular layers. The different destinations of these prominent α and β spiny neuron projections [which are of type I synaptic morphology and therefore presumed to be excitatory (LeVay 1973)] also suggest a continued separation, in some fashion, of the two streams of visual information entering $4C\alpha$ and $4C\beta$, at least in these first interlaminar relays, even though the streams are interrelated by aspinuous neuron axon relays.

To understand the anatomical substrates for function in laminae 4C, apart from the lamina 6 input, one must consider two other laminae—5A and 4A—that contain smooth dendritic neurons with clear axon projections into $4C\alpha$ and $4C\beta$ (Lund 1986, 1987). Laminae 5A and 4A are also axonally linked in reciprocal fashion to each other by aspinous neuron axon relays, and a weak spiny stellate neuron axon projection passes from lamina 4A to 5A. As we have already discussed, $4C\alpha$ and $4C\beta$ both project to 5A and 4A by a variety of both smooth and spiny neuron axon links. Lamina 5A therefore seems to occupy a special role in terms of the smooth dendrite populations contained within its narrow limits (Figure 7). These 5A neurons are extremely diverse in their morphologies but they include varieties that have clear projections to each of the thalamic input layers—4A, $4C\alpha$, $4C\beta$, and 6; additionally, in 5A, varieties of these neurons project to the superficial laminae 3, 2 and 1 (Lund 1987). Only this last projection

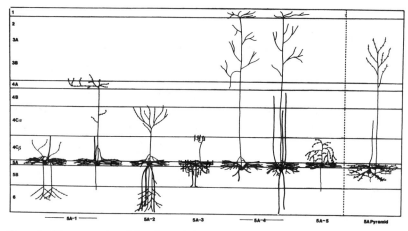

Figure 7   Diagram summarizing the varieties of local circuit neurons (usually GABAergic) found in lamina 5A of monkey cortical area V1. The axon projection of a typical pyramidal neuron of the same region is illustrated for comparison. Note the different patterns of axon projections to other laminae shown by the local circuit neuron populations. (From Lund 1987.)

pattern is shared by the spine-bearing neurons of 5A; these are small pyramids with vestigial, thread-like apical dendrites (Lund 1973).[1]

The axonal links of aspinous lamina 5A neurons (with perhaps GABAergic properties) to thalamic recipient laminae suggest an important role for the 5A layer. A gating function is possible through which the neurons of the lamina determine the relative weighting of outflow from each of the thalamic input zones and coordinate it with input from layer 6. The crucial question is what would drive the layer 5A neurons in order for them to achieve such decisive activity, and what function does the primary excitatory projection from the layer (presumably from the pyramidal neurons) perform in the superficial layers? Another important question concerning this 5A region is whether or not its neurons receive direct thalamic input.

---

[1] It was previously stated (see legend to Figure 15 of Blasdel et al 1985) that smooth dendritic neurons of lamina 5A also contributed axon arbors to lamina 4B. With continued careful examination of Golgi impregnation and reexamination of the cells in question, the author now believes that in fact these smooth dendritic projections from 5A target 4Cα, or upper 4Cα alone in some cases, and that so far, there is no clear evidence that their arbors enter 4B (see Lund 1984). In Golgi Rapid material the 4B borders are not always well defined and upper 4Cα may stain heavily, thus providing a misleading upper boundary, while 4B remains unstained. Perhaps we will find a 5A to 4B projection after all, or perhaps the injections intruded sufficiently on laminae 4A or 4Cα, both recipients of 5A projections, to retrogradely label the cells in lamina 5A.

Figure 1(*A–B*) shows the position and narrow depth of this lamina; interestingly, V1 in the infant monkey shows a narrow, cytochrome-rich band at approximately the same position and with a pale stain band that covers much of 4Cβ. A cytochrome oxidase pattern resembling that of the infant is seen in adult macaque monkeys treated systematically with monomeric acrylamide, which apparently induces degeneration of color opponent ganglion cells in the retina (Eskin & Merigan 1986) and sharply reduces both cytochrome oxidase staining of the parvocellular layers of the LGN and cytochrome oxidase staining in much of 4Cβ. Thus input to this lower narrow cytochrome-rich band may come from other than the parvocellular layers. For the moment, however, whether 5A in fact coincides with the narrow cytochrome-rich band currently placed within the lowermost limit of 4Cβ, or whether 5A lies below this cytochrome-rich band, must be decided by further experiments.

The limits of lamina 4A are, for convenience of description, here defined as the narrow zone of cytochrome-rich thalamic terminations that form a reticular or honeycomb network; this network exhibits circular lacunae, free of thalamic terminals, that vary in size, but generally measure about 100 μm across (see Figure 3). The problem with this definition is that although this region is rich in cell bodies, the dendrites and axons of these neurons are distributed above and below this narrow zone and arborize significantly in the neuropil of laminae 4B and 3B. There are a few exceptions where particular neuron varieties have dendrites that seem stratified within or against this zone [in particular, small, spiny stellate neurons and a very small class of aspinous neurons of the chandelier variety (Lund, unpublished observations)] but, here again, even the local axon distribution of these neurons does not restrict itself to the 4A lamina. It is clear from our earlier discussion that this lamina is also targeted by axonal relays from 4Cβ and α (including both aspinous and spiny neuron axon components). In addition, we have already mentioned layer 6 pyramidal neuron axonal collaterals and their apical dendrite terminal tufts, both of which enter 4A; input to lamina 4A also comes from aspinous neurons of 5A.

Physiologically, lamina 4A shows characteristics of the parvocellular LGN layers that innervate it (Blasdel & Fitzpatrick 1984); its field map resembles in its fine detail that of 4Cβ to which it is related, point by point, by rising 4Cβ axon relays. Lamina 4A is the most superficial layer in the cortex that contains a distinct population of spiny stellate neurons; although pyramidal cell bodies also lie in the layer, their dendrites largely spread above and below the layer (Lund 1973). The predominant axon projection of the small, spiny stellate neurons of the region is upward into layers 3 and 2, although a fine descending process to lamina 5A and on down to lamina 6 can be traced. Much work, both anatomical and physiological, is needed before we can understand lamina 4A. For example, an as yet unanswered question is whether the absence of a rich cytochrome band at this point in the human and ape striate cortex (Horton 1984, Tigges & Tigges 1986) is indicative of the absence of both thalamic input and the elaborate neuropil evident in this region of the macaque monkey striate cortex.

The rich composition of different 4A smooth dendritic neurons and their complex interdigitation with both lamina 4B and the superficial layers in the macaque suggest that lamina 4A may help coordinate the ongoing patterns of activity in 4B with that of the layers above, two regions that we have suggested are driven primarily by different streams of incoming visual information. The insertion of thalamic input from both parvo-cellular and intercalated layers into lamina 4A is interesting, and it will be valuable to learn more about the physiological properties of the LGN cells that give rise to these inputs in comparison to those that project to other laminae. If this region does coordinate the activity of lamina 4B with the more superficial cortex, then the properties of this thalamic input may be expected to reflect particular qualities that would aid in this coordination.

Organization of the Superficial Laminae

We now consider the superficial laminae 3B, 3A-2, and 1. The intrinsic neuronal organization reaches an extreme complexity within these super-ficial layers, with relays from all the other deeper cortical laminae also entering into its neuropil. Within these superficial laminae are found a more diverse population of smooth dendritic neurons than is seen in any of the other layers, and the topography of laterally spreading, intrinsic axonal projections from the pyramidal neurons of the region is extra-ordinarily elaborate. Moreover, the region contains a diverse population of efferent neurons, projecting to cortical areas V2, V3A, V4, and PO (Rockland & Pandya 1979, Lund et al 1981, Tigges et al 1981, Colby et al 1983, Yukie & Iwai 1985, Van Essen et al 1986) that often occur in regularly patterned arrays and at least some of which are apparently spatially segre-gated from each other according to both function and efferent destina-tion (Livingstone & Hubel 1984a).

The principal intrinsic input to lamina 3B is from spiny stellate neurons of 4Cβ, but neurons of lowermost 4Cα and 4A also contribute. We have already seen the pattern of retrograde label if an HRP injection is placed in 3B; such an injection into the superficial laminae (3A, 2, and 1) gives no labelled cells within any of the regions of 4A or 4C that received LGN input. Instead, this more superficial injection labels a cluster of cells in lamina 4B immediately under the injection site, a laterally spread population of retrogradely labelled cells in lamina 5, mainly in 5A, and a few labelled neurons in lamina 6.

The thalamic input from the intercalated layers of the LGN focusses its patchy projections chiefly on the 3B region (Livingstone & Hubel 1982, Fitzpatrick et al 1983a), and lamina 3B is also served by collaterals of the thalamic projection to 4A (Blasdel & Lund 1983). To understand more fully the organization of inputs from other layers to lamina 3B, the interested reader should examine the results of the small injection studies of Blasdel et al (1985) and Fitzpatrick et al (1985) and of the smooth dendritic neuron relays from 4C and 5A detailed by Lund (1987). The 3B region receives its predominant lamina 5 input from 5A rather than from 5B and receives only a narrowly focussed, weak input from lamina 6.

Properties such as binocular fusion, orientation, and end stopping seem to be generated in the neuropil of 3B (Hubel & Wiesel 1968, Hubel et al 1977, Livingstone & Hubel 1984a). I suggest, however, that these properties are generated in more than one lamina. I have already proposed that 4Cα and 4B neurons also achieve some of the same functional characteristics (and see Malpeli et al 1986, Schwark et al 1986, for evidence of such a pattern for cat), but perhaps these properties are kept in vertical alignment between laminae by interlaminar relays—an alignment that may be initiated developmentally from properties developing in a single layer, e.g. 4Cα.

This functional elaboration and the eventual parcellation of different functions to different efferent neuron groups must continue via the prominent and broadly spreading axon projections that rise from lamina 3B neurons into laminae 3A-2 and 1 (Blasdel et al 1985). Joining this projection to 3A-2 and 1 are axon relays from 4B that rise in strong columnar point-to-point fashion; in addition, lamina 5B neurons contribute rising axon projections that spread broadly in the lateral direction on reaching lamina 2-3A (Blasdel et al 1985). Lamina 6 provides a very light projection to this superficial cortex, where it distributes in a point-to-point vertical fashion. These projections from laminae 4B, 5, and 6 include the axons of neurons of both aspinous (Figure 10) and spiny neuron varieties.

Evidence for the complex lateral patterning of intrinsic projections in the superficial layers was first seen in the studies of Rockland & Lund (1983). HRP injected into layers 2 and 3 is found to be transported laterally both retrogradely and orthogradely in elaborate lattice patterns. Close to the injection site in macaque and squirrel monkey striate cortex, the HRP fills processes and cell bodies that form walls around regularly spaced, sparsely labelled, lacunae (Figures 8 and 9). Further distant from the injection site, the HRP label, both retrograde and orthograde, forms regularly spaced clusters about 500 μm apart in layer 2-3 around the site of the injection. While of roughly the same spacing, these labelled patches do not regularly coincide with the cytochrome-rich patches of thalamic input in the same layers (Figure 8*B*). Livingstone & Hubel (1984b) demonstrated, using small injections of HRP, that when this tracer is placed within a cytochrome-rich patch, or "blob", labelling occurs only in neighboring cytochrome-rich patches, and when the injection is in an intervening cytochrome-poor region, cytochrome-poor regions are labelled, but again of similar 500 μm spacing. Curiously, when using WGA-HRP in larger injections (Livingstone & Hubel 1984a), transport was most clearly shown to cytochrome-rich patches. Note, however, that these labelling patterns may reflect the affinity of uptake between specific neuron groups and the tracer substance used, and that activity levels may also play a role in amount of uptake. Nonetheless, these results indicate a lateral parcellation of neuron groups into discrete, interconnected patches in the superficial layers, each set of patches of similar spacing, but each group offset from one another spatially in a series of lattice-like arrays.

The functional role of these lattices has begun to be realized as the results of studies examining the relationship of V1 to V2 have become available. Livingstone & Hubel (1984a) have shown that efferent neurons in the cytochrome-rich patches of V1 project to bands in area V2 characterized by narrow width and rich cytochrome oxidase content. These same bands show a predominance of color-coded neurons and project, in turn, to V4 (DeYoe & Van Essen 1985, Shipp & Zeki 1985), a region thought to be of particular importance for color vision. Patches of efferent neurons in cytochrome-poor regions in V1 superficial layers project to cytochrome-poor bands of V2 that seem to be functionally more concerned with orientation than color properties. Since the superficial layers of V1 also send projections to cortical areas other than V2 [at least to V3A (Zeki 1980, Ungerleider & Mishkin 1982, Van Essen et al 1986), to V4 (Zeki 1978, Yukie & Iwai 1985), and to area PO (Colby et al 1983 and personal communication)], it is likely that the spatial organization of the efferent neurons in this region will be found to hold still further complexities in terms of both anatomy and physiology.

Rockland & Lund (1983) suggested that the connections between these lattice patches are made by the axons of pyramidal neurons, since the retrograde label clearly filled pyramidal neuron somata and proximal dendrites. Indeed, pyramidal neurons in the superficial layers of striate cortex establish patchy terminal fields from laterally spreading axon trunks (Gilbert & Wiesel 1983, Martin & Whitteridge 1984, McGuire et al 1985). The terminal sites of these patchy projections in the monkey seem to be predominantly dendritic spines (McGuire et al 1985, Rockland 1985,

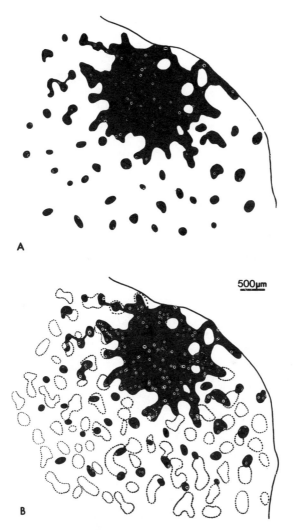

Figure 8(A–B) (*A*) Reconstruction of HRP-labelled loci (*black*) around an injection site in macaque monkey striate cortex, layers 2 and 3, in tangential section. (*B*) The same reconstruction of HRP-labelled loci as seen in *A* but with the position of cytochrome-rich thalamic input patches indicated by hatched profiles. (The alternately stained section series were matched by blood vessel patterns.) The two labels clearly do not coincide, but both may surround common lacunae (see Figure 9). (From Rockland & Lund 1983.)

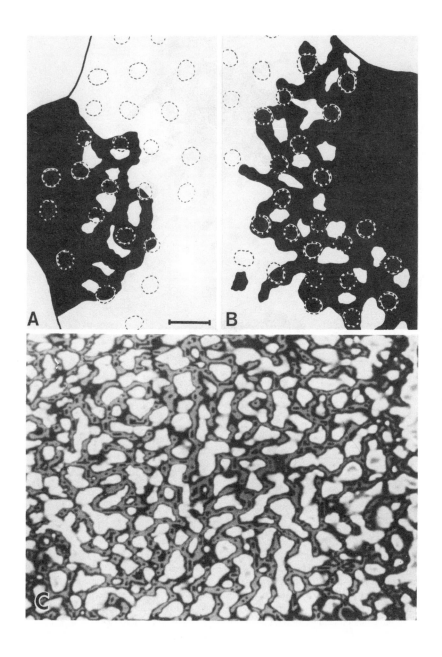

Kisvarday et al 1986), but, so far, the actual weight of these inputs on single postsynaptic pyramidal neurons or smooth dendritic neurons is unknown. The degree to which the axons of GABAergic neurons participate in the long distance connections of these lattice arrays is also unknown.

We have already discussed the fact that patchy lateral connections made by cells within the superficial layers are found also in upper lamina 4Cα and lamina 4B and, evidence for laterally patchy connections by cells intrinsic to the lamina is also found in lamina 6 (Blasdel et al 1985). With injections of HRP that bridge the depth of the cortex, patches of label are found to be in columnar register around the injection site in laminae 2 and 3, 4B, upper 4Cα, 5, and 6 (Rockland & Lund 1983). Patches continue the furthest distance lateral to the injection site in lamina 4B. If the injection is kept to the superficial layers, above lamina 4B, patches of label occur only in laminae 4B, 5, and 6 immediately under patches produced in the superficial layers without further lateral spread.

These patterns of label are difficult to explain. If one thinks of the system of lattices in the superficial layers as creating a continuum of interconnected points across the surface of the cortex, why is it that large or small injections in the superficial layers produce patches around them of the same spacing distance? Since we know that both cytochrome-rich and cytochrome-poor points are each interconnected by axon lattices as separate systems in the superficial layers, this should potentially at least double the number of labelled points if the injection site includes both systems. When injections include all layers, one might suppose that the intrinsic lattices in different layers would be labelled somewhat haphazardly—so why are the labelled patches in alignment through the layers and why do they maintain the roughly 500 μm spacing? One could suggest

---

Figure 9(A–C)   (*A* and *B*) Two maps from tangential sections through HRP injections in squirrel monkey striate cortex, layers 2 and 3. The HRP label adjacent to the injection sites forms solid walls (*indicated in black*) that surround unlabelled lacunae. The dotted circles indicate the position of cytochrome-rich patches of thalamic input in the same layers in sections adjacent to those stained for HRP. The cytochrome-rich patches fit into the HRP-labelled walls but avoid the unlabelled lacunae. (*C*) Computer-derived image from study of orientation preference maps using voltage-sensitive dye monitoring in live macaque monkey V1 cortex. The dark lines mark the position of rapid or abrupt change in orientation sequence, and the white areas mark where orientation change occurs in a continuous and even manner. The cytochrome-rich patches in this same region of cortex lie in the dark regions of rapid change or of break in orientation sequence. It is suggested that the HRP pattern seen above in *A* and *B* may be following the same regions as indicated by dark lines in *C*. (Blasdel & Salama 1986.) (*A* and *B* from Rockland & Lund 1983. *C*, modified from Blasdel & Salama 1986.) Scale bar = 500 μm.

that connections between the laminae only serve one of the lattice systems in the superficial layers, but this reasoning still leaves much unexplained since Livingstone & Hubel (1984b) found that small HRP injections in either cytochrome-rich blobs or in non-blob regions give label in the deeper layers.

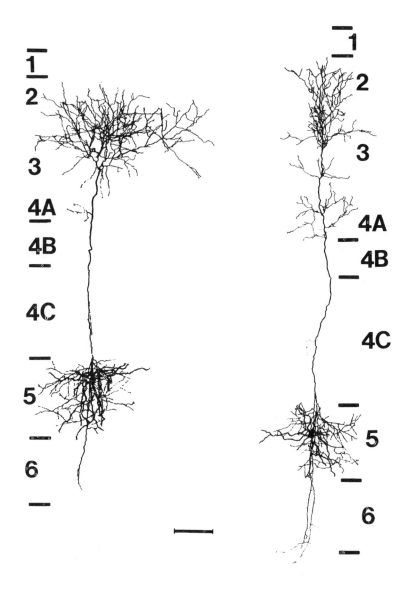

The kinds of theoretical arguments put forward by Mitchison & Crick (1982) in regard to data on patchy connections in tree shrews and by Mitchison (1985) when discussing the primate patchy connections are of considerable interest because of the possibility of obtaining discontinuous patterns of label, when using neuronal tracers such as HRP, in sheet-like systems of neural connections made with continuous but highly ordered patterns of connectivity. Testing models with a continuous lateral connectivity by computer produces discontinuous patterns of simulated label. The discontinuous label patterns that are produced derive from the properties of the connectional rules in the continuum and show the anatomist that discontinuity of label need not necessarily indicate an absence of long-distance connections in lightly labelled regions.

In discussing the form of interlaminar projections, either vertical columnar point-to-point form or spreading and fan-like in macaque cortex, Blasdel et al (1985) suggest that these patterns of projection may reflect the match, or mismatch, of the functional map in the interconnected laminae. For instance, in terms of its neurons' responses, lamina $4C\beta$ has a detailed point-by-point representation of the retinal surface, made up of small, circularly symmetric receptive fields. The lamina $4C\beta$ projection to lamina 4A is made in a vertical point-to-point fashion, and indeed 4A also shares a similarly accurate map of the retina (Blasdel & Fitzpatrick 1984). However, the $4C\beta$ projection to lamina 3B has spreading axon fans, and some collateral processes even take sideways steps on entering 3B (Fitzpatrick et al 1985).

The visual "map" in 3B, as physiologically characterized, has a locally disjointed topographic order in terms of retinal surface and has instead substituted a spatially highly ordered map of directionally selective units (Hubel & Wiesel 1978). The spreading fans and lateral "steps" of the $4C\beta$ axon arbors therefore may represent the redistribution of the retinal map in order to build a sequence of oriented units where retinal locus is repeated several times, each time for a different orientation. The vertical point-to-point projection of lamina 4B upon 3A-2 in turn may reflect a similarity

---

Figure 10 Two examples of local circuit neurons (perhaps GABAergic) with somata and dendritic fields in lamina 5 of macaque monkey V1. Both neurons have a local axon arbor in lamina 5 and a strong projection to the superficial layers, especially to lamina 2-3A. Descending axon trunks are present, and from at least one of these cells enters the white matter. Pyramidal neurons of layer 5 also project to the superficial layers and out of V1. Lamina 5 receives its principal input from neurons of lamina 2-3A, so the reciprocal projection can be presumed to contain both excitatory and inhibitory components. The neurons were drawn from Golgi Rapid preparations of infant monkey. Scale bar = 100 μm. (Lund, work in progress.)

of map in the two regions, since both contain ordered arrays of orientation-specific neurons that seem to be in strict alignment (Hubel & Wiesel 1968, 1978), whereas in both lamina 4B and the 3A-2 region the retinal map is locally disordered. It is of some interest that the layer 5 projection upon lamina 2-3A is fan-like and spreading, even though there is apparently alignment of periodicity in the relationship between lamina 2-3A and 5B. It may be that the individual axons from lamina 5B cell groups search out more than one period in the lamina 2-3A lattices.

The relationship of the supragranular lattice arrays to the circuits responsible for encoding the orientation of linear visual stimuli is poorly understood. The lattice relays probably do not literally reflect the pathways by which orientation signals are generated but rather represent some topographic relationship between the layout of neurons, with different orientation specificities, and the lattice connections. The voltage-sensitive dye study of Blasdel & Salama (1986) shows the overall topography of orientation domains relative to cytochrome-rich patches and to ocular dominance domains in the supragranular layers. The cytochrome-rich patches, which clearly form part of a lattice array (Livingstone & Hubel 1984b), lie in regions where a break in orientation sequence occurred. But the cytochrome-rich patches, known from the work of Livingstone & Hubel (1984a) to lack orientation specific responses, make up only part of the walls around roughly circular regions of smoothly changing sequences of orientation-specific neurons; these walls mark the regions where rapid shifts in orientation (Figure 9C) preference occur. The form of this pattern of gradual versus abrupt change in orientation resembles that of the overall pattern of HRP filling of walls around lacunae (Figure 9A, B), as seen in the Rockland & Lund (1983) study where the cytochrome-rich patches were found to always lie in the HRP-labelled walls of the lattice. This suggests that the break-points or fractures in orientation sequences may form the regions that interconnect heavily in lattice form and that the regions of slow orderly change are less widely connected.

In an attempt to test directly for patterns of connectivity relative to the voltage-sensitive dye patterns, Blasdel and Lund in preliminary studies have made small HRP injections into regions of single orientations or fracture points as visualized in the living cortex using voltage-sensitive dye techniques. The resultant transport of HRP to patches around the injections sites, even when the injection was in a center of slow orientation change, does not appear to match well to repeating regions of the same or even opposite orientation; instead the transported HRP patches appeared to always lie in the walls where the orientation sequences fracture, rather than in regions where smooth orientation change occurs. As in the Livingstone & Hubel (1984b) study, injections in cytochrome-rich patches were found to project to neighboring cytochrome-rich patches. Additional studies of intrinsic connectivity in relation to function are required but undoubtedly the results will further illustrate an exquisite precision of intrinsic cortical connectivity in the superficial layers.

One particularly intriguing and, as yet, unsolved problem in striate cortex is the nature of the anatomical substrate for the generation of orientation specificity. I have suggested that responses to stimulus orientation may be generated in more than one lamina (in 4Cα and in 3B), and there is no reason that layer 6 should not also generate orientation-specific responses within its substance (see Mapeli et al 1986, Schwark et al 1986). It is possible, as suggested by theoretical models such as those proposed by Linsker (1986a,b,c), that we shall never be able to recognize the circuitry anatomically, since the network properties of the relevant connections may not be recognizable in conventional anatomical terms. Still unknown is how, or whether (Bauer et al 1980, 1983), the orientation sequences in each of the laminae showing them are aligned vertically. The multiple vertical interlaminar connections that exist might seem likely pathways for ensuring correlated firing patterns to help alignment (Hubel & Wiesel 1968, 1978) during development, and the work of Bauer et al (1983) shows that the orientation specificities of neurons of lamina 4B and lamina 2-3 seem to be in alignment, even if the deeper layers are not.

Laminae 5 and 6

The deeper layers of the striate cortex, laminae 5B and 6, are the source of subcortical afferent relays and, in the case of layer 6, are also a source of projections to extrastriate regions of visual cortex. Earlier I noted that lamina 5 should be subdivided into 5A and 5B. This subdivision is based on a number of criteria: 5A makes prominent connections to the principal thalamic input zones of the cortex (layers 4C and 4A), whereas 5B projections do not (Blasdel et al 1985); Lamina 5A does not seem to be a promiment source of efferents, whereas lamina 5B contains several efferent cell populations, with strong projections to the pulvinar and superior colliculus (Lund et al 1975) that include cells of a variety of sizes and of both spine bearing and aspinous morphologies and perhaps therefore of different functions.

We have seen that the principle sources of afferents to 5B are from the supragranular laminae, especially lamina 2-3A, and from lamina 4B. Lamina 5B receives only weak projections from layer 6 (Blasdel et al 1985). In contrast, layer 6 receives no strongly marked projections from more superficial layers, but each superficial lamina seems to contribute a vertically focused light projection to which aspinous neurons make considerable contribution (see the studies of Fitzpatrick et al 1985, Blasdel et al 1985; Lund 1987). In these HRP injection studies, the most prominent and spreading input to layer 6 seems to come from lamina 5—but one should be aware that the injection of HRP in layer 5 may encroach upon a specialized border zone at the junction of layers 5 and 6 that contributes both axonal and dendritic elements to lamina 6 in a wide-spreading fashion. This border zone can be recognized in fiber stains as having a distinct and prominent horizontal fiber plexus.

In this same region lie the cell bodies of giant pyramidal neurons (sometimes called Meynert cells), whose basal dendrites sweep down to the base of lamina 6 and spread horizontally for considerable distances (see Lund 1973). These cell bodies usually occur in upper lamina 6, but occasionally also intrude over the border into lowermost lamina 5. In material stained for cell bodies they may be confused in lamina 5 with another population of large pyramidal neurons whose basal dendrites spread purely within lamina 5 and whose apical dendrite, in contrast to that of the cells with dendrites spreading in deep 6, is well developed and presents a vigorous spreading arbor of terminal branches to layers 3A-2 and 1. The giant pyramidal neurons with basal dendrite fields in layer 6 have been shown to project to both cortical area MT and to the superior colliculus (Spatz 1975, Lund et al 1975, Fries 1984, Fries et al 1985); they share these efferent destinations with some smaller neurons that are mostly clustered in upper lamina 6.

The laminae 5-6 border stratum also contains a group of horizontally oriented smooth dendritic neurons whose axons spread laterally over long distances at the junction of laminae 5 and 6 and emit, at intervals, vertically oriented, beaded collaterals into both laminae 5 and 6 (Figure 11) (Lund, work in progress). In their axon morphology and extensive lateral spread, these neurons resemble ones seen in 4Cα whose axons spread horizontally for long distances in upper α and 4B (see Figure 6). Cells with similar axon morphology in cat cortex have been termed "basket" neurons, and their axon terminals have been found to contact cell bodies and initial apical dendritic segments of pyramidal neurons (Somogyi et al 1983b).

This coincidence in form may be related to that fact that lamina 6 and upper lamina 4Cα plus lamina 4B are the two strata of monkey cortex where neurons with markedly direction-selective responses to movement are found (Dow 1974, Livingstone & Hubel 1984a, Movshon et al 1985). Both regions also have populations of cells projecting to the extrastriate cortical area MT (V5) where motion and specific direction of motion are primary driving stimuli. These long, lateral projections by neurons with smooth dendrites, possibly GABAergic and inhibitory in function, may play a role in the generation of direction-selective response properties. It might be that the property of direction selectivity is generated more than once in V1, as was proposed for orientation selectivity, and the system in lamina 6 might therefore have an independent substrate for directionality from that in 4B. Neurons of layer 6, particularly in upper 6 and lowermost 5 (Kennedy & Bullier 1985) as well as those of 4B, project to area V2 where, in the case of the fibers from 4B, it is suggested that the projection terminates in a system of broad, cytochrome-rich, stripes (Van Essen et al 1986). Neurons in these broad stripes are direction selective, and the stripes include neurons efferent to area MT (DeYoe & Van Essen 1985, Shipp & Zeki 1985).

Additional Inputs to V1

A large number of subcortical structures project to the primary visual cortex, apart from the lateral geniculate nucleus. A recent review of these projections—in terms of retrograde cell labelling within subcortical regions following tracer injections within area V1—was carried out by Tigges & Tigges (1984). The strength of these projections and which laminae they target in V1 are in many cases unknown. The pulvinar contributes two, or even three, separate projections to V1 (see Perkel et al 1986 for a recent account), but how each of these inputs contributes to the reported laminar pattern of terminals in layers 1 and 2 is unclear (Ogren & Hendrickson 1976). Information about the laminar patterning of projections to V1 from extrastriate cortex is also incomplete, but a useful review is provided by Van Essen (1984); projections to V1 from V2, V3, and V4 terminate in layer 1, and, in the case of V2, the projection adds

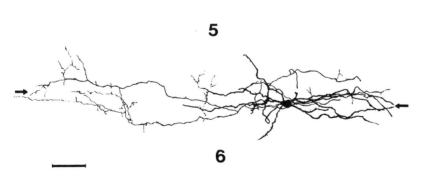

Figure 11   Golgi Rapid impregnation of a local circuit neuron at the border of laminae 5 and 6. This neuron is of the "basket" variety, similar to a population seen in laminae 4Cα (see Figure 6). These neurons influence the neuropil in laminae of V1 (upper 4Cα, 4B, and 6) where neurons with direction-specific responses are located. Scale bar = 50 μm. (Lund, work in progress.)

lighter input to layers 4B and 5B. The projection from V3 and MT to V1 terminates most heavily in lamina 4B and also contributes to layer 6. The function of these inputs is not known.

Summary

I hope that this review of the internal anatomy of the monkey primary visual cortex makes clear the high degree of specialization that exists in each of the cortical laminae and their constituent neurons. Each lamina is driven by different patterns of relays from the LGN and by different patterns of intrinsic interlaminar projections. The elaborate laminar and intralaminar segregation of efferent neuron arrays suggests that the extra-ordinary precision of inter- and intralaminar connectivity provides a unique functional role for each set of efferent neurons. The organization of aspinous (presumed inhibitory) local circuit neurons suggests that they are highly specialized, and within each lamina and via interlaminar relays each variety may only accomplish a single, particular task. The cortex neuropil does not give the immediate impression of "random" networks, and if such exist, they must surely be between very tightly determined subgroups of neurons. Clearly a very detailed physiological exploration of V1 is still needed, with new consideration of thalamic axon function, of efferent neuron characteristics, of laminar differences, and of spatial organization of properties within laminae, in order to match known ana-tomical detail with function. The concept of columnar organization in cortical organization of V1 may eventually be redefined in more complex terms that accurately describe the anatomical and functional parcellation evident in cortical depth and perhaps may link it to a means by which a correlation of different aspects of the visual image is achieved.

ACKNOWLEDGMENTS

I would like to thank Dr. Raymond Lund for his critical comments on the manuscript and Roberta Erickson and Thomas Harper for their help with its preparation. My work was supported by grant EY-05282 from the National Eye Institute.

Literature Cited

Bauer, R., Dow, B. M., Snyder, A. Z., Vautin, R. 1983. Orientation shift between upper and lower layers in monkey visual cortex. *Exp. Brain Res.* 50: 133–45

Bauer, R., Dow, B. M., Vautin, R. G. 1980. Laminar distribution of preferred orientations in foveal striate cortex of monkey. *Exp. Brain Res.* 41: 54–60

Blakemore, C. Vital-Durand, F. 1986. Organisation and post-natal development of the monkey's lateral geniculate nucleus. *J. Physiol.* 380: 453–91

Blasdel, G. G., Fitzpatrick, D. 1984. Physiological organization of layer 4 in macaque striate cortex. *J. Neurosci.* 4: 880–95

Blasdel, G. G., Lund, J. S. 1983. Termination

of afferent axons in macaque striate cortex. *J. Neurosci.* 3: 1389–1413

Blasdel, G. G., Lund, J. S., Fitzpatrick, D. 1985. Intrinsic connections of macaque striate cortex: Axonal projections of cells outside lamina 4C. *J. Neurosci.* 5: 3350–69

Blasdel, G. G., Salama, G. 1986. Voltage-sensitive dyes reveal a modular organization in monkey striate cortex. *Nature* 321: 579–85

Bullier, J., Henry, G. H. 1980. Ordinal position and afferent input of neurons in monkey striate cortex. *J. Comp. Neurol.* 193: 913–35

Colby, C. L., Gattass, R., Olson, C. R., Gross, C. G. 1983. Cortical afferents to visual area PO in the macaque. *Soc. neurosci. Abstr.* 9: 152

Creutzfeldt, O. D., Lee, B. B., Elepfandt, A. 1979. A quantitative study of chromatic organization and receptive fields in the lateral geniculate body of the rhesus monkey. *Exp. Brain Res.* 35: 527–45

deMonasterio, F. M., Schein, S. J., McCrane, E. P. 1981. Staining of blue-sensitive cones of the macaque retina by a fluorescent dye. *Science* 213: 1278–81

DeYoe, E. A., Van Essen, D. C. 1985. Segregation of efferent connections and receptive field properties in visual area V2 of the macaque. *Nature* 317: 58–61

Dow, B. M. 1974. Functional classes of cells and their laminar distribution in monkey visual cortex. *J. Neurophysiol.* 37: 927–46

Dow, B. M., Gouras, P. 1973. Color and spatial specificity of single units in rhesus monkey foveal striate cortex. *J. Neurophysiol.* 36: 79–100

Eskin, T. A., Merigan, W. H. 1986. Selective acrylamide-induced degeneration of color opponent ganglion cells in macaques. *Brain Res.* 378: 379–84

Fitzpatrick, D., Itoh, K., Diamond, I. T. 1983a. The laminar organization of the lateral geniculate body and the striate cortex in the squirrel monkey (*Saimiri sciureus*). *J. Neurosci.* 3: 673–702

Fitzpatrick, D., Lund, J. S., Blasdel, G. G. 1985. Intrinsic connections of macaque striate cortex. Afferent and efferent connections of lamina 4C. *J. Neurosci.* 5: 3329–49

Fitzpatrick, D., Lund, J. S., Schmechel, D. E. 1983b. Glutamic acid decarboxylase immunoreactive neurons and terminals in the visual cortex of monkey and cat. *Soc. Neurosci. Abstr.* 9: 616

Fitzpatrick, D., Lund, J. S., Schmechel, D. E., Towles, A. C. 1987. Distribution of GABAergic neurons and terminals in macaque striate cortex. *J. Comp. Neurol.* 264: 73–91

Freund, T. F., Martin, K. A. C., Somogyi, P., Whitteridge, D. 1985. Innervation of cat visual areas 17 and 18 by physiologically identified X- and Y-type thalamic afferents. II. Identification of postsynaptic targets by GABA immunocytochemistry and Golgi impregnation. *J. Comp. Neurol.* 242: 275–91

Fries, W. 1984. Cortical projections to the superior colliculus in the macaque monkey: A retrograde study using horseradish peroxidase. *J. Comp. Neurol.* 230: 55–76

Fries, W., Keizer, K., Kuypers, H. G. J. M. 1985. Large layer VI cells in macaque striate cortex project to both superior colliculus and prestriate visual area V5. *Exp. Brain Res.* 58: 613–16

Garey, L. J., Powell, T. P. S. 1971. An experimental study of the termination of the lateral geniculo-cortical pathway in the cat and monkey. *Proc. R. Soc. London Ser. B* 179: 41–63

Gilbert, C. D., Wiesel, T. N. 1983. Clustered intrinsic connections in cat visual cortex. *J. Neurosci.* 3: 1116–33

Gouras, P. 1974. Opponent-color cells in different layers of foveal striate cortex. *J. Physiol.* 238: 583–602

Hendrickson, A. E., Wilson, J. R., Ogren, M. P. 1978. The neuroanatomical organization of pathways between the dorsal lateral geniculate nucleus and visual cortex in old world and new world primates. *J. Comp. Neurol.* 182: 123–36

Hornung, J. P., Garey, L. J. 1981. The thalamic projection to cat visual cortex: ultrastructure of neurons identified by Golgi impregnation or retrograde horseradish peroxidase transport. *Neuroscience* 6: 1053–68

Horton, J. C. 1984. Cytochrome oxidase patches: A new cytoarchitectonic feature of monkey visual cortex. *Philos. Trans. R. Soc. London Ser. B* 304: 199–253

Hubel, D. H., Wiesel, T. N. 1968. Receptive fields and functional architecture of monkey striate cortex. *J. Physiol.* 195: 215–43

Hubel, D. H., Wiesel, T. N. 1969. Anatomical demonstration of columns in the monkey striate cortex. *Nature* 221: 747–50

Hubel, D. H., Wiesel, T. N. 1972. Laminar and columnar distribution of geniculocortical fibers in the macaque monkey. *J. Comp. Neurol.* 146: 421–50

Hubel, D. H., Wiesel, T. N. 1978. Functional architecture of macaque monkey visual cortex. Ferrier Lecture. *Proc. R. Soc. London Ser. B* 198: 1–59

Hubel, D. H., Wiesel, T. N., LeVay, S. 1974. Visual field of representation in layer IVC of monkey striate cortex. *Soc. Neurosci. Abstr.* 4: 264

Hubel, D. H., Wiesel, T. N., LeVay, S. 1977. Plasticity of ocular dominance columns in monkey striate cortex. *Philos. Trans. R. Soc. London Ser. B* 278: 377–409

Katz, L. C., Burkhalter, A., Dreyer, W. J. 1984. Fluorescent latex microspheres as a retrograde neuronal marker for *in vivo* and *in vitro* studies of visual cortex. *Nature* 310: 498–500

Kennedy, H. Bullier, J. 1985. A double-labeling investigation of the afferent connectivity to cortical areas V1 and V2 of the macaque monkey. *J. Neurosci.* 5: 2815–30

Kisvarday, Z. F., Cowey, A., Somogyi, P. 1987. Synaptic relationships of a type of GABA-immunoreactive neuron (clutch cell), spiny stellate cells and LGN afferents in layer IVC of the monkey striate cortex. *Neuroscience* 19: 741–61

Kisvarday, Z. F., Martin, K. A. C., Freund, T. F., Magloczky, Zs., Whitteridge, D., Somogyi, P. 1986. Synaptic targets of HRP-filled layer III pyramidal cells in the cat striate cortex. *Exp. Brain Res.* 64: 541–52

LeVay, S. 1973. Synaptic patterns in the visual cortex of the cat and monkey: Electron microscopy of Golgi preparations. *J. Comp. Neurol.* 150: 53–86

LeVay, S. 1986. Synaptic organization of claustral and geniculate afferents to the visual cortex of the cat. *J. Neurosci.* 6: 3564–75

LeVay, S., Wiesel, T. N., Hubel, D. H. 1980. The development of ocular dominance columns in normal and visually deprived monkeys. *J. Comp. Neurol.* 191: 1–51

Leventhal, A. G., Rodieck, R. W., Dreher, B. 1981. Retinal ganglion cell classes in the old world monkey: Morphology and central projections. *Science* 213: 1139–42

Linsker, R. 1986a. From basic network principles to neural architecture: Emergence of spatial opponent cells. *Proc. Natl. Acad. Sci. USA* 83: 7508–12

Linsker, R. 1986b. From basic network principles to neural architecture: Emergence of orientation-selective cells. *Proc. Natl. Acad. Sci. USA* 83: 8390–94

Linsker, R. 1986c. From basic network principles to neural architecture: Emergence of orientation columns. *Proc. Natl. Acad. Sci. USA* 83: 8779–83

Livingstone, M. S., Hubel, D. H. 1982. Thalamic inputs to cytochrome oxidase-rich regions in monkey visual cortex. *Proc. Natl. Acad. Sci. USA* 79: 6098–6101

Livingstone, M. S., Hubel, D. H. 1984a. Anatomy and physiology of a color system in the primate visual cortex. *J. Neurosci.* 4: 309–56

Livingstone, M. S., Hubel, D. H. 1984b.

Specificity of intrinsic connections in primate primary visual cortex. *J. Neurosci.* 4: 2830–35

Lund, J. S. 1973. Organization of neurons in the visual cortex, area 17, of the monkey (*Macaca mulatta*). *J. Comp. Neurol.* 147: 455–96

Lund, J. S. 1984. Spiny stellate neurons. In *Cerebral Cortex*, ed. A. Peters, E. G. Jones, 1: 255–308. New York/London: Plenum

Lund, J. S. 1986. Local circuit neuron projections to divisions of thalamic recipient lamina 4C in the monkey primary visual cortex. *Soc. Neurosci. Abstr.* 12: 129

Lund, J. S. 1987. Local circuit neurons of macaque monkey striate cortex: I. Neurons of laminae 4C and 5A. *J. Comp. Neurol.* 257: 60–92

Lund, J. S., Boothe, R. 1975. Interlaminar connections and pyramidal neuron organization in the visual cortex, area 17, of the macaque monkey. *J. Comp. Neurol.* 159: 305–34

Lund, J. S., Hendrickson, A. E., Ogren, M. P., Tobin, E. A. 1981. Anatomical organization of primate visual cortex area VII. *J. Comp. Neurol.* 202: 19–45

Lund, J. S., Lund, R. D., Hendrickson, A. E., Bunt, A. H., Fuchs, A. F. 1975. The origin of efferent pathways from the primary visual cortex, area 17, of the macaque monkey. *J. Comp. Neurol.* 164: 287–304

Malpeli, J. G., Lee, C., Schwark, H. D., Weyand, T. 1986. Cat area 17. I. Pattern of thalamic control of cortical layers. *J. Neurophysiol.* 56: 1062–73

Martin, K. A. C., Whitteridge, D. 1984. Form, function and intracortical projections of spiny neurones in the striate visual cortex of the cat. *J. Physiol.* 353: 463–504

Mates, S. L., Lund, J. S. 1983. Neuronal composition and development in lamina 4C of monkey striate cortex. *J. Comp. Neurol.* 221: 60–90

McGuinness, E., MacDonald, C., Sereno, M., Allman, J. 1986. Primates without blobs: The distribution of cytochrome oxidase activity in striate cortex in Tarius, Hapalemur, and Cheirogaleus. *Soc. Neurosci. Abstr.* 12: 130

McGuire, B. A., Gilbert, C. D., Wiesel, T. N. 1985. Ultrastructural characterization of long-range clustered horizontal connections in monkey striate cortex. *Soc. Neurosci. Abstr.* 11: 17

McGuire, B. A., Hornung, J.-P., Gilbert, C. D., Wiesel, T. N. 1984. Patterns of synaptic input to layer 4 of cat striate cortex. *J. Neurosci.* 4: 3021–33

Michael, C. R. 1978. Color vision mechanisms in monkey striate cortex: Dual-opponent cells with concentric receptive fields. *J. Neurophysiol.* 46: 587–604

Michael, C. R. 1986. Functional and morphological identification of double and single opponent color cells in layer IVCβ of the monkey's striate cortex. *Soc. Neurosci. Abstr.* 12(pt. 2): 1497

Mitchison, G. J. 1985. Does the striate cortex contain a system of oriented axons? In *Models of the Visual Cortex*, ed. D. Rose, V. G. Dobson, pp. 443–52. Chichester/New York: Wiley

Mitchison, G., Crick, F. 1982. Long axons within striate cortex: Their distribution, orientation and patterns of connections. *Proc. Natl. Acad. Sci. USA* 79: 3661–65

Mitzdorf, U., Singer, W. 1979. Excitatory synaptic ensemble properties in the visual cortex of the macaque monkey: A current source density analysis of electrically evoked potentials. *J. Comp. Neurol.* 187: 71–84

Movshon, J. A., Adelson, E. H., Gizzi, M. S., Newsome, W. T. 1985. The analysis of moving patterns. In *Pattern Recognition Mechanisms*, ed. C. Chagas, R. Gattass, C. Gross. *Exp. Brain Res.* (Suppl. 11). New York: Springer-Verlag

Ogren, M. P., Hendrickson, A. 1976. Pathways between striate cortex and subcortical regions in *Macaca mulatta* and *Saimiri sciureus*: Evidence for a reciprocal pulvinar connection. *Exp. Neurol.* 53: 780–800

Parker, A. J., Hawken, M. J. 1984. Contrast sensitivity and orientation selectivity in lamina IV of the striate cortex of old-world monkeys. *Exp. Brain Res.* 54: 367–73

Perkel, D. J., Bullier, J., Kennedy, H. 1986. Topography of the afferent connectivity of area 17 in the macaque monkey: A double-labelling study. *J. Comp. Neurol.* 253: 359–73

Peters, A., Proskauer, C. C., Feldman, M. L., Kimerer, L. 1979. The projection of the lateral geniculate nucleus to area 17 of the rat cerebral cortex. V. Degenerating axon terminals synapsing with Golgi impregnated neurons. *J. Neurocytol.* 8: 331–57

Rakic, P. 1976. Prenatal genesis of connections subserving ocular dominance in the rhesus monkey. *Nature* 261: 467–71

Ribak, C. E. 1978. Aspinous and sparsely spinous stellate neurons in the visual cortex of rats contain glutamic-acid decarboxylase. *J. Neurocytol.* 7: 461–78

Rockland, K. S. 1985. Intrinsically projecting pyramidal neurons of monkey striate cortex: An EM-HRP study. *Soc. Neurosci. Abstr.* 11: 17

Rockland, K. S., Lund, J. S. 1983. Intrinsic laminar lattice connections in primate visual cortex. *J. Comp. Neurol.* 216: 303–18

Rockland, K. S., Pandya, D. N. 1979. Laminar origins and terminations of cortical connections of the occipital lobe in the rhesus monkey. *Brain Res.* 179: 3–20

Schiller, P. H. 1986. The central visual system. *Vision Res.* 26: 1351–86

Schiller, P. H., Colby, C. L. 1983. The responses of single cells in the lateral geniculate nucleus of the rhesus monkey to color and luminance contrast. *Vision Res.* 23: 1631–41

Schiller, P. H., Malpeli, J. G. 1978. Functional specificity of lateral geniculate nucleus laminae of the rhesus monkey. *J. Neurophysiol.* 41: 788–97

Schwark, H. D., Malpeli, J. G., Weyand, T. G., Lee, C. 1986. Cat area 17. II. Response properties of infragranular layer neurons in the absence of supragranular layer activity. *J. Neurophysiol.* 56: 1074–87

Shapley, R., Perry, V. H. 1986. Cat and monkey retinal ganglion cells and their visual functional roles. *TINS* 9: 229–35

Shipp, S., Zeki, S. 1985. Segregation of pathways leading from area V2 to areas V4 and V5 of macaque monkey visual cortex. *Nature* 317: 322–25

Somogyi, P. 1978. The study of Golgi stained cells and of experimental degeneration under the electron microscope: a direct method for the identification in the visual cortex of three successive links in a neuron chain. *Neuroscience* 3: 167–80

Somogyi, P., Freund, T. F., Smith, A. D. 1983a. The section-Golgi impregnation procedure. 2. Immunocytochemical demonstration of glutamic acid decarboxylase in Golgi impregnated neurons and in their afferent synaptic boutons in the visual cortex of the cat. *Neuroscience* 9: 475–90

Somogyi, P., Kisvarday, Z. F., Martin, K. A. C., Whitteridge, D. 1983b. Synaptic connections of morphologically identified and physiologically characterized large basket cells in the striate cortex of cat. *Neuroscience* 10: 261–94

Spatz, W. B. 1975. An efferent connection of the solitary cells of Meynert. A study with horseradish peroxidase in the marmoset. *Callithrix. Brain Res.* 92: 450–55

Tigges, J., Tigges, M. 1984. Subcortical sources of direct projections to visual cortex. In *Cerebral Cortex, Vol. 3, Visual Cortex*, ed. A. Peters, E. G. Jones, pp. 351–78. New York/London: Plenum

Tigges, J., Tigges, M., Anschel, S., Cross, N. A., Letbetter, W. D., McBride, R. L. 1981. Areal and laminar distribution of neurons interconnecting the central visual cortical areas 17, 18, 19 and MT in squirrel monkey (*Saimiri*). *J. Comp. Neurol.* 202: 539–60

Tigges, M., Tigges, J. 1986. Cytochrome oxidase (CO) staining pattern in striate cortex: Differences between monkeys and apes. *Soc. Neurosci. Abstr.* 12: 129

Ungerleider, L. G., Mishkin, M. 1982. Two cortical visual systems. In *Analysis of Visual Behavior*, ed. D. J. Ingle, M. A. Goodale, R. J. W. Mansfield. Cambridge, MA: Mass. Inst. Technol. Press

Van Essen, D. C. 1984. Functional organization of primate visual cortex. In *Cerebral Cortex, Vol. 3, Visual Cortex*, ed. A. Peters, E. G. Jones, pp. 259–329. New York/London: Plenum

Van Essen, D. C., Newsome, W. T., Maunsell, J. H. R., Bixby, J. L. 1986. The projections from striate cortex (V1) to areas V2 and V3 in the Macaque monkey: Asymmetries, areal boundaries and patchy connections. *J. Comp. neurol.* 244: 451–80

Weber, J. T., Huerta, M. F., Kaas, J. H., Harting, J. K. 1983. The projections of the lateral geniculate nucleus of the squirrel monkey: Studies of the interlaminar zones and the S layers. *J. Comp. Neurol.* 213: 135–45

Wiesel, T. N., Hubel, D. H. 1966. Spatial and chromatic interactions in the lateral geniculate body of the rhesus monkey. *J. Neurophysiol.* 29: 1115–56

Williams, D. R., Macleod, D. I. A., Hayhoe, M. M. 1981. Punctate sensitivity of foveal blue-sensitive cones. *Vision Res.* 21: 1357–75

Winfield, D. A., Powell, T. P. S. 1983. laminar cell counts and geniculo-cortical boutons in area 17 of cat and monkey. *Brain Res.* 277: 223–29

Wolf, W., Hicks, T. P., Albus, K. 1986. Contribution of GABA-mediated inhibitory mechanisms to visual responsible properties of neurons in the kitten's striate cortex. *J. Neurosci.* 6: 2779–95

Wong-Riley, M. T. T. 1979. Changes in the visual system of monocularly sutured or enucleated cats demonstrable with cytochrome oxidase histochemistry. *Brain Res.* 171: 11–28

Yukie, M., Iwai, E. 1985. Laminar origin of direct projection from cortex area V1 to V4 in the rhesus monkey. *Brain Res.* 346: 383–86

Zeki, S. M. 1978. The cortical projections of foveal striate cortex in the rhesus monkey. *J. Physiol.* 277: 227–44

Zeki, S. M. 1980. A direct projection from area V1 to area V3A of rhesus monkey visual cortex. *Proc. R. Soc. London Ser. B* 207: 499–506

Ann. Rev. Neurosci. 1988. 11 : 289–327

FORMATION OF
TOPOGRAPHIC MAPS

Susan B. Udin

Department of Physiology, State University of New York at Buffalo,
Buffalo, New York 14214

James W. Fawcett

Department of Physiology, Cambridge University, Cambridge, England

INTRODUCTION

Most sensory inputs to the central nervous system (CNS) are topographically arranged. For instance, the various retinofugal projections
are topographically arranged in the retinoreceptive nuclei, the auditory
projection connects tonotopically with the cochlear nucleus, and sensory
projections terminate topographically in the primary sensory nuclei. The
topography of these projections is continued through several relay nuclei
as far as the cortex. Working out the mechanisms by which the topography
of these various projections is set up has been a major focus of developmental neurobiology over the past 30 years or so. In this article we
attempt to summarize some of this work and to present a coherent view
of the development of topographic maps in the nervous system.

Historically, the general view held by neurobiologists was that there
must be a single unifying mechanism by which topographic connections
are set up. Four main classes of hypothesis were advanced:

1. The postsynaptic structure possesses a series of "addresses" on it, which
   tell ingrowing axons where to terminate.
2. The presynaptic axon terminals interact with one another on an infor-
   mationless target structure, and arrange themselves topographically,
   by means of either positional labels or patterns of electrical activity.
3. Axons, as they grow, remain neighbors with axons that come from

0147–006X/88/0301–0289$02.00

neighboring neurons in the presynaptic structure, and therefore arrive at the postsynaptic structure already topographically arranged.

4. Axons grow out in a topographically arranged time sequence, and connect to a target generated in a matching temporal pattern.

All these hypotheses now to some extent appear correct, and in fact several different mechanisms cooperate to ensure the topography of sensory maps. (See Willshaw & von der Malsburg 1979, Fraser 1980, Fraser & Hunt 1980, Whitelaw & Cowan 1981 for reviews of theoretical issues.) In the following review we look at the evidence for each of the proposed mechanisms of map formation and try to delineate the role of each. We use as our model mainly the visual system, specifically the retinotectal and retinocollicular projections, since much of the work has been done on them. Much valuable information has also come from the study of the retinogeniculate projection; this subject was reviewed in the previous volume of this series (Shatz & Sretevan 1986), and readers are encouraged to read this article in association with the following.

RETINOTECTAL AND RETINOGENICULATE PROJECTIONS

Topography of Fiber Pathways

If topography is to be transmitted from one structure to another simply by axons from neighboring neurons remaining neighbors all the way to their target, then a topographic arrangement of the axons should be demonstrable at all points in their pathway. The topographic arrangement of axons in the optic nerve, for instance, should be just as precise as it is in the optic tectum. Early results suggested that this might indeed be the case (Attardi & Sperry 1963, Scalia & Fite 1974, reviewed in Horder & Martin 1978), but further examination of the topography of the axons in the retinotectal projections of a number of animals led to somewhat variable results.

THE OPTIC NERVE   Order in the first part of the pathway, the optic nerve, ranges from very precise, as in fish with ribbon-shaped optic nerves (Scholes 1979), fairly precise, as in teleost fish (Rusoff & Easter 1980, Bodick & Levinthal 1980, Easter at al 1981, Bunt 1982) and chicks (Rager & Rager 1978, Rager 1980, Thanos & Bonhoeffer 1983), rather imprecise, as in frogs (Fawcett 1981, Reh et al 1983, Scalia & Arango 1983, Bunt & Horder 1983) and rats (Bunt & Lund 1982), to very imprecise, as in cats (Horton et al 1979). Since the accuracy of the topography of the axons in many species' optic nerves is considerably lower than that of their retinotectal terminations, the conclusion that axons must maintain their neighbor relations in order to make topographic maps seemed unlikely. However, all the above studies suffered from the weakness that axon topography was being studied some time after axonal growth; perhaps the axons had become disarranged after they had grown (e.g. Cima & Grant 1982). Williams & Rakic (1985), however, examined the anatomy of axons and growth cones in serial EM sections of the embryonic monkey optic nerve, and found that growth cones frequently change the axons to which they adhere. One end of a growth cone might have its processes adherent to one set of axons, while the other end of the growth cone would be associated with a quite different group of axons. These findings collectively make untenable the idea that topography in the retinofugal projections is entirely a product of the maintenance of neighbor relations between axons.

THE OPTIC TRACT   Studies of the optic tract in frogs and fish have revealed that the topography of axons in this part of the visual pathway is generally fairly precise, and that the topography of this order is different from that found in the optic nerve, thus indicating that a rearrangement of axon topography must occur in the region of the chiasma (Scalia & Fite 1974, Scholes 1979, Steedman 1981, Easter et al 1981, Fujisawa et al 1981a,b, Fawcett & Gaze 1982, Bunt 1982, Reh et al 1983, Fawcett et al 1984b). The topography is present from the earliest embryonic stages (Holt 1984, Holt & Harris 1983), and is particularly precise as axons grow onto the optic tectum itself (Fawcett & Gaze 1982, Reh et al 1983, Stuermer & Easter 1984a, Rusoff 1984). The optic tract is arranged so that circumferential retinal order is represented across one axis of the tract, and time of axon growth across the other, with the latest growing axons positioned most superficially. A similar ordering is found in the chick (Thanos & Bonhoeffer 1983). In mammals, time of axon arrival is represented across the depth of the tract as in lower vertebrates (Walsh et al 1983, Walsh & Polley 1985, Guillery et al 1982, Walsh & Guillery 1985, B. E. Reese, personal communication, Reese & Guillery 1987), and in the cat there is a map of circumferential retinal position across the other axis; however, this is a complex map, with each type of ganglion cell appearing to make a separate, overlapping map (Aebersold et al 1981, Torrealba et al 1982, Mastronarde 1984). As in the lower vertebrates, there is a change in fiber order in the region of the chiasma.

How is the mapping of circumferential order achieved? In frogs there is evidence for some form of active fiber guidance. In *Xenopus*, eyes can be reconstructed at embryonic stages to contain two nasal, ventral, or temporal half retinas. Despite such eyes producing a roughly normal number of axons, the nerve fibers only travel in the part of the optic tract that would contain axons from the appropriate half of a normal eye (Straznicky et al 1979, Steedman 1981, Fawcett & Gaze 1982). The mechanism behind this active fiber guidance may involve axons specifically following axons from neighboring regions of the eye that have already reached the tectum, since some behavior of this kind is demonstrated by regenerating retinotectal axons (Gaze & Fawcett 1983; Fawcett 1985, Taylor & Gaze 1985). In the chick, however, mechanical factors may largely govern the pathway that axons take as far as the optic tectum. Injection of anti N-CAM (Neural Cell Adhesion Molecule) into the developing chick eye deranges the pathway of axons through the retina and into the optic nerve, and the axons displaced by this treatment remain displaced right into the optic tectum (Thanos et al 1984). Only when they reach the tectum do a few of the axons correct their pathway.

The observation of topographic ordering of axonal growth in the optic tract led to the hypothesis that ordered ingrowth of axons alone might be sufficient to order the retinofugal projections; the topography might be cleaned up later by axon-axon interactions (see below). However, a necessary prediction of such a hypothesis is that axons growing into the tectum from a completely anomalous route should be unable to form a normal topographic map. In the axolotl, Harris (1982) was able to make animals in which a single eye innervated an otherwise "virgin" tectum via a completely abnormal pathway, and such animals had normally ordered retinotectal projections. One is therefore led to the conclusion that the precise topography of axons as they grow onto the optic tectum is not a critical factor in establishing the topography of the axon terminations. There must be some form of addresses on the tectum that tell the axons where to terminate. (See *Chemoaffinity Labels* below.)

Ordering in the deep-to-superficial axis of the tract is clearly related to time of axon growth; the most recently grown axons are found superficially. This is presumably achieved by an affinity of growing axons for the glia limitans, or by mechanical factors. Could time of axon arrival be a topographic ordering factor? In frogs this is unlikely, since delaying axon growth does not affect the final topography of the retinotectal map (Feldman et al 1971, Holt 1984).

In the chick and frog there is evidence for specific interactions between axons from temporal retina, which tend to adhere to one another, as opposed to fibers from nasal retina, which do not (Gaze & Fawcett 1983, Bonhoeffer & Huf 1985, Halfter et al 1981). The role that these interactions play in the ordering of axonal growth is not clear.

Is there any functional reason that the optic tract should be so precisely ordered? One possible advantage of such an arrangement is that axons as they grow onto their target are immediately guided to the correct area in which to terminate. If axons were to grow in randomly, they would have to search much of the target structure for an appropriate termination site; as it is they are guided there directly.

Chemoaffinity Labels

The chemoaffinity hypothesis postulates the existence of chemical labels, distributed in a graded fashion across the retina and tectum, which allow each retinal cell to recognize its proper tectal termination site (Sperry 1944). In this section, we present evidence that argues compellingly that retinal ganglion cells do behave as though they bear quite finely graded "identities" that reflect their positions of origin. We also present evidence for positional labels of a much weaker and more elusive nature in the tectum.

RETINAL LOCUS SPECIFICITIES   The most compelling evidence that retinal ganglion cells bear some label or characteristic by which they can be distinguished from one another comes from studies of compound eyes in *Xenopus*. Surgical techniques can be used in embryos to construct eyes by fusing various fragments of host and donor eye rudiments. For example, a nasal quadrant or half of a donor's right eye can be excised, rotated, and implanted in place of the temporal quadrant or half of a host's eye. (See Figure 1A.) As the compound eye grows, the grafted cells and their progeny can retain their original nasal-pole characteristics; the graft's projections to the tectum can make connections in register with the host's nasal retinal projections (see Figure 1B). Thus the graft continues to behave as though it were still in its original nasal position (Gaze et al 1963, Feldman & Gaze 1975, Gaze & Straznicky 1980, Straznicky 1981, Cooke & Gaze 1983, Willshaw et al 1983). Similar results can be obtained with sectors taken from any part of the eye; projections from wedges as small as about 30° can be resolved with electrophysiological techniques. Whether the mosaic of specificities is so fine-grained that each ganglion cell is distinguishable from its neighbors is not yet known. These experiments demonstrate that not only the ganglion cells, but also their axons, carry these specificities, and that the optic axons must be able to interact with one another in such a way as to ensure that axons bearing the same positional labels innervate the same area of the tectum. These interactions are independent of the actual position that the ganglion cells occupy in the compound eye. We emphasize that this form of axon-axon interaction is not likely to be mediated by a process of correlations of firing patterns (see below). Indeed, in compound eyes, ganglion cells that have the same embryological identity may well be situated at opposite poles of the retina and therefore have minimal similarity in their firing patterns. Thus, it seems reasonable to propose that there is a mechanism of axon-axon interaction, such as graded adhesion, that is independent of activity-mediated axon-axon interactions.

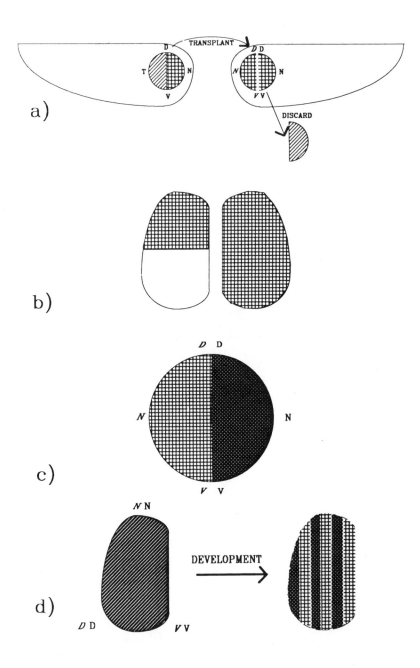

How and when do ganglion cells acquire their identities? To test whether retinal ganglion cells inherit their identities or whether they acquire them from neighboring tissue, whole eyes or parts of eyes may be inserted into new positions or new orientations in *Xenopus* embryos. Later, one may assess whether the optic axons form a map as though their cell bodies were still in their original locations or whether they form maps that are consistent with the cells' new positions (Hunt & Jacobson 1972, Sharma & Hollyfield 1974, Feldman et al 1983). In the previous paragraph, we cited some experiments that suggest that retinal cell identities are stable, but many data seem to suggest the opposite (Jacobson 1967, Hunt & Jacobson 1973, 1974, Cooke & Gaze 1983, Willshaw et al 1983, O'Rourke & Fraser 1986b).

Some of these apparent contradictions stem from the fact that cellular movements have a major influence on the final pattern of projections formed by a rotated retina or retinal graft and may confound the experimenter's intent in constructing a particular anatomical arrangement. Holt (1980) has shown that neural precursors can migrate from the optic stalk into the eye cup and reform an essentially normal retina after a surgical graft or rotation has been performed; and Ide et al (1984) have found that cells within a surgically halved eye can move to new positions and reconstitute a duplicate half-eye that maps as the mirror-image of the original half. In such cases, the final polarity of the maps reflects the original specificities of cells that have moved to positions unintended by the experimenters. Perhaps such cell movements within retinal grafts explain why some grafts seem to give completely or partially respecified maps whereas others yield maps that are strictly consistent with the original location and polarity of the grafted segment (Hunt & Jacobson 1973, Ide et al 1979, Cooke & Gaze 1983, Willshaw et al 1983).

We do not yet know how long during development such rearrangements can occur. A prolonged period of malleability might help to explain the intriguing results of O'Rourke & Fraser (1986b), who have studied the projections from an eye containing a temporal half-eye graft in dorso-ventrally inverted orientation. Their new vital-dye labeling techniques

---

Figure 1   (*a*) Construction of a double-nasal eye. Nasal half of right eye replaces temporal half of left eye to produce compound eye with normal dorso-ventral polarity. (*b*) When compound eye innervates a tectum lacking other retinal input (*right side*), the projection from each half-eye spreads out to cover the whole rostrocaudal extent of the tectum. When the compound eye innervates a tectum that also has input from a normal eye, the double-nasal eye's projection maps in register with the nasal half of the normal eye. (*c*, *d*) The two nasal halves are shown with different patterns to illustrate that their projections initially overlap but later segregate into bands. N: nasal; T: temporal; D: dorsal; V: ventral.

reveal that the graft initially projects in an inverted manner, which reflects its embryological origin; nevertheless, the projection from the subsequently produced cells is not inverted. Does this switch in polarity result from cell movements, from respecification of graft cell identities, or even from axon-axon interactions between graft and host tissue, or from other mechanisms, discussed below, that can influence the topography of the map quite independently of positional labels?

The physical nature of retinal "labels" or "specificities" is not understood. Do differentially distributed cell adhesion molecules constitute the labels? Recent experiments to examine this question have tested the adhesive behavior of cultured embryonic chick retinal axons. Bonhoeffer & Huf (1985) have shown that temporal axons strongly prefer to grow along other temporal axons and not upon nasal axons: nasal axons showed no preference. A complementary nasal-temporal distinction was shown by Halfter et al (1981), who found that tectal membranes adhere avidly to nasal retinal neurites but not to temporal neurites. Neither group found any differential behavior of dorsal versus ventral retina. Thus, these adhesion assays show dramatic nasal-temporal distinctions but do not reveal any of the fine-grain distinctions that are implied by other experiments on retinal specificities.

Another approach to searching for the basis of retinal specificities is to try to identify some cellular component that is differentially distributed across the retina. Liu et al (1983) have found that peanut agglutinin binds to chick retinal plexiform layers with a spatiotemporal gradient that begins near the center of the temporoventral quadrant and progresses toward the periphery. Trisler et al (1981) identified an antigen, dubbed TOP (toponymic), that is distributed in a stable gradient from ventronasal (low) to temporodorsal (high) retina. Constantine-Paton et al (1986) have discovered an antigen, probably a ganglioside, that is more densely distributed in the dorsal half of the rat retina than in the ventral half; the gradient first becomes apparent in the retina at embryonic day 17, the same age at which retinal axons begin to terminate in the tectum. The involvement of any of these gradients with the mechanisms underlying retinal specificities is still unknown.

TECTAL LOCUS SPECIFICITIES   Experiments in which optic axons are forced to reach the tectum via abnormal routes demonstrate that the tectum bears markers that can align the retinal map to conform to an intrinsic tectal polarity. For example, ablation of one tectal lobe on the day of birth in hamsters induces the axons appropriate to the ablated tectum to recross the midline between the tecta; even though they enter via this abnormal pathway, they form a map that is "normal" in the sense that nasal axons innervate caudal tectum, temporal axons innervate rostral tectum, and so forth (Finlay et al 1979). Similarly, axons from surgically implanted eyes can reach the tectum by a variety of abnormal routes but still form maps of proper orientation with respect to retinal landmarks (Sharma 1972b, Constantine-Paton 1981, Harris 1982). In addition, early rotations of the tectal primordium often lead to the establishment of ectopic optic tracts, but the retinotectal maps still become oriented normally with respect to the tecto-diencephalic border, wherever that happens to be (Chung & Cooke 1978).

The above experiments demonstrate that the tectum contains, at minimum, enough information to entrain the retinotectal map to the proper *polarity*. The following experiments indicate that the tectum actually bears *positional* markers rather than simply polarity markers. In other words, some physico-chemical property of tectal cells exists that serves not simply to orient the map but also to help position that individual components of the map. One such class of experiments involves the translocation of grafts of tectal tissue by exchanging rostral and caudal pieces of tectum; in many cases, cut retinotectal axons regrow to innervate the piece of tectum that they would normally innervate, despite the fact that this action brings them to the wrong place relative to the tectal borders and to the axons occupying the intact tissue (Levine & Jacobson 1974, Sharma 1975, Yoon 1980, Gaze & Hope 1983).

Examination of the distribution of retinotectal projections from partial or compound eyes also reveals that optic axons of a given retinal origin preferentially innervate the correct tectal location when they first reach the tectum (although rearrangements may occur later) (Crossland et al 1974, Straznicky et al 1981, Holt & Harris 1983, Holt 1984, O'Rourke & Fraser 1986a). Even when retinal axons are prevented from firing by application of tetrodotoxin, they are able to find approximately correct target regions with the tectum (Harris 1980, 1984, Meyer 1983, Schmidt & Edwards 1983) and lateral geniculate nucleus (Archer et al 1982, C. J. Shatz and M. P. Stryker, personal communication).

What is the nature of the interaction between retinal axons and tectal cells that underlies this selective innervation? The most popular hypotheses have focused on adhesion properties and have motivated experiments to test whether a given set of retinal cells, axons, or membranes adheres preferentially to the appropriate tectal membranes (Barbera 1975, Marchase 1977, Meyer 1982b, Fraser 1985). Intriguing evidence for such a mechanism comes from the experiments of Bonhoeffer & Huf (1982), who found that chick axons from temporal retina prefer to grow upon rostral (appropriate) rather than caudal (inappropriate) tectal cell monolayers. This preference is graded; for example, temporal axons grow more readily on cells from the rostral 20% of the tectum rather than from cells taken from the 20% immediately caudal to that rostral strip. However, these results have not yet been replicated for nasal, dorsal, or ventral retinal axons. A transient gradient of TOP molecules (see above) has recently been found in the developing chick tectum (Trister & Collins 1987). The molecule is most abundant in ventral tectum and least abundant dorsally. This distribution is complementary to the retinal TOP gradient; thus, parts of the retina with high TOP concentrations connect to the parts of the tectum which initially have high TOP concentrations, while corresponding regions with low amounts of TOP also become connected. These results suggest, but do not prove, that preferential adhesion is the mechanism underlying topographic innervation of target tissue by afferent axons.

Plasticity in Retinotectal and Retinocollicular Projections

As explained in the preceding section, evidence is now good that the optic tectum bears positional labels that guide retinofugal fibers to an appropriate termination site. However, there is also much evidence that these labels are not absolute determinants of the termination sites of axons.

MISMATCH EXPERIMENTS   One way of testing whether retinofugal axons are rigidly matched to a target region of the tectum is to create a size mismatch between retina and tectum by surgically reducing the size of either the retina or the tectum, so creating a situation in which half the tectum is denervated, or half the retina has its target removed. The fish and frog retinotectal projections show considerable plasticity following these manipulations. If half the tectum is ablated, the "dispossessed" axons that had innervated the ablated area of tectum generally gradually colonize the remaining half tectum, and the incumbent fibers retreat to make space for them. This leads to a "compressed" retinotectal map, with the entire retina represented over half a tectum (Gaze & Sharma 1970, Yoon 1971, Sharma 1972a, Cook 1979, Jacobson & Levine 1975, Ingle & Dudek 1977, Meyer 1977, Schmidt 1983) [although in *Rana* this only happens if the optic nerve also has been cut and allowed to regenerate (Udin 1977)]. Similarly, when half the retina is removed, thereby denervating half the tectum, the axons from the remaining half retina "expand" to cover the entire tectum (Horder 1971, Schmidt et al 1978, Udin & Gaze 1983).

Experiments in *Xenopus* yield the same type of result. Surgically constructed compound eyes contain only half the usual complement of retinal addresses but a roughly normal number of retina ganglion cells (see Figure 1a). Such compound eye maps are not confined to the "appropriate" half of the tectum, but expand to cover its entire surface (Figure 1b, right) (Gaze et al 1963, Straznicky et al 1974, 1981). Similarly, when half an eye is removed at embryonic stages, and the remaining half rounded up to prevent regrowth, the projection from this half-sized eye, when mapped in adult frogs, covers the entire tectum (Straznicky et al 1980a).

Both expansion and compression are seen in the mammalian reti-nocollicular projection, but only during the first ten days or so of life; after this period plasticity is no longer demonstrable. Thus ablation of half the newborn hamster colliculus leads to a compressed map of the whole retina on the remaining colliculus (Schneider & Jhaveri 1974, Finlay et al 1979) [although it is comprised of a lower number of axons than normal (Udin & Schneider 1981)]. Similarly, ablation of part of the retina allows remaining axons to fill in vacated space on the colliculus (Frost & Schneider 1979).

In contrast, the chick is unusual in that its retinotectal projection appears to have little plasticity: when part of the eye is ablated at early embryonic stages, the remaining retina only connects to the positionally appropriate region of the tectum, leaving much of the tectal surface uninnervated (Crossland et al 1974, McLoon 1985).

A form of developmental plasticity known as "shifting connections" is seen in the fish and frog retinotectal projection. As an animal grows, both the retina and optic tectum grow with it, and new retinofugal fibers constantly arrive at and connect with the tectum. However, retina and tectum grow in very different ways, the retina by the addition of cells all round the ciliary margin, and the tectum by addition of cells to its caudomedial edge (Straznicky & Gaze 1971, 1972, Johns & Easter 1977, Beach & Jacobson 1979, Tay et al 1982, Raymond & Easter 1983). Gaze et al (1974) argued that in order to maintain a retinotopic map on the tectal surface, the earlier arriving axons would gradually have to shift their initial functional connections to a site more caudally in the tectum. It has since been demonstrated in *Xenopus* (Gaze et al 1979b, Fraser 1983), *Rana* (Reh & Constantine-Paton 1984, Fraser & Hunt 1986), goldfish (Easter & Stuermer 1984), and perciform fish (Rusoff 1984) that this process occurs. Birds and mammals develop over a much shorter period of time, and the retinofugal fibers innervate a target that is nearly fully developed, so one would not expect shifting connections to be so important in development.

These results point toward the existence of competitive interactions between optic fibers on the tectal surface. If each axon terminal competes to occupy as much space on the tectum as possible, then retinofugal axons will always occupy all the available tectal space. Many workers therefore attempted to develop theories of retinotectal map formation that did not rely on tectal labels at all, but used a combination of axonal guidance (see first section) and axonal competition. This period of theoretical uncertainty continued until (*a*) experimental evidence demonstrated that there must be positional labels on the tectum and that axonal pathway guidance was not the main mechanism whereby the retinotectal map was set up, and (b) the significance of the phenomenon of ocular dominance stripes was understood.

OCULAR DOMINANCE STRIPES   In fish and frogs, the retinotectal axons from one eye all project to the contralateral optic tectum. However, a variety of maneuvers can cause axons from both eyes to innervate the same tectum. When this is done, the projections from the two eyes, instead of intermixing, separate out into mutually exclusive ocular dominance stripes (Levine & Jacobson 1975, Law & Constatine-Paton 1980, Straznicky et al 1980b, Willshaw & Gaze 1986). Ocular dominance stripes may also be created in chicks (Fawcett & Cowan 1985). The axons from the two half eyes that constitute a "compound" eye also separate into stripes (Fawcett & Willshaw 1982, Ide et al 1983). (See Figure 1c,d.) In mammals, a proportion of the axons from each eye innervate the ipsilateral colliculus: initially this ipsilateral projection is spread diffusely, but during the first few days of life it segregates into eye-specific patches, which are probably equivalent to ocular dominance stripes (Graybiel 1975, Harting & Guillery 1976, Land & Lund 1979, Frost et al 1979, Williams & Chalupa 1982, Godement et al 1984).

A single unifying mechanism cannot explain both topographic mapping and ocular dominance stripe formation. Instead, at least two mechanisms must be invoked, which have been brought into conflict on the doubly innervated tectum. For instance, in the case of a double nasal compound eye (see Figure 1), the fibers from the two nasal poles of the retina grow to the caudal tectum, and initially their terminals overlap (J. W. Fawcett, D. J. Willshaw, and J. S. H. Taylor, unpublished results). This part of their behavior must be controlled by a system of retinal positional addresses, since the axons from the graft nasal pole behave like the normal nasal retina. However, the axon terminals from the two "nasal" poles now separate out into eye-specific stripes. On the basis of retinal positional labels they should be indistinguishable; therefore, another mechanism must be responsible. If there were a mechanism that caused each cell's axon to associate preferentially with the axons of its present retinal neighbors, then axons from the two nasal poles would subdivide their tectal territory, forming dominance areas. Similarly, embryologically equivalent axons from two separate eyes would occupy the same general region of the tectum, but, within that region, their tendency to associate with axons from neighboring ganglion cells would promote the formation of ocular dominance areas. Activity cues, discussed below, are likely to underlie this behavior.

REFINEMENT OF THE RETINOTECTAL MAP   When the optic nerve of a frog or fish is cut, the axons regenerate back to the tectum and make connections. Initially, the topographic accuracy of those connections is low, and an electrophysiologically recorded retinotectal map shows some degree of disorganization, as reflected in the large, diffuse morphology of the ter- minal arbors. However, in time the topography of the map improves, until it is as accurate as a normal one (Gaze & Jacobson 1963, Udin 1978, Meyer 1980, Humphrey & Beazley 1982, Schmidt & Edwards 1983, Rankin & Cook 1986), and the terminal arbors become smaller and more focused (Meyer 1980; Fujisawa et al 1982, Schmidt et al 1984, Stuermer 1984, Stuermer & Easter 1984b). There is also evidence that the accuracy of the fish and frog retinotectal map is rather low early in normal development, and improves with time (Gaze et al 1974; O'Rourke & Fraser 1986a). Again, the morphology of the terminal arbors reflects this topographic refinement; in young frogs, arbors are large relative to the size of the tectum, particularly in the rostro-caudal axis, but as the tectum grows the proportion of its area occupied by each arbor diminishes (Sakaguchi & Murphey 1985, Lázár 1973, Piper et al 1980, Fujisawa 1987).

In the newborn rodent, the topography of the retinocollicular projection is imprecise. A proportion of ganglion cells connect to completely inap- propriate regions of the colliculus, and the terminals of the retinocollicular axons may extend over much of the length of the collicular surface. Over the first few days of life, the terminal arbors become focused to a localized area, and most of the ganglion cells that have sent their axons to inap- propriate parts of the colliculus die. These mechanisms cooperate to refine the topography of the map substantially (Cowan et al 1984, Schneider & Jhaveri 1984, Fawcett & O'Leary 1985, Sachs et al 1986, O'Leary et al 1986). Similarly, in the developing chick, some axons make inappropriate projections that are not demonstrable later on (McLoon 1982, 1985).

How might the refinement of the map be achieved? Theoretically, it could be refined by axons gradually searching out correct tectal addresses with more and more precision. The available evidence, however, suggests that tectal labels are imprecise guides, at least in fish and frogs. Moreover, the ocular dominance stripe phenomenon suggests a mapping mechanism that is independent of retinal and tectal positional labels. The obvious alternative is a mechanism that acts by ensuring simply that neighboring ganglion cells project to neighboring tectal sites. How might information on the relative contiguity of ganglion cells be transmitted to the axon terminals? The most likely possibility is that neighboring ganglion cells have relatively synchronous patterns of electrical activity, and a mech- anism exists that ensures that neighboring terminal arbors that fire synchronously are stable, whereas nonsynchronously firing terminals are unstable—a class of mechanism originally proposed by Stent (1973). Some evidence indicates that neighboring ganglion cells do have correlated activity (Arnett 1978, Mastronarde 1983, Ginsburg et al 1984). This hypothesis, if correct, predicts that abolishing electrical activity in the retinofugal axons should disable both ocular dominance stripe formation and the refinement of a disorganized retinotectal map. Both these predictions turn out to be correct. Electrical activity in the eye can be blocked by injections of tetrodotoxin (TTX), and if this block is maintained over the relevant period, stripe formation (Meyer 1982a, Boss & Schmidt 1984, Reh & Constatine-Paton 1985), ocular dominance patch formation in hamsters (Holt & Thompson 1984), and map refinement (Meyer 1983, Schmidt & Edwards 1983) are prevented. The refinement of the rat retinocollicular projection by the preferential death of erroneously projecting cells is also disabled; retinal ganglion cell death becomes random in the presence of a TTX block (Fawcett & O'Leary 1985, O'Leary et al 1986).

The hypothesis that correlated firing leads to synaptic stabilization implies that simultaneous converging input produces a qualitatively different effect than the same amount of nonsimultaneous input. Recent research on glutamate receptors has revealed a possible mechanism whereby simultaneous activity could trigger events, such as increased calcium influx, which would not occur in response to noncorrelated activity: NMDA (N-methyl-D-aspartate) receptors conduct calcium currents when the receptors bind NMDA or glutamate and, at the same time, the membrane is depolarized by several excitatory inputs (MacDermott et al 1986, Ascher & Nowak 1987, Cull-Candy & Usowicz 1987, Jahr & Stevens 1987). Blocking these receptors with the antagonist APV (aminophosphonovaleric acid) by chronic application to the tectum of three-eyed tadpoles does not block optic nerve firing but does induce desegregation of retinotectal stripes (Cline et al 1987). This exciting result raises the question of whether NMDA receptors also mediate the contribution of activity to the formation of other topographic maps, such as those described below.

Mapping refinement in fish is prevented when all the ganglion cells are made to fire synchronously by rearing animals under stroboscopic illumination (Schmidt & Eisele 1985, Cook & Rankin 1986). However, simple competition for terminal space continues in the absence of electrical activity (Meyer & Wolcott 1984), and a retinotectal map possessing at least a crude degree of order can be formed (Harris 1980), presumably based entirely on tectal positional labels. Moreover, apart from the electrically controlled mechanism, terminal arbors can interact with one another on the basis of positional labels (see above).

The formation of experimentally induced ocular dominance stripes in chicks occurs while the animal is still in the egg, usually in the dark, and possesses a retina that is anatomically immature and unable to respond to light stimula (Rager 1976). Such stripes are formed in rodents during the period when the ipsilateral projection is becoming restricted into patches, topographic targeting errors are being corrected, the eyelids are closed, and the retinal circuitry is incomplete (Weidman & Kuwabura 1968). Assuming that these processes are all electrically driven, as the TTX experiments suggest, then spontaneous activity in the ganglion cells must be responsible. Such spontaneous activity has been observed in newborn rabbit optic nerve fibers, at an equivalent stage of retinal development (Masland 1977, McArdle et al 1977), and in dissociated rat retinal ganglion cells (Lipton & Harcourt 1984).

TERMINAL ARBORS   A topographic map is made up of many terminal arbors, and the behavior of these determines the dynamic properties of the map as a whole. What do we know about the forces that shape terminals? The simplest model would be one in which terminal arbors constantly compete to occupy tectal space; success or failure in this competition would be governed by an impulse activity controlled mechanism, which would give a competitive advantage to the terminals making the most topographically appropriate connections. However, this picture appears to be oversimplified. Some evidence suggests that at a certain stage in their evolution, terminal arbors will withdraw their more far-flung branches even in the absence of competition. For instance, even in the presence of a TTX block, regenerated fish terminal arbors change in time from a diffuse to a more focused morphology (J. T. Schmidt, personal communication), and some evidence suggests that this is also the case in the developing rat retinocollicular projection (Fawcett & O'Leary 1985, O'Leary et al 1986). Similarly, Sretevan & Shatz (1986) report that terminal arbors in the cat geniculate are of normal size even when competition for terminal space is reduced by removing one eye; and in the peripheral nervous system, neuromuscular junction terminals reduce the number of muscle fibers they innervate, and therefore their size, even when competition is considerably reduced by decreasing the number of axons innervating the muscle (Brown et al 1976). Some terminals, however, increase their size in development rather than contract (Sur et al 1984). Although little evidence is available on the subject, it may be that the natural history of most terminal arbors is to focus themselves by retracting their more far-flung branches, regardless of competition from other arbors.

MATCHING OF SENSORY MAPS IN THE VISUAL SYSTEM

Virtually all visual nuclei receive multiple sets of inputs, which are aligned to bring the individual representations of external space into register. For example, the superior colliculus receives input from the contralateral eye directly via the optic nerve and also, indirectly, via structures such as the striate cortex and the parabigeminal nucleus; the ipsilateral eye also sends both direct and indirect inputs; and the auditory and somatosensory systems additionally contribute topographic maps (Stein 1984). These inputs may be interdigitated or intermingled in the same laminae or may be superimposed in adjacent laminae, but in all cases, the maps are in topographic register with one another such that a single region in the visual field is represented in a small region of the nucleus. At least two mechanisms can contribute to the establishment of such congruence. First, chemospecific cues may be equally accessible to all populations and all of the afferent axons are equipped to read the cues properly. Second, visually-evoked firing patterns may provide the information that aligns one group of axons with respect to another. As we shall discuss in this section, the relative roles of these mechanisms can differ markedly in different nuclei and from species to species, with activity cues dominant in some cases and chemospecific cues dominant in others.

Isthmotectal Maps

This point is dramatically illustrated by examining the relation between the tectum (superior colliculus) and the midbrain structure known as the nucleus isthmi in amphibians (Grobstein et al 1978, Gruberg & Udin 1978), parabigeminal nucleus in mammals (Graybiel 1978, Sherk 1978), or nucleus isthmi pars parvocellularis in birds (Hunt & Künzle 1976). This nucleus receives a topographic input from the ipsilateral tectum (superior colliculus); the nucleus then sends a matching topographic map back to the ipsilateral tectum and, in amphibians and mammals, also sends a topographic map to the contralateral tectum. The uncrossed isthmotectal projection is quite capable of forming a topographically normal map in the absence of any visual input, as experiments using bilaterally enucleated frogs (Constantine-Paton & Ferrari-Eastman 1981) and birds (O'Leary & Cowan 1983) demonstrate.

Similarly, the role of visual input on the crossed isthmotectal/parabigeminotectal projection has been tested in several species. This projection relays visual input to the tectum from the ipsilateral eye, forming a map in register with the map from the contralateral eye (Keating & Gaze 1970, Glasser & Ingle 1978, Graybiel 1978, Gruberg & Udin 1978, Sherk 1978). In order to test whether visual information is used to bring these two maps into register, one may rear an animal with one eye rotated. This manipulation does not change the anatomical connectivity of the isthmic/parabigeminal axons in *Rana pipiens* or *Hyla moorei* frogs (Jacobson & Hirsch 1973, Beazley 1979, Kennard & Keating 1985) or in cats (H.

Sherk, C. Peck, and S. B. Udin, unpublished observations); the two eyes' maps thus develop out of register.

In contrast, early eye rotations in *Xenopus* lead to dramatically different results (Gaze et al 1970, Keating 1974). Even though one eye is rotated while the other eye is normally oriented, the isthmotectal map on each tectum comes to be in register with the contralateral map on that tectum. For example, if the right eye is rotated 90° clockwise and the left eye is unrotated, both maps to the left tectum will be rotated 90° while both maps to the right tectum will be normal. Horseradish peroxidase tracing experiments have shown the morphological correlate of this arrangement to be the rerouting of isthmotectal axons; the axons enter the tectum normally and grow toward their normal termination sites but then develop abnormal trajectories that eventually enable them to form a reoriented map (Udin & Keating 1981, Udin 1983). Probably a given axon terminal becomes stabilized at a new position because there is an optimum correlation between its visually evoked activity and that of the adjacent retinotectal axons from the opposite eye. Ultrastructural evidence suggests that isthmotectal and retinotectal axons both terminate on tectal cell dendrites (Székeley & Lazar 1976, Udin & Fisher 1986). Presumably, a visual stimulus at the appropriate receptive field location will activate the retinotectal synapse and will somehow "ready" the dendrite; if the isthmotectal axon fires after an appropriate delay (due in part to the polysynaptic nature of the ipsilateral pathway), a further change occurs that helps to stabilize the recently activated isthmic axon.

Further evidence for the role of visual input in aligning the two maps during normal development is the fact that dark-rearing prevents the isthmotectal axons from establishing a normal topographic ipsilateral map (Keating & Feldman 1975).

Callosal Visual Connections

Another system in which visual activity influences the formation of retinotopic registration between two separate sets of input is the callosal projection in the mammalian visual cortex. In normal adult mammals, areas 17 and 18 in each hemisphere contain maps of all of the contralateral hemifield and a few degrees of the ipsilateral hemifield. These half-maps are "tied together" by callosal axons that relay information about positions along the vertical meridian from the 17/18 border in one hemisphere to the 17/18 border in the opposite hemisphere.

The adult callosal connections are sculpted out of the much more diffuse distribution that has been observed in newborn rodents (Lund et al 1984, Mooney et al 1984) and kittens (Innocenti 1981) [but not monkeys (Dehay et al 1986)]. Callosal axons originate from cells distributed throughout areas 17 and 18; as normal development proceeds, most of these axons are eliminated from the callosum, leaving predominantly those axons originating from cells near the 17/18 border (Olavarria & Van Sluyters 1985). Despite the widespread origin of most early callosal axons, the axons normally arborize only in the appropriate area, the 17/18 border, from the earliest stages at which they enter the cortical plate (Innocenti 1981, Lund et al 1984, Mooney et al 1984, Olavarria & Van Sluyters 1984). One may hypothesize that visual input preferentially stabilizes connections between cells with the same receptive field locations. Under normal circumstances, this mechanism would ensure that cells with receptive fields near the vertical meridian remain interconnected via the corpus callosum but that other cells, with receptive fields farther from the vertical meridian, would fail to find corresponding targets and would die or withdraw their axons (Stanfield et al 1982, Innocenti & Clarke 1983, Stanfield & O'Leary 1985, Innocenti et al 1986), perhaps because of competition from topographically appropriate geniculocortical axons.

If visual input is a factor in stabilizing connections between cells with matching receptive field locations, then misalignment of the two eyes' visual fields as a result of strabismus should induce anomalous connections. For example, if convergent squint is induced in the one eye, then the cells at the 17/18 border will receive input from two different visual field positions from the two eyes. (See Figure 2.) Moreover, cells at some distance from the 17/18 border will receive input that normally should be relayed to the 17/18 border. Thus, one might expect to find more widespread zones of callosal cells and terminals, with their limits being separated from the 17/18 border by a distance corresponding to the degree and direction of strabismus. This prediction is only partially borne out. Misalignment of the eyes in kittens does lead to an expansion of the zone of callosal cells (Innocenti & Frost 1979, Berman & Payne 1983, Elberger et al 1983, but see Lund et al 1978), but there is no obvious correlation of the degree of expansion or the specific location of the ectopic cells with the magnitude of the strabismus or with the animals' fixation strategies. In many cases, the callosal terminal zone also is expanded, again with no clear correlation with the magnitude of the strabismus (Lund et al 1978, Lund & Mitchell 1979, Berman & Payne 1983). Much of the variability of results no doubt is due to poorly controllable factors such as variable degrees of convergence throughout the critical period and differing degrees of "suppression" of input from the strabismic eye. Firmer conclusions cannot be made, however, until more precise information about the detailed topography of the projections becomes available.

Monocular enucleation in neonatal rats leads to an ectopic band of callosal connections in the hemisphere ipsilateral to the remaining eye

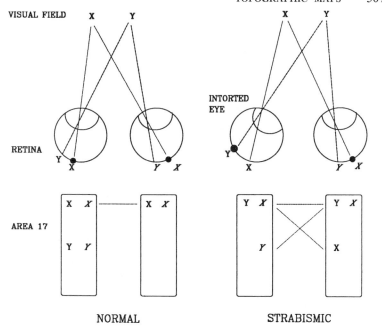

Figure 2   Left: In normal adult cats, the area centralis of each eye sees position X on the vertical meridian and transmits this input bilaterally via the lateral geniculate to the cortex near the border between areas 17 and 18. Callosal axons connect the 17/18 border regions but do not connect with areas with receptive fields distant from the vertical meridian (position Y). *Right*: In strabismic cats different connections are predicted. In this example, the left eye is deviated such that position Y rather than X falls on the area centralis. Thus the left eye relays input about position Y to the 17/18 border, while position X is represented in the right hemisphere. Callosal axons that connect cells with the same receptive field locations should connect more sites than normal.

(Olavarria et al 1979); a double representation of the vertical meridian, resulting from anomalous retinogeniculocortical connections, may underlie this extra band of callosal projections. As with studies of strabismic animals, clear interpretation of those results await studies of the projections' topography.

Direct Ipsilateral Retinofugal Maps

In contrast to the marked and obvious influence of visual input on the formation of binocular maps relayed by callosal fibers and by the nucleus isthmi, vision and activity seem to play little, if any, role on the formation of ipsilateral maps relayed directly from the retina. For example, the direct ipsilateral retinothalamic projections in *Xenopus* do not compensate for an early eye rotation (Kennard 1981) even though the isthomotectal axons in the same animals can fully compensate for the same rotation. The ipsilateral retinothalamic axons instead continue to form connections consistent with their embryonic origin rather than their visual fields.

This lack of influence of visual activity on ipsilateral retinotectal axons is further borne out by examining the response of retinotectal axons to monocular enucleation in neonatal rodents. Some ipsilateral axons establish an expanded projection that retains the normal ipsilateral polarity, with the nasotemporal retinal axis mapped rostrocaudally on the tectum (Finlay et al 1979, Thompson 1979, Fukuda et al 1984, Reese 1986), while other axons may form a second projection of opposite polarity (Thompson 1979, Fukuda et al 1984); this latter projection presumably originates from cells with a "contralateral identity" that normally grow to the ipsilateral tectum and are then eliminated (Jeffery & Perry 1982, Martin et al 1983, Insausti 1984, Fawcett et al 1984a). Thus, the same eye, and even cells in the same region of that eye, can give rise to two opposite maps despite the mismatch of visually evoked firing that results. Some developmental process thus confers drastically different positional specificities upon neighboring retinal ganglion cells.

Little is known about the process that establishes ipsilateral versus contralateral identities in ganglion cells. Briefly, we may mention that ipsilaterally projecting cells are normally born according to a different schedule than contralaterally projecting cells (reviewed in Dräger 1985) and that hormonal (Hoskins & Grobstein 1985) and genetic abnormalities, such as the Siamese mutation in cats, can alter these schedules (Kliot & Shatz 1985) and can change the proportions of cells that project ipsilaterally versus contralaterally.

Registration of Visual and Somatic Maps in the Tectum

The tectum contains not only visual maps but also maps from receptors on the body surface (Wickelgren 1971, Stein 1984). The possibility that the visual maps help to establish the polarity or topography of these other maps has prompted several groups to test whether abnormal visual input to the tectum interferes with establishment of orderly somatosensory maps. Thus far, the evidence for interaction is scanty. Early eye enucleation allows the normally deep somatosensory inputs to innervate more superficial layers that have lost their retinal input (Rhoades 1980, Rhoades et al 1981, Harris 1982, 1983) but does not interfere with the formation of otherwise normal somatotopic maps. Similarly, abnormal visual input induced by a rotated eye does not cause any change in the orientation of the underlying somatosensory map (Harris 1983). However, if optic axons in hamsters are induced to form a mirror image map, the somatosensory map undergoes partial rearrangement, with formation of a partial mirror image map of the ipsilateral body surface superimposed on the normally organized contralateral body map (Mooney et al 1979).

Registration of Visual and Auditory Maps in the Tectum

The superior colliculus contains auditory and multimodal cells with auditory receptive fields that are aligned, at least crudely, with the corresponding maps of visual space when the eyes are centered in the orbit (Dräger & Hubel 1975, Chalupa & Rhoades 1977, L. R. Harris et al 1980, Knudsen 1982, Palmer & King 1982, Middlebrooks & Knudsen 1984, Meredith & Stein 1986). Auditory receptive fields tend to be quite large, and their geometrical centers are often not well-aligned with the visual fields of the same cells, but a better match of visual and auditory maps is achieved by considering auditory "best areas," the most responsive regions of the auditory receptive fields (Jay & Sparks 1984, Middlebrooks & Knudsen 1984).

An example of plasticity of auditory topography comes from the dramatic rearrangements that Knudsen (1983) has demonstrated in the developing barn owl, which displays an exceptionally well-developed ability to localize sound sources (Knudsen 1981). The barn owl is unusual in having multimodal auditory/visual units in all layers of the tectum, not just the intermediate and deep layers (Knudsen & Knudsen 1983). If one ear is plugged in an adult barn owl, the binaural imbalance produces a systematic shift in auditory best areas such that they are out of register with the visual fields; but if one ear is plugged in a young owl, then the auditory receptive fields shift to restore the registration of the auditory with the visual maps (Knudsen 1983). Behavioral tests indicate that the auditory map shifts only if visual input is provided during development; moreover, the magnitude and direction of the shift can be manipulated by rearing the owls with prisms that skew the visual input (Knudsen & Knudsen 1985). Thus, the auditory map organizes with respect to the visual map.

This situation seems reminiscent of the readjustments seen in the tectum of *Xenopus*, in which the ipsilateral eye's map, relayed via the nucleus isthmi, organizes with respect to the contralateral eye's map (Gaze et al 1963). In that system, developing isthmic axons change their termination sites in the tectum when they encounter changes in the visual map coming from the other eye (Udin & Keating 1981). By analogy, one might predict that auditory axons projecting from the owl's inferior colliculus would change their termination sites when they encounter a mismatched visual map in the tectum (Knudsen 1983). Instead, anatomical mapping in one owl reared with a plugged ear indicates that there is no anatomical shift in this connection (Knudsen 1985). At least two possible explanations can be offered for this surprising result. One suggestion is that the auditory map becomes rearranged at some earlier stage in the auditory pathway (Knudsen 1985), but how visual information would be able to guide such a shift at such locations, where visual input is not available to guide the reorganization of the auditory map (Knudsen & Knudsen 1983), is unclear. Another possibility is that a physiological change in the excitatory and inhibitory processes that construct tectal auditory fields could underlie the observed auditory plasticity; the fact that many auditory fields are extremely large relative to the imposed auditory-visual shifts (Knudsen 1982) raises the possibility that intratectal processes could stably alter best-field locations, much as may occur transiently during eye position shifts in primates (Jay & Sparks 1987). The drawback of this model is that the auditory inputs from the inferior colliculus in owls already have best field maps (Knudsen & Knudsen 1983), and these maps would have to be suppressed in favor of newly constructed maps in the superior colliculus.

Not all species show comparable influences of the visual map upon the developing auditory map. King et al (1985) tested the effects of removing one eye and rotating the other eye by 180° in ferrets at the time of eye opening; the auditory map did not compensate for this discrepancy. Another way to test the possible instructive role of vision upon auditory maps is to rear animals in the dark. Rauschecker & Harris (1983) found that visually deprived kittens showed a shift of some auditory units to superficial layers of the superior colliculus and also noted that auditory fields of multimodal units were abnormally large, indicating that visual input normally plays a role in restricting auditory receptive field size in the colliculus.

TOPOGRAPHIC MAPS OUTSIDE THE VISUAL SYSTEM

The phenomenon of topographic mapping is by no means confined to the visual system. In the following section we examine present evidence on how topography is set up in some other parts of the brain.

Auditory System

The tonotopic maps of the auditory system provide opportunities to address some of the same questions discussed above concerning the mechanisms that establish topographic maps. The technical problems posed by the complicated, delicate, and poorly accessible sensory epithelium of the cochlea and by the difficulty of completely controlling auditory input have slowed progress in studying auditory development, but several systems are beginning to yield intriguing data.

The auditory system is characterized by a plethora of tonotopically organized structures, with representations of frequency mapped along one axis (Aitkin et al 1984). That many of these structures also show clear cytogenetic gradients has raised the question of whether gradients of neuron age provide cues to assist the formation of tonotopic maps (Altman & Bayer 1981). This mechanism seems to be plausible for some, but not all, sets of auditory nuclei. For example, in the chick, the primary brainstem nucleus (nucleus magnocellularis, NM) and the secondary brainstem nucleus (nucleus laminaris, NL) develop with corresponding gradients: the first-born cells in each area eventually mediate high frequencies whereas the last-born cells mediate low frequencies (Rubel et al 1976, Smith & Rubel 1979, Jackson et al 1982). Young & Rubel (1986) have examined the development of the map from NM to the ipsilateral NL. Both of these structures derive from the same ventricular epithelium, with NM arising before NL. The axons from NM initially grow dorsally toward the ventricular epithelium, where they contact and follow the appropriate NL cells as the cells migrate ventrally. Thus, the relative position or relative maturation of an NM axon and its NL target cells may underlie the striking precision with which the NM axons initially establish their tonotopic projections in NL.

In contrast, most developing or regenerating sets of connections among auditory nuclei are initially rather crudely organized and undergo a period of refinement that may entail restriction of arbor size and/or shifting of axonal connections (Jhaveri & Morest 1982, Jackson & Parks 1982, Zakon 1983, Young & Rubel 1986). Neuronal activity may well be an important factor in the process of refinement of tonotopic maps. For example, Young & Rubel (1986) find that refinement of the tonotopic organization of contralateral inputs to NL occurs in large part after the NL has become responsive to auditory input (Jackson et al 1982). In addition, Sanes & Constantine-Paton (1985) demonstrated that abnormal auditory input (clicks or 11 KHz and 14 KHz tones) interferes with the refinement of tonotopy in the inferior colliculus of developing mice; it is not yet known whether the abnormally broad tuning curves are the result of abnormally widespread axon arborization or of improper balance of inhibitory versus excitatory connections.

Somatosensory Systems

SPINAL CORD SENSORIMOTOR PROJECTIONS   At first viewing, the "map" from muscle sensory afferents to motor neurons in the ventral horn seems rather crude, there being extensive overlap in the terminal zones of both sensory axons and motor dendrites corresponding to flexor and extensor muscles (Smith 1983, Lichtman et al 1984). However, physiological studies show that the actual synaptic connections are highly precise (see review by Mendell 1984); for example, intracellular recordings demonstrate that a given spindle afferent makes remarkably few detectable connections onto antagonist motor neurons in either developing or adult frogs (Frank & Westerfield 1983, Lichtman & Frank 1984). Light microscopic comparisons of normal developing dorsal root ganglion (DRG) processes in rats and chicks also are consistent with the early establishment of a very orderly pattern of connections, both peripherally and centrally, without a preceding stage of more widespread or disorderly connections (Honig 1982, Scott 1982, Smith 1983, Tosney & Landmesser 1985a, Dodd & Jessell 1986).

Axon-axon guidance mechanisms may play a role in setting up these orderly connections. Sensory axons enter the limb in intimate association with emerging motor axons (Tosney & Landmesser 1985b), and the sensory axons require guidance by motor axons in order to reach muscles (Landmesser & Honig 1986, Scott 1986). The sensory axon acquires its identity, it has been suggested, as a result of contacting a given muscle; this imposed identity then enables the central process to synapse upon appropriate targets within the cord (Miner 1956, Hollyday & Mendell 1975, Honig 1982, Smith 1983, Frank & Jackson 1986). Thus, if one were to change the axons' peripheral connections during development, one would expect to find altered central connections. One way to alter peripheral connections is to delete some of the dorsal root ganglia and to allow the spared ganglia to expand into the vacated periphery. In bullfrog tadpoles, removal of DRG2, which provides most of the sensory input to the forelimb, induces DRG3 or DRG4 to innervate the arm (Frank & Westerfield 1983, Smith 1986). Anatomical (Smith 1986) and physiological methods (Frank & Westerfield 1983) demonstrate that some of the sensory cells in those ganglia innervate spindles in the arm and are able to synapse upon the newly appropriate motor neurons. In *Rana pipiens* tadpoles, removal of caudal DRGs leads to an expansion of the spared DRGs' peripheral territories but not to an anatomically visible change centrally (Davis & Constantine-Paton 1983); nevertheless, the extensive overlap of fields of adjacent DRGs may perhaps still allow for formation of physiologically appropriate connections. Partial removal of neural crest in chicks results in a limited expansion of afferents from the spared dorsal root ganglia (Scott 1984) and to a similarly limited rearrangement of connections within the spinal cord (Eide et al 1982). Although these deletion experiments show varying degrees of peripheral plasticity, all are consistent with the interpretation that afferents from a given muscle connect to motoneurons that innervate that muscle and not to antagonists or other inappropriate motoneurons.

Grafting experiments also lead to the conclusion that inducing a sensory cell's peripheral axon to innervate a given peripheral target somehow allows that cell's central process to find the appropriate targets in the cord. In bullfrog tadpoles, Smith & Frank (1987) replaced DRG2 with thoracic DRGs; they found that some of the grafted thoracic cells could innervate limb muscles and make highly specific monosynaptic connections onto corresponding motor neurons.

Does the sensory axon's central process find its appropriate motor neuron targets by using cues based on related activity patterns? In an effort to test this hypothesis, Frank & Jackson (1986) altered sensory input from the triceps during development by tenotomizing the muscle, or by suturing it to an antagonist tendon; in both cases, the inappropriate spindle activity did not interfere at all with establishment of precise connections of triceps spindle afferents onto triceps motoneurons.

The identities that DRG axons acquire during some critical period of development seem to remain stable throughout life. Frank & Westerfield (1982) found that the crtical period probably ends at about stage 9, when limb muscles normally become innervated; DRG3 axons could replace ablated DRG2 axons in the limb and spinal cord only if DRG2 was removed prior to stage 9. Using older tadpoles as well as postmetamorphic frogs, Sah & Frank (1984) found that transected DRG2 axons regenerate and reestablish connections onto proper motoneurons with almost as much accuracy as normal.

THE PROJECTION OF RODENT VIBRISSAE TO THE CORTEX   Rats, mice, hamsters, and guinea pigs have on their snouts rows of vibrissae, which together constitute the animal's most important tactile sensory organ. The importance of the vibrissae to the rat is reflected in the fact that their projection to the sensory cortex occupies a large proportion of its area. In layer IV of the cortex in the region of the vibrissal projection, the cortical cells are divided into groups (barrels), the pattern of which exactly corresponds to the pattern of the vibrissae on the snout (Woolsey & Van der Loos 1970). Each vibrissa can be mapped functionally to the corresponding barrel in the cortex (Welker 1976, Simons 1978). This cortical map is a third-order topographic projection; the axons innervating the hair follicles first project to the principal sensory nucleus and three subnuclei in the trigeminal nucleus in the medulla (Nord 1967, Arvidsson 1982); the neurons here project to the ventrobasal thalamus (Emmers 1965, Waite 1973, Shipley 1974, Verley & Onnen 1981), which in turn projects to the cortex. At each stage of this pathway, cytoarchitectonic or histochemical patches corresponding to the pattern of the vibrissae on the snout can be detected (Van der Loos 1976, Belford & Killackey 1979a,b, Woolsey et al 1979, Durham & Woolsey 1984).

Development   The vibrissal projection develops from periphery to center. The trigeminal nucleus first receives innervation from the trigeminal ganglion before birth (Erzurumlu & Killackey 1983), whereas axons from the thalamus do not grow into layer IV of the cortex until after birth (Wise & Jones 1978, Crandall & Caviness 1984). The vibrissa-related organization of patches stained with succinate dehydrogenase (SDH) is first detected in the trigeminal nuclei around the time of birth, then in the thalamus on days 1–4, and in the cortex on days 3–6 (Killackey & Belford 1979, Erzurumlu & Killackey 1983). Similar results are reported for 2-deoxy-glucose mapping of the vibrissal projection (Melzer et al 1986). However, a topographically organized set of connections between thalamus and cortex is present before the appearance of the anatomical specialization into whisker barrels (Dawson & Killackey 1985, Crandall & Caviness 1984), and indeed the thalamocortical projection is topographically organized in cortical layer VI before it has invaded layer IV. The projection from the vibrissae to the trigeminal nucleus is also topographically organized before barrels can be detected there by SDH histochemistry (Erzurumlu & Killackey 1983). The formation of barrels seems, therefore, to be the final phase of differentiation of an already topographically arranged projection. There is also some evidence to suggest that the topography of the projection to the cortex is less precise at birth than in adult animals, a notion that implies that some topographic refinement occurs during the early postnatal period (Crandall & Caviness 1984, Kossut & Hand 1984b, Melzer et al 1986). Electrical activity may play some role in this refinement. Removing the whiskers at birth reduces electrical activity in the trigeminal nerve; this does not prevent barrel formation, but it does lead to neurons in the barrel field having abnormal response properties, and to a decrease in the precision of the map (Durham & Woolsey 1978, Simons & Land 1986).

A precise topography exists in the pathway of axons from the vibrissae to the trigeminal nucleus; the axons from the whisker follicles grow out parallel to one another and fasciculate into bundles, each of which contains axons from a localized area of the whisker field (Erzurumlu & Killackey 1983). Whether this topographic arrangement of axons is necessary for axons to make topographic connections in the V nucleus has not been fully tested. However, if the overall number of axons innervating the whisker field is reduced prenatally by anti-NGF treatment, the fascicular pattern of the nerve is abnormal, whereas the barrel pattern in the cortex is unaffected (Sikich et al. 1986). This suggests that topographic mapping in the whisker projection does not depend on the maintenance of topography in the fiber pathway.

Plasticity   The vibrissal projections show a very dramatic and precise form of plasticity. If one or more vibrissae are lesioned in a newborn animal, the corresponding barrels in the cortex and thalamus fail to develop, and the surrounding barrels expand to fill the vacated space (Van der Loos & Woolsey 1973, Weller & Johnson 1975, Killackey et al 1976, Jeanmonod et al 1977, Belford & Killackey 1979b, 1980, Woolsey et al 1979, Durham & Woolsey 1984, 1985). This plasticity is only present in the few days after birth. In order to have an effect on the barrels in the thalamus, lesions must be made before P4, whereas lesions up to P6 will still affect the barrel pattern in the cortex (Weller & Johnson 1975, Woolsey & Wann 1976, Killackey & Belford 1979, Woolsey et al 1979, Jeanmonod et al 1981). Electrophysiological studies give essentially the same results as the anatomical experiments, and also reveal that areas of cortex that have been denervated by whisker lesioning may receive an abnormal input from the fur surrounding the lesion (Waite & Taylor 1978, Pidoux et al 1979). Functional mapping with 2-deoxyglucose shows that the connections from an unlesioned vibrissa can expand to occupy territory vacated by lesioning (Kossut & Hand 1984a) and that the functional mapping corresponds closely to the altered barrel pattern (Durham & Woolsey 1985). The vibrissal projection exhibits a different form of expansion when the overall population of axons innervating the vibrissae is reduced by treating developing guinea pigs in utero with anti-NGF. The barrel pattern in such animals is quite normal, both in size and shape (Sikich et al 1986).

The cortical barrel field can also show compression. If an area of the cortex is lesioned, a normal number of barrels will develop, but they will skirt around the damaged area. The axons that would have innervated the damaged area must, therefore, have displaced axons from the neighboring region of the cortex, thus leading to a readjustment of the topography of the barrel field (Ito & Seo 1983).

SUMMARY

The catalogue of data presented here for many systems demonstrates that multiple mechanisms are involved in the formation of topographic maps. We are not yet in a position to explain why a particular mechanism appears to dominate in some situations and not in others. Certain generalizations can be made, however. First, at least some form of chemospecificity can be invoked to help explain connectivity in all of the experiments we have cited. Often, the differential identities of a population of neurons can be reflected in an orderly pattern of axon outgrowth and in the actively maintained preservation of neighbor relations as the axons grow toward their targets; such orderly arrangements are not obligatory, but, where present, they facilitate the speedy establishment of orderly maps when the axons reach their target nuclei. Within a terminal zone, chemospecific cues may dominate and constrain a given axon to terminate in a specific location, but axon-axon interactions commonly supercede chemospecific matching. At least two forms of axon-axon interactions occur, one based on some sort of biochemical properties related to the axons' embryological identity and another based on the axons' electrical activity.

Tasks for the future are to identify the cellular bases of each of these mechanisms and to understand the situations in which each is manifested.

ACKNOWLEDGMENTS

We thank Dr. Jane Dodd and Dr. Sheryl Scott for their assistance with our discussion of sensory inputs to the spinal cord, and we also thank Mrs. Joan Holland and Mrs. Jean Seiler for secretarial services.

Literature Cited

Aebersold, H., Creuzfeld, O. D., Kühnt, U., Sanides, D. 1981. Representation of the visual field in the optic tract and the optic chiasma of the cat. *Exp. Brain Res.* 42: 127–45

Aitkin, L. M., Irvine, D. R. F., Webster, W. R. 1984. Central neural mechanisms of hearing. *Hand. Physiol.* 3(1): 675–737

Altman, J., Bayer, S. A. 1981. Time of origin of neurons of the rat inferior colliculus and the relations between cytogenesis and tonotopic order in the auditory pathway. *Exp. Brain Res.* 42: 411–23

Archer, S. M., Dubin, M. W., Stark, L. A. 1982. Abnormal development of kitten retino-geniculate connectivity in the absence of action potentials. *Science* 217: 743–45

Arnett, D. W. 1978. Statistical dependence between neighboring retinal ganglion cells in the goldfish. *Exp. Brain Res.* 32: 49–53

Arvidsson, J. 1982. Somatotopic organization of vibrissae afferents in the trigeminal sensory nuclei of the rat studied by transganglionic transport of HRP. *J. Comp. Neurol.* 211: 84–92

Ascher, P., Nowak, L. 1987. Electrophysiological studies of NMDA receptors. *Trends Neurosci.* 10: 284–88

Attardi, D. G., Sperry, R. W. 1963. Preferential selection of central pathways by regenerating optic fibers. *Exp. Neurol.* 7: 46–64

Barbera, A. J. 1975. Adhesive recognition between developing retinal cells and the optic tecta of the chick embryo. *Devel. Biol.* 46: 167–91

Beach, D. B., Jacobson, M. 1979. Patterns of cell proliferation in the retina of the clawed frog during development. *J. Comp. Neurol.* 183: 603–14

Beazley, L. D. 1979. Intertectal connections are not modified by visual experience in developing *Hyla moorei. Exp. Neurol.* 63: 411–19

Belford, G. R., Killackey, H. P. 1979a. Vibrissae representation in subcortical trigeminal centers of the neonatal rat. *J. Comp. Neurol.* 183: 305–22

Belford, G. R., Killackey, H. P. 1979b. The development of vibrissae representation in subcortical trigeminal centers of the neonatal rat. *J. Comp. Neurol.* 188: 63–74

Belford, G. R., Killackey, H. P. 1980. The sensitive period in the development of the trigeminal system of the neonatal rat. *J. Comp. Neurol.* 193: 335–50

Berman, N., Payne, B. R. 1983. Alterations in connections of the corpus callosum following convergent and divergent strabismus. *Brain Res.* 274: 201–12

Bodick, N., Levinthal, C. 1980. Growing optic nerve fibers follow neighbors during embryogenesis. *Proc. Natl. Acad. Sci. USA* 77: 4374–78

Bonhoeffer, F., Huf, J. 1982. *In vitro* experiments on axon guidance demonstrating an anterior-posterior gradient on the tectum. *EMBO J.* 1: 427–31

Bonhoeffer, F., Huf, J. 1985. Position-dependent properties of retinal axons and their growth cones. *Nature* 315: 409–10

Boss, V. C., Schmidt, J. T. 1984. Activity and the formation of ocular dominance patches in dually innervated tectum of goldfish. *J. Neurosci.* 4: 2891–2905

Brown, M. C., Jansen, J. K. S., Van Essen, D. 1976. Polyneuronal innervation of skeletal muscle in newborn rats and its elimination during maturation. *J. Physiol.* 261: 387–422

Bunt, S. M. 1982. Retinotopic and temporal organization of the optic nerve and tracts in the adult goldfish. *J. Comp. Neurol.* 206: 209–66

Bunt, S. M., Horder, T. J. 1983. Evidence for an orderly arrangement of optic axons within the optic nerves of the major nonmammalian verterbrate classes. *J. Comp. Neurol.* 213: 94–114

Bunt, S. M., Lund, R. D. 1982. Optic fiber arrangements in the visual pathways of Long-Evans hooded rats. *Soc. Neurosci. Abstr.* 8: 451

Chalupa, L. M., Rhoades, R. W. 1977. Responses of visual, somatosensory, and auditory neurones in the golden hamster's superior colliculus. *J. Physiol.* 270: 595–626

Chung, S.-H., Cooke, J. 1978. Observations on the formation of the brain and of nerve connections following embryonic manipulation of the amphibian neural tube. *Proc. R. Soc. London Ser. B* 201: 335–73

Cima, C., Grant, P. 1982. Development of the optic nerve in *Xenopus laevis*. II: Gliogenesis, myelination and metamorphic remodelling. *J. Embryol. Exp. Morphol.* 72: 255–78

Cline, H. T., Debski, E. A., Constantine-Paton, M. 1987. *N*-methyl-D-aspartate receptor antagonist desegregates eye-specific stripes. *Proc. Natl. Acad. Sci. USA* 84: 4342–45

Constantine-Paton, M. 1981. Induced ocular-dominance zones in tectal cortex. In *Organization of the Cerebral Cortex*, ed. F. O. Schmitt, F. G. Worden, G. Adelman, S. G. Dennis, pp. 47–67. Cambridge: MIT Press

Constantine-Paton, M., Blum, A. S., Mendez-Otero, R., Barnstable, C. J. 1986. A cell surface molecule distributed in a dorsoventral gradient in the perinatal rat retina. *Nature* 324: 459–62

Constantine-Paton, M., Ferrari-Eastman, P. 1981. Topographic and morphometric effects of bilateral embryonic eye removal on the optic tectum and nucleus isthmus of the leopard frog. *J. Comp. Neurol.* 196: 645–61

Cook, J. E. 1979. Interactions between optic fibres controlling the locations of their terminals in the goldfish optic tectum. *J. Embryol. Exp. Morphol.* 52: 89–103

Cook, J. E., Rankin, E. C. C. 1986. Impaired refinement of the regenerated retinotectal projection of the goldfish in stroboscopic light: A quantitative WGA-HRP study.

Exp. Brain Res. 63: 421–30

Cooke, J., Gaze, R. M. 1983. The positional coding system in the early eye rudiment of *Xenopus laevis*, and its modification after grafting operations. *J. Embryol. Exp. Morphol.* 77: 53–71

Cowan, W. M., Fawcett, J. W., O'Leary, D. D. M., Stanfield, B. B. 1984. Regressive events in neurogenesis. *Science* 225: 1258–65

Crandall, J. E., Caviness, V. S. 1984. Thalamocortical connections in newborn mice. *J. Comp. Neurol.* 228: 542–56

Crossland, W. J., Cowan, W. M., Rogers, L. A., Kelly, J. P. 1974. The specification of the retino-tectal projection in the chick. *J. Comp. Neurol.* 155: 127–64

Cull-Candy, S. G., Usowicz, M. M. 1987. Multiple-conductance channels activated by excitatory amino acids in cerebellar neurons. *Nature* 325: 525–28

Davis, M. R., Constantine-Paton, M. 1983. Central and peripheral connectivity patterns. *J. Comp. Neurol.* 221: 453–65

Dawson, D. R., Killackey, H. P. 1985. Distinguishing topography and somatotopy in the thalamocortical projections of the developing rat. *Dev. Brain Res.* 17: 309–13

Dehay, C., Kennedy, H., Bullier, J. 1986. Callosal connectivity of V1 and V2 in the newborn monkey. *J. Comp. Neurol.* 254: 20–33

Dodd, J. Jessell, T. M. 1986. Cell surface glycoconjugates and carbohydrate-binding proteins: Possible recognition signals in sensory neurone development. *J. Exp. Biol.* 124: 225–38

Dräger, U. C. 1985. Birth dates of retinal ganglion cells giving rise to the crossed and uncrossed optic projections in the mouse. *Proc. R. Soc. London Ser. B* 224: 57–77

Dräger, U. C., Hubel, D. H. 1975. Physiology of visual cells in mouse superior colliculus and correlation with somatosensory and auditory input. *Nature* 253: 203–4

Durham, D., Woolsey, T. A. 1978. Acute whisker removal reduces neuronal activity in barrels of mouse SmI cortex. *J. Comp. Neurol.* 178: 629–44

Durham, D., Woolsey, T. A. 1984. Effects of neonatal whisker lesions on mouse central trigeminal pathways. *J. Comp. Neurol.* 223: 424–47

Durham, D., Woolsey, T. A. 1985. Functional organization in cortical barrels of normal and vibrissae-damaged mice: A (^{3}H)2-deoxyglucose study. *J. Comp. Neurol.* 235: 97–110

Easter, S. S. Jr., Rusoff, A. C., Kish, P. E. 1981. The growth and organization of the optic nerve and tract in juvenile and adult goldfish. *J. Neurosci.* 1: 793–811

318   UDIN & FAWCETT

Easter, S. S. Jr., Stuermer, C. A. O. 1984. An evaluation of the hypothesis of shifting terminals in the goldfish optic tectum. *J. Neurosci.* 4: 1052–63

Eide, A-L., Jansen, J. K. S., Ribchester, R. R. 1982. The effect of lesions in the neural crest in the formation of synaptic connextions in the embryonic chick spinal cord. *J. Physiol.* 324: 453–78

Elberger, A. J., Smith, E. L., White, J. M. 1983. Spatial dissociation of visual inputs alters the origin of the corpus callosum. *Neurosci. Lett.* 35: 19–24

Emmers, R. 1965. Organization of the first and second somesthetic regions (SI and SII) in the rat thalamus. *J. Comp. Neurol.* 124: 215–27

Erzurumlu, R. S., Killackey, H. P. 1983. Development of order in the rat trigeminal system. *J. Comp. Neurol.* 213: 365–80

Fawcett, J. 1981. How axons grown down the *Xenopus* optic nerve. *J. Embryol. Exp. Morphol.* 65: 219–33

Fawcett, J. W. 1985. Factors guiding regenerating retinotectal fibres in the frog *Xenopus laevis. J. Embryol. Exp. Morphol.* 90: 233–50

Fawcett, J. W., Cowan, W. M. 1985. On the formation of eye dominance stripes and patches in the doubly innervated optic tectum of the chick. *Devel. Brain Res.* 17: 149–63

Fawcett, J. W., Gaze, R. M. 1982. The retinotectal fibre pathways from normal and compound eyes in *Xenopus. J. Embryol. Exp. Morphol.* 72: 19–37

Fawcett, J. W., O'Leary, D. D. M. 1985. The role of electrical activity in the formation of topographic maps in the nervous system. *Trends Neurosci.* 8: 201–6

Fawcett, J. W., O'Leary, D. D. M., Cowan, W. M. 1984a. Activity and the control of ganglion cell death in the rat retina. *Proc. Natl. Acad. Sci. USA* 81: 5589–93

Fawcett, J. W., Taylor, J. S. H., Gaze, R. M., Grant, P., Hirst, E. 1984b. Fibre order in the normal *Xenopus* optic tract, near the chiasma. *J. Embryol. Exp. Morphol.* 83: 1–14

Fawcett, J. W., Willshaw, D. J. 1982. Compound eyes project stripes on the optic tectum on *Xenopus. Nature* 296: 350–52

Feldman, J. D., Gaze, R. M. 1975. The development of the retinotectal projection in *Xenopus* with one compound eye. *J. Embryol. Exp. Morphol.* 33: 775–87

Feldman, J. D., Gaze, R. M., Keating, M. J. 1971. Delayed innervation of the optic tectum in *Xenopus laevis. Exp. Brain Res.* 14: 16–23

Feldman, J. D., Gaze, R. M., Keating, M. J. 1983. Development of the orientation of the visuo-tectal map in *Xenopus. Devel.*

Brain Res. 6: 269–77

Finlay, B. L., Schneps, S. E., Schneider, G. E. 1979. Orderly compression of the retinotectal projection following partial tectal ablation in the newborn hamster. *Nature* 280: 153–55

Finlay, B. L., Wilson, K. G., Schneider, G. E. 1979. Anomalous ipsilateral retinotectal projections in Syrian hamsters with early lesions: Topography and functional capacity. *J. Comp. Neurol.* 183: 721–40

Frank, E., Jackson, P. C. 1986. Normal electrical activity is not required for the formation of specific sensory-motor synapses. *Brain Res.* 378: 147–51

Frank, E., Westerfield, M. 1982. The formation of appropriate central and peripheral connexions by foreign sensory neurons of the bullfrog. *J. Physiol.* 324: 495–505

Frank, E., Westerfield, M. 1983. Development of sensory-motor synapses in the spinal cord of the frog. *J. Physiol.* 343: 593–610

Fraser, S. E. 1980. A differential adhesion approach to the patterning of nerve connections. *Devel. Biol.* 79: 453–64

Fraser, S. E. 1983. Fiber optic mapping of the *Xenopus* visual system: Shift in the retinotectal projection during development. *Devel. Biol.* 95: 505–11

Fraser, S. E. 1985. Cell interactions involved in neuronal patterning: An experimental and theoretical approach. In *Molecular Bases of Neural Development*, ed. G. M. Edelman, W. E. Gall, W. M. Cowan, pp. 481–507. New York: Neurosci. Res. Found.

Fraser, S. E., Hunt, R. K. 1980. Retinotectal specificity: Models and experiments in search of a mapping function. *Ann. Rev. Neurosci.* 3: 319–52

Fraser, S. E., Hunt, R. K. 1986. A physiological measure of shifting connection in the *Rana pipiens* retinotectal system. *J. Embryol. Exp. Morphol.* 94: 149–61

Frost, D. O., Schneider, G. E. 1979. Plasticity of retinofugal projections after partial lesions of the retina in newborn Syrian hamsters. *J. Comp. Neurol.* 185: 517–67

Frost, D. O., So, K.-F., Schneider, G. E. 1979. Postnatal development of retinal projections in syrian hamsters: A study using autoradiographic and anterograde degeneration techniques. *Neuroscience* 4: 1649–77

Fujisawa, H. 1987. Mode of growth of retinal axons within the tectum of *Xenopus* tadpoles, and implications in the ordered neuronal connection between the retina and the tectum. *J. Comp. Neurol.* 260: 127–39

Fujisawa, H., Tani, N., Watanabe, K., Ibata, Y. 1981a. Retinotopic analysis of fibre pathways in amphibians. II. The frog *Rana nigromaculata*. *Brain Res.* 206: 21–26

Fujisawa, H., Tani, N., Watanabe, K., Ibata, Y. 1982. Branching of regenerating retinal axons and preferential selection of appropriate branches for specific neuronal connections in the newt. *Devel. Biol.* 90: 43–57

Fujisawa, H., Watanabe, K., Tani, N., Ibata, Y. 1981b. Retinotopic analysis of fiber pathways in amphibians I. The adult newt, *Cynops pynogaster*. *Brain Res.* 206: 9–20

Fukuda, Y., Hsiao, C.-F., Sawai, H., Wakakuwa, K. 1984. Retinotopic organizations of the expanded ipsilateral projection to the rat's superior colliculus: Variations along its rostrocaudal axis. *Brain Res.* 321: 390–95

Gaze, R. M., Fawcett, J. W. 1983. Pathways of *Xenopus* optic fibres regenerating from normal and compound eyes under various conditions. *J. Embryol. Exp. Morphol.* 73: 17–38

Gaze, R. M., Feldman, J. D., Cooke, J., Chung, S.-H. 1979a. The orientation of the visuotectal map in *Xenopus*: Developmental aspects. *J. Embryol. Exp. Morphol.* 53: 39–66

Gaze, R. M., Hope, R. A. 1983. The visuotectal projection following translocation of grafts within an optic tectum in the goldfish. *J. Physiol.* 344: 257–75

Gaze, R. M., Jacobson, M. 1963. A study of the retinotectal projection during regeneration of the optic nerve in the frog. *Proc. R. Soc. London Ser. B* 157: 420–48

Gaze, R. M., Jacobson, M., Székely, G. 1963. The retinotectal projection in *Xenopus* with compound eyes. *J. Physiol.* 165: 484–99

Gaze, R. M., Keating, M. J., Chung, S.-H. 1974. The evolution of the retinotectal map during development in *Xenopus*. *Proc. R. Soc. London Ser. B* 185: 301–30

Gaze, R. M., Keating, M. J., Ostberg, A., Chung, S.-H. 1979b. The relationship between retinal and tectal growth in larval *Xenopus*. Implications for the development of the retinotectal projection. *J. Embryol. Exp. Morphol.* 53: 103–43

Gaze, R. M., Keating, M. J., Székely, G., Beazley, L. 1970. Binocular interaction in the formation of specific intertectal neuronal connexions. *Proc. R. Soc. London Ser. B* 175: 107–47

Gaze, R. M., Sharma, S. C. 1970. Axial differences in the reinnervation of the goldfish optic tectum by regenerating optic nerve fibres. *Exp. Brain Res.* 10: 171–81

Gaze, R. M., Straznicky, C. 1980. Regeneration of optic nerve fibres from a compound eye to both tecta in *Xenopus*: Evidence relating to the state of specification of the eye and the tectum. *J. Embryol. Exp. Morphol.* 60: 125–40

Ginsburg, K. S., Johnsen, J. A., Levine, M. W. 1984. Common noise in the firing of neighbouring ganglion cells in goldfish retina. *J. Physiol.* 351: 433–51

Glasser, S., Ingle, D. 1978. The nucleus isthmus as a relay station in the ipsilateral visual projection to the frog's optic tectum. *Brain Res.* 159: 214–18

Godement, P., Salaun, J., Imbert, M. 1984. Prenatal and postnatal development of retinogeniculate and retinocollicular projections in the mouse. *J. Comp. Neurol.* 230: 552–75

Graybiel, A. M. 1975. Anatomical organization of retinotectal afferents in the cat: An autoradiographic study. *Brain Res.* 96: 1–24

Graybiel, A. M. 1978. A satellite system of the superior colliculus: The parabigeminal nucleus and its projections to the superficial collicular layers. *Brain Res.* 145: 365–74

Grobstein, P., Comer, C., Hollyday, M., Archer, S. M. 1978. A crossed isthmotectal projection in *Rana pipiens* and its involvement in the ipsilateral visuo-tectal projection. *Brain Res.* 156: 117–23

Gruberg, E. R., Udin, S. B. 1978. Topographic projections between the nucleus isthmi and the tectum of the frog *Rana pipiens*. *J. Comp. Neurol.* 179: 487–500

Guillery, R. W., Polley, E. H., Torrealba, F. 1982. The arrangement of axons according to fiber diameter in the optic tract of the cat. *J. Neurosci.* 2: 714–21

Halfter, W., Claviez, M., Schwarz, U. 1981. Preferential adhesion of tectal membranes to anterior embryonic chick retina neurites. *Nature* 292: 67–70

Harris, L. R., Blakemore, C., Donaghy, M. 1980. Integration of visual and auditory space in the mammalian superior colliculus. *Nature* 288: 56–59

Harris, W. A. 1980. The effects of eliminating impulse activity on the development of the retinotectal projection in salamanders. *J. Comp. Neurol.* 194: 303–17

Harris, W. A. 1982. The transplantation of eyes to genetically eyeless salamanders: Visual projections and somatosensory interactions. *J. Neurosci.* 2: 339–53

Harris, W. A. 1983. Differences between embryos and adults in the plasticity of somatosensory afferents to the axolotl tectum. *Devel. Brain Res.* 7: 245–55

Harris, W. A. 1984. Axonal pathfinding in the absence of normal pathways and impulse activity. *J. Neurosci.* 4: 1153–62

Harting, J. K., Guillery, R. W. 1976. Organization of the retinocollicular pathways in the cat. *J. Comp. Neurol.* 166: 133–44

Hollyday, M., Mendell, L. 1975. Area specific reflexes from normal and supernumerary hind limbs of *Xenopus laevis*. *J. Comp. Neurol.* 162: 205–20

Holt, C. E. 1980. Cell movements in *Xenopus* eye development. *Nature* 287: 850–52

Holt, C. E. 1984. Does timing of axon outgrowth influence initial retinotectal topography in *Xenopus? J. Neurosci.* 4: 1130–52

Holt, C. E., Harris, W. A. 1983. Order in the initial retinotectal map in *Xenopus*: A new technique for labelling growing nerve fibres. *Nature* 301: 150–52

Holt, C. E., Thompson, I. A. 1984. The effects of tetrodotoxin on the development of hamster retinal projections. *J. Physiol.* 357: 24P

Honig, M. G. 1982. The development of sensory projection patterns in embryonic chick hind limb. *J. Physiol.* 330: 175–202

Horder, T. J. 1971. Retention, by fish optic nerve fibres regenerating to new terminal sites on the tectum, of chemospecific affinity for their original sites. *J. Physiol.* 216: 53–55P

Horder, T. J., Martin, K. A. C. 1978. Morphogenetics as an alternative to chemnospecificity in the formation of nerve connections. In *Cell-Cell Recognition, 32nd Symp. Soc. for Exp. Biol.*, ed, A. S. G. Curtis, pp. 275–358. Cambridge: Cambridge Univ. Press

Horton, J. C., Greenwood, M. M., Hubel, D. 1979. Non-retinotopic arrangement of fibres in the cat optic nerve. *Nature* 282: 720–22

Hoskins, S. G., Grobstein, P. 1985. Development of the ipsilateral retinothalamic projection in the frog *Xenopus laevis*. I. Retinal distribution of ipsilaterally projecting cells in normal and experimentally manipulated frogs. *J. Neurosci.* 5: 911–19

Humphrey, M. F., Beazley, L. D. 1982. An electrophysiological study of early retinotectal projection patterns during optic nerve regeneration in *Hyla moorei*. *Brain Res.* 239: 595–602

Hunt, R. K., Jacobson, M. 1972. Development and stability of positional information in *Xenopus* retinal ganglion cells. *Proc. Natl. Acad. Sci. USA* 69: 780–83

Hunt, R. K., Jacobson, M. 1973. Neuronal locus specificity: Altered pattern of spatial deployment in fused fragments of embryonic *Xenopus* eyes. *Science* 180: 509–11

Hunt, R. K., Jacobson, M. 1974. Development of neuronal locus specificity in *Xenopus* retinal ganglion cells after surgical eye transection or the fusion of whole

eyes. *Devel. Biol.* 40: 1–15

Hunt, S. P., Künzle, H. 1976. Observations on the projections and intrinsic organization of the pigeon optic tectum: An autoradiographic study based on anterograde and retrograde, axonal and dendritic flow. *J. Comp. Neurol.* 170: 153–72

Ide, C. F., Fraser, S. E., Meyer, R. L. 1983. Eye dominance columns from an isogenic double-nasal frog eye. *Science* 221: 293–95

Ide, C. F., Kosofsky, B. E., Hunt, R. K. 1979. Control of pattern duplication in the retinotectal system of *Xenopus*: Suppression of duplication by eye-fragment interactions. *Devel. Biol.* 69: 337–60

Ide, C. F., Reynolds, P., Tompkins, R. 1984. Two healing patterns correlate with different adult neural connectivity patterns in regenerating embryonic *Xenopus* retina. *J. Exp. Zool.* 230: 71–80

Ingle, D., Dudek, A. 1977. Aberrant retinotectal projections in the frog. *Exp. Neurol.* 55: 567–82

Innocenti, G. M. 1981. Growth and reshaping of axons in the establishment of visual callosal connections. *Science* 212: 824–27

Innocenti, G. M., Clarke, S. 1983. Multiple sets of visual cortical neurons projecting transitorily through the corpus callosum. *Neurosci. Lett.* 41: 27–32

Innocenti, G. M., Clarke, S., Kraftsik, R. 1986. Interchange of callosal and association projections in the developing visual cortex. *J. Neurosci.* 6: 1384–1409

Innocenti, G. M., Frost, D. O. 1979. Effects of visual experience on the maturation of the efferent system to the corpus callosum. *Nature* 280: 231–33

Insausti, R., Blakemore, C., Cowan, W. M. 1984. Ganglion cell death during development of ipsilateral retino-collicular projection in golden hamster. *Nature* 308: 362–65

Ito, M., Seo, M. L. 1983. Avoidance of neonatal cortical lesions by developing somatosensory barrels. *Nature* 301: 600–2

Jackson, H., Hackett, J. T., Rubel, E. W. 1982. Organization and development of brain stem auditory nuclei in the chick: Ontogeny of postsynaptic responses. *J. Comp. Neurol.* 210: 80–86

Jackson, H., Parks, T. N. 1982. Functional synapse elimination in the developing avian cochlear nucleus with simultaneous reduction in cochlear nerve axon branching. *J. Neurosci.* 2: 1736–43

Jacobson, M. 1967. Retinal ganglion cells: Specification of central connections in larval *Xenopus laevis*. *Science* 155: 1106–8

Jacobson, M., Hirsch, H. V. B. 1973. Development and maintenance of connectivity in the visual system of the frog. II. The effects of eye rotation and visual deprivation. *Brain Res.* 49: 47–65

Jacobson, M., Levine, R. 1975. Plasticity in the adult frog brain: Filling in the visual scotoma after excision or translocation of parts of the optic tectum. *Brain Res.* 88: 339–45

Jahr, C. E., Stevens, C. F. 1987. Glutamate activates multiple single channel conductances in hippocampal neurons. *Nature* 325: 522–25

Jay, M. F., Sparks, D. L. 1984. Auditory receptive fields in primate superior colliculus shift with changes in eye position. *Nature* 309: 345–47

Jay, M. F., Sparks, D. L. 1987. Sensorimotor integration in the primate superior colliculus. II. Coordinates of auditory signals. *J. Neurophysiol.* 57: 35–55

Jeanmonod, D., Rice, F. L., Van der Loos, H. 1977. Mouse somatosensory cortex: Development of the alterations in the barrelfield which are caused by injury to the vibrissal follicles. *Neurosci. Lett.* 6: 151–56

Jeanmonod, D., Rice, F. L., Van der Loos, H. 1981. Mouse somatosensory cortex: Alterations in the barrelfield following receptor injury at different early postnatal ages. *Neuroscience* 6: 1503–35

Jeffery, G., Perry, V. H. 1982. Evidence for ganglion cell death during development of the ipsilateral retinal projection in the rat. *Devel. Brain Res.* 2: 176–80

Jhaveri, S., Morest, D. K. 1982. Sequential alterations of neuronal architecture in nucleus magnocellularis of the developing chicken: A Golgi study. *Neuroscience* 7: 837–53

Johns, P. R., Easter, S. S. Jr. 1977. Growth of the adult goldfish eye II. Increase in retinal cell number. *J. Comp. Neurol.* 176: 331–42

Keating, M. J. 1974. The role of visual function in the patterning of binocular visual connections. *Br. Med. Bull.* 30: 145–51

Keating, M. J., Feldman, J. 1975. Visual deprivation and intertectal neuronal connections in *Xenopus laevis. Proc. R. Soc. London Ser. B* 191: 467–74

Keating, M. J., Gaze, R. M. 1970. The ipsilateral retinotectal pathway in the frog. *Q. J. Exp. Physiol.* 55: 284–92

Kennard, C. 1981. Factors involved in the development of ipsilateral retinothalamic projections in *Xenopus laevis. J. Embryol. Exp. Morphol.* 65: 199–217

Kennard, C., Keating, M. J. 1985. A species difference between *Rana* and *Xenopus* in the occurrence of intertectal neuronal plasticity. *Neurosci. Lett.* 58: 365–70

Killackey, H. P., Belford, G. R. 1979. The formation of afferent patterns in the somatosensory cortex of the neonatal rat. *J. Comp. Neurol.* 183: 285–304

Killackey, H. P., Belford, G., Ryugo, R., Ryugo, D. K. 1976. Anomalous organization of thalamocortical projections consequent to vibrissae removal in the newborn rat and mouse. *Brain Res.* 104: 309–15

King, A. J., Hutchings, M. E., Moore, D. R., Blakemore, C. 1985. The effect of neonatal eye rotation on the representation of auditory space in the ferret superior colliculus. *Soc. Neurosci. Abstr.* 11: 450

Kliot, M., Shatz, C. J. 1985. Abnormal development of the retinogeniculate projection in Siamese cats. *J. Neurosci.* 10: 2641–53

Knudsen, E. I. 1981. The hearing of the barn owl. *Sci. Am.* 245: 112–25

Knudsen, E. I. 1982. Auditory and visual maps of space in the optic tectum of the owl. *J. Neurosci.* 2: 1177–94

Knudsen, E. I. 1983. Early auditory experience aligns the auditory map of space in the optic tectum of the barn owl. *Science* 222: 939–42

Knudsen, E. I. 1985. Experience alters the spatial tuning of auditory units in the optic tectum during a sensitive period in the barn owl. *J. Neurosci.* 5: 3094–3109

Knudsen, E. I., Knudsen, P. F. 1983. Space-mapped auditory projections from the inferior colliculus to the optic tectum in the barn owl (*Tyto alba*). *J. Comp. Neurol.* 218: 187–96

Knudsen, E. I., Knudsen, P. F. 1985. Vision guides the adjustment of auditory localization in young barn owls. *Science* 230: 545–48

Kossut, M., Hand, P. 1984a. Early development of changes in cortical representation of C3 vibrissa following neonatal denervation of surrounding vibrissa receptors: A 2-deoxyglucose study in the rat. *Neurosci. Lett.* 36: 7–12

Kossut, M., Hand, P. 1984b. The development of the vibrissal cortical column: A 2-deoxyglucose study in the rat. *Neurosci. Lett.* 46: 1–6

Land, P. W., Lund, R. D. 1979. Development of the uncrossed retinotectal pathway, and its relation to plasticity studies. *Science* 205: 698–700

Landmesser, L., Honig, M. G. 1986. Altered sensory projections in the chick hind limb following the early removal of motoneurons. *Devel. Biol.* 118: 511–31

Law, M. I., Constantine-Paton, M. 1980. Right and left eye bands in frogs with unilateral tectal ablations. *Proc. Natl. Acad. Sci. USA* 77: 2314–18

Lázár, G. 1973. The development of the

optic tectum in the frog *Xenopus laevis*: A Golgi study. *J. Anat.* 116: 347–55

Levine, R., Jacobson, M. 1974. Deployment of optic nerve fibers is determined by positional markers in the frog's tectum. *Exp. Neurol.* 43: 527–38

Levine, R. L., Jacobson, M. 1975. Discontinuous mapping of retina with tectum innervated by both eyes. *Brain Res.* 98: 172–76

Lichtman, J. W., Frank, E. 1984. Physiological evidence for specificity of synaptic connections between individual sensory and motor neurons in the brachial cord of the bullfrog. *J. Neurosci.* 4: 1745–53

Lichtman, J. W., Jhaveri, S., Frank, E. 1984. Anatomical basis of specific connections between sensory axons and motor neurons in the brachial spinal cord of the bullfrog. *J. Neurosci.* 4: 1754–63

Lipton, S. A., Harcourt, P. 1984. The effect of tetrodotoxin on the death of mammalian retinal ganglion cells. *Soc. Neurosci. Abstr.* 10: 1081

Liu, L., Layer, P. G., Gierer, A. 1983. Binding of FITC-coupled peanut-agglutinin (FITC-PNA) to embryonic chicken retina reveals developmental spatio-temporal patterns. *Devel. Brain Res.* 8: 223–29

Lund, R. D., Chang, F.-L. F., Land, P. W. 1984. The development of callosal projections in normal and one-eyed rats. *Devel. Brain Res.* 14: 139–42

Lund, R. D., Mitchell, D. E. 1979. Asymmetry in the visual callosal connections of strabismic cats. *Brain Res.* 167: 176–79

Lund, R. D., Mitchell, D. E., Henry, G. H. 1978. Squint-induced modification of callosal connections in cats. *Brain Res.* 144: 169–72

MacDermott, A. B., Mayer, M. L., Westbrook, G. L., Smith, S. J., Barker, J. L. 1986. NMDA-receptor activation increases cytoplasmic calcium concentration in cultured spinal cord neurones. *Nature* 321: 519–22

Marchase, R. B. 1977. Biochemical investigations of retinotectal adhesive specificity. *J. Cell Biol.* 75: 237–57

Martin, P. R., Sefton, A. J., Dreher, B. 1983. The retinal location and fate of ganglion cells which project to the ipsilateral superior colliculus in neonatal albino and hooded rats. *Neurosci. Lett.* 41: 219–26

Masland, R. H. 1977. Maturation of function in the developing rabbit retina. *J. Comp. Neurol.* 175: 275–86

Mastronarde, D. N. 1983. Correlated activity of cat retinal ganglion cells. I. Spontaneously active inputs to X and Y cells. *J. Neurophysiol.* 49: 303–15

Mastronarde, D. N. 1984. Organization of the cat's optic tract as assessed by single axon recordings. *J. Comp. Neurol.* 227: 14–22

McArdle, C. B., Dowling, J. E., Masland, R. H. 1977. Development of outer segments and synapses in the rabbit retina. *J. Comp. Neurol.* 175: 253–74

McLoon, S. C. 1982. Alterations in the precision of the crossed retinotectal projection during tectal development. *Science* 215: 1418–20

McLoon, S. C. 1985. Evidence for shifting connections during the development of the chick retinotectal projection. *J. Neurosci.* 5: 2570–80

Melzer, P., Welker, E., Dorfl, J., Van der Loos, H. 1986. Development of order in the whisker-to-barrel pathway of the mouse: Neuronal metabolic responses to whisker stimulation. *Soc. Neurosci. Abstr.* 12: 953

Mendell, L. M. 1984. Modifiability of spinal synapses. *Physiol. Rev.* 64: 260–324

Meredith, M. A., Stein, B. E. 1986. Visual, auditory and somatosensory convergence on cells in superior colliculus results in multisensory integration. 1986. *J. Neurophysiol.* 56: 640–62

Meyer, R. L. 1977. Eye-in-water electrophysiological mapping of goldfish with and without tectal lesions. *Exp. Neurol.* 56: 23–41

Meyer, R. 1980. Mapping the normal and regenerating retinotectal projection of goldfish with an autoradiographic method. *J. Comp. Neurol.* 189: 273–89

Meyer, R. L. 1982a. Tetrodotoxin blocks the formation of ocular dominance columns in goldfish. *Science* 218: 589–91

Meyer, R. L. 1982b. Ordering of retinotectal connections: A multivariate operational analysis. *Curr. Top. Devel. Biol.* 17: 101–45

Meyer, R. L. 1983. Tetrodotoxin inhibits the formation of refined retinotopography in goldfish. *Devel. Brain Res.* 6: 293–98

Meyer, R. L., Wolcott, L. L. 1984. Retinotopic expansion and compression in the retinotectal system of goldfish in the absence of impulse activity. *Soc. Neurosci. Abstr.* 10: 467

Middlebrooks, J. C., Knudsen, E. I. 1984. A neural code for auditory space in the cat's superior colliculus. *J. Neurosci.* 4: 2621–34

Miner, N. 1956. Integumetal specification of sensory fibers in the development of cutaneous local sign. *J. Comp. Neurol.* 105: 161–70

Mooney, R. D., Klein, B. G., Rhoades, R. W. 1987. Effects of altered visual input upon the development of the visual and somatosensory representations in the hamster's superior colliculus. *Neurosci.* 20: 537–55

Mooney, R. D., Rhoades, R. W., Fish, S. E. 1984. Neonatal superior collicular lesions alter visual callosal development in hamster. *Exp. Brain Res.* 55: 9–25

Nord, S. G. 1967. Somatotopic organization in the spinal trigeminal nucleus, the dorsal column nuclei and related structures in the rat. *J. Comp. Neurol.* 130: 343–56

Olavarria, J., Malach, R., van Sluyters, R. C. 1987. Development of visual callosal connections in neonatally enucleated rats. *J. Comp. Neurol.* 260: 321–48

Olavarria, J., Van Sluyters, R. C. 1985. Organization and postnatal development of callosal connections in the visual cortex of the rat. *J. Comp. Neurol.* 239: 1–26

O'Leary, D. D. M., Cowan, W. M. 1983. Topographic organization of certain tectal afferent and efferent connections can develop normally in the absence of retinal input. *Proc. Natl. Acad. Sci. USA* 80: 6131–35

O'Leary, D. D. M., Fawcett, J. W., Cowan, W. M. 1986. Topographic targeting errors in the retinocollicular projection and their elimination by selective ganglion cell death. *J. Neurosci.* 6: 3692–3705

O'Rourke, N. A., Fraser, S. E. 1986a. Dynamic aspects of retinotectal map formation revealed by a vital-dye fiber-tracing technique. *Devel. Biol.* 114: 265–76

O'Rourke, N. A., Fraser, S. E. 1986b. Pattern regulation in the eyebud of *Xenopus* studied with a vital-dye fiber-tracing technique. *Devel. Biol..* 114: 277–88

Palmer, A. R., King, A. J. 1982. The representation of auditory space in the mammalian superior colliculus. *Nature* 299: 248–49

Pidoux, B., Verley, R., Farkas, E., Scherrer, J. 1979. Projections of the common fur of the muzzle upon the cortical area for mystacial vibrissae in rats dewhiskered since birth. *Neurosci. Lett.* 11: 301–6

Piper, E. A., Steedman, J. G., Stirling, R. V. 1980. Three-dimensional computer reconstruction of cobalt stained optic fibres in whole brains of *Xenopus* tadpoles. *J. Physiol.* 300: 13P

Rager, G. 1976. Morphogenesis and physiogenesis of the retino-tectal connection in the chicken. *Proc. R. Soc. London Ser. B* 192: 331–52

Rager, G. 1980. Development of the retinotectal projection in the chicken. *Adv. Anat. Embryol. Cell Biol.* 63: 1–90

Rager, G., Rager, U. 1978. Systems matching by degeneration. 1. A quantitative electron microscopic study of the generation and degeneration of ganglion cells in the chicken. *Exp. Brain Res.* 33: 65–78

Rankin, E. C. C., Cook, J. E. 1986. Topographic refinement of the regenerating retinotectal projection of the goldfish in standard laboratory conditions: A quantitative WGA-HRP study. *Exp. Brain Res.* 63: 409–20

Rauschecker, J. P., Harris, L. R. 1983. Auditory compensation of the effects of visual deprivation in the cat's superior colliculus. *Exp. Brain Res.* 50: 69–83

Raymond, P. A., Easter, S. S. Jr. 1983. Postembryonic growth of the optic tectum in goldfish. 1. Location of germinal cells and numbers of neurons produced. *J. Neurosci.* 3: 1077–91

Reese, B. E. 1986. The topography of expanded uncrossed retinal projections following neonatal enucleation of one eye: Differing effects in dorsal lateral geniculate nucleus and superior colliculus. *J. Comp. Neurol.* 250: 8–32

Reese, B. E., Guillery, R. W. 1987. Distribution of axons according to diameter in the monkey's optic tract. *J. Comp. Neurol.* 260: 453–60

Reh, T. A., Constantine-Paton, M. 1984. Retinal ganglion cell terminals change their projection sites during larval development of *Rana pipiens*. *J. Neurosci.* 4: 442–57

Reh, T. A., Constantine-Paton, M. 1985. Eye specific segregation requires neural activity in 3 eyed *Rana pipiens*. *J. Neurosci.* 5: 1132–43

Reh, T. A., Pitts, E. C., Constantine-Paton, M. 1983. The organization of the fibers in the optic nerve of normal and tectum-less *Rana pipiens*. *J. Comp. Neurol.* 218: 282–96

Rhoades, R. W. 1980. Effects of neonatal enucleation on the functional organization of the superior colliculus in the golden hamster. *J. Physiol.* 301: 383–99

Rhoades, R. W., Della Croce, D. D., Meadows, I. 1981. Reorganization of somatosensory input to superior colliculus in neonatally enucleated hamsters: Anatomical and electrophysiological experiments. *J. Neurophysiol.* 46: 855–77

Rubel, E. W., Smith, D. J., Miller, L. C. 1976. Organization and development of brain stem auditory nuclei of the chicken: Ontogeny of the *N. magnocellularis* and *N. laminaris*. *J. Comp. Neurol.* 166: 469–90

Rusoff, A. C. 1984. Patterns of axons in the visual system of perciform fish and implications of these paths for rules governing axonal growth. *J. Neurosci.* 4: 1414–28

Rusoff, A. C., Easter, S. S. 1980. Order in the optic nerve of goldfish. *Science* 208: 311–12

Sachs, G. M., Jacobson, M., Caviness, V. S. 1986. Postnatal changes in arborization patterns of murine retinocollicular axons. *J. Comp. Neurol.* 246: 395–408

Sah, D. W. Y., Frank, E. 1984. Regeneration of sensory-motor synapses in the spinal cord of the bullfrog. *J. Neurosci.* 4: 2784–91

Sakaguchi, D., Murphey, R. 1985. Map formation in the developing *Xenopus* retinotectal system: An examination of ganglion cell terminal arborizations. *J. Neurosci.* 5: 3228–45

Sanes, D. H., Constantine-Paton, M. 1985. The sharpening of frequency tuning curves requires patterned activity during development in the mouse, *Mus musculus. J. Neurosci.* 5: 1152–66

Scalia, F., Arango, V. 1983. The antiretinotopic organization of the frog's optic nerve. *Brain Res.* 266: 121–26

Scalia, F., Fite, K. 1974. A retinotopic analysis of the central connections of the optic nerve in the frog. *J. Comp. Neurol.* 158: 455–78

Schmidt, J. T. 1983. Regeneration of the retinotectal projection following compression onto a half tectum in goldfish. *J. Embryol. Exp. Morphol.* 77: 39–51

Schmidt, J. T., Buzzard, M. J., Turcotte, J. 1984. Morphology of regenerated optic arbors in goldfish tectum. *Soc. Neurosci. Abstr.* 10: 667

Schmidt, J. T., Cicerone, C. M., Easter, S. S. 1978. Expansion of the half retinal projection to the tectum in goldfish: An electrophysiological and anatomical study. *J. Comp. Neurol.* 177: 257–78

Schmidt, J. T., Edwards, D. L. 1983. Activity sharpens the map during the regeneration of the retinotectal projection in goldfish. *Brain Res.* 209: 29–39

Schmidt, J. T., Eisele, L. E. 1985. Stroboscopic illumination and dark rearing block the sharpening of the regenerated retinotectal map in goldfish. *Neurosci.* 14: 535–46

Schneider, G. E., Jhaveri, S. R. 1974. Neuroanatomical correlates of spared or altered function after brain lesions in newborn hamster. In *Plasticity and Recovery of Function in the Central Nervous System*, ed. D. Stein, J. Rosen, N. Butters, pp. 65–109. New York: Academic

Schneider, G. E., Jhaveri, S. 1984. Rapid postnatal establishment of topography in the hamster retinotectal projection. *Soc. Neurosci. Abstr.* 10: 467

Scholes, J. H. 1979. Nerve fibre topography in the retinal projection to the tectum. *Nature* 278: 620–24

Scott, S. A. 1982. The development of the segmental pattern of skin sensory innervation in embryonic chick hind limb. *J. Physiol.* 330: 203–20

Scott, S. A. 1984. The effects of neural crest deletions on the development of sensory innervation patterns in embryonic chick hind limb. *J. Physiol.* 352: 285–304

Scott, S. A. 1986. Skin sensory innervation patterns in embryonic chick hindlimb following dorsal root ganglion reversals. *J. Neurobiol.* 17: 649–68

Sharma, S. C. 1972a. Reformation of retinotectal projections after various tectal ablations in adult goldfish. *Exp. Neurol.* 34: 171–82

Sharma, S. C. 1972b. Retinotectal connexions of a heterotopic eye. *Nature New Biol.* 238: 286–87

Sharma, S. C. 1975. Visual projection in surgically created 'compound' tectum in adult goldfish. *Brain Res.* 93: 497–501

Sharma, S. C., Hollyfield, J. G. 1974. Specification of retinal central connections in *Rana pipiens* before the appearance of the first post-mitotic ganglion cells. *J. Comp. Neurol.* 155: 395–408

Shatz, C. J., Sretevan, D. W. 1986. Interaction between retinal ganglion cells during the development of the mammalian visual system. *Ann. Rev. Neurosci.* 9: 171–207

Sherk, H. 1978. Visual response properties and visual field topography in the cat's parabigeminal nucleus. *Brain Res.* 145: 375–79

Shipley, M. T. 1974. Response characteristics of single units in the rat's trigeminal nuclei to vibrissal displacement. *J. Neurophysiol.* 37: 73–90

Sikich, L., Woolsey, T. A., Johnson, E. M. Jr. 1986. Effect of uniform partial denervation of the periphery on the peripheral and central vibrissal system in guinea pigs. *J. Neurosci.* 6: 1227–40

Simons, D. J. 1978. Response properties of vibrissae units in rat SI somatosensory cortex. *J. Neurophysiol.* 41: 798–820

Simons, D. J., Land, P. W. 1986. Sensory deprivation by neonatal whisker trimming alters functional organization in SmI "barrel" cortex. *Soc. Neurosci. Abstr.* 12: 1435

Smith, C. L. 1983. The development and postnatal organization of primary afferent projections to the rat thoracic spinal cord. *J. Comp. Neurol.* 220: 24–43

Smith, C. L. 1986. Sensory neurons with rerouted peripheral axons make appropriate central connections. *Soc. Neurosci. Abstr.* 12: 452

Smith, C. L., Frank, E. 1987. Peripheral specification of sensory neurons transplanted to novel locations along the neuraxis. *J. Neurosci.* 7: 1537–49

Smith, D. J., Rubel, E. W. 1979. Organization and development of brain stem auditory nuclei of the chicken: Dendritic gradients in nucleus laminaris. *J. Comp. Neurol.* 186: 213–40

Sperry, R. W. 1944. Optic nerve regeneration with return of vision in anurans. *J. Neurophysiol.* 7: 57–69

Sretevan, D. W., Shatz, C. J. 1986. Prenatal development of cat retinogeniculate axon arbors in the absence of binocular interactions. *J. Neurosci.* 6: 990–1003

Stanfield, B. B., O'Leary, D. D. 1985. The transient corticospinal projection from the occipital cortex during the postnatal development of the rat. *J. Comp. Neurol.* 238: 236–48

Stanfield, B. B., O'Leary, D. D. M., Fricks, C. 1982. Selective collateral elimination in early postnatal development restricts cortical distribution of rat pyramidal tract neurones. *Nature* 298: 371–73

Steedman, J. G. 1981. *Pattern formation in the visual pathways of* Xenopus laevis. PhD thesis, Univ. London, London

Stein, B. E. 1984. Development of the superior colliculus. *Ann. Rev. Neurosci.* 7: 95–125

Stent, G. S. 1973. Physiological mechanism for Hebb's postulate of learning. *Proc. Natl. Acad. Sci. USA* 70: 997–1001

Straznicky, C. 1981. Mapping retinal projections from double nasal and double temporal compound eyes to dually innervated tectum in *Xenopus. Devel. Brain Res.* 1: 139–52

Straznicky, C., Gaze, R. M. 1971. The growth of the retina in *Xenopus laevis*: An autoradiographic study. *J. Embryol. Exp. Morphol.* 26: 67–79

Straznicky, C., Gaze, R. M. 1972. The development of the tectum in *Xenopus laevis*: An autoradiographic study. *J. Embryol. Exp. Morphol.* 28: 87–115

Straznicky, C., Gaze, R. M., Horder, T. J. 1979. Selection of appropriate medial branch of the optic tract by fibres of ventral retinal origin during development and regeneration: An autoradiographic study in *Xenopus. J. Embryol. Exp. Morphol.* 50: 253–67

Straznicky, K., Gaze, R. M., Keating, M. J. 1974. The retinotectal projection from a double ventral compound eye in *Xenopus. J. Embryol. Exp. Morphol.* 31: 123–37

Straznicky, K., Gaze, R. M., Keating, M. J. 1980a. The retinotectal projection from rounded up half eyes in *Xenopus. J. Embryol. Exp. Morphol.* 58: 79–91

Straznicky, C., Gaze, R. M., Keating, M. J. 1981. The development of the retinotectal projections from compound eyes in *Xenopus. J. Embryol. Exp. Morph.* 62: 13–35

Straznicky, C., Tay, D., Hiscock, J. 1980b. Segregation of optic fibre projections into eye-specific bands in dually innervated tecta in *Xenopus. Neurosci. Lett.* 19: 131–36

Stuermer, C. A. O. 1984. Rules for retinotectal terminal arborizations in the goldfish optic tectum: A wholemount study. *J. Comp. Neurol.* 229: 214–32

Stuermer, C. A. O., Easter, S. S. Jr. 1984a. Rules of order in the retinotectal fascicles of goldfish. *J. Neurosci.* 4: 1045–51

Stuermer, C. A. O., Easter, S. S. Jr. 1984b. A comparison of normal and regenerated retinotectal pathways of goldfish. *J. Comp. Neurol.* 223: 57–76

Sur, M., Weller, R. E., Sherman, S. M. 1984. Development of X and Y cell retinogeniculate terminations in kittens. *Nature* 310: 246–49

Székely, G., Lázár, G. 1976. Cellular and synaptic architecture of the optic tectum. In *Frog Neurobiology*, ed. R. Llinas, W. Precht. pp. 407–34. New York: Springer-Verlag

Tay, D., Hiscock, J., Straznicky, C. 1982. Temporo-nasal asymmetry in the accretion of retinal ganglion cells in late larval and postmetamorphic *Xenopus. Anat. Embryol.* 164: 75–84

Taylor, J. S. H., Gaze, R. M. 1985. The effects of fibre environment on the paths taken by regenerating optic nerve fibres in *Xenopus. J. Embryol. Exp. Morph.* 89: 383–401

Thanos, S., Bonhoeffer, F. 1983. Investigations on the development and topographic order of retinotectal axons: Anterograde and retrograde staining of axons and perikarya with rhodamine in vivo. *J. Comp. Neurol.* 219: 420–30

Thanos, S., Bonhoeffer, F., Rutishauser, U. 1984. Fiber-fiber interaction and tectal cues influence the development of the chicken retinotectal projection. *Proc. Natl. Acad. Sci. USA* 81: 1906–10

Thompson, I. D. 1979. Changes in the uncrossed retinotectal projection after removal of the other eye at birth. *Nature* 279: 63–66

Torrealba, F., Guillery, R. S., Eysel, U., Polley, E. H., Mason, C. A. 1982. Studies of retinal representations within the cat's optic tract. *J. Comp. Neurol.* 211: 377–96

Tosney, K. W., Landmesser, L. T. 1985a. Specificity of early mononeuron growth cone outgrowth in the chick embryo. *J. Neurosci.* 5: 2336–44

Tosney, K. W., Landmesser, L. T. 1985b. Growth cone morphology and trajectory in the lumbosacral region of the chick embryo. *J. Neurosci.* 5: 2345–58

Trisler, D., Collins, F. 1987. Corresponding spatial gradients of TOP molecules in the developing retina and optic tectum. *Science* 237: 1208–9

Trisler, G. D., Schneider, M. D., Nirenberg, M. 1981. A topographic gradient of molecules in retina can be used to identify neuron position. *Proc. Natl. Acad. Sci. USA* 78: 2145–49

Udin, S. B. 1977. Rearrangement of the retinotectal projection in *Rana pipiens* after unilateral caudal half-tectum ablation. *J. Comp. Neurol.* 173: 561–83

Udin, S. B. 1978. Permanent disorganization of the regenerating optic tract in the frog. *Exp. Neurol.* 58: 455–70

Udin, S. B. 1983. Abnormal visual input leads to development of abnormal axon trajectories in frogs. *Nature* 301: 336–38

Udin, S. B., Fisher, M. D. 1986. Electron microscopy of crossed isthmotectal axons in *Xenopus laevis* frogs. *Soc. Neurosci. Abstr.* 12: 1028

Udin, S. B., Gaze, R. M. 1983. Expansion and retinotopic order in the goldfish retinotectal map after large retinal lesions. *Exp. Brain Res.* 50: 347–52

Udin, S. B., Keating, M. J. 1981. Plasticity in a central nervous pathway in *Xenopus*: Anatomical changes in the isthmo-tectal projection after larval eye rotation. *J. Comp. Neurol.* 203: 575–94

Udin, S. B., Schneider, G. E. 1981. Compressed retinotectal projection in hamsters: Fewer ganglion cells project to tectum after neonatal tectal lesions. *Exp. Brain Res.* 43: 261–69

Van der Loos, H. 1976. Barreloids in mouse somatosensory thalamus. *Neurosci. Lett.* 2: 1–6

Van der Loos, H., Woolsey, T. A. 1973. Somatosensory cortex: Structural alterations following early injury to sense organs. *Science* 179: 395–97

Verley, R., Onnen, I. 1981. Somatotopic organization of the tactile thalamus in normal adult and developing mice and in adult mice dewhiskered since birth. *Exp. Neurol.* 72: 462–74

Waite, P. M. E. 1973. Somatotopic organization of vibrissal responses in the ventrobasal complex of the rat thalamus. *J. Physiol.* 228: 527–40

Waite, P. M. E., Taylor, P. K. 1978. Removal of whiskers in young rats causes functional changes in cerebral cortex. *Nature* 274: 600–2

Walsh, C., Guillery, R. W. 1985. Age related fiber order in the optic tract of the ferret. *J. Neurosci.* 5: 3061–69

Walsh, C., Polley, E. H. 1985. The topography of ganglion cell production in the cat's retina. *J. Neurosci.* 5: 741–50

Walsh, C., Polley, E. H., Hickey, T. L., Guillery, R. W. 1983. Generation of cat retinal ganglion cells in relation to central pathways. *Nature* 302: 611–14

Weideman, T. A., Kuwabara, T. 1968. Postnatal development of the rat retina. An electron microscopic study. *Arch. Ophthalmol.* 79: 470–84

Welker, C. 1976. Receptive fields of barrels in the somatosensory neocortex of the rat. *J. Comp. Neurol.* 166: 173–90

Weller, W. L., Johnson, J. I. 1975. Barrels in cerebral cortex altered by receptor disruption in newborn but not in five day old mice. *Brain Res.* 83: 504–8

Whitelaw, V. A., Cowan, J. D. 1981. Specificity and plasticity of retinotectal connections: A computational model. *J. Neurosci.* 1: 1369–87

Wickelgren, B. G. 1971. Superior colliculus: Some receptive field properties of bimodally responsive cells. *Science* 173: 69–72

Williams, R. W., Chalupa, L. M. 1982. Prenatal development of the retinocollicular projection in the cat: An anterograde tracer transport study. *J. Neurosci.* 2: 604–22

Williams, R. W., Rakic, P. 1985. Dispersion of growing axons within the optic nerve of the embryonic monkey. *Proc. Natl. Acad. Sci. USA* 82: 3906–10

Willshaw, D. J., Fawcett, J. W., Gaze, R. M. 1983. The visuotectal projections made by *Xenopus* 'pie slice' compound eyes. *J. Embryol. Exp. Morphol.* 74: 29–45

Willshaw, D. J., Gaze, R. M. 1986. The discontinuous visual projections on the *Xenopus* optic tectum following regeneration after unilateral nerve section. *J. Embryol. Exp. Morphol.* 94: 121–37

Willshaw, D. J., von der Malsburg, C. 1979. A marker induction mechanism for the establishment of ordered neural mappings: Its application to the retinotectal problem. *Philos. Trans. R. Soc. B* 287: 203–43

Wise, S. P., Jones, E. G. 1978. Developmental studies of thalamocortical and commissural connections in rat somatic sensory cortex. *J. Comp. Neurol.* 178: 187–208

Woolsey, T. A., Anderson, J. R., Wann, J. R., Stanfield, B. B. 1979. Effects of early vibrissae damage on neurons in the ventrobasal (VB) thalamus of the mouse. *J. Comp. Neurol.* 184: 363–80

Woolsey, T. A., Van der Loos, H. 1970. The structural organization of layer IV in the somatosensory region (SmI) of the mouse cerebral cortex. *Brain Res.* 17: 205–42

Woolsey, T. A., Wann, J. R. 1976. Areal changes in mouse cortical barrels following vibrissal damage at different postnatal ages. *J. Comp. Neurol.* 170: 53–66

Yoon, M. G. 1971. Reorganization of retinotectal projection following surgical operations on the optic tectum in goldfish. *Exp. Neurol.* 33: 395–411

Yoon, M. G. 1980. Retention of topographic addresses by reciprocally translocated tectal re-implants in adult goldfish. *J. Physiol.* 308: 197–215

Young, S. R., Rubel, E. W. 1986. Embryogenesis of arborization pattern and topography of individual axons in N. laminaris of the chicken brain stem. *J. Comp. Neurol.* 254: 425–59

Zakon, H. H. 1983. Reorganization of connectivity in amphibian central auditory system following VIIIth nerve regeneration: Time course. *J. Neurophysiol.* 49: 1410–27

Ann. Rev. Neurosci. 1988. 11 : 329–52

BEHAVIORAL STUDIES OF PAVLOVIAN CONDITIONING

Robert A. Rescorla

Department of Psychology, University of Pennsylvania, Philadelphia, Pennsylvania 19104

The purpose of this review is to bring to the attention of neuroscientists some of the current thinking of psychologists who study learning processes from a behavioral perspective. The past 20 years have seen enormous changes in the ways that psychologists conceptualize and study elementary learning processes; however, many of those changes are poorly appreciated by the neuroscience community at large.

The review is organized into two parts. First, I described a framework in which to think about the conduct of learning experiments (Rescorla & Holland, 1976). Although that framework will seem simple minded to some, it provides an overall structure within which to place the study of learning, a structure that helps one avoid certain distressingly common conceptual pitfalls. Second, I describe some properties of a carefully studied example of learning that has dominated modern theorizing: Pavlovian conditioning. That description emphasizes not only basic data but also certain key notions in modern theories of pavlovian associative learning. Of course, it is impossible to provide an exhaustive review of either the theory or data in this area. Instead, my goal is to give a flavor of current thinking.

A FRAMEWORK FOR THE STUDY OF LEARNING

Learning is a process by which an organism benefits from experience so that its future behavior is better adapted to its environment. Consequently, any study of a learning process must attend to two times in the life of the organism. At one time (t1) the organism is exposed to some experience, some opportunity to learn. Then at a later time (t2) some assessment

0147–006X/88/0301–0329$02.00

procedure reveals the modification that the t1 experience produced. The question of interest is always whether a particular t1 experience produces an outcome at t2 that is absent without that experience. Consequently, studies of learning typically compare the behavior of two organisms at t2, those who have been exposed to the t1 experience of interest and those who have been spared that exposure and instead had some "control" experience at that time. One can organize many of the commonly studied elementary learning paradigms in terms of the different types of experience that they arrange for the animal at t1 and the different techniques that they use for assessing that learning at t2.

Type of Experience

Organisms might be expected to benefit from a rich array of types of experience. However, psychologists studying nonhuman learning have conventionally categorized experience into three classes. These can all be described in terms of the different modes in which the organism is presented with a stimulus, call it S1.

1. The most primitive thing that can be learned about a stimulus is that it exists. A procedure that is presumably adequate for such learning is the simple presentation of S1 to the organism without any other explicit constraint. The question is whether simple exposure to a stimulus at t1 leaves an after-effect that can be measured at a later time, t2. Of course, the most commonly studied example is habituation, in which the after-effect measured is a decreased response to the stimulus at t2. But, as described below, many other consequences of simple stimulus presentation are detected by other assessment techniques. Whatever the assessment technique, one can think of a simple presentation procedure as affording the animal the opportunity to learn of the existence and properties of the stimulus.

2. In order to survive, organisms must learn more than the existence of individual stimuli in the world; they must learn something of the structure of the environment, the relations among the stimuli. Hence an important second mode of presentation arranges a relation between S1 and some other stimulus, S2. The question is whether exposure to that *relation* modifies the organism in a manner that can be detected at t2. The most common example is Pavlovian conditioning, in which S2 signals the occurrence of S1. In that context, the S2 is termed the conditioned stimulus (CS) and S1 the unconditioned stimulus (US). The conventional result is that the organism shows an augmented response to the CS during the t2 assessment. This sort of treatment and outcome form the basis of the assertion that an association has formed between CS and US. The learning

of relations among stimuli has been extensively studied at the behavioral level and is discussed at length below.

3. Finally, a successful organism might be expected to learn about the impact that its own actions have on the world. A t1 experience adequate to produce such learning would consist of exposure to a relation between S1 and the organism's own behavior. In that context, S1 is commonly labeled a "reinforcer," or "reward," or "punisher." When such an arrangement of a relation between a response and reinforcer produces a change in the likelihood of that response during the subsequent t2 assessment, one normally speaks of "instrumental learning" or "operant conditioning."

A useful example of such presentation modes can be given with food as the S1 for a hungry animal such as a rat. Such an animal will show learning of all three types about a food S1. It learns about the existence of food, as well as the relation of that food to other stimuli and to its own behavior. First, like many animals, rats are initially hesitant to consume an unfamiliar food substance. But they will show habituation of this initial aversive reaction simply as a result of repeated exposure to the substance. Animals tested for their response at t2 after exposure to food at t1 will consume more than do animals not so exposed (e.g. Rozin, 1976). Second, if at t1 some other event, such as increased illumination, signals the coming of food, then the animal will learn that relation, as exhibited in various ways, such as subsequently increased general excitement and salivation during the light. Finally, if one arranges a relation between some aspect of the rat's behavior (such as depressing a lever) and the occurrence of food, the animal will show instrumental learning, as exhibited by increased likelihood of pressing that lever.

When described in an abstract way, ignoring the identities of the stimuli and organism, these three types of t1 experience encompass the vast majority of behavioral studies of simple learning processes. They involve teaching the organism about the existence of a stimulus, about the relation of that stimulus to other stimuli in its environment, and about the relation of that stimulus to the animal's own behavior. One might argue that if we can understand how organisms learn these three things about a stimulus, we will have close to a complete characterization of how they learn about events in their environment. It is just such a characterization that the modern psychology of learning takes as its job.

Types of Assessment: Consequences for Behavior

For each of the different types of S1 experience at t1, there is a range of different consequences in the animal's behavior and therefore a range of techniques that can be used to assess learning at t2. Although some changes in behavior at t2 are studied more often than others, many different types of change can be used to infer the occurrence of learning. It is useful to classify the changes in behavior at t2 into three categories: changes in response evocation, changes in learnability, and changes in value.

These distinctions are most readily illustrated for the case of simple exposure to S1. Of course, there are many instances in which simple exposure to some stimulus seems to produce learning, but the way in which that learning expresses itself at t2 can be quite varied.

1. The most common technique is to assess the likelihood that S1 will evoke a response. A frequent result is "habituation," in which as a result of its presentation at t1, the stimulus S1 will evoke a smaller response at t2. That outcome is widespread among stimuli and species (e.g. Peeke & Hertz 1973), but it can also happen that simple exposure can augment the response to S1; in that case it is conventional to speak of "sensitization" occurring. More generally, learning at t1 can be indexed by any change in the form of the response elicited at t2.

2. Another very useful procedure assesses the ability of S1 to enter into new learning at t2. For instance, it is well documented that repeated presentation of a stimulus at t1 greatly interferes with the animal's subsequent ability to learn new Pavlovian associative relations about that stimulus. The interference shows up whether the exposed stimulus is subsequently used as the signal or as the consequence in a Pavlovian paradigm. Thus if in a t2 assessment period a rat receives the pairing of a light with a shock, the learning of that relation is substantially retarded by the simple exposure at t1 of either the light or the shock. Retardation produced by exposure to the light has been termed "latent inhibition" (Lubow 1973) whereas that produced by exposure to the shock is called the "US-pre-exposure effect" (Randich & LoLordo 1979). In both instances learning occurs because of simple presentation of a stimulus at t1; but that learning is assessed at t2 not in terms of response evocation but instead in terms of the ability for new learning to occur.

3. Finally, learning as a result of simple exposure to a stimulus can be measured in terms of changes in the value of that stimulus. By "value" I mean the ability of that stimulus to serve as a reinforcer in an instrumental learning situation. A frequent occurrence is that exposure to a reinforcer at t1 reduces its ability to produce instrumental learning at t2. For instance, repeated exposure to an electric shock will reduce the ability of that shock to punish a level press.

One can make a similar point about the range of assessment procedures for the other modes of presentation at t1. For instance, the Pavlovian conditioning that results from the signaling of US by CS is commonly assessed by the increased ability of CS to evoke a response; however, other changes in the CS sometimes provide a better measure. The ability of a CS to interfere with learning about other stimuli (so-called "blocking," see below) or to produce second-order conditioning is an excellent alternative measure of its prior conditioning. Consider, for instance, second-order conditioning as an index of a CS-US association. A CS that has signaled a US also takes on the ability to establish Pavlovian conditioning to other events by which it is signaled itself. As Pavlov (1927) first demonstrated, a light paired with food not only evokes salivation but also becomes capable of conditioning salivation to a tone. Rescorla (1980) has argued that this ability of the light to "second-order" condition a tone is often the best procedure for measuring the fact that the light has become conditioned by the US. The advantages of this procedure are particularly clear when one wants to compare the amounts of conditioning to qualitatively different CSs. An important fact about Pavlovian conditioning that is still poorly appreciated is that different CSs often evoke quite different response forms after conditioning with the same US (Holland, 1984). One way to overcome the consequent difficulty in comparing the amounts of conditioning these CSs control is to measure learning not in terms of the responses that they evoke but in terms of their success at second-order conditioning a common target stimulus. Associations between neutral stimuli or between discrete signals and the contexts in which they occur are also often best studied using techniques like second-order conditioning (e.g. Rescorla 1984).

The important point is that a variety of techniques is available for use at t2 to assess the learning that occurred at t1. That learning can be assessed by changes in response evocation, stimulus conditionability, or reinforcement value. Modern behavioral conceptions of learning do not view it as the modification of a particular response pattern. Instead, learning is thought of as an internal change as a result of experience, a change that can be exhibited in various ways. How any learning will be exhibited depends on the demands that the world, and the experimenter, subsequently make upon the organism. It follows that although some means of detecting learning are more conventional than others, there is no a priori reason to favor one technique. Rather one should be eclectic and pick the detection technique that is best suited to exposing learning in the particular preparation one wishes to study. Such eclecticism has greatly facilitated the behavioral study of learning in the last several decades.

An Implication: Avoid Acquisition Curves

I have described learning in terms of differences in t2 behavior as a function of differences in t1 experiences. It is important to note that I have not characterized learning as a change in behavior from t1 to t2. The former description encourages one to study learning by exposing different animals to different experiences at one time and then assessing this learning by a common test procedure at a later time. That in turn leads naturally to a distinction between learning (and the conditions that obtain at the time of learning) and performance (and the conditions that obtain at the time of test). Describing learning as a change in behavior instead encourages one to study learning by tracking the changes that take place over the course of exposure to a treatment. Unfortunately, that often results in interpretative difficulties. One problem is that the behavior of an organism may differ at t1 and t2 for many reasons, such as developmental changes over time. Consequently, a before/after assessment of a treatment is rarely informative by itself; it must be compared with a similar assessment in animals given a different (control) treatment at t1. A less widely recognized problem is that comparison of behavior at t1 and t2 (or in the course of an ongoing treatment) can inadvertently lead to overlooking the important distinction between learning and performance. This can sometimes lead to a profound misinterpretation of the nature of the learning process under study (see Rescorla & Holland 1976).

The habituation paradigm provides a convenient example of the second problem. It is common in studies of habituation to track the course of decrease in response when a stimulus is repeatedly applied. That course can be studied as a function of such primitive variables as stimulus intensity, modality, and spacing (e.g. Thompson & Spencer 1966). For instance, one commonly observed result is that more response decrement occurs when the presented stimulus is less intense, an observation that has led to the inference that habituation is greater for weaker stimuli. But such tracking of the decrement entirely confounds the conditions of learning with the conditions of assessing that learning. When we observe differences in responding on trial n, we do not know whether to attribute them to differences in the stimulus intensity to which the animals were exposed on the previous $n-1$ trials or to differences in the stimulus intensity to which they are responding on trial n. The animals differ both in the learning experience they have had and in the conditions under which we assess that learning. In order to separate those alternatives, we need to expose different animals to different intensities for the first $n-1$ trials (at what we might identify as t1) and then test them with a common intensity, a common assessment procedure (at t2). If lower intensity stimuli indeed lead to more habituation, then we would expect $n-1$ trials of low intensity to result in a lower response to a common test intensity at t2. However, when Davis & Wagner (1968) performed quite an elegant study of that sort with the startle response in rats, they found just the opposite outcome. When assessed by the response to a common test intensity, greater habituation had resulted from exposure to more, rather than less, intense stimuli. This then is a case in which failure to be explicit about the logic of different learning conditions at t1 measured by common assessment procedures at t2 has led to important interpretive difficulties. Although this empirical outcome may vary from preparation to preparation, the conceptual point is important to remember. The assessment of the impact of an independent variable on learning can only be made when a common test procedure is employed.

This conceptual point is not confined to studies of habituation. In fact, any time that one attempts to study the effects of any independent variable on any learning process by plotting an acquisition function, one risks confusing the conditions of learning with those of assessment. During acquisition itself, we cannot safely attribute response differences to the conditions of learning over trials n − 1 as distinct from the conditions of performance on trial n. Indeed, one could easily defend the position that the worst way to study learning is to look at the acquisition curve. It almost of necessity will confound the conditions of learning with those of performance. The importance of having an explicit test procedure that guarantees comparable assessment conditions for animals subjected to different learning conditions is a very elementary point. But it is so regularly overlooked, often resulting in considerable conceptual confusion, that it must be emphasized again and again.

The same logic, of course, points to the inappropriateness of studying extinction by tracking the decrement observed under nonreinforcement. The shift from training to extinction commonly involves many changes, such as deletion of all deliveries of the US. These changes may be expected to have important effects on performance quite aside from any impact on associative learning. Consequently, the simple observation that behavior declines during extinction is very difficult to interpret. A superior procedure for studying extinction would subject some animals (or some equivalently trained stimuli within the same animal) to an extinction treatment, spare others that treatment, and then give both groups (or stimuli) a common test session. This procedure would allow one to assess the experiential consequences of extinction independently of any local performance effects.

In summary, it is quite useful to keep deliberately in mind an elementary structure for studying learning that acknowledges a time of input and a time of assessment. This not only provides a structure in which to think about different learning procedures, it also provides a certain measure of protection against some important conceptual pitfalls.

PAVLOVIAN ASSOCIATIONS

The study of learning at the behavioral level has been dominated by associative paradigms. In the last 20 years considerable energy has gone into the collection of data and development of theories for the particular instance of Pavlovian conditioning, in which the organism is asked to learn the relation between two stimuli. In this section, I discuss four key notions in modern thinking about Pavlovian associations: contiguity, information, inhibition, and salience. My intention is to characterize current thinking and to give an overview of some of the facts that seem relatively secure and general.

Contiguity

I described Pavlovian conditioning above as the arranging of a relation between two stimuli, S1 and S2 (the US and CS). But until quite recently, the only relation that one took seriously was temporal contiguity. One of the oldest ideas in the study of learning is that contiguity is the key notion in the formation of associations (see Warren 1921). Certainly there is broad agreement with the general statement that two events that are contiguous in time are more likely to become associated than are two that are separated. But what can we say beyond that global statement? Here I try to make three more subtle points about the notion of contiguity.

The first point is an old one but requires some elaboration: In many situations strict simultaneity of occurrence between CS and US does not in fact optimally produce associative learning. Although there are well-controlled reports of substantial conditioning with simultaneous conditioning (e.g. Heth & Rescorla 1973, Rescorla 1981), conditioning appears almost universally to be best when there is a slight temporal asynchrony in which the CS precedes the US by a small time interval. That point is made graphically in Figure 1, which is elaboration of one presented by Mackintosh (1983). The figure displays the amounts of responding obtained when the CS-US interval is varied in six representative, and frequently employed, Pavlovian conditioning preparations. These preparations differ in the species investigated (rat, pigeon, and rabbit), the nature of the US (food, water, shock, LiCl), the nature of the CS (auditory, visual, gustatory), and how the learning is assessed (latency, likelihood, and magnitude of an elicited response, as well as ability to serve as a reinforcer).

There are two points to notice about this figure. First, despite wide procedural differences, the results share a common functional relation. In all cases, the success of conditioning is a nonmonotonic function of the amount of time by which the CS precedes the US. When the precedence

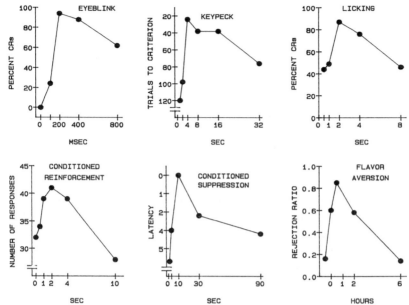

Figure 1   Conditioning as a function of the CS-US interval in six different paradigms: (*a*) eyelid responding over 640 acquisition trials, from Smith et al (1969); (*b*) trials to an acquisition criterion in autoshaping, from Gibbon et al (1977); (*c*) percentage of trials with a lick response during test trials in the course of 270 acquisition trials from Boice & Denny (1965); (*d*) number of bar-presses during a test session in which responding earned the CS, from Bersh (1951); (*e*) latency of responding during a test trial after conditioning, from Yeo (1974): (*f*) rejection of a flavor during a 24 hr test, from Barker & Smith (1974).

is either too short or too long, conditioning is relatively poor. But for each preparation there is some intermediate set of intervals that is successful. Second, these preparations show variation over widely different time scales. For instance, the eyelid preparation shows an optimum when the asynchrony is a few hundred milliseconds. But the time scales and optima are plotted in seconds and even hours for the other preparations. Indeed, because the results in Figure 1 were chosen to show the nature of the overall functions, they fail to display the fact that the preparations also differ widely in the maximum interval over which they will sustain conditioning. In the eyelid preparation, conditioning typically fails when the interval exceeds a few seconds, whereas in conditioned suppression, learning is routinely obtained with intervals of 5 min, and in flavor-aversion conditioning has sometimes been obtained at intervals of 12 hr. Of course, detailed comparisons of this sort across diverse preparations are fraught with technical difficulties (Mackintosh 1983), but the data suggest that there is no absolute temporal interval that produces the best conditioning across the ra ıge of preparations. What conditioning preparations share is the general form of the function, not the absolute values. Although this has been known for many years, a surprising number of workers still apparently expect all conditioning preparations to adhere to a mythical 0.5 sec optimum interval. But there is no reason to celebrate the finding that a particular CS-US interval is especially effective. It is not the parameter values but rather the sharing of functional form that encourages the view that a common process may be involved across a range of preparations.

The second point to be made about contiguity is that there is some evidence that within a preparation it is relative, rather than absolute, time that matters. In particular, the success with which a given CS-US temporal interval produces an association may depend heavily on the temporal interval between trials. Some of the best evidence for this comes from the autoshaping preparation, which has recently been studied in some quantitative detail (e.g. Gibbon et al 1977). Figure 2 shows some representative data from a study that varied both the CS-US interval and the time between trials. The left-hand portion of the figure plots the number of trials to an acquisition criterion as a function of the interval between trials. The data points are joined for the groups sharing a CS-US interval. The negative slope of the functions indicates that the more widely spaced the CS-US trials, the faster the learning. The ordering of the lines within the figure indicates that, within the range studied, the shorter the CS-US interval the faster the learning. Neither of these findings is very surprising.

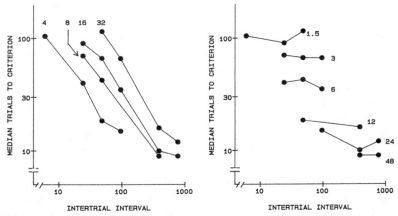

Figure 2 Conditioning as a joint function of the CS-US interval and the intertrial interval (ITI) in autoshaping. Data are plotted as the number of trials to reach a performance criterion. In the *left panel*, data from the same CS-US interval are connected. In the *right panel*, data from the same ratio of ITI to CS-US interval are connected. The value of that ratio is indicated next to each curve. Data are from Gibbon et al (1977).

However, the right-hand panel of Figure 2 highlights a feature of these data that is surprising and important. It reproduces the same data, but joins the points from the groups that share the same *ratio* of intertrial interval to CS-US interval. The striking result is that these curves are nearly flat. This indicates that over quite a broad range of absolute values, the speed of acquisition is relatively constant when that ratio is constant. That suggests that the CS-US contiguity is evaluated not absolutely but in relation to the intertrial interval. Sufficient data are not available to decide the range of conditioning preparations over which this regularity holds, but the results of early experiments by Stein et al (1958) suggest that it may also apply to conditioned suppression in the rat. A constancy of this sort clearly places important constraints on the potential mechanisms that underlie conditioning. Gibbon (1981) discusses several alternative theoretical interpretations of this constraint.

The third point to appreciate is that the contiguity between CS and US can be important for quite different reasons. It is most natural to interpret variation in the CS-US interval as producing variation in the degree or nature of the association between the CS and US. However, in some instances the CS-US contiguity matters not because it promotes an association between the US and the CS but rather because it gives the US the opportunity to modulate *other* learning about the CS. Two examples are worth mentioning. The first involves the habituation that the CS might undergo because of its repeated presentation. It is now well-documented that the period immediately after a stimulus presentation is importantly involved in the development of habituation. Most relevant here, salient events that occur immediately after a stimulus presentation can disrupt habituation to that stimulus (e.g. Wagner 1978). Consequently, arranging for a US to be contiguous with a CS can affect subsequent responding to a CS because that contiguity modulates the success of habituation to the CS. Indeed, it is frequently difficult to separate this modulation of a nonassociative process from the operation of an associative process linking the CS and US. A second example is more subtle but equally important. Sometimes the contiguity of S1 to S2 apparently acts to help establish associations between S2 and another stimulus, S3. In this case, S1 plays a true catalytic role, inducing an association between S2 and S3 by virtue of its contiguity to each (see Rescorla 1982a). The fact that contiguity between two stimuli can have important learning effects other than to promote their association, of course, can greatly complicate the study of associative learning. Rescorla (1982b) discusses techniques for separating these alternative modes of action of contiguity. One might note that classical S-R reinforcement theories of instrumental learning appeal to just such a catalytic process. According to such theories the reinforcer's proximity to the response encourages the development of an association between that response and antecedent stimuli.

These comments make it clear that contiguity is actually quite complex in its effects on learning. Across conditioning situations there is not a common time interval that optimizes conditioning, and within a situation it may be relative rather than absolute time that determines conditioning. Moreover, contiguity can have its effects for multiple reasons, both associative and nonassociative. Any successful search for the neural mechanisms of contiguity's action in learning must be aware of these complexities.

Information

The second key notion in modern thinking about conditioning is that of information. Many have shared the intuition that the functional significance of the CS is to serve as a signal of the US. For instance, that intuition has often been used to explain why the CS should shortly precede the US in order to produce conditioning. Only CSs that precede USs can provide useful information about them. It is now widely agreed that an informational intuition is a useful heuristic. Indeed, several authors have noted that the CS/US relations required for conditioning are very similar to those that a rational scientist would demand to conclude that the CS is the cause of the US (e.g. Dickinson 1980, Rescorla 1985b).

This intuition appears in modern thinking as a further constraint on the operation of contiguity. Current evidence indicates that the simple contiguity between CS and US produces poor conditioning unless that CS also bears an informational relation to the US. I describe below two now-classic experiments that lead to this conclusion, experiments that have had a profound impact on our thinking about conditioning. Both experiments reveal instances in which an otherwise excellent CS-US contiguity fails to produce associative learning because of the absence of an informational relation between CS and US.

The first example is the phenomenon of "blocking," initially reported by Kamin (1968). Kamin's experiments involved fear conditioning in which rat subjects received a 3-min CS preceding the occurrence of a mild footshock US. In such procedures a few pairings are sufficient to produce a CS-US association that expresses itself as the interruption of ongoing activity during subsequent presentation of the CS. The comparison of interest was between two groups, both of which received a compound CS consisting of a light and a noise presented simultaneously prior to the shock. Interest focused on the amount of conditioning gained by the noise, as assessed when that stimulus was presented alone in a subsequent test phase. The groups differed only in their history prior to the conditioning of the compound. One group had received repeated light-shock pairings and the other had been spared that treatment. Kamin found that the group without pretraining on the light showed substantial fear of the noise; however, the group with a history of light-alone conditioning showed almost no conditioning of the noise. Pretraining on the light had blocked the noise conditioning that would ordinarily have occurred when the light/noise compound was followed by shock. The important point to notice is that both groups received the same noise-shock contiguity, yet the levels of conditioning differed substantially. Intuitively, the reason is obvious: The noise provided quite different amounts of information about the shock in the two groups. For the pretrained animals, the light already signaled the coming of shock and the noise was redundant, adding no new information. Apparently, a demonstrably effective contiguity can be rendered ineffective in the absence of an informational relation.

The second example comes from experiments demonstrating the importance of the CS/US contingency, as distinct from the CS/US contiguity, in producing conditioning (Rescorla 1968). The typical Pavlovian experiment intends, of course, to arrange for the CS and US to be contiguous. But it commonly arranges a much stronger relation, in which the CS is informative about the US. This is because the likelihood of obtaining a US is substantially greater given the occurrence of the CS than it is given the absence of the CS. However, one may disentangle contiguity from information by systematically varying the probability of the US given a CS $[p(US/CS)]$ and the probability of a US given no CS $[p(US/\sim CS)]$. Figure 3 shows the results of one experiment that did this in a conditioned suppression paradigm (Rescorla 1968). The figure plots the amount of

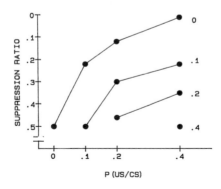

Figure 3   Fear conditioning in rats as a function of the probability of a US in the presence of the CS $[p(US/CS)]$ and the probability of a US in the absence of the CS $[p(US/\sim CS)]$. A ratio value of 0 indicates strong conditioning whereas a ratio of 0.5 indicates little conditioning. From Rescorla (1968).

conditioning, indexed by the degree to which the CS interupts ongoing behavior, against the probability of the US in the presence of the CS. The parameter in the figure is the probability of the US in the absence of the CS. When the conventional procedure of holding $p(US/ \sim CS)$ at 0 is used (*uppermost curve*), conditioning increases monotonically with increases in the value of $p(US/CS)$. In fact, conditioning is an increasing function of $p(US/CS)$ whatever the value of $p(US/ \sim CS)$. However, the interesting finding is that for any given degree of CS-US contiguity [$p(US/CS)$], conditioning depends heavily on the likelihood of the US in the absence of the CS. The functions in Figure 3 are ordered by $p(US/ \sim CS)$; the greater that probability the lower the conditioning. Consequently in order to predict the success of conditioning, one needs to know both probabilities; it is their relationship that orders the data. Conditioning occurs to the degree that $p(US/CS)$ exceeds $p(US/ \sim CS)$. Even when $p(US/CS)$ is high, there is little evidence of conditioning when $p(US/ \sim CS)$ is equally high, as in the left-most point of each function in Figure 3. Stated differently, conditioning depends on the correlation of two events, not their contiguity. Again, the informational intuition is easy to apply. When the CS signals no change in the ongoing likelihood of the US, it is not informative no matter how good the contiguity; under those circumstances one would not intuitively expect much conditioning.

Both of these basic experiments have now been replicated in a variety of conditioning preparations. While the details of the results vary across preparations, the basic phenomena have considerable generality (Mackintosh 1983). As a result they have had a profound effect on modern thinking about the development of Pavlovian associations.

Of course, simple appeal to the information intuition is not a satisfactory explanation of such phenomena. Rather, these results have provided the occasion for rethinking the notion of contiguity. In both kinds of experiment the problem is that an apparently adequate contiguity failed to produce learning. Consequently, the issue is what was wrong with those contiguities. Why did the embedding of joint CS/US occurrences in these conditioning procedures prevent them from yielding conditioning? Two classes of theories have been suggested, both of which retain the primacy of contiguity, each pointing to a failure of processing of one of the events in the contiguity. According to these theories, the failure lies in the fact that either the CS or the US was inadequately processed and hence the nominal contiguity did not actually obtain. The variables that are intuitively identified with information value in fact matter because they affect the processing of either the CS or the US and hence actually destroy the contiguity intended.

These two accounts can best be understood by looking at how they deal with the phenomenon of blocking. Deficits in CS processing are perhaps the most obvious account of blocking (e.g. Sutherland & Mackintosh 1971, Mackintosh 1975). The intuitive idea is that of attention: As a result of the initial conditioning of the light, the animal attends to that light. When the noise is then added on compound trials, the attention to the light adversely affects that to the noise, resulting in less successful processing of the noise. Because the noise is not well processed, it does not benefit from the actual contiguity that the experimenter has arranged with the US. A less obvious alternative is a deficit in US processing (e.g. Rescorla & Wagner 1972). The intuitive notion here is that an expected US is less effective in engaging the learning mechanism. Hence the problem on light-noise compound trials is that the shock is less effective at producing new learning because it is well signaled by the light.

Although these two classes of theories are commonly described as competitors, in fact they share a good many features. For instance, both view contiguity as ultimately responsible for conditioning. They offer ways in which apparently informational relations can be reduced to a somewhat more complex version of contiguity. Both argue that learning occurs when there is a discrepancy between the current level of conditioning that a stimulus controls and the level appropriate to the US that follows that CS. Put casually, both view the organism as only learning when it is "surprised" by the outcome. In that sense, both types of theory are relativistic: USs are evaluated not in absolute terms but in relation to the USs anticipated by the CS. Moreover, both are elementaristic models in which complex stimuli are viewed as composed of elements and total conditioning to a complex is understood in terms of the conditioning of those elements. Finally, both types of theories see the associations among those elements as interacting in important ways. They see the animal as operating within a limited capacity system in which CSs compete either for processing or for the amount of association that they develop. It is that competition that lies at the heart of their ability to explain the animal's sensitivity to information, rather than simple contiguity.

Although I describe these accounts quite casually here, in fact substantial formal versions have been developed for both the CS-processing (e.g. Mackintosh 1975, Moore & Stickney 1980, Pearce & Hall 1980) and the US-processing (e.g. Rescorla & Wagner 1972) models. Such models have enjoyed both striking successes and humbling failures in describing available data and generating new findings (see Mackintosh 1983, Rescorla & Holland 1982). Both have begun to influence the thinking of those studying the neural basis of learning (e.g. Farley & Alkon 1985, Gluck & Thompson 1987, Hawkins & Kandel 1984). Conventional wisdom is that both kinds of theories make a contribution to learning as studied at the behavioral level. As a result there have been some attempts to provide an integration of the two approaches (e.g. Wagner 1981).

At the gross behavioral level the animal is much more sophisticated than classical accounts of Pavlovian conditioning anticipate. A simple-minded contiguity mechanism clearly fails to capture the animal's associative learning. Many organisms are very demanding of the evidence that they obtain from experience before they learn an association between the CS and US. The phenomena briefly described above give only a rough feeling for that sophistication. In fact, those results are but a small portion of the large literature that modern behavioral theories have generated (see Dickinson 1980, Rescorla & Holland 1982, Mackintosh 1983). Phenomena such as overshadowing, overexpectation, unblocking, superconditioning, learned irrelevance, and conditioned inhibition are mainstays of modern thinking about conditioning. They all represent sophisticated features of Pavlovian conditioning whose discovery was prompted by an informational perspective; none of them is readily consistent with a simple contiguity theory. The issue now is what more molecular mechanisms described at both the formal and neural level allow this sophisticated behavior. Considerable analytic progress has been made at the behavioral level. It makes sense to build on that progress when one attempts a neural analysis.

Inhibition

The third important notion in modern behavioral studies of learning is that of inhibition. Of course, inhibition is an old notion at many levels of analysis. It has had many different meanings and within psychology it has had a rather checkered history. But it is currently used within the behavioral study of learning to refer to the learning of a particular associative relation among stimuli. As noted above, an informational view has modified our thinking about how contiguity generates learning about positive relations—what is sometimes called excitatory conditioning. Modern thinking about inhibition is a natural extension of that view to the learning of negative relations. In recent years it has become clear that many organisms can learn negative correlations between CS and US and that such learning constitutes more than simply the failure to learn a positive relation (see Miller & Spear 1985).

The meaning of these assertions can be clarified by describing two procedures that are routinely used to generate conditioned inhibition. The first procedure is a logical extension of the correlation experiment described above. That experiment demonstrated that excitatory conditioning depends on the arrangement of a positive correlation between CS and US. One can similarly show that the arrangement of a negative correlation endows the CS with inhibitory properties. Under such circumstances, the CS signals a decrease in the likelihood of the US, a relation that many organisms are capable of learning. The second procedure frequently used to generate a conditioned inhibitor involves intermixing two types of trials; those on which one stimulus, A, is presented alone and followed by the US and those on which A is accompanied by another stimulus, X, and the US is withheld. Under those circumstances, A becomes a conditioned excitor, typically capable of preventing A from evoking that response.

The existence of an inhibitory association can be assessed in a variety of ways. On occasion, an inhibitory CS will evoke a behaviorally detectable response, commonly the opposite of the response that same CS would have evoked had it been trained as an excitor (e.g. Hearst & Franklin 1977). More often, behaviors observed during an inhibitor do not distinguish it from a stimulus that has never undergone conditioning. Consequently, a variety of other assessment techniques have developed. For instance, a "summation" procedure compares the response to a known excitor when it is presented alone and in combination with the inhibitor. To the degree that the excitor evokes a smaller response in the presence of a stimulus, that stimulus can be identified as an inhibitor; to the degree that this effect depends on a particular training history with the stimulus, it can be identified as a conditioned inhibitor. Alternatively, one could use a "retardation" test in which the putative inhibitor is embedded in a Pavlovian paradigm known to establish a conditioned excitor. To the degree that it develops that excitation less successfully than does a control stimulus, it can be identified as an inhibitor. LoLordo & Fairless (1985) review the results of these and other assessment procedures.

Behavioral studies of conditioned inhibition have focused on two questions. The first issue is how one characterizes more fully the circumstances that produce conditioned inhibition. This question was originally motivated by the observation that conditioned inhibition does not accord well with a simple notion that all associative learning depends on contiguity. Inhibition is a case in which contiguity with the US is blatantly the wrong relation for producing learning. However, it is important to realize that one cannot adequately characterize the circumstances producing inhibition as the simple presentation of the CS in the absence of the US. Of course, as noted above, repeated separate presentation of a CS does produce various kinds of learning, such as habituation; but it does not endow a stimulus with conditioned inhibition (Rescorla 1971, Reiss & Wagner 1972). Attempts to deal with inhibitory phenomena, such as extinction, solely in terms of decremental mechanisms that depend only on the CS presentation, and not its relation to the US, lead to conceptual difficulties. Gleitman et al (1954) made this point about early Hullian theory; a similar point has recently been noted by Gluck & Thompson (1987) about the Hawkins & Kandel (1984) model. Rather, as in the case of conditioned excitation, the intuition of information is useful: in order for a stimulus to become a conditioned inhibitor it must provide information about the US. Stated casually, the stimulus must signal the deletion of an otherwise expected US. Put somewhat more formally, in order to become an inhibitor, a stimulus must be paired with the absence of a US in the presence of an excitor previously conditioned by that US (see Konorski 1948, LoLordo & Fairless 1985). Just as the evaluation of a US was previously seen to be made relativistically, so the evaluation of the absence of a US is made relative to what might have occurred. As in the case of excitatory conditioning, this proposition can be given a mathematical formulation (e.g. Wagner & Rescorla 1972).

Modern work on conditioned inhibition has also focused on a second question: What is learned? The answer to that question now seems likely to depend on the conditioning preparation and the procedures used to establish inhibition. We do not have sufficient data to reach a firm decision, but I can indicate some of the possible learned bases for conditioned inhibition. Those possibilities are best seen if we schematize the excitatory association in terms of a learned connection between the CS and US, such that CS is able to activate a representation of the US and hence evoke an observable response, the CR. Within that schematization, four possible loci of action have been proposed (Holland 1985, Rescorla 1975). The inhibitor might act to prevent the response, to prevent activation of the US representation, to interfere with processing of the CS, or to interfere with the functioning of the CS-US association. Interruption at any of these loci would prevent the excitatory CS from evoking its customary response. The primary tool that is used for choosing among these alternatives is the transfer test, assessing the range of stimuli and behaviors that are affected by an inhibitor. For instance, if a conditioned inhibitor, established in conjunction with a particular CS-US pair, transfers its action to other CSs paired with that US, it cannot function simply to interfere with the processing of a particular CS or the action of a particular CS-US association. Rather its action must be on the US representation or on the response production. Similarly, if the transfer excitor happens to evoke a different response form and yet the inhibitor also suppresses that response, it is unlikely that the action is at the point of response production.

A possibility that has received considerable recent attention is that some instances of inhibition involve a stimulus that interferes with a specific CS-US association. That possibility is of particular interest because it would involve a stimulus being associated not with another stimulus but rather with an association between two other stimuli. If in fact an associative relation can itself enter into further associations, this would provide a way for a system built from pair-wise associations to construct more complex relations (cf Mill 1869). Holland (1985) has suggested, in the context of conditioned inhibition, that a stimulus might interfere with the functioning of another association. Moreover, Holland (1983) and Rescorla (1985a) have both presented evidence for an analogous but opposite result in which a stimulus apparently promotes the functioning of an association between two other stimuli.

Many of these possibilities remain to be worked out at the behavioral level. But it is clear that conditioned inhibition is a well-established and pervasive notion in modern thinking about Pavlovian conditioning. No attempt to deal with conditioning at a neural level will be judged adequate if it does not provide an associative mechanism for inhibitory phenomena.

Salience

A final notion that is current in theories of associative learning is that of stimulus salience or "associability." Some stimuli seem to enter into associations more readily than do others. This is true whether one looks at stimuli that serve as signals or those that serve as consequences. Certain facts about stimulus salience are well established but not very surprising. For instance, the relative salience of stimuli varies considerably across species. Moreover, within a broad range, more intense stimuli enter more readily into associations. But there are two aspects of stimulus salience that are of more interest.

First, stimulus salience can itself be modified by experience. This can come about either because a stimulus is presented alone or because it bears an associative relation to other stimuli. For instance, I mentioned above that repeated exposure to a stimulus in the absence of a relation to other stimuli can reduce its salience; the phenomena of latent inhibition and the US-preexposure effect are striking instances of this. Conversely, having a stimulus regularly signal some outcome can not only give it the ability to evoke a CR, it can also modify its salience. In fact, such a salience modification is at the heart of several modern theories of blocking. I noted above that conditioning may enable a stimulus to better engage the animal's attention, making it better processed.

In fact, some conditioned responses that one observes may be a manifestation of that improved processing of the CS. In some instances, a major consequence of Pavlovian conditioning may be to allow a CS to elicit more successfully its original response. Although such enhancement of the original response to the CS has sometimes been rejected as not being "true" conditioning, little basis exists for such an attitude (cf Farley & Alkon 1985). If a CS comes to evoke its original response more adequately because of its pairing with the US, but fails to do so when an appropriate control condition is arranged, then the performance is demonstrably dependent upon the arranging of a CS/US relation. That the learning of that relation is exhibited as the enhancement of an original response may be informative about its nature, but it surely is not grounds for banishing it from the hallowed realm of "true" conditioning. One would not want to claim that just because a stimulus is not initially neutral, it cannot serve as a CS entering into new associations. In any case, it seems clear that stimulus salience can be modified by experience.

The second point of interest about salience is that, like many other conditioning parameters, it is not an absolute property of the stimulus independent of the rest of the conditioning experiment. In particular, there is evidence that the salience of a stimulus may vary widely depending on what it is signaling. Some signals are better for some outcomes and others are better for other outcomes. The most well-known case of this comes from the seminal work of Garcia and his collaborators (e.g. Garcia & Koelling 1966). For instance, if a rat is presented with an external pain, such as a footshock, it is more likely to associate that shock with an antecedent auditory-visual stimulus than with an antecedent flavor. On the other hand, when exposed to internal malaise the rat is more likely to associate that with an antecedent flavor than an antecedent auditory-visual stimulus. The conditionability of flavors and auditory-visual events is not an absolute property of those stimuli but partly depends on what is being used to condition them. Although there was some concern about the adequacy of the original demonstrations of this phenomenon, subsequent work has placed it on sound ground (e.g. Miller & Domjan 1981). Moreover, several other instances have come to light and have been investigated with more sophisticated designs (e.g. LoLordo 1979). Some instances make contact with such historically important associative variables as stimulus similarity or spatial contiguity (Rescorla 1980).

Such findings are of considerable importance because they suggest a preexisting bias on the part of the animal making it susceptible to certain CS-US pairings. Behavioral theories of learning have historically emphasized the arbitrariness of the associative process, that the organism can learn any relations among events with which the world presents it. Certainly, many organisms have considerable flexibility in this regard, but clearly for many species repeated generations have faced a stable world that makes it likely that a particular individual will have to learn some associations rather than others. Hence it would be surprising if biases had not developed favoring the learning of certain associations.

It is less clear how this bias is accomplished. A variety of mechanisms have been proposed and it seems likely that each has applicability to some instance of bias (LoLordo 1979). It is convenient to describe the alternatives in the context of flavor-aversion learning, although several accounts can be rejected for that particular instance. One possibility is that the bias is not based on the association at all, but instead represents differential sensitization. For instance, receiving a toxin might especially sensitize the animal to flavors rather than to audio-visual stimuli; similarly, shock might sensitize it to audio-visual stimuli. Such sensitization could take a variety of forms, some of which are difficult to rule out in the absence of quite subtle experimental designs (Rescorla & Holland 1976). Such a nonassociative mechanism is almost surely used on occasion and often would stand the animal in good stead in dealing with the world.

However, there are also a variety of associative alternatives. For instance, one simple possibility is that the animal may come to the experiment with a preexisting association between certain pairs of stimuli. It may not be that some associations form more readily but rather that they form on top of a better base. Such a "head start" on the association might be either genetically derived or based on part experience with the actual organization of the world (e.g. Mackintosh 1973). Some authors have gone so far as to claim that only stimuli having such a preexisting relation can become associated with each other. Alternatively, it may not be that some associations form better or even get stronger on top of a better base, but rather that they are more readily exhibited in performance. For instance, an association between a CS and US would be especially easy to see if the response that a US conditions to the CS is compatible with that which the CS evokes on its own. However, if they were incompatible, a similar association might form but have more difficulty being exhibited. It is, of course, precisely for such cases that it is important to use an assortment of assessments of the association. Finally, of course, it might be that the organism is predisposed to learn better the relation between certain stimuli.

These alternatives are often difficult to separate and sometimes quite sophisticated experimental designs are necessary. But fussing over those designs is often worth the trouble, because these different bases for the relativity of salience have quite different implications for neural mechanisms of learning.

CONCLUSION

I have had two aims in review. First, I have described a particular framework within which to view learning processes. That framework forces explicit separation of the learning and assessment phases of learning experiments. The separation emphasizes that the same learning can be measured by an assortment of techniques. Second, I have described four key notions in current thinking about Pavlovian conditioning: contiguity, information, inhibition, and salience. I have described some of the functional relations that obtain in each case. But equally I have pointed to some of the alternative ways in which an instance of learning can occur and some of the tools that are useful for separating those alternatives.

Two points should be emphasized in conclusion. First, how modern theories view Pavlovian conditioning differs substantially from the historical view in which many of us were trained. Conditioning was often historically described as a procedure in which a US that regularly evokes a UR transfers its response to an originally neutral CS by virtue of a CS-US contiguity. The modern view of conditioning as the mechanism by which the organism learns relations among events in its world preserves virtually no aspect of that historical description. Properly assessed, associations can form between USs that either do or do not regularly evoke a response and CSs that are behaviorally neutral or elicit their own responses. Moreover, the learning may show up in a variety of behavioral changes, often as assessed by procedures in which the response bears little relation to that originally evoked by the US. Perhaps most importantly the relation of a simple contiguity completely fails to capture the richness of the stimulus relations that Pavlovian conditioning allows the animal to encode.

Second, one must anticipate that similar learning problems will be solved in different ways by different species. Although many examples of Pavlovian conditioning among mammals seem to share important common properties, it would be naive to think that there is a single associative process by which all animals put together a representation of this environment. It would be misguided to establish rigid criteria such as a list of phenomena that a mechanism must meet in order to be dubbed "true" conditioning. Of course any conditioning mechanism that allows the animal to represent its world adequately will be highly constrained by the properties of the world itself. That constraint will force even divergent conditioning mechanisms to share many gross properties. But in the end, progress at the neurobiological level will depend on prior behavioral analysis delineating how in fact the problem of learning stimulus relations is solved for the species under investigation. What the psychology of learning can offer is the conceptual tools, some alternative theories, and a methodological sophistication for the conduct of that analysis. The long history of behavioral analysis will surely make the path of the informed neurobiologist easier both empirically and conceptually.

Literature Cited

Barker, L. M., Smith, J. C. 1974. A comparison of taste aversions induced by radiation and lithium chloride in CS-US and US-CS paradigms. *J. Comp. Physiol. Psychol.* 87: 644–54

Bersh, P. J. 1951. The influence of two variables upon the establishment of a secondary reinforcer for operant responses. *J. Exp. Psychol.* 41: 62–73

Boice, R., Denny, M. R. 1965. The conditioned licking response in rats as a function of the CS-UCS interval. *Psychon. Sci.* 3: 93–94

Davis, M., Wagner, A. R. 1968. Startle responsiveness after habituation to different intensities of tone. *Psychon. Sci.* 12: 337–338

Dickinson, A. 1980. *Contemporary Animal Learning Theory*. Cambridge: Cambridge Univ. Press, 177 pp.

Farley, J., Alkon, D. L. 1985. Cellular mechanisms of learning, memory, and information storage. *Ann. Rev. Psychol.* 36: 419–94

Garcia, J., Koelling, R. A. 1966. Relation of cue to consequence in avoidance learning. *Psychon. Sci.* 4: 123–24

Gibbon, J. 1981. The problem of contingency. In *Autoshaping and Conditioning Theory*, ed. L. C. Locurto, H. S. Terrace, J. Gibbon, pp. 219–53. New York: Academic. 313 pp.

Gibbon, J. Baldock, M. D., Locurto, C., Gold, L., Terrace, H. S. 1977. Trial and intertrial durations in autoshaping. *J. Exp. Psychol. Anim. Behav. Proc.* 3: 264–84

Gleitman, H., Nachmias, J., Neisser, U. 1954. The S-R reinforcement theory of extinction. *Psychol. Rev.* 61: 23–33

Gluck, M. A., Thompson, R. F. 1987. Modeling the neural substrates of associative learning and memory: A computational approach. *Psychol. Rev.* 94: 176–91

Hawkins, R. D., Kandel, E. R. 1984. Is there a cell-biological alphabet for simple forms of learning? *Psychol. Rev.* 91: 375–91

Hearst, E., Franklin, S. 1977. Positive and negative relation between a signal and food: Approach-withdrawal behavior to the signal. *J. Exp. Psychol. Anim. Behav. Proc.* 3: 37–52

Heth, C. D., Rescorla, R. A. 1973. Simultaneous and backward fear conditioning in the rat. *J. Comp. Physiol. Psychol.* 82: 434–43

Holland, P. C. 1983. "Occasion-setting" in conditional discriminations. In *Quantitative Analyses of Behavior: Discrimination Processes*, ed. M. L. Commons, R. J. Hernstein, A. R. Wagner, 4: 267–97.

New York: Ballinger

Holland, P. C. 1984. Origins of behavior in Pavlovian conditioning. In *The Psychology of Learning and Motivation*, ed. G. H. Bower. 18: 129–174. New York: Academic, 381 pp.

Holland, P. C. 1985. The nature of conditioned inhibition in serial and simultaneous feature negative discriminations. In *Information Processing in Animals: Conditioned Inhibition*, ed. R. R. Miller, N. E. Spear, pp. 267–97. Hillsdale, NJ: Erlbaum. 404 pp.

Kamin, L. J. 1968. Attention-like processes in classical conditioning. In *Miami Symp. Predictability, Behavior and Aversive Stimulation*, ed. M. R. Jones, pp. 9–32. Miami: Univ. Miami Press. 145 pp.

Konorski, J. 1948. *Conditioned Reflexes and Neuron Organization*. Cambridge: Cambridge University Press. 277 pp.

LoLordo, V. M. 1979. Selective associations. In *Mechanisms of Learning and Motivation*, ed. A. Dickinson, R. A. Boakes. pp. 367–98. Hillsdale, N. J.: Erlbaum, 468 pp.

LoLordo, V. M., Fairless, J. L. 1985. See Holland, 1985, pp. 267–97

Lubow, R. E. 1973. Latent inhibition. *Psychol. Bull.* 79: 398–407

Mackintosh, N. J. 1973. Stimulus selection: Learning to ignore stimuli that predict no change in reinforcement. In *Constraints on Learning*, ed. R. A. Hinde, J. Stevenson-Hinde, pp. 75–100. New York: Academic. 488 pp.

Mackintosh, N. J. 1975. A theory of attention: Variations in the associability of stimuli with reinforcement. *Psychol. Rev.* 82: 276–98

Mackintosh, N. J. 1983. *Conditioning and Associative Learning*, Oxford: Oxford Univ Press. 316 pp.

Mill, J. 1869. *The Analysis of the Phenomena of the Human Mind*. London: Longmans, Green, Reed, & Dyer

Miller, R. R., Spear, N. E., eds. 1985. *Information Processing in Animals: Conditioned Inhibition*. Hillsdale, N.J.: Erlbaum. 404 pp.

Miller, V., Domjan, M. 1981. Specificity of cue to consequence in aversion learning in the rat: Control for US-induced differential orientations. *Anim. Learn. Behav.* 9: 339–45

Moore, J. W., Stickney, K. J. 1980. Formation of attentional-associative networks in real time: Role of the hippocampus and implications for conditioning. *Physiol. Psychol.* 8: 207–17

Pavlov, I. P. 1927. *Conditioned Reflexes*. London: Oxford Univ Press. 430 pp.

Pearce, J. M., Hall, G. 1980. A model for Pavlovian learning: Variations in the effectiveness of conditioned but not of unconditioned stimuli. *Psychol. Rev.* 87: 532–52

Peeke, H. V. S., Hertz, M. F. 1973. *Habituation, Vol. 1, Behavioral Studies.* New York: Academic. 290 pp.

Randich, A., LoLordo, V. M. 1979. Associative and nonassociative theories of the UCS pre-exposure phenomenon: Implications for Pavlovian conditioning. *Psychol. Bull.* 86: 523–48

Reiss, S. Wagner, A. R. 1972. CS habituation produces a "latent inhibition" but no active "conditioned inhibition." *Learn. Motiv.* 3: 237–45

Rescorla, R. A. 1968. Probability of shock in the presence and absence of CS in fear conditioning. *J. Comp. Physiol. Psychol.* 66: 1–5

Rescorla, R. A. 1971. Summation and retardation tests of latent inhibition. *J. Comp. Physiol. Psychol.* 75: 77–81

Rescorla, R. A. 1975. Pavlovian excitatory and inhibitory conditioning. In *Handbook of Learning and Cognitive Processes,* ed. W. K. Estes. 2: 7–35. Hillsdale, N.J.: Erlbaum. 373 pp.

Rescorla, R. A. 1980. *Pavlovian Second-Order Conditioning: Studies in Associative Learning.* Hillsdale, N.J.: Erlbaum. 120 pp.

Rescorla, R. A. 1981. Simultaneous associations. In *Advances in Analysis of Behavior,* ed. P. Harzem, M. Zeiler, 2: 47–80. New York: Wiley. 417 pp.

Rescorla, R. A. 1982a. Comments on a technique for assessing associative learning. In *Quantitative Analysis of Behavior: Volume III, Acquisition,* ed. M. L. Commons, R. J. Herrnstein, A. R. Wagner. pp. 41–63. Cambridge, Mass.: Ballinger. 468 pp.

Rescorla, R. A. 1982b. Effect of a stimulus intervening between CS and reinforcer in autoshaping. *J. Exp. Psychol. Anim. Behav. Proc.* 8: 131–41

Rescorla, R. A. 1984. Associations between Pavlovian CSs and context. *J. Exp. Psychol. Anim. Behav. Proc.* 10: 195–204

Rescorla, R. A. 1985a. Conditioned inhibition and facilitation. See Holland 1985, pp. 299–326

Rescorla, R. A. 1985b. Pavlovian conditioning analogues to Gestalt perceptual principles. In *Affect, Conditioning, and Cognition: Essays on the Determinants of Behavior,* ed. F. R. Brush, J. B. Overmier, pp. 113–30. Hillsdale, N.J.: Erlbaum. 387 pp.

Rescorla, R. A., Holland, P. C. 1976. Some behavioral approaches to the study of learning. In *Neural Mechanisms of Learning and Memory,* ed. M. R. Rosenzweig, E. L. Bennett, pp. 165–92. Cambridge, Mass.: MIT Press. 637 pp.

Rescorla, R. A., Holland, P. C. 1982. Behavioral studies of associative learning in animals. *Ann. Rev. Psychol.* 33: 265–308

Rescorla, R. A., Wagner, A. R. 1972. A theory of Pavlovian conditioning: Variations in the effectiveness of reinforcement and nonreinforcement. In *Classical Conditioning II,* ed. A. H. Black, W. F. Prokasy, pp. 64–99. New York: Appleton-Century-Crofts. 497 pp.

Rozin, P. 1976. The selection of foods by rats, humans, and other animals. In *Advances in the Study of Behavior,* ed. J. Rosenblatt, R. A. Hinde, C. Beer, E. Shaw. 6: 21–76. New York: Academic. 284 pp.

Smith, M. C., Coleman, S. R., Gormezano, I. 1969. Classical conditioning of the rabbits nictitating membrane response in backward, simultaneous, and forward CS-US intervals. *J. Comp. Physiol. Psychol.* 69: 226–31

Stein, L., Sidman, M., Brady, J. V. 1958. Some effects of two temporal variables on conditioned suppression. *J. Exp. Anal. Behav.* 1: 153–62

Sutherland, N. S., Mackintosh, N. J. 1971. *Mechanisms of Animal Discrimination Learning.* New York: Academic. 539 pp.

Thompson, R. F., Spencer, W. A. 1966. Habituation: A model phenomenon for the study of the neuronal substrates of behavior. *Psychol. Rev.* 73: 16–43

Wagner, A. R. 1978. Expectancies and the priming of STM. In *Cognitive Processes in Animal Behavior,* ed. S. H. Hulse, H. Fowler, W. K. Honig, pp. 177–209. Hillsdale, N.J.: Erlbaum. 345 pp.

Wagner, A. R. 1981. SOP: A model of automatic memory processing in animal behavior. In *Information Processing in Animals: Memory Mechanisms,* ed. N. E. Spear, R. R. Miller. pp. 5–44. Hillsdale, N.J.: Erlbaum. 390 pp.

Wagner, A. R., Rescorla, R. A. 1972. Inhibition in Pavlovian conditioning: Application of a theory. In *Inhibition and Learning,* ed. R. A. Boakes, S. Halliday, pp. 301–16. New York: Academic. 568 pp.

Warren, H. C. 1921. *A History of the Association Psychology.* New York: Scribner's. 328 pp.

Yeo, A. G. 1974. The acquisition of conditioned suppression as a function of interstimulus interval duration. *Q. J. Exp. Psychol.* 26: 405–16

Ann. Rev. Neurosci. 1988. 11 : 353–72

TRANSGENIC MICE:
Applications to the Study
of the Nervous System

*Michael G. Rosenfeld, E. Bryan Crenshaw III,
and Sergio A. Lira*

Howard Hughes Medical Institute, Eukaryotic Regulatory Biology
Program, School of Medicine, University of California at San Diego,
La Jolla, California 92093

Larry Swanson

Howard Hughes Medical Institute, Neural Systems Laboratory,
The Salk Institute, La Jolla, California 92037

Emiliana Borrelli, Richard Heyman, and Ronald M. Evans

Howard Hughes Medical Institute, Gene Expression Laboratory,
The Salk Institute, La Jolla, California 92037

INTRODUCTION

The need to explore the molecular basis of development and function of
the brain has motivated neurobiologists to search for new technologies.
In recent years, the advent of elegant experimental approaches has allowed
insights into how neural circuits are established and how they interact. In
this review, we discuss the application of the emerging transgenic mouse
technology to the study of the nervous system.

Recognition of mutant phenotypes has provided the basis for identification of genes critical for developmental events in lower eukaryotes.
Unlike lower eukaryotes, mammalian genetic systems have produced a
very limited number of these developmental mutants. This severe limitation

0147–006X/88/0301–0353$02.00

can now be circumvented by the ability to introduce genes into the germ line of higher organisms. Based on advances in developmental embryology, it is possible to introduce a cloned gene into a fertilized egg and generate animals containing the transferred sequences (Brinster 1972, Capecchi 1980, Gurdon & Melton 1981, Brinster et al 1981a,b, Constantini & Lacy 1981, Stewart et al 1982, Brinster et al 1985; Gordon & Ruddle 1981, Hogan et al 1986). The resultant animals may express the introduced or "trans" gene, and are referred to as transgenic animals.

To manipulate the development and physiology of the nervous system, strategies must be developed that target gene expression to specific cell types. Although numerous genes are expressed in a number of different cell types and tissues, many neural and endocrine genes exhibit highly restricted and often complex patterns of expression. Identification of regulatory elements that lead to such precise spatially and temporally limited patterns of gene expression are the subject of intensive investigation. Because transgenic technology allows the introduction of cloned genes into every cell of the mouse, this technology provides an in vivo assay system for identifying putative regulatory sequences and for determining their activity in every cell and tissue. Identification of such regulatory sequences will provide a means to accurately target specific gene products to the predicted regions in the mature animal. For example, directing the expression of an oncogene to a specific cell type holds the promise of generating new cell lines. Further, expression of specific gene products can be used to generate models of disease and identify molecular correlates of neural function.

GENERATION OF TRANSGENIC ANIMALS

The basic experimental strategy for application of transgenic animals to biological questions has been to introduce a gene whose product is conveniently detected and/or functional in the recipient animal. Although genes have been successfully microinjected into fertilized oocytes of a number of species (Hammer et al 1985), generation of transgenic mice has been the model used for all experiments reviewed here.

The introduced DNA should generate a unique, detectable product. One strategy is to introduce a linear DNA fragment that contains sequences required for directing correct transcriptional initiation and encodes a gene product that can be distinguished from the potential expression of any comparable endogenous mouse gene. This can be accomplished by using a gene derived from a heterologous species with detectable sequence variation or, if required, by modifying the introduced gene. A second approach has been to create chimeric genes, in which putative regulatory sequences are fused to DNA encoding a detectable reporter product. Reporter function could encode a biologically active molecule, such as an oncogene or regulatory peptide, a specific antigen, or an enzyme with activity distinct from that of endogenous mouse enzymes. The stability of the RNA or protein product of the transgene is a major consideration in choosing the type of analysis necessary for detection. Whereas a stable reporter molecule is crucial to successful analysis for immunohistochemistry, it is less important for sensitive enzymatic detection methods.

Once the appropriate recombinant clone is constructed, the DNA is microinjected into the male pronucleus of fertilized oocytes, and the oocytes are transferred to the oviducts of pseudopregnant females (Gordon & Ruddle 1983, Brinster at al 1985, Hogan et al 1986). The efficiency of generating transgenic mice varies between laboratories and for the injected DNA, but percentages of > 60% viable oocytes after microinjection and > 25% of resultant births as transgenic animals have been obtained (Brinster et al 1985, Hogan et al 1986). Although the precise mechanisms of genomic integration events remain unknown, even in the usual case when more than one copy of the transgene integrates, they do so as tandem arrays at a single genomic locus (Gordon & Ruddle 1983). There is no evidence for site specific integration; rather, in a series of transgenic animals, integration will occur at many, apparently random, sites (Brinster et al 1985, Lacy et al 1983). Because pedigrees will exhibit different integration sites, any tissue-specific gene expression observed would depend on information present in the transgene rather than chromosomal position.

As integration occurs at a stage of embryogenesis prior to DNA replication, most (> 60%) of transgenic mice contain the introduced gene in all somatic and germ cells. The remaining animals are mosaic, with only some cells containing the transgene. The transgene is usually stably integrated into the genome of the mouse, and is normally segregated in the progeny. Thus, Mendelian patterns of inheritance occur, except in the case of mosaic animals lacking germ line expression of the transgene. Gene rearrangement and amplification events can involve the transgene (Palmiter et al 1983b).

The litters from the pseudopregnant recipient are initially screened to determine whether the transgene DNA is present in the genome. Conventionally, this is done by hybridization analysis using DNA extracted from a piece of mouse tail. Once the transgenic founder animals are identified, appropriate matings are used to generate pedigrees for analysis. Analysis of gene expression in each pedigree is accomplished by assessment of specific RNA expression in each tissue, or by measurement of the expressed proteins. RNA expression can be quantified by hybridization techniques or by procedures that confirm correct RNA structure (e.g. S1 nuclease protection assays). The advantages of RNA analysis is that both qualitative and quantitative gene expression are determined irrespective of mRNA translation or product stability.

The protein product can be assayed by enzymatic assay, or by immunological means. In situ hybridization and immunohistochemical procedures provide the ability to analyze RNA and protein expression, respectively, at a cellular level. Based on the availability of these analytic methods, both the temporal and spatial patterns of transgene expression can be determined quantitatively. Finally, because the expressed product may affect cell function, a variety of morphological or physiological alterations may result from transgene expression.

TISSUE-SPECIFIC GENE EXPRESSION

To use the transgenic technology effectively to study questions in the central nervous system, gene expression must be accurately targeted to precise subsets of neurons. We briefly review recent advances in understanding of the genomic determinants of cell-specific gene expression before describing evidence of correct developmental targeting in transgenic animals.

Genomic Determinants of Developmental Specificity

Since the discovery of DNA as the genetic material, one of the central mysteries has been the nature of the code that dictates tissue-specific gene expression. A detailed analysis of this question has been facilitated by the ability to clone and mutagenize developmentally regulated genes, and the ability to assess expression in homologous and heterologous cells by the technique of DNA-mediated gene transfer.

Based on these analyses for a large number of transcription units, specific cis-active genomic sequences clearly are critical for the qualitative and quantitative features of accurate gene transcription (Gluzman & Shenk 1983). The promoter of a gene is a region of DNA the precedes the coding sequence and that regulates transcription precisely. A typical promoter contains a TATA box located 25–30 base pairs (bp) upstream of the mRNA transcription start site that orients RNA polymerase at the initiation site. In addition, single or multiple upstream elements located 40 to 110 base pairs upstream of the transcription initiation site are involved in the frequency of the initiation of transcription. Additional cis-active sequences that can be found throughout the gene are referred to as "enhancers," as they strongly enhance the transcription of genes to which they are linked. Enhancer elements stimulate transcription irrespective of orientation and position, and they can exert their effect over distances of kilobases (Gluzman & Shenk 1983, Luciw et al 1983, Payvar et al 1983, Stuart et al 1985).

The initial definition of regions with roles in tissue-specific gene expression was obtained from transfection of fusion genes containing the suspected regulatory sequences into cells that either normally express, or fail to express, the gene. Based on deletion and mutation analyses, sequences enhancing cell-specific expression have been identified for several genes (e.g. Gruss et al 1981, Gillies et al 1983, Hen et al 1983, Walker et al 1983). Considerable evidence has suggested that these cis-active sequences intereact with DNA-binding proteins and that the DNA-protein complexes modulate gene activity (e.g. Ephrussi et al 1985, Scheidereit et al 1983, Sassone-Corsi et al 1985). Genes under complex regulation likely have multiple enhancers and binding proteins. Thus, the activity of genes may depend on the combined actions of several trans-acting factors (e.g. Edlund et al 1985, Nelson et al 1986, Sassone-Corsi & Borrelli 1986). Conversely, evidence has been presented suggesting that repressor or silencer sequences might act to prevent expression in particular cells (Gorman et al 1985, Borrelli et al 1984, Brand et al 1985). Such repressors could either modify the positive *trans*-acting factors or bind to *cis*-active regulatory sequences. thus, both position and negative regulation may ultimately dictate the pattern of gene expression.

The techniques used for the in vitro study of gene expression, such as DNA-mediated gene transfer into cells, have obvious limitations for the analysis of gene expression during development for two reasons. First, the number of available cell lines limits the analysis of many genes that have already been cloned. Second, clonal cell lines usually represent a particular stage of cellular development and therefore do not allow analysis of the process of differentiation. For these reasons the transgenic technology has been essential for evaluating models of cell-specific gene expression.

Genes Developmentally Restricted to a Single Cell Type

One example of a gene product that provides a model of tissue-specific gene expression in terminally differentiated cells is provided by elastase I, a digestive enzyme synthesized and secreted by the acinar cells of the exocrine pancreas. Initially, a 27 kilobase (kb) genomic fragment containing 7 kb of upstream flanking sequences, 11 kb of exon and intron sequences, and 5 kb of downstream flanking sequences was used to generate transgenic mice (Swift et al 1984). The integrated gene exhibited the physiological pattern of tissue-specific expression. The regulatory information contained in the gene appeared sufficient to specifically direct the normal quantitative expression of the gene product in the appropriate tissue. The transgenic animal technique was used further to delineate the region that directed the pancreas-specific expression of the elastase I gene. Analysis of a series of deletions showed that a 213 bp sequence (between positions −205 and +8) was sufficient to insure the specific expression of the elastase I gene in the pancreas of transgenic mice (Ornitz et al 1985). A cis-acting element(s) present in the 213 bp fragment is likely recognized by trans-acting factors specifically present in pancreatic cells. Thus, in the case of the elastase I gene, a small genomic region contains the information required for the tissue-specific regulation of gene expression in vivo.

Similar results have been observed in the case of other genes, including β-globin (Magram et al 1985, Townes et al 1985), immunoglobulin genes (Storb et al 1984, Grosschedl et al 1984, Yamamura et al 1986), insulin (Burki & Ullrich 1982, Bucchini et al 1986), α-crystallin genes (Overbeek et al 1985), and α2 type 1 collagen gene (Khillan et al 1986).

Another example is provided by analysis of genes encoding pituitary-specific hormones. In the case of the rat prolactin and growth hormone genes, sequences that appear to be important for cell-specific expression have been identified using DNA-mediated gene transfer into cultured cell lines (Nelson et al 1986). Introduction of fusion genes containing the prolactin and growth hormone promoters, respectively, resulted in pituitary-specific gene expression (E. B. Crenshaw et al, S. A. Lira et al, unpublished observations).

Genes Expressed in Several Tissues

Although many genes analyzed in transgenic mice thus far have contained only one defined tissue-specific enhancer, some genes expressed in several tissues may contain multiple enhancers. Expression of the alpha feto-protein (AFP) gene is directed by no less than three enhancers with overlapping tissue specificities. AFP is expressed during fetal development in the visceral endoderm of the yolk sac, in fetal liver, and in the fetal gastrointestinal tract. AFP mRNA expression constitutes about 20% of polyadenylated transcripts in yolk sac, about 5% in fetal liver, and about 0.1% in the fetal gastrointestinal tract. To examine the tissue-specific expression, transgenic mice were created carrying a construct consisting of 6.6 kb of 5′ flanking sequences (Krumlauf et al 1985, Hammer et al 1987). This transcription unit directed expression to all three tissues that normally express the endogenous AFP gene, but the liver overexpressed the transgene several-fold over endogenous expression, while in the gastrointestinal tract expression was often greater than 5000-fold higher than normal. This expression level was seen even when more 5′ flanking information was introduced, suggesting that any repressing elements are not within 14 kb upstream of the AFP gene. When a gene consisting of only

1 kb of 5′ flanking sequences was introduced, no expression of the transgene was observed in the resultant animals.

Three enhancer regions that were previously defined by transient transfection into hepatoma cells were then attached to the inactive 1 kb 5′ flanking region, and appropriate expression in the three tissues was obtained. The enhancer in Region I conferred specificity predominantly to the fetal gut, but also directed weak expression to the fetal liver and yolk sac. Regions II and III directed expression predominantly to the fetal liver and yolk sac. This analysis revealed that all three enhancers were capable of directing expression to the three target tissues, but that each of the enhancers demonstrated a characteristic tissue preference. Based on their results, Hammer et al (1987) propose that these elements interact to confer the multiple tissue specificity observed with the alpha fetoprotein. In contrast, the *Drosophila* yolk protein gene contains an enhancer that confers specificity to fat bodies and one that confers specificity to the ovary (Garabedian et al 1986). The yolk protein enhancers work independently to direct the tissue specificity of the yolk protein. These data exemplify two mechanisms by which multiple enhancers can direct gene expression to different tissues.

Complexity of the developmental signals affecting gene expression is further illustrated by analysis of fusion genes in transgenic mice. For example, fusion of the mouse metallothionein-I (MT-I) promoter to the growth hormone gene produces widespread expression (Palmiter et al 1982). This pattern of expression parallels the tissue specificity of the MT-I promoter. In the brain, though, the fusion gene is expressed in neurons that do not detectably express either the endogenous metallothionein or the growth hormone (GH) genes (Swanson et al 1985). In addition, chimeric genes containing the metallothionein promoter fused to either rat growth hormone or somatostatin sequences result in unexpected expression in pituitary gonadotrophs (Swanson et al 1985, Low et al 1986). These data are consistent with the operation of combinatorial determinants of developmental patterns of gene expression.

Post-transcriptional Developmental Strategies

Alternative RNA processing is a developmental strategy that determines the phenotype in a variety of cells (reviewed in Leff et al 1986). In neurons and in the endocrine system, examples of neuropeptide splicing choices include the calcitonin/calcitonin gene-related peptide (CGRP) (Amara et al 1982), prekininogen (Kitamura et al 1983), and substance P/substance K (Nawa et al 1984) genes. These alternative RNA processing events would appear to exert important effects on development and function of complex systems such as the brain. In the case of the calcitonin/CGRP gene, the peptides calcitonin and CGRP are encoded by alternatively spliced transcripts from the same gene (Amara et al 1982). Calcitonin transcripts are primarily observed in thyroid C cells whereas CGRP transcripts are expressed in specific neurons in the central and peripheral nervous systems, as well as the adrenal medulla (Rosenfeld et al 1983).

To gain insight into developmentally regulated RNA processing, a mouse metallothionein I-calcitonin/CGRP fusion gene was introduced into fertilized mouse ova. The fusion gene was widely expressed and it was thus possible to ask what processing choices each tissue made. Most visceral and muscle tissues expressed the gene, and calcitonin mRNA was the primary product (Crenshaw et al 1987). In contrast, CGRP transcripts were the predominant neuronal product in the brain, and expressed in many neurons, which do not express the endogenous gene. Interestingly, a small fraction of neurons appear to produce calcitonin transcripts. Based on these results, calcitonin expression seems to be the default or "null" choice, while a specific RNA processing machinery, expressed primarily in neurons, is required for processing of CGRP transcripts.

Transgenic animals also provide a means for evaluating tissue-specific machinery for the specific protein processing events required for generation of mature neuropeptides. For example, analysis of the expression of a metallothionein-rat somatostatin fusion gene showed that the pre-somatostatin transgene product was correctly processed in the pituitary gland but not in the liver (Low et al 1985). This observation extends the well documented examples of cell-type specific proteolytic processing of hormone precursors (Lynch & Snyder 1986).

TUMORIGENESIS IN TRANSGENIC MICE

One of the more powerful aspects of the transgenic mouse technique is the ability to manipulate the physiology of the mouse to create models for pathological conditions. Tumorigenesis is a well-studied example of the usefulness of this technique. Several groups have directed expression of oncogenes to many tissues in the mouse, including endocrine and neuronal structures. Tumors have been produced in the lymphoid system with *myc* fusion genes (Adams et al 1985), in mammary gland with mouse mammary tumor virus/*myc* fusion genes (Stewart et al 1984), in the skin with bovine papilloma virus (Lacy et al 1986), and in exocrine pancreas with a mutant *ras* gene that was isolated from EJ bladder carcinoma (Palmiter & Brinster 1986). We discuss tumors produced by transgenes in the brain and in the endocrine system to illustrate how this type of analysis can be applied to neurobiological investigations. These studies demonstrate two general uses for oncogene expression in transgenic mice: (*a*) using a tumor phenotype as an assay for tissue-specific expression of cloned genes, and (b) creating cell lines from these tumors that provide novel cell culture systems for studying the cloned gene product.

One of the initial examples of tumorigenesis in transgenic mice was found in the brain of animals that had been microinjected with a fragment of the SV40 genome (Brinster et al 1984, Palmiter et al 1985, Messing et al 1985). The injected fragment consisted of the early region, which encodes for two viral proteins—large T and small t (tumor) antigens—that have been shown to induce tumors in vivo and in cell culture systems. Transgenic mice that inherit the SV40 sequences develop choroid plexus papillomas at approximately five months of age. The tumors are characterized by expression of the T antigen, while normal tissues express only low levels of T antigen. Often the T antigen sequences are amplified in the tumor and in cell lines isolated from these animals. Similarly, when the viral particles are injected into the ventricles of the brain, the virus induces choroid plexus papillomas, reminiscent of the pathology seen with the transgenic mouse model.

Early regions from other papovaviruses can induce tumorigenesis in several organ systems; other pathological effects are discussed separately. The distribution of cancers reflects the tissue tropism associated with viral infection. The early region of the human papovavirus, JC, was introduced into transgenic mice (Small et al 1986). These viruses can induce a number of tumors of neural origin after injection into newborn hamsters. The JC virus can induce glioblastomas, neuroblastomas, and meningiomas (Small et al 1986). When the early region of JC virus is introduced into mice, the resultant animals develop primarily adrenal neuroblastomas. As seen with mice microinjected with the SV40 early region, tumorigenesis in transgenic mice containing either human papavovirus is associated with the activation of the T antigen expression in the tumors, whereas the normal tissues show low levels of T antigen expression.

The SV40 early region has been used as an oncogenic reporter molecule for the tissue-specific expression of the rat insulin-II gene (Hanahan 1985). This early region was ligated to a fragment of the insulin promoter (-660 to $+8$ bp) that had previously been shown by transfectional analyses to contain the tissue-specific enhancer element (Walker et al 1983). Mice that inherit the chimeric construction die from pancreatic tumors at 9–12 weeks of age. The insulin promoter directs the expression of the SV40 region exclusively to the pancreas, since no tumors arise in other tissues and the T antigen protein is detected only in the pancreas. Although all of the islets express T antigen in young animals, only a few islets give rise to tumors. These facts indicate that secondary events are necessary for the development of tumors in these animals. T antigen expression, found early in the development of the pancreas long before the development of tumors in these animals, has a profound effect on the development of the islets of Langerhans, because it disrupts the normal distribution and number of α, β, and δ cells within the islets. This observation has important implications for the study of the cell-cell interactions that are necessary for the normal development of the pancreas, but more importantly, it has broader implications about the usefulness of this strategy to probe development in other organ systems, such as the nervous system.

SV40 T antigen has been used as a marker with other promoter/enhancer elements. Palmiter and his colleagues have directed tumorigenesis to pancreatic acinar cells with the elastase gene (Ornitz et al 1985) and to the choroid plexus with the SV40 early promoter (Brinster et al 1984). We have used 3 kb of 5′ flanking information of the prolactin gene to direct tumorigenesis to lactotrophs (E. B. Crenshaw et al, manuscript in preparation). A major advantage of using the SV40 early region is the high penetrance of the tumorigenic phenotype with a number of gene promoters; other oncogenes, such as *myc* and *v-src*, have had limited or no transforming potential with the cell types tested (see Palmiter & Brinster 1986).

Another major advantage of the use of oncogenes for reporter molecules is that it allows the isolation of cell lines from tissues that express cloned gene products. Cell lines could conceivably be isolated by transformation of rare cell types, thereby creating novel cell culture systems. This approach has been used to isolate cell lines from exocrine pancreas (R. D. Palmiter, personal communication), choroid plexus tumors (Small et al 1985), and lactotrophs (E. B. Crenshaw et al, manuscript in preparation). There are several potential problems with this approach. First, the cell type that is being isolated must be susceptible to the oncogenic product being expressed. This point is illustrated by the fact that widespread expression of the *myc* gene results in pathology in only a subset of tissues in which it is expressed (Leder et al 1986). Widespread expression of a mouse mammary tumor virus (MMTV)/*myc* fusion gene induces pathology in only a subset of tissues. Also, *myc* does not transform acinar cells when it is expressed from the elastase promoter (see Palmiter & Brinster 1986). This restriction of tumor-promoter activity has not been observed with the SV40 early region, but only a limited number of cell types have been examined. The second potential problem is that cell lines must be isolated and maintained in cell culture. Third, the maintenance of the differentiated phenotype in vitro may prove difficult. Cell lines isolated from choroid plexus tumors show a progressively more transformed phenotype in cell culture (Small et al 1985). We have also observed an increase in the expression of T antigen in lactotroph cell lines maintained in cell culture (unpublished observations). This increase in T antigen expression seems to be correlated with a progressive loss of the lactotroph phenotype. Overall, this strategy may provide some very important new cell lines, but the strategy must be refined to produce more consistent and stable results.

MODELS OF HUMAN DISEASE

Production of Disease Models

The introduction of functional genes into the germline of animals provides an opportunity to develop animal models of particular human diseases. Progressive multifocal leukoencephalopathy (PML) is characterized by degenerative demyelination in the central nervous system (CNS) in which oligodendrocytes and astrocytes are abnormal and focal lesions with demyelination are present (Astrom et al 1958, ZuRhein 1969). Human papovavirus particles have been isolated from brain lesions of patients with PML and thus have been implicated as the causative agent of this disease (ZuRhein 1969). In addition, JC virus is thought to be involved in the formation of multiple glial tumors in some patients with PML (Giarusso & Koeppen 1978).

Transgenic mice that expressed the entire JC virus early region displayed at least two distinct phenotypes in transgenic mice (Small et al 1986). Four lines of mice developed neuroblastomas of the adrenal medulla, as discussed above. Two other lines of mice exhibited a neurological disorder, characterized by a shaking phenotype that was manifested at 2 to 4 wk of age. These tremors became progressively worse; at 3 wk of age the mice developed tonic seizures and at 4-6 wk of age they died. Neuropathological analysis revealed an abnormal formation of the myelin sheath in the CNS. The myelin deficiency in the CNS as well as the shaking phenotype correlated with the level of expression of the JC tumor-antigen in brain tissue. These observations suggest that demyelination may be a direct result of JC tumor-antigen expression in the brain. The similar neuropathology associated with the human disease PML, as well as the isolation of JC viral particles from these focal lesions, suggests that JC virus-containing transgenic mice provide an animal model for studying this disease.

Gene Therapy

Mutant mice that carry phenotypes that closely resemble those displayed by humans with inherited diseases have long been the subject of active research. Although the precise molecular lesion of many of these mutants is unknown, in a few circumstances they have been defined and animal strains established. Recently, investigators have taken advantage of the existence of animal models and of the expanding availability of cloned genes to conduct gene therapy experiments by germ-line incorporation of new genes. Partial or total correction of inherited diseases such as a murine growth disorder (Hammer et al 1984), immune response deficiency (Le Meur et al 1985, Yamamura et al 1985), and a murine β thalassemia (Constantini et al 1986) have been reported.

The feasibility of correcting hereditary disorders in the nervous system by gene replacement has been elegantly demonstrated by Mason and co-workers (1986a,b). Reproductive function was restored in the hypogonadal (*hpg*) mouse line, which contains a mutation affecting the neuroendrocrine control of reproduction. Mice affected by this autosomal recessive mutation fail to develop the gonads and reproductive organs postnatally, and characteristically they have low or absent levels of gonadotropin-releasing hormone (GnRH) in their hypothalamus. This defect precludes normal regulation of the gonadotrophs, which in turn severely hinders sexual and reproductive development.

To restore the reproductive function in the *hpg* mice it was first necessary to define precisely the nature of the genetic lesion. Utilizing the rat GnRH cDNA probe, a single gene encoding a precursor protein for GnRH and GnRH-associated peptide (GAP) was identified. Comparison of the wild-type GnRH gene locus with that of the mutant animals established that the *hpg* GnRH gene had suffered a deletion of approximately 33.5 kb. Although a major portion of the GAP-coding regions was affected, both the promoter and the coding region for GnRH were intact. Finally, low level of GnRH mRNA sequences were detected by in situ hybridization analysis in the *hpg* hypothalamus but no GnRH peptide was detected by immunohistochemistry. The presence of nontranslated GnRH transcript confirmed the existence of the GnRH-neuronal pathway and suggested that the genomic deletion alone was responsible for the mutant phenotype.

An intact GnRH gene was introduced as a 13.5 kb fragment (the wild-type GnRH gene plus 5 kb of upstream and 3 kb downstream sequences) into wild-type mouse zygotes, and two transgenic animals were generated. Mating of *hpg*/+ heterozygotes with wild-type transgenic mice carrying more than 20 copies of the transgene generated an *hpg/hpg* homozygous mouse that expressed the transgene in the hypothalamus. The transgenic *hpg/hpg* homozygotes displayed normal reproductive function, exhibited neural specific expression of GnRH at levels comparable to unaffected mice, and were capable of mating and raising healthy litters. Furthermore, pituitary and serum levels of luteinizing hormone, follicle-stimulating hormone, and prolactin were restored to those of the wild-type mouse. Although the expression was not absolutely specific (low levels of expression were observed in the liver as well as in the paraventricular nucleus of the thalamus) regulation sufficient to restore reproductive function was attained.

In another example, gene augmentation was attempted to correct a genetic defect associated with the somatotroph cells of the pituitary (Hammer et al 1984). A mutant inbred strain of dwarf mice called *little* appear to have a genetic defect in pituitary function. Because intact human and rat growth hormone genes failed to express in unaffected transgenic animals, a heterologous fusion gene, metallothionein–rat growth hormone (MT/GH), was used to restore growth to the *little* mice. These mice had vastly elevated levels of growth hormone compared to their wild-type controls and greatly increased body size (1.5 times larger than normal mice). Reduced fertility was observed in females, analogous to unaffected transgenic mice expressing the MT/GH fusion genes (Palmiter et al 1982).

These examples clearly demonstrate the potential use of targeted (*hpg* mice) and nontargeted (*lit/lit* mice) gene replacement to rescue genetic mutations. A recent extension of this approach has been used to modify the phenotype of the *shiverer* mouse mutant, which appears to correlate with defective myelin basic protein gene expression. In this case the cloned myelin basic protein gene, expressed in transgenic mice, corrected the abnormal phenotype (Readhead et al 1987).

CONSEQUENCES OF TRANSGENE EXPRESSION

The growth and development of an organism is the result of complex interactions among genetic, hormonal, nutritional, and environmental factors. One approach to examine the genetic basis of growth, development, or differentiation is to study the physiological consequences of the expression of a transgene believed to be involved in the regulation of these processes.

Two genes encoding peptides that control somatic growth—growth hormone-releasing factor (GRF) and growth hormone (GH)—have been introduced into the germline of animals under the control of the mouse MT-I promoter (Palmiter et al 1982, 1983a, Hammer et al 1985). GRF is a hypothalamic neuropeptide that functions to increase the expression and secretion of GH in the somatotroph cells of the pituitary. GH is secreted by the anterior pituitary and has marked effects on growth and development in animals. Transgenic mice containing the MT-GH gene had an accelerated rate of growth that began at 3 wk of age and leveled off at about 12 wk of age, at which time the mice were as much as twice the size of their control littermates (Palmiter et al 1982, 1983a). These mice had high levels of expression of the fusion mRNA in their liver and very elevated GH in their serum. Histological examination of transgenic pituitaries revealed decreased numbers of somatotrophs, which most probably was caused by the chronically elevated GH levels.

Analysis of MT-GRF transgenic mice demonstrated detectable GRF levels in serum and elevated levels of serum GH (Hammer et al 1985). The mice showed significant increase in growth and were 25–50% larger than control littermates. The MT-GRF and the MT-GH fusion genes were expressed with a tissue specificity similar to the MT-I gene; i.e. high levels in the liver, kidney, gut, and pancreas, with lower levels in the brain and spleen. One exception, however was the high levels of MT-GRF expression in transgenic pituitaries, in which the levels of the transgene RNA in the pituitary were as high as those seen in the liver. These transgenic MT-GRF mice had hyperplastic pituitaries routinely 3–10 fold greater (and up to 60-fold) in size than control mice. The hyperplastic pituitaries were the result of the selective proliferation of the somatotrophs. This suggests that GRF exerts an effect on the development of a particular cell type by promoting their proliferation. The trophic effects of GRF are likely to be prototypic for other neuropeptides in the central nervous system.

Insertional Mutagenesis

Because the integration site of microinjected DNA is probably random, integration of DNA into the host genome could disrupt the function of an endogenous gene, thereby producing an insertional mutation. The inserted DNA fragments then become markers for genes associated with the mutant phenotype, and facilitate the isolation and characterization of these mutant genes.

Insertional mutations in transgenic mice have created defects affecting limb development (Woychik et al 1985). In a line of mice bearing the MMTV-myc gene, a mutation was isolated that affected the pattern of limb formation in the developing mouse. The transgene and adjoining DNA sequences were cloned. About 1 kb of DNA was deleted during integration. The insertion was shown to be on chromosome 2. It is phenotypically identical and does not complement two previously isolated limb deformity mutations. However, the normal gene has not yet been identified.

A second limb deformity has been identified resulting from insertion of a gene containing the long terminal repeat of Rous sarcoma virus linked to the bacterial gene chloramphenicol acetyltransferase (Overbeek et al 1986). The insertion resulted in mice that has fused toes in the hind and fore feet. A second line of these transgenic animals resulted in embryonic lethality. Other mutations have been identified in transgenic mice. In certain cases the insertional mutation disrupts development prior to implantation (Mark et al 1985), whereas in others insertional events alter normal development following implantation (Wagner et al 1983).

An interesting insertional mutation affecting adults was found in transgenic mice containing an inverted repeat of a MT-I/thymidine kinase fusion gene (Palmiter et al 1983b). The transgene is transmitted normally by females but never by males. Because the males are fertile, it has been proposed that the inserted gene disrupts a gene that must be expressed during the haploid stages of spermatogenesis.

An alternative means of creating insertional mutations has been accomplished by retroviral insertion. In one case, the α2 collagen gene, in which the insertion occurred, has been identified (Schnieke et al 1983, Harbers et al 1981, Lohler et al 1984, Jaenisch et al 1985, Hartung et al 1986).

Although technically demanding, interesting phenotypes produced by insertional mutagenesis could prove a means to identify genes involved in specific developmental and regulatory systems.

CONCLUSION

Understanding the developmental mechanisms responsible for neuronal structure and phenotype is a prerequisite for insights into the origins of complex higher mental functions and the molecular basis of underlying disorders of the nervous system. Insight into these issues in vertebrate models has been slow, in part, because of the absence of a significant genetic approach. One of the potentially most important aspects of the transgenic animal technology is that it provides an approach to characterize genetic function in the absence of any genetic selections. In the near future this technology will probably be used to identify the specific sequence elements required for tissue-specific expression in the nervous system. This will provide the first insights into the generation of the complex developmental specificities that are apparently central to neuroendocrine gene function. Already, this technology has implications beyond simply being able to identify important regulatory elements associated with developmental control of gene expression. The introduction of viral genes into germ lines of animals has led to a number of outcomes, including tumors of the choroid plexus as well as a demyelinating syndrome associated with the JC virus. The potential for generating models of neuronal disease is likely to increase tremendously as genes encoding regulatory proteins and growth factors are isolated.

Similarly, transgenic animal technology may make the difficult investigation of the molecular basis of cell recognition a more tractable problem. In conjunction with advanced histochemical techniques, immunohistochemistry, and in situ hybridization, it holds promise for new under-standing of the organization and function of the vertebrate nervous system. As we have presented, introduction of functional molecules have already provided insights into neuropeptide function and the actions of oncogenes during development.

Several critical advances would accelerate the applicability of the transgenic approach to neurobiology. First, the identification and engineering of regulatory sequences that allow conditional expression of protein products would permit correlation of transgene function with specific functional or developmental events. Gene products with regulatory or cytotoxic functions could then be used to probe functional and/or morphological aspects in the nervous system. Alternatively, functional impairment of a neural circuit could be achieved by inactivating a single endogenous transcript or protein at any given time. In this regard, technologies are being explored that cause a potential disruption of protein or RNA function by cell-specific expression of antibodies or "anti-sense" RNAs. Second, if site-specific insertion were possible, specific gene deletions would be a powerful genetic approach to assigning functional importance to specific gene products and would have obvious advantages for gene therapy.

The relatively few transgenic animals analyzed thus far represent first steps in working out the technical and practical approaches to addressing these important issues. Even at this early stage, it appears that transgenic animal technology is likely to have a fundamental impact on our under-standing of the development and function of the vertebrate nervous system.

ACKNOWLEDGMENTS

E. B. Crenshaw is a predoctoral fellow in the Department of Biology, University of California, San Diego. S. A. Lira is on leave of absence from the Universidade Federal de Pernambuco, Recife, Brazil. Emiliana Borrelli is on leave from the Unite 184 de Biologie Moleculaire et du Genie Genetique de l'INSERM, Strasbourg, France. Richard Heyman is a post-doctoral fellow at the Salk Institute for Biological Studies.

Literature Cited

Adams, J. M., Harris, A. W., Pinkert, C. A., Corcoran, L. M., Alexander, W. S., Cory, S., Palmiter, R. L. 1985. The c-myc oncogene driven by immunoglobulin enhancers induces lymphoid malignancy in transgenic mice. *Nature* 318: 533–38

Amara, S. G., Jonas, V., Rosenfeld, M. G., Ong, E. S., Evans, R. M. 1982. Alternative RNA processing in calcitonin gene expression generates mRNAs encoding different polypeptide products. *Nature* 298: 240–44

Astrom, K. E., Mancall, E. L., Richardson, E. P. 1958. Progressive multifocal leuko-encephalopathy. *Brain* 81: 93–111

Brand, A. H., Breeden, L., Abraham, J., Sternglanz, R., Nasmyth, K. 1985. Characterization of a "silencer" in yeast: A DNA sequence with properties opposite to those of a transcriptional enhancer. *Cell* 41: 41–48

Brinster, R. L. 1972. Cultivation of the mammalian embryo. In *Growth Nutrition and Metabolism of Cells in Culture*, ed. G. Rothblat, V. Cristafalo. 2: 251–86. New York: Academic

Brinster, R. L., Chen, H. Y., Trumbauer, M. E. 1981a. Mouse oocytes transcribe injected *Xenopus* 5S RNA genes. *Science* 211: 396–98

Brinster, R. L., Chen, H. Y., Trumbauer, M. E., Senear, A. W., Warren, R., Palmiter, R. D. 1981b. Somatic expression of herpes thymidine kinase in mice following injection of a fusion gene into eggs. *Cell* 27: 223–31

Brinster, R. L., Chen, H. Y., Messing, A., Van Dyke, T., Levine, A. L., Palmiter, R. D. 1984. Transgenic mice haboring SV40 T antigen genes develop characteristic brain tumors. *Cell* 37: 367–80

Brinster, R. L., Chen, H. Y., Trumbauer, M. E., Yagle, M. K., Palmiter, R. D. 1985. Factors affecting the efficiency of introducing foreign DNA into mice by microinjecting eggs. *Proc. Natl. Acad. Sci. USA* 82: 4438–42

Borrelli, E., Hen, R., Chambon, P. 1984. Adenovirus-2 E1A products repress enhancer-induced stimulation of transcription. *Nature* 312: 608–12

Bucchini, D., Ripochi, M-A., Stinnakre, M-G., Debois, P., Lores, P., Monthioux, E., Absil, J., Lepesant, J-A., Pictet, R., Jami, J. 1986. Pancreatic expression of human insulin gene in transgenic mice. *Proc. Natl. Acad. Sci. USA* 83: 2511–15

Burki, K., Ullrich, A. 1982. Transplantation of the human insulin gene into fertilized mouse eggs. *EMBO J.* 1: 127–31

Capecchi, M. R. 1980. High efficiency transformation by direct microinjection of DNA into cultured mammalian cells. *Cell* 22: 479–88

Cattanach, B. M., Iddon, C. A., Charlton, H. M., Chiapa, S. A., Fink, G. 1977. *Nature* 269: 388–40

Chada, K., Magram, J., Raphael, K., Radice, G., Lacy, E., Constantini, F. 1985. Specific expression of a foreign β-globin gene in erythroid cells of transgenic mice. *Nature* 314: 377–80

Constantini, F., Lacy, E. 1981. Introduction of a rabbin β-globin gene into the mouse germ line. *Nature* 294: 92–94

Constantini, F., Chada, K., Magram, J. 1986. Correction of murine β-thalassemia by gene transfer into the germ line. *Science* 233: 1192–94

Crenshaw, E. B., Russo, A. F., Swanson, L. W., Rosenfeld, M. G. 1987. Neuron-specific alternative RNA processing in transgenic mice expressing a metallothionein-calcitonin fusion gene. *Cell* 49: 389–98

Edlund, T., Walker, M. D., Barr, P. J.,

Rutter, W. J. 1985. Cell-specific expression of the rat insulin gene: Evidence for role of two distinct 5′ flanking elements. *Science* 230: 912–16

Ephrussi, A., Church, G. M., Tonegawa, S., Gilbert, W. 1985. Lineage-specific interactions of an immunoglobulin enhancer with cellular factors in vivo. *Science* 227: 134–40

Garabedian, M. J., Shepherd, B. M., Wensink, P. C. 1986. A tissue-specific transcription enhancer from the *Drosophila* yolk protein 1 gene. *Cell* 45: 859–67

Giarusso, M. H., Koeppen, A. H. 1978. Atypical progressive multifocal leukoencephalepathy and primary cerebral malignant lymphoma. *J. Neurol. Sci.* 35: 391–98

Gillies, S. D., Morrison, S. L., Oi, V. T., Tonegawa, S. 1983. A tissue-specific transcription enhancer element is located in the major intron of a rearranged immunoglobulin heavy chain gene. *Cell* 33: 717–28

Gluzman, Y., Shenk, T., eds. 1983. *Enhancers and Eukaryotic Gene Expression*. Cold Spring Harbor, NY: Cold Spring Harbor Lab.

Gordon, J. W., Ruddle, F. H. 1981. Integration and stable germ line transmission of genes injected into mouse pronuclei. *Science* 214: 1244–46

Gordon, J. W., Ruddle, F. H. 1983. Gene transfer into mouse embryos: Production of transgenic mice by pronuclear injection. *Meth. Enzymol.* 101: 411–33

Gorman, C. M., Rigby, P. W. J., Plane, D. 1985. Negative regulation of viral enhancers in undifferentiated embryonic stem cells. *Cell* 42: 519–26

Grosschedl, R., Weaver, D., Baltimore, D., Constantini, F. 1984. Introduction of a Mu immunoglobulin gene into the mouse germ line: Specific expression in lymphoid cells and synthesis of functional antibody. *Cell* 38: 647–58

Gurdon, J. B., Melton, D. A. 1981. Gene transfer in amphibian eggs and oocytes. *Ann. Rev. Genet.* 15: 189–218

Gruss, P., Dhar, R., Khoury, G. 1981. Simian virus 40 tandem repeated sequences as an element of the early promoter. *Proc. Natl. Acad. Sci. USA* 78: 943–47

Hammer, R. E., Palmiter, R. D., Brinster, R. L. 1984. Partial correction of murine hereditary disorder by germ-line incorporation of a new gene. *Nature* 311: 65–67

Hammer, R. E., Brinster, R. L., Rosenfeld, M. G., Evans, R. E., Mayo, K. E. 1985. Expression of human growth hormone-releasing factor in transgenic mice results in increased somatic growth. *Nature* 315: 413–16

Hammer, R. E., Krumlauf, R., Lamper, S.

A., Brinster, R. L., Tilghman, S. M. 1987. Diversity of α-fetoprotein gene expression in mice is generated by a combination of separate enhancer elements. *Science* 235: 53–58

Hanahan, D. 1985. Heritable formation of pancreatic β-cell tumors in transgenic mice expressing recombinant insulin-simian virus 40 oncogenes. *Nature* 315: 115–22

Harbers, K., Jahner, D., Jaenisch, R. 1981. Microinjection of cloned retroviral genomes into mouse zygotes: Intergration and expression in the animal. *Nature* 293: 540–42

Hartung, S., Jaenisch, R., Breindl, M. 1986. Retrovirus insertion inactivates mouse α1(I) collagen gene by blocking initiation of transcription. *Nature* 320: 365–67

Hen, R. E., Borrelli, E., Sassone-Corsi, P., Chambon, P. 1983. Far upstream sequences are required for efficient transcription from the adenovirus-2 E1A transcription unit. *Nucl. Acids Res.* 11: 8735–45

Hogan, B. L. M., Constantini, F., Lacy, E. 1986. *Manipulation of the Mouse Embryo: A Laboratory Manual*. Cold Spring Harbor, NY: Cold Spring Harbor Lab.

Jaenisch, R., Breindl, M., Harbers, K., Jahner, D., Lohler, J. 1985. Retroviruses and insertional mutagenesis. *Cold Spring Harbor Symp. Quant. Biol.* 50: 439–45

Khillan, J. S., Schmidt, A., Overbeek, P. A., deCrombrugghe, B., Westphal, H. 1986. Developmental and tissue-specific expression by the α₂ type 1 collagen promoter in transgenic mice. *Proc. Natl. Acad. Sci. USA* 83: 725–29

Kitamura, N., Takagaki, Y., Furoto, S., Tanaka, T., Nawa, H., Nakanishi, S. 1983. A single gene for bovine high molecular weight kinogens. *Nature* 305: 545–49

Krumlauf, R., Hammer, R. E., Tilghman, S. M., Brinster, R. L. 1985. Developmental regulation of α-fetoprotein genes in transgenic mice. *Mol. Cell. Biol.* 5: 1639–48

Lacy, M., Alpert, S., Hanahan, D. 1986. Bovine papilloma virus genome elicits skin tumors in transgenic mice. *Nature* 322: 609–12

Lacy, E., Roberts, S., Evans, E. P., Burtenshaw, M. D., Constantini, F. D. 1983. A foreign β-globin gene in transgenic mice: Intergration at abnormal chromosomal positions and expression in inappropriate tissues. *Cell* 34: 343–58

Leder, A., Pattengale, P. K., Kuo, A., Stewart, T. A., Leder, P. 1986. Consequences of widespread deregulation of the c-myc gene in transgenic mice multiple neoplasms and normal development. *Cell* 45: 485–96

Leff, S. E., Rosenfeld, M. G., Evans, R. M.

1986. Complex transcription units: Diversity in gene expression by alternative RNA processing. *Ann. Rev. Biochem.* 55: 1091–1117

Le Meur, M., Gerlinger, P., Benoist, C., Mathis, D. 1985. Correcting an immune-response deficiency by creating Eα gene transgenic mice. *Nature* 316: 38–42

Lohler, J. R., Timpl, R., Jaenisch, R. 1984. Embryonic lethal mutation in mouse collagen I gene causes rupture of blood vessels and is associated with erythropoietic and mesenchymal cell death. *Cell* 38: 597–605

Low, M. J., Hammer, R. E., Goodman, R. H., Habener, J. F., Palmiter, R. D., Brinster, R. L. 1985. Tissue-specific post-translational processing of pre-prosomatostatin encoded by a metallothionein-somatostatin fusion gene in transgenic mice. *Cell* 41: 211–19

Low, M. J., Lechran, R. M., Hammer, R. E., Brinster, R: L., Habener, J. F., Mandel, G., Goodman, R. H. 1986. Gonadotroph-specific expression of metallothionein fusion genes in pituitaries of transgenic mice. *Science* 231: 1002–4

Luciw, P. A., Bishop, J. M., Varmus, H. E., Capecchi, M. R. 1983. Location and function of retroviral and SV40 sequences that enhance biochemical transformation after microinjection of DNA. *Cell* 33: 705–16

Lynch, D. R., Snyder, S. H. 1986. Neuropeptides: Molecular forms, metabolic pathways, and receptors. *Ann. Rev. Biochem.* 55: 773–79

Mark, W. H., Signorelli, K., Lacy, E. 1985. An insertational mutation in a transgenic line results in developmental arrest at day 5 of gestation. *Cold Spring Harbor Symp. Quant. Biol.* 50: 453–63

Mason, A. J., Hayflick, J. S., Zoeller, R. T., Young, W. S. III, Phillips, H. S., Nikolics, K., Seeburg, P. H. 1986a. A deletion truncating the gondadotropic-releasing hormone gene is responsible for hypogonadism in the *hpg* mouse. *Science* 234: 1366–71

Mason, A. J., Pitts, S. L., Nikolics, K., Szonyi, E., Wilcox, J. N., Seeburg, P. H., Stewardt, T. A. 1986b. The hypogonadal mouse: Reproductive functions restored by gene therapy. *Science* 234: 1372–78

Magram, J., Chada, K., Costantini, F. 1985. Developmental regulation of a cloned β-globin gene in transgenic mice. *Nature* 315: 338–40

Messing, A., Chen, H. Y., Palmiter, R. D., Brinster, R. L. 1985. Peripheral neuropathies, hepatocellular carcinomas and islet cell adenomas in transgenic mice. *Nature* 316: 461–63

Nawa, H., Kotani, H., Nakanishi, S. 1984. Tissue-specific generation of two pre-protachikynin mRNAs from one gene by alternative RNA splicing. *nature* 312: 729–34

Nelson, C., Crenshaw, E. B. III, Franco, R., Lira, S. A., Albert, V. R., Evans, R. M., Rosenfeld, M. G. 1986. Discrete *cis*-active genomic sequences dictate the pituitary cell type-specific expression of rat prolactin and growth hormone genes. *Nature* 322: 557–62

Ornitz, D. M., Palmiter, R. D., Hammer, R. E., Brinster, R. L., Swift, G. H., MacDonald, R. J. 1985. Specific expression of an elastase-human growth hormone fusion gene in pancreatic acinar cells of transgenic mice. *Nature* 313: 600–3

Overbeek, P. A., Chepelinsky, A., Khillan, J. S., Piatigorsky, J., Westphal, H. 1985. Lens-specific expression and developmental regulation of the bacterial chloramphenicol acetyltransferase gene driven by the murine αA-crystalline promoter in transgenic mice. *Proc. Natl. Acad. Sci. USA* 82: 7815–19

Overbeek, P. A., Lai, S-P., van Quill, K. R., Westphal, H. 1986. Tissue-specific expression in transgenic mice of a fused gene containing RSV terminal sequences. *Science* 231: 1574–77

Palmiter, R. D., Brinster, R. L., Hammer, R. E., Trumbauer, M. E., Rosenfeld, M. G., Birnberg, N. C., Evans, R. M. 1982. Dramatic growth of mice that develop from eggs microinjected with metallothionein growth hormone fusion genes. *Nature* 300: 611–15

Palmiter, R. D., Norstedt, G., Gelinas, R. E., Hammer, R. E., Brinster, R. L. 1983a. Metallothionein-human GH fusion genes stimulate growth of mice. *Science* 222: 809–14

Palmiter, R. D., Wilkie, T. M., Chen, H. Y., Brinster, R. L. 1983b. Transmission distortion and mosaicism in an unusual transgenic mouse pedigree. *Cell* 36: 869–77

Palmiter, R. D., Chen, H. Y., Messing, A., Brinster, R. L. 1985. SV40 enhancer and large-T antigen are instrumental in development of choroid plexus tumors in transgenic mice. *Nature* 316: 457–60

Palmiter, R. D., Brinster, R. L. 1986. Germ line transformation of mice. *Ann. Rev. Genet.* 20: 465–99

Payvar, F., DeFranco, D., Firestone, G. L., Edgar, B., Wrange, O., Okret, S., Gustafsson, J-A., Yamamoto, K. R. 1983. Sequence-specific binding of glucocorticoid receptor of MTV DNA at sites within and upstream of the transcribed region. *Cell* 35: 381–92

Readhead, C., Popko, B., Takahashi, N., Shine, H. D., Saavedtra, R. A., Sidman, R. L., Hood, L. 1987. Expression of a myelin basic protein gene in transgenic *shiverer* mice: Correction of the dysmyelinating phenotype. *Cell* 48: 703–12

Rosenfeld, M. G., Mermod, J-J., Amara, S. G., Swanson, L. W., Sawchenko, P. E., Rivier, J., Vale, W. W., Evans, R. M. 1983. Production of a novel neuropeptide encoded by the calcitonin gene via tissue-specific RNA processing. *Nature* 304: 129–35

Sassone-Corsi, P., Wildeman, A., Chambon, P. 1985. A *trans*-acting factor is responsible for the simian virus 40 enhancer activity in vitro. *Nature* 313: 458–63

Sassone-Corsi, P., Borrelli, E. 1986. Transcriptional regulation by *trans*-acting factors. *Trends Genet.* 2: 215–19

Scheidereit, C., Geisse, S., Westphal, H. M., Beato, M. 1983. The glucocorticoid receptor binds to defined nucleotide sequences near the promoter of mouse mammary tumor virus. *Nature* 304: 749–52

Schnieke, A., Harbers, K., Jaenisch, R. 1983. Embryonic lethal mutation in mice induced by retrovirus insertion into the α1(I) collagen gene. *Nature* 304: 315–20

Small, J. A., Blair, D. G., Showalter, S. D., Scangos, G. A. 1985. Analysis of a transgenic mouse containing simian virus 40 and v-myc sequences. *Mol. Cell. Biol.* 5: 642–48

Small, J. A., Scangos, G. A., Cork, L., Jay, G., Khoury, G. 1986. The early region of human papovavirus JC induces dysmyelination in transgenic mice. *Cell* 46: 3–18

Stewart, T. A., Pattengale, P. K., Leder, P. 1984. Spontaneous mammary adenocarcinomas in transgenic mice that carry and express MTV/myc fusion genes. *Cell* 38: 627–37

Stewart, T. A., Wagner, E. F., Mintz, B. 1982. Human β-globin gene sequences injected into mouse eggs, retained in adults, and transmitted to progeny. *Science* 217: 1046–48

Stuart, G. W., Searle, P. F., Palmiter, R. D. 1985. Identification of multiple metal regulatory elements in mouse metallothionein-I promoter by assaying synthetic sequences. *Nature* 317: 828–31

Storb, U., O'Brien, R. L., McMullen, M. D., Gollahon, K. A., Brinster, R. L. 1984. High expression of cloned immunoglobulin k gene in transgenic mice is restricted to B lymphocytes. *Nature* 310: 238–41

Swanson, L. W., Simmons, D. M., Arriza, J., Hammer, R., Brinster, R., Rosenfeld, M. G., Evans, R. M. 1985. Novel developmental specificity in the nervous system of transgenic animals expressing growth hormone fusion genes. *Nature* 317: 363–66

Swift, G. H., Hammer, R. E., MacDonald, R. J., Brinster, R. L. 1984. Tissue-specific expression of the rat pancreatic elastase 1 gene in transgenic mice. *Cell* 38: 639–46

Townes, T. M., Lingrel, J. B., Chen, H. Y., Brinster, R. L., Palmiter, R. D. 1985. Erythroid-specific expression of human β-globin genes in transgenic mice. *EMBO J.* 4: 1715–23

Wagner, E. F., Covarrublas, L., Stewart, T. A., Mintz, B. 1983. Prenatal lethalities in mice homozygous for human growth hormone gene sequences integrated in the germ line. *Cell* 35: 647–55

Walker, M. D., Edlund, T., Boulet, A. M., Rutter, W. J. 1983. Cell-specific expression controlled by the 5′ flanking region of insulin and chymotrypsin genes. *Nature*

306: 557–61

Woychik, R. P., Stewart, T. A., Davis, L. G., D'Eustachio, P., Leder, P. 1985. An inherited limb deformity created by insertional mutagenesis in a transgenic mouse. *Nature* 318: 36–40

Yamamura, K., Kikutani, H., Folsom, V., Clayton, L. K., Kimoto, M., Akira, S., Kashiwamura, S., Tonegawa, S., Kishimoto, T. 1985. Functional expression of a microinjection Eα gene in C57BL/6 transgenic mice. *Nature* 316: 67–69

Yamamura, K., Kudo, A., Ebihara, T., Kamino, K., Araki, K., Kumahara, Y., Watanabe, T. 1986. Cell-type specific and regulated expression of a human λ1 immunoglobulin gene in transgenic mice. *Proc. Natl. Acad. Sci. USA* 83: 2152–56

ZuRhein, G. M. 1969. Association of papova-virus with a human demyelinating disease (progressive multifixas leukoencephalopathy). *Progr. Med. Virol.* 11: 185–247

Ann. Rev. Neurosci. 1988. 11:373–93

MUTATIONS AND MOLECULES INFLUENCING BIOLOGICAL RHYTHMS

Jeffrey C. Hall and Michael Rosbash

Department of Biology, Brandeis University, Waltham,
Massachusetts 02254

INTRODUCTION

Rhythmically oscillating phenomena in biological systems have been
analyzed in many ways: behaviorally (reviews: Moore-Ede et al 1982,
Saunders 1982), neurobiologically (reviews: Turek 1985, Jacklet 1985), and
biochemically (reviews: Takahashi & Menaker 1984, Johnson & Hastings
1986). Over the last four years, another angle from which these rhythms
has been approached—"clock genetics"—appears to have also come to
the fore. Yet, this area of rhythm studies has been around for at least 20
years (previous summaries: Feldman et al 1979, Konopka 1981).

In the present review we update several of the recent genetic advances
and also discuss how such investigations are moving into the molecular
area in studies of two organisms, *Drosophila* and *Neurospora*.

The biological clocks underlying rhythmic fluctuations have seemed to
be an eternal mystery. Whether "clock molecular genetics" will allow us
soon to peer into the core mechanisms of the pertinent oscillators remains
to be seen.

Interpretation of the studies of these clocks is not always simple. That
is, a fair amount of jargon surrounds rhythm analysis. Consequently we
begin by listing some of the properties of true clocks, which are deeper
than simply that they control various "parameters" that exhibit some kind
of regular fluctuations having a given periodicity, such as a "*circadian
tau*," or cycle durations in the *ultradian* realm (ca. 10^{-1} to 10^{-3} of a day;
reviews: Shultz & Lavie 1985). We note these additional attributes of

373

clocks because most of the "clock mutations" to be discussed have turned out to affect more than overt circadian or ultradian periodicities:

1. Thus, clocks not only produce such taus, but also a clock has a *phase*, meaning, for instance, that the start of a given cycle usually occurs in a particular relationship to an environmental cycle. For example, peaks of eclosion (pupal-to-adult transition) in *Drosophila* tend to occur just after dawn (or "subjective dawn," see below).
2. When a clock is started (e.g. by a pulse of light) or *entrained* (e.g. by alternating cycles of light and darkness, such as 12 hr : 12 hr "LD") it will subsequently *free run* in constant conditions (whereby the lights do not come on again after the pulse or after the cycles of LD entrainment).
3. Thus, rhythms are *endogenous*, not mere responses to cyclical environmental cues—meaning that the cycles continue during free-running conditions.
4. A free-running rhythm can be *phase-shifted*, e.g. reset by application of the same kind of light pulses used to set the phase in the first place.
5. The phase *advances* and *delays* that occur under these circumstances define a *phase-response curve* (PRC) that has quite similar properties in many species of microbial and metazoan eukaryotes; i.e. delays are induced when the resetting signal is delivered during the early *subjective night* (portion of a free-running cycle that extrapolates back to when it was dark during entrainment), advances occur during late subjective night, and the clock is insensitive to resetting cues during the *subjective day*.
6. Clocks are *temperature-compensated*: circadian, and even ultradian (e.g. Lloyd & Edwards 1984), taus are much the same over rather wide temperature ranges. This seems most meaningful in poikilothermic organisms, such as *Drosophila* (e.g. Zimmerman et al 1968) plus *Neurospora* (e.g. Gardner & Feldman 1981). Temperature compensation is also meaningful in certain vertebrates (such as hibernating species, e.g. Menaker 1959, Rawson 1960), but it does not occur so well in others (e.g. Richter 1979, Gibbs 1983).

RHYTHM MUTANTS

Mutations that alter or seemingly abolish circadian rhythms have been systematically isolated in two *Drosophila* species, *D. melanogaster* and *D. pseudoobscura* (reviews: Konopka 1979, 1981, Hall 1984), and in the microbes *Neurospora crassa* (reviews: Feldman & Dunlap 1983, Feldman 1985) plus *Chlamydomonas reinhardi* (reviews: Feldman 1982, 1983; also see Mergenhagen 1984). The relevant mutants, defining on the order of a half-dozen "clock genes" in most of these species (exception: $N = 2$ in $D.$ *pseudoobscura*), were recovered in brute force screens, whereby the rather heroic efforts of R. J. Konopka and colleagues plus F. R. Jackson (*Dm*), C. S. Pittendrigh (*Dp*), J. F. Feldman and co-workers (*Nc*), and V. G. Bruce (*Cr*) led to mutants with either longer or shorter than normal taus, or to strains that apparently do not exhibit any circadian rhythmicity. In addition, there are two recently isolated mutants, which were encountered in the course of rhythm experiments as opposed to being the results of deliberate mutagenesis efforts: an arrhythmic mutant in the blowfly *Lucilia cuprina* (Smith 1987) and a short-period mutant in hamster (Ralph & Menaker 1987). Finally, experiments on the "polygenic" control of rhythms have occasionally been performed, not only in *Drosophila* (Pittendrigh 1967) but also in certain mammals (Possidente & Hegmann 1980, Büttner & Wollnik 1984, Hotz et al 1987, Wollnik et al 1987).

Detailed analysis of the phenotypes affected by these kinds of mutations—and, just as important, of the *genes* defined by them—have been extensive and deep only in *D. melanogaster* and *N. crassa*, however. The genic studies *per se* have paved the way for molecular analysis of certain of the most interesting genetic factors that participate in the control of the several clock attributes listed above. We discuss how these genetic studies have specifically led to the molecular work in the latter sections of this review. First, however, we review several recent results concerning the "phenogenetics" of the main clock mutants and mention certain characteristics of the less analyzed variants.

Drosophila *Clock Mutants*

Most of the rhythm mutations in *D. melanogaster* have been induced, by chemical mutagens, at one genetic locus, the X-chromosomal *period* (*per*) gene (Konopka & Benzer 1971, Konopka 1987). Some of these *per* mutants have circadian taus far outside the normal range: per^s, 18–20 hr; per^{L1} and per^{L2}, 28–30 hr; the majority (per^{o1} through *per* ^{o4}) are arrhythmic in both eclosion and locomotor activity (re "sleep/wake" cycles of the adult flies). Two other clock mutations on the X chromosome turned up in these screens: *Clock* (*Clk*, activity cycles ca. 22–23 hr) and *Andante* (*And*, eclosion and activity taus, ca. 25–26 hr). Jackson (1983) induced three rhythm-altering mutations on the autosomes of this species: *phase-angle-2* (*psi-2*) and *phase-angle-3* (*psi-3*), each of which cause "too-early" (i.e. pre-dawn) emergence during LD cycles. Each of these mutations turned out, as well, to control free-running periods that are, as in the *And* mutant, slightly longer than normal (cf Pittendrigh 1967). The other autosomal mutation, *gate* (*gat*), is closely linked and could be allelic to *psi-2*. *gat* causes eclosion to be poorly gated, such that the flies do not emerge during the usual narrow windows of time (Jackson 1983).

Further phenotypic studies of the *per* mutants showed that the *per⁰* cases appear to be arrhythmic, in eclosion, even during LD cycling (Konopka 1981; also see Bargiello et al 1984). Yet, these mutants are, in LD, pseudo-rhythmic in their locomotor activity, i.e. they regularly become active when the lights come on or go off (Hamblen et al 1986). This, though, is a mere response to environmental changes as opposed to the expression of an endogenous oscillator. Thus, *per⁰*s are, by definition, not rhythmic in free-running conditions. And, more subtly, they have no apparent clock running during entrainment, whereas the rhythm's phase for wild-type (*per⁺*) flies is such that they anticipate lights-on by becoming active ca. 2 hr in advance of this environmental change (Hamblen et al 1986). This kind of "phase lead" is a common feature of vertebrate rest/activity cycles (e.g. Fuchs 1983).

The *per* results noted above have some parallels with effects of ar-rhythmic mutations in *L. cuprina* and *D. pseudoobscura*. In the former dipteran, the *ary* mutant was found on the basis of arrhythmic eclosion; it also exhibits this phenotype in free-running locomotor activity, although the adults are cyclically active and inactive in LD (Smith 1987). In *D. pseudoobscura*, there are five of these, defining two X-chromosomal genes by complementation tests. Three of the mutants, only, eclose arrhyth-mically in LD; the other two, which are not allelic to each other, exhibit "forced" rhythmic eclosion under these conditions (Pittendrigh 1974). It has also been determined that all five of these mutants are arrhythmic in free-running locomotor activity (Konopka 1979), though further details (e.g. LD vs "DD" comparisons) were not reported. It was shown, near the time of these mutants' isolations, that the genetic variants are "semi-dominant"; specifically, for these *D. pseudoobscura* mutants, females "doubly" heterozygous for a given pair of nonallelic mutations are rhythmic but with longer than normal periods and with a substantially altered phase (Pittendrigh 1974). We mention below other examples of this kind of incomplete recessivity, i.e. for clock mutations in general.

Multiply mutant flies have been produced in *D. melanogaster* as well to determine whether putatively "synergistic" rhythm defects would occur. So far, however, only additive effects on circadian periodicity (Orr 1982; *And* plus *per* and *And* plus *Clk* combinations) or phase (F. R. Jackson, unpublished: *psi-2*, *psi-3* double mutants) have been found. These pheno-types are analogous to those exhibited by *Chlamydomonas* when it is doubly or multiply mutant for rhythm variants (Bruce 1974). (See below for somewhat contrasting results from *Neurospora*.)

The gene dosage of *per*'s normal allele has been manipulated. Increased and decreased copies of per^+ lead to shortened and lengthened circadian periods, respectively (Smith & Konopka 1981, 1982). Not only do females heterozygous for the normal allele and a deletion show longer than normal periods, but per^o mutations are similarly semidominant (Smith & Konopka 1981, 1982). Nearly the same phenotype results when a per^L mutation is made heterozygous with per^+ (Smith & Konopka 1981), thus suggesting that this class of variant is a severe "under-expressor" of information encoded within this genetic locus, with per^os being hypothetically "null" (i.e. equivalent to a deletion). Additional force to these suggestions has come from recent molecular studies (see below). The short-period per^s mutant is a most interesting case. Like per^o and per^L, per^s is semidominant to per^+ (Konopka & Benzer 1971). The aforementioned hamster mutation (tau) is quite similar, shortening circadian periods by ca. 1.5 hr in $tau/+$ animals and by ca. 4 hr when tau is homozygous (Ralph & Menaker 1987). The analogous mutation in Drosophila is formally an "over-producer" of some kind, in the sense that per^s leads to the same kind of shortened circadian period as do genotypes involving extra copies of the normal allele (Smith & Konopka 1982). Further analysis of the relevant behavioral data indicated that circadian periodicity is a logarithmic function of per activity, and led to the inference that per^s's activity level is ca. 35 times that of per^+ (Coté & Brody 1986). The guess, then, was that molecules produced by this particular mutated form of the gene are intrinsically abnormal, as opposed to a hypothesis whereby transcription would occur at a vastly increased rate. This, too, has been confirmed molecularly (see below).

This per^s mutant is also interesting in its responses to phase-shifting light stimuli: The PRC has a normal (12 hr) subjective night, but the subjective day is shortened to ca. 7 hr (Konopka 1979). Also, phase advances and delays of eclosion peaks, induced during subjective night, are larger than in wild-type (Konopka 1979). These data led to a "membrane model" of circadian rhythm control by the product of this Drosophila gene (Konopka & Orr 1980); the ideas here are similar to the "ionic flux" formalisms developed more generally to model circadian clocks (Engelmann & Schrempf 1980). However, the molecular data extant on the per gene product (see below) do not yet give further force to this kind of clock model.

Temperature compensation is somewhat abnormal in the rhythmic per mutants: Whereas the wild-type was always 24 hr, taus were a bit shorter or longer than normal, respectively, in per^s and per^{L1} adults tested for activity rhythms at relatively high temperatures. Both mutants gave closer-to-normal periodicities in low-temperature experiments (Konopka 1979). In contrast, the And and Clk mutants (Orr 1982) and the Chlamydomonas rhythm mutants that were tested appropriately yielded flat (compensated) plots of period vs temperature.

The cases of arrhythmicity induced by per^o mutations or deletions of the locus have been reanalyzed. It turns out that these genotypic changes do not really lead to all-out aperiodic behavior in locomotor activity tests. Instead, ultradian rhythms are routinely found lurking in the data. These periodicities (mean values, ca. 10–15 hr) are extracted from 50–70% of the per^o or per-minus (per^-) flies whose activity was monitored, followed by application of rather special algorithms designed to "pull rhythms out" of noisy data (Dowse et al 1987, Dowse & Ringo 1987). The per^- type used here could be only suggested to be really deleted of the locus (Smith & Konopka 1981); that this genotype is a true deletion of the per gene turned out to be true at a higher level of resolution (see below).

The significance of per's action has entered the ultradian realm. Genetic variants involving this gene affect rhythms whose normal taus are markedly shorter than the cyclic durations with respect to which the original per mutants were isolated. First, there are ca. one-minute periodicities in the courtship song of $D. melanogaster$ males which are shortened by at least 25% in per^s and lengthened by ca. 50% in per^{L1}; per^{o1} seems to eliminate the song rhythms (Kyriacou & Hall 1980, 1986). These oscillations are temperature-compensated over a ca. 20°C range (Kyriacou & Hall 1980). Second, the heart in $Drosophila$ larvae beats regularly, as if this phenotype is an ultradian rhythm (frequency, ca. 3 Hz), and Livingstone (1981) observed that heartbeating is erratic in late-stage larvae expressing per^o. This finding has recently been confirmed and quantified by C. P. Kyriacou (unpublished) and independently by H. D. Dowse (unpublished).

The high-frequency courtship song rhythms are behaviorally significant in terms of their effects on female responses to a male's eventual mating attempts (Kyriacou & Hall 1982, 1984, 1986). These results might have been expected, given that the males from several closely related $Drosophila$ species sing with different taus [ranging from 30–40 s in $D. simulans$ (Kyriacou & Hall 1980, 1986) to 70–80 s in $D. yakuba$ (Kyriacou & Hall 1987)].

Fly rhythmicity, including that involving courtship, has been examined in some neurogenetic experiments. Action potential mutations (that cause paralysis and turn off nerve conduction at high temperatures; review: Tanouye et al 1986) were shown to stop the "song clock" that is inferred to underlie this behavioral rhythm (Kyriacou & Hall 1985). In contrast, TTX treatments applied to known neural oscillators in molluscs (Eskin 1977) and mammals (Schwartz et al 1987) do not arrest the relevant circadian clock: The physiological or behavioral oscillations are shut down during the treatment; but when rhythmicity returns after washing the toxin away, the phase is the same as it was before the treatments. Thus, the song clock in *Drosophila* and circadian clocks—at least in these other species— seem to have a different neural basis (though this notion is admittedly rather vague). Other quasineural studies of the fly's rhythms involved genetic mosaics (Hall 1984), i.e. flies that were part *per*-mutant and part-normal (within each animal). The effects of *per*^{s} on the oscillator controlling the song rhythm was thus localized to the thorax, probably within the ventral ganglia, whereas the effects of this mutation of circadian periods "mapped" to the brain (cf Konopka et al 1983).

There are further, albeit now somewhat old and not-followed-up, findings in the area of *per* neurobiology: Transplantation of an adult *per*^{s} brain to the abdomen of a *per*^{0} fly can in some cases "correct" the defect associated with the host genotype, whereby rhythmic and short-period locomotor activity cycles are manifested (Handler & Konopka 1979). However, this finding does not necessarily implicate some sort of circulating factor—perhaps under the control of this gene's action—as a component of the clock system of intact flies (i.e. including the brain in its usual location). Still, other "neuro-humoral" findings should be considered: Cells of a neurosecretory cluster are in somewhat ectopic locations in the adult brains of *per*^{0} in *D. melanogaster* and in the two types of arrhythmic mutants in *D. pseudoobscura* (Konopka & Wells 1980). Could these brain abnormalities be related to the deficit in octopamine synthesis in the *melanogaster per* mutants (Livingstone & Tempel 1983)?

Neurospora *Clock Mutants*

The clock variants in this lower eukaryote have, as for the *Drosophila* mutants, been analyzed in many ways since they began to be isolated at about the same time that studies of the fly's clock genetics began. Moreover, the results gathered from studies of these fungal mutants in several ways parallel the findings from *Drosophila*, at least in the sense that the multiplicity of clock-functional abnormalities caused by certain mutations in the two organisms are interestingly similar.

Mutations at six genetic loci in *N. crassa* have been induced (by chemical mutagenesis or UV treatment) and isolated with respect to defective circadian rhythms (Feldman & Dunlap 1983). The *frequency* (*frq*) gene has been "hit" the most times, in that there are eight independently isolated alleles. Only one mutation has been found for the other five loci, called *period* (*prd-1* through *prd-4*) or *chrono*. Most of these mutations cause periodicities of conidial banding patterns (associated with growth) to be shorter or longer than the normal 21.5 hr.

In addition, a number of metabolic and biochemical mutants have been found—subsequent to their isolation on "nonclock" criteria—to have altered taus. The effects can be mild when such variants are tested under standard conditions (review: Feldman & Dunlap 1983) or quite dramatic under certain nutritional conditions. For example, the partial fatty acid auxotrophic *cel⁻* strain exhibits normal cycle durations at ca 22°C, but supplementing the medium with certain unsaturated fatty acids lengthened periods by 5–19 hr (Brody & Martins 1979).

The most intensively studied clock gene in *N. crassa* is the *frequency* (*frq*) locus on chromosome VII. This gene, like *per*, has been mutated to yield short-period and long-period variants (Feldman 1985). In addition, what amounts to an arrhythmic type has recently been induced. Thus, the *frq⁹* allele causes growing cultures to exhibit essentially continuous conidiation under standard growth conditions (Loros & Feldman 1986). In other media, rhythmicity can be seen, though it can require a week of growth to be initiated and then has highly variable periods and phases among different cultures (Loros & Feldman 1986). Thus, *frq⁹* can be regarded as analogous to *D. melanogaster*'s *per⁰*s, which are only quasi-arrhythmic (see above).

Another important property of *frq⁹* is that tau is strikingly temperature dependent, to the extent that this mutant seems to have lost temperature compensation (Loros et al 1986). More generally, it turns out that all the long-period *frq* mutants give periods that are inversely related to temperature (though with Q_{10}s less than the values of 2 found for *frq⁹*), whereas short-period mutations at this locus leave temperature compensation intact (Gardner & Feldman 1981). Somewhat of a parallel is found in the aforementioned *cel⁻* mutant, which loses its temperature compensation when grown below 22°C on medium unsupplemented with fatty acids (Mattern et al 1982).

Included in the temperature-compensation-defective *frequency* mutants is the long-period *frq⁷* allele (Dunlap & Feldman 1987). In addition, this mutation leads to an abnormality in phase-shifting—not by light, which can of course strongly reset *Neurospora*'s clock, but in regard to advances and delays that are caused by protein synthesis inhibitors (cf Nakashima et al 1981). Indeed, such treatments generate PRCs, which are non-superimposable on the standard ones elicited by light-pulses, in a variety of systems (e.g. Dunlap et al 1980, Lotshaw & Jacklet 1986). This might mean that protein synthesis is important for the clock's operation only at certain times within a cycle (i.e. these PRCs almost by definition include insensitive phases). In experiments on *frq⁷*, then, pulsatile cycloheximide applications essentially eliminated this PRC, although the mutant is by all means still rhythmic when treated in this manner (Feldman & Dunlap 1987). The strong implication here is that protein synthesis *per se* is not a part of the "oscillatory feedback loop" that underlies this organism's rhythmicity.

Other experiments involving resetting *Neurospora*'s clock (in these cases, by light stimuli) have shown that the subjective day is the portion of a cycle nonrandomly altered by *frq* mutations. For example, essentially the entirety of the tau-shortening effects of the frq^2, frq^4, and frq^6 mutations can be accounted for by the fact that the middle-to-late subset of the light-insensitive half of a free-running cycle is but 65% of normal (Feldman 1985). A similarity to *Drosophila*'s per^s mutant is evident (cf Konopka & Orr 1980).

Most of the clock mutations in *Neurospora* are *semidominant* in their effects on tau, as determined in "heterokaryon" experiments involving the production of "forced" diploids with syncitia containing nuclei of different genotypes (Feldman & Dunlap 1983). Recall that the rhythm variants in two *Drosophila* species and in a hamster are superficially similar in this genetic property (see above), as are the mutants of *Chlamydomonas* (Bruce & Bruce 1978). An exception, in *N. crassa*, is frq^9; this mutation is recessive to the wild-type allele (Loros & Feldman 1986). Such a result has been interpreted to mean that frq^9 is a null mutation that may completely inactivate the locus. Whether or not this is so, the recessivity of this allele is extremely useful in designing "molecular gene rescue" experiments aimed at identifying cloned material that must include the entirety of a genetic locus like this one (see below).

Further usage of heterokaryons has shown that the relative proportions of nuclei—i.e. carrying alleles associated with shortened, lengthened, or normal periods—allow good predictions of the tau that will be exhibited by a given genetically mixed type (Feldman & Dunlap 1983). Again, dosages of clock genes matter (cf Smith & Konopka 1981, 1982).

Many multiply mutant strains of *N. crassa*, involving variants at the various rhythm-affecting loci, have been tested for tau (Lakin-Thomas & Brody 1985). In many such genetic combinations, it appeared as though the resultant periods may best be explainable by "multiplicative" predictions, as opposed to merely additive effects on tau of the relevant single mutations. Thus, one might further infer that the products of these genes *interact* in some manner (perhaps even in terms of physical associations of the hypothetical proteins). In other multiple mutants, a given factor was sometimes epistatic to another; for example, the effects of cel^- on period and temperature-compensation are not seen in a *prd-1* genetic background (Lakin-Thomas & Brody 1985). Hence, the influences of these two genes on the clock could be viewed in terms of sequentially acting products.

Whether or not these kinds of genetical formalisms are themselves forceful, or have specific heuristic value, it is clear that well more than one putative clock gene must be studied, in a given species, before one can begin to understand how the organism constructs and operates its biological clocks. In this regard, the experiments involving "only certain" of the relevant genes, which are now being studied at more concrete levels and are discussed below, must be viewed as inadequate. But they are a start—perhaps toward understanding the molecular mechanisms of living oscillators.

MOLECULAR BIOLOGY OF CLOCK GENES

The per *Locus of* Drosophila melanogaster

Isolation of DNA sequences that were eventually identified as encompassing all of this clock gene required an extensive array of background "cytogenetic" information on the locus (e.g. a narrow localization of the *per* mutations' site, by mapping them with chromosomal aberrations such as deletions; Young & Judd 1978, Smith & Konopka 1981, 1982).

Thus, *per* was cloned by chromosomal "walking and jumping" procedures (see Bender et al 1983 for a general description of this method) by Bargiello & Young (1984) and by microexcision experiments (Reddy et al 1984). These approaches worked because *per* had been pinned down to a tiny region, probably an interval between two small chromosome "bands" on the fly's X chromosome (see above). What could not work, of course, is the commonly applied cloning strategy of working backwards from a known protein product to the relevant genomic sequences.

The boundaries of *per* were rather precisely delimited by molecular mapping of nearby chromosomal "breakpoints" (e.g. ends of deletions). These data were in fact obtained in conjunction with the walking/jumping efforts and the experiments immediately proceeding from the microexcision of DNA that included the gene. An important bonus in the results was proof that the aforementioned per^- genotype is indeed homozygously deleted of ca. 10 kb of DNA (Bargiello & Young 1984, Reddy et al 1984). Subsequent work (see below) showed that this deletion is devoid of the entirety of the gene, i.e. the critical component of the locus's molecular expression.

RNA transcripts complementary to *per*-locus clones were detected in the initial molecular experiments (Bargiello & Young 1984, Reddy et al 1984). Attention was focused primarily on two of them: a 4.5 kb RNA and a ca. 1 kb species, both of which are missing in the aforementioned per^- flies. The smaller of these two RNAs was immediately intriguing, because it oscillates in its abundance over the course of a given day, in wild-type flies that are in LD entrainment or in free-running conditions

(Reddy et al 1984). This molecular circadian rhythm (there is one peak per day, in the middle of the day or subjective day) seems tied into clock functions, because not only does the 1 kb RNA originate from at least the vicinity of the *per* locus, but the fluctuation is also dampened in the *per°* mutants (this transcript's abundance stays at the low, night-time level; Reddy et al 1984, Hamblen et al 1986). Loosely analogous findings have come from other systems: Gross RNA levels can fluctuate in microbial organisms (either in circadian or ultradian domains, respectively; Walz et al 1983, Lloyd & Edwards 1984). Higher up the evolutionary scale, a transcript encoding vasopressin in rat oscillates in its abundance, as determined by in situ hybridization to brain regions. Moreover, the fluctuations can be detected only in the key clock center in this mammal, the suprachiasmatic nucleus (SCN; Uhl & Reppert 1986, cf Turek 1985).

Subsequent experiments on the fly's fluctuating RNA have indicated, perhaps ironically, that this transcript is not the critical component of the *per* locus's expression. The relevant experiments involved construction of germ-line transformants (or "transgenics"), whereby various DNA fragments from the environs of the gene were transduced into *per°* or *per⁻* "hosts." The aggregate pattern of "arrhythmicity rescue" showed that clones homologous only to the 4.5 kb transcript (see above), but not those that could encode the 1 kb species, are able to restore rhythms to these mutants (Bargiello et al 1984, Zehring et al 1984, Hamblen et al 1986). One important inference from these results is that the effects of *per°* mutations on the smaller RNA's abundance oscillation are indirect (e.g. a given *per°* is not directly mutated in the genomic source of the 1 kb transcript). Also note that, whereas "4.5-covering" clones rescue both circadian and courtship song arrhythmicity, those which cover the fluctuating RNA have no influence on either kind of mutant phenotype (Hamblen et al 1986).

Perhaps the most important result of these transformation experiments has been to allow investigators to focus on the 4.5 kb, *per*-derived transcript. Thus, from nucleotide sequencing of the relevant geonomic clones, and cDNA clones complementary to the genomics, the entire informational content of this portion of the locus has been determined (Jackson et al 1986, Yu et al 1987b, Citri et al 1987). Before considering the nucleotide sequencing details (and, for example, the amino acid sequence that was inferred; see below), one first sees that the 4.5 kb RNA's gross structure consists of eight specific exons. Yet, this refers to the major transcript, because it turns out that there is heterogeneity in the primary expression of this clock gene: At least two additional (and less abundant) RNA species are detected by cNDA analysis (Citri et al 1987), and they would encode three different proteins (differing in amino acid composition and sequence in regions relatively near the C-terminal end of the conceptual proteins). These different, but of course related, RNA species are produced by "alternative RNA splicing" events. Their biological significance is not yet known, but it is possible that the different *per* proteins (each?) play separate roles, i.e. in terms of the different kinds of rhythmic phenotypes reported to be influenced by this gene (Konopka & Benzer 1971, Kyriacou & Hall 1980, Livingstone 1981, Weitzel & Rensing 1981). This idea has begun to be examined via creation of "mini-gene transformants" involving transduction of two separate types of intron-less cDNA "cassettes" transduced into the genome (Citri et al 1987). Either cassette allows for circadian locomotor rhythmicity, but the phenotypes are not entirely normal, i.e. in comparison to the transformants involving the gene in its entirety (introns plus exons). Other *per*-influenced phenotypes are now being examined in these transformants. It may be necessary for all three of the different types of cDNA casettes to be combined in a given fly for fully normal circadian rhythms, and the other cycling phenomena just listed, to take place.

Looking at the conceptual *per* protein(s) (Shin et al 1985, Jackson et al 1986, Reddy et al 1986) led immediately to scrutiny of a certain small part of the amino acid sequence: A perfectly alternating series of threonine-glycine repeats (summing to about 40 residues). This rather striking region, which would be in common to all the known alternatively spliced forms (Citri et al 1987), is qualitatively and quantitatively similar to a series of serine-glycine pairs in the core protein of a chondroitin sulfate proteoglycan from rat (Bourdon et al 1985, 1987). Indeed, *per*'s product does have proteoglycan-like characteristics, as determined by application of antibodies used to "track" the material in test-tube assays (Reddy et al 1986, Bargiello et al 1987). A word of caution is that these preliminary results do not imply that the clock gene here produces a protein that is solely some kind of "extracellular matrix" material—which one might, in turn, believe to be significant only for the structure of some kind of fly tissues. Whereas vertebrate proteoglycans often have such attributes, they can in certain cases be found inside cells where they may play "physiological" roles (Evered & Whelan 1986).

Where in fact are *per* products expressed (cf discussion of mosaic experiments, above, that also asked this question)? First, there are results from in situ localization of the 4.5 kb transcript in the embryonic CNS (James et al 1986), at a stage when this RNA can be detected, as well, in "Northern blotting" experiments (Young et al 1985, James et al 1986). This expression, which was seen in all of the segmental brain plus ventral ganglia, was found to wane during late embryogenesis (James et al 1986). Other embryonic tissues in which the *per* gene product is present are the paired salivary glands (Bargiello et al 1987). In larvae, as well, *per*-produced protein is detectable in the salivary glands (and the material indeed appears to be at cell boundaries; see above)—although mRNA transcribed from this gene is at very low abundance during these intermediate developmental stages (Young et al 1985, James et al 1986). During metamorphosis, more robust *per* expression reappears and reaches even higher levels in adults (Young et al 1985, James et al 1986). At the latter stage, "head vs body Northerns" revealed an enrichment in the anterior tissues (vs what obtains for expression of many other mRNAs, including a transcript which "neighbors" the 4.5 kb species; James et al 1986). A further series of experiments, using an antibody against a portion of the *per* protein, has revealed this gene's expression in the central brain of adults and also in the eye plus optic ganglia of the visual system (Siwicki et al 1987).

These results on the clock gene's spatial expression are intriguing from two angles: First, one cannot assert an "exquisitely local" significance to *per*'s action, which would have been possible if the transcript had been detectable only in a narrowly defined portion of the brain (a hypothetical analog, in *Drosophila*, of the vertebrate SCN?). Second, *per* expression early in the life cycle could mean that the gene has developmental significance per se. That is, might it contribute to cell differentiation and pattern formation in the CNS that, at least in part, has to do with the construction of neural oscillators? Furthermore, might there be no influence of the gene product(s) on the actual "ticking" (ongoing physiological operation) of the fly's clocks? These questions are raised, because couplings among "high-frequency" oscillators have long been suggested (e.g. Klevecz et al 1984) to be involved in the formation of lower-frequency, i.e. circadian, clocks. Such a possibility can be viewed in the context of the fact that ultradian oscillators are still "running" in flies whose *per* gene is gone or hypothetically nonfunctional (Dowse et al 1987, Dowse & Ringo 1987). Therefore, such genetic variants may not really have stopped the clock but instead could result in a formation failure of the relevant interneuronal connections—the literal correlates of formalistic couplings. The involvement of *per* in communication among cells, albeit non-neural ones, has now been addressed in less abstract terms: Electrical couplings, as well as dye movements, between larval salivary gland cells (see above) are strikingly affected by *per* mutations, such that coupling strengths vary directly with the known or inferred levels of the gene product's function (Bargiello et al 1987).

The preceding discussion tacitly asks some questions about mutated forms of *per* (also see section on the phenogenetics of *Drosophila* clock mutants). Specifically, are *per*[o]s really null at the level of gene action? And what about the long- or short-period alleles? These matters have been approached by DNA sequencing of *per* genes cloned from the mutants. In one set of studies (Yu et al 1987b), the portions of the mutant-derived clones that could be said to account completely for the aberrant rhythms were identified, so that "brute force" sequencing of the entire mutated genes was not necessary. This was done by transforming per^o hosts with "chimeric" fragments, each derived partly from mutant clones, with the remainder of each cassette coming from per^+. In this way, the locations of the mutated sites in per^{o1} and per^s were narrowed down to a specific 1.7 kb subset of the locus (Yu et al 1987b). A "modest" amount of sequencing then revealed that per^{o1} is caused by a nonsense, translation chain-terminating codon, which is rather centrally located in the gene and "upstream" of the Thr-Gly repeat; per^s's difference from wild-type was found to be a serine-to-asparagine missense mutation, located between the mutated site in per^{o1} and the repeat region.

This strategy for molecularly pinpointing a given mutation is of some general interest, because it is more than a labor-saving device. In addition, it can show that a given alteration is (or is not, as the case may be) the *only* genetic etiology of the phenotypic abnormality. In other words, what if, say, three "site differences" were found in per^s vs per^+? Two of them might have been outside the confines of the aforementioned 1.7 kb subset of the locus. Yet, these two would, in light of transformation results, be concluded as irrelevant to the "fast clocks" associated with this mutant. By the way, it was speculated that the per^s-defined serine, given the specific amino acids neighboring it (see Yu et al 1987b, for background literature), might be a site for phosphorylation of *per* proteins in the wild-type. [Note: This site would be included in all the known *per* products derived by alternative splicing (cf Citri et al 1987).] Such a covalent modification could "down regulate" the products' activities (cf Cochet et al 1984), thus modulating the normal timing functions such that they contribute to proper circadian and ultradian taus; in contrast, the nonphosphorylatable asparagine residue at this site in per^s could cause the timers to run out of control.

The intragenic sites that are altered in per^{o1} and per^s have been confirmed by the sequencing data of Baylies et al (1987). In addition, these workers determined a nucleotide substitution in per^{L1} that results in a valine-to-aspartate change, upstream of the other two mutated sites. One would hypothesize, then, that this nonconservative substitution in the long-period mutant causes hypoactivity or perhaps relative instability of the protein molecules (see above).

There is additional force to the notion that abnormally low expression of *per*—perhaps at any level of gene action—causes clocks to run more slowly. Baylies et al (1987) determined circadian periodicities for a series of transformant strains, each carrying a per^+ fragment transduced to a different genomic location. The taus ranged from a bit longer than normal to more than 35 hr, and these phenotypes were inversely correlated with abundance levels of the 4.5 kb RNA transcribed from the transduced DNA insert. [Note: The genetic background of these transformants had to be per^-, because per^os do not cause a change in this transcript's level (Bargiello & Young 1984, Hamblen et al 1986). For that matter, per^{L1}, as well, has a normal abundance of the 4.5 kb RNA (Bargiello & Young 1984), so one cannot conclude at this level of analysis that it is hypoactive.]

Thus, much is known about per's molecular biology. Yet, and even though the extant data are obviously not enough, it will be important to expand these kinds of investigations to other clock genes in the two *Drosophila* species for which the rhythm mutants exist. But not many of the five to six clock genes in *D. melanogaster*, and none of the two in *D. pseudoobscura*, are readily clonable: The only other locus for which the necessary, fine-level cytogenetic mapping data exist is *Andante* (Smith 1982). This "moderately slow" clock mutant is, by the way, interesting in that it lengthens song-rhythm periods as well as circadian taus (C. P. Kyriacou, unpublished; cf Jackson 1983). One important exception is that males expressing the X chromosomal *Clk* mutation sing with normal, not shortened, periodicities (C. P. Kyriacou, unpublished; cf Konopka 1987). This is perhaps as it should be: The circadian and song clocks could— and, given these genetic results, do—share components. But they are not the self-same thing, so one would expect them to be independently changeable in certain instances.

A further example of this kind comes from in vitro mutagenesis experiments on *per*. The Thr-Gly repeat region, whose discovery was of heuristic value for biochemical reasons, has also proved to be most interesting from the behavioral standpoint. When the repeat is removed from transformation cassettes that are then introduced into the fly's genome (per^o genetic background), *circadian rhythms were almost entirely normal* (Yu et al 1987a). This somewhat unexpected result is not miraculous: A similar kind of gene deletion, created by recombinant DNA methods in the low density lipoprotein (LDL) receptor, led to no detectable abnormalities in LDL receptor function, when the mutated factor was transduced into cultured cells (Davis et al 1986). The deletion in question was in fact rather analogous to that produced for *per*: Each was about the same size, and that for the LDL receptor gene encodes a domain of the protein that is a substrate for "O-linked" glycosylations (Davis et al 1986), as is also inferred to occur for the threonine residues in *per*'s Thr-Gly repeat (Reddy et al 1986, Bargiello et al 1987).

When the courtship song rhythms were examined, in the transformants carrying repeat-deleted DNA inserts, striking changes were found: The cycle durations were 20–25 s shorter than in the control transformants (Yu et al 1987a), thus implying that this domain of the *per* gene is somehow much more concerned with this ultradian rhythm than with the "main" (i.e. circadian) clock function.

Clocks and Molecules in Neurospora crassa

Studies of clocks and molecules in *Neurospora crassa* are not yet as mature as those of *Drosophila*'s *per*. Interestingly, some of the early data in this area, from the fungal system, started with *per* clones: There is preliminary evidence for *per*-homologous factors in *N. crassa*; two clones from a genomic library, produced from this fungal genome, were found to hybridize to sequences from separate subsegments of *per* (Feldman et al 1986). Neither *per* region included the Thr-Gly repeat. This is mentioned because interspecific homologies between *per* and cloned material from several other organisms have been detected; but the only cases analyzed in detail—whereby the isolated *per* homologs from mouse and from *Acetabularia* were sequenced—indicated that the only regions in common between the clones referred to *per*'s Thr-Gly repeat (Shin et al 1985, Li-Weber et al 1987). Therefore, it could be that families of "generalized" proteoglycan-encoding mouse genes were identified in these experiments. Yet, perhaps one or more of these factors is somehow involved in murine and/or algal clocks.

The *per*-homologous *Neurospora* clones have not yet been sequenced, but it is known that one of them maps, via application of Restriction Fragment Length Polymorphism principles and techniques, tantalizingly near a clock locus (J. F. Feldman, unpublished). It remains to be seen if connections between both formalistic and concrete genetics in these two very different organisms is to occur, i.e. after the necessary further experiments on the *N. crassa per* homologs are performed.

In the meantime, a genuinely exciting advance has occurred in the *Neurospora* system. It began when investigators noticed that the *frequency* gene is genetically quite near a locus for which there is a molecular "handle": The *oli* gene on chromosome VII. This locus is defined by oligomycin-resistant mutants, which actually are abnormal in their circadian taus (review: Feldman & Dunlap 1983). In fact, *frq* and *oli* were once thought to be the same gene (Dieckmann & Brody 1980). Therefore, the mitochondrial ATP synthetase known to be encoded by *oli* was suggested to have an important relationship to rhythm control in this organism, which it indeed might (Brody et al 1985). But it is now known that *frq* and *oli* are separate genetic loci, located approximately two map units apart (Loros et al 1986). Nevertheless—and this is the critical factor—*oli* has been cloned (Viebrock et al 1982), and such material can be used as the starting point for a chromosomal walk.

This search for *frq* clones has been performed by J. C. Dunlap and R. McClung (unpublished), by deliberately walking in *both* directions from *oli* (because there are no physical landmarks in *Neurospora*, such as the chromosome breaks near the fly's *per* locus, to orient the walk). After ca. 70–90 kb of walking in the two directions—and knowing that one genetic map unit is about 25 kb in *N. crassa*—transformation techniques were applied (see Dhawale et al 1984 for an example of the method). The results were that a series of overlapping clones, transduced into the frq^9 strain (see above), succeeded in rescuing the effects of this arrhythmic mutant and ultimately pinned down the *frq* locus to an 8 kb subset of chromosome VII. This, then, will permit a host of clone and gene product characterizations, similar to those performed for *per*. The data that should be rapidly forthcoming may have greater force than those from *Drosophila*. For example, *Neurospora* does not develop any kind of intercellular "oscillatory circuitry" to make its circadian clock. Therefore, the information encoded within this fungus's *frq* gene is likely to be right at the core of its oscillator, whereas what *per* is and does may or may not be (see above).

A final fillip to the emerging *Neurospora* story is that two oscillating RNAs have been recently discovered, which, as for the transcript encoded near *per*, show peaks of abundance during the day or subjective day (J. J. Loros, unpublished). The DNA sequences encoding these RNAs have been cloned, and the strategy for their isolation (which was intimately connected to the discovery of the abundance fluctuations) has indicated that the kind of periodic gene expression that can be inferred is relatively rare in this fungus. Additionally, these molecular rhythms are tied in to the circadian clock of *Neurospora*, because the peaks of RNA abundance were farther apart in a long-period *frq* mutant than in the wild-type (J. J. Loros, unpublished). It remains to be seen (*a*) whether the functions encoded by these genes will be tractable and in some way connectable to the clock mechanism (the same sort of question that remains for the oscillating "near-*per*" RNA from *D. melanogaster*), and (*b*) whether one or more of these *N. crassa* clones in fact defines a locus that, when mutated, leads to an abnormal rhythm.

Many questions—those posed here and others—are going to suggest themselves, as geneticists and chrono-biologists delve deeper into the appropriate genes and the normal vs abnormal rhythmic phenotypes in the bread mold and the fruit fly. As these investigators continue to make progress in their molecular studies, it may well be that the working of biological clocks, at least in these two systems, will become less and less mysterious.

ACKNOWLEDGMENTS

Several of the results reviewed here have come from collaborative efforts involving our laboratories and, especially, that of C. P. Kyriacou. Further collaborations with R. J. Konopka, H. D. Dowse, and J. M. Ringo are gratefully acknowledged. We thank the National Institutes of Health (grant GM-33205) for research support, Stuart Brody for discussions, and Jay Dunlap for comments on the manuscript.

Literature Cited

Bargiello, T. A., Jackson, F. R., Young, M. W. 1984. Restoration of circadian behavioural rhythms by gene transfer in *Drosophila. Nature* 312: 752–54

Bargiello, T. A., Young, M. W. 1984. Molecular genetics of a biological clock in *Drosophila. Proc. Natl. Acad. Sci. USA* 81: 2142–46

Bargiello, T. A., Saez, L., Baylies, M. K., Gasic, G., Young, M. W., Spray, D. C. 1987. The *Drosophila* gene *per* affects intercellular junctional communication. *Nature* 328: 686–91

Baylies, M. K., Bargiello, T. A., Jackson, F. R., Young, M. W. 1987. Changes in abundance or structure of the *per* gene product can alter periodicity of the *Drosophila* clock. *Nature* 326: 390–92

Bender, W., Spierer, P., Hogness, D. S. 1983. Chromosomal walking and jumping to isolate DNA from the *Ace* and *rosy* loci and the bithorax complex in *Drosophila melanogaster. J. Mol. Biol.* 168: 17–33

Bourdon, M. A., Oldberg, A., Pierschbacker, M., Rouslahti, E. 1985. Molecular cloning and sequence analysis of a chondroitin sulfate proteoglycan cDNA. *Proc. Natl. Acad. Sci. USA* 82: 1321–25

Bourdon, M. A., Shiga, M., Ruoslahti, E. 1987. Gene expression of the chondroitin sulfate proteoglycan core protein PG19. *Mol. Cell Biol.* 7: 33–40

Brody, S., Dieckmann, C., Mikolojoczyk, S. 1985. Circadian rhythms in *Neurospora crassa*: The effects of point mutations on the proteolipid portion of the mitochondrial ATP synthetase. *Mol. Gen. Genet.* 200: 155–61

Brody, S., Martins, S. A. 1979. Circadian rhythms in *Neurospora crassa*: Effects of unsaturated fatty acids. *J. Bacteriol.* 137: 912–15

Bruce, V. G. 1974. Recombinants between clock mutants of *Chlamydomonas reinhardi. Genetics* 77: 221–30

Bruce, V. G., Bruce, N. C. 1978. Diploids of clock mutants of *Chlamydomonas reinhardi. Genetics* 89: 225–33

Büttner, D., Wollnik, F. 1984. Strain-differentiated circadian and ultradian rhythms in locomotor activity of the laboratory rat. *Behav. Genet.* 14: 137–52

Citri, Y., Colot, H. V., Jacquier, A. C., Yu, Q., Hall, J. C., Baltimore, D., Rosbash, M. 1987. A family of unusually spliced and biologically active transcripts is encoded by a *Drosophila* clock gene. *Nature* 326: 42–47

Cochet, C., Gill, G. N., Meisenhelder, J., Cooper, J. A., Hunter, T. 1984. C-kinase phosphorylates the epidermal growth factor receptor and reduces its epidermal growth factor-stimulated tyrosine protein kinase activity. *J. Biol. Chem.* 259: 2553–58

Coté, G. G., Brody, S. 1986. Circadian rhythms in *Drosophila melanogaster*: Analysis of period as a function of gene dosage at the *per* (period) locus. *J. Theoret. Biol.* 121: 487–503

Davis, C. G., Elhammer, A., Russell, D. W., Schneider, W. J., Kornfeld, S., Brown, M. S., Goldstein, J. L. 1986. Deletion of O-linked carbohydrates does not impair function of low density lipoprotein receptor in transfected fibroblasts. *J. Biol. Chem.* 261: 2828–38

Dhawale, S. S., Paietta, J. V., Marzluff, G. A. 1984. A new, rapid and efficient transformation procedure for *Neurospora. Curr. Genet.* 8: 77–79

Dieckmann, C., Brody, S. 1980. Circadian rhythms in *Neurospora crassa*: Oligomycin-resistant mutations affect periodicity. *Science* 207: 896–98

Dowse, H. B., Hall, J. C., Ringo, J. M. 1987. Circadian and ultradian rhythms in *period* mutants of *Drosophila melanogaster*. *Behav. Genet.* 17: 19–35

Dowse, H. B., Ringo, J. M. 1987. Further evidence that circadian rhythms in *Drosophila* are a population of coupled ultradian oscillators. *J. Biol. Rhythms* 2: 65–76

Dunlap, J. C., Feldman, J. F. 1987. On the role of protein synthesis in the circadian clock of *Neurospora crassa*. *Proc. Natl. Acad. Sci. USA*. In press

Dunlap, J. C., Taylor, W., Hastings, J. W. 1980. The effects of protein synthesis inhibitors on the *Gonyaulax* clock. I. Phase-shifting effects of cycloheximide. *J. Comp. Physiol.* 138: 1–8

Engelmann, W., Schrempf, M. 1980. Membrane models of circadian rhythms. *Photochem. Photobiol. Rev.* 5: 49–86

Eskin, A. 1977. Neurophysiological mechanisms involved in photo-entrainment of the circadian rhythm from the *Aplysia* eye. *J. Neurobiol.* 8: 273–99

Evered, D., Whelan, J., eds. 1986. *Functions of the Proteoglycans.* Chichester, UK: Wiley. 299 pp.

Feldman, J. F. 1982. Genetic approaches to circadian clocks. *Ann. Rev. Plant Physiol.* 33: 583–608

Feldman, J. F. 1983. Genetics of circadian clocks. *BioScience* 33: 426–31

Feldman, J. F. 1985. Genetic and physiological analysis of a clock gene in *Neurospora crassa*. In *Temporal Order*, ed. L. Rensing, N. I. Jaeger, pp. 238–45. Berlin: Springer-Verlag

Feldman, J. F., Dunlap, J. C. 1983. *Neurospora crassa*: A unique system for studying circadian rhythms. *Photochem. Photobiol. Rev.* 7: 319–68

Feldman, J. F., Gardner, G., Denison, R. 1979. Genetic analysis of the circadian clock of *Neurospora*. In *Biological Rhythms and Their Central Mechanisms*, ed. M. Suda, D. Hayaishi, H. Nakagawa, pp. 57–66. New York: Elsevier/North-Holland

Feldman, J. F., Parker, E., Levy, R., O'Donnell, C. 1986. Sequence homology between a *Drosophila* clock gene and *Neurospora crassa* DNA. *Genetics* 113: s29 (Abstr.)

Fuchs, J. L. 1983. Effects of pinealectomy and subsequent melatonin implants on activity rhythms in the house finch (*Carpodacus mexicanus*). *J. Comp. Physiol.* 153: 413–19

Gardner, G. F., Feldman, J. F. 1981. Temperature compensation of circadian periodicity in clock mutants of *Neurospora crassa*. *Plant Physiol.* 68: 1244–48

Gibbs, F. P. 1983. Temperature dependence of the hamster circadian pacemaker. *Am.*

J. Physiol. 244: R607–10

Hall, J. C. 1984. Complex brain and behavioral functions disrupted by mutations in Drosophila. *Dev. Genet.* 4: 355–78

Hamblen, M., Zehring, W. A., Kyriacou, C. P., Reddy, P., Yu, Q., Wheeler, D. A., Zwiebel, L. J., Konopka, R. J., Rosbash, M., Hall, J. C. 1986. Germ-line transformation involving DNA from the *period* locus in *Drosophila melanogaster*: Overlapping genomic fragments that restore circadian and ultradian rhythmicity to *per⁰* and *per⁻* mutants. *J. Neurogenet.* 3: 249–91

Handler, A. M., Konopka, R. J. 1979. Transplantation of a circadian pacemaker in *Drosophila*. *Nature* 279: 236–38

Hotz, M. M., Connolly, M. S., Lynch, C. B. 1987. Adaptation to daily meal timing and its effects on circadian temperature rhythms in two inbred strains of mouse. *Behav. Genet.* 17: 37–51

Jacklet, J. W. 1985. Neurobiology of circadian rhythm generators. *Trends Neurosci.* 8: 69–73

Jackson, F. R. 1983. The isolation of biological rhythm mutations on the autosomes of *Drosophila melanogaster*. *J. Neurogenet.* 1: 3–15

Jackson, F. R., Bargiello, T. A., Yun, S.-H., Young, M. W. 1986. Product of *per* of *Drosophila* shares homology with proteoglycans. *Nature* 320: 185–88

James, A. A., Ewer, J., Reddy, P., Hall, J. C., Rosbash, M. 1986. Embryonic expression of the *period* clock gene in the central nervous system of *Drosophila melanogaster*. *EMBO J.* 5: 2313–20

Johnson, C. H., Hastings, J. W. 1986. The elusive mechanism of the circadian clock. *Am. Sci.* 74: 29–36

Klevecz, R. R., Kaufmann, S. A., Shymko, R. M. 1984. Cellular clocks and oscillators. *Int. Rev. Cytol.* 86: 97–128

Konopka, R. J. 1979. Genetic dissection of the *Drosophila* circadian system. *Fed. Proc.* 38: 2602–5

Konopka, R. J. 1981. Genetics and development of circadian rhythms in invertebrates. *Handb. Behav. Neurobiol.* 4: 173–81

Konopka, R. J. 1987. Neurogenetics of *Drosophila* circadian rhythms. In *Evolutionary Genetics of Invertebrate Behavior*, ed. M. D. Huettel, pp. 215–21. New York: Plenum

Konopka, R. J., Benzer, S. 1971. Clock mutants of *Drosophila melanogaster*. *Proc. Natl. Acad. Sci. USA* 68: 2112–16

Konopka, R. J., Orr, D. 1980. Effects of a clock mutation on the subjective day—implications for a membrane model of the *Drosophila* clock. In *Development and Neurobiology of* Drosophila, ed. O.

Siddiqi, P. Babu, L. M. Hall, J. C. Hall, pp. 409–16. New York: Plenum

Konopka, R. J., Wells, S. 1980. *Drosophila* clock mutations affect the morphology of a brain neurosecretory cell group. *J. Neurobiol.* 11: 411–15

Konopka, R. J., Wells, S., Lee, T. 1983. Mosaic analysis of a *Drosophila* clock mutant. *Mol. Gen. Genet.* 190: 284–88

Kyriacou, C. P., Hall, J. C. 1980. Circadian rhythm mutations in *Drosophila* affect short-term fluctuations in the male's courtship song. *Proc. Natl. Acad. Sci. USA* 77: 6929–33

Kyriacou, C. P., Hall, J. C. 1982. The function of courtship song rhythms in *Drosophila. Anim. Behav.* 30: 794–801

Kyriacou, C. P., Hall, J. C. 1984. Learning and memory mutations impair acoustic priming of mating behaviour in *Drosophila. Nature* 308: 62–65

Kyriacou, C. P., Hall, J. C. 1985. Action potential mutations stop a biological clock in *Drosophila. Nature* 314: 171–73

Kyriacou, C. P., Hall, J. C. 1986. Interspecific genetic control of courtship song production and reception in *Drosophila. Science* 232: 494–97

Kyriacou, C. P., Hall, J. C. 1987. Genetic and molecular analysis of behavioral rhythms in *Drosophila. J. Neurogenet.* 4: 147–48 (Abstr.)

Lakin-Thomas, P. L., Brody, S. 1985. Circadian rhythms in *Neurospora crassa*: Interactions between clock mutations. *Genetics* 109: 49–66

Livingstone, M. S. 1981. Two mutations in *Drosophila* affect the synthesis of octopamine, dopamine and serotonin by altering the activities of two different amino-acid decarboxylases. *Neurosci. Abstr.* 8: 384 (Abstr.)

Livingstone, M. S., Tempel, B. L. 1983. Genetic dissection of monoamine synthesis in *Drosophila. Nature* 303: 67–70

Li-Weber, M., de Groot, G. J., Schweiger, H.-G. 1987. Sequence homology to the *Drosophila per* locus in higher plant nuclear DNA and in *Acetabularia* chloroplast DNA. *Mol. Gen. Genet.* 209: 1–7

Lloyd, D., Edwards, S. W. 1984. Epigenetic oscillations during the cell cycles of lower eukaryotes are coupled to a clock. In *Cell Cycle Clocks*, ed. L. N. Edmund, pp. 27–46. New York: Marcel Dekker

Loros, J. J., Feldman, J. F. 1986. Loss of temperature compensation of circadian period length in the *frq-9* mutant of *Neurospora crassa. J. Biol. Rhythms* 1: 187–98

Loros, J. J., Richman, A., Feldman, J. F. 1986. A recessive circadian clock mutant at the *frq* locus of *Neurospora crassa. Genetics* 114: 1095–1110

Lotshaw, D. P., Jacklet, J. W. 1986. Involvement of protein synthesis in the circadian clock of *Aplysia* eye. *Am. J. Physiol.* 250: R5–17

Mattern, D. L., Forman, L. R., Brody, S. 1982. Circadian rhythms in *Neurospora crassa*: A mutation affecting temperature compensation. *Proc. Natl. Acad. Sci. USA* 79: 825–29

Menaker, M. 1959. Endogenous rhythms of body temperature in hibernating bats. *Nature* 184: 1251–52

Mergenhagen, D. 1984. Circadian clock: Genetic characterization of a short-period mutant of *Chlamydomonas reinhardi. Eur. J. Cell Biol.* 33: 13–18

Moore-Ede, M. C., Sulzman, F., Fuller, C. A. 1982. *The Clocks that Time Us.* Cambridge, Mass: Harvard Univ. Press

Nakashima, H., Perlman, J., Feldman, J. F. 1981. Genetic evidence that protein synthesis is required for the circadian clock of *Neurospora crassa. Science* 212: 361–62

Orr, D. P.-Y. 1982. Behavioral neurogenetic studies of a circadian clock in *Drosophila melanogaster.* PhD thesis. Calif. Inst. Technol., Pasadena. 198 pp.

Pittendrigh, C. S. 1967. Circadian rhythms. I. The driving oscillator and its assay in *Drosophila pseudoobscura. Proc. Natl. Acad. Sci. USA* 58: 1762–67

Pittendrigh, C. S. 1974. Circadian oscillations in cells and the circadian organization of multicullar systems. In *The Neurosciences Third Study Program*, ed. F. O. Schmitt, F. G. Worden, pp. 437–58. Cambridge, Mass: MIT Press

Possidente, B., Hegmann, J. P. 1980. Circadian complexes: Circadian rhythms under common gene control. *J. Comp. Physiol.* 139: 121–25

Ralph, M. R., Menaker, M. 1987. A genetic mutation of the circadian system in golden hamsters. *Neurosci. Abstr.* 13: 213 (Abstr.)

Rawson, K. S. 1960. Effects of tissue temperature on mammalian activity rhythms. *Cold Spring Harbor Symp. Quant. Biol.* 25: 105–13

Reddy, P., Jacquier, A. C., Abovich, N., Petersen, G., Rosbash, M. 1986. The *period* clock locus of D. melanogaster codes for a proteoglycan. *Cell* 46: 53–61

Reddy, P., Zehring, W. A., Wheeler, D. A., Pirrotta, V., Hadfield, C., et al. 1984. Molecular analysis of the *period* locus in Drosophila melanogaster and identification of a transcript involved in biological rhythms. *Cell* 38: 701–10

Richter, C. P. 1975. Deep hypothermia and its effects on the 24-hour clock of rats and hamsters. *Johns Hopkins Med. J.* 136: 1–10

Saunders, D. S., ed. 1982. *Insect Clocks.* Oxford: Pergamon. 409 pp. 2nd ed.

Schulz, H., Lavie, P., eds. 1985. *Ultradian Rhythms in Physiology and Behavior.* Berlin: Springer-Verlag. 340 pp.

Schwartz, W. J., Gross, R. A., Morton, M. T. 1987. The suprachiasmatic nuclei contain a tetrodotoxin-resistant pacemaker. *Proc. Natl. Acad. Sci. USA* 84: 1694–98

Shin, H.-S., Bargiello, T. A., Clark, B. T., Jackson, F. R., Young, M. W. 1985. An unusual coding sequence from a *Drosophila* clock gene is conserved in vertebrates. *Nature* 317: 445–48

Siwicki, K. K., Rosbash, M., Hall, J. C. 1987. *Drosophila* circadian clock probed with antibodies to the *period* gene product. *Neurosci. Abstr.* 13: 213 (Abstr.)

Smith, P. H. 1987. Naturally occurring arrhythmicity in eclosion and activity in *Lucilia cuprina*: Its genetic basis. *Physiol. Entomol.* 12: 99–107

Smith, R. F. 1982. Genetic analysis of the circadian clock system of *Drosophila melanogaster.* PhD thesis. Calif. Inst. Technol., Pasadena. 134 pp.

Smith, R. F., Konopka, R. J. 1981. Circadian clock phenotypes of chromosome aberrations with a breakpoint at the *per* locus. *Mol. Gen. Genet.* 183: 243–51

Smith, R. F., Konopka, R. J. 1982. Effects of dosage alterations at the *per* locus on the circadian clock of *Drosophila. Mol. Gen. Genet.* 185: 30–36

Takahashi, J. S., Menaker, M. 1984. Circadian rhythmicity: Regulation in the time domain. In *Biological Regulation and Development*, ed. R. F. Goldberger, K. R. Yamamoto, Vol. 3B, pp. 285–303. New York: Plenum

Tanouye, M. A., Lamb, C. A., Iverson, L. E., Salkoff, L. 1986. Genetics and molecular biology of ionic channels in *Drosophila. Ann. Rev. Neurosci.* 9: 255–76

Turek, F. W. 1985. Circadian neural rhythms in mammals. *Ann. Rev. Physiol.* 47: 49–64

Uhl, G. R., Reppert, S. M. 1986. Suprechiasmatic nucleus vasopressin messenger RNA: Circadian variation in normal and Brattleboro rats. *Science* 232: 390–93

Viebrock, A., Perz, A., Sebald, W. 1982. The imported preprotein of the proteolipid subunit of the mitochondrial ATPase synthase from *Neurospora crassa*: Molecular cloning and sequencing of the mRNA. *EMBO J.* 1: 565–71

Walz, B., Walz, A., Sweeney, B. M. 1983. A circadian rhythm in RNA in the dinoflagellate, *Gonyaulax polyedra. J. Comp. Physiol.* 151: 207–13

Weitzel, G., Rensing, L. 1981. Evidence for cellular circadian rhythms in isolated fluorescent dye-labelled salivary glands of wild type and an arrhythmic mutant of *Drosophila melanogaster. J. Comp. Physiol.* 143: 229–35

Wollnik, F., Gärtner, K., Büttner, D. 1987. Genetic analysis of circadian and ultradian locomotor activity rhythms in laboratory rats. *Behav. Genet.* 17: 167–78

Young, M. W., Jackson, F. R., Shin, H.-S., Bargiello, T. A. 1985. A biological clock in *Drosophila. Cold Spring Harbor Symp. Quant. Biol.* 50: 865–75

Young, M. W., Judd, B. H. 1978. Nonessential sequences, genes, and the polytene chromosome bands of *Drosophila melanogaster. Genetics* 88: 723–42

Yu, Q., Colot, H. V., Kyriacou, C. P., Hall, J. C., Rosbash, M. 1987a. Behaviour modification by *in vitro* mutagenesis of a variable region within the *period* gene of *Drosophila. Nature* 326: 765–69

Yu, Q., Jacquier, A. C., Citri, Y., Colot, H. V., Hamblen, M., Hall, J. C., Rosbash, M. 1987b. Molecular mapping of point mutations in the *period* gene that stop or speed up biological clocks in *Drosophila melanogaster. Proc. Natl. Acad. Sci. USA* 84: 784–88

Zehring, W. A., Wheeler, D. A., Reddy, P., Konopka, R. J., Kyriacou, C. P., Rosbash, M., Hall, J. C. 1984. P-element transformation with *period* locus DNA restores rhythmicity to mutant, arrhythmic Drosophila melanogaster. *Cell* 39: 369–76

Zimmerman, W. F., Pittendrigh, C. S., Pavlidis, T. 1968. Temperature compensation of the circadian oscillator in *Drosophila pseudoobscura* and its entrainment by temperature cycles. *J. Insect Physiol.* 14: 669–84

Ann. Rev. Neurosci. 1988. 11 : 395–421

SOME ASPECTS OF LANGUAGE PROCESSING REVEALED THROUGH THE ANALYSIS OF ACQUIRED APHASIA: The Lexical System

Alfonso Caramazza

Cognitive Neuropsychology Laboratory, The Johns Hopkins University, Baltimore, Maryland 21218

INTRODUCTION

Acquired aphasia is the loss of some aspect of language processing consequent to brain damage. The specific form of aphasia observed in a patient is determined by the locus of cerebral insult. However, given the complexity of the language processing system, involving as it does complex linguistic mechanisms—phonological, lexical, syntactic, semantic, and pragmatic—as well as associated cognitive systems (e.g. working memory), a vast number of different forms of aphasia may be observed. Each form of aphasia observed is presumed to result from the particular type of damage to a component or combination of components of the language processing system. It is unrealistic, therefore, in the limited space available here, to attempt a review of the full range of possible language deficits in aphasia. A more manageable task is to focus the review on just one subsystem of the language faculty. This review focuses on the lexical system.

Normally in a review I would at this point move directly to a presentation of the main theoretical and empirical developments in the area of lexical processing and the analysis of diverse forms of aphasia involving lexical deficits. However, developments over the past decade have led to a reconsideration of the theoretical and methodological underpinnings of the once dominant approach in neuropsychological research, with the result that

395

the approach has been challenged and the interpretability of most of its empirical findings questioned. The full implications of this challenge are only now becoming apparent. It is necessary, therefore, to consider briefly the nature of this critique so as to motivate the selection of material reviewed here (as well as the exclusion of certain materials).

The organization of this chapter is as follows. First I present a brief critique of the classical approach in neuropsychological research and a discussion of the theoretical and methodological assumptions of a new approach identified as *cognitive neuropsychology*. I then review the major empirical and theoretical developments in the area of lexical processing and lexical deficits in aphasia. A brief discussion of the implications of these results for a functional neuroanatomy of language concludes the review.

METHODOLOGICAL AND THEORETICAL ASSUMPTIONS FOR A COGNITIVE NEUROPSYCHOLOGY OF LANGUAGE

The modern study of acquired language disorders is based on a set of theoretical and methodological principles that distinguish it from, and even put it in opposition to, the classical study of aphasia. This latter approach is primarily concerned with establishing clinico-pathological correlates for different forms of aphasia. By contrast, the modern study of acquired aphasia has as its objective that of specifying the computational structure of normal language processing. Within this framework, those relationships between the cognitive/linguistic mechanisms comprising the language faculty and brain structures that may emerge from the analysis of aphasia, while very important, do not constitute the principal objective of research. That is, although research on aphasia will undoubtedly serve to provide an important source of constraints on a functional neuro-anatomy for language processing, it need not, and in much recent work it appears not to, be explicitly committed to such a goal. This does not mean that cognitive neuropsychology is unconcerned with the problem of relating cognitive mechanisms to the brain. To the contrary, the relationship of cognition to the brain is one of its objectives, but such a goal cannot take precedence over that of specifying the nature of the cognitive mechanisms that must be neurally implemented. To state the problem differently, the objective of cognitive neuropsychology is to articulate and to attempt to answer the correct type of empirical questions about brain/cognition relationships—questions that can only be formulated through an explicit theory of cognitive functioning. Thus, a "neuro-scientific" theory of cognitive abilities will not be formulated by directly relating behavior to neural events but through the mediation of cognitive operations. Cognitive neuropsychology rejects as prejudicial the eliminative materialism of some neuroscientists [and philosophers; cf Churchland (1986)] and operates instead with the assumption that a neuroscientific theory of cognition will be a theory about cognitive mechanisms and not directly about behavior; cognitive descriptions of mental events will not be replaced by neural descriptions but may be reduced to this latter level of description, should such a day arrive.

Classical neuropsychological research operated within a medical model framework mostly uninformed by cognitive or linguistic theory, and certainly unconcerned with the objective of developing a computationally explicit account of language processing. The syndromes that were correlated to anatomical sites for clinico-pathological analyses were based on impoverished notions of language processing using clinically derived, common sense classification schemes for language impairments (e.g. Benson 1985, Damasio 1981, Kertesz 1985). The symptoms that comprised the syndromes were grossly nonanalytic behavioral categories such as poor repetition, poor auditory language comprehension, poor naming ability, and so forth—behavioral conglomerates that are subserved by highly complex sets of cognitive and linguistic mechanisms. There are several reasons for rejecting this approach as a framework within which to explore the structure of the cognitive/linguistic mechanisms that subserve language processing and their relationship to the brain. However, before briefly presenting the details of this critical analysis I should like to consider one major accomplishment that has been achieved through research carried out within this framework.

Despite the serious limitations of this approach, what little is known about the functional neuroanatomy for language has come to us principally through clinico-pathological correlations for the aphasias. Although it has been known at least since the time of Hippocrates (ca 400 B.C.) that insult to the brain may result in disturbances of the language faculty, it was not until the detailed analysis of Broca, Wernicke, Charcot, Lichtheim, Dejerine, and others in the second half of the nineteenth century that a firm foundation was laid for relating language processes to the brain. Indeed by the end of that century French and German neurologists had described all the major aphasia syndromes at a level of detail that seemed to allow little opportunity for improvement. These investigators, under the influence of Gall's phrenological hypothesis, which proposed that distinct areas of the cerebral cortex subserve different cognitive faculties, set out to chart the functional burden of distinct parts of the cortex; that is, to localize the language faculty and its principal subcomponents in particular areas of the brain. Their labors and those of other students of aphasia since that time have not gone unrewarded. Neuropsychologists have amassed a systematic body of observations relating locus of brain damage to patterns of language dysfunction. These observations have estabished not only that language processing is subserved by neural structures in the left hemisphere (in most people) but that there is a highly articulated functional organization within this hemisphere, with different parts assumed to subserve different components of language processing. These results are well known and have been reviewed many times (e.g. Caplan 1987, Caramazza & Berndt 1982, Damasio & Geschwind 1984).

The general picture to emerge from this research program may be summarized thus: The linguistic components of language processing—syntactic, morphological, lexical-semantic, and phonological—are subserved by neural structures in the perisylvian region of the left hemisphere (see Figure 1); other regions of the brain, notably the right hemisphere, play a less important, supportive role in language processing. Thus, there is now considerable evidence that an intact right hemisphere may be needed for subtle interpretation of language, such as the appreciation of irony, metaphor, and humor as well as the emotional content of a linguistic act, but not for strictly linguistic processing (e.g. Brownell et al 1984, Gardner et al 1983).

This general view of the neural representation of language processes has received considerable support from neuropsychological research with other methodologies and techniques. Research with split-brain patients (patients whose two cerebral hemispheres have been disconnected for medical reasons), where the capacities of the two hemispheres may be

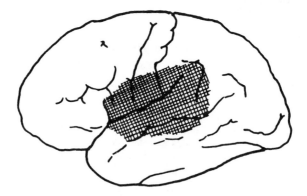

Figure 1   Schematic representation of the lateral surface of the left hemisphere with shading of the perisylvian region.

investigated in relative isolation, has confirmed that linguistic capacities are represented exclusively in the left hemisphere. A similar conclusion has been reached through electrical stimulation research of the exposed cerebral cortex during neurosurgical procedures. This latter research has shown that electrical stimulation of the cortex results in temporary linguistic impairments only when the stimulation is applied to the perisylvian region of the left hemisphere. And, finally, studies of regional cerebral blood flow with emission tomography during language activities in normal subjects have also led to a similar conclusion. These studies, which measure the metabolic (blood flow) activity in different regions of the brain during the performance of language tasks, have shown that it is the perisylvian region of the left hemisphere that is most directly implicated in language processing (cf Caplan 1987).

Without in the least intending to minimize the importance of that which has been learned about the neuropsychology of language through the methods currently at our disposal, I emphasize, nevertheless, that we have only succeeded in providing a gross, nonanalytic mapping of the language faculty onto the brain—at best a gross functional neuroanatomy. Is this the most that may be achieved through the analysis of language disorders consequent to brain damage? This question receives different answers depending on whether we place the focus on the neural or the cognitive part of the brain/cognition equation. Let us consider first the brain part of the equation.

The answer here is not an entirely encouraging one. The effort to relate functional disorders of language to locus of brain damage, no matter how fine-grained an analysis, can only result in a "modern phrenology." This is not to say that such an achievement would be insignificant. Quite the contrary: A fine-grained mapping of component parts of the language processing system onto neural structures would place important constraints on theories of the neurophysiological bases for language. But the type of observation at our disposal cannot lead to a neurophysiology of language. Furthermore, the fact that natural language is a uniquely human ability severely restricts the range of experimental opportunities for exploring the neurophysiological mechanisms for language processing—for example, we cannot use those experimental procedures currently within the armamentarium of the neurophysiologist for the analysis of neural activity in nonhuman animals. Does this mean that we must abandon the hope for a neurophysiology of language? Although current opportunities are limited, there is the hope that technological developments will eventually make it possible to investigate neural activity in humans directly with ethically acceptable means. In the meantime, we are not completely disarmed. We could rely on a bootstrap strategy that exploits whatever brain/cognition principles might emerge from the analysis of diverse cognitive processes in various nonhuman species to develop a computational neurophysiology of language processing; that is, a theoretical neurophysiology that relies on principles of neuronal functioning to develop neuronal-net models of specific linguistic processes—an approach that has received considerable attention in recent years (Arbib et al 1982, Hinton & Anderson 1981, Rumelhart & McClelland 1986). I return to this general issue in the concluding section of this review.

By contrast to the less-than-optimistic conclusion about the possibility of an experimental neurophysiology of language, the outlook for progress in developing a detailed functional theory of language processing through the analysis of different forms of acquired aphasia is very encouraging. The pragmatic motivation for using language deficits to inform and constrain theories of normal language processing comes from the observation that brain damage does not result in undifferentiated loss of language ability but in the selective loss of some ability in the face of otherwise normal performance. Thus, for example, brain damage may selectively impair language processes while sparing other perceptual and cognitive abilities. However, if brain damage were to result in dissociations of functions that are no finer than global cognitive systems (e.g. language, calculation, etc), the resulting patterns of impaired performance would be of little value in determining the processing structure of these systems. Fortunately for our enterprise, brain damage may result in highly specific patterns of dysfunction, presumably reflecting the componential structure of cognitive systems. We can use these highly articulated patterns of impaired performance to evaluate and develop models of normal language processing. However, such an enterprise cannot be carried out within the framework of classical neuropsychology. To fully appreciate this claim we must consider, albeit very briefly here, the assumptions that motivate the possibility of drawing meaningful inferences about normal language processing from patterns of language disorders (see Caramazza 1986a, for detailed discussion).

As already noted, the object of cognitive neuropsychology is to develop a theory of cognitive functioning through the analysis of patterns of cognitive dysfunction consequent to brain damage. The theoretical assumption that motivates the use of impaired performance as the basis for inferring the structure of normal processes is that the transformations of the normal system under conditions of damage are not indefinite or random but, instead, obey precise constraints determined by the intrinsic structure of the normal system: A pattern of impaired performance reflects a discoverable (and specifiable) transformation of the normal cognitive system (what I have called elsewhere the assumption of "transparency";

Caramazza 1984, 1986a). In this framework, a pattern of impaired performance is taken as support for a theory of the processing structure of a cognitive system (over some alternative theory) if it is possible to specify a transformation—a functional lesion—in the proposed theory (but not in some alternative theory) of the cognitive system such that the transformed system may account for the observed pattern of performance. This procedure allows a precise criterion for the empirical evaluation of a cognitive theory through the analysis of the performance of cognitively impaired brain-damaged patients.

The role played by "functional lesions" in the proposed framework for research is analogous to that played by "experimental conditions" in a typical experimental paradigm; that is, in a regular experiment the relationship between data and theory is mediated by specific experimental conditions, and in research with brain-damaged patients it is mediated by functional lesions (as well as experimental conditions). However, the two situations are disanalogous in one crucial respect: Whereas experimental conditions are under the control of the experimenter (and therefore known a priori), functional lesions are not known a priori but must themselves be inferred from the performance of patients. Thus, although we may consider a brain-damaged patient as constituting an "experiment of nature," where the functional lesion represents some of the experimental conditions of the experiment, these latter conditions are not known a priori, as would be the case in a regular experiment, and therefore they raise particular problems whose solution has important methodological consequences. Specifically, given that functional lesions may only be specified a posteriori—that is, once all the relevant patterns of performance for inferring a functional lesion in a cognitive system are available—there can be no theoretical merit in a classificatory scheme of patients' performance that is based on any arbitrary subset of a patient's performance. Two important consequences follow from these observations: (a) patient classification cannot play any significant role in cognitive neuropsychological research and (b) patient-group studies do not allow valid inferences about the structure of normal cognitive processes.

On the issue of patient-classification-based research, not only are there methodological arguments against its validity but, in addition, there are theoretical and practical considerations that undermine its usefulness (see Badecker & Caramazza 1985, Caramazza 1984, Caramazza & Martin 1983, Marshall 1982, 1986). The great majority of classification-based research has used theoretically uninformed behavioral categories for patient classification. Patients are classified as being of a particular type on the basis of criteria such as the following: whether or not a patient has poor repetition performance, or poor language comprehension performance, and so forth. However, since performance of such complex tasks as repetition or comprehension involves many cognitive mechanisms, impaired performance on these tasks may be due to damage to any one or combination of the cognitive mechanisms implicated in the performance of the task as a whole. Thus, poor performance in such tasks does not guarantee a theoretically useful homogeneity of the patients classified by these criteria. Furthermore, there is little value in reviewing classification-based research on aphasia for strictly pragmatic reasons. This research has led to little if any insight into the structure of normal language processes despite over a century of work.

The second major consequence of recent analyses of the logic of research in cognitive neuropsychology is that valid inferences about the structure of cognitive systems from patterns of cognitive dysfunctions are only possible for single-patient studies (Caramazza 1984, 1986a, Caramazza & McCloskey 1988, Shallice 1979). The arguments for this contention are straightforward but too long to present here. Suffice it to say that the principal argument is based on the observation that functional lesions can only be postulated a posteriori—that is, on the basis of all the relevant evidence needed to fix a functional lesion in a cognitive system.

Thus far I have focused on some negative conclusions of recent methodological and theoretical developments in cognitive neuropsychology; that is, I have presented recent conclusions concerning the impossibility of using the clinically based, classical methods of research on aphasia for learning about the structure of normal language processes and their neural correlates. A focus on these negative conclusions has been found necessary because of the need for clearly identifying the type of theoretical questions that may be profitably addressed through investigations of patients with cognitive deficits and for specifying the attendant methodology for addressing these issues. These developments may also be viewed positively, however: They offer us a theoretically coherent basis for a productive cognitive neuropsychology that increasingly interacts with other subdisciplines of the cognitive and neural sciences.

THE LEXICAL SYSTEM

Even though the focus of this review has been resricted to just a single subsystem of the language processing system, the ground to be covered is still quite extensive. The lexical system is very complex involving many linguistic and cognitive dimensions as well as being implicated in many different types of cognitive functions such as sentence comprehension and production, reading, writing, and naming. Consequently a further restriction of focus is necessary. The primary focus will be on single-word processing tasks, although an effort will be made to link the account of the lexicon that emerges from the review to the broader issue of sentence processing. Three sets of issues will be dealt with in this review: the general architecture of the lexical system; the representational content in different lexical processing components; and the processing structure within components. Although these issues are not entirely independent, it is useful to draw these distinctions for purposes of exposition.

The Functional Architecture of the Lexical System

The dominant view of the functional architecture of the lexical system is that it consists of a distributed but interconnected set of lexical components (e.g. Allport & Funnell 1981, Caramazza 1986b, Morton 1981, Shallice 1981). Over the past 10 to 15 years an impressive range of theoretical arguments and empirical evidence has been amassed in support of this view. The modal model that has emerged has the following structure. A major distinction is drawn between input and output lexical components; that is, lexical components involved in the comprehension (recognition) or production of words, respectively. A second major distinction is drawn between modality-specific input or output lexical components: The orthographic input lexicon, those mechanisms involved in processing written words, is distinguished from the phonological input lexicon, those mechanisms involved in processing spoken words. These modality-specific input lexicons are distinguished from their corresponding output lexicons, those mechanisms involved in the production of written and spoken words. It is further assumed that modality-specific lexical components are interconnected through a lexical-semantic system that stores the semantic representations for words. A schematic representation (as a visual aid) of these processing components is shown in Figure 2.

The evidence in favor of this view of the architecture of the lexical system is quite compelling. On strictly theoretical grounds the distinction between modality-specific components is unimpeachable—the mechanisms involved in processing visual and acoustic signals and the orthographic and phonological lexical representations these give rise to, are computationally independent. In one case—reading—the computational problem involves computing a lexical representation on the basis of visual information and subsequently letters or graphemes; in the other—listening—the computational problem involves computing a lexical representation on the basis of acoustic information and subsequently phonetic and phonemic information. Obviously, the computed representations must be different objects—orthographic or phonological lexical representations. A similar argument may be made for the input/output lexicon distinction.

The available empirical evidence is no less compelling. Two types of

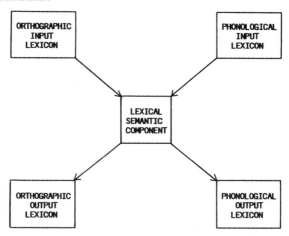

Figure 2 Schematic structure of the lexical system.

evidence have been reported: patients who present with selective damage to one or another lexical component (e.g. Basso et al 1978, Hier & Mohr 1977, Miceli et al 1985, Michel 1979) and patients who present with different patterns of impairments to different components (e.g. Beauvois & Dérouesné 1981, Goodman & Caramazza 1986a). Both kinds of evidence may be taken as support for a distributed view of the lexical system. Thus, for example, Goodman & Caramazza (1986b) have reported a patient who presents with damage to the output graphemic lexicon but who has normal access to other components of the lexical system (e.g. the output phonological lexicon and the lexical-semantic component). This pattern of performance is strong evidence against any theory of the lexicon that assumes a nondistributed, unitary lexical system. Equally supportive of the distributed theory of the lexicon are those patterns of performance in which different types of dysfunctions are found for different components of the lexicon. To give just one example of this type of result, Beauvois & Dérouesné (1979) have reported a patient whose impaired reading performance was radically different from his impaired spelling performance. This patient's impairment in reading involved only those cognitive mechanisms required for converting print-to-sound for novel or unfamiliar words—the patient could not read nonwords but essentially had no difficulty in reading words (see also Goodman & Caramazza 1986a). Thus the graphemic input lexicon must be intact, as must be the phonological output lexicon. By contrast, this patient's spelling impairment resulted from damage to the graphemic output lexicon that spared those mechanisms involved in converting sound-to-print and the phonological input lexicon (see also Goodman & Caramazza 1986a). This pattern of dissociation of deficits can only be explained by assuming selective damage to different components of a distributed lexical system. There is now a vast cognitive neuropsychological literature that demonstrates differential patterns of impairment for different parts of the lexical system (see also Allport & Funnell 1981, Shallice 1981 for reviews).

Lexical Representations

The distributed lexical system under consideration here distinguishes between modality-specific lexical components. These distinctions capture the most salient (perceptual) features of lexical information—phonological and orthographic information are represented in distinct processing components. However, there are other important lexical features that must be accounted for in a more general theory of lexicon. These include form class or categorial information (i.e. noun, verb, etc), morphological structure (root/stem and affixes), thematic structure (the argument structure of predicates), and semantic information (the meaning of words and morphemes). This information must be captured at some level of the lexical system. In this section I review some of the experimental evidence in favor of these representational distinctions. I also briefly review theoretical arguments and empirical evidence that bear on the issue of which lexical components may be assumed to capture the hypothesized lexical features.

FORM CLASS Although words have an independent status, their primary function is to convey meaning in sentential contexts. It is only when words are used in sentences that the full range of their syntactic, semantic, morphological, and phonological properties become apparent. Thus, for example, the word "jump" may be used as a noun (I watched the jump with trepidation) or as a verb (I watched him jump with trepidation). The two uses of "jump" have distinct grammatical roles (noun vs verb), different meanings, and accept different inflectional affixes (the noun accepts -s for plural, the verb accepts -s, -ing, and -ed to mark person and tense). The grammatical class of a word and its subcategorization features (e.g. transitive/intransitive) also determine the type of derivational affixes it accepts (e.g. only verbs accepts the -able derivational affix as in "enjoyable" but not "*windowable" and, furthermore, this applies only for transitive verbs as in "enjoyable" but not "*appearable"). Clearly, then, the lexicon must represent not only the phonological and orthographic structure of words but also their syntactic, semantic, and morphological properties (e.g. Chomsky 1965).

As indicated above, a crucial property of lexical items is their form (grammatical) class; that is, whether a word is (functions as) a noun, a verb, an adjective, an adverb, or a function word. These lexical properties play a determining role in the organization of the lexicon. Already in the classical literature there were clear indications for the dissociability of impairment of different form classes of words. The strongest evidence was for the dissociability of function words (articles, auxiliaries, prepositions, etc) from other form classes (nouns, verbs, and adjectives) (e.g. De Villiers, 1978, Goodglass 1976, Stemberger 1984; see Berndt & Caramazza 1980, Lesser 1978 for reviews). Patients clinically classified as agrammatic aphasics—that is, patients whose spontaneous speech is characterized by the relative omission of function words—could be argued to have a selective impairment in lexical access of function words (this position has been argued most forcefully by Bradley et al 1980). However, this type of deficit does not allow us to distinguish between a deficit at some level of sentence production (where a syntactic frame for sentence production is specified) and a pure lexical access deficit (Caramazza & Berndt 1985, Miceli & Caramazza 1988). Nonetheless, independently of whether the deficit in any one patient who makes errors with (or omissions of) function words in sentence production is ultimately found to be at the level of specifying a sentence frame or in lexical access, such patterns of language impairment are prima facie evidence for a representational distinction between function words and other word classes, and hence for a particular form of organization of the lexicon.

More direct evidence exists for the selective impairment of lexical access of function words. There are reports of patients whose performance in single word processing—reading, writing, or repetition of single words—is either relatively poor or relatively good when compared to other form classes (e.g. Bub & Kertesz 1982, papers in Coltheart et al 1980, Friederici & Schoenle 1980, Nolan & Caramazza 1982, 1983). Although it initially appeared that a deficit in function word processing was associated with a more general morphological processing impairment (Beauvois & Dérouesné, 1979, DeBastiani et al 1983, Patterson 1982), it is now clear that these two types of deficits are dissociable (Caramazza et al 1985, Funnell 1983). Thus, we have evidence for at least one type of organizational distinction at some level of the lexicon. (I take up the issues of the level at which these distinctions may be represented, below.)

Evidence for other organizational distinctions within the lexicon has also been obtained. There are numerous reports of patients whose reading or writing performance is differentially affected for nouns, verbs, and adjectives (see papers in Coltheart et al 1980). The typical result reported is for better performance for nouns relative to verbs and adjectives. The systematicity of this result raises the possibility that a lexical dimension other than form class is responsible for this ordering of performance difficulty. Indeed, patients who present with the form class effect described, typically show greater difficulties in processing abstract than concrete (or high imageability) words. This association of deficits allows the possibility that the relevant dimension affected by brain damage in the patients in question is not form class but concreteness/abstractness. However, an effect of form class has been obtained even when concreteness/abstractness is controlled for (e.g. Baxter & Warrington 1985, Shallice & Warrington 1975). Furthermore, there is evidence in the literature on naming disorders for a double dissociation in naming difficulty for verbs and nouns (Baxter & Warrington 1985, McCarthy & Warrington 1985, Miceli et al 1984). Some patients have considerably greater difficulty naming nouns than verbs, other patients present with the reverse pattern in naming difficulty for nouns and verbs. This result suggests that, at least in some patients, the underlying cause of their naming impairment is selective damage to different subsets of the lexicon—subsets defined by form-class member-ship. The double dissociation of processing difficulty for nouns and verbs has also been documented for a word comprehension task (Miceli et al 1988). In this latter case the reported dissociation for form class also concerned a dissociation by modality of use. That is, some patients pre-sented with selective impairment in comprehension of verbs without a corresponding difficulty in naming of this class of words.

The results reviewed in this section are unequivocal in one regard: they support the view that the lexicon is organized by grammatical class. They do not provide as compelling a basis, if at all, for determining where in the lexical system form class information is represented. Nonetheless, I propose that presently our best answer to this latter question is that form class information is represented in each modality-specific lexicon (i.e. in the phonological input and output lexicons, and in the orthographic input and output lexicons). Although empirical support for this position—modality specific form class effects in lexical processing (e.g. Baxter & Warrington 1985)—is scanty, there are good theoretical reasons for adopt-ing it. Basically, the argument is that since morphological structure is strictly dependent on form class information, this latter information must be represented at the same lexical level as that at which morphological structure is represented. And, as we will see below, morphological structure is represented in modality-specific lexicons.

MORPHOLOGICAL STRUCTURE Words are not unanalyzable units—they have phonological (and orthographic) and morphological structure. The word "nationalized" is considered to be composed of the verb stem "nationalize" plus the inflectional affix, -ed (past tense). In turn, the verb "nationalize" is derived from the adjective "national" by the addition of the derivational affix, -ize, which is itself derived from the noun "nation" by the addition of the adjectival derivational affix, -al. Thus, we may analyze words into stems (or roots), derivational affixes [affixes that serve to specify the form class of the derived word; e.g. nation (noun) → national (adjective)], and inflectional affixes that mark the tense, number, and gender of a word (see Scalise 1984 for review). A crucial issue for a theory of the lexicon is whether morphological structure is explicitly represented in the lexicon and how it is represented and used in language processing.

Various theoretical positions have been taken on this issue. A major contrast is between the view that words are represented in the lexicon in morphologically decomposed form (e.g. Taft 1985) versus the view that words are represented as nondecomposed wholes (e.g. Butterworth 1983). A second distinction, relevant only for the case of morphologically decomposed lexical representations, is whether lexical access is only possible after a word stimulus is parsed into its morphological components (stems or roots and affixes) (e.g. Taft 1979) or whether lexical access may proceed through both whole-word and morphemic access procedures (e.g. Caramazza et al 1985). Other issues concern the proper relationship between derivational and inflectional morphology and whether inflectional morphology is represented in the lexical or syntactic system (e.g. Anderson 1982). In the limited space available here I consider only the general issue of morphological decomposition as it emerges through the analysis of the word-processing performance in brain-damaged patients.

Various reports in the literature have dealt with morphological processing in brain-damaged patients. Some of this research has focused on the patterns of omissions (e.g. De Villiers 1978, Gleason 1978, Goodglass 1976, Goodglass & Berko 1960) or substitutions (e.g. Miceli et al 1983) of inflectional affixes in patients clinically classified as agrammatic aphasics. These reports have clearly documented a dissociation in processing inflectional affixes (impaired) versus word stems ("intact"). The reverse pattern of dissociation, impaired stem production and spared inflectional affix production, has also been reported (e.g. Caplan et al 1972). These patterns of results would appear to be prima facie evidence for morphological decomposition in the lexicon. However, as in the case of function word omission (or substitution) in spontaneous sentence production (discussed above), these results are ambiguous with respect to the locus of deficit: A patient may fail to produce (or fail to produce correctly) an inflectional affix because of damage to the inflectional component of a morphologically decomposed representation or because of damage to a component of the syntactic frames computed in the course of sentence production. The relevant data needed to resolve this issue involves patterns of selective impairment in single word processing. Such data are available.

An important source of evidence comes to us from the oral reading errors in patients with acquired dyslexia. An often noted feature in dyslexic patients in the presence of morphological errors—that is, errors such as reading "walked" for "walking" (inflectional error) or "kindness" for "kindly" (derivational error). Errors of this type were first most clearly documented in patients clinically classified as deep dyslexic. These are patients who in addition to morphological errors also make semantic (read "priest" for "minister") and visual (read "bear" for "fear") errors as well as presenting with other processing impairments [i.e. a form class effect, a concreteness/abstractness effect, a frequency effect, and disproportional difficulties in reading nonwords; see Coltheart et al (1980) for review and discussion]. Although the presence of morphological errors as part of the complex clinical picture in these patients may be suggestive, it does not permit an unequivocal conclusion regarding the issue in question; namely, whether or not lexical representations are morphologically decomposed. After all, the putative morphological errors may be no more than visual or semantic errors. However, the existence of "morphological" reading errors may be used in more focused analyses to address the question of concern here.

Patterson (1980, 1982) and Job & Sartori (1984) have described in some detail patients whose reading errors were almost exclusively of the morphological type. These authors interpreted the highly selective impairment in their patients (essentially restricted to the production of morphological paralexias) as evidence for a selective deficit to the morphological component of the lexicon. This conclusion has been challenged, however. Badecker & Caramazza (1988) have argued that the mere production of "morphological" paralexic errors is not sufficient grounds for concluding that the basis for the impairment is a deficit to the morphological component of the lexicon. Equally plausibly these errors could be considered to be highly similar visual errors or highly similar semantic errors. The ambiguity of interpretation could be resolved only if it turned out that a pattern of errors is only explicable by appeal to a morphological and no other lexical (semantic) or perceptual dimension. Note that this objection does not imply that the cases described by Patterson and Job & Sartori may not, after all, truly be cases of selective deficit to a morphological processing component. All that is asserted is that the presented evidence is not sufficient to unambiguously decide the issue. Fortunately, there is at least one case report of a patient whose impaired lexical processing performance is unequivocally the result of a selective deficit to the morphological component of the phonological output lexicon.

Miceli & Caramazza (1988) have described a patient, F. S., who makes morphological errors in spontaneous sentence production and in repetition of single words. The great majority of this patient's single word repetition errors were morphologically related to the target response. Crucially, these morphologically related responses were almost all inflectional errors (97%). The massive presence of morphological errors restricted to the inflectional category is only explicable by appeal to a morphological principle—a distinction between inflectional and derivational morphology: the evidence for a true morphological processing impairment. The highly selective deficit for inflectional morphology in a single-word processing task reported for F. S. allows the conclusion that lexical entries are represented in morphologically decomposed form—stems (or roots) are represented independently of their inflectional and derivational affixes, which, in turn, constitute independent components within the lexicon.

In this section I have reviewed evidence in support of the view that the lexical system represents words in morphologically decomposed form. As a final issue in this area I argue that morphological structure is represented directly in modality-specific lexicons. However, the evidence for this conclusion is, at best, indirect.

Caramazza et al (1985) have described a patient with a selective deficit in reading nonwords. The patient could read all types of words but made on the order of 40% errors in reading nonwords. However, when his reading performance for "morphologically legal" nonwords (e.g. "walken," composed of the inappropriately combined morphemes, walk- and -en) was assessed, it was found that he read these nonwords much better than comparable nonwords that did not have any morphological structure (e.g. "wolkon"). Since we may safely assume that nonwords do not have permanent entries in the lexical system, the better performance for the "morphologically legal" nonwords must be due to the activation of morphemic representations (e.g. walk- and -en) in the orthographic input lexicon. If this argument is correct, we must conclude that morphological structure is represented in modality specific lexicons.

In conclusion, the evidence from the analysis of language impairments in brain-damaged patients taken together with results in the literature on normal word processing (e.g. Stemberger 1985, Taft 1985) and linguistics (e.g. Scalise 1984) strongly argues for the autonomous representation of morphological structure in the lexical system.

LEXICAL SEMANTICS   That of various features of a word its meaning is the most important is quite obvious. Despite this and despite the fact that word meaning is increasingly seen as playing a determining role in linguistic theory (e.g. Chomsky 1981, Wasow 1985), we do not have the detailed theory of lexical meaning that would be commensurate with the crucial role of this dimension of lexical items. The absence of theory has left empirical work in this area in disarray so that we do not have anything like a coherent research program in the analysis of disorders of lexical meaning. Consequently, in this section I focus on an interesting empirical phenomenon concerning semantic organization of the lexicon, without attempting to provide a general model of this component of the lexicon (in contrast to what I have attempted to do for other components of the lexical system). The phenomenon I consider here is that of category-specific deficits.

We have seen in preceding sections that brain damage may result in highly specific deficits. The patient with a selective deficit of inflectional morphology or the patients with selective deficit in processing function words are cases in point. Results such as these allow us to articulate the functional architecture of the modality-specific lexicons. In recent years Warrington and her colleagues (Warrington 1975, 1981, Warrington & McCarthy 1983, Warrington & Shallice 1984), following an earlier observation by Goodglass et al (1966), have described a number of patients with selective deficits to specific semantic categories. These results provide evidence relevant to the organization of the semantic lexicon.

Goodglass et al (1966) provided a quantitative analysis of a large number of patients in which they show that different patients present with different patterns of relative difficulty in auditory comprehension of semantic categories. Warrington and her colleagues in a series of detailed single-patient analyses have documented selective dissociations between concrete (impaired) and abstract words (spared) (the reserve pattern is commonly reported), inanimate (impaired) and animate words (spared), and living things and foods (impaired) and inanimate words (spared). Perhaps the most striking result in this domain is one reported by Hart et al (1985). The patient, M. D., presented with a very selective disturbance of the ability to name items from two related semantic categories. Despite normal naming performance with the items from many different semantic categories, the patient showed a striking and consistent naming deficit for the categories "fruits" and "vegetables." Thus, as can be seen in Table 1, the patient performed poorly in naming fruits and vegetables in the face of spared ability to name items from other categories.

The patient's difficulties in processing the members of the categories fruits and vegetables extended to a number of other tasks. Thus, the patient presented with difficulties in sorting pictures of fruits and vegetables into the appropriate categories, i.e. sorting together fruits separately from vegetables; he had difficulties in generating the names of members of the two categories when given the category, i.e. producing apple, orange, peach, etc in response to the category "fruits"; and he showed a selective difficulty in naming fruits and vegetables from definition as well as from

Table 1   Number of correct naming responses[a]

	Fruit	Semantic category Vegetables	Other[b]
Line drawings	5/11	7/11	11/11
Colored drawings	4/6	5/7	18/18
Photographs	11/18	12/18	222/229
Real objects	10/13	13/23	11/11
TOTAL	30/48 (0.63)	37/59 (0.63)	262/269 (0.97)

[a] From Hart et al (1985).
[b] The "other" category includes vehicles, toys, tools, animals, body parts, food products, school, bathroom, kitchen and personal items, clothing, colors, shapes, and trees.

tactile presentation. By contrast, he showed normal performance with these categories in a word-picture matching task and in judgments of category, size, texture, and shape when given the name of individual fruits and vegetables. Normal performance on these latter tasks demonstrates that the patient's knowledge of these categories is intact but can only be accessed from the lexicon.

Although the absence of a well-developed theory of lexical semantics makes it difficult to provide a systematic interpretation of these category (semantic)-specific deficits, these latter results provide a provocative source of data on which far-reaching speculations about the structure of lexical organization may be based. Thus, at the very least, these results strongly argue for a highly structured lexical organization based on semantic categories. The implication of these results for neural organization is considered below.

Processing Principles

The material reviewed thus far has allowed us to address issues concerning the architecture of the lexical system and the types and organization of information represented in lexical components. I turn now to a consideration of the processing principles that govern the access of this information.

Two general classes of lexical processing models have been proposed: serial search models and passive, parallel activation models (activation, for short). Of these two classes, the activation models have clearly emerged dominant over the past decade. The basic assumption of activation models is that a stimulus (or input at some level of the lexical system) activates in parallel all stored representations. The degree of activation of any representation is proportional to the overall similarity between the input and the stored representation. Thus, for example, the stimulus word "car" will activate the representations "cat," "tar", "cart," "cord," etc to different degrees. In this example, "car" will be activated most strongly and "cat" will be activated more than "cord" and so forth. When the level of activation of a representation reaches a set, threshold value, the representation becomes available for further processing to other components of the processing system. Models of this type are known as *serial stage models*. If we relax the assumption that only the representation that reaches a threshold value can serve to activate subsequent stages of processing and we allow all representations that reach a minimal level of activation to activate representations in other components of the system, we have what are called *cascade models of processing* (McClelland 1979). Here I assume, for the sake of simplicity, a serial stage model (although it is quite likely that the cascading principle is a more realistic characterization of the processing sequence).

A distributed model of the lexical system, as that discussed in this review, which operates on the principle of passive, parallel activation, provides a natural framework for considering various features of impaired language performance. Two such features are the ubiquitous frequency effect (words of high usage frequency are in many cases relatively spared in comparison to words of lower frequency) and certain types of error responses produced by patients in single-word processing tasks.

It is a well-established phenomenon in the psychological literature that reaction time to recognize a word or to decide that a string of letters forms a word (lexical decision) is inversely proportional to the frequency of usage of a word (and, similarly, for error rates) (see Gordon 1983 for review). Activation models account for this effect by assuming that the activation threshold of a representation is lowered with repeated presentations of the stimulus or input (Morton 1970). Thus, high-frequency words have lower thresholds than low-frequency words and, therefore, can be activated more easily, resulting in lower reaction times (RTs) and lower error rates, than low-frequency words. This differential effect of word frequency is also found in aphasic patients' performance (see Gordon & Caramazza 1982). To give just one example, many dyslexic patients make more errors in reading low-frequency words than in reading high-frequency words. What is important for our present concern, however, is that the presence of a frequency effect may be associated with certain types of error responses, thus allowing us to identify the locus of deficit responsible for a patient's impaired performance. That is, we may take the presence of a word frequency effect as an indication of a deficit to the lexical system and the type of error (e.g. visual or semantic) as an indication of a deficit at a specific level within the lexical system.

I indicated above that two types of errors produced by dyslexic patients are visual and semantic paralexias. Various accounts have been offered as the basis for these types of errors (e.g. Caramazza 1986b, Marshall & Newcombe 1973, Morton & Patterson 1980, Nolan & Caramazza 1982, Shallice & Warrington 1980). I argue that, at least in some cases, these errors arise from independent deficits to the graphemic input lexicon and the phonological output lexicon for visual and semantic paralexic errors, respectively.

Recall that visual paralexic errors are errors such as reading "bead" for "head" and semantic paralexic errors are errors such as reading "airplane" for "ship." In an indepth investigation of a single patient, F. M., Gordon et al (1987) asked the patient to read several thousand words in order to obtain a reliable data base of errors for detailed analysis. The patient's responses were scored either as correct or as an error of one of the following types: visual, semantic, inflectional, derivational, or other—where this last category consists of ambiguous errors, visual-to-semantic errors or word responses that could not be classified in any of the previously listed error categories. Here I first wish to focus on the evidential role of visual and semantic errors to constrain a model of the lexical system.

A priori it is unlikely that these two types of errors have a common basis: A semantic error can only occur if the correct lexical entry has been activated; that is, in order to produce "minister" for "bishop," the lexical entry for "bishop" had to be activated. There is no such constraint for visual errors. This latter type of error most likely arises from damage to the input graphemic lexicon, where an inappropriate lexical representation is activated. To explore this issue consider the following argument. A word that is read correctly is one that successfully activates a lexical entry in the input graphemic lexicon and the output phonological lexicon. By contrast, a word that gives rise to a visual error is one that fails to activate its lexical entry in the input graphemic lexicon and instead activates a visually similar entry in this lexicon. Similarly, a word that gives rise to a semantic error is one that successfully activates a correct lexical entry in the input graphemic lexicon, but fails to activate its lexical entry in the output phonological lexicon, and instead activates a semantically related entry. Note that this argument makes two obvious, but important assumptions: (a) The access procedure for the input graphemic lexicon is orthographically based; (b) the access procedure for the output phonological lexicon is semantically based.

This proposed architecture of the lexical system and, more specifically, the assumptions we have made about the address procedures for the input graphemic lexicon and the output phonological lexicon (i.e. parallel activation), allows us to make a precise prediction about F. M.'s performance on re-reading words read correctly, incorrectly produced responses, words to which he made visual errors, and words to which he made semantic errors on the first reading. The prediction is that he should read very well words he read correctly the first time as well as the incorrectly produced responses but should read poorly words to which he previously made errors. Furthermore, the new errors for words that gave rise to visual errors should be predominantly visual whereas those for words that gave rise to semantic errors should be predominantly semantic. These predictions were borne out.

To further substantiate the claim that visual and semantic errors arise due to difficulties in addressing lexical representations in the input graphemic lexicon and the output phonological lexicon, respectively, we assessed F. M.'s ability to comprehend words that were on a previous occasion read correctly or had resulted in visual or semantic errors. The model of the lexical system proposed here leads to the prediction that F. M. should understand both the words he previously read correctly and those with which he made semantic errors, but he should fail to comprehend the words with which he had made visual errors. This prediction too was borne out.

The implication of these results for claims concerning the processing structure of the hypothesized lexical components is clear-cut. It would appear that a visual error is made when a particular lexical entry in the graphemic input lexicon cannot reach threshold and instead a visually similar representation reaches threshold. Similarly, a semantic error occurs when a representation in the phonological output lexicon cannot reach threshold and instead a semantically related response reaches threshold. This interpretation of the basis for F. M.'s visual and semantic errors is only possible if we assume that lexical representations are activated in parallel and in proportion to the similarity between the input and the stored representation.

CONCLUSION

In this all too brief and highly condensed review I have dealt with three aspects of the structure of the lexical system: the general architecture of the system, the types of representational content in each hypothesized component, and the processing principles that allow access of the information stored in the lexicon. The evidence reviewed not only provides empirical support for the model but, in addition, the model serves as a guide for the interpretation and analysis of cognitive/linguistic disorders. The discussion has focused, however, entirely on functional (cognitive) aspects of the process. We may wish to ask, therefore, whether or not the types of observations available to us from the analysis of cognitive deficits will be relevant to the formulation of a truly *neuro*psychological theory of cognitive functioning. Is a *neuro*psychology of language possible? In a previous section of this review I sounded a pessimistic note with respect to this question. Here, by way of conclusion, I would like to take up this issue in a little more detail.

The classical study of aphasia has failed to lead to any significant insights into the structure of language processing mechanisms and their neural instantiation, other than the gross clinical-pathological mapping already available at the end of the last century. This work clearly established the importance of the perisylvian region of the left hemisphere for language processing but could not go beyond this general phrenological statement. Theoretical and methodological developments over the past decade have introduced the possibility for significant progress for one part of the brain/cognition equation. We have seen that we now have a clearly articulated justification for drawing inferences about normal cognitive processing from the analysis of patterns of cognitive dysfunction, as well as a powerful theoretical and methodological basis for the analysis of cognitive dysfunctions. This development, by itself, is not sufficient to lead to any significant insights into the nature of the neural mechanisms that subserve language processing. It may be sufficient, however, to provide a set of principled constraints on the possible form of a neuropsychological theory of language processing.

Recent work (some of it reviewed here) in cognitive neuropsychology has provided an impressive set of results on the nature of language dysfunctions. It has been possible to demonstrate that language dysfunction may be highly selective, affecting a single component (e.g. Miceli & Caramazza 1988) or even a single representational dimension within a component (e.g. Warrington 1981). Such observations provide a natural set of constraints for a theory of language processing as amply demonstrated above. However, since the observations that enter into this theory-construction process consist of brain/behavior pairs, we may use them to constrain the formulation of a neuropsychological theory of language. Thus far little use has been made of this opportunity. But we may already state an important constraint that has emerged from this research: Given the highly selective and systematic dissociations of function observed in brain-damaged patients, we may conclude that there is a high degree of specialization of cognitive function in the brain; that is, the observations reported support a strong localizationist view of brain organization. This conclusion needs some elaboration.

We have seen that brain pathology may selectively damage one or another component of a distributed lexical system. These results support the modular theory of lexical components presented above. They also suggest, however, that distinct neural structures subserve the hypothesized lexical components. Indeed, the evidence on hand shows a fine-grained localization of function well beyond the level of gross lexical component all the way down to single representational dimensions. This does not necessarily mean (although such may be the case) that distinct neuroanatomical loci are associated with different components of the lexicon. All that is asserted is that a distinct neural process is associated with different cognitive mechanisms and that these neural processes may be selectively damaged. What is clear, however, is that the neuropsychological data do not support an indefinitely plastic, nonlocalizationist model of neural functioning. This is a nontrivial conclusion about neural processing that has emerged from cognitive neuropsychological research.

Cognitive neuropsychological analyses may also be used to provide a fine-grained mapping of cognitive mechanisms to neural structures or processes. That is, we may be able to go beyond the level of merely specifying general constraints for a neural theory of language processing. This is not possible, however, without a profound transformation of the social organization of scientific investigation in this area.

We have seen that valid inferences about the structure of normal cognition are only possible for single-patient studies. The highly detailed investigation of single patients allows us to infer a functional lesion to a model of a cognitive system and thereby provide support for that model. Although the analysis of single patients is well-suited for drawing conclusions about cognitive structure, this methodology is not sufficient for drawing conclusions about brain/cognition relationships. For this latter purpose we need to accumulate enough cases with "identical" functional lesions in order to correlate the identified cognitive mechanisms with the neural structures that support the identified functions. This entails the accumulation of large numbers of cases. However, since the most useful, clear information is likely to come from patients with highly selective deficits and since such cases are relatively rare, it is extremely unlikely that any single investigator or laboratory will have enough cases to carry out the correlational analysis needed for this purpose. This limitation of the cognitive neuropsychological method in relating cognitive mechanisms to neural structures is not an in-principle limitation of the method but only a practical one that may be overcome if adequate measures are taken. Specifically, as I have argued elsewhere (Caramazza & Martin 1983), cognitive neuropsychologists will have to create research consortia, as have done high energy physicists and astronomers in their respective domains. This step will permit the accumulation of cases with the desired characteristics for the needed correlational analysis. (It must be emphasized here, if there is any need, that this proposal in no way implies an indirect justification for the group-study methodology. The frequency analysis proposed here is based on single-patient analyses and does not require the averaging of patients' performance, a methodologically invalid procedure.)

In this concluding section I have identified a procedure for relating language processing mechanisms to brain structures within the methodology of cognitive neuropsychology. We should note, however, that even in the best of all possible worlds this methodology can only lead to a fine-grained, modern phrenology—it will not provide information directly relevant to a neurophysiology of language. This latter goal may be unattainable even with technological developments. The most promising avenue open to us at this time is the development of a computational neuropsychology; that is, the development of neural network models of language processing (e.g. Arbib et al 1982). It is not difficult to imagine how the interaction of increasingly detailed, neurally constrained models of language processing that emerge from cognitive neuropsychological research with neural network models of language processes may lead to a theoretical neurophysiology of language.

ACKNOWLEDGMENTS

The research reported here was supported in part by National Institute of Health grants NS23836 and NS22201, as well as The Seaver Institute and The Lounsbery Foundation. I would like to thank Marie-Camille Havard, Kathy Yantis, and Olivier Koenig for their help in the preparation of this manuscript.

Literature Cited

Allport, A., Funnell, E. 1981. Components of the mental lexicon. *Philos. Trans. R. Soc. London* 295: 397–410

Anderson, S. 1982. Where's morphology? *Ling. Inquiry* 13: 571–612

Arbib, M. A., Caplan, D., Marshall, J. F., eds. 1982. *Neural Models of Language Processes.* NY: Academic

Badecker, W., Caramazza, A. 1985. On considerations of method and theory governing the use of clinical categories in neurolinguistics and cognitive neuropsychology: The case against agrammatism. *Cognition* 20: 97–115

Badecker, W., Caramazza, A. 1988. The analysis of morphological errors in a case of acquired dyslexia. *Brain Lang.* In press

Basso, A., Taborelli, A., Vignolo, L. A. 1978.

Dissociated disorders of speaking and writing in aphasia. *J. Neurol. Neurosurg. Psychiatry* 41: 6, 556

Baxter, D. M., Warrington, E. K. 1985. Category-specific phonological dysgraphia. *Neuropsychologia* 23: 653–66

Beauvois, M.-F., Dérouesné, J. 1979. Phonological alexia: Three dissociations. *J. Neurol. Neurosurg. Psychiatry* 42: 1111–24

Beauvois, M.-F., Dérouesné, J. 1981. Lexical or orthographic agraphia. *Brain* 104: 21–49

Benson, D. F. 1985. Aphasia. In *Clinical Neuropsychology*, ed. K. M. Heilman, E. Valenstein. New York: Oxford Univ. Press

Berndt, R. S., Caramazza, A. 1980. A redefinition of the syndrome of Broca's aphasia: Implications for a neuropsychological model of language. *Applied Psycholinguist.* 1: 225–78

Bradley, D., Garrett, M., Zurif, E. 1980. Syntactic deficits in Broca's aphasia. In *Biological Studies of Metal Processes*, ed. D. Caplan. Cambridge, Mass.: MIT Press

Brownell, H. H., Potter, H. H., Michelow, D., Gardner, H. 1984. Sensitivity to lexical denotation and connotation in brain-damaged patients: A double dissociation? *Brain Language* 22: 253–65

Bub, D. N., Kertesz, A. 1982. Deep agraphia. *Brain Langauge* 17: 147–66

Butterworth, B. 1983. Lexical representation. In *Language Production*, ed. B. Butterworth, Vol. 2. New York: Academic

Caplan, D. 1987. *Neurolinguistics and Linguistic Aphasiology: Introduction.* Cambridge: Cambridge Univ. Press

Caplan, D., Keller, L., Locke, S. 1972. Inflection of neologisms in aphasia. *Brain* 95: 169–72

Caramazza, A. 1984. The logic of neuropsychological research and the problem of patient classification in aphasia. *Brain Language* 21: 9–20

Caramazza, A. 1986a. On drawing inferences about the structure of normal cognitive systems from the analysis of patterns of impaired performance: The case for single-patient studies. *Brain Cognit.* 5: 41–66

Caramazza, A. 1986b. The structure of the lexical system: Evidence from acquired language disorders. *Proc. Clinical Aphasiol. Conf.* 16: 291–301

Caramazza, A., Berndt, R. S. 1982. A psycholinguistic assessment of adult aphasia. In *Handbook of Applied Psycholinguistics*, ed. S. Rosenberg, pp. 477–535. Cambridge: Cambridge Univ. Press

Caramazza, A., Berndt, R. S. 1985. A multicomponent deficit view of agrammatic Broca's aphasia. In *Agrammatism*, ed. M.-L. Kean, Orlando, Fla.: Academic

Caramazza, A., Martin, R. 1983. Theoretical and methodological issues in the study of aphasia. In *Cerebral Hemisphere Asymmetry: Method, Theory and Application*, ed. J. B. Hellige, pp. 18–45. New York: Praeger

Caramazza, A., McCloskey, M. 1988. The case for single-patient studies. *Cognit. Neuropsychol.* In press

Caramazza, A., Miceli, G., Silveri, M., Laudanna, A. 1985. Reading mechanisms and the organization of the lexicon: Evidence from acquired dyslexia. *Cognit. Neuropsychol.* 2: 81–114

Chomsky, N. 1965. *Aspects of the Theory of Syntax.* Cambridge, Mass.: MIT Press

Chomsky, N. 1981. *Lectures on Government and Binding.* Dordrecht, Nertherlands: Foris

Churchland, P. 1986. *Neurophilosophy: Toward a Unified Science of the Mind/Brain.* Cambridge, Mass.: MIT Press

Coltheart, M., Patterson, K., Marshall, J., eds. 1980. *Deep Dyslexia.* London: Routledge & Kegan Paul

Damasio, A. 1981. The nature of aphasia: Signs and syndromes. In *Acquired Aphasia*, ed. N. T. Sarno. New York: Academic

Damasio, A. R., Geschwind, N. 1984. The neural basis of language. *Ann. Rev. Neurosci.* 7: 127–47

DeBastiani, P., Barry, C., Carreras, M. 1983. *Mechanisms for reading nonwords: Evidence from a case of phonological dyslexia in an Italian reader.* Presented at the 1st European Workshop on Cognitive Neuropsychol. Bressanone, Italy

De Villiers, J. G. 1978. Fourteen grammatical morphemes in acquisition and aphasia. In *Language Acquisition and Language Breakdown: Parallels and Divergences*, ed. A. Caramazza, E. B. Zurif, pp. 121–44. Baltimore: Johns Hopkins Univ. Press

Friederici, A. D., Schoenle, P. W. 1980. Computational dissociation of two vocabulary types: Evidence from aphasia. *Neuropsychologia* 18: 11–20

Funnell, E. 1983. Phonological processes in reading: New evidence from acquired dyslexia. *Br. J. Psychol.* 74: 159–80

Gardner, H., Brownell, H. H., Wapner, W., Michelow, D. 1983. Missing the point: The role of the right hemisphere in the processing of complex linguistic materials. In *Cognitive Processes in the Right Hemisphere*, ed. E. Pereceman. NY: Academic

Gleason, J. B. 1978. The acquisition and dissolution of the English inflectional system. See De Villiers 1978, pp. 109–20

Goodglass, H. 1976. Agrammatism. In *Studies in Neurolinguistics*, ed. H. Whitaker, H. A. Whitaker, Vol. 1. New York: Academic

Goodglass, H., Berko, J. 1960. Agrammatism and inflectional morphology in English. *J. Speech Hearing Res.* 3: 257–67

Goodglass, H., Klein, B., Carey, P., Jones, K. J. 1966. Specific semantic word categories in aphasia. *Cortex* 2: 74–89

Goodman, R. A., Caramazza, A. 1986a. Dissociation of spelling errors in written and oral spelling: The role of allographic conversion in writing. *Cognit. Neuropsychol.* 3(2): 179–206

Goodman, R. A., Caramazza, A. 1986b. Aspects of the spelling process: Evidence from a case of acquired dysgraphia. *Lang. Cognit. Processes* 1(4): 263–96

Gordon, B. 1983. Lexical access and lexical

420   CARAMAZZA

decision: Mechanisms of frequency sensitivity. *J. Verbal Learning Verbal Behav.* 22: 146–60

Gordon, B., Caramazza, A. 1982. Lexical decision for open-and closed-class items: Failure to replicate differential frequency sensitivity. *Brain Language* 15: 143–60

Gordon, B., Goodman-Schulman, R. A., Caramazza, A. 1988. Separating the stages of reading errors. *Brain.* Submitted

Hart, J., Berndt, R. S., Caramazza, A. 1985. Category-specific naming deficit following cerebral infarction. *Nature* 316: 439–40

Hier, D. B., Mohr, J. P. 1977. Incongruous oral and written naming: Evidence for a subdivision of the syndrome of Wernicke's aphasia. *Brain Language* 4: 115–26

Hinton, G. E., Anderson, J. A., eds. 1981. *Parallel Models of Associative Memory.* Hillsdale, NJ: Erlbaum

Job, R., Sartori, G. 1984. Morphological decomposition: Evidence from crossed phonological dyslexia. *Q. J. Exp. Psychol.* 36A: 435–58

Kertesz, A. 1985. Aphasia. In *Handbook of Clinical Neurology*, ed. P. J. Vinken, J. W. Bruyn, H. L. Klawans. New York: Elsevier

Lesser, R. 1978. *Linguistic Investigations of Aphasia.* New York: Elsevier North-Holland

Marshall, J. 1982. What is a symptom-complex? In *Neural Models of Language Processes*, ed. M. A. Arbib, D. Caplan, J. F. Marshall, New York: Academic

Marshall, J. C. 1986. The description and interpretation of aphasic language disorder. *Neuropsychologia* 24(1): 5–24

Marshall, J. C., Newcombe, F. 1973. Patterns of paralexia: A psycholinguistic approach. *J. Psycholinguist. Res.* 2: 175–99

McCarthy, R., Warrington, E. K. 1985. Category-specificity in an agrammatic patient: The relative impairment of verb retrieval and comprehension. *Neuropsychologia* 23: 709–27

McClelland, J. L. 1979. On the time-relations of mental processes: An examination of systems of processes in cascade. *Psychol. Rev.* 86: 287–330

Miceli, G., Caramazza, A. 1988. Dissociation of inflectional and derivational morphology. *Brain Language.* In press

Miceli, G., Mazzuchi, A., Menn, L., Goodglass, H. 1983. Contrasting cases of Italian agrammatic aphasia without comprehension disorder. *Brain Language* 19: 65–97

Miceli, G., Silveri, M. C., Nocentini, U., Caramazza, A. 1988. Patterns of dissociation in comprehension and production of nouns and verbs. *Aphasiology.* In press

Miceli, G., Silveri, M. C., Villa, G., Caramazza, A. 1984. On the basis for the agrammatic's difficulty in producing main verbs. *Cortex* 20: 207–20

Miceli, G., Silveri, M. C., Caramazza, A. 1985. Cognitive analysis of a case of pure dysgraphic. *Brain Language* 25: 187–212

Michel, F. 1979. Préservation du langage écrit malgré un deficit majeur du langage oral. *Lyon Méd* 241(3): 141–49

Morton, J. 1970. A functional model of memory. In *Models of Human Memory*, ed. D. A. Normal. New York: Academic

Morton, J. 1981. The status of information processing models of language. *Philos. Trans. R. Soc. London* 295: 387–96

Morton, J., Patterson, K. E. 1980. A new attempt at an interpretation, or, an attempt at a new interpretation. See Coltheart et al 1980

Nolan, K. A., Caramazza, A. 1982. Modality-independent impairments in word processing in a deep dyslexic patient. *Brain Language* 16: 237–64

Nolan, K. A., Caramazza, A. 1983. An analysis of writing in a case of deep dyslexia. *Brain Language* 20: 305–28

Patterson, K. 1980. Derivational errors. See Coltheart et al 1980

Patterson, K. 1982. The relation between reading and phonological coding: Further neuropsychological observation. In *Normality and Pathology in Cognitive Functions*, ed. A. W. Ellis. London: Academic

Rumelhart, D. E., McClelland, J. L. 1986. *Parallel Distributed Processing: Explorations in the Microstructure of Cognition*, Vol. 1: *Foundations.* Cambridge, Mass.: Bradford Books/MIT Press

Scalise, S. 1984. *Generative morphology.* Dordrecht, Netherlands: Foris

Shallice, T. 1979. Case study approach in neuropsychological research. *J. Clin. Neuropsychol.* 1: 183–211

Shallice, T. 1981. Phonological agraphia and the lexical route in writing. *Brain* 104: 413–29

Shallice, T., Warrington, E. K. 1975. Word recognition in a phonemic dyslexic patient. *Q. J. Exp. Psychol.* 27: 187–99

Shallice, T., Warrington, E. K. 1980. Single and multiple component central dyslexic syndromes. See Coltheart et al 1980

Stemberger, J. P. 1984. Structural errors in normal and agrammatic speech. *Cognit. Neuropsychol.* 1: 281–313

Stemberger, J. P. 1985. An interactive activation model of language production. In *Progress in the Psychology of Language*, ed. A. W. Ellis, 1: 143–83. London: LEA

Taft, M. 1979. Recognition of affixed words and the word frequency effect. *Memory Cognit.* 7: 263–72

Taft, M. 1985. The decoding of words in lexical access: A review of the morphographic approach. In *Reading Research: Advances in Theory and Practice*, ed. D. Besner, T. Waller, G. Mackinnon, Vol. 5. New York: Academic

Warrington, E. K. 1975. The selective impairment of semantic memory. *Q. J. Exp. Psychol.* 27: 635–57

Warrington, E. K. 1981. Concrete word dyslexia. *Br. J. Psychol.* 72: 175–96

Warrington, E. K., McCarthy, R. 1983. Category-specific access dysphasia. *Brain* 106: 859–78

Warrington, E. K., Shallice, T. 1984. Category-specific semantic impairments. *Brain* 107: 829–54

Wasow, T. 1985. Postscript. In *Lectures on Contemporary Syntactic Theories: An Introduction to Government-binding Theory. Generalized Phrase Structure Grammar, and Lexical-functional Grammar*, by P. Sells. Stanford: Cent. for the Study of Language and Information

Ann. Rev. Neurosci. 1988. 11:423–53

THE CONTROL OF NEURON NUMBER

Robert W. Williams and Karl Herrup

Section of Neuroanatomy and Department of Human Genetics,
Yale University School of Medicine, New Haven, Connecticut 06510

INTRODUCTION

How an animal senses, perceives, and acts depends on the organization and number of elements that make up its nervous system. Of several kinds of neural elements, ranging in size from ion channels to cytoarchitectonic divisions, the neuron is the fundamental building block. Understanding processes that control numbers of neurons in the brains of different animals at different stages of development is therefore of great importance. Two approaches can be taken to the problem of neuron number. The first is a cellular and molecular approach that focuses on final effects and final causes that regulate this variable in single species. The second approach has a broader focus on evolutionary, ecological, and bioenergetic reasons for, and consequences of, different strategies used to control the size of neuron populations in different species. Here the aim is to understand the diversity of strategies used to modify neuron number in response to natural selection. This review is divided into three sections. The first section provides an analysis of neuron number in adults of different species. The second and third sections examine the two principal processes that control neuron number during development—neuron production and neuron elimination.

NEURON NUMBER AT MATURITY

Total Neuron Number

The total number of neurons in the central nervous system ranges from under 300 for small free-living metazoans such as rotifers and nematodes (Martini 1912, Bullock & Horridge 1965), through about 30 million for

423

0147–006X/88/0301–0423$02.00

the common octopus and small mammals such as shrews (Young 1971, Campbell & Ryzen 1953), to well over 200 billion for whales and elephants. Estimates for the human brain range between 10 billion and 1 trillion. The imprecision in these estimates is due almost entirely to uncertainty about the number of granule cells in the cerebellum, a problem that can be traced back to a study by Braitenberg & Atwood (1958). More recent work by Lange (1975) makes a reasonably accurate estimate possible: The average human brain (1350 gm) contains about 85 billion neurons; of these, 12 to 15 billion are telencephalic neurons (Shariff 1953), 70 billion are cerebellar granule cells (Lange 1975), and fewer than 1 billion are brainstem and spinal neurons.

Behavioral complexity is not a function of body size. It follows that an increase in body mass alone does not require a matched increase in numbers of cells. Neurons could simply be larger and could branch more widely (Tower 1954, Purves et al 1986). Nonetheless, larger individuals and larger species generally do have larger brains that do contain more neurons. Although neurons are larger and packed more loosely in the brains of large species (Holloway 1968, Lange 1975), the increase in brain weight more than offsets the lower density. A 6000-g elephant brain has two to three times as many neurons as does a 1350-g human brain.

In contrast to the paucity and imprecision of data on total neuron number, a great deal is known about the brain weight of vertebrates. This information has been collected in an attempt to provide insight into animal intelligence and evolutionary status (reviewed in Jerison 1985, Martin & Harvey 1985). That effort has met with little success. As has been repeatedly stressed (Sholl 1948, Mangold-Wirz 1966, Mann et al 1986) and repeatedly ignored, brain mass is a compound variable, and little insight can be gained by reducing brain weight to a simple expression made up of one constant, one variable (body weight, surface area, metabolic rate, and even life span) and one coefficient.[1] However, if we can be content with a procrustean generalization, this work does demonstrate that large individuals and large species tend to have large brains. Brain weight is proportional to body weight raised by a power that ranges widely—from 0.1 to 0.8 (Lapicque 1907, Sholl 1948, Stephan 1958, Lande 1979, Ricklefs & Marks 1984). The exponent is less than 1 in all cases; this demonstrates that the increase in brain weight lags behind the increase in body weight. The particular value of the exponent depends in part on the taxonomic level at which comparisons are made. Higher exponents characterize com-

---

[1] For a particularly vivid example of the limitation of this single variable approach, see the controversy surrounding the correlation between brain size and longevity (Mallouk 1975, 1976, Calder 1976).

parisons across orders and classes; lower exponents characterize comparisons at the species level. Exponents between 0.1 and 0.4 are typically derived when comparisons are limited to individuals of the same or closely related species (Wingert 1969, Holloway 1980). However, even within closely related families of species the exponent may vary all the way from 0.2 to 0.6 (Mann et al 1986). This variation is due to differences in ontogenetic, ecological, and evolutionary factors that influence how big a brain a species needs and how big a brain a species can afford.

Importance of Total Neuron Number

The number of neurons and their relative abundance in different parts of the brain is a determinant of neural function and, consequently, of behavior. Phyla whose members have larger brains and more neurons respond to environmental change with a greater range and versatility of behavior (Jerison 1985). Orders of mammals with big brains, such as cetaceans and primates, are more clever than those with little brains, such as insectivores and marsupials. However, the correlation breaks down as we narrow the focus and compare allied species and even individuals within species—the exceptions obscure any trend. No generally valid equation relates neuron number to behavioral complexity. For instance, humans with brains which weigh only half the average of 1350 gm, and in which there is no evidence for any compensatory increase in neuron density, can have normal intelligence (Hechst 1932, Cobb 1965).

Despite this reservation, there is some experimental evidence of a relation between neuron number and intelligence. Triploid and tetraploid newts have the same brain mass as diploid controls, but their neurons are larger and there are only 50–70% of the normal number (Fankhauser et al 1955, Vernon & Butsch 1957). Although the locomotion of these polyploid newts is indistinguishable from normal, they take two to three times as many trials to learn a maze as do normal newts. Thus, a reduction in neuron number in this case lowers performance markedly. Seasonal oscillations in neuron number in the song nuclei of canaries correlates well with singing ability (Goldman & Nottebohm 1983); this observation lends further support to a notion that many of us have been willing to take on faith.[2]

Conversely, in some instances increased neuron number has been shown to result in improved performance. Exposure of immature frogs and rats

---

[2] The only instance we are aware of in which an increase in neuron number may be maladaptive is the case of a mutant mouse, *quaking*. Maurin et al (1985) have reported that the number of noradrenergic neurons in the locus coeruleus of *quaking* is about 20% above normal. This increase is associated with convulsion.

to excess growth hormone can boost neuron number 20 to 60% according to Zamenhof and colleagues (1941, 1966, 1971), and in several instances the hyperplasia or hypertrophy is correlated with improved performance on single-trial avoidance conditioning tasks (Clendinnen & Eayrs 1961, Block & Essman 1965). Similarly, the number of visual cortical neurons excited by one eye has been experimentally increased two-fold in both cats and monkeys, and this increase is associated with smaller receptive fields in visual cortex (Shook et al 1984). Preliminary work supports the idea that such experimental animals are able to resolve smaller differences in the offset between two lines than normal monkeys (M. G. MacAvoy, P. Rakic, and C. Bruce, personal communication).

Structure of Neuron Subpopulations

The nervous system of all vertebrates and many invertebrates is a mosaic composed of hundreds to thousands of neuron subassemblies. To begin systematic study of the cellular demography of the nervous system we need reliable, objective methods to define groups of neurons. One way is to group neurons by common descent. The clone of all labeled cells descended from a single precursor cell, whatever the mixture of types, defines the group (Jacobson & Hirose 1978, Weisblat et al 1978, Jacobson & Moody 1984, Sanes et al 1986, Turner & Cepko 1987). The strength of this method in terms of examining relations between lines of neuronal descent, pattern of cell divisions, and the final fate of neurons is considered in the section on neuron proliferation.

The second, more common way to classify neurons is by shared properties: position, shape, size, axonal projections, electrophysiology, and biochemistry. Neurons of similar type are classed together regardless of whether they are homologous or analogous (Rodieck & Brening 1983, Sternberg & Horvitz 1984). Within a group, neurons may differ enormously in size (Stone 1983), but if the variation is continuous, as is often the case, distinct subpopulations are not recognized. Granule cells of the cerebellar cortex are an example—these cells constitute a huge and nearly homogeneous class that accounts for roughly 7/8 of all neurons in the human brain. However, in the majority of cases, neuron properties vary noncontinuously, and populations may thus be broken down repeatedly into a complex hierarchy of sets that share certain features but not others. Retinal ganglion cells of vertebrates have been subjected to such a detailed breakdown (Kolb et al 1981, Stone 1983). These cells are united in being the only neuron population to project out of the retina, but they can be differentiated across other dimensions into as many as 20 to 30 distinct subclasses. If the characteristics of a large enough number of ganglion cells are plotted in a multidimensional space, clusters of cells become evident. These clusters provide an objective method to classify neuron phenotype (Rodieck & Brening 1983).

Some neurons do not belong to groups. These neurons are unique and can be recognized individually on the basis of size, cytochemistry, axonal projections, and physiological properties in almost all members of a species. In contrast to the large populations of phenotypically equivalent neurons in vertebrates, the nervous systems of many invertebrates are made up of collections of unique neurons. Essentially all neurons in rotifers (Ware 1971 cited in Ware et al 1975), nematodes, and leeches are of this type (Goldschmidt 1909, Sulston 1976, Müller et al 1982). The case has even been made that the complex nervous system of grasshoppers, which contain up to 200,000 neurons, may be made up almost entirely of unique neurons (Goodman 1976). In contrast, unique neurons are rare in vertebrates; the Mauthner neuron of fish and amphibians is certainly the best characterized of these (Faber & Korn 1978).

Small clusters of two to six equivalent neurons are also particularly common in the nervous system of invertebrates, for example, the oculomotor neuron clusters in crayfish (Mellon et al 1976) and the heart accessory cells of the leech (Gao & Macagno 1986). While neurons that are part of a cluster are not unique, they are thought to arise simply by the addition of one to two extra rounds of division at the end of the cell lineage (Goodman 1977, Chalfie et al 1981, Ambros & Horvitz 1984). Identified neurons and neuron clusters are treated only briefly in this review. When there are only one or two cells of a particular type, the most interesting questions center around determination and neuron differentiation. In contrast, when populations of neurons are large and variable, questions center around the size and composition of the precursor pool, the rate of proliferation, the number of cell divisions, and the patterns of cell death among young neurons. The shift in emphasis is from differentiation to kinetics of proliferation and death.

Variation in Neuron Number

Variation in the numbers of neurons in a nucleus or ganglion of a group of animals belonging to the same species is often regarded as an experimental nuisance, and in many cases it is not possible to separate genuine intraspecific diversity from technical artifact (Konigsmark 1970, Williams & Rakic 1986, 1988a,b). Unfortunately, this problem masks a fundamental issue. Not only must the absolute number of neurons be regulated during development, but within a population of interbreeding individuals (a deme) there must also be mechanisms that ensure the production and maintenance of adequate variation in numbers of neurons. Without this variation no evolutionary change in either total or relative numbers of neurons would occur (DeBrul 1960, Armstrong 1982). It is important to appreciate that mechanisms that control neuron number also control the range of variation and, ultimately, rates of brain evolution (Mayr 1963). Thus it is not surprising that whenever and wherever it has been looked for with any persistence, natural variation in the numbers of neurons has been found, even in the "invariant" nervous systems of arthropods and annelids (Lubbock 1858, Hertweck 1931, Goodman 1979, Macagno 1980).

Naturally, there is a balance between the degree of variation and the lowered fitness of individuals at either end of the distribution. Variation will be least in those neuronal systems most exposed to natural selection, and greatest in those systems least exposed to natural selection (Yablokov 1974, Wright 1978). Not only is variation substantial, but the degree of variation is itself highly variable! There can be no blanket statements to the effect that variation is ± 5–10%.

Do small neuron populations show less variation than large populations? If we are willing to settle for the distant view, the answer is, yes. Because each neuron of a small population makes a greater contribution to fitness, it follows that the fewer neurons there are of a given type, the less the tolerable variation in number. Every one of several hundred zebrafish examined by Kimmel & Eaton (1976) had precisely two Mauthner neurons and every one of 50 crayfish examined by Mellon et al (1976) had precisely nine neurons in each ocular motor neuron cluster. In four species of leech studied by Macagno (1979), the variation in neuron number in segmental ganglia was typically under 2%. However, variation in small populations may in some cases be substantial. Kollros & Thiesse (1985) in a careful study of neurons in the mesencephalic V nucleus of *Xenopus* frogs found that neuron numbers ranged from 185 to 284 at stage 65/66. Left-right differences are also substantial, ranging from 5 to 20% (Kollros & McMurray 1955, Williams & Rakic 1988a). Nonetheless, variation in numbers appears to be greatest in particularly large populations of neurons organized in parallel arrays. The range of variation in primate neocortex is greater than ± 50% (van Essen et al 1984, 1986, Williams & Rakic 1986).

In terms of brain evolution it is more important to establish the source and degree of variation in neuron number than the average population size. Is the variation developmental noise, is it experimental noise, or is it a consequence of interactions with the environment? How much of the variation is genetic? How much of the genetic variation is heritable (additive genetic variance) and how much is not heritable (variance due to epistasis and dominance)? Studies of groups of individuals who have virtually the same genome (isogenic or homozygous offspring) could in theory provide insight into these issues. Up to the present, however, isogenic individuals have been used almost exclusively to examine the degree to which genetic and epigenetic mechanisms are able to specify neuron phenotype (Macagno et al 1973, White et al 1976, Goodman 1978). The results in some cases reveal remarkably detailed specification. But there have been surprises. For example, some individuals of a single clone of grasshoppers have extra neurons, whereas other members from the same clone do not (Goodman 1977). Similar variations in meristic traits (traits that can be counted: neurons, scales, hairs, etc) have been examined closely in isogenic mammals, and here too appreciable variation has been found (Newman & Patterson 1911, Storrs & Williams 1968). Unfortunately, no concerted effort has yet been made to rear isogenic organisms under a wide variety of conditions. Such an experiment could reveal how stable neuron number is in the face of developmental and environmental stress (Waddington 1942, Tåning 1950, Katz & Lasek 1978).

Importance of Ratios of Neurons

It is well accepted that the ratio of different types of neurons within single nuclei and the ratio of neurons in interconnected parts of the nervous system are key determinants of neuronal performance (Wimer et al 1976, Katz & Grenander 1982). However, the limits around which ratios may vary without loss of function are in some cases substantial. For instance, in primates, ratios of neurons in the richly interconnected cytoarchitectonic zones of the occipital lobe (visual areas 17, 18, and 19) vary by more than a factor of 2 (van Essen et al 1984, 1986, Williams & Rakic 1986), and an equal magnitude of variation is found in ratios of neurons in the peripheral nervous system (Ebbesson 1965). Great variation is also found between ratios of homologous populations of neurons in different species. Here the differences have an easily recognized basis; brains are customized for the body and behavior of each species and consequently ratios of cells will often be radically different. For example, the ratio between granule cells and Purkinje cells rises from less than 200 to 1 in mice to 3000 to 1 in humans (Wetts & Herrup 1983, Lange 1975).

The substantial variations in cell ratios may often be offset by changes in the efficacy and distribution of synapses and the size of axons and dendrites. Consequently, convergence and divergence ratios may be regulated more precisely than are cell ratios. For example, the ratio of preganglionic to postganglionic neurons in the sympathetic nervous system of mammals drops with increasing body size, but the drop is compensated for by an increase in the size and complexity of axons and dendrites (Purves et al 1986).

Brain Metabolism and Neuron Number

A variety of studies have shown that a surplus of neurons does not give rise to maladaptive behavior (e.g. Hollyday & Hamburger 1976, Chalupa et al 1984, Ellis & Horvitz 1986); on the contrary, any supernumerary neurons or glial cells can be sequestered or integrated into existing neuronal circuitry and may easily result in more adaptive behaviors. This is as true for roundworms (Ellis & Horvitz 1986) as for rats (Zamenhof 1942). A 20% surplus of neurons in *Caenorhabditis elegans* caused by a mutation that blocks cell death results in no behavioral deficits in moving, mating, or egg laying (Ellis & Horvitz 1986). What then are the selective pressures that set an upper limit?

Neurons are greedy cells. In resting humans, 18% of circulating oxygen is consumed by brain tissue even though brain makes up only 2% of the body mass (Kety & Schmidt 1948). In infancy the brain is proportionally much larger (Blinkov & Glezer 1968), a situation that is only partially offset by lower metabolic requirements (Kreisman et al 1986). Even during hibernation and estivation, brain metabolism is not reduced appreciably (Meyer & Morrison 1960). Thus neurons represent a high and fixed metabolic expense, and the cost of feeding neurons and associated glial cells is undoubtedly one of the principal factors that constrain the total size of the neuron population.[3]

The relative cost of neurons depends on an animal's size, its functional metabolic rate (Martin 1981, Eisenberg & Wilson 1982, Armstrong 1983, Armstrong & Bergeron 1985), and the specific metabolic rate of its brain tissue (Krebs 1950). This cost is proportionally highest in small carnivores, birds, and primates (e.g. *Cebus albifrons*), in which the brain may represent up to 10% of the adult body weight (Spitzka 1903). The nervous system of the 1-mm-long roundworm *C. elegans* takes up an even larger fraction of the body—in the males 381 of 1000 cells are neurons. In terms of body mass devoted to neurons this is the brainiest creature of all. For animals

---

[3] This may be particuarly true at early stages of development. As Martin (1981) has argued persuasively, the amount of energy that can be transferred from parent to offspring may set an upper limit to brain size. The advent of fully functional lactation in mammals before the end of the Triassic period was undoubtedly a key innovation that enabled mammals to sustain relatively large populations of neurons and glial cells (Pond 1977, Lillegraven 1979), and this in turn may have contributed significantly to the rapid radiation of mammals during the early Cenozoic. Speciation rates in mammalian genera are estimated to be about five times higher than rates in lower vertebrate genera (Bush et al 1977). The rate of growth of mammalian young that possess particularly large brains is often slow, and gestation and maturation take a long time (Sacher & Staffeldt 1974). We see this pattern clearly in primates (Gould 1977, p. 367), but it is also apparent in other mammals. For instance, the nectar-feeding bat, *Glossophaga soricina*, has a very high brain/body weight ratio, and Eisenberg (1981, p. 307) points out that the young may develop slowly principally because the milk contains little fat. This strategy allows the mother to defray the cost of building a new brain over a longer period. In general, animals with big brains have small litters, slow development, and intense parental care.

such as these, feeding the brain is a challenge, and we would predict that fecundity in nematodes that must support an extra complement of neurons would be reduced substantially. In fact, brood size in mutants with a 20% increase in neuron number (due to a failure of normal cell death) is 15–30% below normal (Ellis & Horvitz 1986).

The high cost of neurons in small species leads to an interesting prediction: If particular populations of neurons function only seasonally—during mating or while raising young—then it follows that these populations might atrophy or die back when not needed. Such a pattern of loss and replacement has recently been reported in the adult canary—a passerine bird with a small body, a relatively large brain (Goldman & Nottebohm 1983), and an extremely high resting metabolic rate. The neuron population in parts of the hyperstriatal complex involved in song production and recognition fluctuates from about 40,000 in spring to 25,000 in fall and winter. Similar periodic regression might also be found in other small, highly encephalized species such as *Mustela putorius* and *Tamias striatus*.

In contrast, large species may be able to afford a superabundance of neurons, even though they may have no pressing need for these cells. The metabolic demands of excess neurons may represent a minor addition to the total energy consumption of a whale. Alleles that contribute to the production of surplus neurons may be retained in the population because the behavioral and metabolic consequences of doing so are negligible. Although this extra baggage does have a small cost, the capacity to retain more neurons than needed increases the range of variation both in neuron number and in patterns of deployment of neurons in different individuals. Maximizing variation increases the evolutionary plasticity of the lineage, an idea analogous to the concept of genetic load—the preservation of alleles in the population that reduce mean fitness but that compensate by providing a reservoir of heritable variation (Mayr 1963). In contrast, an equivalent fractional increase in total neuron number in a small animal will represent a greater increase in the metabolic load and may consequently be strongly selected against (Ricklefs & Marks 1984). Thus, in a mole, canary, or nematode in which the fraction of metabolism devoted to neurons is great, individuals with superfluous neurons will be weeded out quickly.

It follows that the nervous system of small species may be more efficiently designed—that individual neurons may work harder. This is certainly true at a crude level of analysis. At one extreme, the deletion of a single neuron in nematodes lowers fitness to zero by making egg laying impossible (Trent et al 1983), while at the other extreme, the loss of more than 90% of the retinal ganglion cell or Purkinje cell population may result in no detectable behavioral change (Quigley 1982, Wetts & Herrup 1983).

However, the analysis should not, and probably cannot, be taken much farther than this—the functional contributions of particular neurons in different parts of the brain are likely to vary widely and depend on much more than just number (Holloway 1968). To take one surprising example, a drop in neuron number in the optic medulla of the "rigidly specified" fruit fly from 40,000 to 20,000 does not cause a clear-cut behavioral effect (Fischbach & Heisenberg 1981).

CONTROL OF NEURON NUMBER BY ADDITION

In most tissues a deficit of a particular cell type can be corrected by generating new cells, by coaxing cells into the depleted region from surrounding tissues, or by modifying existing cells. As a rule, this is not the case in the nervous system. Precursor cells capable of producing neurons are either few in number or simply do not exist at late stages of development (Miale & Sidman 1961, Rakic 1985). And neurons themselves cannot divide—they are postmitotic cells unable to reenter a cycle of cell division. Consequently, the number of neurons within each part of the nervous system is, with some interesting exceptions, determined at early stages of development, often without the benefit of any exposure to the environment—an odd situation for an organ system designed to deal with the environment. Mature neurons have only limited capabilities to change type or move around. Thus, in contrast to cells in most other tissues, the number, type, and distribution of neurons cannot be regulated about some optimum at maturity. The production and deployment of neurons has to be done the right way the first time, not an easy job given the extraordinary complexity of the nervous system.

The number of neurons that are generated can be controlled in two ways—by changing the number of stem cells or by changing the number of descendants each stem cell produces. Events that alter numbers of stem cells operate early in embryogenesis, but events that change the number of descendants can operate at any time during the expansion of the lineages—the earlier the change, the more dramatic and significant the effects. Changes in stem cell number and descendant number can be controlled either by mechanisms that are intrinsic to the population or by mechanisms that are extrinsic. Intrinsic mechanisms are those that affect only cells of a specified lineage or events that are not easily altered by changing the developmental environment. Extrinsic mechanisms include the influence of heterologous cell interactions, hormones, and environment on the kinetics, commitment, and survival probabilities of stem cells, as well as the recruitment of still uncommitted cells to differentiate into one of several potential subtypes (as during neural crest cell development). Since vertebrates and invertebrates seem to emphasize different areas of this 2 × 2 matrix of possibilities (addition/subtraction × intrinsic/extrinsic), we have found it convenient to consider the groups separately.

Intrinsic Factors

In leeches and nematodes, neuron number is under tight control of an inflexible pattern of cell lineage. Shankland & Weisblat (1984) have shown in the leech that if one of a bilateral pair of precursor cells is eliminated, descendents of the contralateral homologue will occasionally cross the midline. However, this action depletes the donor side to the benefit of the operated side. Up-regulation of cell number either by proliferation or recruitment evidently is not possible, and consequently the question of control of neuron number is rarely raised explicitly. In additional studies, the stereotypic patterns of cell lineage were perturbed either by mutation (e.g. Chalfie et al 1981) or by the deletion of cells (Sulston & Horvitz 1977, Weisblat & Blair 1982). These manipulations have revealed that for the great majority of cells lineage constrains their developmental potential to a single fate, while for a few other neuroblasts cell-cell interactions specify which of two or three alternative fates—frequently arranged hierarchically—is expressed. In these experiments, the manipulations alter the morphological and numerical fate of the lineage lower on the hierarchy. Thus, again, there is no up-regulation of cell number.

The evidence that cell production in vertebrates is controlled by intrinsic processes is most compelling for nonneuronal tissue. In frog, the midblastula transition occurs precisely 12 rounds of division after cleavage. Newport & Kirschner (1982a,b) have shown that, as in ascidians, it is apparently the ratio of DNA to cytoplasm that is important in triggering the transition. Several other systems also illustrate the point that precursor cells must go through a set number of divisions before differentiation begins. Quinn and colleagues (1983a,b, 1984, 1985, 1986) have argued persuasively that precisely four cell divisions are needed between the commitment of a cell to the myogenic lineage and the end of division. Similar evidence has been presented in the case of epidermal stem cells (Potten et al 1982), erythropoietic stem cells (Guesella et al 1976), diploid fibroblasts (Angelo & Prothero 1985), and glial progenitor cells (Temple & Raff 1986).

An elegant analysis of the genesis of the *Xenopus* lateral line system by Winklbauer & Hausen (1983) has shown that the number of cells in one lateral line organ is always a multiple of 7 times a small but variable integer plus a constant of 8. Their interpretation is that the small number of founder cells of each organ is variable, but once they are committed as founders, a stereotypic series of cell divisions invariably ensues—six asymmetric divisions, each leading to one nonmitotic cell and one new stem cell, followed by one symmetric division that produces two nonmitotic cells. The consequence of this pattern is that neuron number, as well as phenotype, is tightly regulated by the lineage that generates the lateral line.

Evidence for intrinsic mechanisms specifying neuron number during development is implicit in much early work on transplantation of pieces of embryonic central nervous system (reviewed in Cooke 1980). However, the methods of analysis are qualitative and only hint at intrinsic mechanisms. As noted by Schoenwolf (1985), no numerical analyses of cell populations that make up the neural plate and tube have yet been undertaken. It would be of considerable theoretical interest to establish whether the number of çells in the human neural plate is any greater than that in the mouse.

SINGLE GENE EFFECTS   Some of the best evidence that neuron number in mammals can be regulated by mechanisms intrinsic to single neurons comes from developmental studies of chimeric mice. These animals are made by aggregating two embryos in vitro. Usually a wild-type and a mutant blastocyst are pushed together at the eight-cell stage creating a mosaic of mutant and wild-type cells. Because single cells have either two mutant alleles or two wild-type alleles, the genotype of the chimera is denoted *mutant/mutant* ↔ + / + . The double-sized blastocyst is then transplanted into a host mother in which it develops normally. Chimeras have been used to study mutations that cause specific reductions in neuron number (reviewed by Mullen & Herrup 1979). In the *purkinje cell degeneration (pcd)* mutant, the entire Purkinje cell population is eliminated (Mullen et al 1976). Mullen (1977) created *pcd/pcd* ↔ + / + chimeras and found that only the *pcd/pcd* neurons died. This landmark study revealed that, as in invertebrates, the mammalian central nervous system has genes required to produce and/or maintain the proper number of neurons.

A more intriguing, and in some ways more compelling case for the involvement of specific genes in the regulation of neuron number is that of the *staggerer (sg)* mutant. This autosomal recessive gene, when homozygous, leads to a substantial reduction in the size of the cerebellum and has marked qualitative and quantitative effects on the development of the Purkinje cell population (Sidman 1968, Bradley & Berry 1978, Herrup & Mullen 1979a). In *sg/sg* ↔ + / + chimeras the cerebellum is intermediate in size between mutants and wild-types. Analysis of the chimeras has demonstrated that both the qualitative defects (small cell size, ectopias, regional variation in morphology) and the severe quantitative defect (a 75% reduction in neuron number in homozygous mutants) are only expressed in cells of the *sg/sg* lineage (Herrup & Mullen 1979, 1981). For example, as shown in Figure 1, if neuron number were reduced non-selectively in chimeras, without regard to genotype, the number of remaining *sg/sg* genotype neurons would fall somewhere on the curve marked *Extrinsic*. On the contrary, if only the *staggerer* lineage were affected by this mutation, then the number of *sg/sg*-type neurons in chimeras would fall on the line marked *Intrinsic* (see Herrup & Mullen 1981). As is clear from the fit of the points (Herrup & Mullen 1981), the data strongly favor an intrinsic over an extrinsic model. The implication of this result is that the mutation at the *staggerer* gene locus disrupts patterns of cell acquisition or survival within the Purkinje cell lineage in a manner reminiscent of several of the *C. elegans* lineage mutations.

LINEAGE-RELATED CONTROL OF NEURON NUMBER   These findings suggest that specific lineages of neurons require specific genes for the regulation

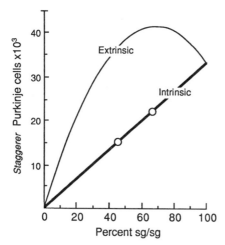

Figure 1   Two theoretical curves that illustrate the expected number of *sg/sg* genotype Purkinje cells (PC) in *sg/sg* ↔ +/+ chimeras, assuming two different modes of gene action (see Herrup & Mullen 1981 for the assumptions and calculations on which these curves are based). The *lower line*, labeled *intrinsic*, represents the function expected if the number of *sg/sg* Purkinje cells is determined by mechanisms that are entirely lineage autonomous. The *upper curve* shows the function expected if the number of *sg/sg* Purkinje cells is determined by factors extrinsic to the cells themselves (e.g. hormones, target-related cell death, etc). This *extrinsic* model predicts that many chimeras should have more *sg/sg* cells than a nonchimeric *sg/sg* mouse. The two *open circles* represent the actual counts of two *sg/sg* ↔ +/+ animals reported by Herrup & Mullen (1981). The close fit to the intrinsic model suggests that Purkinje cell number in the *staggerer* is controlled by factors intrinsic to the lineage. The value "percent *sg/sg*" is determined by the following formula:

$$\frac{\text{PC in the chimera} - \text{PC in } sg/sg}{\text{PC in } +/+ - \text{PC in } sg/sg}.$$

of cell number. This raises the question of whether other lineage-specific factors—perhaps involving groups of genes—also regulate neuron number. Studies of *lurcher* ↔ wild-type chimeras provide support for this idea. Wetts & Herrup (1982) counted the number of Purkinje cells in a series of *lurcher* chimeras. The cell-autonomous action of the mutant gene destroys all *lurcher* Purkinje cells, and consequently only the progeny of wild-type stem cells survive. Because the estimates of Purkinje cell number fall close to integral multiples of 10,000, Wetts & Herrup suggested that this number of Purkinje cells arises from each stem cell, and that the quantal distribution of counts (10,000, 20,000, 30,000, etc) is caused by chance variation in the initial ratio of *lurcher* and wild-type stem cells in different individuals early in development.

Using a mean estimate of 8.5 hr for cell cycle time of neuroblasts (see references in Jacobson 1978), using 12.5 embryonic days as the time of the last Purkinje cell division (Miale & Sidman 1961), and assuming the pattern of cell division is geometric as opposed to asymmetric, it can be estimated that stem cells that give rise to Purkinje cell clones are founded early in development—around the neural plate to early neural tube stages. This is the period during which experimental analysis suggests that cells of the neural ectoderm become unalterably established as nervous system precursor cells (e.g. Chan & Tam 1986, Beddington 1981, 1983). It has not been possible to determine whether the small pool of precursor cells gives rise only to Purkinje cells, but it has been possible to rule out that other stem cells give rise to Purkinje cells.

Extending these studies to the facial nerve nucleus, Herrup et al (1984a,b) were able to show that there is a distinct, clonal organization to this neural structure as well. However, the total number of facial nucleus neurons differed substantially among chimeras, and the number of neurons of any one genotype showed no evidence for being an integral multiple of some unit clone size. Instead, the fraction of cells of any one genotype was always divisible by 1/12. The best explanation of this binding is that the facial nucleus is made up of 12 clones (of mixed genotype) and that these clones are reduced in number uniformly during the period of cell death (Ashwell & Watson 1983).

Recently, Herrup (1986, 1987) and Herrup & Sunter (1986) have demonstrated the pivotal role cell lineage relationships play in establishing neuron number. Struck by the fact that the size of a Purkinje cell quantum did not change in *lurcher* chimeras over a wide range of *lurcher* to wild-type ratios, they examined the clonal relationships of two different inbred strains. The first was the background strain for the *lurcher* mutant itself, C57BL/6; the second was a different wild-type strain of mouse, AKR/J. In the first instance, the chimeras were mosaics of cells that had identical genotypes at every locus except *lurcher*. In the second, the cells of the chimeras differed at many loci throughout the genome (Taylor 1971). Each analysis led to the conclusion that the number of Purkinje cells in these animals occurred in integral multiples of a single value. In the case of C57BL/6, this value was 9200; in the case of AKR/J, however, this value was 7850 Purkinje cells. In each analysis, the quantal nature of the counts was maintained over a wide range of mutant/wild-type ratios. This indicates that the size of quanta (i.e. number of Purkinje cells in a clone) is not influenced by developmental environment: an AKR/J quantum remains the same size whether it occurs in an animal that is mostly AKR/J or in an animal that is predominantly C57BL/6. Since the latter strain has a different intrinsic clone size, the conclusion can be drawn that the size of a Purkinje cell clone is intrinsic to the lineage. In other words, the number of Purkinje cells that a given progenitor will produce is as much a characteristic of that progenitor as is the morphological type of the cell— precisely the same conclusion reached by the analysis of single gene effects.

Extrinsic Factors

The evidence cited above suggests that intrinsic, lineage-autonomous regulation of neuron number occurs during normal development. This evidence, however, does not preclude a role for extrinsic factors. Few factors, however, have been identified as either mitogens or mitotic inhibitors that might serve as hormonal or other extrinsic regulators of neuronal cell division. Nerve growth factor was once thought to be a mitogen of neurons in the peripheral nervous system, but is now evident that it exerts its effects primarily through prevention of cell death and through concomitant hypertrophy of cells (Thoenen & Barde 1980, Thoenen et al 1985). Growth hormone and sex hormones are both candidates for substances with neuronal mitogenic effects. Perhaps the best studied compound of this class is the thyroid hormone, thyroxine.

As early as 1922, Champy showed that factors in the thyroid gland produce a substantial rise in the number of mitotic figures in tadpole retina. Subsequent studies have proved that thyroxine treatment causes an increase in mitotic activity and an increase in numbers of neurons in retina, telecephalon, and midbrain when given to larval frogs (Zamenhof 1941, Beach & Jacobson 1979, Cline & Constantine-Paton 1986). Zamenhof and colleagues (1942, 1966) have examined the effect of growth hormone in mammals and report a 20% increase in cell density in the cortex of rats treated with growth hormone. Obvious questions are whether the hormone is speeding up the rate of cell division or whether additional cells are being recruited in response to growth hormone. Beach & Jacobson (1979) have shown that in the retina of metamorphosing frogs, the effect is due entirely to an increase in the number of dividing cells.

In rodents, administration of thyroxine accelerates the development of the cerebellar cortex and leads to the premature disappearance of the external granule cell layer—the site of synthesis of cerebellar granule cells. Conversely, administration of propylthiouracil, a compound that depletes thyroxine, retards cerebellar development, thereby resulting in the persistence of the external granule cell layer, larger cerebellar volumes, and increased foliation (Nicholson & Altman 1972, Lauder et al 1974). These results suggest that by speeding up development, thyroxine can paradoxically decrease total numbers of neurons generated and retained.

Testosterone was initially thought to increase the mitotic activity of the precursor cells that give rise to the ventral hyperstriatum (a song control nucleus), but recent results reveal that the effect is largely attributable to a sex difference in the severity of neuron death (Konishi 1985). Serotonin has also been cited as a potential mitogen. Depleting serotonin in fetal rats prolongs proliferation in those brain regions that ultimately receive serotonergic input (Lauder & Krebs 1976, 1978). 6-Hydroxydopamine has been shown to retard cerebellar growth (Allen et al 1981), suggesting a possible role for noradrenaline uptake systems in the regulation of granule cell genesis.

Cell-cell interactions that appear to have a mitogenic effect on neuroblasts are also surprisingly rare. There is some evidence that peripheral input modulates the duration or intensity of the proliferation of central neuroblasts (Kollros 1953, 1982, Chiarodo 1963, Macagno 1979, Anderson et al 1980). Extirpation of a leg rudiment in blowfly larvae reduces the central neuron population by 30%, an effect that Anderson et al (1980) attribute to a decrease in the production of small local circuit neurons. Similarly, removal of an eye in frogs is reported to reduce mitotic activity in the ventricular zone underlying the tectum (Kollros 1982). The supposition is that ingrowing axons exert a mitogenic or permissive effect on nearby neuroblasts.

In mammalian cerebellum, several lines of circumstantial evidence suggest that Purkinje cells exert a mitogenic influence on granule cell precursors in the external granule cell layer. This idea is based on observations of granule cell kinetics in mutant mice such as *reeler* (Mariani et al 1977) and *staggerer* (Mallet et al 1976, Landis & Sidman 1978, Herrup & Mullen 1979, Sonmez & Herrup 1984) and on the regeneration patterns observed after administration of ENU (Das & Pfaffenroth 1977). Finally, the recent work of Raymond and colleagues (1986) and Reh & Tulley (1986) in fish and frog suggests that retinal neurons respond to local cues when making their decision to enter or leave the mitotic cycle.

Certain populations of neurons continue to proliferate at maturity. Arthropod sensory receptor neurons, vertebrate olfactory receptors (Graziadei & Monti Graziadei 1979a,b), and avian telencephalic neurons (Goldman & Nottebohm 1983) are three examples. There are also small residual populations of neuroblasts in the optic lobes of insects and in the retina and tectum of fish (Nordlander & Edwards 1969, Raymond et al 1986, Reh & Tulley 1986). In these systems, rates of neuron proliferation appear to be controlled via a feedback loop. Breaking the loop by sectioning the olfactory nerve, for instance, results in a new burst of neuron proliferation (Graziadei & Monti Graziadei 1979b).

There are a few other, less well-substantiated instances of production of neurons in adult life. Tritiated thymidine studies of Kaplan (1981) and of Bayer et al (1982) indicate that neurons may be produced, albeit in limited number, in adult rats. Even more difficult to interpret are careful quantitative studies in which counts of neuron populations that are presumably postmitotic are observed to increase during postnatal growth. The populations that rise in number include inferior olivary neurons (Delhaye-Bouchaud et al 1985, Caddy & Biscoe 1979), principal neurons in the dorsal lateral geniculate nucleus (Satorre et al 1986), and Purkinje cells (Caddy & Biscoe 1979, Diglio & Herrup 1982). Most of the authors cited here explain their results as caused either by error of identification of neurons in young animals or delayed differentiation of subsets of neurons. Tritiated thymidine studies in most of these systems reveal no apparent genesis of neurons. Nonetheless, the findings are intriguing, if only for the cautionary notes they raise about cell counts in developing systems.

Much more is known about the mitogenic effects of neurons and axons on glial cells. DeLong & Sidman (1962) initially showed that removing an eye at birth reduced the number of glial cells in the adult colliculus mainly as a result of decreased rates of glial proliferation. The partial dependence of glial cell proliferation on neurons and on mitogenic factors is a particularly active and interesting topic (Raff et al 1978, Perkins et al 1981, Salzer et al 1980, Lemke & Brokes 1983, Nieto-Sampedro et al 1985).

CONTROL OF NEURON NUMBER BY CELL DEATH

One of the most counterintuitive and seemingly wasteful processes in brain development is the death of an often large fraction of the initial complement of neurons and glial cells. First noted by Studnicka (1905), Collin (1906), von Szily (1912), and others in the vertebrate central nervous system, serious attention to this process and proof that many dying cells are young neurons was delayed until catalytic studies by Glücksmann (1940), Romanes (1946), Hamburger & Levi-Montalcini (1949), Beaudoin

(1955), Källén (1955), and Hughes (1961). The literature has been particularly well reviewed (Källén 1865, Hughes 1968, Prestige 1970, Cowan 1973, Silver 1978, Jacobson 1978, Cunningham 1982, Berg 1982, Beaulaton & Lockshin 1982, Hamburger & Oppenheim 1982, Horvitz et al 1982, Truman 1984, Finlay et al 1987, Oppenheim 1988).

Variation

Neuron death is by no means universal throughout the animal kingdom. The process is unknown in *Aplysia californica* (Jacob 1984) and is also unknown or rare in elasmobranchs, teleosts, and reptiles (but see Fox & Richardson 1982). There is no evidence of motor neuron loss in zebrafish, dogfish, or stingray (reviewed in Mos & Williamson 1986). With some important exceptions (mainly homeotherms), the elimination of neurons is most pronounced in species that undergo metamorphosis and is least pronounced in vertebrates and invertebrates that do not metamorphose and that grow throughout life.

The incidence of death among different populations of neurons also shows great variation in single species—there is no meaningful average incidence. Some populations are eliminated completely (Studnicka 1905, Hughes 1957, Munk 1966, Stewart et al 1987), while other populations live happily ever after (e.g. Armstrong & Clarke 1979, Mos & Williamson 1986, Oppenheim 1986). Even within groups of neurons the distribution of dying cells is uneven. For instance, in the mammalian neocortex, degeneration is restricted mostly to the upper layers (Finlay & Slattery 1983; Williams et al 1987), and in spinal cord and dorsal root ganglia, degeneration is most severe in segments that do not innervate the limbs (Levi-Montalcini 1950, Hamburger 1975). The magnitude of neuron death within homologous populations in different species is also remarkably variable. For instance, 80% of retinal ganglion cells die in cat (Williams et al 1986); 60–70% in rat (Crespo et al 1985), rhesus monkey, and human (Rakic & Riley 1983, Provis et al 1985); 40% in chicken (Rager 1980); and none whatsoever in fish or amphibia (Wilson 1971, Easter et al 1981).

Sex-specific variation in patterns of neuron death has also recently been demonstrated in warm-blooded vertebrates (Konishi 1985, Nordeen et al 1985). In rats, for instance, the pool of motor neurons innervating penis muscles contains 160–200 neurons whereas a homologous pool in females contains only 35–60 small neurons (Breedlove & Arnold 1980, Nordeen et al 1985). The three- to four-fold difference is brought about by a more severe loss in females during the first ten postnatal days (Nordeen et al 1985). Compare this strategy with that used by annelids: In leeches, the two segmental ganglia associated with the sex organs contain twice the baseline number of neurons found in the other ganglia. The abrupt regional difference is not brought about by modulation of the severity of cell loss across the entire series of segments (this really would be wasteful). Instead the neurons are added selectively to the sex ganglia late in development (Ogawa 1939, Stewart et al 1986).

Comparative Perspective on Roles of Neuron Death

In nematodes neuron death is rarely contingent on the status of other cells. Death is an inflexible autonomous fate that serves to eliminate neurons that are apparently unwanted byproducts of patterns of cell lineage. In *C. elegans* one common mechanism, involving the action of at least four *ced* (cell death) gene products, catalyzes the death of most cells—neurons included. Mutations of the genes *ced-3* and *ced-4* prevent degeneration and death entirely (Hedgecock et al 1983, Ellis & Horvitz 1986). The most probable reason that neurons die in nematodes is not to remove potentially detrimental neurons, but to rid the organisms of unneeded, metabolically demanding cells. In this animal, neuron elimination appears to be strictly a means to regulate number, not to regulate ratios of cells or to improve neuronal performance. However, in many, if not most, other phyla, the process of cell elimination in any one population of neurons is regulated by several different kinds of interactions, each designed to optimize cell number or cell performance along a different parameter. Here, neuron death is just the final common pathway of disparate processes having disparate causes. The relative contributions of these processes will differ from animal to animal, from neuron population to neuron population.

Even in insects, in which neuron death often appears to be under the same rigid genomic control noted in nematodes (Whitington et al 1982), there are several interesting exceptions in which the loss is contingent on cell environment and even animal behavior (Truman 1983). For instance, the severe necrosis in the optic lobes of the eyeless fruit fly mutant, *so*, is triggered by the degeneration of the eye imaginal disc cells (Hofbauer & Campos-Ortega 1976, Fischbach & Technau 1984). This is an important advance—neuron death in holometabolous insects is no longer an inexorable fate of certain young neurons.

In chordates, neuron loss appears to be catalyzed in large part by the hormonal, vascular, and cellular environments in which neurons and their processes are situated. Thyroxine, for example, catalyzes the radical transformations in body architecture in amphibians that are associated with a rapid and focal loss of select neuron populations (Hughes 1961, Torrence 1983). All Rohan-Beard neurons are lost in frogs over a short period at the climax of metamorphosis (Hughes 1957). In warm-blooded vertebrates, naturally occurring neuron death is often very substantial and is in most cases at least partially explicable in terms of the size and status of the target (see reviews cited above), the density of synaptic input onto dendrites and cell bodies (Linden & Perry 1982, Okado & Oppenheim 1984), nutritional status (Dobbing & Smart 1974), and hormone levels (Nordeen et al 1985). In some cases, neuron death rids the brain of neurons that fail to make correct connections (e.g. Clarke & Cowan 1976, McLoon 1982, Jacobs et al 1984, O'Leary et al 1986). More to the point of this chapter, neuron death appears to have a major role in the final adjustments of ratios of interconnected neurons.

Neuron Death and Neuron Ratios

The rate and duration of neuron production in distant populations that are ultimately interconnected may not be well matched. This will produce quantitative imbalances.[4] Cell elimination provides a way to adjust neuron ratios after links between populations have been established by axons or dendrites. The larger the nervous system and the more disjointed the production of interconnected cell populations (e.g. motor neurons and muscle fibers, pre- and post-ganglion sympathetic neurons, receptor neurons and central target cells), the more important this secondary fine-tuning will be. For example, motor neurons in vertebrates are lost shortly after their axons have reached target musculature. If the target is removed early enough in development (before or during the period of cell death), then the severity of neuron loss is increased. Cell death can wipe out the entire population if the target is removed early enough (e.g. Hughes & Tschumi 1958, Lanser & Fallon 1984). Complementary studies in which the size and number of target cells are increased have been carried out in several systems (e.g. Hamburger 1939, Hollyday & Hamburger 1976, Pilar et al 1980, Boydston & Sohal 1979, Narayanan & Narayanan 1978, Lamb 1979, Chalupa et al 1984, Maheras & Pollack 1985). In all cases, an increase in number of target cells causes an appreciable increase in the percentage of surviving neurons, but the increase is never as great as would be required to maintain a normal ratio between interconnected cells. Thus, neuron elimination even in these systems is not regulated entirely by extrinsic factors. Some fraction of the loss is evidently intrinsic and thus cannot be regulated during ontogeny to bring about the most adaptive ratio between cells. However, long-term selective pressures on a species may result in a change in the importance of intrinsic control, and as pointed out by Prestige (1970), the relative weight of intrinsic and extrinsic control varies

[4] Hamburger (1975) has pointed out, however, that the overproduction of spinal motor neurons cannot be ascribed to an imprecise programming of the number of mitotic cycles. He observed that the total number of spinal motor neurons produced before the onset of neuron death in 5.5- and 6-day chicken embryos was relatively constant (ranges from 18,900 to 21,600, $n = 11$).

substantially among vertebrate classes. For example, the loss of thoracic motor neurons in the anuran spinal cord is heavily dependent on the size of the target, whereas the loss of the same population of neurons in chick and mouse is largely under intrinsic control. This class difference may be related to the fact that amphibians metamorphose, whereas birds and mammals develop directly.

We remarked above that there is only limited experimental evidence that the number of the neurons in any one nucleus or the ratio of neurons in interconnected populations is important. But natural selection is the final judge, and natural selection can work effectively with small variation over many generations—variation that we might even fail to recognize, and for which there would be little prospect of demonstrating a direct functional effect. The remarkable precision of regulation of number apparent during the development of some neuron populations strengthens the proposition that neuron number matters a great deal. For instance, there is now evidence that ratios of interconnected cells are regulated precisely by naturally occurring cell death. Tanaka & Landmesser (1986a,b) examined spinal motor neurons in chick, quail, and chick ↔ quail chimeras. The ratios of neurons and primary myotubes for each condition was constant. Lanser & Fallon (1987) have extended this work and shown that the number of surviving motor neurons in *wingless* mutant chicks (a mutation with variable penetrance) is a linear function of the mass of the limb musculature. A direct correlation between neuron number and target size has also been demonstrated in the central nervous system. Wetts & Herrup (1983) observed a linear relation between Purkinje cells and granule cell numbers in several wild-type strains of mouse, and experimental studies of *staggerer* ↔ wild-type mouse chimeras provided further proof that ratios are regulated precisely. *Staggerer* granule cells are not intrinsically programmed to die (Herrup 1983). Thus in a *staggerer* chimera more granule cells are produced than can be supported by the healthy, wild-type Purkinje cells. This imbalance is corrected by an increased incidence of cell death (Sonmez & Herrup 1984), and the result is a linear relationship between granule cell and Purkinje cell numbers (Herrup & Sunter 1987). Achieving an optimal match in numbers of interconnected cells is undoubtedly one important role of neuron elimination.

Neuron loss can be interpreted as contributing to an individual's fitness by improving either the efficiency of information processing or the efficiency of metabolism. At a higher level of analysis, the combination of overproduction and loss of neurons may also buffer and conserve heritable variation and thereby increase the evolutionary plasticity of a species (Katz & Lasek 1978, Williams et al 1986, Finlay et al 1987). However, experimental demonstrations that the combination of neuron overproduction and neuron elimination makes a particular contribution to fitness of individuals and species will continue to be hard to obtain. We will almost certainly understand the cell biology of neuron elimination before we understand its underlying cause and utility.

SUMMARY

In comparing strategies used to control neuron number, we find it useful to view nervous system development as occurring in three phases. The phases overlap—each is a process, not an event. The first phase is the development of a genetic nervous system. This is a nervous system of simple genetic intention, not a blueprint or pile of bricks. Its characteristics are abstract: How many neurons is this stem cell destined to produce, which cells are programmed to die, how much target does this cell need to survive. At the level of the individual, the genetic nervous system is essentially fixed, but over generations it is fluid. As this genetic intent interacts within a world of cells, a real brain appears. A new set of rules only tacitly present in the genome is expressed—the embryonic nervous system emerges. The interaction of its parts defines the shape and size of the nervous system. Neuron numbers are adjusted interactively by changes in proliferative potential and the severity of cell death. Cell fates are established in part through interactions with other cells and with hormones. Glial cell numbers are adjusted to match the neuron populations. This phase of brain development is what the embryologist sees under the microscope. The third and final phase of development begins when the brain starts to function and the animal starts to deal with its world. Small changes in neuron number may occur during this period, but these changes are generally of minor functional importance. At this point, the smaller elements of neuronal organization are refined in shape, number, and distribution. Axons are lost or rearranged; dendrites grow, branch, and retract; synapses are fine-tuned; and finally, receptor densities and transmitter titres are adjusted. Some of these interactions and numerical adjustments continue until the animal dies.

ACKNOWLEDGMENT

We thank Kathryn Graehl for editorial help. Supported in part by grants NS20591, NS18381, and the March of Dimes 1-763.

Literature Cited

Allen, C., Sievers, J., Berry, M., Jenner, S. 1981. Experimental studies on cerebellar foliation. II. A morphometric analysis of cerebellar fissuration defects and growth retardation after neonatal treatment with 6-OHDA in the rat. *J. Comp. Neurol.* 203: 771–83

Ambros, V., Horvitz, H. R. 1984. Heterochronic mutants of the nematode *Caenorhabditis elegans. Science* 226: 409–16

Anderson, H., Edwards, J. S., Palka, J. 1980. Developmental neurobiology of invertebrates. *Ann. Rev. Neurosci.* 3: 97–139

Angelo, J. C., Prothero, J. W. 1985. Clonal attenuation in chick embryo fibroblasts. Experimental data, a model and computer simulations. *Cell Tissue Kinet.* 18: 27–43

Armstrong, E. 1982. Mosaic evolution in the primate brain: Differences and similarities in the hominoid thalamus. In *Primate Brain Evolution. Methods and Concepts,* ed. E. Armstrong, D. Falk, pp. 131–61. New York: Plenum

Armstrong, E. 1983. Relative brain size and metabolism in mammals. *Science* 220: 1302–4

Armstrong, E., Bergeron, R. 1985. Relative brain size and metabolism in birds. *Brain Behav. Evol.* 26: 141–53

Armstrong, R. C., Clarke, P. G. H. 1979. Neuronal death and the development of the pontine nuclei and inferior olive in the chick. *Neuroscience* 4: 1635–47

Ashwell, K. W., Watson, C. R. R. 1983. The development of facial motoneurones in the mouse—Neuronal death and the innervation of the facial muscles. *J. Embryol. Exp. Morphol.* 77: 117–41

Beach, D. H., Jacobson, M. 1979. Influence of thyroxine on cell proliferation in the retina of the clawed frog at different ages. *J. Comp. Neurol.* 183: 615–24

Beaudoin, A. R. 1955. The development of lateral motor column cells in the lumbosacral cord in *Rana pipiens* I. Normal development and development following unilateral limb ablation. *Anat. Rec.* 121: 81–95

Beaulaton, J., Lockshin, R. A. 1982. The relation of programmed cell death to development and reproduction: Comparative studies and an attempt at classification. *Int. Rev. Cytol.* 79: 215–35

Beddington, R. S. P. 1981. A developmental analysis of the potency of embryonic ectoderm in 8th day postimplantation mouse embryos. *J. Embryol. Exp. Morphol.* 69: 265–85

Beddington, R. S. P. 1983. Histogenic and neoplastic potential of different regions of mouse embryonic egg cylinder. *J. Embryol. Exp. Morphol.* 75: 189–204

Berg, D. K. 1982. Cell death in neuronal development. Regulation by trophic factors. In *Neuronal Development,* ed. N. C. Spitzer, pp. 297–331. New York: Plenum

Blinkov, S. M., Glezer, I. I. 1968. *The Human Brain in Figures and Tables: A Quantitative Handbook.* New York: Basic

Block, J. B., Essman, W. B. 1965. Growth hormone administration during preg-

nancy: A behavioural difference in offspring rats. *Nature* 205: 1136–37

Boydston, W. R., Sohal, G. S. 1979. Grafting of additional periphery reduces embryonic loss of neurons. *Brain Res.* 178: 403–10

Bradley, P., Berry, M. 1978. Purkinje cell dendritic tree in mutant mouse cerebellum. A quantitative Golgi study of *weaver* and *staggerer* mice. *Brain Res.* 142: 135–41

Braitenberg, V., Atwood, R. P. 1958. Morphological observations on the cerebellar cortex. *J. Comp. Neurol.* 109: 1–33

Breedlove, S. M., Arnold, A. P. 1980. Hormone accumulation in a sexually dimorphic motor nucleus of the rat spinal cord. *Science* 210: 564–66

Bullock, T. H., Horridge, G. A. 1965. *Structure and Function in the Nervous System of Invertebrates.* San Francisco: Freeman

Bush, G. L., Case, S. M., Wilson, A. C., Patton, J. L. 1977. Rapid speciation and chromosomal evolution in mammals. *Proc. Natl. Acad. Sci. USA* 74: 3942–46

Caddy, K. W. T., Biscoe, T. J. 1979. Structural and quantitative studies on the normal C3H and Lurcher mutant mouse. *Philos. Trans. R. Soc. London Ser. B* 287: 167–201

Calder, W. A. 1976. Aging in vertebrates: Allometric considerations of spleen size and lifespan. *Fed. Proc.* 35: 96–97

Campbell, B., Ryzen, M. 1953. The nuclear anatomy of the diencephalon of *Sorex cinereus. J. Comp. Neurol.* 99: 1–22

Chalfie, M., Horvitz, H. R., Sulston, J. E. 1981. Mutations that lead to reiterations in the cell lineage of *C. elegans. Cell* 24: 59–81

Chalupa, L. M., Williams, R. W., Henderson, Z. 1984. Binocular interaction in the fetal cat regulates the size of the ganglion cell population. *Neuroscience* 12: 1139–46

Champy, C. 1922. L'action de l'extrait thyroidien sur la multiplication cellulaire. Caracter electif de cette action. *Arch. Morphol. Gen. Exper.* 4: 1–58

Chan, W. Y., Tam, P. P. L. 1986. The histogenetic potential of neural plate cells of early somite-stage mouse embryos. *J. Embryol. Exp. Morphol.* 96: 183–93

Chiarodo, A. T. 1963. The effects of mesothoracic leg disc extirpation on the postembryonic development of the nervous system of the blowfly *Sarcophaga bullata. J. Exp. Zool.* 153: 263–77

Clarke, P. G. H., Cowan, W. M. 1976. The development of the isthmo-optic tract in the chick, with special reference to the occurrence and correction of developmental errors in the location and connection of isthmo-optic neurons. *J. Comp. Neurol.* 167: 143–64

Clendinnen, B. G., Eayrs, J. T. 1961. The anatomical and physiological effects of prenatally administered somatotrophin on cerebral development. *J. Endocrinol.* 22: 183–93

Cline, H. T., Constantine-Paton, M. 1986. Thyroxine effects on the development of the retinotectal projection. *Soc. Neurosci. Abstr.* 12: 437

Cobb, S. 1965. Brain size. *Arch. Neurol.* 12: 555–61

Collin, R. 1906–1907. Recherches cytologiques sur le développement de la cellule nerveuse. *Névraxe* 8: 181–303

Cooke, J. 1980. Early organization of the central nervous system: Form and pattern. *Curr. Top. Devel. Biol.* 15: 373–407

Cowan, W. M. 1973. Neuronal death as a regulative mechanism in the control of cell number in the nervous system. In *Development and Aging in the Nervous System*, ed. M. Rockstein, pp. 19–41. New York: Academic

Crespo, D., O'Leary, D. D. M., Cowan, W. M. 1985. Changes in the number of optic nerve fibers during late prenatal and postnatal development in the albino rat. *Dev. Brain Res.* 19: 129–34

Cunningham, T. J. 1982. Naturally occurring neuron death and its regulation by developing neural pathways. *Internat. Rev. Cytol.* 74: 163–86

Das, G. D., Pfaffenroth, M. J. 1977. Experimental studies on the postnatal development of the brain. III. Cerebellar development following localized administration of ENU. *Neuropath. Appl. Neurobiol.* 3: 191–212

DeBrul, L. 1960. Structural evidence in the brain for a theory of the evolution of behavior. *Perspect. Biol. Med.* 4: 40–57

Delhaye-Bouchaud, N., Geoffroy, B., Mariani, J. 1985. Neuronal death and synapse elimination in the olivocerebellar system. I. Cell counts in the inferior olive of developing rats. *J. Comp. Neurol.* 232: 299–308

DeLong, G. R., Sidman, R. L. 1962. Effects of eye removal at birth on histogenesis of the mouse superior colliculus: An autoradiographic analysis with tritiated thymidine. *J. Comp. Neurol.* 118: 205–23

Diglio, T., Herrup, K. 1982. A significant fraction of the adult number of mature cerebellar Purkinje cells first appears between postnatal days 16 and 30 in the mouse. *Soc. Neurosci. Abstr.* 8: 636

Dobbing, J., Smart, J. L. 1974. Vulnerability of developing brain and behaviour. *Br. Med. Bull.* 30: 164–68

Easter, S. S. Jr., Rusoff, A. C., Kish, P. E. 1981. The growth and organization of the optic nerve and tract in juvenile and adult goldfish. *J. Neurosci.* 1: 793–811

Ebbesson, S. O. E. 1965. Quantitative studies of superior cervical sympathetic ganglia in a variety of primates including man. I. The ratio of preganglionic fibers to ganglionic neurons. *J. Morphol.* 124: 117–32

Eisenberg, J. F. 1981. *The Mammalian Radiations. An Analysis of Trends in Evolution, Adaptation, and Behavior.* Chicago: Univ. Chicago Press

Eisenberg, J. F., Wilson, D. E. 1982. Relative brain size and feeding strategies in the Chiroptera. *Evolution* 32: 740–51

Ellis, H. M., Horvitz, H. R. 1986. Genetic control of programmed cell death in the nematode *C. elegans*. *Cell* 44: 817–29

Faber, D. S., Korn, H., eds. 1978. *Neurobiology of the Mauthner Cells.* New York: Raven

Fankhauser, G., Vernon, J. A., Frank, W. H., Slack, W. V. 1955. Effect of size and number of brain cells on learning in larvae of the salamander, *Triturus viridescens*. *Science* 122: 692–93

Finlay, B. L., Slattery, M. 1983. Local differences in the amount of early cell death in neocortex predict adult local specializations. *Science* 219: 1349–51

Finlay, B. L., Wikler, K. C., Sengelaub, D. R. 1987. Regressive events in brain development and scenarios for vertebrate brain evolution. *Brain Behav. Evol.* 30: 102–17

Fischbach, K. F., Heisenberg, M. 1981. Structural brain mutant of *Drosophila melanogaster* with reduced cell numbers in the medulla cortex and with normal optomotor yaw response. *Proc. Natl. Acad. Sci. USA* 78: 1105–9

Fischbach, K. F., Technau, G. 1984. Cell degeneration in the developing optic lobes of the sine oculis and small-optic-lobes mutants of *Drosophila melanogaster*. *Devel. Biol.* 104: 219–39

Fox, G. Q., Richardson, G. P. 1982. The developmental morphology of *Torpedo marmorata*: Electric lobe-electromotoneuron proliferation and cell death. *J. Comp. Neurol.* 207: 183–90

Gao, W.-Q., Macagno, E. R. 1986. Extension and retraction of axonal projections by existence of neighboring homologues. I. The HA cells. *J. Neurobiol.* 18: 43–59

Glücksmann, A. 1940. Development and differentiation of the tadpole eye. *Br. J. Ophthal.* 24: 153–78

Glücksmann, A. 1951. Cell deaths in normal vertebrate ontogeny. *Biol. Rev.* 26: 59–86

Goldschmidt, Rr. 1909. Das Nervensystem von *Ascaris lumbridoides* und *megalodephala*. II. *Z. Wiss. Zool.* 92: 306–57

Goldman, S. A., Nottebohm, F. 1983. Neuronal production, migration, and differentiation in a vocal control nucleus of the adult canary brain. *Proc. Natl. Acad. Sci. USA* 80: 2390–94

Goodman, C. 1976. Constancy and uniqueness in a large population of small interneurons. *Science* 193: 502–4

Goodman, C. S. 1977. Neuron duplications and deletions in locust clones and clutches. *Science* 197: 1384–86

Goodman, C. S. 1978. Isogenic grasshoppers: Genetic variability in the morphology of identified neurons. *J. Comp. Neurol.* 182: 681–706

Goodman, C. S. 1979. Isogenic grasshoppers: Genetic variability and development of identified neurons. In *Neurogenetics: Genetic Approaches to the Nervous System*, ed. X. O. Breakefield, pp. 101–51. New York: Elsevier-North Holland

Gould, S. J. 1977. *Ontogeny and Phylogeny.* Cambridge, Mass: Belknap Press/Harvard Univ. Press

Graziadei, P. P. C., Monti Graziadei, G. A. 1979a. Neurogenesis and neuron regeneration in the olfactory system of mammals. I. Morphological aspects of differentiation and structural organization of the olfactory sensory neurons. *J. Neurocytol.* 8: 1–18

Graziadei, P. P. C., Monti Graziadei, G. A. 1979b. Neurogenesis and neuron regeneration in the olfactory system of mammals. II. Degeneration and reconstitution of the olfactory sensory neurons after axotomy. *J. Neurocytol.* 8: 197–213

Guesella, J., Geller, R., Clarke, B., Weeks, V., Housman, D. 1976. Commitment to erythroid differentiation by Friend erythroleukemia cells: A stochastic analysis. *Cell* 9: 221–29

Hamburger, V. 1939. Motor and sensory hyperplasia following limbbud transplantations in chick embryos. *Physiol. Zool.* 12: 268–84

Hamburger, V. 1975. Cell death in the development of the lateral motor column of the chick embryo. *J. Comp. Neurol.* 160: 535–46

Hamburger, V., Levi-Montalcini, R. 1949. Proliferation, differentiation and degeneration in the spinal ganglia of the chick embryo under normal and experimental conditions. *J. Exp. Zool.* 111: 457–501

Hamburger, V., Oppenheim, R. W. 1982. Naturally occurring neuronal death in vertebrates. *Neurosci. Comment.* 1: 39–55

Hechst, B. 1932. Über einen Fall von Mikroencephalie ohne geistigen Defekt. *Arch. Psychiat. Nervenk.* 97: 64–76

Hedgecock, E., Sulston, J. E., Thompson, N. 1983. Mutations affecting programmed cell deaths in the nematode *Caenorhabditis elegans. Science* 220: 1277–80

Herrup, K. 1983. Role of *staggerer* gene in determining cell number in cerebellar cortex. I. Granule cell death is an indirect consequence of *staggerer* gene action. *Devel. Brain Res.* 11: 267–74

Herrup, K. 1986. Cell lineage relationships in the development of the mammalian CNS: Role of cell lineage in control of cerebellar Purkinje cell number. *Devel. Biol.* 115: 148–54

Herrup, K. 1987. Roles of cell lineage in the developing mammalian brain. *Curr. Top. Devel. Biol.* 21: 65–97

Herrup, K., Letsou, A., Diglio, T. J. 1984b. Cell lineage relationships in the development of the mammalian CNS: The facial nerve nucleus. *Devel. Biol.* 103: 329–36

Herrup, K., Mullen, R. J. 1979. *Staggerer* chimeras: Intrinsic nature of Purkinje cell defects and implications for normal cerebellar development. *Brain Res.* 178: 443–57

Herrup, K., Mullen, R. J. 1981. Role of the *staggerer* gene in determining Purkinje cell number in the cerebellar cortex of mouse chimeras. *Devel. Brain Res.* 1: 475–85

Herrup, K., Sunter, K. 1986. Cell lineage dependent and independent control of Purkinje cell number in the mammalian CNS: Further quantitative studies of *Lurcher* chimeric mice. *Devel. Biol.* 117: 417–27

Herrup, K., Sunter, K. 1987. Numerical matching during cerebellar development: Quantitative analysis of granule cell death in *staggerer* mouse chimeras. *J. Neurosci.* 7: 829–36

Herrup, K., Wetts, R., Diglio, T. J. 1984a. Cell lineage relationships in the development of the mammalian CNS. II. Bilateral independence of CNS clones. *J. Neurogenet.* 1: 275–88

Hertweck, H. 1931. Anatomie und Variabilität des Nervensystems und der Sinnesorgane von *Drosophila melanogaster* (Meigen). *Z. Wiss. Zool.* 139: 559–663

Hofbauer, A., Campos-Ortega, J. A. 1976. Cell clones and pattern formation: Genetic eye mosaics in *Drosophila melanogaster. Wilhelm Roux Arch. Devel. Biol.* 179: 275–89

Holloway, R. L. 1968. The evolution of the primate brain: Some aspects of quantitative relations. *Brain Res.* 7: 121–72

Holloway, R. L. 1980. Within-species brain-body weight variability: A reexamination of the Danish data and other primate species. *Am. J. Phys. Anthropol.* 53: 109–21

Hollyday, M., Hamburger, V. 1976. Reduction of the naturally occurring motor neuron loss by enlargement of the periphery. *J. Comp. Neurol.* 170: 311–20

Horvitz, H. R., Ellis, H. M., Sternberg, P. W.

1982. Programmed cell death in nematode development. *Neurosci. Comment.* 1: 56–65

Hughes, A. F. 1957. The development of the primary sensory system in *Xenopus laevis* (Daudin). *J. Anat.* 91: 323–38

Hughes, A. F. 1961. Cell degeneration in the larval ventral horn of *Xenopus laevis*. *J. Embryol. Exp. Morphol.* 9: 269–84

Hughes, A. 1968. *Aspects of Neural Ontogeny.* New York: Academic

Hughes, A., Tschumi, P. A. 1958. The factors controlling the development of the dorsal root ganglia and ventral horn in *Xenopus laevis* (Daud.). *J. Anat.* 92: 498–527

Jacob, M. H. 1984. Neurogenesis in *Aplysia californica* resembles nervous system formation in vertebrates. *J. Neurosci.* 4: 1225–39

Jacobs, D. S., Perry, V. H., Hawken, M. J. 1984. The postnatal reduction of the uncrossed projection from the nasal retina in the cat. *J. Neurosci.* 4: 2425–33

Jacobson, M. 1978. *Developmental Neurobiology.* New York: Plenum. 2nd ed.

Jacobson, M., Hirose, G. 1978. Origin of the retina from both sides of the embryonic brain: A contribution to the problem of crossing at the optic chiasma. *Science* 202: 637–39

Jacobson, M., Moody, S. A. 1984. Quantitative lineage analysis of the frog's nervous system. I: Lineages of Rohon-Beard neurons and primary motoneurons. *J. Neurosci.* 4: 1361–69

Jerison, H. J. 1955. Quantitative analysis of evolution of the brain in mammals. *Science* 133: 1012–14

Jerison, H. J. 1973. *Evolution of the Brain and Intelligence.* New York: Academic

Jerison, H. J. 1985. Animal intelligence as encephalization. *Philos. Trans. R. Soc. London Ser. B* 308: 21–35

Källén, B. 1955. Cell degeneration during normal ontogenesis of the rabbit brain. *J. Anat.* 89: 153–62

Källén, B. 1965. Degeneration and regeneration in the vertebrate central nervous system during embryogenesis. *Prog. Brain Res.* 14: 77–96

Kaplan, M. S. 1981. Neurogenesis in the 3-month-old rat visual cortex. *J. Comp. Neurol.* 195: 323–38

Katz, M. J., Lasek, R. J. 1978. Evolution of the nervous system: Role of ontogenetic mechanisms in the evolution of matching populations. *Proc. Natl. Acad. Sci. USA* 75: 1349–52

Katz, M. J., Grenander, U. 1982. Developmental matching and the numerical matching hypothesis for neuronal cell death. *J. Theor. Biol.* 98: 501–17

Kety, S. S., Schmidt, C. F. 1948. The effects of altered arterial tensions of carbon dioxide and oxygen on cerebral blood flow and cerebral oxygen consumption of normal young men. *J. Clin. Invest.* 27: 484–92

Kimmel, C. B., Eaton, R. C. 1976. Development of the Mauthner cell. In *Simpler Networks and Behavior*, ed. J. C. Fentress, pp. 186–202. Sunderland, Mass: Sinauer

Kolb, H., Nelson, R., Mariani, A. 1981. Amacrine cells, bipolar cells and ganglion cells of the cat retina: A Golgi study. *Vision Res.* 21: 1081–1114

Kollros, J. J. 1953. The development of the optic lobes in the frog. I. The effects of unilateral enucleation in embryonic stages. *J. Exp. Zool.* 123: 153–87

Kollros, J. J. 1982. Peripheral control of midbrain mitotic activity in the frog. *J. Comp. Neurol.* 205: 171–78

Kollros, J. J., McMurray, V. M. 1955. The mesencephalic V nucleus in anurans. *J. Comp. Neurol.* 102: 47–61

Kollros, J. J., Thiesse, M. L. 1985. Growth and death of cells of the mesencephalic fifth nucleus in *Xenopus laevis* larvae. *J. Comp. Neurol.* 233: 481–89

Konigsmark, B. W. 1970. The counting of neurons. In *Contemporary Research Methods in Neuroanatomy*, ed. W. J. H. Nauta, S. O. Ebbesson, pp. 315–40. New York: Springer-Verlag

Konishi, M. 1985. Birdsong: From behavior to neuron. *Ann. Rev. Neurosci.* 8: 125–70

Krebs, H. A. 1950. Body size and tissue respiration. *Biochim. Biophys. Acta* 4: 249–69

Kreisman, N. R., Olson, J. E., Horne, D. S., Holtzman, D. 1986. Developmental increases in oxygen delivery and extraction in immature rat cerebral cortex. *Neurosci. Abstr.* 12: 451

Lamb, A. H. 1979. Evidence that some developing limb motoneurons die for reasons other than peripheral competition. *Devel. Biol.* 71: 8–21

Lande, R. 1979. Quantitative genetic analysis of multivariate evolution, applied to brain: body size allometry. *Evolution* 33: 402–16

Landis, D. M. D., Sidman, R. L. 1978. Electron microscopic analysis of postnatal histogenesis in the cerebellar cortex of staggerer mutant mice. *J. Comp. Neurol.* 179: 831–63

Lange, W. 1975. Cell number and cell density in the cerebellar cortex of man and some other mammals. *Cell Tiss. Res.* 157: 115–24

Lanser, M. E., Fallon, J. F. 1984. Development of the lateral motor column in the *limbless* mutant chick embryo. *J. Neurosci.* 4: 2043–50

Lanser, M. E., Fallon, J. F. 1987. Development of the branchial lateral motor column in the *Wingless* mutant chick embryo: Motoneuron survival under varying degrees of peripheral load. *J. Comp. Neurol.* 261: 423–34

Lapicque, L. 1907. Le poids encéphalique en fonction du poids corporel entre individus d'une même espèce. *Bull. Mem. Soc. Anthropol. Paris* 8: 313–45

Lauder, J. M., Altman, J., Krebs, H. 1974. Some mechanisms of cerebellar foliation: Effects of early hypo- and hyperthyroidism. *Brain Res.* 76: 33–40

Lauder, J. M., Krebs, H. 1976. Effects of *p*-chlorophenylalanine on time of neuronal origin during embryogenesis in the rat. *Brain Res.* 107: 638–44

Lauder, J. M., Krebs, H. 1978. Serotonin as a differentiation signal in early neurogenesis. *Devel. Neurosci.* 1: 15–30

Lemke, G. E., Brokes, J. P. 1983. Identification and purification of glial growth factor. *J. Neurosci.* 4: 75–83

Levi-Montalcini, R. 1950. The origin and development of the visceral system in the spinal cord of the chick embryo. *J. Morphol.* 86: 253–83

Linden, R., Perry, V. H. 1982. Ganglion cell death within the developing retina: A regulatory role for retinal dendrites? *Neuroscience* 7: 2813–27

Lillegraven, J. A. 1979. Reproduction in Mesozoic mammals. In *Mesozoic Mammals: The Fist Two-Thirds of the Mammalian History*, ed. J. A. Lillegraven, Z. Kielan-Javorowska, W. A. Clemens, pp. 259–76. Berkeley: Univ. Calif. Press

Lubbock, J. 1858. Variability of the nervous system. *Proc. R. Soc. London* 9: 480–86

Macagno, E. R., Lopresti, V., Levinthal, C. 1973. Structure and development of neuronal connections in isogenic organisms: Variation and similarities in the optic system of *Daphnia magna. Proc. Natl. Acad. Sci. USA* 70: 57–61

Macagno, E. R. 1979. Cellular interactions and pattern formation in the development of the visual system of *Daphnia magna* (Crustacea, Branchiopoda). I. Interactions between embryonic retinular fibers and laminar neurons. *Devel. Biol.* 73: 206–38

Macagno, E. R. 1980. Number and distribution of neurons in leech segmental ganglia. *J. Comp. Neurol.* 190: 283–302

Maheras, H. M., Pollack, E. D. 1985. Quantitative compensation by lateral motor column neurons in response to four functional hindlimbs in a frog tadpole. *Devel. Brain Res.* 19: 150–54

Mallet, J., Huchet, M., Pougeois, R., Changeux, J. P. 1976. Anatomical, physiological and biochemical studies on the cerebellum from mutant mice. III. Protein differences associated with the *weaver*, *staggerer* and *nervous* mutation. *Brain Res.* 103: 291–312

Mallouk, R. S. 1975. Longevity in vertebrates is proportional to relative brain weight. *Fed. Proc.* 34: 2102–3

Mallouk, R. S. 1976. Author's reply (to: Aging in vertebrates: Allometric considerations of spleen size and lifespan by W. A. Calder III). *Fed. Proc.* 35: 97–98

Mangold-Wirz, K. 1966. Cerebralisation und Ontogenesemodus bei Eutherien. *Acta Anat.* 63: 449–508

Mann, M. D., Towe, A. L., Glickman, S. E. 1986. Relationship between brain size and body size among Myomorph rodents. *Soc. Neurosci. Abstr.* 12: 111

Mariani, J., Crepel, F., Mikoshiba, K., Changeux, J. P., Sotelo, C. 1977. Anatomical, physiological and biochemical studies of the cerebellum from *reeler* mutant mouse. *Philos. Trans. R. Soc. London Ser. B* 281: 1–28

Martin, R. D. 1981. Relative brain size and basal metabolic rate in terrestrial vertebrates. *Nature* 293: 57–60

Martin, R. D., Harvey, P. H. 1985. Brain size allometry. Ontogeny and phylogeny. In *Size and Scaling in Primate Biology*, ed. W. L. Jungers, pp. 147–73. New York: Plenum

Martini, E. 1912. Studien über die Konstanz histologischer Elemente. III. *Hydatima senta. Z. Wiss. Zool.* 102: 425–645

Maurin, Y., Berger, B., Le Saux, F., Gay, M., Baumann, N. 1985. Increased number of locus ceruleus noradrenergic neurons in the convulsive mutant *Quaking* mouse. *Neurosci. Lett.* 57: 313–18

Mayr, E. 1963. *Animal Species and Evolution*. Cambridge, Mass: Belknap Press/Harvard Univ. Press

McLoon, S. C. 1982. Alteration in precision of the crossed retinotectal projection during chick development. *Science* 215: 1418–20

Mellon, D. Jr., Tufty, R. H., Lorton, E. D. 1976. Analysis of spatial constancy of oculomotor neurons in the crayfish. *Brain Res.* 109: 587–94

Meyer, M. P., Morrison, P. 1960. Tissue respiration and hibernation in the thirteen-lined ground-squirrel, *Spermophilus tridecemlineatus. Bull. Museum Comp. Zoo.* 124: 405–21

Miale, I. L., Sidman, R. L. 1961. An autoradiographic analysis of histogenesis in the mouse cerebellum. *Exp. Neurol.* 4: 277–96

Mos, W., Williamson, R. 1986. A quantitative analysis of the spinal motor pool and its target muscle during growth in the

Dogfish, *Scyliorhinus canicula*. *J. Comp. Neurol.* 248: 431–40

Mullen, R. J. 1977. Site of gene action and Purkinje cell mosaicism in *pcd* ↔ normal chimeric mice. *Nature* 270: 245–47

Mullen, R. J., Eicher, E. M., Sidman, R. L. 1976. Purkinje cell degeneration. A new neurological mutant in the mouse. *Proc. Natl. Acad. Sci. USA* 73: 208–12

Mullen, R. J., Herrup, K. 1979. Chimeric analysis of mouse cerebellar mutants. In *Neurogenetics: A Genetic Approach to the Central Nervous System*, ed. X. O. Breakefield, pp. 271–97. New York: Elsevier-North Holland

Müller, K. J., Nicholls, J. G., Stent, G. S. 1982. *Neurobiology of the Leech*. New York: Cold Spring Harbor Press

Munk, O. 1966. Ocular degeneration in deep-sea fishes. *Galathea Report, Scientific Results of the Danish Deep-sea Expedition Round the World 1950–52* 8: 22–31

Narayanan, C. H., Narayanan, Y. 1978. Neuronal adjustments in developing nucleus centers of the chick embryo following transplantation of an additional optic primordium. *J. Embryol. Exp. Morphol.* 44: 53–70

Newman, H. H., Patterson, J. T. 1911. The limits of hereditary control in armadillo quadruplets: A study of blastogenic variation. *J. Morphol.* 22: 855–926

Newport, J., Kirschner, M. 1982a. A major developmental transition in early *Xenopus* embryos. I. Characterization and timing of cellular changes at the midblastula stage. *Cell* 30: 675–86

Newport, J., Kirschner, M. 1982b. A major developmental transition in early *Xenopus* embryos. II: Control of the onset of transcription. *Cell* 30: 687–96

Nicholson, J. L., Altman, J. 1972. The effects of early hypo- and hyperthyroidism on the development of rat cerebellar cortex. I. Cell proliferation and differentiation. *Brain Res.* 44: 13–23

Nieto-Sampedro, J., Saneto, R. P., De Vellis, J., Cotman, C. W. 1985. The control of glial populations in brain: Change in astrocyte mitogenic and morphogenic factors in response to injury. *Brain Res.* 343: 320–28

Nordeen, E. J., Nordeen, K. W., Sengelaub, D. R., Arnold, A. P. 1985. Androgens prevent normally occurring cell death in a sexually dimorphic spinal nucleus. *Science* 229: 671–73

Nordlander, R. H., Edwards, J. S. 1969. Postembryonic brain development in the monarch butterfly. II. The optic lobes. *Wilhelm Roux Arch. Devel. Biol.* 163: 197–220

Ogawa, F. 1939. The nervous system of earthworm (*Pheretima communissima*) in different ages. *Sci. Rep. Tohoku Univ.* [*Med*]. 13: 395–488

Okado, N., Oppenheim, R. W. 1984. Cell death of motoneurons in the chick embryo spinal cord. IX. The loss of motoneurons following removal of afferent input. *J. Neurosci.* 4: 1639–52

O'Leary, D. D. M., Fawcett, J. W., Cowan, W. M. 1986. Topographic targeting errors in the retinocollicular projection and their elimination by selective ganglion cell death. *J. Neurosci.* 6: 3692–3705

Oppenheim, R. W. 1986. The absence of significant postnatal motoneuron death in the brachial and lumbar spinal cord of the rat. *J. Comp. Neurol.* 246: 281–86

Oppenheim, R. W. 1988. Cell death during neural development. In *Handbook of Physiology, Neuronal Development*, Vol. 1, ed. W. M. Cowan. Washington, DC: Am. Physiol. Soc. In press

Perkins, C. S., Aguayo, A. J., Bray, G. M. 1981. Schwann cell multiplication in *Trembler* mice. *Neuropathol. Appl. Neurobiol.* 7: 115–26

Pilar, G., Landmesser, L., Burstein, L. 1980. Competition for survival among developing ciliary ganglion cells. *J. Neurophysiol.* 43: 233–54

Pond, C. M. 1977. The significance of lactation in the evolution of mammals. *Evolution* 31: 177–99

Potten, C. S., Wichmann, H. E., Loeffler, M., Dobek, K., Major, D. 1982. Evidence for discrete cell kinetic subpopulations in mouse epidermis based on mathematical analysis. *Cell Tiss. Kinet.* 15: 305–29

Prestige, M. C. 1970. Differentiation, degeneration, and the role of the periphery: Quantitative considerations. In *The Neurosciences, Second Study Program*, ed. F. O. Schmitt, pp. 73–82. New York: Rockefeller Univ. Press

Provis, J., van Driel, D., Billson, F. A., Russel, P. 1985. Human fetal optic nerve: Overproduction and elimination of retinal axons during development. *J. Comp. Neurol.* 238: 92–101

Purves, D., Rubin, E., Snider, W. D., Lichtman, J. 1986. Relation of animal size to convergence, divergence, and neuronal number in peripheral sympathetic pathways. *J. Neurosci.* 6: 158–63

Quigley, H. A., Addicks, E. M., Green, W. G. 1982. Optic nerve damage in human glaucoma. III. Quantitative correlation of nerve fiber loss and visual field defect in glaucoma, ischemic neuropathy, papilledema, and toxic neuropathy. *Arch. Ophthalmol.* 100: 135–46

Quinn, L. S., Holtzer, H., Nameroff, M. 1984. Age dependent changes in myogenic precursor cell compartment sizes. Evidence for the existence of a stem cell. *Exp. Cell Res.* 154: 65–82

Quinn, L. S., Holtzer, H., Nameroff, M. 1985. Generation of chick skeletal muscle cells in groups of 16 from stem cells. *Nature* 313: 692–94

Quinn, L. S., Nameroff, M. 1983a. Analysis of the myogenic lineage in chick embryos. III. Quantitative evidence for discrete compartments of precursors cells. *Differentiation* 24: 111–23

Quinn, L. S., Nameroff, M. 1983b. Analysis of the myogenic lineage in chick embryos. IV. Effects of conditioned medium. *Differentiation* 24: 124–30

Quinn, L. S., Nameroff, M. 1986. Evidence for a myogenic stem cell. In *Molecular Biology of Muscle Development. UCLA Symp. Molec. Cell. Biol.*, ed. C. Emerson, D. A. Fischman, B. Nadal-Ginard, M. A. Q. Siddiqui, 29: 35–45. New York: Liss

Raff, M. C., Hornby-Smith, A., Brockes, J. P. 1978. Cyclic AMP as a mitogenic signal for cultured rat Schwann cells. *Nature* 273: 672–73

Rager, G. 1980. Development of the retinotectal projection in the chicken. *Adv. Anat. Embryol. Cell Biol.* 63: 1–92

Rakic, P. 1985. Limits of neurogenesis in primates. *Science* 227: 1054–56

Rakic, P., Riley, K. P. 1983. Overproduction and elimination of retinal axons in the fetal rhesus monkey. *Science* 219: 1441–44

Raymond, P. A., Reifler, M. J., Rivlin, P. K., Clendening, B. 1986. Progenitor cells specific for rods lose their specificity in regenerating goldfish retina. *Soc. Neurosci. Abstr.* 12: 118

Reh, T. A., Tully, T. 1986. Regulation of tyrosine hydroxylase-containing amacrine cell number in larval frog retina. *Devel. Biol.* 114: 463–69

Ricklefs, R. E., Marks, H. L. 1984. Insensitivity of brain growth to selection of four-week body mass in Japanese quail. *Evolution* 38: 1180–85

Rodieck, R. W., Brening, R. K. 1983. Retinal ganglion cells: Properties, types, genera, pathways and trans-species comparisons. *Brain Behav. Evol.* 23: 121–64

Romanes, G. J. 1946. Motor localization and the effects of nerve injury on the ventral horn cells of the spinal cord. *J. Anat.* 80: 117–31

Sacher, G. A., Staffeldt, E. F. 1974. Relation of gestation time to brain weight for placental mammals: Implications for the theory of vertebrate growth. *Am. Natur.* 108: 593–615

Salzer, J. L., Bunge, R. P., Glaser, L. 1980. Studies of Schwann cell proliferation. III. Evidence for the surface localization of the neurite mitogen. *J. Cell Biol.* 88: 767–78

Sanes, J. R., Rubenstein, J. L. R., Nicolas, J.-F. 1986. Use of a recombinant retrovirus to study post-implantation cell lineage in mouse embryos. *EMBO J.* 5: 3133–42

Satorre, J., Cano, J., Reinoso-Suárez, F. 1986. Quantitative cellular changes during postnatal development of the rat dorsal lateral geniculate nucleus. *Anat. Embryol.* 174: 321–27

Schoenwolf, G. C. 1985. Shape and bending of the avian neuroepithelium: Morphometric analyses. *Devel. Biol.* 109: 127–39

Shankland, M., Weisblat, D. A. 1984. Stepwise commitment of blast cell fates during the positional specification of the O and P cell lines in the leech embryo. *Devel. Biol.* 106: 326–42

Shariff, G. A. 1953. Cell counts in the primate cerebral cortex. *J. Comp. Neurol.* 98: 381–400

Sholl, D. 1948. The quantitative investigation of the vertebrate brain and the applicability of allometric formulae to its study. *Proc. R. Soc. London* 135: 243–58

Shook, B. L., Maffei, L., Chalupa, L. M. 1984. Functional organization of the cat's visual cortex after prenatal interruption of binocular interactions. *Proc. Natl. Acad. Sci. USA* 82: 3901–5

Sidman, R. L. 1968. Development of interneuronal connections in brains of mutant mice. In *Physiological and Biochemical Aspects of Nervous Integration*, ed. F. D. Carlson, pp. 163–93. Englewood Cliffs, NJ: Prentice Hall

Silver, J. 1978. Cell death during development of the nervous system. In *Handbook of Sensory Physiology*, ed. M. Jacobson, 9: 419–36. New York: Springer-Verlag

Sonmez, E., Herrup, K. 1984. Role of *staggerer* gene in determining cell number in cerebellar cortex. II. Granule cell death and persistence of the external granule cell layer in young mouse chimeras. *Devel. Brain Res.* 12: 271–83

Spitzka, E. A. 1903. Brain weights of animals with special reference to the weight of the brain in macaque monkey. *J. Comp. Neurol.* 13: 9–17

Stephan, H. 1958. Vergleichend-anatomische Untersuchungen an Insektivorengehirnen. *Morphol. Jahrb.* 99: 853–80

Sternberg, P. W., Horvitz, H. R. 1984. The genetic control of cell lineage during nematode development. *Ann. Rev. Genet.* 18: 489–524

Stewart, R. R., Spergel, D., Macagno, E. R. 1986. Segmental differentiation in the leech nervous system. The genesis of cell number in the segmental ganglia of *Haemopis marmorata. J. Comp. Neurol.* 253: 253–59

Stewart, R. R., Gao, W.-Q., Peinado, A., Zipser, B., Macagno, E. R. 1987. Cell death during gangliogenesis in the leech: Bipolar cells appear and then degenerate in all ganglia. *J. Neurosci.* 7: 1919–27

Stone, J. 1983. *Parallel Processing in the Visual System. The Classification of Retinal Ganglion Cells and its Impact on the Neurobiology of Vision.* New York: Plenum

Storrs, E. E., Williams, R. J. 1968. A study of monozygous quadruplet armadillos in relation to mammalian inheritance. *Proc. Natl. Acad. Sci. USA* 60: 910–14

Studnicka, F. K. 1905. Die Parietalorgane. In *Lehrbuch der vergleichende mikroskopischen Anatomie der Wirbeltiere*, Vol. 5, ed. A. Oppel. Jena: Fischer

Sulston, J. E. 1976. Post-embryonic development in the ventral cord of *Caenorhabditis elegans. Philos. Trans. R. Soc. London Ser. B* 275: 287–97

Sulston, J. E., Horvitz, H. R. 1977. Post-embryonic cell lineages of the nematode *Caenorhabditis elegans. Devel. Biol.* 56: 110–56

Tanaka, H., Landmesser, L. T. 1986a. Interspecies selective motoneuron projection patterns in chick-quail chimeras. *J. Neurosci.* 6: 2880–88

Tanaka, H., Landmesser, L. T. 1986b. Cell death of lumbosacral motoneurons in chick, quail, and chick-quail chimera embryos. A test of the quantitative matching hypothesis of neuronal cell death. *J. Neurosci.* 6: 2889–99

Tåning, Å. V. 1950. Influence of the environment on number of vertebrae in teleostean fishes. *Nature* 165: 28

Taylor, B. A. 1971. Genetic relationships between inbred strains of mice. *J. Heredity* 63: 83–86

Temple, S., Raff, M. C. 1986. Clonal analysis of oligodendrocyte development in culture: Evidence for a developmental clock that counts cell divisions. *Cell* 44: 773–79

Thoenen, H., Barde, Y. A. 1980. Physiology of nerve growth factor. *Physiol. Rev.* 60: 1284–1335

Thoenen, H., Korsching, S., Heumann, R., Acheson, A. 1985. Nerve growth factor. *Ciba Found. Symp.* 116: 113–28

Tower, D. B. 1954. Structural and functional organization of mammalian cerebral cortex: The correlation of neurone density with brain size. *J. Comp. Neurol.* 101: 19–52

Torrence, S. A. 1983. *Ascidian larval nervous system: Anatomy, ultrastructure, and metamorphosis.* PhD thesis, Univ. Washington, Seattle

Trent, C., Tsung, N., Horvitz, H. R. 1983. Egg-laying defective mutants of the nematode *Caenorhabditis elegans. Genetics* 104: 619–47

Truman, J. W. 1983. Programmed cell death in the nervous system of an adult insect. *J. Comp. Neurol.* 216: 445–52

Truman, J. W. 1984. Cell death in invertebrate nervous systems. *Ann. Rev. Neurosci.* 7: 171–88

Turner, D. L., Cepko, C. L. 1987. A common progenitor for neurons and glia persists in rat retina late in development. *Nature* 328: 131–36

van Essen, D. C., Newsome, W. T., Maunsell, J. H. R. 1984. The visual field representation in striate cortex of the macaque monkey: Asymmetries, anisotropies, and individual variability. *Vision Res.* 24: 426–48

van Essen, D. C., Newsome, W. T., Maunsell, J. H. R., Bixby, J. L. 1986. The projections from striate cortex (V1) to areas V2 and V3 in the macaque monkey: Asymmetries, areal boundaries and patchy connections. *J. Comp. Neurol.* 244: 451–80

Vernon, J. A., Butsch, J. 1957. Effect of tetraploidy on learning and retention in the salamander. *Science* 125: 1033–34

von Szily, A. 1912. Über die einleitenden Vorgänge bei der ersten Entstehung der Nervenfasern im N. opticus. *Albrecht von Graefe's Archiv. f. Ophthal.* 81: 67–86, plates 5, 6

Waddington, C. J. 1942. Canalization of development and the inheritance of acquired characters. *Nature* 150: 563–65

Ware, R. W., Clark, D., Crossland, K., Russell, R. L. 1975. The nerve ring of the nematode *Caenorhabditis elegans*: Sensory input and motor output. *J. Comp. Neurol.* 162: 71–110

Weisblat, D. A., Sawyer, R. T., Stent, G. S. 1978. Cell lineage analysis by intracellular injection of a tracer enzyme. *Science* 202: 1295–98

Weisblat, D. A., Blair, S. S. 1982. Cell lineage in leech neurogenesis during normal development and after the ablation of identified blastomeres. *NRP Bull.* 20: 783–93

Wetts, R., Herrup, K. 1982. Cerebellar Purkinje cells are descended from a small number of progenitors committed during early development: Quantitative analysis of lurcher chimeric mice. *J. Neurosci.* 2: 1494–98

Wetts, R., Herrup, K. 1983. Direct correlation between Purkinje and granule cell number in the cerebella of lurcher chimeras and wild-type mice. *Dev. Brain Res.* 10: 41–47

White, J., Southgate, E., Thomson, J. N., Brenner, S. 1976. The structure of the ventral nerve cord of *Caenorhabditis elegans*. *Philos. Trans. R. Soc. London Ser. B* 275: 327–48

Whitington, P. M., Bate, M., Seifert, E., Ridge, K., Goodman, C. S. 1982. Survival and differentiation of identified embryonic neurons in the absence of their target muscles. *Science* 215: 973–75

Williams, R. W., Bastiani, M. J., Lia, B., Chalupa, L. M. 1986. Growth cones, dying axons, and developmental fluctuations in the fiber population of the cat's optic nerve. *J. Comp. Neurol.* 246: 32–69

Williams, R. W., Rakic, P. 1986. Pronounced architectonic differences between monocular and binocular segments of the monkeys striate cortex. *Soc. Neurosci. Abstr.* 12: 1498

Williams, R. W., Rakic, P. 1988a. Elimination of neurons from the lateral geniculate nucleus of rhesus monkeys during development. *J. Comp. Neurol.* In press

Williams, R. W., Rakic, P. 1988b. Direct three-dimensional counting: An inherently accurate method to estimate cell number in sectioned material. Submitted for publication

Williams, R. W., Ryder, K., Rakic, P. 1987. Emergence of cytoarchitectonic differences between areas 17 and 18 in the developing rhesus monkey. *Soc. Neurosci. Abstr.* 13: 1044

Wilson, M. A. 1971. Optic nerve fibre counts and retinal ganglion cell counts during development of *Xenopus laevis* (Daudin). *Q. J. Exp. Physiol.* 56: 83–91

Wimer, R. E., Wimer, C. C., Vaughn, J. E., Barber, R. P., Balvanz, B. A., Chernow, C. R. 1976. The genetic organization of neuron number in Ammon's horns of house mice. *Brain Res.* 118: 219–43

Wingert, F. 1969. Biometrische Analyse der Wachstumsfunktionen von Hirnteilen und Körpergewicht der Albinomaus. *J. Hirnforsch.* 11: 133–97

Winklbauer, R., Hausen, P. 1983. Development of the lateral line system in *Xenopus laevis*. III. Cell multiplication and organ formation in the supraorbital system. *J. Embryol. Exp. Morphol.* 76: 283–96

Wright, S. 1978. *Evolution and the Genetics of Populations, Vol. 4: Variability Within and Among Natural Populations.* Chicago: Univ. Chicago Press

Yablokov, A. V. 1974. *Variability of Mammals.* New Delhi: Amerind

Young, J. Z. 1971. *The Anatomy of the Nervous System of Octupus vulgaris.* Oxford: Clarendon

Zamenhof, S. 1941. Stimulation of the proliferation of neurons by the growth hormone. I. Experiments on tadpoles. *Growth* 5: 123–39

Zamenhof, S. 1942. Stimulation of cortical cell proliferation by the growth hormone. III. Experiments on albino rats. *Physiol. Zool.* 15: 281–92

Zamenhof, S., Mosley, J., Schuller, E. 1966. Stimulation of the proliferation of cortical neurons by prenatal treatment with growth hormone. *Science* 152: 1396–97

Zamenhof, S., Mosley, J., Schuller, E. 1971. Prenatal cerebral development: Effects of restricted diet, reversal by growth hormone. *Science* 174: 954–55

Ann. Rev. Neurosci. 1988. 11:455–95

PROBING THE MOLECULAR STRUCTURE OF THE VOLTAGE-DEPENDENT SODIUM CHANNEL

Robert L. Barchi

David Mahoney Institute of Neurological Sciences, University of
Pennsylvania School of Medicine, Philadelphia, Pennsylvania 19104

INTRODUCTION

A voltage-sensitive sodium channel plays a fundamental role in the generation and propagation of action potentials in most multicellular organisms. This transmembrane protein controls a time- and voltage-dependent conductance to sodium ions that produces the rapid depolarization in membrane potential characterizing the beginning of an action potential.

Because of its central role in signaling and information transfer, the sodium channel has traditionally been the subject of intensive electrophysiological and biophysical research. During the past ten years, advances in biochemistry and molecular biology have led to rapid progress in the molecular characterization of this ion channel, and it is this aspect of sodium channel research that is reviewed in detail here. Since the biochemistry of the sodium channel has been the topic of a number of review articles in the past (c.f. Agnew 1984, Barchi 1982, Catterall 1986), we concentrate here on work carried out since 1982.

OVERVIEW OF SODIUM CHANNEL PHYSIOLOGY

Although the voltage-dependent sodium channel is well known to most neuroscientists, a brief overview of its physiology will serve as an orientation for discussions that follow. Sodium channel physiology has been treated extensively elsewhere and the reader is referred to recent reviews for specific

455

0147–006X/88/0301–0455$02.00

citations to the older literature (Bezanilla 1985, Horn 1984, Yamamoto 1985).

The central role of membrane sodium conductance in the production of an action potential was first shown in the pioneering work of Hodgkin & Huxley (1952). A voltage-dependent increase in sodium channel conductance, with its associated inward sodium current, initiates the transition in membrane potential from a point near the potassium equilibrium potential (-70 to -90 mV in most cells) through zero toward the sodium equilibrium potential ($+30$ to $+50$ mV). This activation process is transient, lasting only a few milliseconds. With progressive depolarization and time, a second process leading to channel inactivation and a reduction in membrane sodium conductance occurs, and the membrane repolarizes as potassium permeability once again predominates. In many excitable membranes, this repolarization process is augmented by the delayed activation of a second channel specific for K^+ ions.

Research using voltage clamp technique in a variety of systems confirmed and refined the basic observations on sodium channel kinetics provided by Hodgkin & Huxley. The smoothly increasing and decreasing conductance curves derived as a function of time with this approach led to a conception of sodium channels as pathways through the membrane that gradually opened with depolarization and then progressively closed again as inactivation set in. Patch clamping, with its ability to resolve current flow through individual ion channels, dramatically changed that view (see Horn 1984). Sodium channels are now known to undergo abrupt transitions from one or more closed states to an open state with a characteristic conductance of about 20 pS (Figure 1). After remaining open for a variable period, the channel abruptly closes, usually not to be seen again until the membrane is taken through a repolarization cycle. After depolarization, delays in closed states leading to channel opening appear to be rate limiting. Once opened, however, channels remain open only briefly relative to the overall duration of macroscopic current flow, and close rapidly to the inactivated state. The time course of inactivation observed in the macroscopic currents reflects predominantly the variability in latency to first opening for the channels in a membrane rather than variability in the duration of channel opening (Aldrich 1986).

Electrophysiological measurements have served to emphasize the strong similarities in kinetic behavior of voltage-dependent sodium channels across the phylogenetic tree, while at the same time cataloging the subtle differences in these properties that make channels from different species unique. One of the major challenges of the biochemist and molecular biologist is to define those aspects of the channel's molecular structure that provide its common but highly complex functional features while also

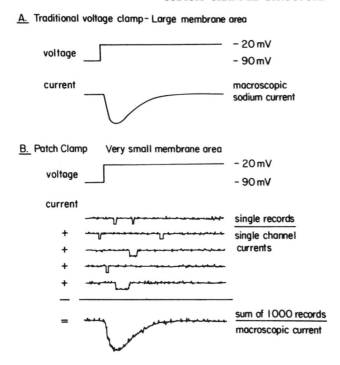

Figure 1 (A) Traditional voltage clamp analysis of sodium channels from nerve and muscle typically demonstrates inward currents that activate progressively after depolarization, reach a peak within a millisecond or so, and then slowly decline toward baseline as inactivation develops. (B) At the level of single ion channels accessible with patch clamp technology, this smoothly increasing and decreasing current is seen to be the result of the statistical averaging of currents through individual sodium channels, each of which abruptly turns on and off with a variable latency after depolarization. Summing thousands of single channel records for a given depolarization step reproduces the form of the macroscopic current.

elucidating those aspects of its architecture that make two types of sodium channel in the same tissue unique.

NEUROTOXINS AND THE SODIUM CHANNEL

In keeping with its central role in information processing, the sodium channel has become the target of a variety of lethal neurotoxins. These toxins have proven valuable in the molecular characterization of the channel protein, since they provide a means of identifying the protein after solubilization and of modulating its conductance state after reconstitution. Sodium channel neurotoxins have been presented in detail elsewhere; for more extensive treatment the reader is referred to the excellent articles by Narahashi (1974), Ritchie & Rogart (1977), Catterall (1980), Lazdunski & Renaud (1982), and Khodorov (1985).

Sodium channel neurotoxins are classified on the basis of the binding sites they occupy on the channel protein. At present, six such binding sites are commonly recognized. Site 1 binds the small, polar, heterocyclic guanidines, tetrodotoxin (TTX) and saxitoxin (STX). A polypeptide toxin from *Conus geographicus* binds at an overlapping site. Toxins binding at site 1 work only from the outside of the membrane, and prevent ion flow through the channel by blocking access to the outer mouth of the pore. Radiolabeled TTX and STX have been especially useful as markers of the sodium channel because of their high affinity for this site (1–10 nM).

Site 2 recognizes a number of lipid-soluble toxins, including batra-chotoxin, veratridine, aconitine, and grayanotoxin. These toxins shift the voltage dependence of channel activation in a hyperpolarizing direction and prevent channel inactivation; this results in persistent channel activation even at normal membrane potentials. These toxins have proven useful for activating purified sodium channels after reconstitution.

A third channel site binds polypeptide toxins from scorpions and sea anemonae that slow or block channel inactivation. Allosteric interactions can be demonstrated between their binding and the binding of toxins at site 2. Their binding is typically voltage-dependent.

Site 4 binds another class of scorpion toxins, designated beta toxins, which affect mainly channel activation. Their binding does not interact with ligands at any of the other sites. Two additional sites, 5 and 6, have recently been defined largely on the basis of single toxins whose properties do not fit into the preceding four categories. Brevetoxin markedly enhances the effects of toxins that bind at site 2, but does not compete for binding with polypeptide toxins at site 3 (Catterall & Gainer 1985, Poli et al 1986); its binding site is designated site 5. Toxins from the coral *Goniopora* markedly slow channel inactivation but again do not compete for binding at site 3 (Gonoi et al 1986). Their binding site is now designated site 6.

PURIFICATION OF SODIUM CHANNEL PROTEINS

Although high-affinity binding sites for tetrodotoxin were solubilized from eel electroplax membranes as early as 1972 (Henderson & Wang 1972, Benzer & Raftery 1973), the first successful purification of the channel protein, identified by high-affinity binding of TTX, was reported by Agnew et al in 1978. In this and most subsequent purification reports, solu-bilization of the channel was accomplished with non-ionic detergents such as Lubrol PX, Triton X-100, or NP-40. Stabilization of high-affinity toxin binding after solubilization, a major problem in early attempts at channel purification, was accomplished by the addition of exogenous phospholipid to all detergent-containing solutions. The phospholipid requirements for this stabilizing effect are broad, and a range of molar ratios of detergent : phospholipid between 3 : 1 and 10 : 1 can be effective (Agnew & Raftery 1979, Barchi et al 1980), thus suggesting that the effect results from the formation of mixed detergent-phospholipid vesicles whose hydrophobic interiors are particularly favorable for the preservation of channel structure, rather than from the satisfaction of a specific lipid requirement of the channel protein itself.

Purification of the eel channel, and subsequent channel purifications from mammalian muscle and nerve (Barchi et al 1980, Hartshorne & Catterall 1981, Norman et al 1983, Kraner et al 1985) and from cardiac tissue (Lombet et al 1981, Lombet & Lazdunski 1984) took advantage of the positively charged nature of the channel with an initial step of ion exchange chromatography. Further enrichment was obtained by iterative molecular sieve chromatography with the eel, and by lectin affinity chromatography followed by molecular sieving or gradient centrifugation with the mammalian preparations. In most of these preparations, purified material has ultimately been obtained that exhibits binding of [3H]-STX or [3H]-TTX of between 0.4 and 0.9 mol per mol channel protein, indicating that a high degree of purification has been achieved.

More recently, partially denatured preparations of sodium channel from eel (Nakayama et al 1982) and from rat skeletal muscle (Casadei et al 1986) have been obtained directly from tissue homogenates by immunoaffinity chromatography using antibodies raised against the purified channel protein.

Physical Properties of Solubilized Channel Proteins

Physical properties for the rat brain and skeletal muscle channels were initially determined with hydrodynamic studies on solubilized but unpurified preparations in which the channel was identified by its toxin-binding characteristics (Hartshorne et al 1980, Barchi & Murphy 1980). These studies suggested an overall molecular weight for the mixed micelle containing the channel protein, lipid, and detergent of between 560,000 and 600,000. Measured Stokes radii for the channel in this form ranged between 80 and 95 Å, with sedimentation values of 9–12 S. By using known values for the partial specific volume of the detergent and phospholipid, values were calculated for the molecular weight of the channel protein itself of between 287,000 and 314,000 for rat, rabbit and human skeletal muscle and 316,000 for the rat brain channel. Electron inactivation studies that specifically examined the rate of loss of the STX, TTX, or polypeptide binding sites suggested a somewhat smaller limiting size of ~ 260 kDa for the components containing these binding sites (Levinson & Ellory 1972, Barhanin et al 1983b, Angelides et al 1985). Subsequent experiments with purified channel preparations have confirmed these initial measurements. Purified channels exhibit sedimentation coefficients of 8–9 S and apparent Stokes radii of 90–95 Å.

The purified channel exists in a mixed micelle with detergent and phospholipid, with the channel protein accounting for about 40–45% of its mass. The unusually high apparent partial specific volume (~ 0.83 ml/gm) and the large Stokes radius for the solubilized protein reflect the contribution of lipid and detergent to the micelle. The overall size of this mixed micelle is considerably larger than of the protein itself and contributes directly to its behavior on preparative columns and gradients.

Subunit Composition

ALPHA SUBUNITS   After some early controversy, a consensus now appears to be emerging on the subunit composition of the purified sodium channel (Table 1). Although initially reported to contain three subunits, the eel channel is now known to consist of only a single large component of $\sim 280,000$ MW (Agnew et al 1980, Miller et al 1983). Probably as the result of proteolytic nicking during membrane isolation, early preparations of the rat skeletal muscle sodium channel exhibited a smaller glycoprotein component of $\sim 150,000$ MW (Barchi 1983); more recent work, including immunoaffinity purification of this channel protein from crude muscle homogenates (Casadei et al 1986), confirms that it too contains a large 260,000 MW component (Figure 2). Preparations from rat brain, rabbit skeletal muscle, and chick cardiac muscle all demonstrate comparable

Table 1 Subunit composition of purified sodium channels

Source	Subunits	Core Peptide	Ref.
Eel	280 kDa	208 kDa	Agnew et al 1983
Rat brain	260 kDa 36 kDa 33 kDa	220 kDa 23 kDa 21 kDa	Messner & Catterall 1985
Rat skeletal muscle	260 kDa 38 kDa		Casadei et al 1986
Rabbit skeletal muscle	260 kDa 38 kDa	214 kDa 26.5 kDa	Roberts & Barchi 1987
Chick heart	230-270 kDa		Lombet & Lazdunski 1984

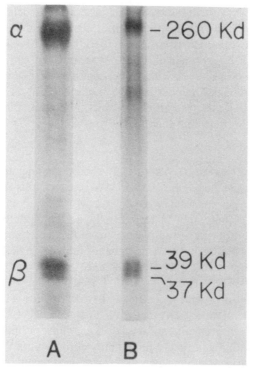

Figure 2 Voltage-dependent sodium channels from rat and rabbit skeletal muscle contain a large 260 kDa α subunit and a smaller 38 kDa β subunit. Comparable results are obtained with channels purified by traditional biochemical means from rabbit muscle (A) or by immunoaffinity column chromatography using monoclonal antibodies directed solely against the α subunit in rat muscle (B). The β subunit occasionally appears as a doublet under reducing conditions on SDS gels. These bands probably represent differential glycosylation, since the core peptides are identical after enzymatic removal of carbohydrate. Detailed analysis of stoichiometry in both systems indicates that each α subunit is associated with only one β subunit (Kraner et al 1985, Casadei et al 1986, Roberts & Barchi 1987).

large α subunits of ~ 260,000 MW (Hartshorne & Catterall 1984, Lombet & Lazdunski 1984, Kraner et al 1985).

The 260 kDa component is glycosylated, as indicated by its ability to bind lectins on blot transfers (Cohen & Barchi 1981, Messner & Catterall 1985) and by its retention on lectin affinity columns (Cohen & Barchi 1981, Moore et al 1982). Microheterogeneity in the extent of glycosylation probably accounts for the diffuse appearance of this band seen on SDS-PAGE (Miller et al 1983).

Ferguson analysis indicates that the anomalous migratory behavior on

SDS-PAGE seen with all these 260 kDa proteins is associated with a high free solution mobility for the protein after denaturation in SDS (Miller et al 1983, Barchi 1983, Hartshorne & Catterall 1984, Messner & Catterall 1985). Although this could in part be due to the presence of a large number of sialic acid residues, work with the eel channel suggests that the protein may bind an unusual number of SDS molecules per unit protein mass (Miller et al 1983). The presence of covalently bound lipid moieties may play a role in this detergent binding (Levinson et al 1987).

β SUBUNITS AND CHANNEL FUNCTION   In the eel channel and the channel isolated from chick heart, the 260,000 MW component is the only protein whose purification consistently correlates with the presence of channel toxin binding activity and, at least for the eel, ultimately of functional channel activity. In preparations from rat brain and from rat and rabbit skeletal muscle, however, one or two smaller subunits in the 30,000–40,000 MW range, designated β subunits, are consistently found (Hartshorne & Catterall 1981, 1984, Kraner et al 1985, Casadei et al 1986).

The β subunit is present in a strict 1 : 1 stoichiometry with alpha in rat and rabbit skeletal muscle (Kraner et al 1985, Casadei et al 1986). This is true whether measurements are made on channels purified by standard biochemical methods or by immunoaffinity chromatography. The muscle β subunit copurifies with α when immunoaffinity techniques are used in which the antibody recognizes only the alpha subunit (Casadei et al 1986).

In the rat brain channel, two β subunits are present with a stoichiometry of $1 : 1 : 1$ $\alpha : \beta_1 : \beta_2$. β_2 is covalently attached to alpha through a disulfide link (Hartshorne et al 1982). Affinity-labeled toxins have been used to support the intimate relationship between the β subunits and α in the rat brain channel. Photoactivatible derivatives of the toxin from the scorpion, *Lieurus quinquistriatus*, will specifically label both α and β_1 in synaptosomes, and the relative labeling of the two varies with the position of the activatible residue on the toxin (Beneski & Catterall 1980, Sharkey et al 1984). Aryl azide derivatives of the beta scorpion toxin from *Centruroides suffusus suffusus* will also label both alpha and β_1 in synaptosomes (Darbon et al 1983). Furthermore, labeling of both alpha and β_1 with *Lieurus* derivatives can be demonstrated in fresh crude homogenates of brain in the presence of a variety of protease inhibitors, thus supporting the contention that β does not derive from α by proteolysis (Sharkey et al 1984).

Although the specific role of the β subunits remains enigmatic, additional information has recently become available detailing the interactions of these subunits with α in the rat brain. Messner & Catterall (1986) have demonstrated that β_1 can be dissociated from the solubilized α-β_1-β_2 complex by treatment with 1 M $MgCl_2$ and subsequent sucrose gradient centrifugation. High-affinity STX binding was lost in proportion to the amount of β_1 dissociated. Both the dissociation of β_1 and the loss of STX binding could be prevented by carrying out the $MgCl_2$ incubation in the presence of TTX.

Conversely, β_2 could be removed by treatment of the solubilized channel with dithiothreotol in the presence of TTX. The loss of the covalently attached β_2 did not result in the parallel loss of STX binding (Messner & Catterall 1986), and the α-β_1 complex alone is capable of functional reconstitution into lipid vesicles (Catterall 1986).

Recently, it has been demonstrated that the α subunit of the rat brain sodium channel contains all the molecular machinery necessary to produce a functional sodium channel (Noda et al 1986b, Stühmer et al 1987). Functional rat brain sodium channels were expressed in oocytes by using mRNA generated by transcription of the cDNA encoding the sequence for rat brain α II. This system, which contained no exogenous information for small β-like subunits, demonstrated voltage-activated sodium currents with spontaneous inactivation; these currents were blocked by TTX with an approximate K_i of 14 nM. Subsequently, Goldin et al (1986) reached a similar conclusion using mRNA for the rat brain α subunit selected by cross-hybridization with antisense cDNA prepared from clones containing various fragments of the channel α subunit message. Taken in conjunction with the reconstitution work reviewed above on the purified 260 kDa protein from eel electroplax, these studies provide strong evidence for the hypothesis that the α subunit alone is capable of forming a functional sodium channel.

Chemical Analysis of Subunits

The 260 kDa component of the eel channel and the α and β subunits of the rat brain and rabbit skeletal muscle channels have been obtained preparatively and subjected to microchemical analysis. In general, the chemical properties of the corresponding subunits from each preparation are very similar.

Amino acid analyses of the 260 kDa proteins from eel, rat brain, and rabbit skeletal muscle contain no surprises (Miller et al 1983, Grishin et al 1984, Elmer et al 1985, Roberts & Barchi 1987). The three proteins appear similar in their relative content of various amino acids, vary little from compositions reported for many soluble proteins, and do not contain an unusually high concentration of hydrophobic amino acids.

Direct determination of the carbohydrate composition of these large proteins bears out the predictions of earlier studies with lectin binding. All three contain more than 20% (w/w) carbohydrate (see Table 2). The predominant sugar present was sialic acid, accounting for nearly 50% of the total carbohydrate present. High levels of n-acetylhexosamines were also present, consistent with the presence of processed complex n-linked carbohydrate moieties on these channel proteins.

Enzymatic deglycosylation of the rat brain and rabbit skeletal muscle α subunits results in an increase in their migration on SDS-PAGE to positions consistent with core peptide molecular weights of 220 kDa and 209 kDa, respectively (Messner & Catterall 1985, Roberts & Barchi 1987).

Levinson et al (1987) have shown that the eel sodium channel contains an unusually large amount of bound lipid. After exhaustive dialysis of the purified protein in SDS followed by extraction of residual associated lipid with chloroform/methanol, an additional 3 to 5% by weight lipid remained, apparently covalently attached to the protein. This lipid could be removed by transesterification in acid methanol, and proved to be predominantly palmitate and stearate.

Measurements of circular dichroism have been made on preparations of sodium channel from rat brain that contained only the α subunit on SDS-PAGE (Elmer et al 1985). The reconstituted preparation exhibited a spectrum reflecting a high α-helical content of approx. 65%. In the solubilized form, however, apparent α-helical content declined to 32% with a concomitant rise in β sheet and random coil configurations. Both preparations retained high-affinity TTX binding.

Amino acid analyses of the isolated β subunits from rat brain and rabbit muscle channels are also unremarkable (Grishin et al 1984, Roberts & Barchi 1987). Both of the rat brain β subunits and the β subunit from rabbit skeletal muscle are heavily glycosylated, each containing more than 30% by weight carbohydrate (Table 2). The predominant carbohydrates are again sialic acid and n-acetylhexhosamines, although the sialic acid content is lower than that found in the α subunits.

Enzymatic deglycosylation of the β subunits can be carried out more easily than with the α subunits. Stepwise increases in mobility of the protein on SDS-PAGE are found with increasing time of incubation with endoglycosidase-H. Core peptide molecular weights of 23 kDa and 21 kDa are reported for the β_1 and β_2 subunits of the brain channel, respectively (Messner & Catterall 1985), whereas the rabbit muscle channel β subunit is reduced to 26.5 kDa (Roberts & Barchi 1987).

Limited peptide mapping has been carried out on the rat brain β subunits (Messner & Catterall 1985). Cleavage with a variety of proteases produces different maps with the two subunits, a finding that suggests they are not closely homologous.

SODIUM CHANNEL STRUCTURE 465

Table 2 Carbohydrate analysis of sodium channel subunits

260 kDa α subunit

	Rabbit muscle[1]		Rat brain[2]		Eel electroplax[3]	
	% Total carb.	% Total weight	% Total carb.	% Total weight	% Total carb.	% Total weight
Fucose	1.5 +/− 0.4	0.4	4.3	0.9	1.5	0.5
Mannose	15.3 +/− 1.3	4.1	13.5	2.8	8.3	2.4
Galactose	17.6 +/− 1.1	4.6	13.8	2.8	5.2	1.5
N-acetyl hexosamine	21.4 +/− 2.2	5.7	22.3	4.6	45.3	13.3
Sialic acid	44.1 +/− 1.9	11.7	46.1	9.5	39.7	11.8
Total		26.5 +/−1.5		20.5		29.5

38 kDa β subunit

	Rabbit muscle[1]		Rat brain[2]		Eel electroplax[3]	
Fucose	trace	trace	3.3	1.0		
Mannose	4.1 +/− 1.9	1.2	22.3	6.4		
Galactose	7.5 +/− 2.4	2.2	13.8	4.0		
N-acetyl hexosamine	56.1 +/− 8.7	16.7	23.4	6.8		
Sialic acid	32.2 +/− 6.7	9.6	37.2	10.8		
Total		29.7 +/−1.9		28.9		

[1] From Roberts & Barchi (1986).
[2] From Grishin et al (1984).
[3] From Miller et al (1983).

RECONSTITUTION OF PURIFIED SODIUM CHANNELS

The voltage-sensitive sodium channel has a number of biophysical and biochemical properties that uniquely define it among membrane ion channels. While neurotoxin binding is one such characteristic, the presence of toxin binding alone does not provide evidence for channel function. The purified protein should also retain the ability to select for sodium among monovalent cations, should exhibit a characteristic maximal rate of ion flow or single channel conductance, and should be capable of characteristic voltage-dependent transitions between conducting and nonconducting states. For the purified channel protein, demonstration of these properties requires its reconstitution into lipid vesicles or planar bilayers. Successful reconstitution has now been demonstrated for purified sodium channels from rat and rabbit skeletal muscle, rat brain, and eel electroplax (for reviews, see Tanaka et al 1986, Hartshorne et al 1986, and Agnew et al 1986, respectively).

Lipid Vesicles and Cation Fluxes

The general approach followed by most groups for the reconstitution of purified sodium channels into lipid vesicles involves supplementation of the purified protein with excess phospholipid (usually egg phosphatidylcholine) and removal of detergent by incubation with polystyrene beads. This usually produces a bimodal population of vesicles, about half of which are large enough (> 1000 Å) to allow the measurement of cation influx.

After reconstitution, sodium channels regain the thermal stability lost when they were first solubilized (Weigele & Barchi 1982, Talvenheimo et al 1982, Rosenberg et al 1984a). Measurements of STX binding indicate that the channels incorporate into vesicles randomly, with about half of the channels in the normal outward facing orientation. Specific sodium influx can be activated by batrachotoxin or veratridine at concentrations corresponding to those activating the channel in its native membrane environment (Figure 3); half of this flux is blocked by externally applied TTX; the remainder is blocked when TTX is present both on the inside and the outside of the vesicles, consistent with the random orientation of channels in these vesicles.

Purified channels incorporated into egg phosphatidylcholine vesicles are blocked by local anesthetics (Talvenheimo et al 1982, Rosenberg et al 1984a), but they usually lose their capacity to bind α scorpion toxins in their characteristic voltage-dependent manner. Tamkun et al (1984) have shown that *Leiurus* toxin binding can be restored when rat brain lipids are

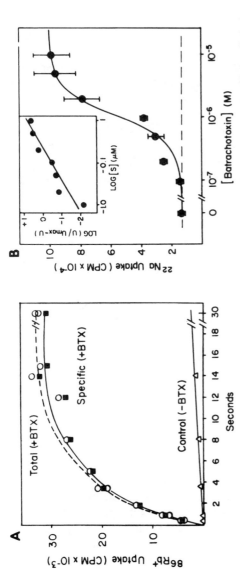

Figure 3 (A) Influx of ^{86}Rb$^+$ into vesicles containing the purified sodium channel from rabbit skeletal muscle. The purified channel was reconstituted into egg phosphatidylcholine vesicles and incubated for 45 min at 36°C with 5 μM batrachotoxin (BTX) dissolved in a small amount of ethanol (*open circles*) or with an equivalent amount of solvent alone (*open triangles*). Specific BTX-activated influx (*solid squares*), defined as total minus control influx, occurred with a half-time of about 3 sec in this example. Fluxes were measured after equilibration of the vesicles at 22°C. (B) Dose-dependent activation of the purified reconstituted channel by BTX. Vesicles containing the purified channel were incubated for 45 min at 36°C with the indicated concentrations of BTX or with an equivalent amount of carrier solvent alone. Influx of ^{22}Na$^+$ was then measured after 5 sec, a period sufficient for all vesicles containing open channels to equilibrate with this cation. (*Insert*) Hill plot of the same data indicating an apparent K_d of 2 μM and the Hill coefficient of 1.2. (Figures adapted from Tanaka et al 1986.)

used for reconstitution. Full voltage-sensitive scorpion toxin binding is also retained by channels reconstituted into a mixture of phosphatidylcholine and phosphatidylethanolamine (65%/35%) (Feller et al 1985).

Cation selectivity has been measured directly in reconstituted preparations of rat and rabbit skeletal muscle channels activated by batrachotoxin by using a quenched flow system to extend temporal resolution to < 100 msec. Both these purified channels demonstrate cation selectivity sequences of $Na^+ > K^+ > Rb^+ > Cs^+$, with relative selectivity among cations comparable to that measured in native channel preparations (Tanaka et al 1983, Kraner et al 1985) (Figure 4). A selectivity sequence of $Na^+ > Rb^+ > Cs^+$ was also reported for purified rat brain channels activated by veratridine (Tamkun et al 1984).

Although batrachotoxin-activated sodium channels do not inactivate, they do exhibit voltage-dependent activation (Huang et al 1982). Saturating concentrations of batrachotoxin shift the activation curve 30 mv in the hyperpolarizing direction. This voltage-dependent activation has been demonstrated in vesicles containing purified sodium channels by using a K^+ concentration gradient and valinomycin to control the transmembrane potential (Furman et al 1986). With rabbit skeletal muscle sodium channels reconstituted into PC vesicles, all channels that could be activated by batrachotoxin and blocked by TTX could also be closed by hyperpolarization. The range of potentials over which increasing activation was seen in these vesicles corresponds to that associated with voltage-dependent activation of batrachotoxin-treated channels in normal membranes.

Reconstitution into Planar Lipid Bilayers

Sodium channel activity normally involves voltage-dependent channel openings that occur on a millisecond timescale. Although voltage-dependent activation can be inferred from population measurements with channels reconstituted into lipid vesicles, direct measurement requires systems in which the activity of single sodium channels can be observed (Furman et al 1986). Either patch-clamping of vesicles containing the purified channels or incorporation of these channels into planar lipid bilayers can provide this resolution. The major limitation of both methods is the extremely small percentage of the reconstituted channel population that is sampled; thus caution is required in generalizing observed single channel behavior as representative of the entire population of purified channel proteins.

Rosenberg et al (1984b) first reported single-channel measurements on rat brain sodium channels by patch-clamping blebs from multilammelar

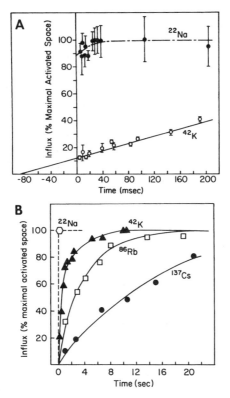

Figure 4   Quenched flow measurements of $^{22}Na^+$ and $^{42}K^+$ influx in BTX-stimulated vesicles at short incubation periods. The data points represent specific BTX-stimulated uptake corrected for control uptake. Each point is the mean $+/-$ S.D. of three or more separate experiments. The effective quenching time for the apparatus was determined by extrapolation of the initial linear uptake rate for $^{42}K^+$ into the BTX-activated vesicles (*open circles*), yielding a value of 90 msec. $^{22}Na^+$ influx on this time scale (*solid circles*) was very rapid, and an initial rate could not be resolved; an upper limit of 50 msec for the half-time for $^{22}Na^+$ can be estimated based on the measured quenching time for the system. (B) The time course for $^{42}K^+$, $^{86}Rb^+$, and $^{137}Cs^+$ uptake into BTX-activated vesicles as determined with the quenched flow apparatus. Specific uptake data from several preparations, corrected for control uptake and plotted as a percent maximal activated space, are shown for $^{42}K^+$ (*solid triangles*), $^{86}Rb^+$ (*open squares*), and $^{137}Cs^+$ (*solid circles*). The maximal space accessible through activated sodium channels was the same for all cations in any given preparation. Maximal $^{22}Na^+$ uptake was already observed at the shortest interval shown in this graph (*open six-sided figures*). All measurements were made at 18°C. Assuming a maximal half-time for sodium uptake of 50 msec, these data yield cation uptake ratios for $^{22}Na^+ : ^{42}K^+ : ^{86}Rb^+ : ^{137}Cs^+$ of $1 : 0.14 : 0.02 : 0.005$. (Adapted from Tanaka et al 1983.)

vesicles containing the purified protein. These investigators recorded voltage-dependent single channel events in the absence of activating neurotoxins that had many of the properties of native sodium channels. This single-channel activity had a conductance of $\sim$ 11 pS in 95 mM Na^+, a mean open time of $\sim$ 1.9 msec, and a seven-fold selectivity for Na^+ over K^+. Using a similar approach with batrachotoxin-activated channels, other investigators recorded voltage-dependent but non-inactivating single channel events from vesicles containing the purified rat skeletal muscle channel with a single channel conductance of 15–18 pS in 100 mM Na^+ and channel openings of up to 50 msec (Barchi et al 1984).

A number of groups have reported the successful incorporation of purified sodium channels into planar lipid bilayers. In the absence of activating toxins, Hanke et al (1984) observed large inactivating voltage-dependent events of $\sim$ 150 pS after insertion of vesicles containing purified rat brain sodium channel into solvent-free bilayers. This channel type was also observed with purified preparations of eel sodium channel. With the brain channel, a second, non-inactivating, class of channels with a conductance of $\sim$ 25 pS was also seen by this group, although the voltage-dependence of channel activation was quite different from that exhibited by sodium channels prior to isolation.

Hartshorne et al (1985), again using purified rat brain channel protein, detected single channel events in bilayers in the presence of batrachotoxin that had properties very similar to those reported for the native channel. Single channel openings were voltage-dependent, with the half-probability of channel opening occurring at -91 mV and open time increasing with depolarization. The relationship between channel activation and voltage indicated an average gating charge of $\sim$ 4. A single channel conductance of 25 pS was determined in 500 mM Na^+. More detailed analysis of the kinetics of channel activation have recently been reported, and again the purified channel exhibits behavior very similar to that observed with unpurified sodium channels (Keller et al 1986).

The eel channel has also been examined in solvent-containing planar bilayers. Results again differed from those reported by Hanke et al. In the presence of batrachotoxin, single channel events with a conductance of 23.7 pS in 500 mM Na^+ and 18.9 pS in 100 mM Na^+ were observed that activated with depolarization (Levinson et al 1987). Average half-maximal activation occurred at -77 mV. The reconstituted channel had a 4.5-fold preference for Na^+ over K^+. No large conductance events were seen.

Purified sodium channels from rabbit skeletal muscle have also been reconstituted into planar bilayers and their single channel properties examined after activation with batrachotoxin (Furman et al 1986) (Figure 5). These channels also exhibit steep voltage-dependent activation with half-

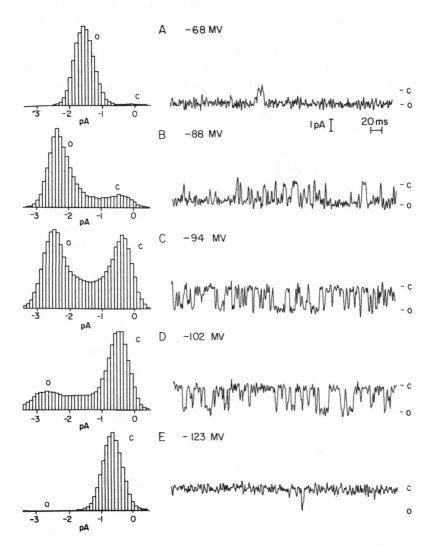

Figure 5 Single channel currents and amplitude histograms from a purified rabbit sodium channel reconstituted into a planar bilayer, demonstrating the voltage-dependence of channel gating. Hyperpolarization (A–E) progressively shifts the channel from the open (O) to the closed state (C). This channel exhibited 50% open activity at about −95 mv and had a single channel conductance of 22 pS after activation by BTX (1 μM) with 500 mM NaCl in the *cis* chamber and 200 mM NaCl in the *trans* chamber. (From Furman et al 1986.)

maximal opening probabilities at potentials between -95 and -115 mV. The calculated gating charge movement was 3.8, and the single-channel conductance averaged 20 pS in 500 mM Na^+.

One possible reason for the differences noted between the more recent studies of the rat brain and muscle and eel channels and the earlier work of Hanke et al (1984) may be the different lipids used to form the solvent-free bilayers in the latter case as compared to those in the Mueller-Rudin bilayers used by other investigators.

THE MOLECULAR ARCHITECTURE
OF THE SODIUM CHANNEL

Channel Primary Sequence

A major advance in the structural characterization of the sodium channel proteins has been the deduction of the complete primary sequence of the eel sodium channel (Noda et al 1984), as well as the sequences of two related sodium channels and a part of a third from rat brain (Noda et al 1986a), through the cloning of DNA complementary to their respective mRNAs. In each case the nucleotide sequence for a message of ~ 6800 base pairs was obtained with an open reading frame for 1820 amino acids in the eel sequence and 2009 and 2005 amino acids for the two complete brain channel sequences. The molecular weights for the core proteins without carbohydrate calculated from these sequences are 208,321 for the eel channel, and 228,758 and 227,840 for the two completely sequenced brain clones.

Homology matrix comparison of these linear sequences demonstrates the presence of four large internal repeats within the primary structure of each channel (Noda et al 1984, 1986a). Interspecies sequence homology is highest within these domains, while regions between domains exhibit lower homology and contain large stretches that are present in one species but not in the other. This is especially true in the region between the first and second internal repeat domains, where there is a 12-residue segment present in eel but absent in the two brain sequences, followed shortly by two stretches totaling 202 residues that are found in the brain sequences but not in the eel channel.

Partial sequence information has also been reported for a gene encoding a putative sodium channel protein in *Drosophila* (Salkoff et al 1987). A genomic library was screened with a complementary probe to the fourth homologous repeat domain of the rat brain II sodium channel message. Sequences encoding four homologous segments were deduced in the fly that corresponded with remarkable precision to the four domains found in the vertebrate sodium channels (Figure 6). Unique sequences that dis-

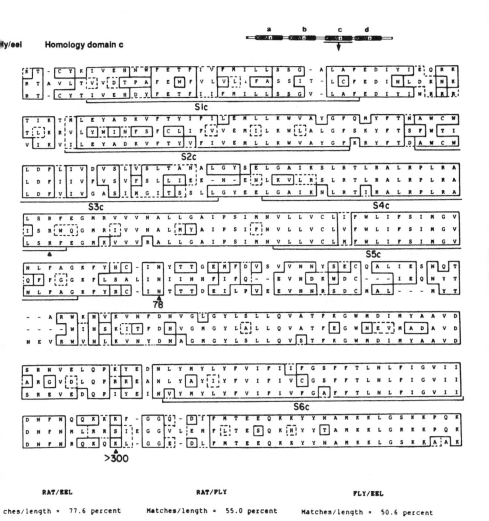

Figure 6 Sequences for each of the four large internal repeat domains found in the voltage-dependent sodium channel show a remarkable degree of conservation among species. Primary sequences for the third internal repeat domain (domain C) of the eel electroplax sodium channel, the rat brain sodium channel, and the putative sodium channel from *Drosophila* are shown here. Areas of complete sequence homology are enclosed in *solid boxes* while conservative substitutions are enclosed with *dotted lines*. This degree of sequence homology is typical of all four repeat domains. With the exception of a conserved link between domains C and D, the stretches of sequence joining the domains are much more divergent. (From Salkoff et al 1987.)

tinguish these homologous domains from each other are absolutely conserved between *Drosophila* and the vertebrate channels, leading the authors to conclude that the sodium channel protein evolved more than 600 million years ago, prior to the divergence of the vertebrate and invertebrate species. In addition, sequences in each domain containing recurrent arginine or lysine residues alternating with two nonpolar residues, which are postulated to be involved in channel gating (S-4 amphipathic helices, see below), are particularly strongly conserved; of a total of 24 positive charges in these regions of the four domains, all are present in identical positions in fly, rat, and eel (see Figure 7).

These observations suggest that the structural features of the channel most important for their common functional characteristics will be contained within the four homologous domains. The extended linear sequence of the sodium channel can be conceived of as four homologous subunits incidently joined within the same primary sequence, a situation functionally similar to the separate homologous subunits that comprise the acetylcholine receptor (for review see Maelicke 1986). It is tempting to speculate that the large insert between domains 1 and 2, unique to the mammalian channel, is related in some way to interactions with the β subunits.

Complete sequence information on the mammalian skeletal muscle sodium channel is not yet available, but recent studies using probes derived from the brain and eel sequences suggest that the muscle message will be closely related. Goldin et al (1986), using cDNA fragments from the rat brain channel message, demonstrated cross-hybridization with a low-abundance 9 kb poly-A mRNA species in rat skeletal muscle. Cooperman et al (1987) have carried out similar experiments with a 1134 base pair probe complementary to residues 1575–1996 of the rat brain II sequence of Noda et al (1986a). This stretch includes the entire fourth internal repeat domain, a region of the primary sequence that is highly conserved between eel and rat. The probe cross-hybridizes with a 9 kb message in rat skeletal muscle, again suggesting that the muscle sodium channel message is about the same size as those from eel and rat brain, with considerable homology with those known sequences.

Nuclease protection studies were carried out with probes derived from genomic clones of the 5′ regions of the rat brain I and II channel sequences. These probes each included stretches of 5′-untranslated regions unique to that channel sequence (Cooperman et al 1987). Although appropriate fragments of both probes were protected by rat brain mRNA, neither was protected by mRNA from innervated or denervated adult rat skeletal muscle, indicating that the sodium channel message(s) expressed in muscle is not the same as either of the principal messages found in the brain.

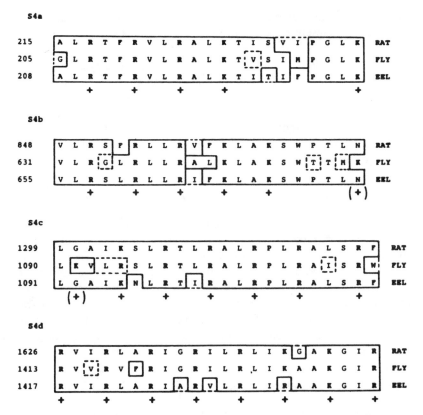

Figure 7 Each of the sodium channel primary sequences that have been reported to date contain four regions in which positively charged residues recur at every third position. These regions, which are capable of forming amphipathic helices and are postulated to be involved in channel voltage-dependent gating, are located in each of the four internal repeat domains. These segments are strongly conserved among species. In most channel models, these stretches of the primary structure are designated the S-4 helices. Here sequences are given for the S-4 helices in each of the repeat domains for the rat brain, *Drosophila*, and eel sodium channels. *Solid lines* enclose areas of complete sequence homology while *dashed lines* indicate conservative substitutions. (From Salkoff et al 1987.)

The recent elucidation of the primary sequence of the dihydropyridine-binding component of the rabbit skeletal muscle calcium channel (Tanabe et al 1987) provides dramatic evidence of the conservation of architecture in voltage-activated ion channels. This large protein, containing 1873 amino acids with a calculated molecular weight of 212,018 without carbohydrate, may well represent a functional calcium channel. It is remarkably similar in size to the sodium channel. It contains four internal repeat domains that closely resemble those found in the sodium channel, and each domain contains an amphipathic helix with lysine or arginine residues at every third position. Sequence homology between this calcium channel protein and the rat brain II sodium channel exceeds 32% in each of the repeat domains, reaching more than 60% if conservative substitutions are included. These two voltage-activated channels, while electrophysiologically quite different, are clearly closely related; the conservation of major structural features in both implies a close coupling between structure and function.

Models of Channel Secondary and Tertiary Structure

The availability of primary sequence information for the sodium channel has lead to a number of models predicting the secondary and tertiary structure of the protein (Noda et al 1984, 1986a, Guy & Seetharamulu 1986, Greenblatt et al 1985, Kosower 1985). Each model was generated on the basis of computer-assisted analysis of the primary sequence by using various approaches that predict secondary structure for short segments within this sequence. Three of the models recognize the presence of the four internally homologous repeat domains in the primary sequence and approach the modeling from the viewpoint of a pseudosymmetric aggregate of these four "subunits." The Kosower model, on the other hand, makes no explicit assumptions regarding these internal repeat domains.

The first model for the sodium channel secondary and tertiary structure was proposed by Noda et al (1984) when the primary sequence was initially presented. Using hydropathy profiles, this group identified six regions, designated S_1 to S_6, within each of four homologous domains, that could potentially form alpha helices. No hydrophobic helical segments were identified outside of these repeat domains.

Two helices, S_5 and S_6, were highly hydrophobic, containing no charged residues in any of the repeats for 24–38 residues; in each case, these nonpolar regions were flanked by charged residues. These helices were assigned a transmembrane orientation. Regions S_1 and S_2 contained several charged residues, but could in general form amphipathic helices; these were also placed within the membrane. The S_4 region was especially interesting in that it contained highly conserved positive residues at every third position in each of the repeat domains (Figure 7). Organized as a 3_{10} helix, a strongly amphipathic structure is formed having a vertical stripe of positive charge with the remaining surface of the cylinder covered with nonpolar residues. The authors felt it unlikely that this positively charged helix would reside within the membrane, and assigned it and helix S_3 a cytoplasmic location.

Since no leader segment was found in the channel message, the N-terminus of the protein was assumed to be cytoplasmic. Subsequent trans-membrane crossings by helices 1, 2, 5, and 6 in each repeat, with helices 3 and 4 on the cytoplasmic membrane surface, form the basic folding pattern for this model. The ion channel would be formed in some manner by the interaction of the weakly amphipathic S_1 and S_2 helices in each domain. This configuration orients 6 of 10 potential n-glycosylation sites on the extracellular surface of the molecule and 8 of 8 potential cAMP-dependent phosphorylation sites on the cytoplasmic side.

Greenblatt et al (1985) subsequently suggested a model that identified two additional alpha helices in each domain, for a total of 8 helices per repeat region. Their model proposed that the amphipathic S_3 helices interact to form the channel lining, and that the positively charged S_4 helices are transmembrane but buried in the interior of each of the pseudo-subunits. Gating involves the movement of the S_4 helix in response to changing potential.

A third model (Kosower 1985) makes no explicit definition of internal repeat domains. A total of 20 transmembrane helical segments are identi-fied throughout the primary sequence, 14 hydrophobic and 6 amphipathic. The stretches of sequence corresponding to the S_4 helices of Noda et al (1984) are assigned a 3_{10} helical conformation and used to form the channel lining along with two other 3_{10} helices containing predominantly negative charges.

The most detailed of the proposed structural models is that of Guy & Seetheramalu (1986). This model is similar to that of Greenblatt et al (1985) in that eight transmembrane segments are postulated in each of four repeat domains. The major difference lies in the assignment of the positively charged S_4 amphipathic helix in each domain as a component of the ion pore. A second short amphipathic helical segment with coun-terbalancing negative charges is identified in domains A, B, and C, while a section of beta-pleated sheet provides balancing charge in domain D. These S_7 helices alternate with S_4 in forming the pore lining. The two extra transmembrane segments proposed in this model are located in stretches of the primary sequence between helices S_5 and S_6 in the original notation of Noda et al (1984).

In this model, gating is envisioned as a screw-like motion of the S_4 helices in one direction and the S_7 segments in the other, such that each positive charge becomes neutralized by a negative side-group one below it on the adjacent helix. With this concept, a 60° rotation of each helix will produce a 4.5 Å translocation of S_4 helices in a direction perpendicular to the plane of the membrane, and will result in the effective net transfer of six charges from one side of the membrane to the other.

This model also allows the location of all eight phosphorylation sites on the cytoplasmic side of the membrane and between 7/10 and 10/10 of the potential glycosylation sites on the external surface, depending on the ultimate conformation of stretches near the C-terminus of the protein.

In their most recent report, Noda et al (1986b) presented a modification of their original structural model. These modifications are based in part on examination of residues that are highly conserved between rat brain and eel. The original six helical segments are identified in each repeat domain, but all are now placed within the membrane. The S_2 helix, which contains highly conserved glutamic acid and lysine residues, is designated as forming the channel lining. As in other models, the S_4 helix is assigned the voltage-sensing function and placed within the interior of each pseudo-subunit in a manner similar to that suggested by Greenblatt et al (1985).

Each of these models is based largely on theoretical analyses of the primary sequence data and on the imaginative insight of the investigators, since very few data are available that can be used to constrain the predicted folding patterns. Although all of these models remain highly speculative, they can now serve as a starting point for further studies probing the topographical organization of the channel in the membrane.

Probes of Sodium Channel Topography

Some insight into the organization of the sodium channel in the membrane can be derived from biochemical studies already reviewed. For example, the α subunit of the brain channel must be a transmembrane protein, since it is accessible to covalent labeling with scorpion toxins and tetrodotoxin derivatives acting only on the external surface of the membrane (Beneski & Catterall 1980, Barhanin et al 1983a, Darbon et al 1983, Sharkey et al 1984, Lombet et al 1983) while at the same time it is susceptible to phosphorylation on the cytoplasmic surface by an internal cAMP-dependent protein kinase (Costa & Catterall 1984a).

Angelides and his colleagues have attempted to map the distance between various neurotoxin binding sites on the channel by measuring the fluorescence energy transfer between toxins labeled with donor or acceptor fluorophores. These workers report a distance of 35 Å between the TTX binding site 1 and the α scorpion site 3 (Angelides & Nutter 1983), 34 Å between site 1 and the β scorpion site 4 (Darbon & Angelides 1984), and 37 Å between site 2, occupied by batrachotoxin, and site 3 (Angelides & Brown 1984). Sites 3 and 4 exhibit blue shifts in probe emission spectra on binding, suggestive of a hydrophobic environment for both. The interpretation of these distances is complicated by intrinsic uncertainties in the absolute location of the fluorophores relative to the point of interaction of the toxins with the channel, and by the lack of a defined location for any one of the sites on the channel structure. However, if site 1 (TTX) is assumed to be on the outer surface of the channel, then the distances would be consistent with a localization of sites 3 and 4 in a hydrophobic region extending partially into the external half of the bilayer.

Other approaches can be used to provide a more detailed analysis of channel topography. Discriminate stretches of primary sequence predicted in various models to be located either on the intracellular or extracellular surface of the membrane can be identified and oligopeptides corresponding to these sequences synthesized. Antibodies raised against these synthetic peptides are then used to determine their location in intact tissue or in vesicles containing oriented channels. This approach can generate independent information about sodium channel topography while providing additional constraints for the testing of channel models.

In all current models, the C-terminus of the protein is on the cytoplasmic side of the membrane. In order to test this hypothesis, Gordon et al (1987a) synthesized a 13-amino-acid oligopeptide corresponding to residues 1781–1794 at the C-terminus of the eel primary channel sequence. Antibodies against this peptide specifically identified the 260 kDa sodium channel protein on western blots of crude eel membrane proteins.

In sealed, right-side-out vesicles prepared from eel electroplax, only low levels of antibody binding were detectable prior to permeabilization. After treatment with 0.05% saponin, however, specific binding increased eight- to ten-fold (Figure 8A), thus suggesting the the channel's C-terminus residues on the cytoplasmic side of the membrane (Gordon et al 1987a).

This localization was subsequently confirmed with immunocytochemical techniques. Voltage-dependent sodium channels are confined to the innervated face of the electroplax cells. Antibodies to peptide 1781–1794 bound only to the innervated face of these cells, and this labeling was specifically blocked by preadsorption of the serum with the purified peptide. Bound antibody, detected at the EM level with colloidal gold-labeled second antibody, was found exclusively on the cytpoplasmic side of the bilayer (Gordon et al 1987a), in agreement with the localization inferred from antibody binding studies in oriented vesicles (Figure 8B).

The C-terminus of the eel sodium channel, including a region containing a potential N-glycosylation site (residue 1806), resides on the cytoplasmic side of the plasma membrane. If the four internal repeat domains are homologous in organization then the backbone of the protein must cross the membrane an even number of times in each domain. Therefore, the N-terminus and the interdomain regions should be cytoplasmic as well.

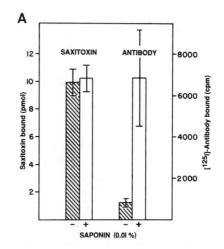

B   INNERVATED

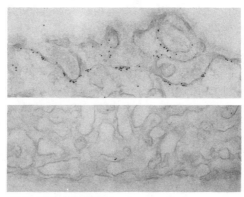

NONINNERVATED

Figure 8 (A) Binding of antibodies directed against the C-terminal region of the eel voltage-dependent sodium channel and of [³H]-saxitoxin (STX) to vesicles formed from electroplax membranes. Sites labeled with [³H]-STX, which binds to a specific site on the external surface of the sodium channel, are nearly all accessible in the intact vesicles with little increase in binding detectable after permeabilization with 0.01% saponin. The epitope recognized by this antiserum is not accessible on the external surface of the intact vesicles. Permeabilization increases bound antibody eight-fold, implying a cytoplasmic location for the 1783–1794 C-terminal peptide. (B) Localization of binding of the C-terminal antibody at the ultrastructural level with colloidal gold-labeled second antibody. Most gold particles are found associated with the innervated membrane with little labeling of the non-innervated membrane, consistent with the known distribution of sodium channels in this tissue. Colloidal gold particles along the innervated membrane are exclusively on the cytoplasmic surface of this membrane and its invaginations, confirming a cytoplasmic localization for the protein's C-terminal region. (From Gordon et al 1987a.)

Analysis of a second peptide corresponding to amino acid residues 927–938, which lies in the primary sequence between homologous domains 2 and 3, has also been reported (Gordon et al 1987b). Antiserum and immunoaffinity purified IgG against this peptide also immunoreacted exclusively with a 260 kDa band on western blots of eel membrane protein and specifically labeled the innervated face of the eel electroplax at the light microscopic level. Electron microscopy indicates that the epitope recognized by this antibody is also cytoplasmic, in agreement with that proposed in the folding models of Noda et al (1984), Guy & Seetharamulu (1986), and Greenblatt et al (1985), but not that of Kosower (1985).

Antibodies have also been raised to a synthetic peptide corresponding to residues 210–223 of the eel primary sequence, a portion of the positively charged S_4 helix in the first repeat domain (Meiri et al 1987). When applied to the external surface of dorsal root ganglion cells in culture, these antibodies specifically shifted the inactivation curve for the sodium channel in a hyperpolarizing direction along the voltage axis by 10–20 mV without modifying the kinetics of channel activation. The authors interpret these results as implying that the involved segment of the primary sequence can be made accessible to antibody binding on the external surface of the membrane, and that this segment may be linked in some manner to channel inactivation. Exposure of the internal surface of the channels to the antibodies was not attempted.

SYNTHESIS AND EXPRESSION OF SODIUM CHANNELS

Biosynthesis and Intracellular Processing

The biosynthesis of the sodium channel alpha subunit has been studied in primary cultures of rat brain neurons (Schmidt et al 1985, Schmidt & Catterall 1986); this system produces a mature sodium channel whose subunit structure is identical to that found in the adult rat brain. When grown in the presence of inhibitors of glycosylation, anti-α antiserum recognized a protein component of 203 kDa in these primary cultures (Schmidt & Catterall 1986). This correlates well with the reported size for the core protein of the α subunit after enzymatic deglycosylation in the rat and the eel (Messner & Catterall 1985, Roberts & Barchi 1987, Miller et al 1983), as well as with the size of the core protein predicted from the known primary sequence for both the rat brain and the eel α components (Noda et al 1984, 1986a,b). When glycosylation is allowed to proceed normally, several intermediate forms of the protein can be identified with pulse-chase techniques (Schmidt & Catterall 1986). These include two components of 224 kDa, one of which contains sufficient n-acetyl-glucosamine to be retained by wheat germ agglutinin, and an additional discrete form of 249 kDa (Figure 9). The latter is then processed, possibly through the addition of further sialic acid residues, to its final size of 260 kDa.

Nearly 70% of the newly synthesized α subunit equilibrates with an intracellular, membrane-associated pool that is not covalently associated with either β_1 or β_2 (Schmidt et al 1985). A smaller percentage is processed to the mature complex ultimately found in the neuronal surface membrane. Schmidt et al speculate that this large intracellular pool may provide the precursor for the mature surface channel, and that the rate-limiting step for surface channel formation may be the association of this subunit with the β components.

Preliminary studies have also been reported for the biosynthesis of eel electroplax sodium channels in frog oocytes after the injection of total eel mRNA (Thornhill & Levinson 1987). In this system, the earliest product recognized by antibodies against the purified mature eel channel is a

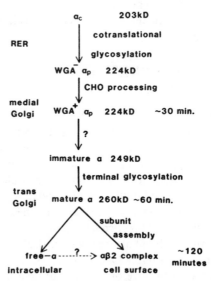

Figure 9   A scheme for the steps in the biosynthesis, co-transitional, and post-transitional processing of the rat brain sodium as proposed by Schmidt & Catterall (1986). At least in the eel, an additional step of post-translational modification involving the covalent addition of acyl chains to the α component may be present (Thornhill & Levinson 1986). (Figure from Schmidt & Catterall 1986.)

protein of $\sim$ 230,000 MW, comparable in size to that seen in the rat neuronal system when cotranslational glycosylation is not inhibited. Eel mRNA translated in vitro with a rabbit reticylocyte lysate system also produces a core protein with an apparent molecular weight of 230 kDa that is recognized by the same antibody (Thornhill & Levinson 1987).

In the eel, biosynthesis may be accompanied by the covalent attachment of acyl chains to the core peptide as an additional component of post-translational modification (Thornhill & Levinson 1987, Levinson et al 1987). The mature eel channel contains considerable bound lipid and exhibits an unusually high free solution mobility in SDS, possibly due in part to the binding of SDS to these lipid moieties. The 230,000 core protein identified immunologically following cell-free synthesis, on the other hand, shows electrophoretic mobility in SDS comparable to that of other standard proteins (Thornhill & Levinson 1987), a result that suggests that these lipid chains are attached at a more distal point in the biosynthetic pathway.

Regulation of Channel Expression

In rat muscle cells grown in culture, several factors influence the density of sodium channels expressed on the cell surface. In general, factors that tend to increase cytoplasmic free Ca^{2+} result in the down-regulation of sodium channels, while factors that increase intracellular cAMP have the opposite effect (Sherman et al 1985). In cultured rat and chick muscle, increased electrical activity decreases sodium channel density, presumably through the activation of voltage-dependent Ca^{2+} channels (Sherman & Catterall 1984, Bar-Sagi & Prives 1985).

Interference with glycosylation affects the expression of functional channels in vivo. Neuroblastoma cells exposed to low concentrations of L-fucose or D-galactose show an inhibition of the usual induction of toxin-stimulated Rb^+ efflux characteristic of sodium channel activity (Giovanni et al 1981), perhaps due to a specific regulatory role played by these monosaccharides in the glycosylation process. In a similar cell line, inhibition of glycosylation with tunicamycin reduces the density of surface sodium channels to $\sim$ 25% of control within 60 hr, and reduces toxin-stimulated influx proportionately (Waechter et al 1983) without affecting the binding properties of the residual channels. Based on the rate of loss of STX-binding activity, Waechter et al (1983) estimate a half-time for sodium channel turnover in these cells of 26 hr. A variety of other inhibitors of glycosylation and glycoprotein processing have a similar effect on sodium channel expression in culture (Negishi & Glick, 1986). Taken together, these studies suggest that the oligosaccharide residues of the sodium channel must be processed to the mature complex type for normal expression of functional channels in the surface membrane.

Phosphorylation of the α Subunit

The α subunit of the rat brain sodium channel can be rapidly and selectively phosphorylated both after purification (Costa et al 1982) and in synaptosomal membranes (Costa & Catterall 1984a). Phosphorylation mediated by the purified catalytic subunit of cAMP-dependent protein kinase incorporates 3–4 mol phosphorous per mol channel in the purified channel; phosphorylation in intact synaptosomes followed by tryptic analysis of the α subunit demonstrates five major reaction sites within the protein's primary structure (Costa & Catterall 1984a). Phosphorylation can also be induced by treatment of intact synaptosomes with 8-Br-cAMP. The α subunit of the rat brain channel can also be phosphorylated by purified C-kinase both in synaptosomal membranes and after channel purification (Costa & Catterall 1984b).

Phosphorylation of sodium channels in synaptosomes has no effect on neurotoxin-binding parameters but has been reported to reduce slightly (16–26%) the magnitude of toxin-induced Na^+ influx in this system (Costa & Catteral 1984a). On the other hand, addition of either dibutyryl-cAMP or inhibitors of phosphodiesterase does not alter sodium currents through the sodium channel in frog nodes of Ranvier under voltage-clamped conditions (Arhem et al 1986). The physiological role for sodium channel phosphorylation in vivo remains to be defined.

Channel Distribution and Mobility

Sodium channels are not uniformly distributed on the surface of nerve or muscle cells. In nerve, concentration of sodium channels at the axon hillock and at the nodes of Ranvier in myelinated axons has long been appreciated. More recently, several groups have reported a nonuniform distribution of channels along the length of adult muscle fibers as well as at more local levels in the sarcolemma.

The concentration of sodium channels at the peripheral node of Ranvier has been confirmed immunocytochemically with antibodies specific for the purified channel protein (Haimovich et al 1984, Meiri et al 1984). Cytochemical evidence has also been reported suggesting a similar although less striking nonhomogeneous distribution of sodium channels in the unmyelinated squid giant axon (Greeff et al 1985).

Channels in myelinated fibers maintain their inhomogeneous distribution even after acute demyelination, thus suggesting the presence of mechanisms restricting their movement longitudinally along the axolemma (Sears & Bostock 1981). Recent studies have shown that channels within the nodal area itself, however, may be capable of local lateral diffusion (Pappone & Cahalan 1985). In addition, long-term demyelination can be associated with the appearance of sodium channels in the internodal regions, possibly playing a role in the reestablishment of functional conduction after pathological demyelination (Sears & Bostock 1981, Meiri et al 1985).

In skeletal muscle, sodium channels are concentrated at the neuromuscular junction; their density falls off rapidly in the perijunctional region to reach a level about ten-fold lower in the sarcolemma some distance away (Beam et al 1985, Caldwell et al 1986, Angelides 1986, Dreyfus et al 1986, Haimovich et al 1987). At the ultrastructural level, sodium channels are distributed throughout the postjunctional folds of the neuromuscular junction; they are not restricted to the apex of the folds as are acetylcholine receptors (Figure 10) (Haimovich et al 1987).

Local sodium channel distribution in the nonjunctional sarcolemma has been mapped at higher resolution with the loose patch clamp technique (Almers et al 1983, Caldwell et al 1986). At this level, channels are organized into bands of higher density running mainly along the fiber axis. Channel density in these bands may be three-fold higher than that in adjacent areas of the sarcolemma. This uneven lateral distribution is not a general phenomenon; potassium channels mapped in the same regions do not show similar distributional patterns.

Using fluorescence photobleaching recovery in conjunction with labeled sodium channel neurotoxins, Angelides (1986) has found that sodium channels at the neuromuscular junction of innervated muscle cells in culture are highly immobilized, while sodium channels in uninnervated muscle fibers show a much higher degree of lateral mobility. Extending the earlier work of Stühmer & Almers (1982), Weiss et al (1986) concluded that sodium channels in the sarcolemma of adult frog muscle are also highly immobilized. Finally, nonjunctional sarcolemmal sodium channels in cultured muscle cells exhibit a higher level of lateral mobility than that seen in mature muscle fibers (Angelides 1986).

The mechanisms by which sodium channels are immobilized in the surface membrane of nerve and muscle, and the factors that control this local organization, remain to be elucidated. This area should be a fruitful one for future research.

SODIUM CHANNEL SUBTYPES

Although much of the preceding discussion has focused on aspects of channel structure that are common to the various purified sodium channels, the subtle differences among the channels may be equally impor-

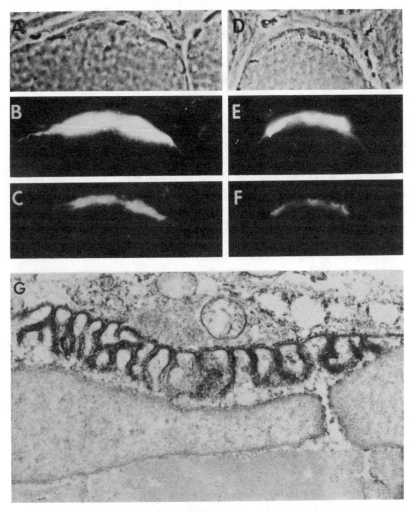

Figure 10 Voltage-dependent sodium channels are present at a higher density in the post-synaptic membrane of the neuromuscular junction in rat skeletal muscle than in the surrounding sarcolemma. In this example, monoclonal antibodies against the sodium channel α subunit are used to localize the channel at the light- and electron-microscopic level. (A, D) Phase contrast images of the endplate regions of two muscle fibers from the rat *tibialis anterior* in transverse section. (B, E) Immunofluorescent labeling with fluorescein-conjugated antibodies against the voltage-dependent sodium channel. (C, F) Simultaneous imaging of the same endplates with rhodamine-labeled α-bungarotoxin. (G) Ultrastructural localization of antibody binding to the sodium channel in the neuromuscular junction. Channels are distributed throughout the postjunction folds and are not selectively localized near their apex as is the case with the acetylcholine receptor. (Adapted from Haimovich et al 1987.)

tant for our future understanding of the regulation of channel expression and distribution. Strong evidence indicates that multiple related subtypes of the voltage-dependent sodium channel exist in the rat, and even that multiple subtypes may be expressed in the same tissue at any given time.

Evidence for the existence of sodium channel subtypes comes from a number of different sources. It has long been recognized that sodium channels that appear at different times during nerve and muscle development have markedly different sensitivity to the neurotoxins STX and TTX (see for example the work of Frelin et al 1981, 1983, 1984; Sherman & Catterall 1982, Strichartz et al 1983, Sherman et al 1983, Baumgold et al 1983, Gonoi et al 1985, Haimovich et al 1986). One population of channels is blocked by nanomolar concentrations of these toxins whereas the other is resistant to block by TTX at concentrations several orders of magnitude higher. Similarly, denervation of mature skeletal muscle induces the appearance of TTX-resistant sodium channels through a process that requires the synthesis of new protein (Thesleff et al 1974, Barchi & Weigele 1979, Rogart & Regan 1985). In the chick, TTX-sensitive sodium channels are found in the brain at the same time that TTX-resistant channels are being expressed in cardiac muscle (Rogart et al 1983), and both TTX-sensitive and TTX-resistant channels can be identified in the same preparations of adult cardiac muscle (Renaud et al 1983).

Voltage-clamp studies of denervated frog skeletal muscle has confirmed the presence of two sodium channel populations and has defined some of their macroscopic kinetic properties (Pappone 1980). Recently, single channel measurements have separated two populations of sodium channels in muscle culture having distinct biophysical properties that might correspond to these TTX-sensitive and TTX-resistant channel subtypes (Weiss & Horn 1986). One class of channels was more resistant to TTX block, had a lower single-channel conductance, and was activated at more hyperpolarizing potentials. The second class was blocked by nanomolar concentrations of TTX, had a higher single channel conductance, and was activated at more depolarizing potentials. The proportion of TTX-resistant to TTX-sensitive channels was higher in myoblasts than in myotubes. TTX-sensitive and TTX-resistant sodium channels also differ in their sensitivity to block by Cd^{2+} and Zn^{2+} (Frelin et al 1986).

Sodium channels from adult rat brain and skeletal muscle, both of which are sensitive to TTX, can be distinguished by the toxins from *Conus geographicus*. These interesting polypeptides neurotoxins specifically block the muscle channel but produce no discernable effect on channels from the brain (Cruz et al 1985, Kobayashi et al 1986, Ohizumi et al 1986a,b). Biophysical measurements with these channels from rat also indicate that although the channels are similar in most respects, the muscle channel has a single-channel conductance about 20% higher than the brain channel.

The binding of other neurotoxins points to the presence of multiple channel subtypes in mature innervated muscle. The toxin from *Cen-truriodes suffusus suffusus* (Jaimovich et al 1982) and *Titius serrulatus* (Barhanin et al 1984), as well as a synthetic derivative of tetrodotoxin (Jaimovich et al 1983), show differential binding to channels in muscle surface membrane as opposed to those in the T-tubular system. Another synthetic derivative of TTX preferentially binds to the T-tubular sodium channel (Jaimovich et al 1983). Differential sensitivity of sodium channels in these contiguous membranes to TTX derivatives has been confirmed with electrophysiological measurements (Jaimovich et al 1982).

Monoclonal and polyclonal antibodies prepared against purified sodium channels also differentiate channel subtypes in nerve and muscle from a given animal as well as within a single mature tissue. Some polyclonal antisera raised against the rat brain channel do not recognize the sodium channel in rat muscle (Wollner & Catterall 1985), while other antisera to the purified skeletal muscle sodium channel show significant cross-reactivity between muscle channels and channels in the peripheral node of Ranvier (Haimovich et al 1984). Monoclonal antibodies against the rat skeletal muscle sodium channel distinguish between epitopes in the surface and the T-tubular membranes, although all apparently recognize the same 260 kDa band on western blots of muscle protein (Haimovich et al 1987). Some of these antibodies discriminate between channels in the T-tubular system of fast and slow muscle fibers within the same muscle cross-section. These antibodies do not cross-react with the channel from rat brain.

Although the skeletal muscle sodium channel has not yet been completely sequenced, the sodium channel expressed in rat skeletal muscle differs from both the rat brain I and rat brain II sequences expressed in rat brain (Cooperman et al 1987). Probes derived from genomic clones directed to the 5′ region of these sequences, which include unique stretches of adjacent 5′-untranslated regions, are protected by rat brain mRNA but not by skeletal muscle mRNA. In the same preparations, a probe complementary to the rat brain II fourth homologous repeat domain hybridizes with a 9.5 kb message in both tissues. Thus, although homologous to brain sodium channels, the muscle channel most likely represents the expression of yet another gene in this family.

The strongest evidence for the presence of multiple subtypes of channel in the same tissue comes from the cloning experiments of Noda et al (1986a). These investigators have obtained the complete sequence for two highly homologous sodium channels and the partial sequence for a third, all of which are expressed simultaneously in the rat brain. Since these sequences were obtained from cDNA prepared to poly-A message derived from brain, the messages for each channel type must be present in this preparation, although possibly not in the same cell type. The sequences for these two channels show the highest degree of homology within the four internal repeat domains, with greater divergence in the interdomain regions.

It will be of considerable interest to determine which of the subtle differences that are found among these related channel subtypes contribute to their differential localization, expression, and function. Indeed, focusing on these regions in the primary sequence may provide additional insight into the relationship of channel structure to function. The structural relationship of these sodium channels to those identified in nonexcitable cells such as Schwann cells (Shrager et al 1985), astrocytes (Bevan et al 1985), fibroblasts (Pouyssegur et al 1980), and lymphocytes (DeCoursey et al 1985) remains a complete enigma.

Literature Cited

Agnew, W. S. 1984. Voltage-regulated sodium channel molecules. *Annu. Rev. Physiol.* 46: 517–30

Agnew, W. S., Levinson, S. R., Brabson, J. S., Raftery, M. A. 1978. Purification of the tetrodotoxin binding component associated with the voltage-sensitive sodium channel from *Electrophorus electricus* electroplax membranes. *Proc. Natl. Acad. Sci. USA* 75: 2606–10

Agnew, W. S., Miller, J. A., Ellisman, M. H., Rosenberg, R. L., Tomiko, S. A., Levinson, S. R. 1983. The voltage-regulated sodium channel from the electroplax of *Electrophorus electricus*. *Cold Spring Harbor Symp. Quant. Biol.* 48: 165–79

Agnew, W. S., Moore, A. C., Levinson, S. R., Raftery, M. A. 1980. Identification of a large molecular weight peptide associated with a tetrodotoxin binding protein from the electroplax of *Electrophorus electricus*. *Biochem. Biophys. Res. Commun.* 92: 860–66

Agnew, W. S., Rosenberg, R. L., Tomiko, S. A. 1986. Reconstitution of the sodium channel from *Electrophorus electricus*. In *Ion Channel Reconstitution* ed. C. Miller, pp. 307–36. New York: Plenum

Agnew, W., Raftery, W. 1979. Solubilized tetrodotoxin-binding component from the electroplax of *Electrophorus electricus*. Stability as a function of mixed lipid-detergent micelle composition. *Biochemistry* 18: 1912–19

Aldrich, R. W. 1986. Voltage-dependent gating of sodium channels: Towards an inte-

grated approach. *Trends Neurosci.* Feb: 82–85

Almers, W., Stanfield, P. R., Stühmer, W. 1983. Lateral distribution of sodium and potassium channels in frog skeletal muscle: Measurements with a patch-clamp technique. *J. Physiol.* 336: 261–84

Angelides, K. J. 1986. Fluorescently labeled Na^+ channels are localized and immobilized to synapses of innervated muscle fibres. *Nature* 321: 63–66

Angelides, K. J., Brown, G. B. 1984. Fluorescence resonance energy transfer on the voltage-dependent sodium channel. Spatial relationship and site coupling between the batrachotoxin and *Leiurus quinquestriatus quinquestriatus* alpha-scorpion toxin receptors. *J. Biol. Chem.* 259: 6117–26

Angelides, K. J., Nutter, T. J. 1983. Mapping the molecular structure of the voltage-dependent sodium channel. Distances between the tetrodotoxin and *Leiurus quinquestriatus quinquestriatus* scorpion toxin receptors. *J. Biol. Chem.* 258: 11958–67

Angelides, K. J., Nutter, T. J., Elmer, L. W., Kempner, E. S. 1985. Functional unit size of the neurotoxin receptors on the voltage-dependent sodium channel. *J. Biol. Chem.* 260: 3431–39

Arhem, P., Lindström, A., Johansson, S. 1986. Ionic channels in the node of Ranvier are not modulated by cyclic adenosine monophosphate. *Brain Res.* 371: 182–86

Barchi, R. L. 1982. Biochemical studies of

the excitable membrane sodium channel. *Int. Rev. Neurobiol.* 23: 69–102

Barchi, R. L. 1983. Protein components of the purified sodium channel from rat skeletal muscle sarcolemma. *J. Neurochem.* 40: 1377–85

Barchi, R. L., Murphy, L. E. 1980. Size characteristics of the solubilized sodium channel saxitoxin binding site from mammalian sarcolemma. *Biochim. Biophys. Acta* 597: 391–98

Barchi, R. L., Weigele, J. B. 1979. Characteristics of saxitoxin binding to the sodium channel of sarcolemma isolated from rat skeletal muscle. *J. Physiol.* 295: 383–96

Barchi, R. L., Cohen, S. A., Murphy, L. E. 1980. Purification from rat sarcolemma of the saxitoxin binding component of the excitable membrane sodium channel. *Proc. Natl. Acad. Sci. USA* 77: 1306–10

Barchi, R. L., Tanaka, J. C., Furman, R. E. 1984. Molecular characteristics and functional reconstitution of muscle voltage-sensitive sodium channels. *J. Cell. Biochem.* 26: 135–46

Barhanin, J., Ildefonse, M., Rougier, O., Sampaio, S. V., Giglio, J. R., Lazdunski, M. 1984. *Tityus γ* toxin, a high affinity effector of the Na$^+$ channel in muscle, with a selectivity for channels in the surface membrane. *Pflugers Arch.* 400: 22–27

Barhanin, J., Pauron, D., Lombet, A., Norman, R. I., Vijverberg, H. P., Giglio, J. R., Lazdunski, M. 1983a. Electrophysiological characterization, solubilization and purification of the *Tityus γ* toxin receptor associated with the gating component of the Na$^+$ channel from rat brain. *EMBO J.* 2: 915–20

Barhanin, J., Schmid, A., Lombet, A., Wheeler, K. P., Lazdunski, M., Ellory, J. C. 1983b. Molecular size of different neurotoxin receptors on the voltage-sensitive Na$^+$ channel. *J. Biol. Chem.* 258: 700–2

Bar-Sagi, D., Prives, J. 1985. Negative modulation of sodium channels in cultured chick muscle cells by the channel activator batrachotoxin. *J. Biol. Chem.* 260: 4740–44

Baumgold, J., Parent, J. B., Spector, I. 1983. Development of sodium channels during differentiation of chick skeletal muscle in culture. I. Binding studies. *J. Neurosci.* 3: 995–1003

Beam, K. G., Caldwell, J. H., Campbell, D. T. 1985. Na channels in skeletal muscle concentrated near the neuromuscular junction. *Nature* 313: 588–90

Beneski, D., Catterall, W. A. 1980. Covalent labeling of protein components of the sodium channel with a photoactivable derivative of scorpion toxin. *Proc. Natl. Acad. Sci. USA* 77: 639–43

Benzer, T. I., Raftery, M. A. 1973. Partial characterization of a tetrodotoxin-binding component from nerve membrane. *Proc. Natl. Acad. Sci. USA* 69: 3634–37

Bevan, S., Chiu, S. Y., Gray, P. T., Ritchie, J. M. 1985. The presence of voltage-gated sodium, potassium and chloride channels in rat cultured astrocytes. *Proc. R. Soc. London Ser. B* 225: 299–313

Bezanilla, F. 1985. Gating of sodium and potassium channels. *J. Membr. Biol.* 88: 97–111

Caldwell, J. H., Campbell, D. T., Beam, K. G. 1986. Na channel distribution in vertebrate skeletal muscle. *J. Gen. Physiol.* 87: 907–32

Casadei, J. M., Gordon, R. D., Barchi, R. L. 1986. Immunoaffinity isolation of Na$^+$ channels from rat skeletal muscle. Analysis of subunits. *J. Biol. Chem.* 261: 4318–23

Catterall, W. A. 1980. Neurotoxins that act on voltage-sensitive sodium channels in excitable membranes. *Ann. Rev. Pharmacol. Toxicol.* 20: 15–43

Catterall, W. A. 1986. Molecular properties of voltage-sensitive sodium channels. *Annu. Rev. Biochem.* 55: 953–85

Catterall, W. A., Gainer, M. 1985. Interaction of brevetoxin A with a new receptor site on the sodium channel. *Toxicon* 23: 497–504

Cohen, S. A., Barchi, R. L. 1981. Glycoprotein characteristics of the sodium channel saxitoxin-binding component from mammalian sarcolemma. *Biochim. Biophys. Acta* 645: 253–61

Cooperman, S. S., Grubman, S. A., Barchi, R. L., Goodman, R. H., Mandel, G. 1987. Modulation of sodium channel mRNA levels in rat skeletal muscle. *Proc. Natl. Acad. Sci. USA* 84: 8721–25

Costa, M. C., Casnellie, J. E., Catterall, W. A. 1982. Selective phosphorylation of the alpha subunit of the sodium channel by cAMP-dependent protein kinase. *J. Biol. Chem.* 257: 7918–21

Costa, M. R., Catterall, W. A. 1984a. Cyclic AMP-dependent phosphorylation of the alpha subunit of the sodium channel in synaptic nerve ending particles. *J. Biol. Chem.* 259: 8210–18

Costa, M. R., Catterall, W. A. 1984b. Phosphorylation of the alpha subunit of the sodium channel by protein kinase C. *Cell. Mol. Neurobiol.* 4: 291–97

Cruz, L. J., Gray, W. R., Olivera, B. M., Zeikus, R. D., Kert, L., Yoshikami, D., Moczydlowski, E. 1985. *Conus geographus* toxins that discriminate between neuronal and muscle sodium channels. *J. Biol. Chem.* 260: 9280–88

Darbon, H., Angelides, K. J. 1984. Structural mapping of the voltage-dependent sodium channel. Distance between the tetrodotoxin and Centruroides suffusus suffusus II beta-scorpion toxin receptors. *J. Biol. Chem.* 259: 6074–84

Darbon, H., Jover, E., Couraud, F., Rochat, H. 1983. Photoaffinity labeling of alpha- and beta-scorpion toxin receptors associated with rat brain sodium channel. *Biochem. Biophys. Res. Commun.* 115: 415–22

DeCoursey, T. E., Chandy, K. G., Gupta, S., Cahalan, M. D. 1985. Voltage-dependent ion channels in T-lymphocytes. *J. Neuroimmunol.* 10: 71–95

Dreyfus, P., Rieger, F., Murawsky, M., Garcia, L., Lombet, A., et al. 1986. The voltage-dependent sodium channel is co-localized with the acetylcholine receptor at the vertebrate neuromuscular junction. *Biochem. Biophys. Res. Commun.* 139: 196–201

Elmer, L. W., O. B. J., Nutter, T. J., Angelides, K. J. 1985. Physicochemical characterization of the alpha-peptide of the sodium channel from rat brain. *Biochemistry* 24: 8128–37

Feller, D. J., Talvenheimo, J. A., Catterall, W. A. 1985. The sodium channel from rat brain. Reconstitution of voltage-dependent scorpion toxin binding in vesicles of defined lipid composition. *J. Biol. Chem.* 260: 11542–47

Frelin, C., Cognard, C., Vigne, P., Lazdunski, M. 1986. Tetrodotoxin-sensitive and tetrodotoxin-resistant Na$^+$ channels differ in their sensitivity to Cd^{2+} and Zn^{2+}. *Eur. J. Pharmacol.* 122: 245–50

Frelin, C., Lombet, A., Vigne, P., Romey, G., Lazdunski, M. 1981. The appearance of voltage-sensitive sodium channels during the in vitro differentiation of embryonic chick skeletal muscle cells. *J. Biol. Chem.* 256: 12355–61

Frelin, C., Vigne, P., Lazdunski, M. 1983. Na$^+$ channels with high and low affinity tetrodotoxin binding sites in the mammalian skeletal muscle cell. Difference in functional properties and sequential appearance during rat skeletal myogenesis. *J. Biol. Chem.* 258: 7256–59

Frelin, C., Vijverberg, H. P., Romey, G., Vigne, P., Lazdunski, M. 1984. Different functional states of tetrodotoxin sensitive and tetrodotoxin resistant Na$^+$ channels occur during the in vitro development of rat skeletal muscle. *Pflugers Arch.* 402: 121–28

Furman, R. E., Tanaka, J. C., Meuller, P., Barchi, R. L. 1986. Voltage-dependent activation in purified reconstituted sodium channels from rabbit T-tubular membranes. *Proc. Natl. Acad. Sci. USA* 83: 488–92

Giovanni, M., Kessel, D., Glick, M. C. 1981. Specific monosaccharide inhibition of active sodium channels in neuroblastoma cells. *Proc. Natl. Acad. Sci. USA* 78: 1250–54

Goldin, A. L., Snutch, T., Lübbert, H., Dowsett, A., Marshall, J., et al. 1986. Messenger RNA coding for only the alpha subunit of the rat brain Na channel is sufficient for expression of functional channels in *Xenopus* oocytes. *Proc. Natl. Acad. Sci. USA* 83: 7503–7

Gonoi, T., Ashida, K., Feller, D., Schmidt, J., Fujiwara, M., Catterall, W. A. 1986. Mechanism of action of a polypeptide neurotoxin from the coral *Goniopora* on sodium channels in mouse neuroblastoma cells. *Mol. Pharmacol.* 29: 347–54

Gonoi, T., Sherman, S. J., Catterall, W. A. 1985. Voltage clamp analysis of tetrodotoxin-sensitive and -insensitive sodium channels in rat muscle cells developing in vitro. *J. Neurosci.* 5: 2559–64

Gordon, R. A., Fieles, W. E., Schotland, D. L., Angeletti, R. A., Barchi, R. L. 1987a. Topographical localization of the C-terminal region of the voltage-dependent sodium channel from *Electrophorus electricus* using antibodies against a synthetic peptide. *Proc. Natl. Acad. Sci. USA* 84: 308–13

Gordon, R. D., Fieles, W. E., Schotland, D L., Barchi, R. L. 1987b. Immunochemical testing of current molecular models of the voltage-dependent sodium channel from *Electrophorus electricus. Biophys. J.* 51: 437a

Greeff, N. G., Akert, K., Sandri, C. 1985. Cytochemical evidence for an inhomogeneous distribution of sodium channels in the squid giant axon. *Neurosci. Lett.* 12: 153–57

Greenblatt, R. E., Blatt, Y., Montal, M. 1985. The structure of the voltage-sensitive sodium channel. Inferences derived from computer-aided analysis of the *Electrophorus electricus* channel primary structure. *FEBS Lett.* 193: 125–34

Grishin, E. V., Kovalenko, E. V., Pashkov, V. N., Shamotienko, O. G. 1984. Purification of rat brain sodium channels. *Membr. Biophys. USSR* 1: 858–67

Guy, H. R., Seetharamulu, P. 1986. Molecular model of the action potential sodium channel. *Proc. Natl. Acad. Sci. USA* 83: 508–12

Haimovich, B., Bonilla, E., Casadei, J., Barchi, R. 1984. Immunocytochemical localization of the mammalian voltage-dependent sodium channel using polyclonal antibodies against the purified protein. *J. Neurosci.* 4: 2259–68

Haimovich, B., Schotland, D. L., Fieles, W. E., Barchi, R. L. 1987. Localization of sodium channel subtypes in adult rat skeletal muscle using channel-specific monoclonal antibodies. *J. Neurosci.* 7: 2957–66

Haimovich, B., Tanaka, J. C., Barchi, R. L. 1986. Developmental appearance of sodium channel subtypes in rat skeletal muscle cultures. *J. Neurochem.* 47: 1148–52

Hanke, W., Boheim, G., Barhanin, J., Pauron, D., Lazdunski, M. 1984. Reconstitution of highly purified saxitoxin-sensitive sodium channels into planar lipid bilayers. *EMBO J.* 3: 509–15

Hartshorne, R. P., Catterall, W. A. 1981. Purification of the saxitoxin receptor of the sodium channel from rat brain. *Proc. Natl. Acad. Sci. USA* 78: 4620–24

Hartshorne, R. P., Catterall, W. A. 1984. The sodium channel from rat brain. Purification and subunit composition. *J. Biol. Chem.* 259: 1667–75

Hartshorne, R. P., Coppersmith, J., Catterall, W. A. 1980. Size characteristics of the solubilized saxitoxin receptor of the voltage-sensitive sodium channel from rat brain. *J. Biol. Chem.* 255: 10572–75

Hartshorne, R. P., Keller, B. U., Talvenheimo, J. A., Catterall, W. A., Montal, M. 1985. Functional reconstitution of the purified brain sodium channel in planar lipid bilayers. *Proc. Natl. Acad. Sci. USA* 82: 240–44

Hartshorne, R. P., Messner, D. J., Coppersmith, J. C., Catterall, W. A. 1982. The saxitoxin receptor of the sodium channel from rat brain: Evidence for two nonidentical beta subunits. *J. Biol. Chem.* 257: 13888–91

Hartshorne, R., Tamkun, M., Montal, M. 1986. The reconstituted sodium channel from rat brain. See Agnew et al 1986, pp. 337–62

Henderson, R. H., Wang, J. H. 1972. Solubilization of a specific tetrodotoxin-binding component from garfish olfactory nerve. *Biochemistry* 11: 4565–69

Hodgkin, A. L., Huxley, A. F. 1952. The components of membrane conductance in the giant axon of *Loligo*. *J. Physiol.* 116: 473–96

Horn, R. 1984. Gating of channels in nerve and muscle: A stochastic approach. In *Ion Channels: Molecular and Physiological Aspects*, ed. W. D. Stein, pp. 53-97. New York: Academic

Huang, L. M., Moran, N., Ehrenstein, G. 1982. Batrachotoxin modifies the gating kinetics of sodium channels in internally perfused neuroblastoma cells. *Proc. Natl. Acad. Sci. USA* 79: 2082–85

Jaimovich, E., Chicheportiche, R., Lombet, A., Lazdunski, M., Ildefonse, M.,

Rougier, O. 1983. Differences in the properties of Na^+ channels in muscle surface and T-tubular membranes revealed by tetrodotoxin derivatives. *Pflugers Arch.* 397: 1–5

Jaimovich, E., Ildefonse, M., Barhanin, J., Rougher, O., Lazdunski, M. 1982. Centruroides toxin, a selective blocker of surface sodium channels in skeletal muscle: Voltage-clamp analysis and biochemical characterization of the receptor. *Proc. Natl. Acad. Sci. USA* 79: 3896–3900

Keller, B. U., Hartshorne, R. P., Talvenheimo, J. A., Catterall, W. A., Montal, M. 1986. Sodium channels in planar lipid bilayers. Channel gating kinetics of purified sodium channels modified by batrachotoxin. *J. Gen. Physiol.* 88: 1–23

Khodorov, B. I. 1985. Batrachotoxin as a tool to study voltage-sensitive sodium channels of excitable membranes. *Prog. Biophys. Mol. Biol.* 45: 57–148

Kobayashi, M., Wu, C. H., Yoshii, M., Narahashi, T., Nakamura, H., Kobayashi, J., Ohizumi, Y. 1986. Preferential block of skeletal muscle sodium channels by geographutoxin II, a new peptide toxin from *Conus geographus*. *Pflugers Arch.* 407: 241–43

Kosower, E. M. 1985. A structural and dynamic molecular model for the sodium channel of *Electrophorus electricus*. *FEBS Lett.* 182: 234–42

Kraner, S. D., Tanaka, J. C., Barchi, R. L. 1985. Purification and functional reconstitution of the voltage-sensitive sodium channel from rabbit T-tubular membranes. *J. Biol. Chem.* 260: 6341–47

Lazdunski, M., Renaud, J. 1982. The action of cardiotoxins on cardiac plasma membrane. *Ann. Rev. Physiol.* 44: 463–73

Levinson, S. R., Ellory, J. C. 1972. Molecular size of the tetrodotoxin binding site estimated by irradiation inactivation. *Nature* 245: 122–23

Levinson, S. R., Duch, D. S., Urban, B. W., Recio-Pinto, E. 1986. The sodium channel from *electrophorus electricus*. *Ann. NY Acad. Sci.* 479: 162–78

Lombet, A., Lazdunski, M. 1984. Characterization, solubilization, affinity labeling and purification of the cardiac Na^+ channel using *Tityus* toxin gamma. *Eur. J. Biochem.* 141: 651–60

Lombet, A., Norman, R. I., Lazdunski, M. 1983. Affinity labelling of the tetrodotoxin-binding component of the Na^+ channel. *Biochem. Biophys. Res. Commun.* 114: 126–30

Lombet, A., Renaud, J.-F., Chicheportiche, R., Lazdunski, M. 1981. A cardiac tetrodotoxin binding component: Biochemical identification, characterization, and properties. *Biochemistry* 20: 1279–85

Maelicke, A., ed. 1986. Nicotinic acetylcholine receptor: Structure and function. *Nato ASI Ser., Ser. H: Cell Biology*, Vol. 3. New York: Springer Verlag

Meiri, H., Pri-Chen, S., Korczyn, A. D. 1985. Sodium channel localization in rat sciatic nerve following lead-induced demyelination. *Brain Res.* 359: 326–31

Meiri, H., Spira, G., Marei, S., Namir, M., Schwartz, H., Komoriya, A., Kosower, E., Palti, Y. 1987. Mapping a region associated with sodium channel inactivation using antibodies to a synthetic peptide corresponding to a part of the channel. *Proc. Natl. Acad. Sci. USA* 84: 5058–62

Meiri, H., Zeitoun, I., Grunhagen, H. H., Lev-Ram, V., Eshhar, Z., Schlessinger, J. 1984. Monoclonal antibodies associated with sodium channel block nerve impulse and stain nodes of Ranvier. *Brain Res.* 310: 168–73

Messner, D. J., Catterall, W. A. 1985. The sodium channel from rat brain. Separation and characterization of subunits. *J. Biol. Chem.* 260: 10597–10604

Messner, D. J., Catterall, W. A. 1986. The sodium channel from rat brain. Role of the beta 1 and beta 2 subunits in saxitoxin binding. *J. Biol. Chem.* 261: 211–15

Miller, J. A., Agnew, W. S., Levinson, S. R. 1983. Principal glycopeptide of the tetrodotoxin-saxitoxin binding protein from *Electrophorus electricus*: Isolation and partial chemical and physical characterization. *Biochemistry* 22: 462–70

Moore, A. C., Agnew, W. S., Raftery, M. A. 1982. Biochemical characterization of the tetrodotoxin binding protein from *Electrophorus electricus. Biochemistry* 21: 6212–20

Nakayama, H., Withy, R. M., Raftery, M. A. 1982. Use of a monoclonal antibody to purify the tetrodotoxin-binding component from the electroplax of *Electrophorus electricus. Proc. Natl. Acad. Sci. USA* 79: 7575–79

Narahashi, T. 1974. Chemicals as tools in the study of excitable membranes. *Physiol. Rev.* 54: 813–89

Negishi, M., Glick, M. C. 1986. Perturbation of glycoprotein processing affects the neurotoxin-responsive Na^+ channel in neuroblastoma cells. *Carbohydr. Res.* 149: 185–98

Noda, M., Ikeda, T., Kayano, T., Suzuki, H., Takeshima, H., et al. 1986a. Existence of distinct sodium channel messenger RNAs in rat brain. *Nature* 320: 188–92

Noda, M., Ikeda, T., Suzuki, H., Takeshima, H., Takahashi, T. 1986b. Expression of functional sodium channels from cloned cDNA. *Nature* 322: 826–28

Noda, M., Shimizu, S., Tanabe, T., Takai, T., Kayano, T. et al. 1984. Primary structure of *Electrophorus electricus* sodium channel deduced from cDNA sequence. *Nature* 312: 121–27

Norman, R. I., Schmid, A., Lombet, A., Barhanin, J., Lazdunski, M. 1983. Purification of binding protein for *Tityus* gamma toxin identified with the gating component of the voltage-sensitive Na^+ channel. *Proc. Natl. Acad. Sci. USA* 80: 4164–68

Ohizumi, Y., Minoshima, S., Takahashi, M., Kajiwara, A., Nakamura, H., Kobayashi, J. 1986a. Geographutoxin II, a novel peptide inhibitor of Na channels of skeletal muscles and autonomic nerves. *J. Pharmacol. Exp. Ther.* 239: 243–248

Ohizumi, Y., Nakamura, H., Kobayashi, J., Catterall, W. A. 1986b. Specific inhibition of [3H] saxitoxin binding to skeletal muscle sodium channels by geographutoxin II, a polypeptide channel blocker. *J. Biol. Chem.* 261: 6149–52

Pappone, P. 1980. Voltage-clamp experiments in normal and denervated mammalian skeletal muscle fibers. *J. Physiol.* 305: 377–410

Pappone, P. A., Cahalan, M. D. 1985. Demyelination as a test for a mobile Na channel modulator in frog node of Ranvier. *Biophys. J.* 47: 217–23

Poli, M. A., Mende, T. J., Baden, D. G. 1986. Brevetoxins, unique activators of voltage-sensitive sodium channels, bind to specific sites in rat brain synaptosomes. *Mol. Pharmacol.* 30: 129–35

Pouyssegur, J., Jacques, Y., Lazdunski, M. 1980. Identification of a tetrodotoxin-sensitive sodium channel in a variety of fibroblast lines. *Nature* 286: 162–64

Renaud, J. F., Kazazoglou, T., Lombet, A., Chicheportiche, R., Jaimovich, E., Romey, G., Lazdunski, M. 1983. The Na^+ channel in mammalian cardiac cells. Two kinds of tetrodotoxin receptors in rat heart membranes. *J. Biol. Chem.* 258: 8799–8805

Ritchie, J. M., Rogart, R. 1977. The binding of saxitoxin and tetrodotoxin to excitable tissue. *Rev. Physiol. Biochem. Pharmacol.* 79: 2–50

Roberts, R., Barchi, R. L. 1987. The voltage-sensitive sodium channel from rabbit skeletal muscle: Chemical characterization of subunits. *J. Biol. Chem.* 262: 2298–2303

Rogart, R. B., Regan, L. J. 1985. Two subtypes of sodium channel with tetrodotoxin sensitivity and insensitivity detected in denervated mammalian skeletal muscle. *Brain Res.* 329: 314–18

Rogart, R. B., Regan, L. J., Dziekan, L. C., Galper, J. B. 1983. Identification of two sodium channel subtypes in chick heart and brain. *Proc. Natl. Acad. Sci. USA* 80: 1106–10

Rosenberg, R. L., Tomiko, S. A., Agnew, W. S. 1984a. Reconstitution of neurotoxin-modulated ion transport by the voltage-regulated sodium channel isolated from the electroplax of *Electrophorus electricus*. *Proc. Natl. Acad. Sci. USA* 81: 1239–43

Rosenberg, R. L., Tomiko, S. A., Agnew, W. S. 1984b. Single-channel properties of the reconstituted voltage-regulated Na channel isolated from the electroplax of *Electrophorus electricus*. *Proc. Natl. Acad. Sci. USA* 81: 5594–98

Salkoff, L., Butler, A., Wei, A., Scavarda, N., Giffen, K., Ifune, C., Goodman, R., Mandel, G. 1987. Genomic organization and deduced amino acid sequence of a putative sodium channel gene in *Drosophila*. *Science* 237: 744–48

Schmidt, J. W., Catterall, W. A. 1986. Biosynthesis and processing of the alpha subunit of the voltage-sensitive sodium channel in rat brain neurons. *Cell* 46: 437–44

Schmidt, J., Rossie, S., Catterall, W. A. 1985. A large intracellular pool of inactive Na channel alpha subunits in developing rat brain. *Proc. Natl. Acad. Sci. USA* 82: 4847–51

Sears, T. A., Bostock, H. 1981. Conduction failure in demyelination: Is it inevitable? In *Demyelinating Disease: Basic and Clinical Electrophysiology*, ed. S. G. Waxman, J. M. Ritchie, pp. 357–76. New York: Raven

Sharkey, R. G., Beneski, D. A., Catterall, W. A. 1984. Differential labeling of the alpha and beta 1 subunits of the sodium channel by photoreactive derivatives of scorpion toxin. *Biochemistry* 23: 6078–86

Sherman, S. J., Catterall, W. A. 1982. Biphasic regulation of development of the high-affinity saxitoxin receptor by innervation in rat skeletal muscle. *J. Gen. Physiol.* 80: 753–68

Sherman, S. J., Catterall, W. A. 1984. Electrical activity and cytosolic calcium regulate levels of tetrodotoxin-sensitive sodium channels in cultured rat muscle cells. *Proc. Natl. Acad. Sci. USA* 81: 262–66

Sherman, S. J., Chrivia, J., Catterall, W. A. 1985. Cyclic adenosine 3′ : 5′-monophosphate and cytosolic calcium exert opposing effects on biosynthesis of tetrodotoxin-sensitive sodium channels in rat muscle cells. *J. Neurosci.* 5: 1570–76

Sherman, S. J., Lawrence, J. C., Messner, D. J., Jacoby, K., Catterall, W. A. 1983.

Tetrodotoxin-sensitive sodium channels in rat muscle cells developing in vitro. *J. Biol. Chem.* 258: 2488–95

Shrager, P., Chiu, S. Y., Ritchie, J. M. 1985. Voltage dependent sodium and potassium channels in mammalian cultured Schwann cells. *Proc. Natl. Acad. Sci. USA* 82: 948–52

Strichartz, G., Bar-Sagi, D., Prives, J. 1983. Differential expression of sodium channel activities during the development of chick skeletal muscle cells in culture. *J. Gen. Physiol.* 82: 365–84

Stühmer, W., Almers, W. 1982. photobleaching through glass micropipettes: Sodium channels without lateral mobility in the sarcolemma of frog skeletal muscle. *Proc. Natl. Acad. Sci. USA* 79: 946–50

Stühmer, W., Methfessel, C., Sakmann, B., Noda, M., Numa, S. 1987. Patch clamp characterization of sodium channels expressed from rat brain cDNA. *Eur. Biophys. J.* 14: 131–38

Talvenheimo, J. A., Tamkun, M. M., Catterall, W. A. 1982. Reconstitution of neurotoxin-stimulated sodium transport by the voltage-sensitive sodium channel purified from rat brain. *J. Biol. Chem.* 257: 11868–71

Tamkun, M. M., Talvenheimo, J. A., Catterall, W. A. 1984. The sodium channel from rat brain. Reconstitution of neurotoxin-activated ion flux and scorpion toxin binding from purified components. *J. Biol. Chem.* 259: 1676–88

Tanabe, T., Takeshima, H., Mikami, A., Flockerzi, V., Takahashi, H., et al. 1987. Primary structure of the receptor for calcium channel blockers from skeletal muscle. *Nature* 328: 313–18

Tanaka, J. C., Eccleston, J. F., Barchi, R. L. 1983. Cation selectivity characteristics of the reconstituted voltage-dependent sodium channel purified from rat skeletal muscle sarcolemma. *J. Biol. Chem.* 258: 7519–26

Tanaka, J. C., Furman, R. E., Barchi, R. L. 1986. Skeletal muscle sodium channels: Isolation and reconstitution. See Agnew et al 1986, pp. 277–306

Thesleff, S., Vyskogil, F., Ward, M. R. 1974. The action potential in the endplate and extrajunctional regions of rat skeletal muscle. *Acta Physiol. Scand.* 91: 196–202

Thornhill, W. B., Levinson, S. R. 1986. Biosynthesis of electroplax sodium channels. *Ann. NY Acad. Sci.* 479: 356–63

Waechter, C. J., Schmidt, J. W., Catterall, W. A. 1983. Glycosylation is required for maintenance of functional sodium channels in neuroblastoma cells. *J. Biol. Chem.* 258: 5117–23

Weigele, J. B., Barchi, R. L. 1982. Functional

reconstitution of the purified sodium channel protein from rat sarcolemma. *Proc. Natl. Acad. Sci. USA* 79: 3651–55

Weiss, R. E., Horn, R. 1986. Functional differences between two classes of sodium channels in developing rat skeletal muscle. *Science* 233: 361–64

Weiss, R. E., Roberts, W. M., Stühmer, W., Almers, W. 1986. Mobility of voltage-dependent ion channels and lectin recep-tors in the sarcolemma of frog skeletal muscle. *J. Gen. Physiol.* 87: 955–83

Wollner, D. A., Catterall, W. A. 1985. Anti-genic differences among the voltage-sen-sitive sodium channels in the peripheral and central nervous systems and skeletal muscle. *Brain Res.* 331: 145–49

Yamamoto, D. 1985. The operation of the sodium channel in nerve and muscle. *Prog. Neurobiol.* 24: 257–91

Ann. Rev. Neurosci. 1988. 11:497–535

NEUROETHOLOGY OF ELECTRIC COMMUNICATION

Carl D. Hopkins

Division of Biological Sciences, Section of Neurobiology & Behavior, Cornell University, Ithaca, New York 14853

INTRODUCTION

Two groups of tropical freshwater fish, one South American and one African, send and receive weak electric signals in social communication. Like other, more familiar communication modalities, electric communication is a highly evolved system with many functions, including: sex- and species-recognition, courtship behavior, mate assessment, territoriality and other forms of spacing behavior, appeasement, alarm, and aggression. Electric communication is more than a zoological curiosity restricted to a few unusual groups of fishes; it has sparked considerable interest among neuroethologists because it is an ideal system to study the neurobiological basis of natural behavior. One reason for this is that the signal repertoires of electric fish are quite rich and varied, and the functions served by these signals are diverse. This makes the system interesting from the ethological perspective. Also, electric communication signals appear to be relatively simple in structure, and therefore are comparatively easy to generate and imitate. And finally, electric fish have a highly specialized electrosensory system that contains, in some species, a subpopulation of receptors and neurons whose sole function is in communication signal sensing. Because these sense organs appear dedicated to communication, the adaptive characteristics of the receptors and of the entire sensory submodality are best understood in the context of communication behavior. In this article I divide the analysis of electric communication mechanisms into three parts: signals, signal transmission, and signal reception.

SIGNALS

Central to any communication system are the signals themselves: What are they like and how are they generated and modulated for social communication?

497

Six phylogenetically diverse groups of fishes have evolved electric organs but only two have developed elaborate use of electric organs for communication: the South American Gymnotiformes (Figure 1) (electric eels and relatives), the African Mormyriformes (Figure 2) (elephant-nose fishes), both teleosts. A good deal is known about the structure and physiology of electric organs in these fish, largely as a result of a series of comparative studies carried out in the 1950s and 1960s by H. Grundfest, R. D. Keynes and H. Martins-Ferreira, M. V. L. Bennett, A. Fessard, T. Szabo, and others (reviews in Fessard 1958, Chagas & de Carvalho 1961, Grundfest 1957, 1966, Grundfest & Bennett 1961, Bennett 1961b, 1970, 1971a, Bass 1986b). Recently, interest in the structure and function of electric organs has revived: first, because there are correlates between the structure of the electric organ and the stereotyped waveform of the electric discharge, and second, because electric organs appear to change their characteristics both during larval development and during sexual maturation (see Kirschbaum 1977, 1981, Kirschbaum & Westby 1975, Bass 1986a, Bass & Hopkins 1985, Bass et al 1986a).

EODs VERSUS SPIs   Electric signals of these fish have a fixed part and a variable part. The electric organ discharge (EOD) is a stereotyped waveform fixed by the anatomy and physiology of the electric organ in the fish's periphery. EODs do not appear to be modulated under voluntary control, although they do change during development. The sequence of pulse intervals (SPI) is controlled by a chain of command interneurons in the midbrain and medulla, which produce patterns of discharges that make up the widely varying repertoire of social signals. In some "wave" species, the SPIs are highly regular and the EOD durations are long compared to the average inter-pulse interval. Among "pulse" species, the SPIs are more irregular, and the EOD duration is short compared to the inter-pulse interval. Both EODs and SPIs play a role in communication.

Generating the EOD (Gymnotiformes)

Although electric fish generate EODs differing widely in waveform and duration, the variation traces back to properties of electrogenic membranes in the individual electrocytes of the electric organs (Bennett 1961b, 1970, 1971a, Grundfest 1957, 1966, Keynes & Martin-Ferreira 1953, Bass 1986a, Bass & Volman 1985, Hagedorn & Carr 1985). Spike-generating membranes produce the transient currents that are the main part of most discharges, while other membranes behave as delayed rectifiers, large capacitors, large polarization sources, and low-resistance inexcitable surfaces (see Bennett 1971a). EOD variation is especially diverse among the gymnotiformes from South America (Figure 3), which systematists now

GYMNOTIFORMES

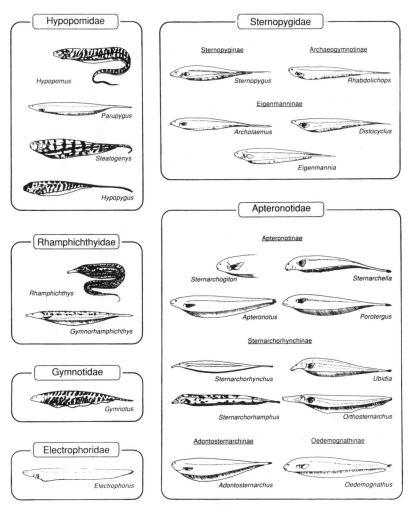

Figure 1   Electric communication is especially well developed among the Gymnotiformes, composed of 23 genera and six families of South American electric eels. There may be more than 100 different species of gymnotiforms, but systematists continue to discover new species with continued exploration. All of the Hypopomidae, Rhamphichthyidae, Gymnotidae, and Electrophoridae produce pulse discharges. All of the Sternopygidae and Apteronotidae produce wave discharges. Based on Mago-Leccia (1978).

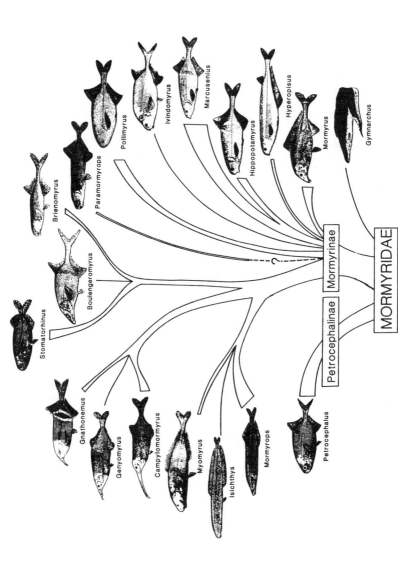

Figure 2 Electric communication is also well developed among the Mormyridae, a primitive African family of about 200 species in the Osteoglossomorpha (bony tongue fishes). The three main groups within the family are the Petrocephalinae, the Mormyrinae, and the Gymnarchinae. All except *Gymnarchus niloticus* produce pulse discharges. The width of each branch on this tree is proportional to the number of living species within the genus. Based on a phylogenetic tree of Taverne (1972). (From Hopkins 1986b.)

divide into six families: Electrophoridae, Gymnotidae, Rhamphichthyidae, Hypopomidae, Sternopygidae, and Apteronotidae (Mago-Leccia 1978, Mago-Leccia & Zaret 1978).

MONOPHASIC EODs   The monotypic species, *Electrophorus*, one of the species of *Hypopomus* (Hypopomidae), and all species in the family Sternopygidae, have comparatively simple electric organs composed of drum- or plate-like electrocytes, arranged in four to six columns along the longi-

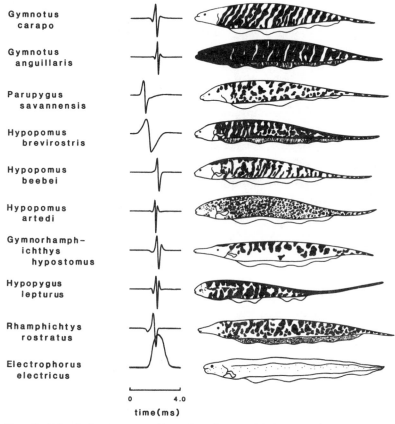

Figure 3   Pulse discharging gymnotiforms from the coastal Guianas of South America differ in the waveforms of their EODs. *Electrophorus electricus* has a simple monophasic discharge generated by an electrocyte with a single active face (posterior). Two of the *Hypopomus* generate biphasic waveforms using electrocytes with two active faces, posterior followed by anterior. The remaining species all have accessory electric organs that fire out of phase with the main organ and produce triphasic and tetraphasic waveforms (from Hopkins & Westby 1986).

tudinal axis of the fish (except in *Electrophorus*, which has three electric organs filling the body cavity). All electrocytes receive cholinergic inputs on their posterior (caudal) surfaces. The electromotor nerve depolarizes the posterior (innervated) face of the electrocyte and causes current to flow rostrally through the fish, through the skin, and caudally in a return path through the water. Voltage-sensing electrodes in the water record a monophasic head-positive potential during the firing of the posterior face of the organ (Figure 3).

All of these species have a simple, *monophasic* EOD (Bennett 1961b, 1971a, p. 400). Only the posterior, or innervated, face is electrically excited during the EOD. The anterior face of the monophasic *Hypopomus* has deep folds and tubular invaginations in the membrane (Schwartz et al 1975), which increase its surface area over 40 times and reduce the series resistance of this face. In *Electrophorus* the noninnervated face lacks sodium channels entirely, whereas the innervated face is densely populated. In both species, the folded surface is electrically inexcitable. In all wave-discharging species (Sternopygidae), the net DC component to the EOD caused by the monophasic pulses is balanced by an opposing, long-time constant, DC baseline voltage generated from extracellular K^+-ion accumulation in the spaces between electrocytes (Bennett 1971a).

BIPHASIC EODs   Many gymnotiforms in the Hypopomidae (e.g. *Hypopomus, Parupygus*) produce biphasic EODs (see Hopkins & Heiligenberg 1978, Hopkins & Westby 1986 and Figure 3, Heiligenberg & Bastian 1980). The externally recorded EOD matches the activity of single electrocytes (Bennett 1961b, 1971a). The first phase is produced when the caudal (innervated) face fires a spike, as before; the second phase is caused by an action potential in the rostral face that is triggered by the depolarizing current from the first phase discharge. Action potentials from the two faces may differ. In one species of *Hypopomus*, the innervated face fires a short spike while the uninnervated face fires a long-lasting spike that outlasts the former (Bennett 1961b). In females of a second species, *Hypopomus occidentalis*, the spikes produced by the two faces of the electrocytes are nearly equal in duration and amplitude, but slightly delayed with respect to one another so that the EOD is biphasic and symmetric (Hagedorn & Carr 1985). This contrasts with the breeding males where the electrocytes are asymmetric, apparently having been enlarged by effects of testosterone, and a short first-phase spike is followed by a long second-phase discharge.

COMPLEX EODs   All other pulse gymnotiforms produce more complicated EODs composed of three or four phases. Since the basic electrocyte produces only biphasic discharges, these fish have had to evolve a number and variety of "accessory electric organs" whose effects combine with the main electric organ. Accessory electric organs are found on the head, under the chin, and in the abdomen. The first case of this was noted in *Steatogenes elegans* (Hypopomidae) by Steindachner (1880), Lowry (1913), and Bennett (1961a).

Originally called "submental filaments," these accessory electric organs are composed of a single column of electrocytes innervated on their rostral surfaces. They lie on the underside of the head surrounding the urogenital opening and produce a monophasic, head-negative pulse approximately 1 ms before the main electric organ. A second pair of "postopercular" accessory organs, which lies near the skin surface posterior to the operculum, is innervated on the posterior side and fires a head-positive pulse in synchrony with the submental organs. The net effect of both accessory organs is to add an early head-negative component to the overall EOD (Bennett 1971a), which would have been biphasic in their absence.

Another species, *Hypopygus lepturus* (Figure 3), which is a close relative of *Steatogenys elegans*, carries only "postpectoral" electric organs (Nijssen & Isbrücker 1972), but it also has a complex EOD waveform. *Gymnorhamphichthys hypostomus*, a small sand-dwelling species with a four-phase EOD, also has submental electric organs (Bennett 1971a).

One of the most common gymnotiforms, *Gymnotus carapo*, has a triphasic discharge that is produced by the unusual arrangement of electrocytes in the abdomen. One column of cells is functionally turned around, anterior for posterior, so that the discharge has a reverse polarity in this region (Szabo 1961a, Trujillo-Cenóz et al 1984). The abdominal part of the electric organ has only a medial and a lateral column of electrocytes. The medial row resembles the four columns in the normal caudal electric organ in innervation, polarity, and physiology. The lateral rows of cells, by contrast, are all doubly innervated. The posterior faces are innervated from spinal roots VIII to XXI and the anterior faces are innervated by a shorter path from the pacemaker through roots I to VII. The command spike arrives at the anterior faces first, firing a head-negative spike, and after a short delay, at the posterior faces of both the abdominal electrocytes and caudal electrocytes, where it triggers a head-positive spike. Current through the caudal electrocytes depolarizes the anterior faces to fire, generating the final head-negative phase to the EOD. Although this unusual double innervation pattern was described in 1961 by Szabo, the observation received little attention until recently, when Trujillo-Cenóz et al (1984) isolated the caudal EOD $(+/-)$ from the abdominal EOD $(-/+)$ by cutting the spinal cord at different levels. *Gymnotus anguillaris* probably has a similar electric organ structure, and there may be other examples. Bennett (1971a, p. 402) describes an unidentified *Hypopomus* with a triphasic discharge in which a monophasic, head-negative potential is recorded from the abdominal region.

One can label the lateral row of electrocytes in the abdominal region of *Gymnotus* as an accessory electric organ, since its anatomy, physiology, and innervation are so different from the rest of the electric organ. These specialized, functionally reversed, electrocytes in *Gymnotus* may well be homologues of the accessory electric organs in *Steatogenys*. Both receive innervation from the rostral cord, although in *Steatogenys* the nerve is a recurrent branch that reaches the cephalic or submental region. One can further conclude that complex EOD waveforms are achieved in gymnotiforms only when accessory electric organs are present. Since the EOD waveform appears to be important in species and sex recognition in electric fish, some advantage may be gained, in terms of signal design, by the evolution of accessory organs.

NEURALLY DERIVED ELECTRIC ORGANS   Electric organs from fish in the family Apteronotidae are derived from nerve, not muscle. Spinal electromotor neurons first emerge from the cord and enter a lateral column with other similar electromotor neurons, run anteriorly for a few segments, and finally turn posteriorly to return to end at the level near where they started (Bennett 1971a, Waxmann et al 1972). The axons are enlarged up to 100 μm in diameter within the electric organ, and there are specialized enlarged nodes at the rostral-most and caudal-most ends of the axons that permit current to flow longitudinally along the electric organ axis as the discharge travels along the electric organ.

All of the known Apteronotids generate wave discharges, usually in the 600 Hz to 1800 Hz frequency range. The EOD waveform is usually biphasic but sometimes triphasic (see Kramer et al 1981), thus indicating a reversal in polarity as the impulse travels from the anterior part of the electric organ to the posterior part.

One species, *Adontosternarchus sachsii*, has a unique accessory electric organ in the chin region below the mouth, which is derived from an electrosensory nerve. Lacking motor control, this accessory electric organ fires at a different frequency and phase from the main organ. The discharge near the chin is monophasic (Bennett 1971a) and restricted to the head region.

LARVAL ELECTRIC ORGANS   A distinct class of larval electric organs arises in young gymnotiforms (and mormyrids), but the significance of these organs in behavior is unknown (see Kirschbaum 1981, 1983, Kirschbaum & Westby 1975, Westby & Kirschbaum 1977, 1978, Denizot et al 1978; review in Bass 1986b).

Generating the EOD (*Mormyridae*)

The African family Mormyridae, the largest group of electrogenic fishes, is made up of over 200 species, belonging to 26 genera (Figures 2 and 4). The mormyrids are primitive Osteoglossomorphs ("bony tongued fishes") (see Hopkins 1986b). All of the mormyrids produce pulse-discharges except one species, *Gymnarchus niloticus*, which produces a wave discharge. The EODs of mormyrids appear far more complex than those of their South American counterparts. By adding a penetrating stalk system to the electrocyte, the mormyrids generate complex EOD waveforms without accessory electric organs.

One of the best ways to appreciate the diversity of EOD waveforms is through field studies where one can survey a population of sympatric species and examine the diversity of their signals. Hopkins (1980, 1981, 1983b, 1986b) and Hopkins & Bass (1981) have studied a community of mormyrids in Gabon West Africa and found over 20 species of mormyrids with a diversity of EOD waveforms (Figure 4). The EODs of mormyrids are so stereotyped for a given species, and yet variable between different species, that the EOD serves as a good taxonomic character (Hopkins 1986b). EODs can be monophasic, biphasic, biphasic with plateaus and inflection points, triphasic, triphasic with plateaus and inflection points, or tetraphasic. The duration may range from 100 μs to well over 10 ms. Sex differences in EODs add to the complexity of a given community of species.

The structure of the individual electrocytes of mormyrid electric organs correlates with the EOD waveform (Szabo 1958, Bruns 1971, Bennett & Grundfest 1961, Bennett 1971a, Bass 1986a,b). Electrocytes are flat disk-shaped structures arranged in four longitudinal columns in the caudal peduncle like stacks of dinner plates on edge. Numerous small stalks emerge from one surface of the electrocyte disk like stilt roots, and coalesce into larger and larger branches. The stalks may penetrate through tight-fitting tunnels in the disk, or may not, but eventually join and receive cholinergic input from the electromotor neuron. Spikes generated in the posterior and anterior faces of the electrocyte generate head-positive and head-negative potentials as with the gymnotiforms, but the stalk also fires a spike, and this can add another wrinkle to the EOD waveform if the current generated there is allowed to pass through the electrocytes via the penetrating stalk system. Some species have doubly penetrating stalks, while others have nonpenetrating stalks. Each variant has its own distinctive EOD waveform.

Bass (1986a) has recently examined the structure of the Gabon mormyrids and has found six different types of basic plans for electrocytes.

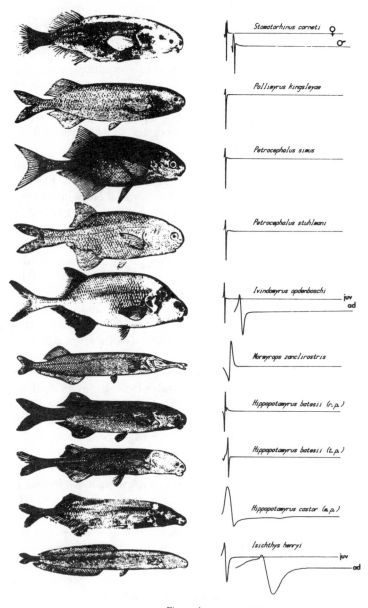

Figure 4

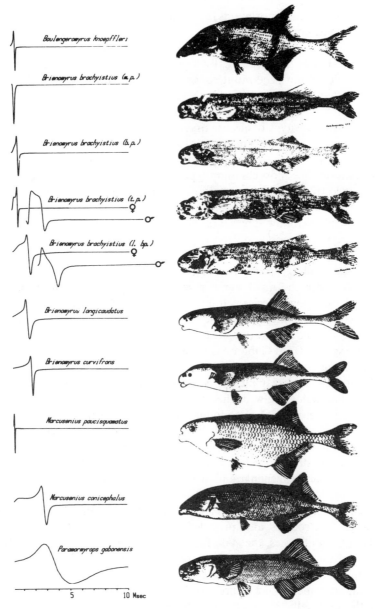

Figure 4   The mormyrids from the Ivindo River district of Gabon in West Africa produce a diversity of EOD waveforms. The fish are arranged by genus, and each species' EOD is illustrated with head positivity upward. In cases where the sexes differ, two traces are presented. One may conclude from examination of EOD diversity that EODs are important in species and sex recognition (from Hopkins 1986b).

The most primitive mormyrids in the genus *Petrocephalus* have electrocytes with a single nonpenetrating stalk. All examples of this type receive innervation on the posterior side of the electrocyte, and all species generate biphasic EODs that are initially head-positive. Species with electrocytes with multiple non-penetrating stalks are similar to those with a single stalk except that the nerve branches before arriving at the electrocyte and innervates each electrocyte stalk in several places. Two of the Gabon species with this type of electric organ have notably long duration EODs, but the waveforms are biphasic.

There are two types of electrocytes with penetrating stalks with single innervation. The most common type has innervation on the anterior side of the electrocyte with penetrations to the posterior side. Most of these species produce a triphasic discharge that starts out with a small head-negative phase followed by a large head-positive/head-negative phase. The second type, seen only in *Mormyrops zanclirostris*, is inverted—anterior for posterior—so that the innervation is on the posterior side with penetrations reaching through to fuse with the anterior side. The EOD is biphasic, with the main signal going head-negative then head-positive (Bass 1986a). One species is found to have multiple penetrating stalks in which the stalks, oriented like the majority of the single penetrating stalks, receive multiple innervation. The EOD is very long in duration.

One species from Gabon has both double penetrating stalks and non-penetrating stalks, and the EOD is tetraphasic (see Denizot et al 1978, 1982 for a second example in *Pollimyrus isidori*). It is tempting to make a functional analogy between the double system of stalks here and the double innervation in the lateral abdominal electrocytes in *Gymnotus* described above. With slight changes in the relative lengths of doubly penetrating vs nonpenetrating stalks (or in *Gymnotus*, between the direct spinal nerve and the recurrent nerve), the timing of the spikes in the stalk and those in the innervated face will change and thereby change the summed potential that makes up the EOD (see Westby 1984).

Bennett & Grundfest (1961) first examined the physiology of these unusual electrocytes and provided a reasonable model of how they function. In all cases, the stalk fires a spike with a longer duration than either of the two electrocyte faces, possibly because of differences in ion channel kinetics. When the stalks penetrate, radially directed current entering the stalk membrane is directed through the electrocyte and shows up on the externally recorded EOD to contribute the first phase to the waveform (head negative in the case of penetrating stalks innervated on the posterior side). As the stalk potential reaches the anterior surface of the electrocyte, and spreads over that surface, a depolarizing spike is generated (head positive). The spike through this face triggers a spike in the opposite face of the electrocyte, thereby reversing the polarity of the external field for a second time (head negative). The size of the first potential is related to the surface area of the stalk, and therefore to the number of branches on the stalk system. Bass (1986a) finds several species for which this correlation holds. In many species (i.e. *Mormyrus rume*; see Bennett & Grundfest 1961, Bennett 1971a) the two primary faces of the electrocyte have different kinetics, and the durations of the spikes generated there can differ. This model is explained in detail in Bass (1986a).

Sex Differences in EODs and Sexual Dimorphism of Electric Organs

It is not surprising that one finds sex differences in the electric signals of both the mormyrids and the gymnotiforms. What is surprising is that for many species, the sex differences are encoded into the brief EOD waveform. For both the mormyrids and gymnotiforms, recent evidence shows that the sex difference is correlated with a number of morphological differences at the level of the electric organ.

GYMNOTIFORMES   The first case of a sex difference in electric discharge was noted for the wave discharging species, *Sternopygus macrurus* (Hopkins 1972). Mature males discharge at a pacemaker frequency of 50 Hz to 90 Hz, females discharge at 100 to 150 Hz, and juveniles are intermediate. A parallel effect is on the EOD waveform, which becomes elongated in males compared to females, such that the entire waveform retains a species-specific wave shape (i.e. the waveform is isomophic with respect to time stretching or compression) (Mills & Zakon 1987). Natural sex differences in EODs and discharge frequency can be produced by testosterone administration (Meyer 1983, 1984). In *Eigenmannia*, Kramer (1985) found male EOD pulse durations shorter than female's, with no change in pacemaker frequency—a nonisomorphic change.

Hagedorn & Carr (1985) recently found a case of a pulse gymnotiform with a sex difference in EOD. *Hypopomus occidentalis* males from Panama produce biphasic discharges with twice the duration of those of females. Female EODs have two peaks that are nearly symmetrical in amplitude and duration, whereas the first phase of the male EODs is larger and shorter than the second phase. The change in the EOD can be produced in females after treatment with androgens (Hagedorn & Carr 1985). Further, the sex difference in EOD is correlated with a 100 to 300% enlargement of the caudal filament (which is largely composed of electric organ) of males compared to females. This sexual dimorphism is caused by the swelling of individual electrocytes among males. The action potential recorded intracellularly from the caudal (innervated) face of the electrocyte does not differ from males to females. The action potential in the rostral face (noninnervated) is reduced in amplitude and prolonged for males compared to females.

Sex differences in pulse waveforms also occur in *Hypopomus beebei*, *Parupygus savannensis*, and *Hypopomus sp* (Hypopomidae) (Hopkins & Westby 1986, J. Bastian and C. D. Hopkins, unpublished). Sex differences in EODs may be expressed only in the breeding season; this explains why they have been observed mainly in field studies. There are no known records of an EOD sex difference in a species with accessory electric organs.

MORMYRIDAE   Sex differences in EODs of pulse-discharging mormyrids were first noted by Hopkins (1980, 1981) for three species from Gabon (Figure 4). This list of species with sex differences has now increased to six (C. D. Hopkins and J. D. Crawford, unpublished) and it will undoubtedly increase further with additional field work. Westby & Kirschbaum (1982) found a clear sex difference in the EODs of the laboratory-bred, *Pollimyrus isidori*, and it appears to be a very widespread phenomenon among mormyrids. Males produce longer duration EODs than females in all cases, usually by nonisomorphic stretching. Bass & Hopkins (1983) demonstrated that females and juveniles can develop male-like EODs after androgen treatment.

Bass et al (1986a) found a morphological correlate of the sex difference in EODs of mormyrids. Androgens cause a general thickening of the anterior faces of electrocytes from *Brienomyrus brachyistius*, a species with penetrating stalks. In these fish, the anterior face is the side where the nerve terminals meet with the stalk, but the stalk does not fuse on this side but plunges through the electrocyte to the posterior side. The anterior surface, therefore, is the last face to fire an action potential. The thickening response only occurs in those species that have a natural sex difference in EOD. The hormone-mediated thickening is caused by surface invaginations—a series of tube-like penetrations into the anterior surface. Bennett (1971a) speculated that surface proliferation in the anterior face should increase the total capacitance of that surface and decrease its resistance. Bass et al (1986a), Bennett (1961b, 1971a), Bennett & Grundfest (1961), and Schwartz et al (1975) have found that fish with longer duration EODs have surface proliferations on the side that fires its spike last (usually the anterior face). This is especially pronounced among males where sex differences occur. A decrease in surface resistance will reduce the depolarization voltage across the anterior membrane, caused by current from the posterior face, and this will delay the onset of the anterior surface spike. An increase in capacitance will also delay the anterior surface spike by slowing the depolarizing effect of the posterior surface spike. The

Hodgkin-Huxley model would predict that the capacitance increase will not significantly alter the waveform of the spike in the anterior face. It is especially interesting that Bass & Volman (1985) report that the action potentials recorded inside the electrocytes of androgen-treated mormyrids are longer in duration than the action potentials of control animals. Bass et al (1986a) do not exclude possible significant changes in the ion channel characteristics of androgenized electrocyte membranes, since this has not been investigated.

In summary, the diversity of EOD waveforms among pulse gym-notiforms is largely due to the diversity of arrangement of electrocytes in the electric organs—both by the addition of accessory electric organs and by the development of special columns of cells, some with double innervation. In mormyrids, the EODs are diverse because the electrocytes are so complex and varied. In both groups, we commonly find sex differences in waveforms and find larval electric organs.

Control of the Sequence of Pulse Intervals (SPIs)

The electric organs are activated from a spinal motor nucleus, either through a cholinergic synapse in the case of myogenic electric organs, or directly on the axons of the neurogenic organ in the Apteronotids. The spinal motor neurons are activated in turn from the midline medullary relay nucleus in the brainstem (Szabo 1961b, Bennett et al 1967a,b, Szabo & Enger 1964, Libouban et al 1981, Dye & Meyer 1986). The relay nucleus is activated, in turn, from a medullary structure that has been called a pacemaker or command nucleus. Electrotonic coupling between cells in the pacemaker helps to synchronize the discharge.

MORMYRIDAE   The electromotor neurons in the spinal cord reside in a small nucleus within the caudal peduncle at the level of the electric organ (Bass et al 1986b,c). These cells receive descending input from large (15 μm) axons in a dorsallateral tract in the spinal cord (Szabo 1961b). The descending axons arise from the midline "relay nucleus" in the reticular formation, which is composed of only 25 to 30 large diameter (40–50 μm) adendritic cells (Bennett et al 1967a,b). Bell et al (1983) and Grant et al (1986) have identified a second, smaller (c. 30 μm) cell type directly ventral to the large cells, which provide the primary input to the large relay cells. There are only about 16 of these small fusiform cells. They have been called command cells by Grant et al (1986) because they appear to be the initiation site for the EOD motor command. The command cells are all coupled by gap junctions (Elekes & Szabo 1985). The small command cells have extensive dendritic trees, which receive diffuse inputs from many different sources (see Bell et al 1983, Grant et al 1986). Electrophysiological recordings by Grant et al (1986) have demonstrated that the command cells receive synaptic input that is not phase locked to the command to fire a pulse. Hyperpolarization of the command cells causes a decrease in firing rate from the command.

One specific source of inputs to the command nucleus is a so-called precommand nucleus in the mesencephalic tegmentum; another is from the mesencephalic torus ventralis posterior. Output from the command cell branches, with one branch going to the medullary relay nucleus, and the second to some small cells in the bilateral nucleus called the bulbar command associated nucleus (BCA). These neurons send fibers back to the medullary relay. This excitatory/excitatory loop probably provides the positive feedback needed to create a double action potential in the relay cell (see Bennett et al 1967a,b, Bennett 1971a, Bell & Szabo 1986, Bell 1986, Elekes et al 1985).

The diffuse input to the command nucleus is clearly important in determining the pattern of interpulse intervals generated by the command nucleus. There are many patterns, dependent upon social situation (below) or on electrolocation performance situation, yet none of these pathways has been worked out in detail.

GYMNOTIFORMS   The same organization—relay nucleus and command nucleus—occurs in the neural control of gymnotiform electric signals (Dye & Meyer 1986). Large cells in the pacemaker nucleus in the ventral medulla are relay cells that connect to the spinal electromotor neurons in the spinal cord; the small cells in the nucleus are the command cells. In gymnotiforms, the wave species tend to have a homogeneous organization in which relay cells and command cells are intermingled, whereas among pulse species the two cell types are more clearly segregated as they are in mormyrids (Dye & Meyer 1986). Where distinct, the command nucleus lies dorsal to the relay nucleus.

Through lesion studies, Szabo (1961b) demonstrated that the pacemaker nucleus can generate the rhythm to drive the electric organ in the absence of descending inputs from midbrain or higher centers. Meyer (1984) has studied autoactivity of the pacemaker of *Apteronotus* in a slice preparation and has modulated discharge frequency with androgens. The intrinsic connections of the pacemaker nucleus are described in Elekes & Szabo (1981), Bennett et al (1967a,b), and Tokunaga et al (1980).

HRP injections of the pacemaker nucleus of gymnotiforms have demonstrated that inputs to the pacemaker nucleus are derived from a single source: the prepacemaker nucleus in the mesencephalic tegmentum (Heiligenberg et al 1981). The prepacemaker has two cell types: large multipolar cells, which send thick axons to the pacemaker, and smaller, spindle-shaped cells, which also send thinner axons to the pacemaker. Electrical stimulation in the prepacemaker nucleus produces a change in discharge frequency (Bastian & Yuthas 1984). Recently Kawasaki & Heiligenberg (1987) have succeeded in evoking rapid transient frequency increases from the pacemaker of *Eigenmannia* by electrical stimulation in the large multipolar cells of the prepacemaker. The modulations resemble courtship and agonistic "chirps" and "interruptions" described in the behavioral studies of Hopkins (1974c) and Hagedorn & Heiligenberg (1985). The prepacemaker in turn receives inputs from the nucleus electrosensorius, which lies at the border of the diencephalon and pretectum (Bastian & Yuthas 1984). Electrosensorius is a nucleus that responds to electrical stimuli of social significance, the jamming avoidance response (see Heiligenberg 1977, 1986 for review).

In summary, both the mormyrids and the gymnotids have a similar motor control system, composed of a command nucleus and a relay nucleus, for controlling the discharge of the electric organ. Synchrony of the discharge is promoted by electrotonic coupling among the relay cells. The discharge rhythm can be modified through inputs to the command system, but these inputs, and their effects, are known only in a general way, and the role of these structures in the generation of species-specific patterns of discharge remains to be elucidated.

Patterns in the Sequences of Pulse Intervals

The pattern of electric discharges can be described simply as a sequential list of time intervals between EODs. A large number of sequences of pulse intervals (SPIs) have been described and characterized, and many of these function as "displays" in the social behavior of the fish. Patterns of SPIs may be tonic patterns (unmodulated activity), cessations, frequency modulations, or interactions between the discharges of two or more individuals. All four types of patterns have communicative functions.

TONIC DISCHARGE PATTERNS   Electric fish generate tonic discharge patterns when they are resting and when they are active but not interacting socially. This tonic activity is the baseline against which other discharge patterns can be compared. Although the primary function of the tonic activity is for electrolocation, it has a signal function as well.

Wave species use tonic discharge frequency in species recognition, sex recognition, and possibly individual recognition. Wave species living sympatrically sometimes have non-overlapping discharge frequencies (Hopkins 1972, 1974a,b, Kramer et al 1981), and in at least one species, *Sternopygus macrurus*, males and females diverge in frequency. A number of studies have shown agonistic or sexual responses to sinusoidal signals of the appropriate frequency range. In *Eigenmannia virescens*, spawning occurs only when the male has a lower frequency than the female (Hagedorn & Heiligenberg 1985).

Although males and females have widely overlapping frequency ranges, there are few observed cases where a male with a high frequency will mate with a female with a lower frequency. Discharge frequency is an indicator of dominance in *Eigenmannia* in community aquaria: Dominant male individuals have the lowest frequency, subordinates have higher frequencies. Similarly, among females, the dominant female discharges at the highest frequency. If a new dominance order is established in a tank after fighting, the new dominant male gradually lowers its frequency. In *Apteronotus* the effect is reversed: Males spawn only with females of frequency lower than their own and dominant males tend to have the highest frequency in a social group (see Hagedorn & Heiligenberg 1985, Hagedorn 1986). Finally, in a case of possible individual recognition, a small sample of mated pairs of *Sternopygus macrurus*, male and female, adopted frequencies exactly one octave apart (Hopkins 1974a, C. D. Hopkins and K. P. Harpham, unpublished).

The resting discharge of pulse gymnotiforms is usually quite regular but the coefficient of variation can be vastly different between species. *Gymnotus carapo*, a typical pulser, generates intervals that vary by as much as 20 ms (mean interval = 18 ms) when undisturbed, whereas the pulse intervals of *Rhamphichthys rostratus* may vary by only a fraction of a millisecond (mean interval = 31 ms) (Hopkins & Westby 1986). Low frequency pulsers (e.g. *Hypopomus brevirostris*, mean interval = 44 ms) tend to fire more irregularly than high frequency species (e.g. *Hypopomus sp.*, mean = 7.3 ms; Hopkins & Westby 1986). The tonic resting frequency of many species is lower and less variable than the discharges during active swimming nonsocial discharge. Bullock (1969), Black-Cleworth (1970), Hopkins & Heiligenberg (1978), and Hopkins & Westby (1986) all have comparative data on discharge interval of pulse species.

Pulse mormyrids have far more variable tonic discharges than gymnotiforms. The intervals typically range between 10 ms and 500 ms depending on the species and the time of day, and intervals typically cluster into three or more modal peaks (Harder et al 1967, Heiligenberg 1976, 1977, Bauer 1974, Kramer 1976a, Bell et al 1974, Hopkins 1986b). Additional patterns have been noted. Many mormyrids switch from random intervals, for which there is little correlation between adjacent interpulse intervals, to regular intervals (Moller 1970, Hopkins 1986b), although this is undoubtedly a social response that occurs when one fish senses a second one. Tonic resting discharges differ from discharges during swimming, which differ from active electrolocation (Serrier 1973, Teyssèdre & Serrier 1986, Graff 1986, Toerring & Moller 1984).

INTERRUPTIONS   Discharge interruptions are complete breaks in an otherwise continuous discharge. All electric fish, except the Apteronotids, generate them: Short interruptions (< 1 s) make a strong contrast with the continuous discharge and are used both as aggressive "threats" and as courtship displays; long cessations (> 1 s) render the signaller electrically silent and are used to signal submission or alarm (reviewed in Hopkins 1974a, 1977, 1983a). *Eigenmannia virescens*, a well-studied South American wave species, produces stereotyped 20 to 100 ms interruptions in bouts of one to five during fighting. These displays accompany attacks, precede "butts" and charges, and may even cause subordinates to flee (Hopkins 1974c). During courtship, males generate continuous trains of interruptions (called "chirps" by Hagedorn & Heiligenberg 1985), usually of longer duration (100 to 300 ms) when they are near a female in spawning condition. Continuous chirping may continue for several hours on an evening when a pair is spawning (Hagedorn & Heiligenberg 1985). Chirps in *Eigenmannia* can be evoked by stimulating single cells in the prepacemaker nucleus (Kawasaki & Heiligenberg (1987).

The wave mormyrid, *Gymnarchus*, also produces discharge interruptions in agonistic behavior. Short interruptions are used by dominants during threats and attacks; long cessations, including complete electrical silence, are produced by subordinates after an attack. The long interruption serves to prevent or reduce attacks from dominants. This remarkable case of convergent evolution of display form and function deserves special note (Hopkins 1977), for it suggests that selection may shape the form of electrical displays in much the same way as it has shaped vocal displays (see Marler 1959, Morton 1975, Wiley & Richards 1980).

Pulse fish produce cessations with a similar function. *Gymnotus carapo* produces brief cessations (< 1 s) during attacks and threats; long cessations are used by subordinates who probably "hope" to remain undetected. Mormyrids give discharge cessations as well, in very similar contexts (Kramer & Bauer 1976, Bell et al 1974, Hopkins 1986b).

FREQUENCY MODULATION (FM)   Frequency modulations are defined as transient changes in the frequency of the electric discharge. A very common FM is an acceleration in discharge rate followed by a decrease back to the resting rate. *Brief accelerations, bursts, buzzes, SIDs (sharp increase in frequency followed by decrease), scallops, rasps,* and *chirps* are all examples of frequency modulations (see Hopkins 1986b, Hagedorn 1986, and Dye & Meyer 1986 for recent summaries). The functions of FMs are numerous, so I mention only a few here.

Black-Cleworth (1970) provided one of the first detailed descriptions of electric displays in *Gymnotus carapo*. SIDs very often are given prior to an overt attack, during an attack, charge, or head butt, or during predatory attack. SIDs cause subordinates to retreat. Similar displays have been noted for other pulse gymnotids, such as *Hypopomus* (Westby 1974, 1975a–c, 1979).

Mormyrids also produce brief accelerations of their discharge (Bauer 1974, Kramer 1976a,b, Moller 1970, Bell et al 1974, Kramer & Bauer 1976), which have been called bursts, buzzes, and smooth accelerations by various authors. All are relatively similar in function and form. Many of these displays accompany highly aggressive acts such as antiparallel swimming, intense aggression, attack, head butting, and others. These displays are summarized in Hopkins (1986b).

A distinctive modulation, called a "rasp," is composed of a burst of pulses at a relatively constant frequency (rather than a strongly decreasing frequency). Rasps have been noted for *Hypopomus* (Hopkins 1974a, Hagedorn 1986) during courtship, and they have been documented in several species of mormyrids (Hopkins & Bass 1981).

Some of the most remarkable examples of frequency modulations are found in the wave species that were once thought to be invariant. In some wave species, the resting discharge frequency has a coefficient of variation of less than 0.01% (Bullock 1969). Against this constancy, FM signals of only a few Hz change can be significant. Male *Sternopygus macrurus* have elaborate courtship songs composed of frequency increases and decreases that range from 50 to 150 Hz.

The FMs are interposed with cessations in the discharge that last seconds. Breeding male *Sternopygus* frequently congregate in groups of five to six individuals in 5 m^2 and produce these elaborate rises and frequency modulations to passing females. Males respond to playback of sine waves but only when the frequency of the sine is in the 100 to 150 Hz range typical of females (Hopkins 1974b). Females also produce FMs but the frequency excursions recorded from females are not as large as for males. Females often make 1 to 5 Hz increases and decreases in frequency when paired with a male.

Eigenmannia also produces FMs in social interactions. Short rises (less than 1 s) of 10 to 50 Hz are used as aggressive threats. Long rises (5 to 40 s) of 2 to 20 Hz signify submission. Female *Eigenmannia* produce an elaborate warbling FM of 5 to 20 Hz during courtship and spawning (Hopkins 1974c, Hagedorn & Heiligenberg 1985, Hagedorn 1986). Again, in another striking example of convergence, *Gymnarchus* also produces FMs during appeasement behavior, and they resemble those of *Eigenmannia* very closely both in form and in context (see Hopkins 1977).

The most stereotyped and best-studied FM is the jamming avoidance response (JAR) known for some wave species, including *Eigenmannia* and *Gymnarchus*. The JAR is a small frequency shift (<20 Hz) upward or downward given in response to a pacemaker-like stimulus with a frequency near to the fish's own discharge frequency (Watanabe & Takeda 1963). The response is a highly stereotyped reflex, whose circuitry has been the study of intense investigation by Heiligenberg and co-workers (see Heiligenberg 1986).

INTERACTIONS BETWEEN MORE THAN ONE FISH   The SPIs of electric fish are often influenced by electrical stimuli from other fish. Gymnotiform pulse fish tend to shift the phase of their pulses either to avoid coincidences or to synchronize (Langner & Scheich 1978, Hopkins & Westby 1986, Heiligenberg 1974, 1977, Baker 1980, 1981); wave species show active phase coupling (Langner & Scheich 1978) and jamming avoidance (Bullock et al 1972a,b, Scheich 1974, 1977a–c, Heiligenberg 1977, 1986); mormyrids echo each other's discharges at preferred latencies (Bell et al 1974, Heiligenberg 1974, Bauer & Kramer 1974). Interactions between electric signaling and electric sensing may have evolved to prevent electroloca-tion jamming or they may have arisen as a communication strategy (see Hopkins & Westby 1986 for discussion on scan/sampling hypothesis).

Signals, then, are encoded as different EODs and as SPIs. The EOD is a stereotyped carrier or signature for the more variable SPI. Combined, EODs and SPIs function in signaler identification, in conveying intent to attack and flee, in maintaining spacing, and in reproductive behavior.

SIGNAL TRANSMISSION

The physical principles that apply to electrostatics determine, to a large extent, the characteristics of signal transmission in electric fish. From these, one can derive measures of the active space of an electric fish's signal. Physical principles also make predictions about the temporal properties of received signals and about the problems of signal localization.

Active Space

The *active space* of a signal is the area or volume within which the signal can evoke responses from other organisms. Bossert & Wilson (1963) introduced the active space concept in their analysis of diffusion of chemi-cal communication signals, and others have applied the concept to a variety of different sensory and communication modalities (see Brenowitz 1982, Wilczynski 1986, Hopkins 1986b). Active space is determined by four independently varying factors: (*a*) the amplitude of the signal at the source; (*b*) the rate of attenuation of the signal with distance in the natural environ-ment; (*c*) the receiver's masked sensory threshold for the signal in the presence of noise; and (*d*) the nature and the amplitude of the noise in the environment. For the electric modality, empirical measurements demonstrate that the active space is limited to a radius of 50 cm to several meters only (Moller & Bauer 1973, Squire & Moller 1982), although the maximum range varies with the species, sex, and size of the signaler. What factors contribute to this limited range?

AMPLITUDE AT SOURCE   Gymnotiforms and mormyrids both produce discharges measuring several hundred millivolts to tens of volts, peak to peak, in air (see Bennett 1971b, Bell et al 1976), but in water the resistive load of the water reduces the peak voltage by 50% or more. The fish in water can be effectively modeled as a Thevenin equivalent (see Harder et al 1967, Knudsen 1975, Bell et al 1976, Heiligenberg 1975), i.e. a voltage source with skin resistance and water resistance in the current path. When the total resistance through the external path is matched by the skin resistance, then the voltage is half the maximum value.

Bell et al's (1976) measurements of current and voltage in the mormyrid, *Gnathonemus petersii*, indicate that the first phase of the biphasic EOD can be effectively modeled as a voltage source (c. 8 V) in series with a fixed internal source resistance of approximately 4.3 to 4.8 kOhms and a variable load resistance due to the water. The peak current, observed at low load resistances, ranges from 2 to 4 mA. The second phase of the EOD is triggered by the current from the first phase. As the external resistance increases, this current drops, weakening the depolarizing current and ultimately causing a drop in the peak voltage of the second phase. With this comes a change in the shape of the EOD. Since the anterior faces of the electrocytes receive no innervation of their own, they are normally depolarized only by current from the posterior faces. Interestingly, significant distortions of EODs only occur at water resistivities of 100 kOhm·cm or higher, a value that is much higher than that found in most natural creek and river waters in Africa. Knudsen (1975) and Heiligenberg (1975) showed a number of gymnotiforms to be impedance matched to the resistance of their natural water. It would be interesting to explore this general phenomenon in extreme environments, namely in the Rio Negro of Brazil where the resistivity of the water can reach 200 kOhm·cm.

The amplitude of the peak current generated by an electric fish is directly affected by the fish's size (Knudsen 1975). In *Eigenmannia* (monophasic discharge) and *Apteronotus* (neuronal electric organ), the EOD waveform does not change in water with different conductances. The empirically measured dipole moment of the discharge varies widely from 41 to 1240 μA·cm for a number of small gymnotiforms. Plots of dipole moment as a function of fish size revealed a linear increase of

75 μA·cm per cm standard length. The electric eel's voltage output in air is similarly related to length (Brown & Coates 1952).

Species differences in EOD amplitude can be explained simply by the number of electrocytes in the electric organ column. The voltage from individual electrocytes sum like batteries in series (see Bell et al 1976); another factor is the cross-sectional area of the organ, which affects the internal source resistance. Whereas *G. petersii* has between 97 and 120 electrocytes per column, each contributing approximately 100 mV to the total voltage, Bass (1986a) found that *Paramormyrops gabonensis* from Gabon, a species with a very weak, long duration EOD, has only 20 to 22 electrocytes in a column; while *Isichthys henryi*, another weak-discharging species, has only 41 to 48 electrocytes in a column. Some mormyrids such as *Mormyrus rume* with numerous electrocytes have discharges that are strong enough to give a weak electric shock when handled. These fish evidently have a considerably larger signal amplitude that will carry further in communication.

Although physiological factors control the current and voltage of the discharge at the source, elementary electrostatic theory predicts the magnitude and direction of current at a distance from the source.

RATE OF ATTENUATION   The rate at which signals are attenuated with distance is probably the most significant determinant of active space of electric communication, and geometric spreading factors are of primary importance. Electric fish are close to an ideal dipole, for which current declines in proportion to the inverse of the cube of the distance from the source density in a homogeneous environment.

Knudsen's (1975) empirical measurements support the dipole field predictions, at least for distances greater than one to two body lengths. When individuals of two species were tethered at the surface of a pool of water 1 meter deep and log-log plots of the magnitude of the electric field as a function of distance were compiled, the plots had a slope of -3, as expected for a dipole. For distances less than a body length, in the electric "near field," the body shape and conductivity and inhomogeneities of the skin surface cause deviations from perfect dipole. Comparable empirical data are lacking that measure the rate of decrease of the electric fields in real environments like shallow streams, swamps, or other more complex natural bodies of water.

The inverse cube reduction of electric currents by geometric spreading is more severe than inverse square or inverse first power attenuation, as found for sound intensity or chemical concentration. This, more than any other factor, will limit electric communication to short ranges. Electric signalers must increase the amplitude of their signals eight-fold in order to double their active space radius (as compared to a four-fold increase in sound intensity, or a simple doubling of chemical concentration).

ELECTRORECEPTOR THRESHOLD   Active space is determined not only by the signal, and by signal transmission, but also by the receiver sensitivity. Electroreceptor thresholds have been determined both physiologically and behaviorally, and the results have been recently reviewed by Bullock (1982), Zakon (1986), and Hopkins (1983a). I mention only two points here. First, behavioral measures of sensitivity to EOD stimuli indicate that communication ranges are on the order of 1 m for wave discharging gymnotiforms (Knudsen 1974). Second, electroreceptor sensitivity measures aid in the identification of Knollenorgans as the primary communication sensor in mormyrids, but similar measures do not distinguish between receptor types in the pulse gymnotiforms.

Knudsen (1974) measured absolute thresholds to sinusoidal electrical stimuli in *Eigenmannia* and *Apteronotus* using a conditioning paradigm with food reward. Both species were behaviorally tuned to sinewave frequencies near their own discharge frequency, although *Apternotus* was also extremely sensitive to low frequencies (presumably due to ampullary electroreceptors). At the best frequency, the peak sensitivity of both *Eigenmannia* and *Apteronotus* was found to be 0.25 μV/cm in 2.0 kOhm·cm water. The threshold increased to approximately 1.0 μV/cm in 30 kOhm·cm water. Using this measure, and using data on the rate of attenuation of signals from fish, Knudsen (1975) computes the range of communication to be between 40 and 116 cm in 2 kOhm cm water (50 to 160 cm in 10 kOhm·cm water), depending upon the size and species of fish. The range is greatest at 0° and 180° with respect to the signaler, smallest at 90°. These distance estimates are consistent with the empirically determined measurements of 30 cm for mormyrids (Moller & Bauer 1973, Russell et al 1974). Squire & Moller (1982) used the unconditioned EOD-cessation response to evaluate the distance at which one mormyrid fish, passively restrained, could sense another restrained fish moving toward it. The maximum distances were greatest when the conductivity of the water was low (resistivity high) and greatest when the fish were parallel to each other.

A clear difference is found in mormyrids in threshold and tuning of the Knollenorgan electroreceptors and the mormyromast electroreceptors. Knollenorgans are tuned to the peak of the power spectrum of the species-specific EOD, and they are sensitive to sinusoidal stimuli on the order of 0.1 mV/cm longitudinal stimulation. Mormyromast receptors are tuned to frequencies well below the peak of the power spectrum, and the threshold is an order of magnitude higher (Bullock 1982, Hopkins 1980).

Gymnotiform pulse fish have two classes of tuberous electroreceptors, pulse marker units and burst duration coders. The relative sensitivity of the two receptor classes to externally generated fields has been questioned in the literature. Szabo & Fessard (1974) and Hopkins & Heiligenberg (1978) report that pulse marker (Type I of Szabo) units are more sensitive than burst duration coders, whereas Bastian (1976) reports that pulse markers are less sensitive than burst duration coders. Recently, J. Bastian (personal communication) has studied receptor sensitivities in pulse markers and burst duration coders from *Hypopomus beebei* in an attempt to resolve the differences in these three reports. If stimuli are applied so that the current flows transverse to the body axis, then burst duration coders have the same threshold values as pulse markers within 10 dB of each other. When the stimulus current flows longitudinal to the body axis, from electrodes positioned in the head-to-tail orientation, then pulse marker units appear to be 10 to 15 dB more sensitive than burst duration coders. This apparent difference can be explained by simple geometry. Pulse markers are located primarily on the head region, where the strongest current density is found when stimulus currents run parallel to the body axis. Changing the direction of the current flow from longitudinal to transverse changes the distribution of currents, but not the sensitivity of receptors. Bastian concludes that the two classes of receptors have approximately the same absolute sensitivity.

If an electroreceptor is to function exclusively in the detection of conspecifics, it is best placed on the head, or in the caudal abdominal regions where current density will be greatest. This is exactly where most mormyrids have the greatest number of Knollenorgans (Quinet 1971, Harder 1968). Bennett (1965, 1971b) noted a marked increase in Knollenorgan sensitivity on either the head or the most caudal patches of receptors compared to those located at mid-body when these receptors were stimulated with head-to-tail electric fields. They did not differ in sensitivity when local fields were applied across the skin (similar to the transverse stimulus result of Bastian, above).

NOISE   The fourth factor influencing active space is the electrical noise in the communication channel that has the potential of masking communication signals for the receiver. The most significant source of non-biological electrical noise in tropical environments comes from distant lightning storms, which create broadband clicks and tweeks (Hopkins 1973, 1980, 1986b). Noise from lightning is more intense at night when electric fish are active because of earth to ionosphere reflections of distant lightning flashes. The frequency of above-threshold noise may be as high as several events per second (Hopkins 1973). It would be desirable to observe signaling behavior and to determine electrosensory thresholds to biologically significant signals in the presence of actual noise from lightning or artificial noise with similar characteristics.

Electric Signals are Nonpropagating

Unlike sound and light, which propagate as waves, electric signals from fish exist as quasi-stationary electrostatic fields (see Hopkins 1986a). When sounds propagate through the atmosphere or water, they are subject to reflections (echoes, reverberations), refraction (bending of sound rays, generation of shadow zones), absorption by the molecules of water or air, dispersion (differential propagation velocities for different frequencies) (review in Wiley & Richards 1983, Hopkins 1986a). All of these factors tend to distort sound signals during transmission in a largely unpredictable way. By contrast, nonpropagating electric signals from fish appear to be immune to many sources of signal distortion that affect sounds. Much of the distortion introduced into sound signals by the environment concerns the temporal characteristics of the received signal. Hopkins (1986a) has suggested two important consequences of the electric signals' relative freedom from wave-propagation distortions during transmission:

1. Temporal characteristics of waveforms are likely to be far more important in the electric modality than they are in the acoustic modality simply because electrical time domain cues are much less affected by the environment than are acoustic cues. Electric fish may be attuned to subtle time-domain cues in signal waveforms.
2. Electroreceptors may be specialists at time-domain processing of signals, and may show adaptations to take advantage of the detail of information that is potentially available even after transmission through the environment.

Electric Signal Localization

When one electric fish detects and localizes a distant electric field generated by another electric fish, it is called *passive electrolocation* to distinguish it from *active electrolocation* in which the fish senses distortions in the self-generated electric field produced by objects in the environment. Active electrolocation is a well-studied phenomenon (Bastian 1986), whereas passive electrolocation is relatively unexplored. Recently, Schluger & Hopkins (1987) studied passive electrolocation in the mormyrid, *Brienomyrus brachyistius*, and found that these fish are capable of finding a nonmoving electrical dipole by approaching it along the current lines. The fish maintain a precise alignment between their body axis and the local electric field vector while swimming. This leads them to the current source in a reliable, although indirect, pathway. Additional studies by Davis & Hopkins (1988) indicate that gymnotids adopt the same behavioral strategy when attempting to find other electric fish. Unlike organisms engaged in the passive detection of sound or optical signals, electric fish are deprived of a number of important physical cues related to the location of distant signalers. Since electric signals are essentially non propagating, there are negligible time delays associated with distance and direction of signals. Further, electric currents follow curved lines, thus making extrapolation of locally perceived vectors useless for predicting the source of a current. Alignment to current lines and following them is a relatively simple yet reliable strategy for locating conspecifics in space.

SIGNAL RECEPTION

A crucial question posed by the finding of EOD and SPI diversity among the mormyrids and gymnotiformes is whether or not these signals have communicative significance, i.e. whether receivers respond to the signals and make discriminations among them. The question is especially interesting for EODs that are only 0.5 to 1.0 ms in duration—far shorter than most other time-varying signals that the nervous system is capable of dealing with. Can electric fish process EODs and extract meaningful information from the fine structure of waveforms?

Behavioral Studies

A number of ethological studies have demonstrated a clear relationship between the emission of an electrical signal by one electric fish and the evocation of a response by a second (Valone 1970, Black-Cleworth 1970, Hopkins 1972, 1974b,c, Hagedorn & Heiligenberg 1985). Among these are quantitative studies of sequences of electrical and motor acts of social interactions. These studies have demonstrated that electric displays can influence receivers to respond in predictable ways even in the absence of any other observable behavior (Black-Cleworth 1970, Hopkins 1974c). Playback experiments have also been particularly valuable in demonstrating that electrical cues are responsible for eliciting responses, and have been used to isolate salient cues for evoking natural responses in communication.

Hopkins & Bass (1981) and Hopkins (1983b, 1986b) have used field playback of synthetic EODs with a number of SPIs to test the importance of EOD waveform for species and sex recognition among mormyrids. These studies show that males of the Gabon species, *Brienomyrus brachyistius*, will electrically "court" playback electrodes introduced into their territories that play female EODs. The males will not court male EODs played to them or the EODs of any other sympatric species, even if they were delivered with the SPI that was effective when paired with female EODs. In all cases, the courtship signals were easily recognized as high-frequency bursts of EODs called "rasps," given predominantly by males.

To demonstrate that the SPI are of secondary importance in evoking courtship responses, Hopkins (1983b) scrambled the natural SPI and showed that they are as effective as the natural SPI in evoking rasps. Interval sequences derived from random number tables also served well in this context.

TEMPORAL CODING OF EODs   Hopkins & Bass (1981) used playback experiments and found that *Brienomyrus brachyistius* (*triphasic*) males were attending to the temporal features of the female's EOD waveform, rather than the power spectrum characteristics of the pulses. When they played female EODs that had been reversed in time, or female EODs subjected to unity-gain, constant phase-shift digital filtering, they evoked fewer courtship rasps from males than when normal female EODs were played (Hopkins & Bass 1981). Males presented with rectangle-wave stimuli courted the electrodes only when the duration of the rectangle matched the duration of the female's EOD. The square wave was thus a good "sign stimulus," in the ethological sense, for the female discharge. This behavioral result strongly suggests that mormyrids are sensitive to time-domain cues occurring within the EOD waveform.

EOD RECOGNITION IN PULSE GYMNOTIFORMS   Pulse gymnotiforms produce EODs with complex and often species-specific waveforms (Hopkins 1972, 1974c, Hopkins & Heiligenberg 1978, Heiligenberg & Bastian 1980). Because most of these EODs are extremely brief by comparison with the duration of a single nerve impulse, it is of interest to consider the sensory mechanisms by which one fish distinguishes one EOD from another. Hopkins & Westby (1986) summarized three possible mechanisms for EOD recognition, including *spectral coding*, which refers to the possible encoding of EODs by electroreceptors tuned to different frequencies as cross fiber patterns. *Temporal coding* refers to the possibility that EODs are encoded as a characteristic volley of nerve spikes patterned in the time domain, as appears to be the case for mormyrids. *Scan sampling*, a novel third method, applies only to the pulse gymnotiforms. According to this notion, a receiver senses a signaler's EOD as an amplitude modulation or "beat" set up by the combination of its own discharge with the signaler's. The receiver might use the modulation envelope to access the signaler's EOD waveform. Hopkins & Westby (1986) found support for the scan sampling hypothesis by demonstrating that one pulse species, *Hypopomus beebei*, was able to discriminate very similar EODs as long as the stimulus was free-running with respect to the test fish's discharge. When stimulus pulses were clamped to the firing of the fish's own discharge, the captives did not discriminate the EODs being tested.

THE ROLE OF SPIs   Various approaches have been taken in the study of sequences of pulse intervals (Harder et al 1967, Lissmann 1958, Black-Cleworth 1970, Graff 1986, Kramer 1976a, Hopkins 1972, 1974a–c, Hopkins & Bass 1981, Heiligenberg & Altes 1978, Westby 1974, 1979, Teyssèdre & Serrier 1986, Hopkins & Westby 1986). Some of the first playback experiments done on electric fish tested the role of discharge frequency in species and sex recognition in *Sternopygus* (Hopkins 1972, 1974b).

Studies of the importance of SPIs in the pulse species in evoking specific responses by mormyrids have been more difficult to perform and analyze. Because the SPIs are continuous and highly variable, demonstration of clear responses to a specific segment of an SPI has been difficult. Hopkins (1974c) played artificial discharge interruptions to *Eigenmannia* and found a reduction in attacks directed at the model. Increasing numbers of interruptions caused a reduction in attack intensity. Kramer (1979) found that mormyrids discriminate SPIs recorded from attacking fish from those recorded from resting fish. Similar studies by Teyssèdre & Serrier (1986) have also shown quantitative differences in responses to different SPIs. Much remains to be learned about specific SPI patterns and their role in eliciting qualitatively different actions from repondents (see review in Hopkins 1986b).

Physiology and Anatomy of Signal Reception

Specialized electroreceptors have evolved for the reception of electric communication signals, the third step in the communication process. Bullock (1982), Heiligenberg & Bastian (1984), and Bullock & Heiligenberg (1986) recently assembled authoritative reviews on many different aspects of electroreception discovered during the past two decades, so here I only discuss a few points of key importance to communication.

KNOLLENORGANS IN MORMYRIDS ARE COMMUNICATION SENSORS   Mormyrids have three classes of electroreceptors—ampullary, mormyromast, and Knollenorgan—that can be recognized anatomically and physiologically (Szabo 1965, 1974, Bennett 1967, 1970, 1971b, Moller & Szabo 1981, Hopkins 1983a, Bass & Hopkins 1984, Bullock 1982). The Knollenorgans are ideally suited to receive electric communication signals: They are putative *communication sensors*. Knollenorgans are frequency-tuned to the peak of the power spectrum of the species-typical EOD (Hopkins & Bass 1981, Bass & Hopkins 1984), whereas ampullary receptors are tuned to 1–10 Hz, a value well outside the frequency range of energy in the EOD. Mormyromasts are also tuned too low to receive weak EODs. Knollenorgans are two orders of magnitude more sensitive than mormyromast electroreceptors, so they alone are suited for detecting the attenuated discharges from other fish (Hopkins 1981a). As mentioned, they are found prominently on the head, where they are most sensitive.

The strongest evidence for the Knollenorgan's communication role is that the relay of sensory information from Knollenorgans to midbrain is blanked by a corollary discharge from the EOD command nucleus for approximately 1 ms at the instant that the fish expects to receive its own EOD. Because of this the fish never hears itself on the Knollenorgan pathway. Thus the receptor must function solely in receiving other fish's discharges (Zipser & Bennett 1976; review in Bell & Szabo 1986; see also Hopkins 1986b, Bell 1986, Bell & Szabo 1986).

Hopkins & Bass (1981) also demonstrated that Knollenorgans respond to species-specific EOD waveforms by firing in phase with the outside-positive-going transients in the stimulus waveform. Complex waveforms like EODs evoke spikes on positive-going edges, but since the EODs are short, and the Knollenorgan has a refractory period, usually only one or two spikes are produced in response to an EOD waveform. In the natural situation, where one fish transmits an EOD and a second fish receives it, the receiver gains information about the positive-going and the negative-going transients within the waveform over different parts of its body. When sender and receiver are perpendicular to each other, current flows inward across receptors on the left side of the body and outward through the receptors on the right side. Knollenorgans on the left will respond to the head-positive parts of the sender's waveform, and those on the right will respond to the head-negative transients in the same EOD. The CNS must gather information about both polarities—thereby retaining the timing of the spike-code for the EOD stimulus, even though any one set of receptors might have been refractory and would not have been able to encode such a rapid EOD with a temporal pattern of spikes.

Knollenorgan afferents project to the nucleus of the electrosensory lateral line lobe (nELLL). Knollenorgan afferents project to large, round, adendritic cell bodies making up the nucleus of the lobe (Szabo & Ravaille 1976, Bell 1979, Szabo et al 1983, Mugnaini & Maler 1986). Two to five large club endings with gap junctions from afferents terminate on each of the large cells in the nucleus. These five terminals may represent more than one Knollenorgan in the periphery, but the exact anatomy of convergence is unknown. The large cells on nELLL receive a second type of synaptic input in the form of a smaller (chemical) synaptic ending that covers much of the cell body and axon hillock not already occupied by the large club endings. The small terminals are the most probable source of the descending corollary discharge. The synapses are rich in GABA and are presumed inhibitory (Mugnaini & Maler 1987). The output from nELLL is through the large cells that project, via the lateral lemniscus, to the midbrain nucleus, ELa, anterior part of the exterolateralis nucleus of the mesencephalon (Enger et al 1976a,b).

The ELa is composed of two types of cells: medium (c. 15 μm) and small (c. 4 μm) (Szabo et al, 1975, Haugedé-Carré 1979, Mugnaini & Maler 1987). The axons from the nELLL terminate on the medium cells, again with large club endings with gap junction morphology. Collaterals of axons from nELLL branch to terminate on the cell bodies of the small cells, again with gap junctions. The axons of the medium cells do not leave the nucleus; they have never been labeled by HRP injections in surrounding structures (Bell et al 1981, Finger et al 1981).

Mugnaini & Maler have recently shown (1987) that the medium cells send axons to other parts of the ELa, but never leave the ELa. The terminal fields end either on the small cells, where they appear to make inhibitory (GABAergic) contact (Mugnaini & Maler 1987), or on other medium cells (located at some distance), where they again make inhibitory contact. Whether ipsilateral and contralateral inputs to ELa terminate on the same or different medium cells is unknown. Nothing is known about the topographic relationships of the intrinsic connections within ELa.

The small cells in the ELa send axons that leave the nucleus (the only output of the ELa) and make a massive projection to the posterior exterolateral nucleus of the torus (ELp) via small axons. The ELp receives inputs from a number of structures besides ELa, including cells in the torus longitudinalis, perilemniscal cells in the caudal mesencephalon, and cells in the isthmic granule nucleus (review in Bell & Szabo 1986). Commissural connections do not seem to be present between the ELa's or the ELp's left and right sides. The ELp is composed of many cell types. Systematic studies of this nucleus have not been undertaken.

The ELp sends a large output to the isthmic granule nucleus, both contralaterally and ipsilaterally (Haugedé-Carré 1979). This nucleus projects, in turn, to the Knollenorgan-processing area of the valvula of the cerebellum. ELp also projects to the inferior olive and to the ipsilateral medial ventral nucleus of the torus (an area that receives direct input from the nELLL). Both of these areas send outputs to the valvula of the cerebellum. ELp also projects to nucleus preeminentialis (ipsi and contra), and probably finally back to the molecular layer of the electrosensory lateral line lobe cortex, where it might have a descending influence on ampullary receptor input and mormyromast input.

Although this pathway has been described in broad outline, much work remains to determine what each of the structures is doing in the analysis of communication signals.

GYMNOTIFORMES   The South American electric fish have evolved a rapid timing circuit for electroreceptors similar to the pathway described above. The parallels between the two circuits are striking at the medullary and receptor levels, although the precise connectivity within the torus is considerably different.

In the periphery of "wave" gymnotiformes, electroreceptors appear to fall into two physiological categories, T-units (phase coders) and P-units (probability coders) (Scheich et al 1973, Hopkins 1976). T units are the most sensitive, and fire 1 to 1 with the animal's own EOD at or near the zero crossing with minimal jitter. In the pulse gymnotiformes, pulse marker units fire a single spike with a short latency in response to low-threshold stimuli, while burst duration coders fire bursts of spikes of varying duration and latency. The T-system of the wave species is very well known anatomically and physiologically (Carr et al 1986a,b, Heiligenberg & Rose 1985). T-receptors project to the ELLL and converge upon large spherical cells in the ELLL via electrotonic synapses. The spherical cells send their axons to layer VI (nucleus magnocellularis) of the torus semicircularis of the midbrain via the lateral lemniscus.

Carr et al (1986a,b) have employed HRP and electronmicroscopy to identify terminals of the axons of spherical cells in the ELLL as they end in the contralateral torus layer VI of *Eigenmannia*. The spherical cells synapse on large (so-called "giant") cells of layer VI (20–40 μm diameter cell body) with large, club-like endings. The afferents also terminate on the distal dendrites of smaller cells in layer VI (6–12 μm diameter) with electrotonic synapses. The small cells receive additional input from gap junctions on their cell bodies from different giant cells. The axons of the giant cells travel within layer VI to contact small cells from all possible locations on the body surface. Because the small cells receive inputs to their distal dendrites from T-receptors and also receive direct inputs from the giant cells, Carr et al (1986a,b) suggest that the small cell dendrite serves as a type of delay line that looks for differences in receptor firing times for different parts of the body. If the small cells were to fire only in response to simultaneous arrival of inputs from the distal dendrites (delayed input) and from the cell body itself (nondelayed input), then the length of the dendrite and its diameter could affect the optimal difference in spike timing from the two inputs, thereby permitting time comparison with a precision determined by the dendritic delay line.

A second electroreceptor system, based on the P-system or the burst duration coder electroreceptors, runs in parallel to the T-system (pulse marker) from the ELLL to the torus. The anatomy of these structures is reviewed in Carr & Maler (1986), and the physiology, mainly concerned with electrolocation performance, is reviewed in Bastian (1986). At all levels, P- and T-systems interact with each other to extract information about the external stimulus. These interactions are especially interesting, and will provide much fertile ground for future work on the sensory processing of communication signals in electric fish.

ACKNOWLEDGMENTS

I thank W. Heiligenberg and C. Shumway for comments on this review, and S. Warsham, M. Marchaterre, and B. Baldwin for the figures.

Literature Cited

Baker, C. L. 1980. Jamming avoidance behavior in Gymnotoid electric fish with pulse-type discharges: sensory encoding for a temporal pattern discrimination. *J. Comp. Physiol.* 136: 165–81

Baker, C. L. 1981. Sensory control of pacemaker acceleration and deceleration in Gymnotiform electric fish with pulse-type discharges. *J. Comp. Physiol.* 141: 197–206

Bass, A. H. 1986a. Species differences in electric organs of mormyrids: Substrates for species-typical electric organ discharge waveforms. *J. Comp. Neurol.* 244: 313–30

Bass, A. H. 1986b. Electric organs revisited: Evolution of a vertebrate communication and orientation organ. See Bullock & Heiligenberg 1986, 1: 13–70

Bass, A. H., Denizot, J. P., Marchaterre, M. A. 1986a. Ultrastructure features and hormone-dependent sex differences of mormyrid electric organs. *J. Comp. Neurol.* 254: 511–28

Bass, A. H., Hopkins, C. D. 1983. Hormonal control of sexual differentiation: Changes in electric organ discharge waveform. *Science* 220: 971–74

Bass, A. H., Hopkins, C. D. 1984. Shifts in frequency tuning of electroreceptors in androgen-treated mormyrid fish. *J. Comp. Physiol.* 155: 713–24

Bass, A. H., Hopkins, C. D. 1985. Hormonal control of sex differences in the electric organ discharge of mormyrid fish. *J. Comp. Physiol.* 156: 587–604

Bass, A. H., Segil, N., Kelley, D. B. 1986b. Androgen binding in the brain and electric organ of a mormyrid fish. *J. Comp. Physiol.* 159: 535–44

Bass, A. H., Volman, S. 1985. Steroid-induced changes in action potential waveforms of an electric organ. *Neurosci. Abstr.* 11: 159

Bastian, J. 1976. Frequency response characteristics of electroreceptors in weakly electric fish (Gymnotoidei) with a pulse discharge. *J. Comp. Physiol.* 112: 165–90

Bastian, J. 1986. Electrolocation: Behavior, anatomy, and physiology. See Bullock & Heiligenberg 1986, 1: 577–612

Bastian, J., Yuthas, J. 1984. The jamming avoidance response of *Eigenmannia*: Properties of a diencephalic link between sensory processing and motor output. *J. Comp. Physiol.* 154: 895–908

Bauer, R. 1974. Electric organ discharge activity of resting and stimulated *Gnathonemus petersii*. *Experientia* 28: 699–70

Bauer, R., Kramer, B. 1974. Agonistic behaviour in mormyrid fish: Latency relationship between electric discharges of *Gnathonemus petersii* and *Mormyrus rume*. *Experientia* 30: 51–52

Bell, C. 1979. Central nervous system physiology of electroreception, a review. *J. Physiol.* 75. 361–79

Bell, C. C. 1986. Electroreception in mormyrid fish. See Bullock & Heiligenberg 1986, 1: 423–52

Bell, C. C., Bradbury, J., Russell, C. J. 1976. The electric organ of a mormyrid as a current and voltage source. *J. Comp. Physiol.* 110: 65–88

Bell, C. C., Finger, T. E., Russell, C. J. 1981.

Central connections of the posterior lateral line lobe in mormyrid fish. *Exp. Brain Res.* 42: 9–22

Bell, C. C., Libouban, S., Szabo, T. 1983. Neural pathways related to the electric organ discharge command in mormyrid fish. *J. Comp. Neurol.* 216: 327–38

Bell, C. C., Myers, J. P., Russell, C. J. 1974. Electric organ discharge patterns during dominance related behavioral displays in *Gnathonemus petersii. J. Comp. Physiol.* 92: 201–28

Bell, C. C., Szabo, T. 1986. Electroreception in mormyrid fish. See Bullock & Heiligenberg 1986, 1: 375–421

Bennett, M. V. L. 1961a. Electric organs of the knifefish Steatogenes. *J. Gen. Physiol.* 45: 590A

Bennett, M. V. L. 1961b. Modes of operation of electric organs. *Ann. NY Acad. Sci.* 94: 458–509

Bennett, M. V. L. 1965. Electroreceptors in Mormyrids. *Cold Spring Harbor Symp. Quant. Biol.* 30: 245–62

Bennett, M. V. L. 1967. Mechanisms of electroreception. In *Lateral Line Detectors*, ed. P. Cahn, 1: 313–93. Bloomington: Indiana Press

Bennett, M. V. L. 1970. Comparative physiology: Electric organs. *Ann. Rev. Physiol.* 32: 471–528

Bennett, M. V. L. 1971a. Electric organs. In *Fish Physiology*, ed. W. S. Hoar, D. J. Randall, 5: 347–491. New York: Academic

Bennett, M. V. L. 1971b. Electroreception. In *Fish Physiology*, ed. W. S. Hoar, D. J. Randall, 5: 493–574. New York: Academic

Bennett, M. V. L., Grundfest, H. 1961. Studies on the morphology and electrophysiology of electric organs. III. Electrophysiology of electric organs in mormyrids. See Chagas & de Carvalho 1961, 1: 113–35

Bennett, M. V. L., Pappas, G. D., Aljure, E., Nakajima, Y. 1967a. Physiology and ultrastructure of electrotonic junctions. II. Spinal and medullary electromotor nuclei in mormyrid fish. *J. Neurophysiol.* 30: 180–208

Bennett, M. V. L., Pappas, G. D., Gimenez, M., Nakajima, Y. 1967b. Physiology and ultrastructure of electrotonic junctions. IV. Medullary electromotor nuclei in gymnotid fish. *J. Neurophysiol.* 30: 236–300

Black-Cleworth, P. 1970. The role of electric discharges in the non-reproductive social behavior of *Gymnotus carapo. Anim. Behav. Monogr.* 3: 1–77

Bossert, W. H., Wilson, E. O. 1963. The analysis of olfactory communication among animals. *J. Theoret. Biol.* 5. 443–69

Brenowitz, E. A. 1982. The active space of red-winged blackbird song. *J. Comp. Physiol.* 147: 511–22

Brown, M. V., Coates, C. W. 1952. Further comparisons of length and voltage in the electric eel, *Electrophorus electricus. Zoologica* 37: 191–97

Bruns, V. 1971. Elektrisches organ von Gnathonemus (Mormyridae). *Z. Zellforsch.* 122: 538–63

Bullock, T. H. 1969. Species differences in effect of electroreceptor input on electric organ pacemakers and other aspects of behavior in Gymnotid fish. *Brain Behav. Evol.* 2: 85–118

Bullock, T. H. 1982. Electroreception. *Ann. Rev. Neurosci.* 5: 121–70

Bullock, T. H., Hamstra, R. H. Jr., Scheich, H. 1972a. The jamming avoidance response of high frequency electric fish. I. General features. *J. Comp. Physiol.* 77: 1–22

Bullock, T. H., Hamstra, R. H. Jr., Scheich, H. 1972b. The jamming avoidance response of high frequency electric fish. II. Quantitative aspects. *J. Comp. Physiol.* 77: 23–48

Bullock, T. H., Heiligenberg, W., eds. 1986. *Electroreception.* New York: Wiley

Carr, C. E., Heiligenberg, W., Rose, G. 1986a. A time comparison circuit in the electric fish midbrain. I. behavior and physiology. *J. Neurosci.* 6: 107–19

Carr, C. E., Maler, L., Taylor, B. 1986b. A time-comparison circuit in the electric fish midbrain. II. functional morphology. *J. Neurosci.* 6: 1372–83

Carr, C. E., Maler, L. 1986. Electroreception in gymnotiform fish. Central anatomy and physiology. See Bullock & Heiligenberg 1986, 1: 319–73

Chagas, C., de Carvalho, A. P., eds. 1961. *Bioelectrogenesis.* Amsterdam: Elsevier

Davis, E., Hopkins, C. D. 1988. Behavioral analysis of electric signal localization in the electric fish, *Gymnotus carapo* (Gymnotiformes). *Anim. Behav.* Submitted

Denizot, J. P., Kirschbaum, F., Westby, G. W. M., Tsuji, S. 1978. The larval electric organ of the weakly electric fish *Pollimyrus* (Marcusenius) *isidori* (Mormyridae, Teleostei). *J. Neurocytol.* 7: 165–81

Denizot, J. P., Kirschbaum, F., Westby, G. W. M., Tsuji, S. 1982. On the development of the adult electric organ in the mormyrid fish *Pollimyrus isidori* (with special focus on the innervation). *J. Neurocytol.* 11: 913–34

Dye, J. C., Meyer, J. H. 1986. Central control of the electric organ discharge in weakly electric fish. See Bullock & Heiligenberg 1986, 1: 71–102

Elekes, K., Ravaille, M., Bell, C. C., Libouban, S., Szabo, T. 1985. The mormyrid brainstem. II. The medullary electromotor relay nucleus: An ultrastructural horseradish peroxidase study. *Neuroscience* 15: 417–29

Elekes, K., Szabo, T. 1981. Comparative synaptology of the pacemaker nucleus in the brain of weakly electric fish (Gymnotidae). In *Sensory Physiology of Aquatic Lower Vertebrates*, ed. T. Szabo, G. Czéh, pp. 107–27. Budapest: Akadémiai Kiadó

Elekes, K., Szabo, T. 1985. The mormyrid brainstem. III. Ultrastructure and synaptic organization of the medullary "pacemaker" nucleus. *Neuroscience* 15: 431–43

Enger, P. S., Libouban, S., Szabo, T. 1976a. Fast conducting electrosensory pathways in the mormyrid fish, *Gnathonemus petersii*. *Neurosci. Lett.* 2: 133–36

Enger, P. S., Libouban, S., Szabo, T. 1976b. Rhombo-mesencephalic connections in the fast conducting electrosensory system of the mormyrid fish, *Gnathonemus petersii*. An HRP study. *Neurosci. Lett.* 3: 239–43

Fessard, A. 1958. Les organes électriques. In *Traité de Zoologie*, ed. P. P. Grassé, 13: 1143–1238. Paris: Masson

Finger, T., Bell, C., Russell, C. J. 1981. Electrosensory pathways to the valvula cerebelli in mormyrid fish. *Exp. Brain Res.* 42: 23–33

Graff, C. 1986. *Signaux électriques et comportement social du poisson à faibles décharges*, Marcusenius macrolepidotus (*Mormyridae, Teleostei*). PhD thesis, Universite de Paris-Sud

Grant, K., Bell, C. C., Clausse, S., Ravaille, M. 1986. Morphology and physiology of the brainstem nuclei controlling the electric organ discharge in mormyrid fish. *J. Comp. Neurol.* 245: 514–30

Grundfest, H. 1957. The mechanisms of discharge of electric organs in relation to general and comparative electrophysiology. *Prog. Biophys.* 7: 1–85

Grundfest, H. 1966. Comparative electrobiology of excitable membranes. *Adv. Comp. Physiol. Biochem.* 2: 1–116

Grundfest, H., Bennett, M. V. L. 1961. Studies on the morphology and electrophysiology of electric organs. I. Electrophysiology of marine electric fishes. See Chagas & de Varvalho 1961, 1: 57–95

Hagedorn, M. 1986. The ecology, courtship, and mating of gymnotiform electric fish. See Bullock & Heiligenberg 1986, 1: 497–525

Hagedorn, M. M., Carr, C. 1985. Single elec-

trocytes produce a sexually dimorphic signal in South American electric fish, *Hypopomus occidentalis* (Gymnotiformes, Hypopomidae). *J. Comp. Physiol.* 156: 511–23

Hagedorn, M. M., Heiligenberg, W. 1985. Court and spark: Electric signals used during the courtship and mating of a gymnotid electric fish. *Anim. Behav.* 33: 254–65

Harder, W. 1968. Die bezeihungen zwischen elektrorezeptoren, elektrischem organ, seitenlinienorganen und nervensystem bei den mormyridae (Teleostei, Pisces). *Z. Vergl. Physiol.* 59: 272–318

Harder, W., Schief, A., Uhlemann, H. 1967. Zur empfindlidikeif des schunch elektrischen fisches *Gnathonemus petersii* (Mormyriformes, Teleostei) gegenuber elektrischen feldern. *Z. Vergl. Physiol.* 54: 89–108

Haugedé-Carré, F. 1979. The mesencephalic exterolateral posterior nucleus of the mormyrid fish, *Brienomyrus niger*: Efferent connections studied by the HRP method. *Brain Res.* 178: 179–84

Heiligenberg, W. F. 1974. Electrolocation and jamming avoidance in a *Hypopygus* (Rhamphichthyidae, Gymnotoidei), an electric fish with pulse-type discharges. *J. Comp. Physiol.* 91: 223–40

Heiligenberg, W. F. 1975. Theoretical experimental approaches to the spatial aspects of electroreception. *J. Comp. Physiol.* 103: 247–72

Heiligenberg, W. 1976. Electrolocation and jamming avoidance in the mormyrid fish, *Brienomyrus*. *J. Comp. Physiol.* 109: 357–72

Heiligenberg, W. 1977. Principles of electrolocation and jamming avoidance in electric fish. In *Studies of Brain Function*, ed. H. B. Barlow, E. Florey, O. J. Grussner, H. van der Loos, 1: 1–85. Berlin/Heidelberg/New York: Springer-Verlag

Heiligenberg, W. 1986. Jamming avoidance responses. See Bullock & Heiligenberg 1986, 1: 613–49

Heiligenberg, W., Altes, R. A. 1978. Phase sensitivity in electroreception. *Science* 199: 1001–4

Heiligenberg, W. F., Bastian, J. 1980. Species specificity of electric organ discharges in sympatric gymnotoid fish of the Rio Negro. *Acta Biol. Venez.* 10: 187–203

Heiligenberg, W., Bastian, J. 1984. The electric sense of weakly electric fish. *Ann. Rev. Physiol.* 46: 561–83

Heiligenberg, W., Finger, T., Matsubara, J., Carr, C. 1981. Input to medullary pacemaker nucleus in the weakly electric fish, *Eigenmannia* (Sternopygidae, Gymnotiformes). *Brain Res.* 211: 418–23

Heiligenberg, W., Rose, G. 1985. Phase and amplitude computations in the midbrain of an electric fish: Intracellular studies of neurons participating in the jamming avoidance response (JAR) of Eigenmannia. *J. Neurosci.* 5: 515–31

Hopkins, C., D. 1972. Sex differences in electric signaling in an electric fish. *Science* 176: 1035–37

Hopkins, C. D. 1973. Lightning as a background noise for communication among electric fish. *Nature* 242: 268–70

Hopkins, C. D. 1974a. Electric communication in fish. *Am. Sci.* 62: 426–37

Hopkins, C. D. 1974b. Electric communication in the reproductive behavior of *Sternopygus marcus* (Gymnotoidei). *Z. Tierpsychol.* 35: 518–35

Hopkins, C. D. 1974c. Electric communication: Functions in the social behavior of *Eigenmannia virescens*. *Behaviour* 50: 270–305

Hopkins, C. D. 1976. Stimulus filtering and electroreception: Tuberous electroreceptors in three species of Gymnotid fish. *J. Comp. Physiol.* 111: 171–207

Hopkins, C. D. 1977. Electric communication. In *How Animals Communicate*, ed. T. A. Sebeok, 1: 263–89. Bloomington: Indiana Univ. Press

Hopkins, C. D. 1980. Evolution of electric communication channels of mormyrids. *Behav. Ecol. Sociobiol.* 7: 1–13

Hopkins, C. D. 1981a. On the diversity of electric signals in a community of mormyrid electric fish in West Africa. *Am. Zool.* 21: 211–22

Hopkins, C. D. 1983a. Functions and mechanisms in electroreception. In *The Central Nervous System of Teleost Fishes*, ed. G. Northcutt, R. Davis, 1: 216–59. Ann Arbor: Univ. Michigan Press

Hopkins, C. D. 1983b. Neuroethology of species recognition in electroreception. *Adv. Vertebrate neuroethol.* 56: 871–81

Hopkins, C. D. 1986a. Temporal structure of non-propagated electric communication signals. *Brain Behav. Evol.* 28: 43–59

Hopkins, C. D. 1986b. Behavior of Mormyridae. See Bullock & Heiligenberg 1986, 1: 527–76

Hopkins, C. D., Bass, A. H. 1981. Temporal coding of species recognition signals in an electric fish. *Science* 212: 85–87

Hopkins, C. D., Heiligenberg, W. 1978. Evolutionary designs for electric signals and electroreceptors in Gymnotid fishes of Surinam. *Behav. Ecol. Sociobiol.* 3: 113–34

Hopkins, C. D., Westby, G. W. M. 1986. Time-domain processing of electric organ discharge waveforms by pulse-type electric fish. *Brain Behav. Evol.* 29: 77–104

Kawasaki, M., Heiligenberg, W. 1987. Individual prepacemaker neurons can modulate the pacemaker cycle of the gymnotiform electric fish, *Eigenmannia*. *J. Comp. Physiol.* 161: In press

Keynes, R. D., Martins-Ferreira, H. 1953. Membrane potentials in the electroplates of the electric eel. *J. Physiol.* 119: 315–51

Kirschbaum, F. 1977. Electric-organ ontogeny: Distinct *larval* organ precedes the *adult* organ in weakly electric fish. *Naturwissenschaften* 64: 387–88

Kirschbaum, F. 1981. Ontogeny of both larval electric organ and electromotoneurones in *Pollimyrus isidori* (Mormyridae, Teleostei). *Adv. Physiol. Sci.* 31: 129–57

Kirschbaum, F. 1983. Myogenic electric organ precedes the neurogenic organ in apteronotid fish. *Naturwissenschaften* 70: 205–6

Kirschbaum, F., Westby, G. W. M. 1975. Development of the electric discharge in mormyrid and gymnotid fish (*Marcusenius sp.* and *Eigenmannia virescens*). *Experientia* 15: 1290–93

Knudsen, E. I. 1974. Behavioral thresholds to electric signals in high frequency electric fish. *J. Comp. Physiol.* 91: 333–53

Knudsen, E. I. 1975. Spatial aspects of the electric fields generated by a weakly electric fish. *J. Comp. Physiol.* 99: 103–18

Kramer, B. 1976a. Flight associated discharge pattern in a weakly electric fish *Gnathonemus petersii* (Mormyridae, Teleostei). *Behaviour* 59: 88–95

Kramer, B. 1976b. The attack frequency of *Gnathonemus petersii* toward electrically silent (denervated) and intact conspecifics, and toward another Mormyrid (*Brienomyrus niger*). *Behav. Ecol. Sociobiol.* 9: 425–46

Kramer, B. 1985. Jamming avoidance in the electric fish *Eigenmannia*: Harmonic analysis of sexually dimorphic waves. *J. Exp. Biol.* 119: 41–69

Kramer, B., Bauer, R. 1976. Agonistic behaviour and electric signalling in a mormyrid fish, *Gnathonemus petersii*. *Behav. Ecol. Sociobiol.* 1: 45–61

Kramer, B., Kirschbaum, F., Markl, H. 1981. Species specificity of electric organ discharges in a sympatric group of gymnotid fish from Manaus (Amazonas). In *Advances in Physiological Sciences*, ed. T. Szabo, G. Czéh, 1: 195–220. Budapest: Akadémia Kiadó

Langner, G., Scheich, H. 1978. Active phase coupling in electric fish: Behavioral control with microsecond precision. *J. Comp. Physiol.* 128: 235–40

Libouban, S., Szabo, T., Ellis, D. 1981. Comparative study of the medullary command (pacemaker) nucleus in species of the four weakly electric fish families. In *Sensory Physiology of Aquatic Lower Vertebrates*, ed. T. Szabo, G. Czéh, 1: 95–106. Budapest: Akadémia Kiadó

Lissmann, W. H. 1958. On the function and evolution of electric organs in fish. *J. Exp. Biol.* 35: 156–91

Lowrey, A. 1913. A study of the submental filament considered as probable electric organs in the Gymnotid eel, *Steatogenes elegans* (Steindachner). *J. Morphol.* 24: 685–694

Mago-Leccia, F. 1978. Los peces de la familia Sternopygidae de Venezuela. *Acta Cient. Venez.* 29: 1–89

Mago-Leccia, F., Zaret, T. 1978. Taxonomic status of *Rhabdolichops troscheli* (Kaup 1856), and speculations on gymnotiform evolution. *Env. Biol. Fishes* 3: 379–84

Marler, P. 1959. Developments in the study of animal communication. In *Darwin's Biological Work*, ed. P. R. Bell. London: Cambridge Univ. Press

Meyer, H. 1983. Steroid influences upon the discharge frequencies of a weakly electric fish. *J. Comp. Physiol.* 153: 29–37

Meyer, H. 1984. Steroid influences upon the discharge frequencies of intact and isolated pacemakers of weakly electric fish. *J. Comp. Physiol.* 154: 659–68

Mills, A., Zakon, H. 1987. Coordination of EOD frequency and pulse duration in a weakly electric wave fish: The influence of androgens. *J. Comp. Physiol.* 161: 417–30

Moller, P. 1970. Communication in weakly electric fish, *Gnathonemus niger* (Mormyridae). I. Variation of electric organ discharge (EOD) frequency elicited by controlled electric stimuli. *Anim. Behav.* 18: 768–86

Moller, P., Bauer, R. 1973. Communication in weakly electric fish, *Gnathonemus petersii* (Mormyridae). II. Interaction of electric organ discharge activities of two fish. *Anim. Behav.* 21: 501–12

Moller, P., Szabo, T. 1981. Lesions in the nucleus mesencephali exterolateralis: Effects on electrocommunication in the mormyrid fish, *Gnathonemus petersii* (Mormyriformes). *J. Comp. Physiol.* 144: 327–33

Morton, E. S. 1975. Ecological sources on avian sounds. *Am. Nat.* 108: 17–34

Mugnaini, E., Maler, L. 1986. Cytology and immunocytochemistry of the nucleus of the lateral line lobe in the electric fish, *Gnathonemus petersii* (Mormyridae): Evidence suggesting that GABAergic synapses mediate an inhibitory corollary discharge. *Synapse* 1: 32–56

Mugnaini, E., Maler, L. 1987. Cytology and immunocytochemistry of the nucleus extralateralis anterior of the mormyrid brain: Possible role of GABAergic synapses in temporal analysis. *Anat. Embryol.* 176: 313–36

Nijssen, H., Isbrücker, I. J. H. 1972. On *Hypopygus lepturus*, a little known dwarf gymnotid fish from South America (Pisces, Gypriniformes, Gymnotoidei),. *Zool. Mededelingen* 47: 160–76

Quinet, P. 1971. Etude systematique des organes sensoriels de la peau des Mormyriformes. *Mus. Roy. de l'Afrique Cent. Terv.* 190: 1–97

Russell, C. J., Myers, J. P., Bell, C. C. 1974. The echo response in *Gnathonemus petersii* (Mormyridae). *J. Comp. Physiol.* 92: 181–200

Scheich, H. 1974. Neuronal analysis of wave form in the time domain: Midbrain units in electric fish during social behavior. *Science* 185: 365–67

Scheich, H. 1977a. Neural basis of communication in the high frequency electric fish, *Eigenmannia virescens* (jamming avoidance response). I. Open loop experiments and the time domain concept of signal analysis. *J. Comp. Physiol.* 113: 181–206

Scheich, H. 1977b. Neural basis of communication in the high frequency electric fish, *Eigenmannia virescens* (jamming avoidance response). II. Jammed electroreceptor neurons in the lateral line nerve. *J. Comp. Physiol.* 113: 207–27

Scheich, H. 1977c. Neural basis of communication in the high frequency electric fish, *Eigenmannia virescens* (jamming avoidance response). III. Central integration in the sensory pathway and control of the pacemaker. *J. Comp. Physiol.* 113: 229–55

Scheich, H., Bullock, T. H., Hamstra, R. H. Jr. 1973. Coding properties of two classes of afferent nerve fibers: High frequency electroreceptors in the electric fish, *Eigenmannia. J. Neurophysiol.* 36: 39–60

Schluger, J., Hopkins, C. D. 1988. Electric fish approach stationary signal sources by following electric current lines. *J. Exp. Biol.* In press

Schwartz, I. R., Pappas, G. D., Bennett, M. V. L. 1975. The fine structure of electrocytes in weakly electric teleosts. *J. Neurocytol.* 4: 87–114

Serrier, J. 1973. Modifications instantanées du rythme de l'activité électrique d'un mormyre, *Gnathonemus petersii*, provoqué par la stimulation electrique artificielle de ses électrorecepteurs. *J. Physiol.* 66: 713–28

534   HOPKINS

Squire, A., Moller, P. 1982. Effects of water conductivity on electrocommunication in the weak-electric fish, *Brienomyrus niger* (Mormyriformes). *Anim. Behav.* 30: 375–82

Steindachner, F. 1880. Fisch-fauna des cauca und der flusse bei guaya quil. *Denkschrift. Kaiser. Akad. Wiss. Wien Math. Klasse* 42: 55–104

Szabo, T. 1958. Structure intime de l'organe électrique de trois mormyrides. *Z. Zellforsch.* 49: 33–45

Szabo, T. 1961a. Les organes électrique de *Gymnotus carapo. Koninkl. Ned. Akad. Wetenschap.* 64: 584–86

Szabo, T. 1961b. Anatomo-physiologie des centres nerveux specifiques de quelques organes électriques. In *Bioelectrogenesis*, ed. C. Chagas, A. P. de Carvalho, 1: 185–201. Amsterdam: Elsevier

Szabo, T. 1965. Sense organs of the lateral line system in some electric fish of the Gymnotidae, Mormyridae and Gymnarchidae. *J. Morphol.* 117: 229–50

Szabo, T. 1974. Anatomy of the specialized lateral line organs of electroreception. *Hand. Sensory Physiol.* 3: 13–56

Szabo, T., Enger, P. 1964. Pacemaker activity of the medullary nucleus controlling electric organs in high-frequency gymnotid fish. *Z. Vergl. Physiol.* 49: 285–300

Szabo, T., Fessard, A. 1974. Physiology of electroreceptors. *Hand. Sensory Physiol.* 3: 59–124

Szabo, T., Ravaille, M. 1976. Synaptic structure of the lateral line lobe nucleus in mormyrid fish. *Neurosci. Lett.* 2: 127–31

Szabo, T., Sakata, H., Ravaille, M. 1975. Electronic coupled pathways in the central nervous system of some teleost fish, Gymnotidae and Mormyridae. *Brain Res.* 95: 459–74

Szabo, T., Ravaille, M., Libouban, S., Enger, P. S. 1983. The mormyrid rhombencephalon. I. Light and EM investigations on the structure and connections of the lateral line lobe nucleus with HRP labeling. *Brain Res.* 26: 1–19

Taverne, L. 1972. Ostéologie des genres *Mormyrus* Linne, *Mormyrops* Muller, *Hyperopisus* Gill, *Isichthys* Gill, *Myomyrus* Boulenger, *Stomatorhinus* Boulenger, et *Gymnarchus* Cuvier. Considérations générales sur la systématique des poissons de l'ordre des Mormyriformes. *Mus. R. Afr. Cent. Terv. Belg. Ann. Ser. Octavo. Sci. Zool.* 200: 1–194

Teyssèdre, C., Serrier, J. 1986. Temporal spacing of signals in communication studied in weakly electric mormyrid fish (Teleostei, Pisces). *Behav. Proc.* 12: 77–98

Toerring, M. J., Moller, P. 1984. Locomotor and electric displays associated with electrolocation during exploratory behaviour in mormyrid fish. *Behav. Brain Res.* 12: 291–306

Tokunaka, A., Akert, K., Sandri, C., Bennett, M. V. L. 1980. Cell types and synaptic organization of the medullary electromotor nucleus in a constant frequency fish, *Sternarchus albifrons. J. Comp. Neurol.* 192: 407–26

Trujillo-Cenóz, O., Echague, J. A., Macadar, O. 1984. Innervation pattern and electric organ discharge waveform in *Gymnotus carapo* (Teleostei, Gymnotiformes). *J. Neurobiol.* 15: 273–81

Valone, J. A. Jr. 1970. Electrical emissions in *Gymnotus carapo* and their relation to social behavior. *Behaviour* 37: 1–14

Watanabe, A., Takeda, K. 1963. The change of discharge frequency by a.c. stimulus in a weakly electric fish. *J. Exp. Biol.* 40: 57–66

Waxman, S. G., Pappas, G. D., Bennett, M. V. L. 1972. Morphological correlates of functional differentiation of nodes of Ranvier along single fibers in the neurogenic electric organ of the knife fish, *Sternarchus. J. Cell Biol.* 53: 210–24

Westby, G. W. M. 1974. Assessment of signal value of certain discharge patterns in the electric fish, *Gymnotus carapo*, by means of playback. *J. Comp. Physiol.* 92: 327–41

Westby, G. W. M. 1975a. Has the latency dependent response of *Gymnotus carapo* to discharge-triggered stimuli a bearing on electric fish communication? *J. Comp. Physiol.* 96: 307–41

Westby, G. W. M. 1975b. Further analysis of the individual discharge characteristics predicting social dominance in the electric fish, *Gymnotus carapo. Anim. Behav.* 23: 249–60

Westby, G. W. M. 1975c. Comparative studies of the aggressive behaviour of two gymnotid electric fish (*Gymnotus carapo* and *Hypopomus artedi*). *Anim. Behav.* 23: 192–213

Westby, G. W. M. 1979. Electrical communication and jamming avoidance between resting *Gymnotus carapo. Behav. Ecol. Sociobiol.* 4: 381–93

Westby, G. W. M. 1984. Simple computer model accounts for observed individual and sex differences in electric fish signals. *Anim. Behav.* 32: 1254–56

Westby, G. W. M., Kirschbaum, F. 1977. Emergence and development of the electric organ discharge in the mormyrid fish, *Pollimyrus isidori. J. Comp. Physiol.* 122: 251–71

Westby, G. W. M., Kirschbaum, F. 1978. Emergence and development of the electric organ discharge in the mormyrid fish, *Pollimyrus isidori*. II. Replacement of the larval by the adult discharge. *J. Comp. Physiol.* 127: 45–59

Westby, G. W. M., Kirschbaum, F. 1982. Sex differences in the waveform of the pulse-type electric fish, *Pollimyrus isidori* (Mormyridae). *J. Comp. Physiol.* 145: 399–403

Wilczynski, W. 1986. Sexual differences in neural tuning and their effect on active space. *Brain Behav. Evol.* 28: 83–94

Wiley, R. H., Richards, D. G. 1983. Adaptations for acoustic communication in birds: Sound transmission and signal detection. In *Acoustic Communication in Birds*, ed. D. Kroodsma, E. H. Miller, 1. 131–81. New York: Academic

Zakon, H. H. 1986. The electroreceptive periphery. See Bullock & Heiligenberg 1986, 1: 103–56

Zipser, B., Bennett, M. V. P. 1976. Interaction of electrosensory and electromotor signals in lateral line lobe of a mormyrid fish. *J. Neurophysiol.* 39: 713–21

Ann. Rev. Neurosci. 1988. 11:537–63

NEUROGENETIC DISSECTION OF LEARNING AND SHORT-TERM MEMORY IN *DROSOPHILA*

Yadin Dudai

Department of Neurobiology, The Weizmann Institute of Science, Rehovot 76100, Israel

Introduction

The ability of an organism to acquire and store new information in the nervous system, and to retrieve it for later use, should be coded by genes. Given the appropriate mutants, one might identify these genes and their products and thus elucidate the molecular and cellular components that make learning and memory possible. This approach is practical only if the effect of certain single gene mutations on learning and memory can be dissociated from the effect of the same mutations on other physiological (or developmental) processes. Otherwise, mutations that disrupt learning and memory will result in sick or even dead organisms, and the transformation of genetic information into biological learning apparatuses will remain cryptic. Luckily, at least in the fruit fly, *Drosophila melanogaster*, mutations exist that cause flies to be stupid or amnesiac, without severely disabling them in other respects. Such mutations shed interesting light on what might be regarded as the elementary biological technologies of learning and memory. And as the current data suggest, these technologies are not restricted to fruit flies; they are very probably shared by other organisms as well.

The study of memory in the brain of a fly was initiated in the laboratory of Seymour Benzer at the California Institute of Technology more than a decade ago. The rationale was anchored in a much broader experimental adventure, namely, the neurogenetic dissection of behavior (Benzer 1967, 1973). Single gene mutations, it was argued, could be used as powerful microdissecting tools to elucidate and analyze the intricate molecular sys-

537

0147–006X/88/0301–0537$02.00

tems that transform a linear genetic code into a four-dimensional behaving organism. This seemed a reasonable experimental risk when innate behaviors, e.g. phototaxis, were considered. Soon scores of nonphototactic, sluggish, trembling, flightless, and other strange-looking flies began to emerge among the progeny of chemically mutagenized *Drosophila*. Yet the mere suggestion that the same approach could be extended to the study of learning evoked sheer skepticism. Can flies learn?! The conviction that they cannot was initially shared by both researchers and editors of popular science columns.

The breakthrough came with the development of the first reliable *Drosophila* learning paradigm by Quinn et al (1974). It became apparent that flies can behave quite intelligently if presented with appropriate tasks. The ability of *Drosophila* to habituate, sensitize, undergo classical conditioning and operant conditioning and to modify their behaviors in paradigms that include elements of all or part of the above, under entirely sterile, artificial laboratory situations, or even in the ethological, romantic context of courtship, is now well documented (Aceves-Pina & Quinn 1979, Booker & Quinn 1981, Corfas & Dudai 1987, Dudai 1977, Duerr & Quinn 1982, Fischbach 1981, Gailey et al 1982, Heisenberg et al 1985, Kyriacou & Hall 1984, Mariath 1985, Menne & Spatz 1977, Platt et al 1980, Quinn et al 1974, Siegel & Hall 1979, Tempel et al 1983, Tully & Quinn 1985). The behaviors, their modifications, and the effect of genetic aberrations on performance have been reviewed by several authors (Aceves-Pina et al 1983, Dudai 1985a, 1986, 1987, Hall 1982, 1984, 1986, Quinn 1984, Quinn & Greenspan 1984, Tully 1984, 1987). Even the possible correlation between phylogenetic fitness and learning ability of a seemingly innately programmed fruit fly was discussed (Hall 1986, Hewitt et al 1983). The behavioral paradigms therefore are not detailed here. Instead, several issues that are central to the experimental approach and to the data obtained by this approach to date are singled out and discussed. Among these are the following: When should a mutant be defined as a learning mutant? How general are the effects of learning mutations on learning? Do the mutations affect acquisition, retention, or retrieval? What is the nature of the molecular lesions, and what is the time window during which these lesions exert their effect on learning and memory? And finally, what contributions have been made by the *Drosophila* studies to our knowledge of elementary learning and memory mechanisms, and what might these studies be expected to contribute in the near future?

What is a Learning Mutant?

Since the availability of proper mutants is a necessary condition for the neurogenetic dissection of learning and memory, the definition of such mutants deserves proper discussion. The original, orthodox, and somewhat ideal definition of learning mutants could be stated as follows: a single gene mutant that is deficient in acquisition and/or storage and/or retrieval but lacks any developmental and/or sensory and/or motor defects that might account for the aberrant behavior.

Soon it became apparent that the criteria must be relaxed and the definition should become more permissive; three factors contributed to this. First, it became clear that psychological and physiological tests available for *Drosophila* are often not yet sophisticated enough (especially when compared to the batteries of tests available for mammals) to permit exclusion of minor, non-learning-related behavioral alterations. Second, it became clear that if one looks carefully enough, additional, pleotropic effects of learning mutations are discovered (see below). And third, it became apparent that useful information regarding elementary learning and memory mechanisms would be disregarded if the identification of pleotropic effects would automatically exclude the mutant in question from further analysis. Therefore, a more correct and realistic definition of a learning mutation would be the following: a single gene mutation that affects acquisition and/or storage and/or retrieval, and whose effect on these processes can be reasonably dissociated, under the appropriate experimental conditions, from the effect on other functions of the organism.

The latter, more relaxed definition permits one to take advantage of restricted experimental situations (e.g. conditioning paradigms that do not require phototaxis for testing learning in nonphototactic flies) and also of unique experimental tools available for *Drosophila* (e.g. the use of temperature-sensitive (*ts*) mutants to overcome a developmental lesion and to test the effect of the mutation on behavior in the adult). But still implicit in the experimental approach is the understanding that each putative mutant must be carefully scrutinized for sensory, motor, or motivational defects that might explain the failure to learn and/or remember in the appropriate paradigms. Alas, detailed, quantified behavioral analysis of putative learning and memory mutants is often not yet available. Relatively detailed behavioral data were published to date mainly for two mutants that were isolated in a deliberate screen for conditioning mutants among the progeny of chemically mutagenized flies, namely, *dunce* (*dnc*) and *rutabaga* (*rut*). It is of interest to see how these mutants fare when their behavior (learning and memory excluded) is compared to normal.

dnc was the first learning mutant to be isolated (Dudai et al 1976, Byers 1980). It was independently rediscovered several times in deliberate screens for learning mutants (references as above), for female sterility mutants (Mohler 1977), and for mutations that affect cAMP-phosphodiesterase (Byers et al 1981, Davis & Kiger 1981, Davis & Kauver 1984). Several alleles are currently known, but behavior was studied mainly in the two that were isolated in the learning screen, namely dnc^1 and dnc^2. The flies were found to be capable of apparently normal motor and sensory activities, including sensing odorants and reinforcements used in learning tests (Dudai et al 1976, Dudai 1979, Byers 1980). The female sterility phenotype is suppressible by genetic elements independent of the defects in memory and in phosphodiesterase (Salz et al 1982, Shotwell 1982). *dnc* can therefore be regarded as a mutant that should tell us something specific about the genetics and biochemistry of learning systems. Nevertheless, even those alleles of *dnc* that at first sight fit the "platonic" definition of a learning mutant do show some aberrations in other behaviors, e.g. wing spread and wing buzzing (Byers 1980; see also below).

The behavior of *rut* was analyzed using the same assays used for *dnc*, namely, tests for phototaxis, olfactory acuity, shock sensitivity, and general motor activity (Dudai et al 1984, 1987b). Again, similar to dnc^1 and dnc^2, *rut* has normal external morphology, can sense odorants and shock, and has normal phototaxis. If there is any sensory or motor defect that might account for its poor performance in learning paradigms (and see below), it is not trivial. *rut* is therefore another mutant that seems to correspond to the original definition of a learning mutant. Nevertheless, some additional phenotypes are detected. The mutant is more resistant than normal to formamidine toxins (Dudai et al 1987a), and has more varicosities in axons of an identified mechanosensory neuron than the CS wild-type from which it was isolated (Corfas & Dudai 1987). Interestingly, *rut* seems to partially suppress female sterility of *dnc*, although the double mutant *dnc rut* still fails to learn (Livingstone et al 1984). This attests again to the dissociation of the developmental effect of *dnc* (see above) from its effect on learning and, together with the aforementioned *rut* phenotypes, suggests that the *rut* gene product does have roles in addition to its role in learning.

Very little information has been published on the physiology and behavior of other mutants that were isolated in a deliberate screen for learning and memory deficits. *turnip* (*tur*) was briefly reported to be affected in motor activity and phototaxis (Hall 1982). No description is yet available for the physiology and behavior of *amnesiac* (*amn*), *cabbage* (*cab*), *zucchini* (reviewed by Aceves-Pina et al 1983, Hall 1982). Some of these mutants are further discussed below with respect to their learning and memory deficits.

The deliberate screen for learning and memory mutants among the progeny of mutagenized flies is only one method of identifying such mutants. It is an efficient method: To date, mainly the X chromosome (containing only ca 20% of the fly genes) was screened, and the probability of identifying an interesting candidate following EMS-mutagenesis could be calculated to be ca 1 in 250. Such a high probability already indicates that learning and memory are very sensitive to genetic manipulations and involve the action of many gene products. The limiting step in mutant isolation is the behavioral screen: Each mutagenized chromosome (i.e. the fly carrying it) has to be expanded into an isogenic population and subjected separately to the conditioning test (e.g. olfactory avoidance conditioning or olfactory classical conditioning).

Another method of identifying appropriate mutants is to search among existing genetic aberrations of *Drosophila*. Indeed, additional mutants that are often referred to in the literature as learning mutants were originally isolated due to the presence of some other morphological, physiological, biochemical, or behavioral defects. In such mutants, by definition, reduced learning or fleeting memory is not the exclusive defect; yet this should not belittle their potential contribution to the analysis of elementary learning mechanisms: *dnc* was also independently isolated as a female sterility and a phosphodiesterase mutant, and *rut* might have been isolated as a formamidine-resistant mutant.

Among the mutants previously isolated in various screens and later found to have interesting effects on memory are *Dopa-decarboxylase* mutants (*Ddc*). Because these mutations also disrupt development, temperature-sensitive alleles, Ddc^{ts1} and Ddc^{ts2}, were used to study the effect on learning. Flies carrying these mutations die if they develop at a high, restrictive temperature (e.g. 29°C) but grow normally at a lower, permissive temperature (e.g. 20°C). If enzyme activity is turned off in these flies, as adults, by shifting them to a restrictive temperature, they survive and can be tested for their behavior. Such flies show marked defects in some forms of learning (Tempel et al 1984), yet, again, detailed behavioral analysis was not reported; some deviations from normal in proboscis extension and response to starvation, which might indicate motivational abnormalities, were detected (Tempel et al 1984). These motivational problems might be intimately associated with the learning defects (Mason 1984).

Several additional classes of mutants were tested for their effects on learning in recent years. These include biological rhythm mutants, ion channel mutants, and neuroanatomical mutants. Several independently isolated mutants that have abnormally long rhythms, e.g. per^1, per^{l2}, *Andante*, *psi-2*, and *psi-3* (but not mutants with abbreviated or chaotic rhythms), were reported to be defective in some aspects of conditioned behavior in the context of courtship (Ackerman & Siegel 1986, Hall 1984, 1986, Jackson et al 1983). The K^+ channel mutant *Shaker* (*Sh*) was reported to be defective in conditioned courtship and in olfactory classical conditioning (Cowan & Siegel 1984, 1986). Another mutation that may be associated with defective K$^+$ channel function, *tetanic* (*tta*), was also found to lesion classical conditioning (Ferrus, personal communication). The presumptive Na$^+$ channel mutant *nap^{ts}* was also thought to be defective in conditioning, but some of the data actually reflect very small deviation from normal (Cowan & Siegel 1986), and reference to *nap* as a conditioning mutant should be treated with reservation.

Learning deficiency was also reported for two mutants that were isolated by virtue of defective brain neuroanatomy, more precisely, lesions in their mushroom bodies (Heisenberg et al 1985). Although the neuroanatomical defects were quite dramatic, no effects were found in perception of the conditioned stimuli.

In conclusion, *Drosophila* mutations reported to affect learning and/or memory could be described as representing a spectrum of specificity in their effect on conditioning. The most specific seem to be some mutations that were identified in a deliberate screen for learning mutants. But even these reveal, upon careful examination, additional defects in physiology or behavior. Other mutations, with gross biochemical, physiological, or neuroanatomical defects, could still contribute interesting information on learning. In all cases, the development, physiology, and behavior must be scrutinized to permit assessment of the specificity of the effects on learning. Regretfully, the behavioral analysis of the innate and plastic behavioral repertoire still lags much behind the mere availability of putative mutants.

How General are the Effects on Learning?

A mutation that affects learning in a variety of conditioning paradigms, preferably paradigms that rely on different sensory modalities and motor responses, might be a better starting point for a search for elementary learning mechanisms than a mutation that affects only one or few learning situations. The latter mutation may still yield interesting information on specific learning systems and might even reveal previously unknown differences between distinct learning mechanisms, but universality is sacrificed. This is why attempts have been made in recent years to develop novel conditioning paradigms and to assay the behavior of learning mutants in different paradigms. The emerging view is that, in general, mutants isolated in a deliberate screen for conditioning deficits behave abnormally in learning paradigms in addition to the one used for their screening. This indicates that the gene products lesioned by the mutations have a widespread role in learning and/or memory. However, differences in behavior do emerge when different conditioning paradigms are used.

The paradigm employed for isolation of *dnc, rut, amn, tur, cab,* and additional X-linked conditioning mutants (reviewed in Aceves-Pina et al 1983) was the original olfactory avoidance conditioning paradigm developed by Quinn et al (1974; for methods see isolation of *dnc*, Dudai et al 1976). *dnc*, *rut*, and *amn* showed defective performance in a classical conditioning version of the olfactory conditioning paradigm (Tully & Quinn 1985). The same mutants also failed in reward learning, in which starved flies are conditioned to prefer a sucrose-coupled odorant (Tempel et al 1983). Abnormal performance was also detected in nonassociative conditioning. When sucrose is applied to the leg of a starved, water-satiated fly, the fly responds by proboscis extension, and this feeding reflex shows habituation; *dnc*, *rut*, and *tur* displayed decreased habituation (Duerr & Quinn 1982). The results for *amn* were ambiguous, probably indicating normal habituation, and those for *cab* were not reproducible enough to merit analysis (Duerr & Quinn 1982). The aforementioned reflex also shows sensitization following application of concentrated sucrose solution to the proboscis; sensitization of *tur* was indistinguishable from normal, while that of *dnc* was first greater than normal followed by abnormally rapid decline, and that of *rut* and *amn* was initially normal, followed again by abnormally rapid decline (Duerr & Quinn 1982). Another reflex that shows habituation and dishabituation is a cleaning reflex. Here, upon mechanical stimulation of thoracic bristles, the fly cleans, with a patterned set of movements, the field covered by the stimulated bristles (Corfas & Dudai 1987). *rut* was found to be capable of habituation and dishabituation of this reflex, but habituation was abnormally short-lived (G. Corfas and Y. Dudai, unpublished). Thus, some of the mutants isolated as defective in associative conditioning were also defective in simpler, nonassociative forms of learning.

Individual *Drosophila* can be trained to alter their leg position (e.g. lift it) over a conducting solution to avoid electric shock. This response is thus an operant paradigm in which mechanosensory cues are employed. *dnc*, *tur*, and *cab* were found to perform poorly in this paradigm (Booker & Quinn 1981). In yet another operant conditioning paradigm, in which individual flies were trained to change their position on a platform in response to heat negative reinforcement, *dnc*, *amn*, and *rut* also displayed defective performance (Mariath 1985).

Gotz & Biesinger (1985) found that *Drosophila* exposed to diethylether narcosis later avoid the center of an arena; this behavioral modification, which involves a change in the search behavior of the fly and its control, is suppressed in *dnc* (Gotz & Biesinger 1985).

Using another sensory modality, i.e. color or light intensity, as the conditioned stimulus (CS) in a classical conditioning paradigm, results were somewhat conflicting. Under certain experimental conditions, the immediate performance of dnc^1 and dnc^2 following training was indistinguishable from normal, although it must be noted that the learning scores of both mutants and CS wild type were low (as compared, for example, to another wild type, AS) (Bicker & Dudai 1978). In another study, the performance of dnc^1, *amn*, *tur*, and *rut* was abnormally low during the first minutes after visual training but became indistinguishable from normal afterwards (Folkers 1982).

Perhaps some of the most intriguing conditioning paradigms developed for *Drosophila* take advantage of behavioral modifications within the ethological context of courtship. Variations on themes borrowed from the intricate love life of the fruit fly have been used to probe the fly's ability to learn from experience. These include the acquired female-avoidance behavior of males following rejection by fertilized females (Siegel & Hall 1979); the avoidance of courtship of immature-males by experienced males (Gailey et al 1982); and acoustic priming of females (i.e. speeding up subsequent mating performance) by the rhythmic pulse component of the male's love song (Kyriacou & Hall 1984).

Most of these behavioral modifications are associative and not merely due to habituation or sensitization (Ackerman & Siegel 1986, Hall 1984, 1986). *amn* was found to be mutant in the female rejection paradigm (Siegel & Hall 1979); *dnc*, *amn*, *rut*, and *tur* (but not *cab*) were mutants in the male-courtship paradigm (Gailey et al 1982); and *dnc*, *rut*, and *amn* were found to be mutant in the love-song priming paradigm (Kyriacou & Hall 1984). Imitating some facets of *Drosophila* natural behavior, an artificial negative reinforcement conditioning paradigm was developed in which female fruit flies were covered with an aversive quinine powder and exposed to courting males. This resulted in temporary avoidance of females by the conditioned males; *dnc* and *amn* were mutants in this situation, too (Ackerman & Siegel 1986).

An entirely different type of naturally occurring experience-dependent neuronal modification has been recently reported by Balling et al (1987). Wild-type flies at eclosion differ in the number of Kenyon cell fibers in their mushroom bodies. The latter are brain structures implicated in associations (see below). During the first week of adult life the number of fibers adjusts to a level that depends on the experience of the flies. *dnc* and *rut* did not show such an experience-dependent modulation of fiber number.

It therefore appears that mutants isolated in intentional screens for learning deficits behave as mutants in very different learning situations. This seems also to be the situation for some, though not all, mutants that were first isolated due to the presence of other physiological or anatomical defects and only later were crowned as learning mutants. Ddc^{ts} flies were reported to be defective in olfactory avoidance conditioning, reward learning, and depression of male courtship following exposure to rejecting females (Tempel et al 1984). Ddc^{ts} flies were not, however, deficient in olfactory classical conditioning (Tully 1987). Data for circadian rhythm mutations was published only for conditioned courtship behavior (Jackson et al 1983, Hall 1984, 1986). In the case of ion channel mutants, *Sh* was reported to affect conditioned male courtship (Cowan & Siegel 1984) and olfactory classical conditioning (Cowan & Siegel 1986); whether it could be regarded as a learning mutant in the olfactory avoidance paradigm is not clear, since the somewhat decreased performance of a *Sh* allele in this paradigm may be due to decreased phototaxis and/or different genetic background (Dudai 1977), and another allele learns normally (Ferrus, personal communication). To date, the presumptive channel-related mutant *tta* has been tested only in olfactory conditioning (Ferrus, personal communication). Another presumptive channel mutant, *nap*, is thought to slightly affect olfactory classical conditioning (Cowan & Siegel 1986), and to depress memory of experience-modified courtship (Cowan & Siegel 1984). Data on learning paradigms using other sensory modalities is still lacking.

In two neuroanatomical mutants, mushroom body deranged (*mbd*) and mushroom body miniature (*mbm*), learning was completely abolished in an olfactory avoidance paradigm and was partially suppressed in a reward olfactory paradigm (Heisenberg et al 1985). *mbd* was not affected in color discrimination tasks and in operant conditioning involving visual cues (cited in Heisenberg et al 1985).

No information is available on the behavior of cuticle-color mutants, *e*, *b*, and *t*, which display low learning scores in an olfactory avoidance paradigm (Dudai 1977), in other conditioning paradigms; it it therefore not clear how much of the decreased performance of these mutants results from specific sensory defects, e.g. decreased phototaxis.

Plasticity of Innate Behaviors

The point that emerges from some of the studies mentioned thus far, is that while the lives of *dnc*, *rut*, and *amn* are reasonably normal, this does not imply that behavioral plasticity is utterly dissociated from the innate behavioral repertoire in *Drosophila*. Two types of innate behaviors mentioned above deserve special consideration in this respect.

The first is courtship, probably the most astonishing aspect of *Drosophila* behavior. It is amazing to discover how learning and memory are interwoven in this complex set of actions and interactions, which not long ago was depicted by the metaphor of a prerecorded tape that is merely being played back by the fly nervous system. Even the sexual life of flies benefits from experience (reviewed in Hall 1984, 1986, Quinn & Greenspan 1984). The ability to learn is associated with evolutionary fitness in *Drosophila* (Hewitt et al 1983); indeed, it has been suggested that learning has evolved in the fruit fly in the first place because it contributes to efficient reproduction (Hall 1986). The correlation with reproductive fitness might contribute, over time, to the accumulation, in homozygous mutant stocks, of genetic modifiers that compensate for the primary learning defect (Tully & Quinn 1985).

Another intimate relationship seems to exist between some aspects of ethological learning (i.e. modified courtship) and biological rhythms (reviewed in Hall 1984, 1986). The involvement of biological clocks in at least some type(s) of neuronal plasticity is still a mystery; is it due to the clock action per se or merely to a gene product that is shared by biological clocks and learning systems but that functions in a different molecular context in each case? Mutations like *per*[1] may provide the answer.

What Processes are Affected: Acquisition, Retention, Retrieval?

Current conditioning paradigms in *Drosophila* can reliably measure memories that last for up to 24 hours. Memories can probably be extended by intensifying training (Quinn et al 1974, Dudai 1977, Dudai et al 1987b, Folkers 1982, Tully & Quinn 1985), but this is not routinely done. On the other hand, even a relatively short memory life-span permits identification of phases in *Drosophila* memory (Quinn & Dudai 1976, Dudai 1977, Dudai et al 1987b, Tempel et al 1983, Tully & Quinn 1985). This means that mutations could be searched for in post-consolidation phases. However, mutations described to date were isolated in screens for defects in learning or short-term memory. Screens intended to specifically isolate mutants in later memory phases are indeed underway (Y. Dudai, unpublished; T. Tully, unpublished).

Given that the mutations affect early events in memory formation, the question can be raised: Do the mutations affect the acquisition process; or do they affect storage; or do they affect retrieval, or, alternatively, combinations of the above? The answer to this question will not be simple. Suppose the initial molecular events in acquisition are normal but a lesion occurs downstream the molecular cascade and prevents storage of data for more than fractions of a second. Although from a mechanistic point of view acquisition does take place and early memory is defective, behavioral tests would probably detect defects in acquisition. Moreover, it is possible that the same molecular complex that acquires information also stores it for the first seconds. Dissociation of acquisition from storage, if at all possible, would ultimately require, in this case, kinetic analysis of the appropriate molecular events. It might therefore be quite risky to conclude, on the basis of behavioral tests, that the mechanism of acquisition is indeed defective. Nevertheless, behavioral experiments intended to unveil very short-lived and feeble memory are worth doing, and they often expose interesting properties of the mutant's plastic behavior.

Sometimes it appears that the answer provided by behavioral analysis of acquisition and short-term memory in mutants depends on the conditioning paradigm employed. A few illustrative examples are presented here.

The most extensive studies were performed on the mutations isolated in a deliberate screen for conditioning mutants, e.g. *dnc* and *rut*. Using various versions of an olfactory avoidance paradigm, it was demonstrated that dnc^1 and dnc^2 can form an almost normal association between odor and shock, but memory is very labile and short lived, i.e. it decays within tens of seconds to a few minutes (Dudai 1979, 1983). In reward learning, too, *dnc* showed essentially normal learning but fleeting memory, although here memory lasts for hours (Tempel et al 1983). From the aforementioned studies it can therefore be concluded that *dnc* is a memory and not an acquisition mutant. Using saving experiments, no evidence was detected for "mute" memory in dnc^1 and dnc^2; thus it is more likely that the problem is in storage and not in retrieval (Dudai 1983). In a sensitization paradigm, *dnc* showed larger than normal behavioral modification when tested 15 sec after presentation of the sensitizing stimulus, but this was followed by an abnormally rapid decline in sensitization within less than a minute (Duerr & Quinn 1982). In contrast, employing classical conditioning to odorants, it was found that *dnc* is affected in acquisition (Tully & Quinn 1985; Dudai, unpublished). One must always keep in mind the possibility that a relatively small reduction in acquisition may cause marked abbreviation of memory (e.g. Dudai et al 1984, 1987). However, in both the olfactory avoidance paradigm (Dudai 1979) and in the reward learning paradigm (Tempel et al 1983), the acquisition scores of *dnc* were very close to normal, whereas in the classical conditioning paradigm they were very much reduced (Tully & Quinn 1985), even if tested within less than a minute after training, under conditions where interference should be minimal (Dudai, unpublished).

The picture seems to be even more complicated. In an operant conditioning paradigm involving alterations in body posture to avoid an aversive heat source, *dnc* showed somewhat reduced performance but no evidence of rapidly declining memory, when compared to normal (Mariath 1985). During training and visual conditioning, *dnc* did not reach the learning level of normal flies, whereas memory was normal 30 min after training and remained normal 90 min afterwards (longer memory curves were not reported) (Folkers 1982). It could thus be concluded that, depending on the paradigm, *dnc* could be regarded either as an acquisition mutant or as a short-term memory mutant, or both.

rut was also subjected to a battery of learning and memory tests, and the results suggest that this mutant is affected in acquisition, in short-term memory, or in both. *rut* displayed normal sensitization 15 sec after a sensitizing stimulus, but the response declined at an abnormally rapid rate (Duerr & Quinn 1982). In olfactory avoidance conditioning, memory, when measured 30 sec after shock, was somewhat lower than normal and decayed very rapidly (Dudai 1983, Dudai et al 1984); the memory of *rut* after extensive training in this paradigm resembled that of normal flies after brief training. Long-term memory did not seem to be affected (Dudai et al 1984). Interestingly, saving experiments suggested the presence of "mute" memory and defective retrieval (Dudai 1983). In classical conditioning to odorants, tests performed immediately after training revealed significantly reduced memory in *rut* (Tully & Quinn 1985, Dudai et al 1987b), and no evidence for normal acquisition could be detected even when the tests were performed within less than a minute after training and involved only minimal interference (Dudai et al 1987b). Here too, longer-term memory did not seem to be affected (Tully & Quinn 1985), and residual memory consolidated into an anesthesia-resistant form (Dudai et al 1987b). In reward learning, performance at the end of training was much reduced and memory very much abbreviated (Tempel et al 1983). As in the case of *dnc*, visual acquisition was impaired but memory at 2 hrs was normal (Folkers 1982).

Likewise, in the case of *amn*, the effect on acquisition and memory depends on the paradigm employed. Acquisition was normal but memory decayed at an abnormally rapid rate in olfactory avoidance conditioning (Quinn et al 1979) and in reward learning (Tempel et al 1983). Acquisition was also close to normal in classical conditioning of odorants, and memory declined rapidly immediately after training but not subsequently (Tully & Quinn 1985). So too was the case in acoustical priming of female mating behavior by the male's love song (Kyriacou & Hall 1984). On the other hand, initial performance in a paradigm in which males avoid females following rejection was lower than normal, and so was memory (Siegel & Hall 1979); in conditioned avoidance of immature males, initial scores were again lower than normal (Gailey et al 1982).

In contrast with the memory decay curves obtained in olfactory avoidance (Quinn et al 1979), the results of olfactory classical conditioning experiments (Tully & Quinn 1985) did not suggest that *amn* is lesioned in a memory phase that is significantly later than the phase in which *dnc* and *rut* are affected. Saving experiments, performed in the olfactory avoidance paradigm, suggested that some memory in *amn* may be stored in a cryptic form, i.e. that *amn* is a retrieval mutant (Quinn et al 1979).

In visual conditioning, as is the case with *dnc* and *rut*, acquisition of *amn* was defective, but later memory was within normal range (Folkers 1982).

Flies homozygous for *tur* learn significantly less than normal when tested immediately after training in olfactory classical conditioning. The mutation appears to act as recessive for learning but as dominant for memory; although immediate performance of *tur*/+ flies was normal, memory decayed more rapidly than normal, as did the memory of the homozygous mutant (Tully & Quinn 1985). In visual conditioning, *tur* acquisition was defective, but memory, tested two hours after training, was not (Folkers 1982). Habituation was decreased, but sensitization, on the other hand, was normal (Duerr & Quinn 1982). *cabbage* was reported to display low memory scores seconds after training; retrieval defects may be involved in the reduced performance (Dudai 1983). Ddc^{ts} seemed to affect acquisition in both olfactory avoidance conditioning and reward conditioning, but later memory decayed at a normal rate (Tempel et al 1984). When tested 5 min after training, *Sh* did not display reduced courtship, normally displayed by males following rejection by females, and was therefore thought to be defective in acquisition; in the same paradigm nap^{ts} was reported to display normal acquisition but reduced memory (Cowan & Siegel 1984). The allele Sh^5 was also reported to affect acquisition (and probably short-term memory), but not later memory, in olfactory classical conditioning (Cowan & Siegel 1986).

In conclusion, all learning and memory mutations studied to date affect early events in memory formation, although the magnitude and the time scale of the effect differ from one mutant to another. Under some circumstances, normal or close-to-normal acquisition can be detected in a few mutants (e.g. *dnc*), but memory is labile and deteriorates rapidly; in other cases, acquisition appears to be affected. The possibility that in such cases rapid loss of memory results from the availability of a weaker memory to start with, and not from inherent defects in storage, must therefore be tested. Clearly, in many instances it is difficult to dissociate the effects on acquisition from the effects on early memory. This may indeed indicate that memory during the first seconds or minutes after training is stored by the same molecular system that acquires it.

The Molecular Lesions

What gene products are defective or missing in learning and memory mutants described to date, and how do they fit together to form a learning apparatus? The answer to this question comprises, no doubt, the ultimate goal of the neurogenetic dissection of learning in *Drosophila*.

In some cases, especially in mutants isolated in a deliberate screen for conditioning mutants, like *amn*, *cab*, and additional, even less characterized putative mutants, the identity of the affected gene product is still a mystery. The recent history of *Drosophila* neurogenetics and molecular biology hints that even here the mystery may vanish before long. In some other cases, however, information on the gene product involved is already available. This information indicates that second messenger systems and phosphorylation cascades play relatively specific roles in acquisition and short-term memory in *Drosophila*.

AMINERGIC NEUROTRANSMITTERS   *Ddc* mutants, as the name implies, are lesioned in the gene that codes for the enzyme dopa-decarboxylase (Wright et al 1982). In *Drosophila*, dopa-decarboxylase converts L-DOPA into dopamine and 5-hydroxytryptophan into serotonin (Livingstone & Tempel 1983). *Ddc* mutations therefore decrease dopamine and serotonin levels. By manipulating the level of *Ddc* in temperature-sensitive mutants (and see above), it is possible to show that these aminergic neurotransmitters play a role in acquisition of some kinds of associative learning. They probably do not play a major role in other, nonplastic behaviors (Tempel et al 1984). Some learning capabilities are also spared (Tully 1987). In *Drosophila*, as in other organisms, both monoamines exert at least part of their physiological effect by activating adenylate cyclase (Uzzan & Dudai 1982). Biogenic amines are, however, also known to initiate intracellular molecular cascades via G-regulatory subunits, which are not coupled to the adenylate cyclase system (Sasaki & Sato 1987). The behavioral effects of *Ddc* may therefore be due to improper initiation of the cAMP cascade, but other cascades may be involved as well. The *Ddc* gene was cloned and found to be developmentally regulated (Scholnick et al 1983).

RECEPTORS FOR BIOGENIC AMINES   Receptors for octopamine and serotonin exist in *Drosophila* in multiple affinity-states, the interconversion of which is regulated by G-regulatory proteins (Dudai & Zvi 1984b,c). Smith and his colleagues found that *tur* eliminates the high affinity serotonin binding-site (Aceves-Pina et al 1983); this however was later found to be secondary to the effect of *tur* on G units, which itself is secondary to the effect on kinase-C system(s) (Smith et al 1986). Again, it is not yet clear whether the serotonin-binding sites revealed by direct binding of [^{3}H]serotonin are coupled to adenylate cyclase or whether another second-messenger system is involved; it is, however, plausible to suggest that the secondary effect of *tur* on aminergic receptors contributes to the effect of the mutation on learning and memory.

ADENYLATE CYCLASE   A subpopulation, or a functional state, of adenylate cyclase is lesioned by the mutation *rut* (Aceves-Pina et al 1983, Dudai 1985, Dudai & Zvi 1984a, 1985, Dudai et al 1983, 1984, 1985; Livingstone 1985, Livingstone et al 1984, Yovell et al 1987). The enzyme that is affected by the mutation is characterized by several kinetic and regulatory properties (Dudai 1985), including stimulation by Ca^{2+}-calmodulin; *rut* lesions this stimulation (Dudai & Zvi 1984a, Livingstone 1985, Livingstone et al 1984, Yovell et al 1987b). The nature of the *rut* gene product is not yet clear, although it seems to be intimately associated with the function of the catalytic unit(s) of adenylate cyclase (Dudai & Zvi 1985, Dudai et al 1984, Livingstone et al 1984).

cAMP PHOSPHODIESTERASE   Mutations in the *dnc* locus reduce or abolish the activity of an isozyme of cAMP-phosphodiesterase, PDE II, which contributes ca 70% to the total cAMP-hydrolyzing activity in the fly and displays high affinity for cAMP but not cGMP (Byers et al 1981, Davis & Kiger 1981, Kauver 1982, Shotwell 1983, Solti et al 1983). The identification of *dnc* as a cAMP phosphodiesterase mutant (Byers et al 1981) was a turning point in the neurogenetic dissection of learning in *Drosophila*. It proved that the molecular lesion in a conditioning mutant can indeed be traced, and it paved the way for identification of other mutations, by turning attention to aberrations in the cAMP cascade.

The region of the X chromosome that contains *dnc* has been cloned and analyzed (Chen et al 1987, Davis & Davidson 1984, 1986). The picture that emerges from the molecular biological studies is much more complex than was probably expected on the basis of biochemical studies of *dnc*. Genomic sequences that code for *dnc* RNA span at least 100 kb and five chromosomal bands. *dnc* exons are separated by a very large intron, and two other genes reside within this intron. Six developmentally regulated RNAs, of 9.6, 7.4, 7.2, 7.0, 5.4, and 4.5 kb were identified as transcripts of the normal *dnc* gene, and they were thought to arise by alternative splicing (Davis & Davidson 1986). Two *dnc* alleles were found to have alterations in the RNA expression patterns (Davis & Davidson 1986). cDNA clones representing portions of the polyA RNAs of *dnc*$^+$ were isolated and cloned (Chen et al 1986). The deduced amino acid sequence of the open reading frame is highly homologous to the amino acid sequence of a cyclic nucleotide phosphodiesterase from bovine brain (Chen et al 1986). Taken together, the genetic, biochemical, and molecular biological data prove that *dnc* is indeed a structural gene for cAMP phosphodiesterase II in *Drosophila*. Interestingly, a weak homology exists between a region of the *dnc*$^+$ gene and the egg-laying hormone precursor of *Aplysia californica*; this might be related to the effect of mutations in *dnc* on female fertility (Chen et al 1986).

PROTEIN KINASE   Mutation in *tur* drastically reduces the activity of Ca^{2+}-phospholipid-dependent protein kinase (protein kinase C) (Smith et al 1986). In vitro phosphorylation of two membrane proteins, of MW 76 and 84 kDa, is dependent on phospholipids and Ca^{2+} in wild type *Drosophila* and is much reduced in *tur*; the 84 kDa polypeptide co-elutes upon ion-exchange chromatography with the peak of C kinase activity (Smith et al 1986). In *tur*/+ flies, C kinase activity is half that of the wild type. But whether *tur* codes for a form of C kinase or for a regulatory protein is not yet established. It was also found that, secondary to the C kinase defect, *tur* reduces activation of adenylate cyclase by guanyl nucleotides (Smith et al 1986) and also reduces the level of high affinity receptors for monoamines (Aceves-Pina et al 1983; and see above).In addition, the affinity of cAMP-dependent protein kinase for an analogue of cAMP in vitro was found to be lower than normal (Buxbaum & Dudai 1987). It is therefore possible that the effect of *tur* on associative learning results from defects in the interaction of the C-kinase cascade with the cAMP cascade. Protein kinase C has been recently implicated in regulation of neuronal excitability in many systems (reviewed in Miller 1986).

In this context it is of interest to note that subcellular translocation of another protein kinase, Ca^{2+}-calmodulin kinase, and alteration in Ca^{2+}-stimulated protein phosphorylation, were implicated in prolonged visual adaptation in normal *Drosophila* (Willmund 1986, Willmund et al 1986). This sensory adaptation may serve as a model for studying some aspects of plasticity.

SUBSTRATES FOR PROTEIN KINASES   Attempts have been made in recent years to identify those proteins that are the subject of altered phosphorylation in the cAMP-cascade mutants *dnc* and *rut*. Several alterations in protein phosphorylation in *dnc* have been reported (Devay et al 1984, 1986), but their relevance to learning processes in vivo is not clear (Buxbaum & Dudai 1987).

ADDITIONAL COMPONENTS OF SECOND MESSENGER SYSTEMS   Additional mutations that affect learning and memory may involve second messenger systems. In preliminary experiments, *amn* slightly increased cAMP levels in fly homogenates (Livingstone et al 1984, Uzzan 1981). Additional experiments failed to establish a consistent effect (Tully 1987). Recently, *amn* was mapped to a small region on the X chromosome, molecular cloning has become feasible and, hopefully, will shed light on the gene product (Tully & Gergen 1986). The mutant *e* has twice the normal level of dopamine (Hodgetts & Konopka 1973), and *t* has abnormal dopamine levels (Konopka 1972); both may affect learning in an olfactory-conditioned avoidance paradigm (Dudai 1977).

Two additional classes of macromolecules have been implicated in learning and memory by the *Drosophila* studies. The first are ion channels. As mentioned above, putative K$^+$ channel (*Sh*) and Na$^+$ channel (*nap*) mutants were reported to affect acquisition (*Sh*) or acquisition and retention (*nap*) (Cowan & Siegel 1984, 1986). At least part of the differences between *nap* and normal are weak (Cowan & Siegel 1986), and the mutation strongly affects other behaviors [e.g. courtship (Cowan & Siegel 1984)]. Further analysis must determine whether these mutations can be regarded as relatively specific conditioning mutants (and see above). The possibility that channel mutations affect learning is indeed appealing; it has been suggested that phosphorylation of ion channels is involved in learning in identified cellular systems (Abrams 1985, Alkon 1984, Byrne 1985, Goelet et al 1986, Kandel & Schwartz 1982). It is also of interest to note that the double mutant *Sh dnc* has an enhanced *Sh* phenotype and *Sh rut* has a depressed *Sh* phenotype, which suggests an interaction between the K$^+$ channel and the cAMP cascade (Ferrus, personal communication).

A second class of macromolecules is especially intriguing, because it hints at a different type of learning model, which in many respects is still terra incognita. As mentioned above, biological clock mutations that bring about longer-period rhythms disrupt learning within the context of courtship (Jackson et al 1983, Hall 1986). Central in this group are *per* alleles. *per* has been cloned (Bargiello et al 1984, Reddy et al 1984), and the gene product shares homology with proteoglycans (Jackson et al 1986, Reddy et al 1986). Recently it has been reported that *per* affects intercellular communication via gap junctions in salivary glands (Bargiello et al 1987). Although the cellular localization of the *per* gene product is not yet clear, proteoglycans are usually found in extracellular compartments and in association with cell surfaces. This raises the possibility that extracellular proteins play a role not only in biological clocks but also in some aspects of neuronal plasticity that may be relevant to learning (and see, in this respect, Shashoua 1985). Further analysis of *per* and *per* homologues may thus shed light on the role of extracellular components in neuronal plasticity.

The Time Window in Which the Molecular Lesions are Expressed

The observation that certain gene products are required for acquisition and memory in *Drosophila* leads to the question, what is the time window during which the normal counterparts of the affected gene products must function in order to ensure normal learning and memory? Three possibilities must be considered:

1. The ongoing activity of the gene product, during acquisition and/or learning, is directly required for successful performance.

2. The implicated gene products per se are not directly required for acquisition and retention, but rather some cellular processes secondary to the activity of the identified genes. These cellular processes are chronically altered, and the defect is observed only when learning occurs, either because the processes relate directly to learning, or merely because learning is very sensitive to metabolic alterations in the cell. For example, suppose the cAMP cascade is involved in regulation of gene expression in the appropriate neuronal components (reviewed in Dudai 1987, Goelet et al 1986, Kondrashin 1985, Nagamine & Reich 1985). In such a case, the altered expression of gene(s), whose function is not yet known (and which might not necessarily relate to phosphorylation cascades), might be the cause of the defective learning and memory.

3. The development of the nervous system is abnormal, and missing cells or altered wiring lead to defective performance.

The first possibility is currently the most plausible and would account for at least a substantial portion of the phenotype. The following arguments could be raised in favor of the possibility:

(*a*) One would expect possibilities 2 and 3 to lead to gross alterations in development and behavior. This often is not the case. Some conditioning mutants display apparently normal sensory and motor behavior. When developmental problems are encountered, they can be dissociated from the effect on learning and memory (Livingstone et al 1984, Shotwell 1983, Tempel et al 1984; see also discussion in Chen et al 1986).

(*b*) At least in the case of dnc^1 and dnc^2, conditions exist under which the flies can display almost normal acquisition but much abbreviated retention (Dudai 1979, Tempel et al 1983). This argues against a wiring defect.

(*c*) In several cases it is possible to simulate the defective phenotype by using drugs. The brief pharmacological manipulations would not be expected to lead to developmental alterations (Cowan & Siegel 1984, Dudai et al 1986, Folkers & Spatz 1984).

In spite of the above, the contribution of developmental defects to the defective conditioning cannot yet be excluded and awaits further analysis. Indeed it seems that the relationship between learning and development, as it appears from the *Drosophila* studies, is quite intimate; under certain circumstances, mutations that affect conditioning may also affect some aspects of development (Bellen & Kiger 1987, Bellen et al 1987, Livingstone et al 1984, Tempel et al 1984). In addition, possibility 2 mentioned above, namely indirect effects of second messenger cascade on expression of other genes, should also be considered. Consolidation in the much widely used olfactory conditioning paradigm starts during training or a few minutes afterwards (Dudai 1977, Dudai et al 1987b, Quinn & Dudai 1976, Tempel et al 1983). Assuming that consolidation may involve protein synthesis and gene expression (Barondes 1970, Goelet et al 1986), one could consider the possibility that short-term memory is reduced because the consolidation process(es) contribute(s) to the observed memory at this early stage. This might be a factor to consider in the study of *rut* (Dudai et al 1987b).

Some Mechanistic Considerations

The data obtained from learning and memory mutants of *Drosophila* do not yet portray a detailed molecular model for an elementary learning apparatus. They do implicate second messenger systems, mainly the cAMP cascade, and cross-talk between these systems. The involvement of second messenger systems in acquisition and retention at the early phases of memory is in accord with results obtained from the study of simple learning in molluscs (Abrams 1985, Alkon 1984, Byrne 1985, Carew & Sahley 1986, Kandel & Schwartz 1982, Schwartz & Greenberg 1987), and neuronal plasticity in vertebrates (e.g. Akers et al 1986). The involvement of the cAMP cascade agrees with the molecular and cellular models suggested for acquisition and short-term memory of both nonassociative and associative behavioral modifications of defensive reflexes in *Aplysia* (Abrams 1985, Byrne 1985, Kandel & Schwartz 1982, Schwartz et al 1983). However, *dnc* or *rut* do not provide a proof for the validity of a specific model. The mutants do provide intriguing insights into the properties of postulated elementary learning mechanisms in general:

1. Some elementary molecular mechanisms that are involved in behavioral plasticity are indeed shared by neuronal systems that execute different behavioral algorithms.

2. The phenotypic manifestation of a molecular defect, and the nature and magnitude of contribution of a molecular mechanism to behavioral plasticity, may depend on the behavioral paradigm used as an assay.

3. Elementary learning and memory systems take advantage of ubiquitous information-mediating systems within the cell. It is tempting to postulate that such systems have evolved specializations (and see below) to enable dissociation of transience from persistence in handling data that passes through the neuronal network. This conclusion must however be taken with a pinch of salt. Identification of components of ubiquitous biochemical cascades as sites of lesions in conditioning mutants is aided

by our conceptual and technical ability to identify these cascades. Some molecular surprises may therefore await us. Extracellular matrix macromolecules are one possible example; transient gene activators might be another. Nevertheless, the fact that a large proportion of the mutants isolated to date are lesioned in second messenger systems is very probably not a coincidence.

4. Some defects in second messenger systems cause specific effects and do not interfere with basic functions of the organism, learning and memory excluded. This implies great heterogeneity within second messenger systems, i.e. the existence of subpopulations and functional states of enzymes, and possibly receptors and substrates, that are recruited within the context of specific cellular programs, e.g. acquisition or storage of new information. *dnc* has several forms of cAMP-phosphodiesterase, and analysis of the gene indicates that heterogeneity is much deeper, as several developmentally regulated transcripts arise from the gene; in *rut*, only a small portion of total adenylate cyclase is missing and only a few regulatory properties of the enzyme are altered or abolished. Such heterogeneity of components of second messenger systems appears to be the rule rather than the exception, as recently revealed by molecular biological studies in other systems as well (e.g. Coussens et al 1986).

5. Very probably there is no single, exclusive molecular mechanism of acquisition and storage, but rather several mechanisms operate in concert. Learning mutants always learn something. *rut* might serve as an interesting example in this respect. The mutant abolishes Ca^{2+}-calmodulin activation of adenylate cyclase; this is the form of the enzyme that was suggested by the *Aplysia* studies to serve as a molecular convergence site for unconditioned and conditioned stimuli during associative learning, and as a storage site for short-term memory (Abrams 1985, Byrne 1985, Eliot et al 1986, Hawkins et al 1983, Schwartz et al 1983, Walters & Byrne 1983, Yovell et al 1987a,b). Yet *rut* can acquire a substantial amount of new information. There must be other processes that run in parallel (Dudai et al 1987b). Moreover, some learning mutants are affected in the crosstalk between different intracellular messengers (cAMP, Ca^{2+}-calmodulin, phospholipids). One may indeed suggest that parallel processing starts in the brain at the subcellular level (Klein et al 1986).

Necessary versus Sufficient Operations of Learning Systems

At its current state of the art, the neurogenetic approach can provide clues to the nature of cellular components that are *necessary* for acquisition or retention; but it cannot indicate what components are *sufficient* for the mechanism to function properly. Even in experimental systems that are easily amenable to cellular analysis and in which cellular components of the learning network have been identified and can be manipulated, only rarely has it been established that an identified cellular change is indeed sufficient to bring about the appropriate behavioral modification (Farley et al 1983). However, if a reductive cellular model exists with discrete molecular steps that are presumed to underlie the behavioral plasticity, then one might heuristically abandon psychological memory, work with reduced cellular or subcellular preparations, and ask whether identified macromolecules can indeed perform the cellular alteration in question, and whether the properties of these macromolecules are sufficient to cause the expected cellular change. For example, suppose that closure of a K^+ channel is an elementary step in learning. Can a given protein kinase close the channel with the appropriate kinetics? Such reductive tests of discrete steps in learning and memory models have been carried out in several systems (Farley & Auerbach 1986, Shuster et al 1985).

It is realistic to suggest that in the foreseeable future, and in concert with cellular research strategies and possibly other experimental organisms, the identification and isolation of *Drosophila* gene products that are necessary for learning and memory will contribute to the identification of sufficient conditions for elementary operations of learning systems. The idea that the neuron learns by recruiting multiple molecular cascades does not exclude the possibility that each of these cascades per se can lead to acquisition and storage of data. Distributed processing might be a safety device or a mechanism to increase efficiency.

One possible method for the identification of sufficient conditions for elementary learning and memory events may involve reconstitution studies, in which identified gene products will be integrated, step by step, into subcellular or cellular preparations, and the ability to execute alterations in identified cellular properties (e.g. ion permeability) in an associative or nonassociative manner will be tested. Recent attempts, which draw from combined studies on *Drosophila* and *Aplysia*, to characterize in vitro the response of adenylate cyclase to transient stimuli which are thought to represent conditioned and unconditioned stimuli, may hint at future trends (Yovell et al 1987b). These studies are guided by a model that suggests that two features of acquisition and short-term memory may reside in properties of the adenylate cyclase complex: Memory during the first minutes after training may reside in persistent activation of the cyclase (Schwartz et al 1983), and temporal specificity (namely, the inefficiency of backward conditioning) may reside in the mode of activation of the cyclase by transient pulses of Ca^{2+} and transmitter (Abrams 1985). The postulates of this specific model might indeed prove naive (Yovell et al 1987b), but other working hypotheses could also be tested using a similar experimental approach, which simulates physiological stimuli by discrete and controlled biochemical stimuli and tests their effects on an identified enzyme or receptor.

Concluding Remarks

A little more than a decade after the adventure was initiated (Quinn et al 1974) and the first mutant isolated (Dudai et al 1976), the neurogenetic dissection of learning in *Drosophila* is illuminating molecular and cellular niches that appear to function during elementary acquisition and short-term retention processes. Additional putative conditioning mutants await analysis. Many more mutants are required; only a small fraction of the *Drosophila* genome has been examined for its effects on learning. Improved behavioral and molecular tools now make it possible to study later memory phases. It will be intriguing to test the role of genes homologous to *dnc*, *rut*, or *tur* in higher organisms, human included. It will be even more interesting to investigate the yet unknown gene products that make *cab*, *amn*, and other mutants forget so quickly what other flies remember.

Drosophila adds its unique experimental advantages to the inter-disciplinary search for elementary learning and memory mechanisms. The idea that second messenger systems play a role in elementary learning did not of course originate with the study of *Drosophila*. But fruit flies make it possible to clearly pinpoint the exact gene products that are involved, even if these gene products are very minute in quantity or operate only transiently and are therefore difficult to detect by cellular research methods. This advantage of the neurogenetic approach becomes even more important when one considers the great heterogeneity in transcripts and in their developmental expression, which is now being revealed in many genes.

An additional benefit of the neurogenetic approach is its ability to tie together apparently remote behaviors and physiological functions and to unravel their hidden common denominators and evolution. The interplay between nature and nurture in *Drosophila* courtship is a beautiful example. Furthermore, the role played by plasticity in mostly-innate behaviors, and the effect of learning and memory mutations on execution of such behaviors, may hint at the evolutionary drives that have brought learning into the fly's life and that have generated novel gene products or extended the cellular role of old ones.

Scholars who attempt to decipher complex behavioral modifications in vertebrates often ask what is the relevance of *Drosophila* (and of *Aplysia*, *Hermissenda*, *Lymax* or *Pleurobranchia*) to our own memories. Clearly, fruit flies are not going to provide clues for the structure of internal representations and distributed memory systems. But they can elucidate biological technologies that later evolved to make even complex memories possible.

ACKNOWLEDGMENTS

I am grateful to Seymour Benzer and Eric Kandel for their warm and continuous encouragement. Tom Abrams and Jack Byrne provided valuable comments on the manuscript. The following agencies support research in my laboratory: The US-Israel Binational Science Foundation (BSF), Jerusalem; The Fund for Basic Research, The Israel Academy of Science, Jerusalem; The US-Israel Agricultural Research and Development Fund (BARD); The Weizmann Institute–Rockefeller University Research Fund; and the Julia and Leo Forchheimer Center for Molecular Genetics.

Literature Cited

Abrams, T. W. 1985. Activity-dependent presynaptic facilitation: An associative mechanism in *Aplysia*. *Cell. Mol. Neurobiol.* 5: 123–45

Aceves-Pina, E. O., Booker, R., Duerr, J. S., Livingstone, M. S., Quinn, W. G., Smith, R. F., Sziber, P. P., Tempel, B. L., Tully, T. P. 1983. Learning and memory in *Drosophila*, studied with mutants. *Cold Spring Harbor Symp. Quant. Biol.* 48: 831–40

Aceves-Pina, E. O., Quinn, W. G. 1979. Learning in normal and mutant *Drosophila* larvae. *Science* 206: 93–96

Ackerman, S. L., Siegel, R. W. 1986. Chemically reinforced conditioned courtship in *Drosophila*: Responses of wild-type and the *dunce, amnesiac* and *don giovanni* mutants. *J. Neurogenet.* 3: 111–24

Akers, R. F., Lovinger, D. M., Colley, P. A., Linden, D. J., Routtenberg, A. 1986. Translocation of protein kinase C activity may mediate hippocampal long-term potentiation. *Science* 231: 587–89

Alkon, D. L. 1984. Calcium-mediated reduction of ionic currents: A biophysical memory trace. *Science* 226: 1037–45

Balling, A., Technau, G. M., Heisenberg, M. 1987. Are the structural changes in adult *Drosophila* mushroom bodies memory traces? Studies on biochemical mutants. *J. Neurogenet.* 4: 65–73

Bargiello, T. A., Jackson, F. R., Young, M. W. 1984. Restoration of circadian behavioral rhythms by gene transfer in *Drosophila*. *Nature* 312: 752–54

Bargiello, T. A., Saez, L., Baylies, M. K., Gasic, G., Young, M. W., Spray, D. C. 1987. The *Drosophila* clock gene *per* affects intercellular junctional communication. *Nature* 328: 686–91

Barondes, S. H. 1970. Cerebral protein synthesis inhibitors block long-term memory. *Int. Rev. Neurobiol.* 12: 177–205

Bellen, H. J., Gregory, B. K., Olsson, C. L., Kiger, J. A. 1987. Two *Drosophila* learning mutants, *dunce* and *rutabaga*, provide evidence of a maternal role for cAMP on embryogenesis. *Devel. Biol.* 121: 432–44

Bellen, H. J., Kiger, J. A. 1987. Sexual hyperactivity and reduced longevity of *dunce* females of *Drosophila melanogaster*. *Genetics* 115: 153–60

Benzer, S. 1967. Behavioral mutants of *Drosophila* isolated by countercurrent distribution. *Proc. Natl. Acad. Sci. USA* 58: 1112–19

Benzer, S. 1973. Genetic dissection of behavior. *Sci. Am.* 229: 24–37

Bicker, G., Dudai, Y. 1978. Comparison of visual and olfactory learning in *Drosophila*. *Naturwissenschaften* 65: 494–95

Booker, R., Quinn, W. G. 1981. Conditioning of leg position in normal and mutant *Drosophila*. *Proc. Natl. Acad. Sci. USA* 78: 3940–44

Buxbaum, J. D., Dudai, Y. 1987. In vitro protein phosphorylation in head preparations from normal and mutant *Drosophila melanogaster*. *J. Neurochem.* 49: 1161–73

Byers, D. 1980. *Studies on learning and cyclic*

AMP phosphodiesterase of the dunce mutant of Drosophila melanogaster. PhD thesis. Calif. Inst. Technol., Pasadena

Byers, D., Davis, R. L., Kiger, J. A. 1981. Defect in cyclic AMP phosphodiesterase due to the *dunce* mutation of learning in *Drosophila melanogaster. Nature* 289: 79–81

Byrne, J. H. 1985. Neural and molecular mechanisms underlying information storage in *Aplysia:* Implications for learning and memory. *Trends Neurosci.* 9: 478–82

Carew, T. J., Sahley, C. L. 1986. Invertebrate learning and memory: From behavior to molecules. *Ann. Rev. Neurosci.* 9: 435–87

Chen, C. N., Denome, S., Davis, R. L. 1986. Molecular analysis of cDNA clones and the corresponding genomic coding sequences of the *Drosophila dunce*+ gene, the structural gene for cAMP phosphodiesterase. *Proc. Natl. Acad. Sci. USA* 83: 9313–17

Chen, C. N., Malone, T., Beckendorf, S. K., Davis, R. L. 1987. At least two genes reside within a large intron of the *dunce* gene of *Drosophila. Nature* 329: 721–24

Corfas, G., Dudai, Y. 1987. A cleaning reflex and its modification by experience in *Drosophila. Soc. Neurosci. Abstr.* 13

Coussens, L., Parker, P. J., Rhee, L., Yang-Feng, T. L., Chen, E., Waterfield, M. D., Francke, U., Ullrich, A. 1986. Multiple, distinct forms of bovine and human protein kinase C suggest diversity in cellular signaling pathways. *Science* 233: 859–66

Cowan, T. M., Siegel, R. W. 1984. Mutational and pharmacological alterations of neuronal membrane function disrupt conditioning in *Drosophila. J. Neurogenet.* 1: 333–44

Cowan, T. M., Siegel, R. W. 1986. *Drosophila* mutations that alter ionic conduction disrupt acquisition and retention of a conditioned odor avoidance response. *J. Neurogenet.* 3: 187–201

Davis, R. L., Davidson, N. 1984. Isolation of the *Drosophila melanogaster dunce* chromosomal region and recombinational mapping of *dunce* sequences with restriction site polymorphisms as genetic markers. *Mol. Cell. Biol.* 4: 358–67

Davis, R. L., Davidson, N. 1986. The memory gene *dunce*+ encodes a remarkable set of RNAs with internal heterogeneity. *Mol. Cell. Biol.* 6: 1464–70

Davis, R. L., Kauver, L. M. 1984. *Drosophila* cyclic nucleotide phosphodiesterase. *Adv. Cyclic Nucleotide Protein Phosphoryl. Res.* 16: 393–402

Davis, R. L., Kiger, J. A. 1981. *dunce* mutants of *Drosophila melanogaster:*

Mutants defective in the cyclic AMP phosphodiesterase enzyme system. *J. Cell Biol.* 90: 101–7

Devay, P., Pinter, M., Yalcin, A. S., Friedrich, P. 1986. Altered autophosphorylation of adenosine 3′,5′-phosphate dependent protein kinase in the *dunce* memory mutant of *Drosophila melanogaster. Neuroscience* 18: 193–203

Devay, P., Solti, M., Kiss, I., Dombradi, V., Friedrich, P. 1984. Differences in protein phosphorylation *in vivo* and *in vitro* between wild type and *dunce* mutant strains of *Drosophila melanogaster. Int. J. Biochem.* 16: 1401–8

Dudai, Y. 1977. Properties of learning and memory in *Drosophila melanogaster. J. Comp. Physiol.* 114: 69–89

Dudai, Y. 1979. Behavioral plasticity in a *Drosophila* mutant, *dunce DB276. J. Comp. Physiol.* 130: 271–75

Dudai, Y. 1983. Mutations affect storage and use of memory differentially in *Drosophila. Proc. Natl. Acad. Sci. USA* 80: 5445–48

Dudai, Y. 1985a. Genes, enzymes and learning in *Drosophila. Trends Neurosci.* 8: 18–21

Dudai, Y. 1985b. Some properties of adenylate cyclase which might be important for learning. *FEBS Lett.* 191: 165–70

Dudai, Y. 1986. Cyclic AMP and learning in *Drosophila. Adv. Cyclic Nucleotide Protein Phosphoryl. Res.* 20: 343–61

Dudai, Y. 1987. The cAMP cascade in the nervous system: Molecular sites of action and possible relevance to neuronal plasticity. *CRC Crit. Rev. Biochem.* 22: 221–81

Dudai, Y., Buxbaum, J., Corfas, G., Ofarim, M. 1987a. Formamidines interact with *Drosophila* octopamine receptors, affect the flies' behavior and reduce their learning ability. *J. Comp. Physiol.* 161: 739–46

Dudai, Y., Buxbaum, J., Corfas, G., Orgad, S., Segal, D., Sher, B., Uzzan, A., Zvi, S. 1986. Defective cAMP metabolism and defective memory in *Drosophila. Acta Biochim. Biophys. Hung.* 21: 177–92

Dudai, Y., Corfas, G., Zvi, S. 1987b. What is the possible contribution of Ca^{2+}-calmodulin stimulated adenylate cyclase in acquisition, consolidation and retention of memory in *Drosophila. J. Comp. Physiol.* In press

Dudai, Y., Jan, Y.-N., Byers, D., Quinn, W. G., Benzer, S. 1976. *dunce,* a mutant of *Drosophila* deficient in learning. *Proc. Natl. Acad. Sci. USA* 73: 1684–88

Dudai, Y., Sher, B., Segal, D., Yovell, Y. 1985. Defective responsiveness of adenylate cyclase to forskolin in the *Drosophila* mutant *rut. J. Neurogenet.* 2: 365–80

Dudai, Y., Uzzan, A., Zvi, S. 1983. Abnormal activity of adenylate cyclase in the *Drosophila* memory mutant *rutabaga*. *Neurosci. Lett.* 42: 207–12

Dudai, Y., Zvi, S. 1984a. Adenylate cyclase in the *Drosophila* memory mutant *rutabaga* displays an altered Ca^{2+} sensitivity. *Neurosci. Lett.* 47: 119–24

Dudai, Y., Zvi, S. 1984b. High-affinity [3H]octopamine binding sites in *Drosophila melanogaster*: Interaction with ligands and relationship to octopamine receptors. *Comp. Biochem. Physiol.* C77: 145–51

Dudai, Y., Zvi, S. 1984c. [3H]Serotonin binds to two classes of sites in *Drosophila* head homogenate. *Comp. Biochem. Physiol.* C77: 305–9

Dudai, Y., Zvi, S. 1985. Multiple defects in the adenylate cyclase of *rutabaga*, a memory mutant of *Drosophila*. *J. Neurochem.* 45: 355–65

Dudai, Y., Zvi, S., Segel, S. 1984. A defective conditioned behavior and a defective adenylate cyclase in the *Drosophila* mutant *rutabaga*. *J. Comp. Physiol.* 155: 569–76

Duerr, J. S., Quinn, W. G. 1982. Three *Drosophila* mutations that block associative learning also affect habituation and sensitization. *Proc. Natl. Acad. Sci. USA* 79: 3646–50

Eliot, L., Dudai, Y., Abrams, T. W., Kandel, E. R. 1986. Activation of adenylate cyclase in *Aplysia* by Ca^{2+}/calmodulin: A possible molecular site of stimulus convergence in associative conditioning. *Soc. Neurosci. Abstr.* 12: 400

Farley, J., Auerbach, S. 1986. Protein kinase C activation induces conductance changes in *Hermissenda* photoreceptors like those seen in associative learning. *Nature* 319: 220–23

Farley, J., Richards, W. G., Ling, L. J., Liman, E., Alkon, D. 1983. Membrane changes in a single photoreceptor cause associative learning in *Hermissenda*. *Science* 221: 1201–3

Fischbach, K. F. 1981. Habituation and sensitization of the landing response of *Drosophila melanogaster*. *Naturwissenschaften* 68: 332

Folkers, E. 1982. Visual learning and memory of *Drosophila melanogaster* wild type C-S and the mutants *dunce[1]*, *amnesiac*, *turnip* and *rutabaga*. *J. Insect Physiol.* 28: 535–39

Folkers, E., Spatz, H. Ch. 1984. Visual learning performance of *Drosophila* is altered by neuropharmaca affecting phosphodiesterase activity and acetylcholine transmission. *J. Insect Physiol.* 30: 957–65

Gailey, D. A., Jackson, F. R., Siegel, R. W. 1982. Male courtship in *Drosophila*: The conditioned response to immature males and its genetic control. *Genetics* 102: 771–82

Goelet, P., Castellucci, V. F., Schacher, S., Kandel, E. R. 1986. The long and the short of long-term memory—a molecular framework. *Nature* 322: 419–22

Gotz, K. G., Biesinger, R. 1985. Centrophobism in *Drosophila melanogaster*. *J. Comp. Physiol.* 156: 319–27

Hall, J. C. 1982. Genetics of the nervous system in *Drosophila*. *Q. Rev. Biophys.* 15: 223–479

Hall, J. C. 1984. Complex brain and behavioral functions disrupted by mutations in *Drosophila*. *Dev. Genet.* 4: 355–78

Hall, J. C. 1986. Learning and rhythms in courting, mutant *Drosophila*. *Trends Neurosci.* 9: 414–18

Hawkins, R. D., Abrams, T. W., Carew, T. J., Kandel, E. R. 1983. A cellular mechanism of classical conditioning in *Aplysia*: Activity dependent amplification of presynaptic facilitation. *Science* 219: 400–5

Heisenberg, M., Borst, A., Wagner, S., Byers, D. 1985. *Drosophila* mushroom body mutants are deficient in olfactory learning. *J. Neurogenet.* 2: 1–30

Hewitt, J. K., Fulker, D. W., Hewitt, C. A. 1983. Genetic architecture of olfactory discriminative avoidance conditioning in *Drosophila melanogaster*. *J. Comp. Psychol.* 97: 52–58

Hodgetts, R. B., Konopka, R. J. 1973. Tyrosine and catecholamine metabolism in wild type *Drosophila melanogaster* and a mutant, *ebony*. *J. Insect Physiol.* 19: 1211–20

Jackson, F. R., Bargiello, T. A., Yun, S. H., Young, M. 1986. Product of *per* locus of *Drosophila* shares homology with proteoglycans. *Nature* 320: 185–88

Jackson, F. R., Gailey, D. A., Siegel, R. W. 1983. Biological rhythm mutations affect an experience-dependent modification of male courtship behavior in *Drosophila melanogaster*. *J. Comp. Physiol.* 151: 545–52

Kandel, E. R., Schwartz, J. H. 1982. Molecular biology of learning: Modulation of transmitter release. *Science* 218: 433–43

Kauvar, L. M. 1982. Defective cyclic adenosine $3':5'$-monophosphate phosphodiesterase in the *Drosophila* memory mutant *dunce*. *J. Neurosci.* 2: 1347–58

Klein, M., Hochner, B., Kandel, E. R. 1986. Facilitatory transmitters and cAMP can modulate accommodation as well as transmitter release in *Aplysia* sensory neurons: Evidence for parallel processing in a single cell. *Proc. Natl. Acad. Sci. USA* 83: 7994–98

562   DUDAI

Kondrashin, A. A. 1985. Cyclic AMP and regulation of gene expression. *Trends Biochem. Sci.* 10: 97–98

Kyriacou, C. P., Hall, J. C. 1984. Learning and memory mutations impair acoustic priming of mating behavior in *Drosophila*. *Nature* 308: 62–65

Livingstone, M. S. 1985. Genetic dissection of *Drosophila* adenylate cyclase. *Proc. Natl. Acad. Sci. USA* 82: 5992–96

Livingstone, M. S., Sziber, P. P., Quinn, W. G. 1984. Loss of calcium/calmodulin responsiveness in adenylate cyclase of *rutabaga*, a *Drosophila* learning mutant. *Cell* 37: 205–15

Livingstone, M. S., Tempel, B. L. 1983. Genetic dissection of monoamine neurotransmitter synthesis in *Drosophila*. *Nature* 303: 67–70

Mariath, H. A. 1985. Operant conditioning in *Drosophila melanogaster* wild-type and learning mutants with defects in the cyclic AMP metabolism. *J. Insect Physiol.* 31: 779–87

Mason, S. T., ed. 1984. *Catecholamines and Behavior*. Cambridge: Cambridge Univ. Press

Menne, D., Spatz, H. Ch. 1977. Colour vision in *Drosophila melanogaster*. *J. Comp. Physiol.* 114: 301–12

Miller, R. J. 1986. Protein kinase C: A key regulator of neuronal excitability? *Trends Neurosci.* 11: 538–41

Mohler, J. D. 1977. Developmental genetics of the *Drosophila* egg. I. Identification of 59 sex-linked cistrons with maternal effects on embryonic development. *Genetics* 85: 259–72

Nagamine, Y., Reich, E. 1985. Gene expression and cAMP. *Proc. Natl. Acad. Sci. USA* 82: 4606–10

Platt, S. A., Holliday, M., Drudge, O. W. 1980. Discrimination learning of an instrumental response in individual *Drosophila melanogaster*. *J. Exp. Psychol. Anim. Behav. Proc.* 6: 301–11

Quinn, W. G. 1984. Work in invertebrates on the mechanisms underlying learning. In *Biology of Learning* ed. P. Marler, H. S. Terrace, pp. 197–246. Berlin: Springer-Verlag

Quinn, W. G., Dudai, Y. 1976. Memory phases in *Drosophila*. *Nature* 262: 576–77

Quinn, W. G., Greenspan, R. J. 1984. Learning and courtship in *Drosophila*: Two stories with mutants. *Ann. Rev. Neurosci.* 7: 67–93

Quinn, W. G., Harris, W. A., Benzer, S. 1974. Conditioned behavior in *Drosophila melanogaster*. *Proc. Natl. Acad. Sci. USA* 71: 708–12

Quinn, W. G., Sziber, P. P., Booker, R. 1979.

The *Drosophila* memory mutant *amnesiac*. *Nature* 277: 212–14

Reddy, P., Jacquier, A. C., Abovich, N., Peterson, G., Rosbash, M. 1986. The *period* clock locus of *D. melanogaster* codes for a proteoglycan. *Cell* 46: 53–61

Reddy, P., Zehring, W. A., Wheeler, D. A., Pirrota, V., Hadfield, C., Hall, J. C., Rosbash, M. 1984. Molecular analysis of the *period* locus in *Drosophila melanogaster* and identification of a transcript involved in biological rhythms. *Cell* 38: 701–10

Salz, H. K., Davis, R. L., Kiger, J. A. 1982. Genetic analysis of chromomere 3D4 in *Drosophila melanogaster*: The *dunce* and *sperm-amotile* genes. *Genetics* 100: 587–96

Sasaki, K., Sato, M. 1987. A single GTP-binding protein regulates K^+-channels coupled with dopamine, histamine and acetylcholine receptors. *Nature* 325: 259–62

Scholnick, S. B., Morgan, B. A., Hirsh, J. 1983. The cloned dopa decarboxylase gene is developmentally regulated when reintegrated into the *Drosophila* genome. *Cell* 34: 37–45

Schwartz, J. H., Bernier, L., Castellucci, V. F., Palazzolo, T., Saitoh, T., Stapelton, A., Kandel, E. R. 1983. What molecular steps determine the time course of the memory for short-term sensitization in *Aplysia*. *Cold Spring Harbor Symp. Quant. Biol.* 48: 811–19

Schwartz, J. H., Greenberg, S. M. 1987. Molecular mechanisms for memory: Second-messenger induced modifications of protein kinases in nerve cells. *Ann. Rev. Neurosci.* 10

Shashoua, V. 1985. The role of extracellular proteins in learning and memory. *Am. Sci.* 73: 364–70

Shotwell, S. L. 1982. *A biochemical and genetic analysis of the cyclic AMP phosphodiesterase defect in* dunce, *a memory mutant of Drosophila*. PhD thesis. Calif. Inst. Technol., Pasadena

Shotwell, S. L. 1983. Cyclic adenosine 3′:5′-monophosphate phosphodiesterase and its role in learning in *Drosophila*. *J. Neurosci.* 3: 739–47

Shuster, M. J., Camardo, J. S., Siegelbaum, S. A., Kandel, E. R. 1985. Cyclic AMP dependent protein kinase closes the serotonin sensitive K channels of *Aplysia* sensory neurons in cell free membrane patches. *Nature* 313: 322–95

Siegel, R. W., Hall, J. 1979. Conditioned responses in courtship behavior of normal and mutant *Drosophila*. *Proc. Natl. Acad. Sci. USA* 76: 3430–34

Smith, R. F., Choi, K.-W., Tully, T., Quinn,

W. G. 1986. Deficient protein kinase C activity in *turnip*, a *Drosophila* learning mutant. *Soc. Neurosci. Abstr.* 12: 399

Solti, M., Devay, P., Kiss, I., Londesborough, J., Friedrich, P. 1983. Cyclic nucleotide phosphodiesterases in larval brain of wild type and *dunce* mutant strains of *Drosophila melanogaster*: Isoenzyme patterns and activation by Ca^{2+}/calmodulin. *Biochem. Biophys. Res. Commun.* 111: 652–58

Tempel, B. L., Bonini, N., Dawson, D. R., Quinn, W. G. 1983. Reward learning in normal and mutant *Drosophila*. *Proc. Natl. Acad. Sci. USA* 80: 1482–86

Tempel, B. L., Livingstone, M. S., Quinn, W. G. 1984. Mutations in the dopa decarboxylase gene affect learning in *Drosophila*. *Proc. Natl. Acad. Sci. USA* 81: 3577–81

Tully, T. 1984. *Drosophila* learning: Behavior and biochemistry. *Behav. Genet.* 14: 527–57

Tully, T. 1987. *Drosophila* learning and memory revisited. *Trends Neurosci.* 10: 330–35

Tully, T., Gergen, J. P. 1986. Deletion mapping of the *Drosophila* memory mutant *amnesiac*. *J. Neurogenet.* 3: 33–47

Tully, T., Quinn, W. G. 1985. Classical conditioning and retention in normal and mutant *Drosophila melanogaster*. *J. Comp. Physiol.* 157: 263–77

Uzzan, A. 1981. *Studies on adenylate cyclase and on protein phosphorylation in normal Drosophila melanogaster and in learning and memory mutants*. M.S. thesis. Weizmann Inst., Rehovot, Israel

Uzzan, A., Dudai, Y. 1982. Aminergic receptors in *Drosophila melanogaster*: Responsiveness of adenylate cyclase to putative neurotransmitters. *J. Neurochem.* 38: 1542–50

Walters, E. T., Byrne, J. H. 1983. Associative conditioning of single sensory neurons suggest a cellular mechanism for learning. *Science* 219: 405–8

Willmund, R. 1986. Long-lasting modulation of protein kinase activity in the brain of *Drosophila melanogaster* induced by visual adaptation. *J. Insect Physiol.* 32: 1–8

Willmund, R., Mitschulat, H., Schneider, K. 1986. Long-term modulation of Ca^{2+}-stimulated autophosphorylation and subcellular distribution of the Ca^{2+}/calmodulin-dependent protein kinase in the brain of *Drosophila melanogaster*. *Proc. Natl. Acad. Sci. USA* 83: 9789–93

Wright, T. R. F., Black, B. C., Bishop, C. P., Marsh, J. L., Pentz, E. S., Steward, R., Wright, E. Y. 1982. The genetics of dopa decarboxylase in *Drosophila melanogaster*. V. *Ddc* and *1(21)amd* alleles: Isolation, characterization and intragenic complementation. *Mol. Gen. Genet.* 188: 18–26

Yovell, Y., Dudai, Y., Kandel, E. R., Abrams, T. W. 1987a. The Ca^{2+}-calmodulin responsiveness of adenylate cyclase: Quantitative analysis of the enzyme from *Aplysia*, *Drosophila*, and rat brain. In preparation

Yovell, Y., Dudai, Y., Kandel, E. R., Abrams, T. W. 1987b. The response of adenylate cyclase from the central nervous system of *Aplysia* to transient stimuli. *Proc. Natl. Acad. Sci. USA*. In press

SUBJECT INDEX

A

Acetylcholine
biosynthesis of
enzymes for, 178-79
potassium conductance and,
129
Acetylcholine receptor
signal transduction by, 184
Acetylcholine receptor/channel
cyclic AMP-dependent protein
kinase and, 121-22
Acid
neuromuscular block due to
calcium and, 7
Aconitine
sodium channels and, 458
Actin
neuritic cytoplasm and, 37-38
Adenosine diphosphate
See ADP
Adenylate cyclase
Drosophila learning mutants
and, 550
5-hydroxytryptamine-sensitive,
48-49
Adenyl cyclase
tachykinins and, 24
Adontosternarchus sachsii
electric organ of, 504
ADP
ATP-sensitive potassium
channels and, 108-9
Aldosterone
5-hydroxytryptamine receptors
and, 54
Alpha-fetoprotein gene
expression of
enhancers of, 358
Alzheimer's disease
N-methyl-D-aspartate receptor
and, 72
Aminergic neurotransmitters
Drosophila learning mutants
and, 549-50
Amino acid receptor
excitatory
brain damage and, 71
Amino acids
excitatory
excitotoxic action of, 71-72
neurotransmission and, 61-
75
9-Aminoacridine
ATP-sensitive potassium
channel inhibition and,
106

γ-Aminobutyric acid
biosynthesis of
enzymes for, 178-79
calcium current inhibition by,
129
4-Aminopyridine
ATP-sensitive potassium
channel inhibition and,
106
Amphibians
body architecture transformations in
thyroxine and, 441
Mauthner neuron of, 427
Androgens
electric organ discharge and,
509
male reproductive behaviors
and, 230
Anisogamy
male courtship and, 227
Annelids
nervous systems of
neuron number in, 438
Aphasia
agrammatic, 406
clinico-pathological correlates
for, 396-97
lexical deficits in, 395
word frequency and, 413
Aplysia californica
neuron death and, 440
Aplysia neurons
potassium conductance in,
129
Apteronotids
electric organs of, 504
Apteronotus
electroreceptor thresholds in,
520
pacemaker of
autoactivity of, 512
Arthropods
nervous systems of
neuron number in, 438
sensory receptor neurons of
maturity and, 439
Association cortex
parallel circuits of, 137-53
Astrocytes
progressive multifocal
leukoencephalopathy and,
363
Atropine
neuromuscular block due to
calcium and, 7

Auditory system
tonotopic maps of, 310-11
Axons
microtubule-associated proteins and, 30
optic nerve
topographic arrangement of,
290-91
optic tract
topographic arrangement of,
291-93

B

Baclofen
potassium currents and, 129
Barium
neuromuscular block due to
calcium and, 7
neuromuscular junction and, 9
Batrachotoxin
sodium channels and, 458
Bay wrens
duetting in, 236-37
Behavior
sexually dimorphic, 225-46
central nervous system and,
238-43
evolution of, 226-29
neuromuscular junction
and, 243-45
parental in rats, 234-35
Beta toxins
sodium channel activation
and, 458
Biogenic amine receptors
Drosophila learning mutants
and, 550
Biological rhythms, 373-90
clock genes, 382-90
ethological learning in *Drosophila* and, 545-46
rhythm mutants, 374-82
Birds
brain metabolism in
relative costs of, 430
song control neurons of
maturity and, 439
song production and recognition in
neuron population in, 431
Bird song
intersexual selection and, 226-
27
sex differences in
central nervous system and,
238

565

CUMULATIVE INDEXES

CONTRIBUTING AUTHORS, VOLUMES 7–11

574

CHAPTER TITLES, VOLUMES 7–11